VENOMOUS ANIMALS
OF THE UNITED STATES AND CANADA

A GUIDE TO
Vertebrates and Invertebrates of Land and Sea

Lawrence L. C. Jones
Southwest Zoologists' League
Tucson, Arizona

RIO NUEVO
PUBLISHER

IMAGE CONTRIBUTORS

The following people and organizations graciously allowed me use of their many fine photographs. Partial proceeds to North American poison control centers will be donated on their behalf, as well as others listed in the Acknowledgments. Many images were acquired through a Creative Commons with Attribution (CCA) license through Flickr and Wikimedia Commons, and if only "handles" (e.g., my "GilaMan" on Flickr) were available for the name of the photographer, the handle is in quotes. If I was uncertain about ownership, images were checked with a reverse image search engine. Unfortunately, due to space constraints I could not include the nearly 2,000 images I considered from all photographers, but others are listed in the Acknowledgments. I appreciate all organizations that helped by supporting their photographers. Many of the people below also helped with other aspects of the project, such as information on symptoms of envenomation, but to save space are not repeated in the Acknowledgments.

Robert Agular, Gary Alpert, Jim Atkinson, Ivy Baremore, Silke Baron, Sam Beebee, Richard Bejarano, Michael Bentley, Joseph Berger, Giacomo Bernardi, Ed Bierman, Ronald Billings, Tom Blackwell, Jerod Blazer, Bloodworm Depot, Mike Bogan, Wayne Boo, Brenda Bowling, Chris Brown, Margarethe Brummermann, "Buddha Dog," Savannah Burgess, John Butler, Howard Byrne, Mike Cardwell, "Carolyn," Janice Haney Carr, Tina Carvalho, A. Catenazzi Center for Disease Control Image Library, Bruce Christman, Michelle Christman, R. C. Clark, Allen Collins, Paul Condon, Douglas Costa, Garrett Craft, Kerry Crowther, Andrew David, Sam Droege, Bernard Dupont, "Eric + F," Bart Everson, Noah Fields, William Flaxington, Bryan Fry, Judy Gallegher, Rich Gassaway, Andy Gluesenkamp, Gilles Gonthier, Annette Govindarajan, Janet Graham, Rachel Graham, Brian Gratwicke, Randy Gray, Greg Green, Ratha Grimes, Vince Guida, Philippe Guillaume, Vidal Haddad, Jr., Bryan Hamilton, Gail Hampshire, Bob Hansen, Marshal Hedin,

◀ Tiger Rattlesnake, the most venomous rattlesnake species. © *Tell Hicks*

Brent Hendrixson, Jeremy T. Hetzel, Terry Hibbitts, Troy Hibbitts, Tell Hicks, Mary Hightower, David Hill, Kyle Hovey, Bryan Hughes, Laszlo Ilyes, "Incidence Matrix," "Insect Unlocked," "Jim G," Hunter Johnson, Sonke Johnson, "Jolt to Horizontal," Steve Jurvetson, "Jynto (talk)," Derek Keats, Mike Keeling, Kirk Kilfolye, Douglas King, Jerry Kirkhart, Douglas Klug, Li Cheng Shih, Don Loarie, Elias Levy, Kevin Lino, Jacinto Lluch-Valero, "Louis," Bill Love, Stephen Mackessy, Andreas Marz, Danny Martin, Jim McCullough, Ryan McMinds, Christian Mehlfuhrer, Stephanie Young Merzel, Allison Miller, Carol Miron, John Murphy, Patrik Neckman, Bengt Nyman, "Oakley Originals (and/or Beekeeper)," Bjorn Ognibenl, Jerry Payne, Gary Peeples, Bob Peterson, John Porter, Terry Poutry, "Wildman" Phil Rakoci, Marian Rice, Kevin Ripka, "Ron W," Sylke Rohrlach, Jim Rorabaugh, Harry Rose, Matthew Run, Findlay Russell, Joseph Ryan, Eli Sarnat, Greg Schechter, Krista Schmidt, Justin Schmidt, Katja Schultz, Justin Schwartz, Susan Sferra, Jackson D. Shedd, "smudge 9000," M. Smyly, "Snapper," Mark Spangler, Spirosk Photography, "Symac," Kim Starr, Kevin Stolezenbach, Diana Stralberg, Ingrid Taylor, Taro Taylor, "That Other Paper," James Tuich, Andrew Walker, Tami Warner-Minton, William Wells, H. Weerman, Alex Wild, Joe Wilson, M. D. Wilson, Daniel Wojcik, Laura Wolf, Ron Wolf, Mark Wolfson, Alexander Wong, Angel Yanagihara, Rickard Zurpre.

Some agencies and organizations that provided gratis images, often in the public domain, without an individual photographer's name include: Animal Diversity Web, Antweb, Barkshire Community College Image Library, Bloodworm Depot, Center for Disease Control (CDC), Clemson University, EcoStinger, Florida Division of Plant Services, Loma Linda University, Mar Alliance, National Institute of Allergy and Infection Diseases, National Natural Toxins Research Center, National Oceanic and Atmospheric Administration (NOAA), University of Arizona VIPER Institute, USDI Bureau of Land Management, USGS Bee Inventory and Monitoring, and Wikimedia Commons.

CONTENTS

IMAGE CONTRIBUTORS v
FOREWORD xii
PREFACE xiv

PART I 1

1 Introduction 1
SIDEBAR: *Aposematism, Mimicry, and Crypsis* 7
SIDEBAR: *Resistance to Venom and the Predator-Prey Arms Race* 10

2 A Venom and Envenomation Primer 12
SIDEBAR: *The Cost of Being Bitten* 38

3 Which Animals Are Considered Venomous for this Book? 40
SIDEBAR: *Venomous Parasites* 42

4 Precautions 45
SIDEBAR: *I Like to Hang Out in Restrooms* 65

5 First Aid 66
SIDEBAR: *The Making of Antivenom* 75
SIDEBAR: *Antivenom in the United States* 76

6 How to Use this Book 78
SIDEBAR: *Pain Indices* 88

PART II: TAXA ACCOUNTS 91

7 Terrestrial Invertebrates 91

Centipedes (Class Chilopoda) 100
Bark Scorpions (Genus *Centruroides*) 108
Giant Hairy Scorpions (Genus *Hadrurus*) 119
Vaejovid Scorpions (Family Vaejovidae) 126
Other Scorpions (Families Chactidae, Scorpionidae, and Superstitionidae) 135
Widow Spiders (Genus *Latrodectus*) 143
SIDEBAR: *Why Do Widow Spiders Have Such a Potent Venom?* 152
Recluse Spiders (Genus *Loxosceles*) 154
SIDEBAR: *Spider Bites: Tracking a Mystery Wrapped in an Enigma* 164
Funnel Weaver and Grass Spiders (Family Agelenidae) 166
Tarantulas (Genus *Aphonopelma*) **and Similar Taxa** 175
Yellow Sac Spiders (Genus *Cheiracanthium*) 186
Other Spiders (Order Araneae) 193
Yellowjackets and Hornets (Subfamily Vespinae) 210
Paper Wasps (Subfamily Polistinae) 224
Tarantula Hawks (Genera *Pepsis* and *Hemipepsis*) 230
Velvet Ants (Family Mutillidae) 238
Fire Ants (*Solenopsis geminata* Group; Primarily *S. invicta* and *S. richteri*) 246
Harvester Ants (Genus *Pogonomyrmex*) 255
European Honey Bee (*Apis mellifera*) 264
Bumblebees (Genus *Bombus*) 272
Large Carpenter Bees (Genus *Xylocopa*) 278
Assassin Bugs (Family Reduviidae) 284
Stinging Caterpillars (Order Lepidoptera) 293
Other Terrestrial Invertebrates (Phylum Arthropoda) 307

8 Terrestrial Vertebrates 318

SIDEBAR: *Self-Immunization* 326
Gila Monster (*Heloderma suspectum*) 327
Copperheads (*Agkistrodon contortrix* and *A. laticinctus*) 335

Cottonmouths (*Agkistrodon conanti* and *A. piscivorus*)   342
Grand Canyon Rattlesnake (*Crotalus abyssus*)   349
Eastern Diamond-backed Rattlesnake (*Crotalus adamanteus*)   354
Western Diamond-backed Rattlesnake (*Crotalus atrox*)   359
Sidewinder (*Crotalus cerastes*)   366
Arizona Black Rattlesnake (*Crotalus cerberus*)   372
Midget Faded Rattlesnake (*Crotalus concolor*)   378
Southern Pacific Rattlesnake (*Crotalus helleri*)   383
Timber Rattlesnake (*Crotalus horridus*)   389
Rock Rattlesnake (*Crotalus lepidus*)   396
Great Basin Rattlesnake (*Crotalus lutosus*)   403
Black-tailed Rattlesnakes (*Crotalus molossus* and *C. ornatus*)   409
Northern Pacific Rattlesnake (*Crotalus oreganus*)   416
Twin-spotted Rattlesnake (*Crotalus pricei*)   421
Southwestern Speckled Rattlesnake (*Crotalus pyrrhus*)   425
Red Diamond Rattlesnake (*Crotalus ruber*)   431
Mojave Rattlesnake (*Crotalus scutulatus*)   435
Panamint Rattlesnake (*Crotalus stephensi*)   441
Tiger Rattlesnake (*Crotalus tigris*)   445
Prairie Rattlesnake (*Crotalus viridis*)   450
Ridge-nosed Rattlesnake (*Crotalus willardi*)   457
Eastern Massasauga (*Sistrurus catenatus*)   464
Pygmy Rattlesnake (*Sistrurus miliarius*)   468
Grassland Massasauga (*Sistrurus tergeminus*)   473
Sonoran Coralsnake (*Micruroides euryxanthus*)   479
Harlequin Coralsnake (*Micrurus fulvius*)   486
Texas Coralsnake (*Micrurus tener*)   490
Other Snakes (Family Colubridae and Their Relatives)   495
American Short-tailed Shrews (Genus *Blarina*)   509

9 Aquatic Invertebrates   515
Octopuses and Squids (Class Cephalopoda)   520
Cone Snails (Family Conidae)   531
Nudibranchs (Order Nudibranchia)   541
Polychaete Worms (Class Polychaeta)   549

True Jellyfishes (Class Scyphozoa) 556
Box Jellyfishes (Class Cubozoa) 569
Clinging Jellyfish (*Gonionemus* sp.) 581
Portuguese Man-of-War (*Physalia physalis*) **and Pacific Bluebottle** (*P. utriculus*) 585
Aquatic Insects (Especially Families Belostomatidae and Naucoridae) 592
Other Aquatic Invertebrates 601

10 Aquatic Vertebrates 615

Chimaeras (Order Chimaeriformes) 619
Horn Shark (*Heterodontus francisci*) 624
Dogfish Sharks (Family Squalidae) **and Their Relatives** 629
Round Stingrays and Similar Taxa (Genera *Urotrygon*, *Hexatrygon*, and *Plesiobatis*) 637
Dasyatid Stingrays (Family Dasyatidae) 646
Pelagic Stingray (*Pteroplatytrygon violacea*) 655
Butterfly Rays (*Gymnura altavela* and *G. marmorata*) 658
Pelagic Eagle Rays (Genus *Aetobatus*) 662
Giant Mobula (*Mobula mobular*) 667
Eagle Rays (Genus *Myliobatis*) 672
Cownose Rays (Genus *Rhinoptera*) 677
Sea Catfishes (Family Ariidae) 682
Freshwater Catfishes (Family Ictaluridae and Others) 687
Typical Scorpionfishes (Subfamily Scorpaeninae) 696
Lionfishes (Subfamily Pteroinae) 704
Rockfishes and Thornyheads (Family Sebastidae) 712
Yellow-bellied Seasnake (*Pelamis platura*) 720
Other Aquatic Vertebrates (i.e., Other Fishes in Class Actinopterygii) 725

PART III 736

11 Venomous Animals of American Territories 736

12 Envenomation Stories: What Is It Like to Be Bitten or Stung? 748

   SIDEBAR: *Categorizing Factors Leading to Envenomation* 749
   Terrestrial Invertebrates 752
   Terrestrial Vertebrates 783
   Aquatic Invertebrates 822
   Aquatic Vertebrates 829

13 Medical and Pharmacological Values of Venom 837

14 Some Institutions Associated with Venomous Animals 840

   SIDEBAR: *Arizona Poison and Drug Information Center* 840
   SIDEBAR: *The Spider Pharm* 842
   SIDEBAR: *Phoenix Herpetological Sanctuary* 845
   SIDEBAR: *Venom Immunochemistry, Pharmacology, and Emergency Response (VIPER) Institute* 848
   SIDEBAR: *Arizona-Sonora Desert Museum* 851
   SIDEBAR: *Cone Snail Lab, University of Utah* 855
   SIDEBAR: *Venom Analysis Lab, University of Northern Colorado, Greeley* 859
   SIDEBAR: *Venom Research Labs and Hospital, Loma Linda University* 862
   SIDEBAR: *Pacific Cnidarian Research Lab (PCRL), University of Hawaii at Manoa* 866
   SIDEBAR: *National Natural Toxins Research Center, Texas A&M University* 870
   SIDEBAR: *Venom I (Miami-Dade County) & II (Lake County), Florida: Venom Response Programs* 873

ACKNOWLEDGMENTS 875
REFERENCES 878
GLOSSARY 879
INDEX 891

FOREWORD

People of all ages and all cultures are fascinated by venomous animals. Why? Is there something innate in our evolutionary history that predisposes us to be frightened by, and infatuated with, venomous animals? The answer is resounding—throughout our recent history and that of our ancient ancestors millions of years ago, our lives depended on knowing venomous animals. One mistake and you could be dead. Curiosity and infatuation are the best teachers for good reasons. Africa, where we originated, boasts of an amazing diversity and abundance of venomous animals, including a variety of snakes, spiders, scorpions, stinging insects, and others. The most dangerous stinging insects are the honeybees that thrive in abundance in Africa. No better example of human fright and fascination exists than in our relationship with honeybees. On the one hand we fear them for good cause given that they readily sting, the stings hurt, and they kill more people than any other venomous insect species. We are also fascinated with them for good cause because they supply us with food, folk medicine, and lessons to pass to our children. Fascination with bees keeps us safe and our tummies happy.

Like Africa, North America shines with a diversity and wealth of venomous animals to please every human instinct, good or bad. Many North Americans dislike the thought of venomous animals, yet at many social events stories regaling harrowing adventures with venomous animals are enjoyed. Love and hate! This love and hate with venomous animals was brought home to me one day when the editor of a prestigious publisher approached me to write a book about stinging insects and the pain they cause. I thought that only I and a few other colleagues liked stinging insects despite the pain they cause and that nobody would care. Wrong, the book has been immensely popular throughout the world. Lesson: for good or bad we all want to know about venomous animals.

Larry Jones charismatically writes about all forms of venomous animals of North America, be they large or small. His knowledge spans across the lands to the top of the mountains and the bottom of the seas, with dips into freshwater. His passion for the subject rings throughout the book, carrying the reader along a wild, exciting series of adventures. Each animal has a true story to tell. Each story tells something not only about the animal, but about us as human beings.

The book's contents range from the somewhat familiar largest rattlesnake, the Eastern Diamond-backed Rattlesnake, and the smallest rattlesnake, the Desert Massasauga, to more obscure venomous snakes including the Sonoran Coralsnake. Oh, and don't forget mammals. We learn that some shrews are venomous. Oceans and freshwaters teem with venomous animals, some familiar like octopuses and scorpionfishes, others more obscure—for example, the freshwater and sea catfishes. Venomous invertebrates abound in waters and range from jellyfish to "toe biters" (giant aquatic water bugs). By far the most important North American venomous animals in terms of causing human deaths are the insects. Stinging insects kill approximately sixty people in the United States per year compared to about five per year for snakes. Honeybees alone cause about half the deaths, with all but about one per year being the result of anaphylactic reactions (the remaining one by massive envenomation by thousands of bees). Even as scary as that sounds, honeybees kill only about one person in 10 million per year, a number smaller than killed by horses. Why do we talk about stinging insects more than horses as a threat? Quite simply because stinging insects sting and cause pain. Therein lies our fascination with venomous animals. They won the biological war of wits, and we lost. That is, until we understand the true biology and beauty of venomous animals. The following pages provide the facts and details. All we need to do as readers is sit back and imbibe the stories deeply. As they say, nature is more fascinating than fantasy.

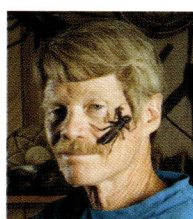

Justin O. Schmidt

Developer of the Insect Sting Pain Scale,
Author of *The Sting of the Wild*

Justin "King of Sting" Schmidt and friend.

PREFACE

Everywhere I Go, There They Are!

A Sonoran Desert moment: a Gila Monster and tarantula. *L.L.C. Jones*

People often ask me when I knew I wanted to be a biologist, and I invariably answer, "When I was in the womb." OK, I don't actually remember that far back, but some of my earliest memories were of animals, some of them venomous. And everywhere I go, I encounter venomous species. One of my first memories was of the terrarium full of spiders I collected when I was four years old. Even then I knew to avoid black widow spiders, but my nanny pointed out I had collected an immature female black widow, which looked different than the adults, so my era of learning the biology of venomous animals had begun. Probably at age six, I started another spider collection with the girl next door and started to learn the species. I became immersed in all things living and I knew I wanted to be a naturalist, just like Charles Darwin. I never achieved Charlie's accomplishments, but my heart was there.

Although I grew up in the quasi-urban environment of the southern California beach cities, there was some local wildlife to be enjoyed. This included some encounters with venomous animals near my home. While collecting ants for an ant farm, I noticed that if you stand barefooted on a harvester ant nest, they will sting you. I also spent many of my beach-bumming days being stung by Purple-striped Sea Nettles. In high school, my buddies and I often ventured off to the desert to look for snakes, tarantulas, scorpions, and anything else that moved. I also became an avid SCUBA diver and learned

You know you live in a cool place when the bike path has this sign.
L.L.C. Jones

about the many mysteries of undersea life in the biologically rich environs of southern California's giant kelp forests. Stingrays, Horn Sharks, sea anemones, sea urchins, and octopuses were frequently encountered.

After high school I went to college. There I took full advantage of courses being offered to further my education in zoology. My classes covered a wide variety of topics, including biochemistry, cytology, cell physiology, electron microscopy, invertebrate biology, entomology, ecology, terrestrial arthropods (spiders and scorpions, etc., including making a collection), herpetology, and ichthyology—all of which apply to venomous animals. I did my graduate work on venomous fishes and I worked my way through college at a marine aquarium store. There I became familiar with tropical lionfishes, Blue-spotted

Ribbon Rays, rabbitfishes, surgeonfishes, plotosid catfishes, and even the deadly Reef Stonefish and blue-spotted octopuses. My first real job was as marine biologist/aquarist at a public aquarium in southern California. Then I moved to northern California and my trade shifted to ecological aspects of terrestrial animals. During the first half of my career, I was a field biologist and researcher in various parts of the Pacific Northwest, where I learned to respect yellowjackets, an omnipresent occupational hazard. Then mid-career I transferred to southern Arizona as a management biologist. The job was not nearly as satisfying as research, but at least I managed to make it to Tucson, which Leslie Boyer of the VIPER Institute calls "Ground Zero for Venomous Animals." And here I remain through retirement, where I am a kid in a candy store. Now I encounter a plethora of venomous species whenever I walk the dogs, hike, or just sit in my backyard.

Now that I am mostly retired, I can chart my own course, and am engaged in a scorpion study, a long-term lizard study, natural history writing, leading herpetology field trips, and working intermittently at a local zoo on amphibians, reptiles, stingrays, and invertebrates. Retirement was the perfect opportunity to bring together my knowledge of marine and terrestrial fauna that are venomous to humans, as well as peruse the literature to get the latest and greatest information to synthesize in this volume. Honestly, I was a bit overwhelmed by the immensity of the task and just how much information is out there. And I am only scratching the surface. To try and paraphrase the complex world of venomous animals and their venoms is a daunting task, because it can't really be done. But this passion gave me an excuse to travel to other parts of the country to look for and photograph animals of land and sea. I hope the readers will enjoy learning about all things venomous as much as I do, and learn to admire and respect them, rather than fear and loathe them.

Larry Jones
Tucson, Arizona (Ground Zero for Terrestrial Venomous Animals)

PART I

1 Introduction

When I first began work on this book, someone asked me why I would include both terrestrial and aquatic organisms in a single volume. To them, it seemed odd combining creatures from completely different environments. My answer is this: virtually all of us live on land, but most of us are also near some kind of body of water; we often venture to the sea for livelihood or recreation; and nearly all of us are affected by venomous creatures from one or both realms. According to the National Oceanic and Atmospheric Administration (NOAA), in 2010, 123 million people in the United States (39%) lived in coastal counties, and the percentage is expected to increase by 8% in 2020. Those who don't live on the coast often seek out shorelines when on vacation. This book is designed to help the reader understand the amazing world of venomous animals of both land and sea, and how to protect oneself from envenomation in both environs. Here's a real-life example of an encounter where creatures from both land and sea made for an interesting camping trip. A small group of friends was spending a Thanksgiving holiday in 2016 on a beach in Sonora, Mexico. On the same 100-meter stretch of beach, within twenty-four hours, there was a stingray sting, two scorpion stings, a Leatherjacket fish sting, and an octopus bite!

Although there are no statistics on the total number of envenomations per year or per capita in the United States and Canada, it is probably safe to say that most of us have been bitten or stung by some sort of creature over the years. Many people have been stung numerous times by a variety of animals. However, there are some statistics and estimates available in the literature. It has been estimated that 150 million people are stung by jellyfish-like organisms every year, worldwide. There are some one-half million annual cnidarian stings in Chesapeake Bay alone, and an estimated 200,000 annual stings occur in Florida waters. Other areas, such as California and

Sonoran Coralsnake, arguably the most beautiful creature on Earth. *L.L.C. Jones*

Texas, are probably similar. According to a 2018 update, there are more than 1,000,000 emergency room visits annually from animal-related injuries, and approximately 220,000 of those visits—including about 60–70 fatalities—are due to stinging insects. The cost of these ER visits is about two billion dollars annually. And those are only the exposures reported by emergency rooms, as most go unreported. For example, one estimate is that there are probably about 24 million fire ant stings annually in the Southeast, and allergies increase from previous exposures every year. Thus, the physical, financial, and emotional cost of envenomation *each year* is staggering. Having said that, I do not wish to instill fear into my readers, but instead create a sense of awe and wonder about these amazing animals, with a strong dose of respect and caution. It is important to put things in perspective. While being bitten or stung can be painful, very few people die from envenomation in the United States. There are about six snakebite deaths annually (compared to 20,000–90,000 deaths from snakebites worldwide), and almost none from other venomous species, except bees and wasps, but deaths are usually attributable to allergy, rather than the venom itself. Only about two people per 10 million die from all venomous species combined, compared to 1,159 per 10 million people from traffic accidents (data from 2016).

While most people avoid venomous animals, a surprising number intentionally seek them out, including researchers, photographers, nature nuts, and exotic pet owners. If you ask them what the attraction is, they will usually answer that they find these animals "fascinating." What is it that is so fascinating about these creatures? We love our top ten lists, so I came up with one to help explain our fascination with venomous animals:

1. **Many are beautiful.** There is no need to dwell on this because you can see the many photographs of stunning animals in this book. Each time I see an Arizona Coralsnake, I am in awe of their glistening, vibrant beauty. Most often, bright colors evolved as aposematic coloration to let potential predators with color vision know to leave them alone.

2. **Some are cryptic (but may still be beautiful).** Unlike most poisonous animals, many venomous species don't advertise their toxic prowess, but instead try to remain hidden to blend in with the substrate. For example, a scorpionfish looks like a rock covered in encrusting organisms and algae. Octopuses are the epitome of crypsis. They change their color, shape, and texture in a flash—and squids actually do flash. Most rattlesnakes are also cryptically colored and blend in with rocks or dappled sunlight on leaf litter. However, if you see a rattlesnake on a plain background, you will see they also have beautiful patterns.

3. **They often have intriguing morphology.** Morphology goes hand in hand with venom delivery. Think of stingrays, which are flattened against the bottom of the sea and armed with a caudal appendage to guard against potential predators. And if you wanted to "create" an alien life-form, you probably can't come up with anything more alien than a scorpion or jellyfish.

4. **Their venom is interesting stuff.** If you are a biochemist or just interested in the topic, the composition and actions of an animal's venom "cocktail" is fascinating. Most chemicals in venoms are organic molecules that interact with the recipient of the venom in specific ways. A small amount of liquid can do some powerful stuff to other organisms. Venom can have a profound effect on human physiology, and untold numbers of researchers are learning about venom compounds to produce medicines to ease our suffering.

5. **They have evolved venom independently countless times over the millennia.** From an evolutionary perspective, venomous animals give us insight into relatedness among taxa, and convergent and parallel evolution. It is interesting how many times animals have independently

evolved a similar venom and venom apparatus. Take the fishes for example. Recent research shows that venom evolved some eighteen times in various taxa independently, and in almost all cases, venom is delivered through various spines. Another example: histamine and histamine-releasing agents are widespread among a broad category of taxa, as are enzymes in the phospholipase and hyaluronidase families.

6. **They have interesting mechanisms for venom delivery.** Humans didn't invent the hypodermic needle; they got the idea from animals, including such distantly related taxa as wasps and rattlesnakes. There is a diverse array of venom-delivery systems, such as the stinging cells of cnidarians, the last abdominal segment of a scorpion, the forelegs of a centipede, the setae of a caterpillar, and the spicules of a sponge.

7. **Some are excellent invaders.** Being cryptic and having venom gives some animals an advantage when it comes to becoming established outside their native range. In their alien habitats they may lack natural predators. Jellyfishes have pretty much spread globally into new areas within their temperature parameters, by means of currents and ship ballast. Lionfishes became popular in the exotic pet industry and they are now invasive in the Atlantic, Gulf of Mexico, and Caribbean, because good-intentioned pet owners thought they should "set Nemo free." Many spiders are excellent stowaways and some now are distributed worldwide.

8. **Some have developed a resistance to venom or evolved mechanisms to avoid envenomation.** There is a predator-prey arms race going on. Some animals that would be potential prey are highly resistant to venom. Some other animals avoid predation by their anatomy (e.g., the thin, scaly legs and fluffy feathers of birds). Some avoid predation by their behavior (e.g., rodents alerting their ilk to the presence of a threat through vocalization or kicking sand).

9. **Some are, or have spawned, mimics.** Scientists have described different types of mimicry in animals. The one we are most familiar with, Batesian mimicry, is where a nonvenomous animal visually mimics a venomous one to avoid being eaten. One case many of us are familiar with is the coralsnake and the harmless coralsnake-mimicking species. Some venomous animals even mimic other venomous animals.

10. **They give us a primal rush when we see them.** Even though I have encountered hundreds of rattlesnakes, it still gets my heart racing a

bit when I hear their alarm. I don't consider myself a thrill-seeker and avoid being bitten at all costs, but I'm a primate and research shows we evolved to be wary of snakes. Whenever I look at a spider under a microscope, I think to myself, "I'm glad these things aren't as big as dogs."

As you read this book, you will notice that I am a safety-first kind of guy, and I recommend precautions and seeking medical attention whenever warranted. I am also a believer in personal responsibility and personal choice. If someone wants to free-handle venomous animals, that is their life and their choice, but there is also a cost to society. For example, if someone handles a rattlesnake for the thrill and gets tagged, the insurance company and taxpayers pick up the $100,000+ tab. I don't want to financially subsidize their foolish behavior. A person who milks rattlesnakes for a living is a whole different story—they are doing that to save lives and bring down medical costs. Many people take calculated risks, such as diving with stingrays in the Caribbean. There is some risk, but accidents rarely happen. I also cannot condone the behavior of many sensational television "biologists." By free-handing venomous animals, they are doing a disservice to our children

While some thrill-seekers think free-handling venomous snakes is their right, it is a dangerous practice, and we all pay the insurance bills. This photograph is from a Rattlesnake Rodeo; most of these are now safe and educational festivals.
"That other paper"/CCA-SA/Flickr

Many people choose to interact with venomous animals that are low-risk for serious envenomation, but Southern Stingrays can sting. *Jeremy T. Hetzel/CCA/Flickr*

by sending the wrong message. For example, I recall one of these television thrill-seekers free-handling wild Tiger Rattlesnakes, which are the most toxic of all rattlesnakes. He was telling us that it was his calm demeanor and his professionalism that made it safe. "Don't try this at home; I am a professional" should translate to "Don't try this at all and I shouldn't even be doing this because I am setting a bad example; if I get bitten it is traumatic and expensive and I could lose a limb." What kind of message is it for your 10-year-old to see their TV hero handling dangerous animals without precautions? Many people have had serious injury and even death from free-handling dangerous animals.

The book is laid out into three major sections: front matter, taxa accounts, and back matter. The front matter gives the reader explanations to help understand the taxa accounts, especially in the chapter called, "A Venom and Envenomation Primer." There are also chapters on precautions and first aid. Initially, this information was embedded within the taxa accounts, but there was so much repetition (e.g., use boots and gloves as precautions, and use hot water as a first aid for marine stings) that they were combined into stand-alone chapters. The bulk of the book is composed of the taxa accounts, divided into sections for terrestrial invertebrates, terrestrial vertebrates, aquatic invertebrates, and aquatic vertebrates. Each of the headings within taxa accounts are discussed in the front matter chapter, "How to Use this Book." The back matter includes a chapter to discuss the additional taxa that may be found in the American territories, one that discusses venom as a potential treasure trove of medicinal uses, and one with novel stories of people who have been envenomated—we can only imagine their pain as we sit back by the fireplace reading of their plight. Also, in the back matter, there are some examples of institutions where researchers deal with venomous animals on a daily basis. Scattered throughout the text are additional sidebars.

As you will see in the following pages, nowhere in North America is completely safe from venomous animals. I hope you will become more aware of these creatures and their habits, take appropriate precautions, and seek medical attention as needed. Most of all, I hope you will learn to embrace all of our splendid life-forms, venomous or otherwise. After all, we are all just animals trying to survive on this sphere we call Earth.

Aposematism, Mimicry, and Crypsis

Besides the obvious benefit of being able to deliver a toxic substance, venomous animals have a variety of other mechanisms for procuring prey and avoiding predation. Many sport bright colors and/or conspicuous patterns, which is known as aposematism. Some nonvenomous animals may look like venomous species to avoid predation, which is known as mimicry. Other venomous animals are just the opposite, appearing dull and blending in with the environment, a phenomenon known as crypsis. Some of these are animals among the most beautiful and interesting creatures on Earth.

In the visual sense, aposematism is also known as warning coloration. Avoidance is sometimes instinctive among predators, but it can also be a learned behavior. An aposematic venomous animal has bright or contrasting colors that warn potential predators

The red wings are warning colors of many tarantula hawks. *L.L.C. Jones*

that it should not be engaged with. Examples include Gila Monsters, coralsnakes, wasps, bees, and many others. There are other types of aposematism, including acoustic, behavioral, and odor, which are also warning signs. Some taxa, such as velvet ants, can have a variety of aposematic attributes among their antipredator arsenal.

This blister beetle has a poisonous exudate, but it also mimics a tarantula hawk. *Jim McCullough/CCA/Flickr*

Mimicry is what a nonvenomous (usually) animal will use to be mistaken for something venomous to avoid predation. Examples include snakes banded with yellow, red, and black, to mimic coralsnakes; beetles that mimic tarantula hawks; and flies that mimic bees and wasps. There are several types of mimicry, including Batesian, Mertensian, Müllerian, and Wasmannian, named after the person who described the nuances of visual mimicry. The most familiar type is Batesian, where a nonvenomous animal looks like a venomous species, so both can avoid common predators. A good example is a Sonoran Shovel-nosed Snake mimicking an Arizona Coralsnake. Müllerian mimicry may involve a similar pattern across a suite of species. Again, the velvet ants of the Southwest are one of the best textbook examples.

Crypsis and camouflage refer to mechanisms to avoid detection from predators or prey. In essence, this is background mimicry, as cryptic species generally match the background substrate, although the term includes species that hide under cover to avoid detection. Examples of visually cryptic species include scorpions, recluse spiders, pit vipers, octopuses, stingrays,

(left) This News Bee is actually a fly that mimics a yellowjacket. They are called News Bees because they fly in your face and "give you the news." That habit is behavioral aposematism, and it mimics the yellowjacket's buzz. (right) The Hornet Moth is a species that mimics a yellowjacket. *L.L.C. Jones*

and scorpionfishes. Scorpions, by the way, are not cryptic to animals that have ultraviolet vision. As an example of supreme visual crypsis, the Speckled Rattlesnake matches its substrate with shocking accuracy, despite the variety of background colors between nearby populations. Perhaps even more cryptic are the typical scorpionfishes, which match nuances of the environment, such as algae on rocks. The world leader in this department, however, goes to the octopuses, which can not only change their color but also their shape and texture, at a moment's notice, to match the background. An unfortunate use of the term crypsis that also applies to biology is taxonomic crypsis—when a species' taxonomic hierarchy or status is "hidden" among other similar species. A taxonomically cryptic species is usually one that has gone undetected as a different species for a long time, because it looks so similar to related species. Taxonomically cryptic species are generally "discovered" when researchers conduct genetic analyses of different populations. In some cases the species are phenotypically identical, but genetically distinct.

This Southwestern Speckled Rattlesnake is cryptic; it matches the granite background supremely. *Alexander Wong*

Resistance to Venom and the Predator-Prey Arms Race

The "arm's race" between predator and prey is a well-known biological phenomenon. A good example that people may be familiar with are big cats and antelope in Africa. The antelope had to get faster and more alert to outrun lions and leopards, while the cats had to become faster also, to keep up, and they had to develop better hunting strategies. The same sort of thing is true for venomous animals, and it has resulted in several independent lines of animals becoming resistant to the venom of their prey or predators. As animals became resistant, stronger and more complex venom evolved.

An example of the venom arms race is that between horned lizards and their primary prey, harvester ants. These ants evolved the most potent venom of any insects on Earth. The venom is highly effective against vertebrate predators that may kill individuals or threaten the nest. Humans, for example, have little resistance to the venom. While there are other predators of harvester ants besides horned lizards, these lizards are fairly specialized, and may feed almost exclusively on them. Horned lizards feed near the entrance of a mound or along one of the major ant foraging trails. As the ants walk by, the lizards will lap them up with their sticky tongue. Unlike many other lizards, they do not chew their prey, but swallow it whole. The feeding behavior has several adaptations that do not expose them to venom. The tongue sticks to the back of the ant, then it is quickly withdrawn into a curled shape, so the biting and stinging parts are out of harm's way. As they are drawn into the mouth and esophagus, they are covered with pharyngeal mucus. In the protective mucus ball, they are incapable of biting or stinging. However, even if ants do sting the lizard, the immune factors in their plasma makes the lizard highly resistant to the venom.

Another well-known case is that of kingsnakes and pit vipers, particularly rattlesnakes. Kingsnakes have long been known to be resistant to various degrees to pit viper venom, while most other snakes lack this ability. Kingsnakes are opportunists that feed on small mammals, birds, and lizards, but will readily feed on other snakes, including pit vipers. Contrary to popular belief, however, kingsnakes are not rattlesnake specialists. Kingsnakes, in general, have natural resistance to pit viper venom, but that depends on the species of kingsnake, the species of pit viper, and whether or not they co-occur. Most kingsnakes have resistance to most tissue-destructive pit vipers, while most (but not all) are not immune to the neurotoxic Mojave toxin, and none are known to be immune to the venom of the Eastern Coralsnake. There is also a tendency of greater venom resistance in snakes that co-occur, such as eastern species of kingsnakes and cottonmouths. Some small mammals also have varying degrees of resistance to snake venom.

Rattlesnakes, like this Sidewinder, lose the battle with kingsnakes, most of which are resistant to venom. *Phil Rakoci*

Some animals have developed behavioral strategies to avoid being eaten, instead of a resistance to venom. For example, Kangaroo Rats have developed an acrobatic body and use behavioral avoidance techniques.

Horned lizards have evolved a mechanism to feed almost exclusively on the most venomous insects on Earth—harvester ants. *LLC Jones*

2 A Venom and Envenomation Primer

The study of venomous animals and their toxins spans many categories of natural sciences, including biology, zoology, ecology, anatomy, physiology, chemistry, biochemistry, genetics, cytology, toxinology, toxicology, venomics, pharmacology, and medicine. Suffice to say it is a complex field with thousands of researchers each doing their part to give us a better understanding of venomous animals and how their toxins affect humans. Although these creatures may be capable of inducing pain in humans, they can also provide us with new drugs to help ease suffering. This chapter is intended to be a brief primer for readers who may not be well-versed in the sciences of venom, venomous animals, and the effects of their toxins on humans. I recommend reading this section before reading the taxa accounts, but you can refer back to it. If you are keenly interested in the topics discussed, there are countless volumes of literature on each subject, available online and in university libraries.

Toxins, or toxic compounds, are chemicals that cause harm to humans. Venom is a type of toxin that is only found in animals, and it is delivered

The Circumtropical Glaucous Nudibranch is a venomous nudibranch with aposematic coloration. *Sylke Rohrlack/CCA-SA/Flickr*

Hopkin's Rose is a poisonous dorid nudibranch with aposematic coloration. *Ron Wolf*

into the prey or victim with some sort of anatomical specialization. A poison is also a toxin, but it can be found in animals, plants, and according to some definitions, inorganic sources, such as household cleaners. Animal poisons are not introduced into the victim by the animal; rather, they must be absorbed by ingestion or contact with the skin or mucous membranes to cause harm. Hence, there are no poisonous snakes, except for a few species of keelback snakes (genus *Rhabdophis*) not found in our area—and they are also venomous. Examples of animal poisons include skin secretions of toads, newts, keelback snakes, and the exudate of blister beetles. If a poisonous caterpillar is held, it is harmless; if ingested, it will cause harm. There are shades of gray. Some animals have toxic skin secretions (crinotoxins) and if there is a poorly developed mechanism for introduction into a victim, it could be construed as either poisonous or venomous. Two such animals included in the book are sponges and soapfishes. If the delivery mechanism of a venomous animal involves mouthparts, it is referred to as a bite. If it involves different body parts, it is called a sting. Examples of venomous animals that bite include snakes, spiders, true bugs, cone snails, and octopuses. Examples of venomous animals that sting include centipedes, scorpions, ants, wasps, bees, stinging caterpillars, stingrays (note they are not called "biterays"), scorpionfishes, and jellyfishes. Some species within animal groups that typically transfer venom through a sting may actually bite, including certain ants that bite, then spray toxic compounds into the wound. The venom-fang

The Western Corsair assassin bug delivers a bite, because mouthparts are used.
L.L.C. Jones

blennies of some American territories are among the few fishes that have a venomous bite, rather than sting. It should be noted that the word "sting" is sometimes referred to as both a noun and a verb. To avoid confusion, I refer to "sting" as the act of envenomation, but never a body part itself, which I call a stinger, spine, or barb.

A bite or a sting can always be used for defense, although many animals are disinclined to do so. In some animals, its primary purpose may be for procuring prey. For example, stingrays and European Honey Bees only sting in defense, but spiders and snakes use their venom both for procuring prey and defense. Many taxa have different venom components specifically used for defense, predation, or both. Many species can control the amount of venom they deliver, which is known as metering. Those taxa that meter their venom, such as rattlesnakes, widow spiders, and fire ants, have muscular control of their venom glands. While a large dose of venom may be injected if an animal feels threatened, species that meter their venom can also deliver a "dry" bite or sting. This is presumably done to warn the intruder without wasting valuable venom needed for feeding, as it takes some time to build up venom reserves. In at least some species, it is energetically expensive for an

The forcipules of a centipede are modified legs, so a venomous pinch can be construed as a sting, but some people consider the forcipules to be mouthparts, so they say centipedes bite. *L.L.C. Jones*

animal to replace expended venom. It is well established that snakes will meter their venom and usually inject a partial dose, so typically do not unload their available volume. A few taxa are now known to have different venom components that they use independently for defense versus predation. Assassin bugs (at least those studied) even have different venom glands that manufacture different venom components; if they receive a stimulus from a threat, they bite with defensive venom, but if they sense prey, they deliver predatory venom. There are some other taxa that may also do this (e.g., cone snails and perhaps some cnidarians), and more research will help show the array of taxa that have such adaptations.

Some animals bite or sting "actively," such as when a snake strikes directly at a perceived threat, while some stinging animals sting "passively." A passive sting is when the threat brushes up against the animal. Jellyfishes and sponges are good examples—most or all cannot actively defend themselves by directing a strike, but instead sting when touched, triggered by a chemical or sensory cue. Hence, there is no such thing as a passive bite, but many

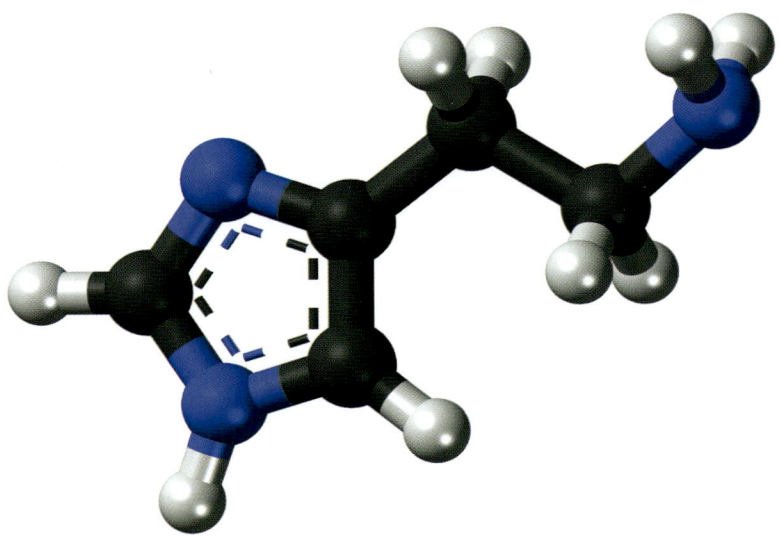

Histamine ball-and-stick model. It is a simple organic compound (amine) that can cause massive changes in the human body after envenomation. Black = carbon, white = hydrogen, and blue = nitrogen. *"Jynto (talk)"/CCA/Wikimedia Commons*

examples of active stings. When venom is injected, it is known as envenomation. A vernacular term that has come into widespread use is that of being "tagged," which is a useful word because there is no technical counterpart. This simply means the victim was bitten or stung, be it a dry or envenoming exposure. The term "exposure" in medical terminology also refers to being bitten or stung, although it is unclear if dry bites and stings constitute exposure. Another unfortunate medical term is "intoxication," which really means the body has reacted to a toxin, but in popular usage refers also to recreational imbibing of adult beverages. Table 1 is an overview of the major groups of venomous animals and their mechanisms of envenomation.

On the topic of unfortunate terminology are several commonly used terms and phrases that are unstandardized or misleading. One is the word "harmless." It literally means without harm, so indicates that a harmless species cannot cause detriment to one's health or comfort. An example of a harmless snake is a threadsnake, because it cannot possibly bite a human. However, many species often thought of as harmless are not. Coachwhip snakes, for example, are usually considered harmless snakes, but they are prone to bite and their long, sharp teeth can inflict painful wounds. They are

TABLE 1 Summary of major taxa groups, with their method of envenomation, in the order shown in this book. Abbreviations: bite (B), sting (S); used in defense (D) or procuring prey (P); active (A) or passive (P). Many animals will bite, but only those with a venomous bite are "B." All have "D" or they wouldn't be in this book. For defense/prey, when both are used, I attempted to put the primary use first. Active means that the animal can strike or actively bite, while passive means you need to rub against it.

Taxa	Bite/ Sting	Defense/ Prey	Active/ Passive	Mechanism
Centipedes	S	P, D	A	Forcipules (forelegs)
Scorpions	S	P, D	A	Stinger on abdomen
Spiders	B	P, D	A	Chelicerae, fangs
Wasps, bees, ants	S	D, P	A	Modified ovipositor in females
Assassin bugs	B	P, D	A	Proboscis harpoon
Caterpillars	S	D	P	Spines
Gila Monster	B	D	A	Teeth
Snakes	B	P, D	A	Teeth
Shrews	B	P, D	A	Teeth
Cephalopods	B	P, D	A	Beak
Cone snails	B	P, D	A	Proboscis harpoon
Aeolid nudibranchs	S	D	P	Sequestered nematocysts in cerata
Polychaetes (biting)	B	P, D	A	Proboscis or jaws
Polychaetes (stinging)	S	D	P	Spines
Cnidarians	S	P, D	P	Nematocysts
Aquatic insects	B	P, D	A	Mouthparts
Sponges	S	D	P	Epidermal spicules
Echinoderms	S	D	P	Spines
Chimaeras	S	D	A	Spines
Fishes	S	D	A	Spines

not venomous, but they are not harmless. Even some venomous animals are frequently deemed harmless in published and online references, including most scorpions and spiders with the ability to penetrate human skin, and dipsadid snakes. I have even read accounts dubbing some stingrays and Arizona Coralsnakes as harmless. They are not. A frequently used phrase is "medically significant" or the synonym, "medically important." This might not be problematic, except that there is no standard definition and there are different interpretations. For example, bees and wasps are usually considered medically significant but they are far less venomous than Maricopa Harvester Ants, which are rarely mentioned as being medically significant. The vast majority of people stung by bees and wasps have a mild, local reaction and do not seek medical attention, but the sheer number of stings results in large numbers of allergic reactions nationally, and this elevates their medical importance. It is often said Arizona Bark Scorpions are the only medically important U.S. scorpion, and yet there are hundreds of calls to poison control centers and ER admissions due to other species. One colleague had a logical explanation for the phrase when he suggested that "medically significant/important species are those that have a reasonable likelihood of requiring medical intervention when one is envenomated." This sounds reasonable, but it is still objective, so I prefer to avoid the use of the phrase. Lastly, there is a suite of terms used in the medical profession that I avoid because they are confusing, misleading, or wrong. These are the "-isms." For example, loxoscelism is a term that applies to symptoms one shows after being bitten by a spider in the genus *Loxosceles*. Some regard it as a suite of symptoms, while others specifically use it to mean the necrotic (or dermonecrotic) lesions that are occasionally observed from recluse spider bites. It is also called "necrotic arachnidism," which of course refers to necrotic lesions caused by any type of arachnid. Similarly, "tegenarism" is the term for symptoms that occur when bitten by a spider of the genus *Tegenaria*. One problem is that it was introduced specifically for the Hobo Spider, which at the time had the binomial *Tegenaria agrestis*. It is no longer in that genus, so does not apply to a Hobo Spider bite, but does apply to some other species in the genus *Tegenaria*. Another problem is that it was used at a time when it was believed that Hobo Spiders also caused dermonecrosis, but based on verified case reports, these symptoms have not been confirmed in humans bitten by a Hobo Spider. To my knowledge, no species currently in the genus *Tegenaria* have even been accused of causing dermonecrosis—at least from confirmed bite cases. There are many other examples. The "-isms" just need to go away from the literature; authors and physicians only need to refer to the source of envenomation and the specific symptoms.

Venom is often described as a chemical cocktail. While a true cocktail can be as simple as a gin and tonic, venom of many animals is composed of dozens to hundreds of different chemical compounds, or fractions. These compounds are usually organic in nature—that is, they have carbon (at least) and/or oxygen, nitrogen, and hydrogen atoms, among others. Some of the organic compounds are small molecules with only a few atoms. This includes serotonin and histamine. Larger organic compounds are more complex, and often have very specific purposes. Peptides are medium-sized organic compounds, while polypeptides are larger stings of peptides, and proteins are larger still. The most complex of the proteins are termed enzymes. Enzymes may be inert by themselves, but if injected into the victim, can cause profound chemical reactions in the human body. Enzymes are catalysts; they speed up chemical reactions. Venom may also cause chemicals that occur naturally in the body to be released, and this action itself may be the source of harm. For example, when histamine is released in the body, it can cause a veritable onslaught of physiological changes and symptoms. Venom also contains water, inorganic compounds, and inert ingredients. When venom is injected into a prey animal or predator, each of the components has a specific task, which usually takes place at the cellular and subcellular level. A good example of specific venom components comes from the peptides of cone snails. A single species may have many neurotoxic peptides, because there are different kinds of nerve targets that behave differently from one another; hence, one neurotoxin doesn't do the trick, and it may take many types to complete the physiological changes required for the cone snail to procure its prey and defend itself. Commonly referenced venom components are summarized in Table 2, although there are many more not shown.

Groups of chemicals and the symptoms they cause are often classified by a medical term with a Latin root. One group of compounds is known as algogenic toxin. These are completely defensive, so have the specific task of causing intense and immediate pain to a perceived threat. These toxins all affect the nervous system. The stimulus to nerve cells triggers the brain to give a sensation of what we know as a pain response. Pain receptors detect heat, cold, or trauma/pressure. In the taxa accounts and Envenomation Stories, you will notice that many people experience severe burning sensations when envenomated, yet there is nothing producing heat (except with inflammation). This shows how the brain can be tricked. Good examples of animals

TABLE 2 Some common and noteworthy venom components, including some examples of biological activity on envenomated humans or prey. The examples of some taxa that possess these venom types may be incomplete and only specific taxa within the group have been shown to possess the compound. There are thousands of unnamed or specifically named toxins not included.

Toxin	Chemical Type	Some Biological Activities	Examples of Taxa
Acetylcholine	Small organic molecule (biogenic amine)	Neurotransmitter--used for nerve to nerve communication and stimulates muscular contraction	Wasps, bees, caterpillars, cephalopods, fishes
Acetylcholinesterase	Enzyme	Depletes neurotransmitters; may cause paralysis	Colubrids and related, elapids
Bradykinin/releasing factors	Peptide	Pain, low blood pressure, prey immobilization	Viperids
Cardiotoxins	Various (mostly peptides)	A type of toxin that affects the heart	Various
CAP [Cysteine-rich secretory proteins (CRISP), Antigen 5 (Ag5), and Pathogenesis-related (PR-1)] proteins	Protein	Activities and biological roles poorly defined, but seem to include prey immobilization and hypothermia	Cephalopods, cone snails, polychaetes, scorpions, insects, colubrids and related, elapids, viperids
Crinotoxins	Various (mostly proteins and peptides)	These are skin poisons, but may become venoms when introduced by spicules (in sponges) or spines (in bony fishes). Cause itching, burning, and rash formation; may also cause blistering and necrosis	Sponges and certain fishes, including toadfishes, catfishes, and soapfishes
Cytotoxins	Various (mostly peptides)	A type of toxin that affects the cells	Various
Disintegrin	Peptide/Protein	Platelet inhibition; encourages hemorrhaging; acts as a "relocator protein" in rattlesnake venoms	Viperids
Esterase	Enzyme	Breaks down esters	Centipedes, recluse spiders, wasps, caterpillars
Hemotoxins and hemolysins	Various (mostly peptides and proteins)	A hemotoxin affects the blood, while hemolysins break down blood cells. "Hemotoxin" is often misused to describe whole venoms of snakes that generally cause tissue damage.	Various, especially well developed in viperids and insects
Histamine	Small organic compound	Involved in immune system. Increases capillary permeability and lowers blood pressure. Causes pain and itching. Massive release during allergic reaction.	Common component in many venoms or is released by venom in humans.

Hyaluronidase	Enzyme	Diffuses other venom components by breaking down hyaluronic acid in interstitial fluids. Called the [venom] spreading factor. Contributes to swelling.	Nearly universal, in centipedes, scorpions, polychaetes, cnidarians, insects, helodermatids, elapids, viperids, cephalopods, fishes
Kallikrein-like	Enzyme	Releases bradykinin; rapid fall in blood pressure; prey immobilization	Wasps, viperids, helodermatids
Kinin	Peptide	Drop in blood pressure	Wasps, ants, caterpillars
Kunitz-type	Peptide	Inhibits serine proteases. May have coagulopathic and neurotoxic activities	Various insects, spiders, scorpions, cone snails, cnidarians, snakes
L-amino acid oxidase	Enzyme	Actual role in envenomation poorly defined, but may cause cell damage and death, hemorrhage, necrosis, and inhibit blood clotting	Elapids, viperids, some colubrids and related
Lectins	Protein	Promotes or inhibits blood clotting	Snakes, fishes, caterpillars
Lipase	Enzyme	Breaks down lipids	Insects, snakes
Metalloprotease/ metalloproteinase	Enzyme	Prey predigestion, hemorrhage, redness, blisters, muscle necrosis	Centipedes, spiders, polychaetes, cnidarians, snakes (especially viperids)
Myotoxins	Various	Destroys normal muscle cell structure and interferes with normal muscle function	Various
Neurotoxins	Various	Toxins that affect the nervous system. The term is often misused to describe whole venoms from snakes that have generally nerve-affecting venom compounds.	Various
Noradrenaline (norepinephrine)	Simple organic molecule	Blood vessel constriction; increased blood pressure	Honey bees, wasps
Nucleases and nucleotidases	Enzyme	Various, including vasodilation; cardiotoxicity; swelling; kidney failure	Viperids, elapids

TABLE 2 Continued

Toxin	Chemical Type	Some Biological Activities	Examples of Taxa
Phosphodiesterase	Enzyme	Low blood pressure, perhaps shock	Insects, stingrays, colubrids and related, elapids, viperids
Phospholipases: especially PLA_2, but also PLA_1, PLB and PLD	Enzyme	There are many phospholipases in venom, especially PLA_2 types. Effects are varied, but can include anticoagulation and interruption of nerve synapses (blocks release of acetylcholine). It causes damage to lipid/cell membranes and can cause muscle necrosis. May cause paralysis. PLD responsible for necrosis from recluse spiders.	Centipedes, scorpions, bees, wasps, ants, spiders, cnidarians, cephalopods, helodermatids, snakes
Porins	protein	These pore-forming proteins disrupt membrane integrity of various cell types leading to prey immobilization, paralysis, and death. Also induce inflammation due to perforation of immune cells leading to cytokine and histamine release.	Hymenoptera, cnidarians, scorpaenids
Protease/Proteinase	Enzyme	A large group of enzymes that breaks down proteins; predigestion, coagulopathy. Many specific enzymes in this table are proteinases.	A broad category of enzymes found in some form in many animal venoms
Purines and pyrimidines	Small organic compound	Low blood pressure, paralysis, cell destruction; prey immobilization	Elapids, viperids
Serotonin	Small organic compound	Neurotransmitter. It occurs naturally in humans. When injected causes pain. Also mediates inflammation, vasodilation.	Many species of plants and animals, including centipedes, wasps, helodermatids, some cnidarians, stinging caterpillars
Three-finger toxins	Peptide/protein	Prey immobilization/paralysis (neurotoxic), or cardiotoxic; may also have anticoagulant properties	Colubrids and related, elapids, some viperids

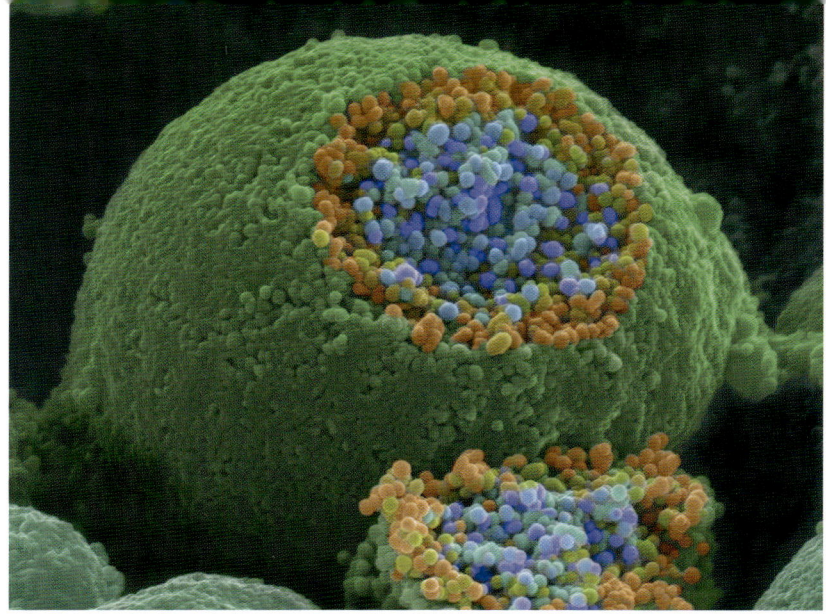

This colorized scanning electron micrograph shows where it all starts in neurotoxin envenomation: the synapse of a neuron. *Tina Carvalho/Public Domain*

with algogenic venom include bees, wasps, stingrays, and Gila Monsters. The message to the victim is clear: leave me alone, clear the area, and let others know what you experienced. Animals with algogenic venom often have bright aposematic coloration or other warning mechanisms. Examples of algogenic compounds include melittin, which is found in European Honey Bees, and serotonin. Serotonin is found naturally in the human body, and it has a variety of functions. Its release can even help create a feeling of well-being, but when injected, say by a stingray, it causes severe pain. There are many other compounds that cause pain indirectly, but are not termed algogenic, because pain-production is not their function and is only an artifact of envenomation. For example, if an enzyme causes hemorrhaging and swelling, the area becomes painful, but that is a response to the physiological reaction associated with swelling, which stimulates pain receptors for pressure/trauma.

Another major class of venoms is termed neurotoxic because the compounds affect the nervous system. Neurons are the cells of the nervous system. A network of these cells conveys messages throughout the body to the brain via chemicals and electrical impulses. The area between the neural cells and the nerve to muscle cells is called the synapse. This is a critical junction where neurotoxins can do considerable harm. Interference of normal regulation of ion (charged particles) flow occurs at the synapse. Some neurotoxic

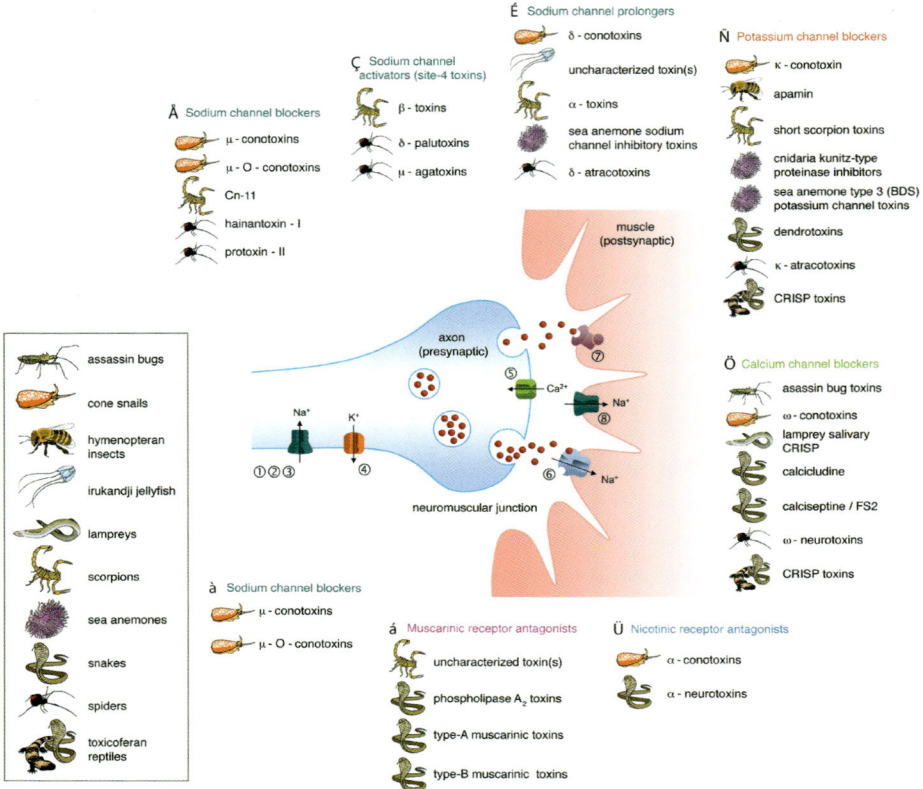

Graphic of ion flow at the neuron-muscle cell junction in neurotoxic animals. Note the diversity of taxa that have similar venom types. *Bryan G. Fry/University of Queensland*

compounds are algogenic, as discussed above. Other neurotoxins can have anesthetic properties or are used to immobilize prey, while others are used defensively. Some raw venoms contain numerous neurotoxins that combined have a variety of functions. Kissing bugs and mosquitoes are examples of venomous animals that have anesthetic qualities to their venom—in essence, these are the exact opposite of algogenic. Anesthetic venom allows parasites to feed without being detected and injured by the host. However, among the more important functions of neurotoxins in nonparasites is to immobilize or kill prey. Many neurotoxins are suited for immobilizing prey because they can cause muscle paralysis. In most cases the prey undergoes flaccid paralysis, wherein the nerves relay or interrupt a chemical message to skeletal muscles to suspend contraction. A paralyzed meal cannot struggle, escape, or injure the predator. Another type of paralysis is called spastic paralysis, which has the opposite effect. In this case, skeletal muscle is stimulated into contracting and the victim can have muscle spasms, fasciculations, or rigidity. Animals that are primarily neurotoxic tend to be more lethal (i.e., a low LD_{50}, see page 28) than those that are not. For example, death may occur in humans when a chemical signal tells their muscles to not contract. If they cannot move their muscles, the signal to move their diaphragm is blocked, so they suffocate. If the signal is to their heart, they have cardiac arrest. Examples of species that have primarily neurotoxic venom includes elapid snakes, widow spiders, scorpions, octopuses, and cubozoans.

Another class of venom is sometimes termed hemotoxic, but this term is misleading. A purely hemotoxic venom, by definition, affects blood. However, the term has also been used erroneously to incorporate a suite of symptoms that cause general tissue damage. In most pit vipers, this type of venom is sometimes referred to as Type I venom (as opposed to Type II venom, which is largely neurotoxic). I usually refer to the actions of this type of venom simply as being tissue-destructive, but "necrotizing venom" is also descriptive. These compounds are characterized by proteases (= proteinases), or enzymes that are designed to break down proteins. Among these are the highly researched metalloproteinases, which have a metal chemical component, in addition to the organic compounds. When a typical pit viper bites its prey, the animal is killed and tissue is destroyed to aid in digestion. When a human is bitten, the action of the venom is the same, although death by pit viper bite is rare these days. Consider a human finger. It is about the same size as a mouse, which is a common prey item of many pit vipers. When a person is bitten on the finger, the venom is literally digesting the flesh by the same mechanism. Many animals do not have the ability to swallow or chew up chunks of prey, so the tissue must be predigested. For example, a spider must liquify its prey and then drink the fluids to obtain nourishment. This is referred to as extraoral digestion. In some taxa, the prey is essentially slurped up through a straw. In addition to most pit vipers, animals that have tissue-destructive venom include recluse spiders, assassin bugs, and giant water bugs.

Categorization may be confusing, however, as some species can have multiple types of venom among their biochemical arsenal. For example, Mojave Rattlesnakes from a certain area are primarily neurotoxic, while those from another area have primarily tissue-destroying enzymes, and those from another area can have both. Also, there are similar terms about the effects of venom that are not categorized as neurotoxic or tissue-destructive, including

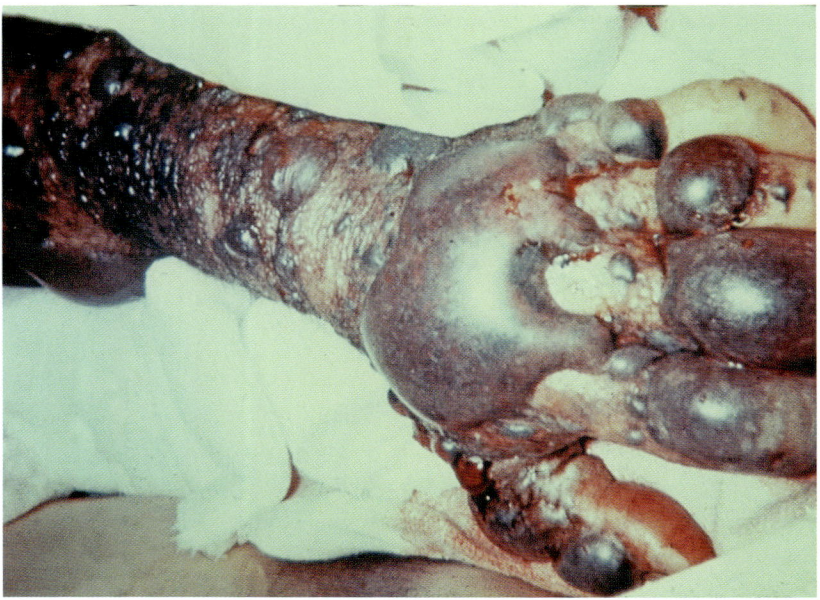

Tissue-destroying, or necrotic venom, literally dissolves flesh. This is five days after a rattlesnake bite. Loss of digits or limbs is not an uncommon side effect of snakebite. *Findlay Russell*

myotoxic (affecting the muscles), cardiotoxic (affecting the heart), cytotoxic (affecting cells), and so on. The bite from some populations of Mojave Rattlesnake are simultaneously neurotoxic, hemotoxic, cytotoxic, myotoxic, and cardiotoxic. This is why emphasis should not necessarily be placed on treating for a particular species or venom type (especially true for snakes). Instead, symptoms need to be monitored and the patient treated accordingly.

Of the thousands of venom fractions found in nature, there are many that have evolved similarly across very different taxa, via convergent or parallel evolution. Some compounds may be chemically similar, but they may have different functions. Hyaluronidase and phospholipase A_2 (PLA_2) are examples of enzymes that are found in most venomous species, and each represents functional families of similar proteins. For example, there are hundreds of different molecules in the PLA_2 family, but all are phospholipases, so as the name implies, they all speed up chemical reactions and interfere with phospholipids, which are molecules in cell membranes. Some PLA_2 are neurotoxic, while others destroy tissues.

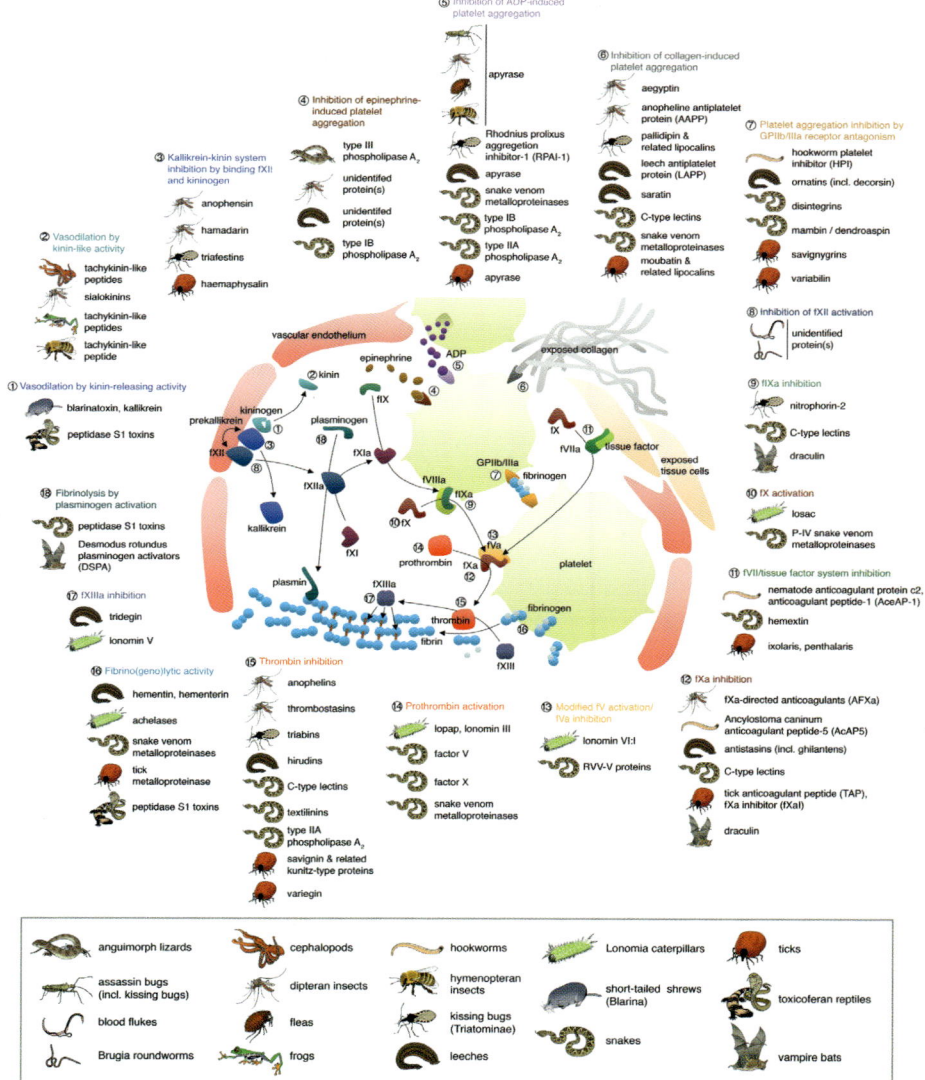

Graphic of some venom actions in animals with tissue-destroying venom. Note the diversity of taxa that have similar venom types. *Bryan G. Fry/University of Queensland*

Hyaluronidase is also widespread across most animal groups because it usually has the function of helping venom disperse in the victim. Hence, it is commonly known as "the spreading factor," although it sometimes has other functions.

A measure I use for comparing venom potency is known as LD_{50}. It refers to the Lethal Dose in 50% of laboratory animals used during toxicity testing, at twenty-four hours post-injection. The test is usually done with whole venom injected into Swiss lab mice, but it may be done with specific venom fractions, and is sometimes done on other species in the lab. In this book, LD_{50} is expressed as mg of dried venom per kg of mouse. It takes many mice to equal one kilogram, so it is usually standardized this way as an aid to comparison of toxicity, and to help us understand how potent it may be in a human. Dried venom is used as the standard because pure venom has a significant and variable amount of water in it, which is inert. LD_{50} tests are not done on humans, of course, although some authorities will venture an estimate of lethality for humans, based on LD_{50} for mice, case reports, and venom yields. In small venomous animals (e.g., spiders), the units are often µg/g, because of tiny venom yield, but the ratio is the same. Also, toxicity to mice may not accurately reflect toxicity to humans, but animals that are highly venomous to mice do seem to be highly venomous to humans, because both mice and humans are not only vertebrates, but also mammals. Injection sites in lab mice can be subcutaneous, intramuscular, intraperitoneal, or intravenous. The injection site can make a huge difference in the reported lethality. Subcutaneous and intramuscular injections tend to be considerably less lethal than intravenous. Most bites and stings to humans are intramuscular or subcutaneous. There can be considerable variation in LD_{50} reports for a given species, due to differences in injection site, population, and age of the animal. An example of a difference for a species is that of the neurotoxic vs. tissue-destructive populations of the Mojave Rattlesnake, the former being decidedly more virulent so has a much lower LD_{50}. Even with the limitations and biases of the method, it is among the best tools we have to gauge relative lethality to mammals.

The lower the LD_{50} value, the more potent the venom is to lab mice, and presumably to humans. An animal with a low LD_{50} is said to be more venomous, more toxic, more lethal, or more virulent, than one with a higher LD_{50}. However, venom yield is an important consideration. This is how much venom the animal has in its stores or can have when the glands are full; it is also measured in dry weight. If a creature has a high venom yield, it can deliver more venom. It is often believed or reported by the media that juvenile rattlesnakes are deadlier than adults, but this is not true. Juvenile snakes sometimes have a lower LD_{50}, but they also have a much lower venom

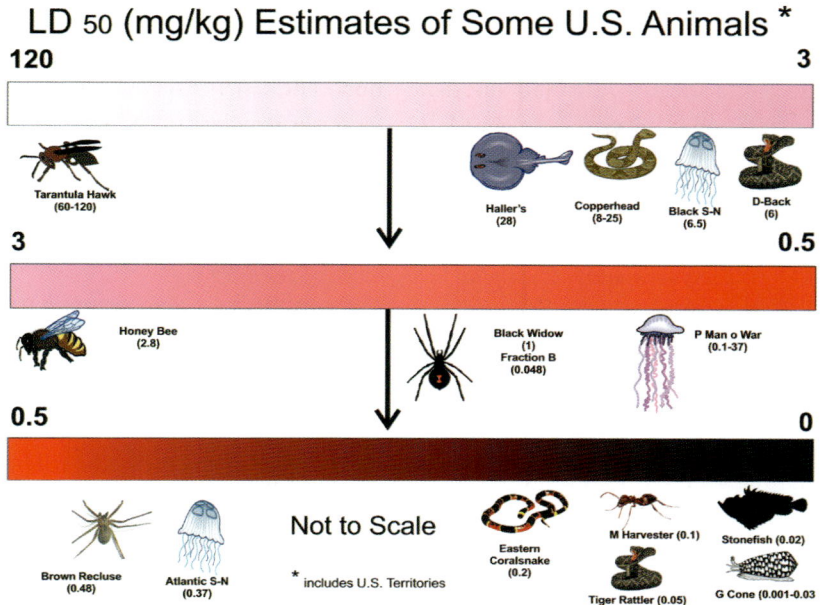

LD_{50} of some taxa described in this book. Low LD_{50} is high toxicity, and does not necessarily relate to pain. *L.L.C. Jones*

yield, so large snakes are more dangerous. Just how dangerous an animal is is somewhat subjective, as LD_{50}, venom yield, ability to cause physical trauma, encounter rate, and behavior can all be taken into consideration. Some examples of LD_{50} reported in this book include:

- **Brown Recluse Spider:** 0.48 mg/kg. This is a very toxic spider, but venom yield is usually less than 0.7 mg, so a fatality for a 68 kg human would be highly unusual.
- **Black Widow Spider (whole venom):** 0.9–1.39 mg/kg. Whole venom is not as potent as that of the Brown Recluse but is still extremely virulent. Widows also have a very small venom yield.
- **Black Widow Spider (fraction B):** 0.048 mg/kg. This is an example of a fraction of the whole venom that has been shown to be highly toxic to mice, so accounts for the high venom virulence in mammals.
- **Harvester Ants:** 0.09–0.7 mg/kg. These are the most toxic insects on Earth. Thank goodness they are small and have a miniscule venom yield!

- **Tarantula Hawk:** 60 to greater than 120 mg/kg. Pain level is not related to LD_{50}. Although these creatures have the second most painful stings of any insect on Earth, they have a very low toxicity. The venom is highly effective to deter predators but would be nonlethal in a healthy human. However, the extreme pain could lead to other outcomes. (Imagine the sensation of having boiling oil poured on your face while you are riding a motorcycle!)
- **Copperhead:** 8–25 mg/kg. This is one of the least toxic pit vipers, plus it has a relatively low venom yield of 40–75 mg. Despite thousands of human envenomations annually, fatal encounters are extremely rare, but not nonexistent.
- **Western Diamond-backed Rattlesnake:** There are many LD_{50} studies, and results are highly variable, but the average for this species seems to be around 6 mg/kg. Because more people are bitten by this species than any other rattlesnake, and it has a wide geographic range, it is often used as a reference point when comparing toxicity to other species. It is a moderate to large snake with a moderate to very large venom yield (average adults have 200–300 mg but has been reported to be as high as 600–1,145 mg or more in large individuals). There are numerous reports of fatalities from this species, but thanks to medical care, that is uncommon these days.
- **Mojave Rattlesnake (neurotoxic):** 0.13–0.54 (average around 0.34) mg/kg. Most Mojave Rattlesnake populations have this type of venom, which is about twenty times as virulent as that of the Western Diamond-backed Rattlesnake. Obviously, this is an extremely dangerous animal to be bitten by, but with prompt medical care, fatalities are rare.
- **Mojave Rattlesnake (tissue-damaging):** 2.29–3.8 mg/kg. Mojave Rattlesnakes with this type of venom are still about twice as toxic as that of Western Diamond-backed Rattlesnake, but much less than neurotoxic populations.
- **Atlantic Sea Nettle:** This species is generally considered more of a nuisance than a serious health risk, but it does have a potent LD_{50} at 0.37 mg/kg. Individual stinging cells are tiny, but if you are wrapped in tentacles by a large individual, it can cause serious envenomation.

- **Black Sea Nettle:** This is a not-so-virulent congener of its Atlantic cousin, with an LD_{50} of 6.5 mg/kg, but they can get huge, so can cause serious injury.
- **Portuguese Man-of-War:** This is the maligned jellyfish-like animal that haunts swimming beaches in the Atlantic and Gulf of Mexico. It has powerful venom—human fatalities have been attributed to it—but by looking at two LD_{50} estimates, we see huge differences in delivery: 0.1 mg/kg (intravenous) to 37 mg/kg (subcutaneous).
- **Haller's Round Stingray:** Hundreds of southern California beach-goers are stung every year by these dinner-plate-sized stingrays. Like tarantula hawks, their venom is extremely algogenic, but not very potent at about 28 mg/kg.
- **Spotted Scorpionfish:** One estimate puts the LD_{50} at 0.28 mg/kg (intravenous).
- **Geographer Cone:** Two estimates of LD_{50} using different methods for humans (not experimenting on humans!) described in greater detail in the American Territories chapter, range from 0.029–0.038 mg/kg, making this possibly the most venomous animal on Earth.

When a person is tagged, it may be a dry bite or sting, or they may have been envenomated. Initial pain is not always an indicator of the degree or seriousness of envenomation. Animals such as recluse spiders and some neurotoxic snakes may have a delayed onset of symptoms. This is part of the reason that recluse spiders are often blamed for dermonecrotic lesions that appear without a known cause. Symptoms are termed "local" if they are confined to the area of the bite or sting and are not usually a cause for concern by themselves. Local symptoms typically involve one or more of the following:

- Pain
- Redness (sometimes bluing, or cyanosis)
- Burning
- Swelling
- Sometimes excessive bleeding limited to the wound.

These are classic symptoms of inflammation, involving the body's own immune system, except for excessive bleeding (although blood flow is increased during inflammation). When a wasp's stinger injects venom, for example,

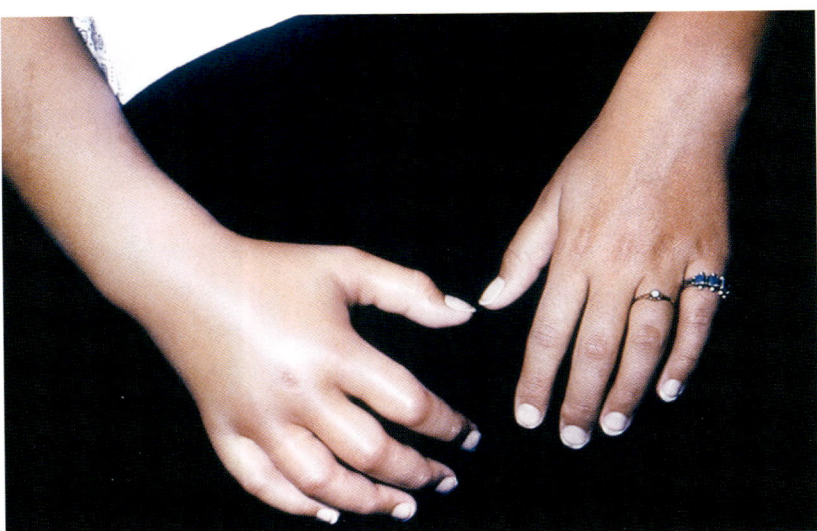

The sting from a yellowjacket causes an inflammatory response.
Margaret A. Parsons/Public Domain/CDC

cells in the body recognize the molecules in the venom as antigens, or foreign invaders that may threaten a person's health. The reaction of redness, swelling, and heat is the body's attempt to rid it of antigens by allowing antibodies (i.e., immunoglobin proteins) to bind with them and render them inert. Histamine, bradykinin, and other chemicals are released in the assault. Swelling is caused by the release of fluids from the bloodstream, while redness is caused by capillaries that are normally closed, to open, which allows blood and the antibodies into the area. In most local reactions, this lasts a few hours, but may linger longer than a few days.

If symptoms extend beyond the area of the envenomation, or cause other physiological maladies, they are termed "systemic," because they affect the anatomical and physiological systems of the human body. These symptoms are a red flag indicating a serious envenomation, so medical intervention is warranted. They are sometimes confused with, or confounded by, allergic reactions. Either way, medical involvement is recommended; the physicians can determine which symptoms are induced by allergy and which are from venom, so can take measures to alleviate symptoms of both. Examples of systemic symptoms include:

- Extensive swelling and discoloration
- Lymph gland involvement (e.g., soreness)

A Venom and Envenomation Primer   33

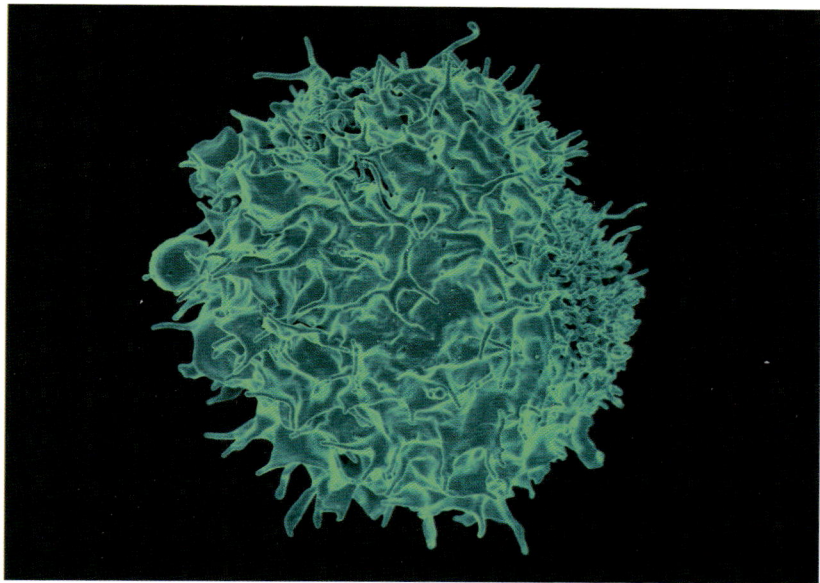

Colorized scanning electron micrograph of a t-cell, a type of lymphocyte in humans that protects against antigens. *National Institute of Allergy and Infectious Diseases (NIAID)/Public Domain*

- Respiratory problems
- Cardiovascular changes (e.g., blood pressure, palpitations, dizziness, fainting)
- Nausea and/or vomiting
- Diarrhea
- Extensive hemorrhaging
- Sensory maladies (e.g., metallic taste, seeing colors, abnormal ringing in the ears)
- Muscle coordination, tremor, or paralysis
- Lack of eye control
- Paresthesias (abnormal sensations such as numbness or tingling away from the sting/bite site, such as tingling of the lips)

Humans are all different. We have different genetic makeup and inherent traits, and different abilities to tolerate and combat envenomation—to a point. Individual variation in the ability to tolerate venom is often referred to as "sensitivity," a term also applied to someone who is allergic to the venom.

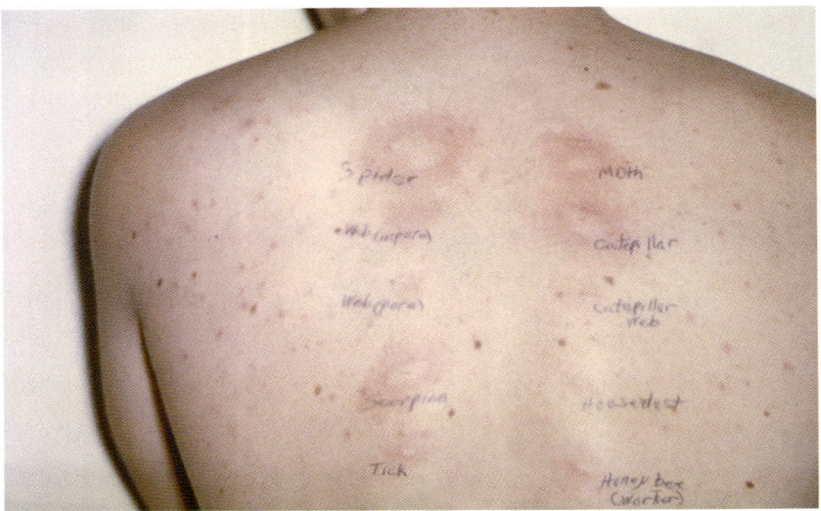

A skin test shows if one is allergic to animal venom. Early diagnosis can be a lifesaver. *CDC/Public Domain*

"Hypersensitivity" is just a term to show extreme sensitivity, and both terms are sometimes used interchangeably. People who are inherently at a greater risk of complications include infants, elderly persons, those with compromised immune systems, and those with allergy to the venom. Medications, alcohol, and recreational drugs can also affect the immune system. Infants lack the weight of an adult so cannot handle the venom doses an adult can, and they do not have a fully developed immune system. This is why an Arizona Bark Scorpion is just an unpleasant ride for a healthy adult, but for a small child, the sting is potentially life-threatening.

When someone is envenomated there will be local and/or systemic symptoms caused by the venom, but it doesn't always end there. There is always a possibility for allergy to the venom and/or antivenom, serum sickness, secondary infection, and sometimes disease transmission. These maladies are not direct effects of the venom itself, but are potential consequences of envenomation, which is why they are often treated independently by physicians.

Allergies to venom may occur when a person has been stung, bitten, or otherwise exposed more than once by the same or similar animal. Exposure can even come from breathing in aerosolized venom, as when a herpetologist is working with snakes in captivity or in the field. Allergy can occur to someone on their second exposure, the hundredth exposure, or never. In the human body, chemicals in venom are recognized as antigens. When someone is first envenomated by a particular animal, antibodies are released to combat the antigens. Antibodies include immunoglobulins (Ig), most notably immunoglobulin E (IgE) and G (IgG). These antibodies are Y-shaped molecules, with the upper tips of the "Y" being the antigen-binding sites. When the antibodies bind with them, antigens can be rendered inert. Specific antibodies begin to build up to combat potential future attacks. However, in some instances on subsequent envenomations, for whatever reason, the body goes overboard and releases an excess of antibodies and chemicals to fight the antigens; this is what causes venom allergy. Ironically, when someone has an allergy to the venom, the human body's attempt to remedy the situation has the opposite effect, causing the body harm.

Symptoms of an allergic reaction are systemic and can include:

- Hives or rash, sometimes swollen and raised
- Skin irritation
- Red, itchy eyes
- Stuffy or runny nose
- Sneezing
- Wheezing and difficulty in breathing.

Anaphylaxis is an extreme allergic reaction that can be life-threatening. It is the cause of most fatalities in the United States and Canada from venomous animals. This is why bees and wasps claim many more lives than rattlesnakes. In the most extreme cases, death from anaphylaxis can occur within minutes, but sometimes it may take several hours. A reaction that is anaphylaxis-like is sometimes called an anaphylactoid reaction, but both are extreme medical emergencies. Anaphylactic shock refers to a sudden drop in blood pressure and the victim loses consciousness. When one or more of the following symptoms (often several) occur, it suggests an anaphylactic reaction.

- Coughing; wheezing; and pain, itching, or tightness in your chest
- Shortness of breath or trouble breathing
- Sudden drop in blood pressure causing fainting, dizziness, confusion, or weakness
- Rapid or irregular heart beat
- Hives; a rash; and itchy, swollen, or red skin

- Runny or stuffy nose and sneezing
- Swollen lips, tongue, throat, face, or hands
- Trouble swallowing, tightness in your throat
- Vomiting, diarrhea, or cramps
- Weak pulse, pallor
- A sense of impending doom

"Cross-sensitivity" is defined as sensitivity to one substance that predisposes an individual to sensitivity to other substances, due to related chemical structure, such as an antigen or allergen. Cross-sensitivity is highest among closely related taxa and unlikely among distantly related taxa. For example, if you are allergic to yellowjackets, there is a good chance you are allergic to paper wasps, but not likely to be allergic to widow spiders. On occasion, however, there may be cross-sensitivity between distantly related taxa. For instance, some people envenomated by Red Imported Fire Ants may become sensitized to Striped Bark Scorpions. Some venomous animals seem to be more allergenic than others. For example, hymenopterans and stinging caterpillars have many known human allergens, so allergy to those taxa is not uncommon.

Serum sickness is similar to an allergic reaction. It occurs when a person reacts to medications, including antivenom. A main difference is that serum sickness can be delayed from 7–10 days after exposure, although it can be sooner. Symptoms may include those of allergy, plus fever, malaise, extreme itching, swollen lymph nodes, joint pain, abdominal cramps, nausea, and vomiting. The patient does not have to have been exposed to antivenom previously, because they are reacting to proteins in the serum (the clear part of the blood). An allergy to horse serum is often the cause. Horses are widely used in the production of certain medicines, including many (but not all) types of antivenom. Some physicians may avoid the use of antivenom because of the relatively high likelihood of complications from allergy or serum sickness. However, others steadfastly believe they need to treat a patient with antivenom, when otherwise warranted. Patients can often be tested for an allergy to specific antivenoms. As one physician put it, "treating patients with allergies is something hospital personnel do every day, so they need to treat for envenomation, and also treat for allergies, should they arise."

Secondary infection is extremely common with envenomation, especially from snakebite and marine stings. A large variety of microorganisms are found virtually everywhere in the environment, and envenomation can

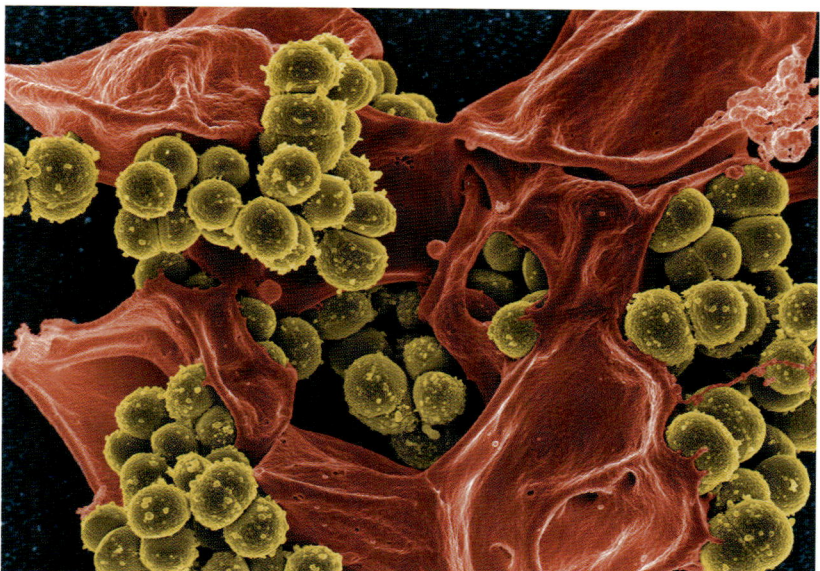

Colorized scanning electron micrograph of MRSA bacteria infecting human tissue. These pathogens sometimes cause secondary infection after envenomation, are resistant to antibiotics, and cause deep-tissue necrosis. NIAID/CCA

be an effective route to infection. Secondary infections can be serious. Tetanus, for example, is rare, but it is a serious and unforgiving disease, caused by the bacterium, *Clostridium tetani*. In untreated cases, mortality rates are high—40–76%. Deep puncture wounds are one of the factors that increases the odds of contracting tetanus, and people have died from tetanus following animal bites and stings. There are numerous cases of human fatalities attributed to venomous animals when the cause of death was actually due to tetanus or massive bacterial infection.

There are many microorganisms that can cause secondary infections following bites and stings. Bacteria are often introduced into the wound immediately, as when one is stung by a marine organism with long spines. Bacteria are common on the skin of marine animals. The most likely culprits of marine infections are bacteria in the genera *Staphylococcus* and *Streptococcus*. Symptoms include pain, redness, and swelling, the signs that the immune system is at work, as well as the formation of lesions and blisters with pus. Infections are often near the surface, but the notorious flesh-eating infections (necrotizing fasciitis) can include deep tissues. Some *Vibrio* bacteria can also cause marine infections, although they are mostly known from consuming seafood. Another bacterium, *Erysipelothrix rhusiopathiae*, is known to cause the so-called fish handler's disease. It does often affect fish handlers who have wounds, because the bacterium is common on the skin of fish, but it can also be introduced through a sting. Symptoms include a purplish swollen area, with a clear area in a circle around it, then another raised, swollen purplish area. The area becomes hot and itchy and is painful to the touch. Like other bacterial infections, it can spread beyond the sting site.

Infection following terrestrial envenomations is similar. Some terrestrial organisms have antimicrobial compounds in their venom, while others can be vectors. For example, bacterial infections from snakebite can include *Staphylococcus* (especially *S. aureus*), *Streptococcus*, *Enterococcus*, *Escherichia coli*, and *Pseudomonas*, among others. Although the bacteria may enter the system with the initial envenomation with the tooth or stinger, it can also develop as the wound caused by venom develops. For example, in snakebite, digestion caused by the venom may go deep into the flesh and provide the perfect opportunity for bacteria to grow. It is not unusual for someone to be envenomated, then when the wound seems to be healing, develop an infection. Infections can be extremely dangerous and can spread systemically, and are rarely, but potentially, life-threatening. Infection can also lead to even more tissue damage, including gangrene. For victims bitten by tissue-destroying snakes, digit and limb loss or disfiguration are not uncommon outcomes.

The Cost of Being Bitten

It is common knowledge among herpetologists working with venomous animals that being bitten can carry a heavy price, not only from the potential damage to tissue (e.g., loss of a digit or limb, or disfigurement), but also from damage to one's bank account. The cost of treatment resulting from a momentary lapse of reason, or accidentally stepping on a snake, can cost more than a house. It is not unusual for the hospital price tag to be more than $100,000. A recent case from Benson, Arizona, made the news because the final bill was about $480,000, due to the cost of sixty-six vials of antivenom and a twelve-day hospital stay. An even costlier case was that of a boy who was bitten in Yosemite National Park (see opening quote for Northern Pacific Rattlesnake and the following photographs)—a whopping $700,000! Indeed,

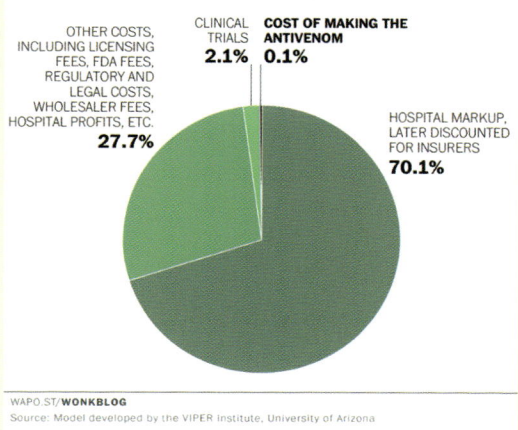

Price components of a vial of antivenom
Percentage of antivenom sticker price, by category

- OTHER COSTS, INCLUDING LICENSING FEES, FDA FEES, REGULATORY AND LEGAL COSTS, WHOLESALER FEES, HOSPITAL PROFITS, ETC. **27.7%**
- CLINICAL TRIALS **2.1%**
- COST OF MAKING THE ANTIVENOM **0.1%**
- HOSPITAL MARKUP, LATER DISCOUNTED FOR INSURERS **70.1%**

WAPO.ST/**WONKBLOG**
Source: Model developed by the VIPER Institute, University of Arizona

The actual cost of making antivenom is miniscule compared to the final cost to the victim and insurance companies. *Courtesy of the VIPER Institute*

many people cannot afford treatment, even with insurance coverage, and at least one death occurred because the patient refused treatment he could not afford. A recent flurry of articles in the news media has begun to alert the public about this. In a recent paper to explain the thousand-fold price markup, Dr. Leslie Boyer of the VIPER Institute cited a *Washington Post* blog titled, "This $153,000 rattlesnake bite is everything wrong with American health care." Her follow-up investigation led to some surprising facts as to why the cost of snakebite (and scorpion stings) is so high in the United States. The same antivenom that costs about $100 in Mexico wholesales for up to $2,300 per vial in the United States, and the retail price is usually much higher.

The actual cost of making the antivenom, which is a tedious process in itself, only accounts for 0.1% of the final retail cost (i.e., what a patient is charged). Clinical trials are another 2.1%. That leaves a whopping 70.1% due to hospital markup, and 27.7% for other fees such as FDA approval, legal fees, wholesaler fees, and hospital profit. Even at $2,300 per vial, the final cost to a patient can be in excess of $10,000 per vial. Part of the markup is intended as a starting point for negotiations with insurance companies, expecting to back down. However, insurance companies do not pay for everything, so the balance, including deductibles, can amount to many thousands of dollars for the consumer. And that is for people who have insurance . . .

(Left) Thirty-six hours after an accidental envenomation by a Northern Pacific Rattlesnake led to tissue damage and a fasciotomy, a procedure that has been hotly debated. (Right) Thirty days and ten surgeries later. The arm and hand were saved. Final bill: $750,000. *Courtesy of Justin Schwartz*

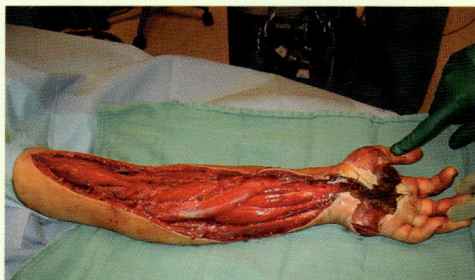
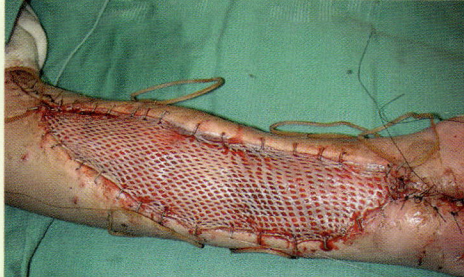

3 Which Animals Are Considered Venomous for this Book?

Technically, there are many more animals considered venomous than I am including in this book, so I made judgment calls on what to include. I include venomous taxa if they meet these criteria:

- They are venomous, rather than just poisonous.
- They are venomous to humans. In general, taxa that are venomous to vertebrates tend to be venomous to mammals, and probably humans. Examples include cone snails that feed on fishes and scorpions that may need to defend themselves against vertebrates.
- They deliver a bite or sting reaching the pain threshold of about that of a European Honey Bee or greater, except as noted below. Hence, most bees and wasps are not even included, although some are mentioned in the "Other Terrestrial Invertebrates" chapter.

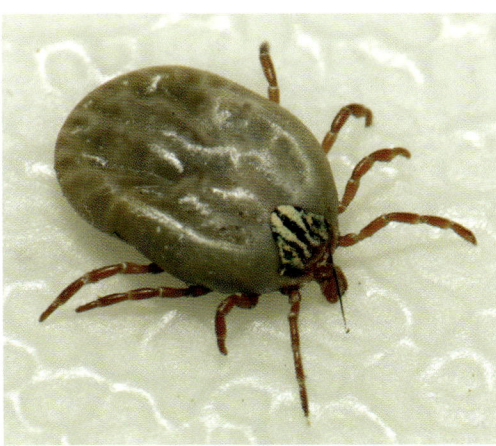

Gulf Coast Tick. Ticks are venomous, but parasitic. This arthropod also causes American Tick Fever in humans. *Katja Schulz/CCA/Flickr*

Which Animals Are Considered Venomous for this Book?   41

Sweat bees have a sting that usually registers a 0.5–1 on the Hymenopteran Sting Pain Index. *Judy Gallagher/CCA/Flickr*

- In the absence of pain, or if it is less than that of a European Honey Bee sting, there are other symptoms that may be disconcerting to the victim. This spans the gamut from a large, itchy rash to life-threatening systemic reactions. Stinging caterpillars and jellyfishes can cause such rashes, and some rattlesnakes may not even hurt much when they bite (save the needle-prick of fangs).
- They deliver a defensive bite or sting, rather than feeding on us. Hence, venomous parasites are not included, except as mentioned in a sidebar. An exception is made for the kissing bugs, because they belong to a venomous group of insects already being covered (assassin bugs), and they sometimes cause venomous reactions.
- There is enough information in the literature on which to build taxa accounts. An example of a venomous animal excluded from a taxon account on these grounds would be the Nemertea, or ribbon worms. The venom is believed to be toxic to humans, but I could find no records of anyone ever having been bitten by one, and natural history information is scant.
- There is a reasonable likelihood that a poorly known taxon is venomous, and it can be included in an existing taxon account.

Some of the more than 300 species of jumping spiders are known to be venomous to humans, but are not medically important, so the whole group is covered in the chapter, "Other Spiders." *L.L.C. Jones*

Venomous Parasites

I would be remiss if I didn't at least mention the parasitic organisms that are venomous to humans. In addition to the kissing bugs, which I do discuss because they belong to a family of mostly nonparasitic species, venomous parasites include: flies (class Insecta, order Diptera), fleas (Insecta, order Siphonaptera), sucking lice (Insecta, order Phthiraptera, suborder Anopleura), bed bugs (Insecta, order Hemiptera, Cimicidae), and mites and ticks (class Arachnida, subclass Acari). There are other internal parasites (e.g., roundworms) and parasitic microorganisms (e.g., swimmer's itch) that cause distress in humans. Most of these do not fall into the "at least as bad as a bee sting" requirement to be in this book, but are technically venomous, and they do sometimes cause distressing symptoms. With many of these parasites, there can be disease transmission, and people can also develop sensitivity and allergy to their bites.

There are many types of parasitic flies, including biting midges (e.g., no-see-ums; Ceratopogonidae), mosquitoes (Culicidae), stable flies (Muscidae), snipe flies (Rhagionidae), sand flies (Psychodidae), black flies (Simuliidae), and horse and deer flies (Tabanidae). Some of these, such as mosquitoes and no-see-ums are not particularly painful when they bite, but do cause

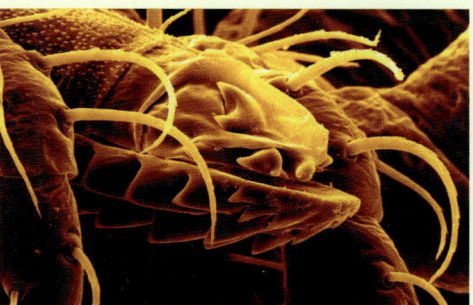

Colorized scanning electron micrograph of the mouthparts of a tick. *NIAID/Public Domain*

Dog Ticks are technically venomous and can transmit disease. *USGS, BLM/Public Domain*

maddening itching, which may be accompanied by welts or rash. The bite may not be noticeable at first because their venomous saliva contains pain suppressants, which allows them to bite and feed before being noticed—advantageous for a parasite. Others, such as some snipe flies, stable flies, and tabanids are quite painful soon after biting. Often, these flies have an anticoagulant in their saliva to allow blood flow for feeding. Flies may also transmit disease. This is well known for mosquitoes, which can carry malaria, yellow fever, dengue, Zika, and West Nile virus. Robber flies bite and are venomous, but not parasitic, which is why there is a chapter in this book that includes them.

Fleas are well-known parasitic biters, and most people know them when their pets get infested. The domestic fleas include dog fleas, cat fleas, and human fleas. These will all bite humans. In addition, there are a variety of other flea species that primarily parasitize wild animals, and these may bite humans and their pets, as well. Wild fleas affecting rodents may carry the infamous plague virus, a disease that is potentially fatal to humans. Fleas inject a hemorrhagic venom in their saliva, which causes pain, redness, itching, and welts. Sucking lice are similar to fleas because they may spend their entire life on a single host, and feed on their blood. Two common species include the head louse and body louse. Bed bugs also bite humans, and their incidence is on the rise, partly because they are becoming resistant to pesticides.

Chiggers are the parasitic life-stage of an otherwise free-living mite. They tend to be most common in areas that are warm and humid, such

This kissing bug, *Triatoma recurva*, is one of the only parasites included in this book—only because some people occasionally report venomous symptoms, and also because it is in a family of insects (assassin bugs) that are otherwise venomous. Kissing bugs sometimes transmit Chagas Disease, though very rarely north of Mexico.
L.L.C. Jones

as the southeastern United States and Midwest. They are mostly limited to summer. Chiggers are another one of the venomous parasites that we would consider a maddening pest. They bite the host, inject digestive enzymes, and then drop off after feeding. There is a commonly held misbelief that they burrow into the skin, but the marks they leave are feeding areas, not entry points. The bite may not be noticed until well after they have exited, and is persistently itchy, often noticed by red bumps in the areas of clothing constriction, such as sock and belt lines. Their larger cousins, ticks, are even more infamous for their parasitic activity on humans and pets. They are known to transmit diseases, including Lyme disease and Rocky Mountain spotted fever. However, there is also a malady called tick paralysis, which is a possible outcome of their venom. When members of the tick genus *Dermacentor* feed, they inject their venom/saliva, which contains a neurotoxin. The more time that a tick is allowed to feed, the more neurotoxin enters the body. Symptoms usually occur within a few days of being bitten, and include numbness and weakness in the legs, followed by paralysis that will spread to the rest of the body. Usually, when the tick is removed, the symptoms will reverse, but not always. In extreme cases, paralysis can result in respiratory failure and death.

4 Precautions

Precaution is a much better option than the expense and trauma of dealing with an envenomation in the first place. There are two things to consider when deciding which precautions to take: the geographic setting and the types of creatures that inhabit the area. For example, when hiking on a trail in a woodland, one would consider threats from terrestrial animals below (e.g., snakes) and above (e.g., stinging caterpillars). Boots and snake gaiters would help guard against snakebite, while a wide-brimmed sun hat and long-sleeve shirt should help against caterpillar stings. When visiting a seashore, there are different potential hazards, such as cnidarians and stingrays. In this case, a stinger (anti-jellyfish) suit is good protection, and doing the stingray shuffle would be in order. Wherever one goes, it always helps to be with others, and to let a responsible party know where you are going and when you intend to return (and when to start making phone calls). One should always have a charged cell phone at the ready, but only satellite phones can reach emergency care numbers in remote areas.

People who begin to develop allergic reactions to venom, even if not serious, should be tested for allergy, and consider immunotherapy, the process of becoming desensitized through a series of antigen injections. For example, a common early sign of becoming sensitized to a particular type of venom may be more swelling than previously experienced, or you may get hives or become dizzy. If you are known to have or suspect an allergy to venom, you should always carry an epinephrine kit (and know how to use it) and have antihistamines in a first-aid kit.

In strictly urban areas, there are relatively few threats from venomous animals, but as one moves to the suburbs and interfaces with natural areas, encounters increase. Your home, yard, garden, garage, shed, woodpile, bird feeder, privy, and greenbelt can be attractive to many kinds of creatures. In northern areas, the diversity of venomous animals is relatively low, although this doesn't necessarily equate to safety. For example, in the coastal areas of northwestern United States and southwestern Canada there are no scorpions, rattlesnakes, black widows, recluses, harvester ants, Africanized Honey Bees, etcetera. However, this is one area where yellowjackets almost make up for the lack of all other venomous species. Suburbanites along the southern U.S. border are more likely to come into contact with a host of potentially dangerous animal taxa. For example, in my suburban home in Tucson, my house and yard list includes centipedes, scorpions (including Arizona Bark Scorpion), Western Black Widow, Arizona Recluse, agelenid spiders, tarantulas, various other spiders, paper wasps, tarantula hawks, velvet ants, harvester ants, Africanized Honey Bee, carpenter bees, assassin bugs (including kissing bugs), Western Diamond-backed Rattlesnake, and Mojave Rattlesnake. If you add my nearby dog-walking trails, I have seen stinging caterpillars, Gila Monster, Tiger and Western Black-tailed Rattlesnake, Sidewinder, and Arizona Coralsnake.

Additional sources of exposure can come from keeping native and exotic animals as pets, traveling to unfamiliar areas, and when you have a workplace hazard due to your occupation. Regardless of where you find yourself, there are three general precautions that apply across the board: (1) Be aware of your surroundings, (2) have some knowledge of the types of venomous animals that occur where your activity is, and (3) use common sense. These three things are kind of the point of this book.

Home

The yard is usually the source of venomous animals inside the home, so a logical precaution is to seal the home from the entry of small creatures. Scorpions, spiders, wasps, bees, kissing bugs, and even snakes are adept at getting in through small cracks and holes, so sealing potential access points is in order. Flying insects may enter the home through openings above ground level, such as an awning. For example, I had Africanized Honey Bees enter through the awning of the roof, and they nested in the attic. Later they swarmed into my living room through a light fixture. Basements and attics are notoriously attractive to any manner of photophobic animals. These areas should be checked periodically. Populations of certain spiders, including Brown Recluse, can be high in these situations.

Keeping venomous creatures out of the home is an important precaution in scorpion country. *L.L.C. Jones*

Despite your best efforts, some creatures will sneak inside anyway, through unseen cracks, open doors, and drains. One commonly employed method that I cannot recommend is the use of sticky traps. If an arachnid or insect steps on it, they will be captured, but so will a plethora of harmless animals. The area they serve is only a needle in a haystack, anyway. For scorpion control, repeated sweeps through the home with a black light at night is effective to detect them when they fluoresce. Scoot the arachnid into a container and do what you will with it. Most domestic spiders are also nocturnal, so searching for them when it is dark can be effective, and one can look for their webs and egg sacs during the day. See the Encounters section of Taxa Accounts to know where they may be. Spiders often enter the home through drainpipes, so it is just a matter of inspecting the basin before washing up.

Perhaps the most effective way to guard against envenomation in the home is by keeping it tidy. Many bites and stings happen when arthropods seek shelter in clothing, towels, sheets, and areas of clutter. Scorpion envenomation often happens when people trod upon them, so shoes are always warranted in homes harboring scorpions—but shoes should not be left on the floor, as scorpions may climb into them. Flip-flops usually (but not always) suffice. Spiders may also seek out the cover of shoe darkness and humidity. Bark scorpions climb very well, so don't allow bedclothes to touch

Sticky traps are environmental nightmares. This trap got one scorpion, but also killed a kingsnake (which eats rattlesnakes) and three lizards. No doubt most scorpions in the area went unharmed. *Bryan Hughes/Rattlesnake Solutions*

Once in the home, spiders and scorpions seek shelter in dark areas. A tidy house is a safe house. *L.L.C. Jones*

the floor. Many exposures from these animals come from rolling over on them in your sleep. Centipedes, spiders, and kissing bugs also climb well.

Yard, Sheds, and Gardens

The risk of envenomation is usually greater in the yard than in the home. Yards are very attractive to a number of animals because of increased water, humidity, and structures to hide in. Some people living in suburbia like to wall off their property, and in some instances, this is a deterrent to certain venomous animals, but not all. Gates may be entryways for animals as large as snakes. Many animals climb walls well, while others can fly or even balloon in (i.e., spiderlings). Somewhat jokingly, rock walls have been called "the preferred habitat of bark scorpions." As in the home, clutter can be an attractant to creatures seeking hidey holes. Structures such as sheds should always be approached cautiously, as any number of venomous vermin may reside in or near them. Woodpiles are virtual apartment complexes for snakes, spiders, scorpions, centipedes, wasps, and other creatures. Woodrats build their own woodpiles and are an important reservoir for kissing bugs and bark scorpions. Gardens provide leafy structure, food, prey, water, shade, and humidity. Bird feeders may seem innocuous, but birds drop seeds on the ground, then feed on the ground—something that rattlesnakes seem to hone in on. Lawn furniture and certain types of walls can be a magnet for widow spiders.

To avoid being tagged by a denizen of the yard, it is always wise to approach with caution, starting before you enter the yard. As an example, one of the Envenomation Stories in this book involves a man who stepped out of his house and onto a rattlesnake. If he had scanned the area before entering his front porch, or worn boots, he likely would not have been bitten. Observing the area you will be in, with eyes and ears, is a necessary first step. Just as an EMT or Good Samaritan would do at a crash scene, look to see if it will be safe for you before proceeding. If you see cobwebs, you know widows may be present; cobwebs are a specific type of web made only by members of the family Theridiidae, so it helps to know the different web types. If you hear buzzing, you should pay attention to flying insects. If threats are detected, you may need to take action. For example, if you see a rattlesnake or honey bee swarm, you should call a professional, if you cannot safely deal with the threat yourself.

If there are no obvious threats detected, you can proceed to do your yardwork, which brings up the yardwork mantra: "gloves and boots." Latex gloves protect against nothing, while leather gloves do. There are also

Snake repellent is really "snake oil," and an environmental hazard for people and pets. L.L.C. Jones

puncture-resistant gloves (none are puncture-*proof*) that can increase safety. A long-sleeved shirt and long pants are usually good for added protection. However, some insects are adept at getting *under* clothing, such as sleeves, to gain a purchase to sting, especially among the Hymenoptera. Stinging caterpillars may fall out of trees and get lodged in the collar. Fire ants and harvester ants may get under pant legs to sting (see the trou-dropping accounts in the Envenomation Stories section). Boots should be made of leather and extend above the ankle. Many people have been spared a traumatic ER visit and horrific medical bill by just wearing boots and jeans. Gloves, socks, and boots should be kept indoors when not in use and elevated beyond the reach of most arthropods. Then they should be shaken out before donning. Some people are tagged when they put boots or gloves on, particularly if they are stored outside or in a garage or shed.

Many homeowners spray the yard and perimeter of their home with pesticides, but there are concerns about the widespread use of environmental poisons. Spiders may not even come into contact with pesticides, as they are elevated above the surface, and flying insects certainly do not. The efficacy of prophylactic application of poisons for scorpion control is poorly known. Of the many people I have spoken to about this, some say it helps, while others notice no difference. Certain products do more to contaminate the area than

Yellowjacket traps attract and kill their target without poisoning the environment.
L.L.C. Jones

controlling pests. For example, so-called snake repellent is more like the infamous "snake oil." There is no scientific data to show it works, although you will find plenty of people who swear by it. To use it, you basically broadcast the hazardous chemical in mothballs (naphthalene or paradichlorobenzene) into the environment where it can cause ecological harm and risk the health of children, pets, and wildlife. Another group of broadcast chemicals are those used to poison ants. Usually, they do not even target ants, but kill any manner of animals that come into contact. These poisons can cause a great deal of ecological harm and risk the health of humans and pets coming into contact. It is truly amazing how the big box stores carry rows of bags of these poisons to protect you against fire ants, even where they don't occur. Repeated use of hot water will actually control fire ants. After the first round of several treatments, the stragglers will start new colonies, and these need to be treated, too. Eventually, they will be eradicated, but it will have to be long-term maintenance, like any control measure, including chemical poisons.

If poisons are to be used, they should target specific taxa and be safe for the environment. For example, insect sprays do not work on spiders. For a spider spray to work, you need to spray the actual spider and its egg sacs, rather than the environment. A single female may construct several egg sacs per season, each with hundreds of eggs. Broadcast spraying may do little good, as the females are often in a safe spot during spraying, and their body does not come into contact with the surface. Brown Recluse Spiders prefer to be indoors, rather than the yard, but the other *Loxosceles* species are more often out-of-doors, where they are virtually never a problem. Other spiders can just be lived with. Keeping scorpion numbers down in the yard will also keep numbers in the house to a minimum. Besides professional control services, there are also traps that target aculeate Hymenoptera; these may use poison bait, but it is not broadcast into the environment.

Reptiles are another group of animals that may pose a risk in yards. If one appears in your yard, contact someone skilled in snake removal. There are services for this. Some fire departments still do this, but many do not, especially considering the number of calls for harmless snake removal by people that do not know what venomous snakes look like, or because they are ophidiophobes. If you choose to remove them yourself, you should be trained on safe practices first. A surprising number of people are bitten when they try to move or kill snakes. Rattlesnakes should not be killed anyway, as they are important predators of rodents, which may carry diseases that can be transmitted to humans. However, translocating them far away is usually a death sentence for the snake. There are numerous studies on this.

Greenbelts, Parks, and Wilderness

The out-of-doors is where most types of creatures are likely to be encountered, including centipedes, hymenopterans, mildly venomous spiders, Gila Monsters, and snakes. Nature enthusiasts may be more aware of their surroundings when hiking than at home, as they must always watch where there boots land, and should always be cautious about where they place their hands. Insects such as yellowjackets, paper wasps, harvester ants, assassin bugs, and stinging caterpillars are always a potential hazard. Yellowjackets may nest in paper nests above ground level, but more often are in subterranean nests. There is always some level of insect buzzing in forests and woodlands, but when the nest is disturbed, they get aggravated and the volume of buzzing increases—a warning before they attack en masse. When the level of buzzing increases, it means it is time to start running. After a couple hundred meters, yellowjackets quit the chase, while Africanized Honey Bees can chase much farther before giving up.

This paper wasp may enter this unprotected beverage. A mouthful of stinging insects is most unpleasant. *L.L.C. Jones*

Africanized Honey Bees are a potentially serious risk, especially when in remote areas. If they think the nest is being threatened, they are persistent, and people can receive a fatal dose of venom from hundreds of stings. If you encounter a wild-bee hive, avoid it with great caution. They are often "set off" by sounds and vibrations, especially motors, but other disturbances will incite them to attack en masse. One thing to remember is there is often a warning sign where some individuals will bump into you. This means leave. If they do attack, you need to get to cover, like inside a car. If you have a hundred bees stinging your face and neck, it may seem counterintuitive to seek shelter, but if you keep running, that could become five hundred to a thousand or more bees. Unless allergic, you will survive a hundred bee stings, but a thousand bees may prove fatal. In one Envenomation Story, a victim survived by being submerged in water for some time, but that is risky, as bees and wasps will hover and wait for you to come up for air. Again, better to make it to your car or a building, if you can. Bees and wasps can also be hazards while picnicking. It is prudent to put covers over your food and drinks.

Gila Monsters and coralsnakes almost never bite anyone that isn't messing with them, but that is not the case for pit vipers. They blend in well with the background, so many exposures happen when snakes are trod upon or

These simple picnic covers can keep wasps and bees out of your food and drink.
L.L.C. Jones

within striking distance. If you try to interact with a snake, your chances of being bitten increase precipitously. Many, and possibly most, venomous snakebites are from someone handling or attempting to kill them. There are many examples of bites from "dead" snakes, as even with the head removed, their nervous system is still functional. The best defense is to stay on trails, where snakes are more visible than in the bush. They are almost impossible to see in dense habitats like prairies, chaparral, and thorn scrub. In these cases, snake boots, chaps, or snake gaiters are recommended. If you are searching for scorpions with a black light, remember, snakes don't glow, so snake protection is essential. If scorpions are out, so are snakes.

Streams, Rivers, and Ponds

In North America, there are few venomous freshwater animals. Among invertebrates, there are only a handful of insects. The main precaution is to not handle them, and most bites are due to an occupational hazard. If walking through streams and ponds, it is always a good idea to wear heavy water shoes or boots. A bigger threat to the feet comes from broken bottles, fish hooks, sharp rocks, and branches, so thick-soled water shoes help anyway. Among vertebrates, are a few more taxa: catfishes and a couple pit vipers.

These nature lovers were well prepared in rattlesnake country with snake gaiters.
L.L.C. Jones

Madtoms are so small, they are functionally like insects, because they are hidden among aquatic plants, rocks, limbs, or debris. Stings usually happen as an occupational hazard, or by unseen individuals. Most stings of other catfishes occur to anglers, who usually learn to avoid the dorsal and pectoral spines. Heavy gloves may help, but also reduce the angler's dexterity. Smaller catfish have sharper spines and are more flexible than larger individuals. On small to medium catfishes, there are a couple options for hook removal. The "lip grip" is when you put your thumb in their mouth, and push against the clenched hand below. Another option is to grab the fish behind the spines, either with your thumb and forefinger around the body, or by grasping their body with the forefinger and index finger straddling the dorsal spine, and thumb on the venter. Large catfishes are not usually such a hazard, because the spines are not sharp, and they can be detained in a net while needle-nosed pliers or fishhook removers do their job. However, the spines are large and stout, so can cause considerable damage. One interesting method of catching catfish is noodling, where anglers reach into overhangs with their hands and pull out a catfish by the mouth. Gloves are in order. Noodling has its unique set of hazards, but to most anglers, the scrapes and stings are usually worth it.

Cottonmouths and Eastern Massasaugas are aquatic to semiaquatic pit vipers, so can be a potential hazard to those entering their environment. They are often encountered near the edge of the water, so that is an area to be particularly aware, and snake protection is warranted along the edge of aquatic systems. There may also be other venomous animals near the shore, so caution is always advised. It is well known that cottonmouths can bite underwater, but as best I can tell it is a rare event and is difficult to prepare for anyway.

Seashore, Estuary, and Ocean

Estuarine and marine waters have many more venomous animals than freshwater. Among invertebrates are mollusks, polychaete worms, and cnidarians. It is easy to avoid being stung or bitten by mollusks (i.e., octopuses, cone snails, and nudibranchs): don't handle them. It is rare that someone would be randomly tagged while just swimming. In our area, the cnidarians are by far our biggest nemeses. Stinging cnidarians of some sort are found in every ocean and bay along all our coastlines, including Hawaii, from Mexico and the Caribbean to the Arctic Sea. Although some polyp-stage cnidarians can sting, the biggest problems are the medusae of jellyfishes, clinging jellyfishes, men-of-war, and box jellyfishes. Thousands of people are stung every year—and sometimes, hundreds or possibly thousands of people can be stung in a single day! This usually happens in the summer when cnidarian blooms coincide with beach recreation. The best precaution is to stay out of the water when there are blooms. If that is not a personal option, then there are two other choices: wearing an anti-jellyfish (stinger) suit or using an anti-sting ointment. Stinger suits are more commonly worn in Australia, where there is a prevalence of dangerous box jellies. They also filter out UV radiation and can be quite fashionable. Anti-sting ointments may be in the form of a sunscreen cream. These work well for some cnidarians, and not so well for others. Be sure to check to see what the latest and greatest product is, and if it has been scientifically proven to be effective. Do not rely on the company's advertisements. Some ointments and sprays are to be used *after* being stung, so make sure to read the label.

Stinger or anti-jellyfish suits haven't really caught on in North America but are required on some tours in Australia. *Courtesy of EcoStinger*

A host of venomous vertebrates are found in all our marine and estuarine waters. Horn Sharks, stingrays, sea catfishes, typical scorpionfishes, and lionfishes are mostly tropical to warm-temperate, while chimaeras, dogfishes, lantern sharks, rockfishes, and thornyheads tend to be cold-temperate or in deep waters. By far the biggest threats to swimmers are the many types of stingrays. The vast majority of stings happen when someone steps on them, although some stingrays are more likely to be encountered when swimming, and one is pelagic. There is a well-known precaution for species that rest or hide on the bottom: "the stingray shuffle." Stingrays feel vibrations in the sand and water, so will avoid contact with humans if given sufficient warning. The stingray shuffle consists of walking slowly and accentuating vibrations on the bottom by pounding the sand with your feet. Anglers in the shallows may use a stick to prod the bottom as they move along. Anglers should be extremely cautious with stingrays that are landed. By using barbless hooks, the line can be cut, and the angler need not deal with a thrashing ray onboard. Anglers may also catch rockfishes and

Dr. Angel Yanagihara (see Pacific Cnidarian Research Lab profile) with the Sting-No-More cream she developed at her lab. She is still in her stinger suit, but as she explains, it is "my safety layer UNDER my 1 ml neoprene wet suit. 'Stinger Suits' were not enough for the work I was doing. I was hauling out these huge lethal stingers with 60, 10-foot-long tentacles, so my level of precaution was extremely high."
Courtesy of Angel Yanagihara

Anglers are at risk of envenomation by any number of marine animals, such as this rockfish, so care should be taken in handling and removing hooks. *"Carolyn"/CCA-EQ/Flickr*

scorpionfishes, which are targeted for food fare, and sometimes sharks, so these all must be treated with respect and handled carefully, if at all. Gloves may help, but not always, as the spines are sharp and stout. On rare occasions, some stingrays may accidentally leap into a boat. If this happens, quickly move away from the animal and give it a wide berth, while you formulate a logical plan. On boats, most stings happen when people try to get them back into the water. They are world-class thrashers and have good aim with their caudal appendages. Extreme caution is warranted if you must evict a stingray onboard.

Yellow-bellied Seasnakes are rare off Hawaii and southern California. If you are one of the lucky few to find one, precautions are simple: don't pick them up. If they have washed up on shore, they are dead or going to die, and local museums should be interested in adding them to their collection. Animals in collections yield a great deal of information on the life and natural history of species. In this case scientists can compile information on sex, diet, migration patterns, reproductive state, parasites, general health, cause of death, and other traits of seasnakes outside their normal range.

Collecting and Keeping Venomous Pets in Captivity

The wild animal pet trade is a multi-billion-dollar business, and smuggling of illegal exotics is second only to the illegal drug trade. However, this is not a forum to discuss poaching, which is indeed an ecologically heinous crime. There are probably tens of thousands of pet owners in the United States and Canada who legally collect and/or keep venomous animals, native or exotic, as pets. There are federal, state, and local regulations about keeping venomous species (especially snakes), but for many animals (especially invertebrates), there are simply no regulations. Popular invertebrates include *Scolopendra* centipedes, scorpions, and large spiders (especially tarantulas). Centipedes and scorpions can be safely handled with tongs, but they must be long enough to keep the stingers away from fingers. Puncture-resistant gloves can be useful. Both of these types of arachnids may be quick to defend themselves, so should not be free-handled. Many collectors and free-handlers pick up scorpions by the telson, but this is a risky technique and many people are stung while doing this bare-handed. Some exotic scorpions have the ability to take human life with one quick jab. Native tarantulas are timid, disinclined to bite, and are not particularly venomous to humans, so can usually be handled with care—although they may be quick to deploy their urticating hairs. However, some exotic tarantulas and other large spiders may be highly venomous, extremely fast, and inclined to bite, so pet owners need to do their homework (e.g., see the baboon spider introductory quote in the Tarantulas account).

In 2007, there were an estimated 586,000 pet snakes in the United States. An unknown percentage of these (certainly the minority) were venomous, called "hots" in the pet trade. Virtually every species of venomous snake is desired as a pet by someone. Especially popular native species seem to be the Arizona and New Mexico sky island rattlesnakes (*Crotalus lepidus, C. pricei*, and *C. willardi*), which require a special permit to collect and also to keep, although many are illegally obtained and kept. Others include diamond-backed rattlesnakes, Southwestern Speckled Rattlesnake, black-tailed rattlesnakes, Timber Rattlesnake, massasaugas, Pygmy Rattlesnake, and copperheads. In addition to these are any number of exotic species, such as vipers and cobras; the latter seem to be particularly attractive to fanciers of exotic hots. There have been hundreds of bites from exotics, but they are not always reported. From 1990–2008, there were 16 deaths by captive snakes in the United States, both by native and exotic species. Gila Monsters and beaded lizards are also esteemed as pets.

Hobbyists who collect and/or maintain venomous snakes and lizards are usually bitten when they let their guard down or are taking unnecessary risks. When animals are collected from the wild, there is always a risk of being tagged, but capture with quality snake tongs and retention in a secure bucket is usually the safest approach. However, snakes can writhe, people can stumble, and accidents can happen. The bucket should have a lid that is easy to secure and has carefully placed air holes. Snake sticks for pinning and lifting snakes are not as trustworthy as tongs. Holding the snake by the neck and head is risky. Even professionals get bitten occasionally. "Tailing" a snake involves holding it by the tail, away from the body, and monitoring its head and movements. This is unhealthy for the snake and risky for the handler, and many people are bitten while attempting it. Some people handle venomous snakes with puncture-resistant gloves, but these are not puncture-proof. The worst possible method of capture is called free-handing, or free-handling, wherein the captor picks up the snake without protection, either in a quick motion, or very slowly to not upset it. Again, many people are envenomated this way. Snake bags are not terribly safe for transporting snakes, as they may bite their captor through the cloth. One thing that I have noticed when compiling the Envenomation Stories is that people have been bitten by snakes during all phases of capture, even using methods that are generally thought of as safe, including tailing, using snake hooks, using tongs, transferring to bags, carrying bags too close to the body, and pinning the head. I have not heard of anyone bitten or sprayed while the snake was in a bucket, although I'm sure it has happened (e.g., lids can come off if not securely locked). Researchers have also been bitten while tubing a snake, a method that is generally thought of as safe.

In the domestic terrarium, people are usually bitten while feeding and watering their pets, cleaning cages, or handling them. Precautions need to be used whenever a hobbyist is interfacing with venomous creatures. Large snakes should be kept in large tanks, where the serpent can be secured in the opposite end by a partition. Even the calmest snake may bite on occasion, particularly if they are hungry.

Home aquaria, particularly marine aquaria, may also house venomous animals. All hobbyists seem to know that some animals are venomous, such as lionfishes, and indeed, many people are stung by them. Unlike hobbyists who keep hot snakes, those with lionfishes in the home are not usually thrillseekers or wanting to live life on the edge. Rather, they are homeowners who think the fish is beautiful. They do not consider having them a risk because they have no intention of handling them. Accidental injury happens when they do not see where the lionfish is when cleaning the aquarium. They may be unaware that a lionfish will quickly dart and jab its spines into an intruder they view as a threat. Precaution: always locate the fish before putting your hand in the tank, and make sure it is at a safe distance. Some other venomous fishes in aquaria include tangs, rabbitfishes, and plotosid catfishes. Less common species that are unregulated may show up in retail aquarium stores, including deadly blue-ringed octopuses, stonefishes, and stingrays. Aquarists should always be careful when aquarium fish must be netted, and they should never be handled.

Occupational Hazards

Certain jobs may put one inherently in harm's way, and precautions for occupational hazards are very similar to those for keeping exotic pets in captivity. Many of the Envenomation Stories in this book document accidental exposure to professionals with occupational hazards. This is partly because I am a biologist and nature buff, so I know many other biologists who do lab and fieldwork or keep venomous animals in captivity. However, people working with venomous animals are more at risk of envenomation than the public.

There are many different occupations that require workers to interface with venomous animals, such as zoo and aquarium personnel, entomologists, beekeepers, fisheries biologists, herpetologists, university biology departments, aquaculturists, marine biologists, field researchers, parks staff, litter control crews, pest management organizations, land management agency field crews, surveyors, agricultural workers, and so on. There are literally hundreds of professions that require their employees to be at some sort of risk of envenomation. For those who are intimately associated with venomous animals, they should be prepared to deal with exposure. For example, researchers and technicians who milk snakes for a living or maintain them in captivity for public display should have emergency procedure protocols in place. This should include a plan to get employees immediate medical attention by a staff trained in snakebite, knowing at any given time where stores of antivenom are, and how to transport them (if needed) in a timely manner. This may seem like a no-brainer, but in my research I have found many instances where there is not a plan in place in the event of envenomation. The fact that there are so many instances of envenomation during occupational activities should demonstrate the urgency of proper planning.

The example of a professional snake milker is at the far end of the risk spectrum; most exposure comes to people who are merely doing their job that does not actually involve their working directly with venomous animals. An example of this comes from my days as a field technician with the USDA Forest Service. As discussed in my Envenomation Story about exposure to yellowjackets, I became very aware of the danger they imposed. In the Pacific Northwest, the number 1 reason for sick leave from fieldwork in the Forest Service was due to yellowjacket stings. Yet, there was never any training on how to avoid being stung, what to do if stung, or the consequences of a severe allergic reaction. At best, there was brief mention of stinging insects and the concept of allergic reactions in a one-day first-aid/CPR course, which was required every few years. To help correct the issue, I organized a brief training program for fellow employees at a scheduled safety meeting. There are many resources available to conduct such training programs and safety protocols. One useful tool we had in the federal government was a Job Hazard Analysis (JHA), which was essentially a safety protocol outline. Anyone can use such a tool. It basically outlines all potential hazards that are likely (or even not-so-likely) to be encountered on the job, and what steps can be taken to avoid exposure. However, it needs to go to the next step with a protocol for what to do if an exposure occurs. The JHA used the concept that "there is no such thing as an accident," but this is unrealistic.

Traveling

In this book, I often mention animals of areas just south of the U.S. border, because people living close to the border often travel to getaways in northern Mexico and the Caribbean. For them, this can be less than a day's drive or a short boat or plane trip. Many of the bites and stings I hear about in Arizona are from friends who have taken short trips to the Gulf of California. I have not written much about trips farther abroad, but whenever people travel to new destinations, especially tropical ones, they should become aware of the types of venomous animals they may encounter in their travels, *before* they go (not to mention diseases and parasites). Read the information in the chapter on the American territories or nearby islands if traveling there. Some of the venomous animals from North America north of Mexico, and bite statistics, pale in comparison to those elsewhere. Here are some highlights, but this is just scratching the surface and each destination is different:

Mexico: One thousand people per year die of scorpion stings; Chagas disease is readily transmitted by kissing bugs; there are more species of rattlesnakes than anywhere else; there are numerous species of large coralsnakes, some of which do not adhere to the "red-touch-yellow, kill a fellow" rhyme.

Central and South America: Similar to Mexico, with some different species. Animals here include the deadly wandering spiders (*Phoneutria* spp.); Chilean Recluse Spider; the world's most dangerous stinging caterpillar (*Lonomia* spp.); Bullet Ants; large lance-head vipers, including the Bushmaster, which may reach 4 m; and many coralsnake species. Yellow-bellied Seasnakes are common. There are numerous dangerous marine creatures and fresh-water stingrays, which are more dangerous than marine counterparts.

Africa, Asia, and Middle East: There are many deadly buthid scorpions. Deadly snakes include Saw-scaled Viper, Russell's Viper, Gaboon Viper, cobras, and mambas. India alone has 11,000–19,000 deaths per year from snakebite. Southeast Asia has land and sea animals similar to Australia (see below), with many of the same species, especially the noteworthy southern West Pacific species.

Australia: Despite its crazy number of venomous animals of land and sea, human fatalities are few. Notable venomous animals include: numerous spiders, including the infamous funnel-webs; Australian Box Jellyfish, the most venomous cubozoan; Irukandji-inducing box jellyfishes; blue-ringed octopuses; Geographer Cone (among other deadly cones); stonefishes; more elapids than harmless snakes, including most of the world's most venomous species, such as Inland Taipan (world's most venomous), Death Adder, Tiger Snake, Brown Snake, and many highly venomous seasnakes.

Blue-ringed octopuses may be common in Australian tidal pools, and can deliver a deadly bite, so should never be handled or kept in aquaria.
Rickard Zurpre/CCA-SA/Flickr

I Like to Hang Out in Restrooms

No, I am not a pervert (unless you consider looking for venomous animals to be a perverse activity). Old buildings and other structures can attract a wide variety of animals that use them for nesting and cover. Outdoor restrooms, be they outhouses or campground facilities, in particular, seem to attract a variety of animals, including venomous ones. Part of the reason is because they may get infrequent use and maintenance, but also because these structures may be lit at night and insects may be attracted to the odor. Insects attract animals that prey on them, which in turn attract animals that prey on those predators. Spiders and scorpions are often seen in restrooms. Awnings of buildings make an excellent area for web-building by spiders and nest-building by paper wasps. The interior of the structure is also attractive, as there may be moderated temperatures during times of extreme weather, and it is likely to be more humid than the surrounding area. Widow spiders are notoriously common in outhouses, and it is unfortunate that they find the underside of the toilet rim an ideal place to catch flies that visit the pit for a bite to eat. This is reflected by the relatively large percentage of people bitten on the genitalia.

So, while looking for photo fodder for this book, I hung out at restrooms. I have seen a variety of interesting animals including countless widow spiders, orb-weaving spiders, tarantulas, scorpions, paper wasps, European Honey Bees, large carpenter bees, *Scolopendra* centipedes, and even a Cottonmouth. So, if you like to view venomous animals, you might consider hanging out at restrooms.

This Florida Cottonmouth was at the entrance of a restroom in the Everglades. *L.L.C. Jones*

Inside I was greeted by a Florida Keys Giant Centipede. *L.L.C. Jones*

5 First Aid

First aid can be done by the victim or a companion at the scene of the incident, or shortly thereafter. Technically, first aid can be considered what a first responder, such as an EMT, will do, but that is beyond the scope of this book. First aid is not the same as treatment, which is done by a medical professional. I will not attempt to tell an EMT or physician how to treat a victim or patient—that is their job. However, I will offer some general input for medical personnel to consider. One thing I will advocate to all readers is having an up-to-date tetanus vaccination.

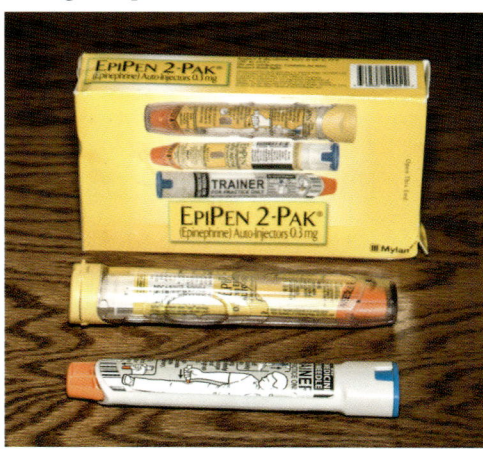

An epinephrine kit can be a lifesaver for people who could go into anaphylaxis. *L.L.C. Jones*

If someone is known to be allergic to venom, they should always carry an epinephrine kit, and use it within a minute of being stung or bitten. Anaphylaxis can have a very quick onset, so it is best to administer epinephrine within a minute of noting symptoms, before you begin to lose cognitive abilities or even consciousness. After using an epinephrine pen, immediately seek medical attention. It is also a good idea to train companions on how to administer it to you ahead of time, and give them written permission to aid you, should you become incapacitated before self-administration.

If someone is in a remote setting and they find themselves in a life-threatening situation (e.g., rattlesnake bite or anaphylaxis), they need rapid medical attention. Someone asked me what to do if they were bitten by a rattlesnake 15 km up a trail and out of cell-phone range—should they hike out, get carried out, or send a runner? I would say none of these is the best option—calling 911 with a satellite phone is a much better way to receive rapid medical care or advice. Short of that, there are no easy answers, so the victim and their party need to make their best-informed decision based on available options. Another key here is having a companion. Hiking alone is riskier than having someone to help you in case of emergency. A person bitten or stung may find themselves incapacitated, so may need someone to call an emergency number or transport them to a medical facility. A golden rule of SCUBA diving is to never dive alone, and the same rule applies on land.

A well-stocked first-aid kit is a good thing to always have at the ready. It should contain disinfectants, analgesics, topical itch cream, topical antibiotics, antihistamines, bandages, gauze, and so on, not to mention specialty items such as prescription medication and an epinephrine kit, as needed. In general, cleaning a wound with soap and water is a good idea, although there may be more pressing issues with serious envenomation. For minor wounds, disinfectants and bandages help keep the wound clean, but keeping major wounds clean and sterile is a more complicated medical issue. In all cases,

First aid for snakebite is a call to 911 and transport to a hospital, despite all of the paraphernalia developed over the years for removing venom from the victim.
L.L.C. Jones

injuries should be monitored for complications, such as infection, allergy, or systemic involvement. If any of these arise, I would recommend immediately calling a poison control center and/or 911, then consider going to a clinic, seeing a physician, or checking into an ER, depending on the severity. Poison control centers are invaluable as they are specialists in the field of poisoning and envenomation, and can also patch you through to 911, help find antivenom stores, arrange transport options, and so on. Also, if you have any doubts about your envenomation, or just need general information about venomous animals, poison control centers are there to help 24/7/365.25.

Arthropods

Spiders, hemipteran insects, and robber flies envenomate by biting, while all other nonparasitic arthropods sting. Aculeate (stinging) Hymenoptera account for the majority of human envenomations across all taxa. The venom is largely algogenic, rather than designed for immobilizing and/or digesting vertebrate prey. If stung by an individual aculeate hymenopteran, the reaction is usually local, and the victim essentially experiences an elevated immune response. However, the social hymenopterans (many wasps, bees, and ants) are unique among all other venomous animals because they can also sting en masse to protect the nest. This is a game changer. While the venom remains algogenic with multiple stings, the venom yield increases, and for species having a high lethality (low LD_{50}), it can be fatal. For example, some harvester ants have venom as lethal as the most toxic rattlesnakes, but the venom yield is much less. First aid for a local envenomation is important, but massive numbers of stings are an emergency requiring rapid medical intervention.

If allergy to the venom is not an issue, and you have not been stung en masse, then there should be no systemic involvement. If you are stung by a species that leaves its venom glands attached to the stinger, like a European Honey Bee, scrape it away gently to keep from getting a full dose of venom. A credit card works well for this. If you are tough, it is just a matter of grinning and bearing the pain till it subsides. If you are not so tough, you probably just need an over-the-counter analgesic, and perhaps an antihistamine to bring down the swelling. An antihistamine can help with inflammation and may also help if you are mildly sensitive. An ice pack may help with the pain and reduce swelling. While cold therapy works for stings that elicit an inflammatory response, such as from arthropods; it is not to be used for marine animal stings.

Something that Hymenoptera have in common is an array of allergens with a certain degree of cross-reactivity among some taxa. Thousands of people are allergic to hymenopteran venom, and the number is always increasing. More people are killed by allergic reactions to Hymenoptera than all other venomous animals combined.

Stings by most scorpions and small centipedes, and mild spider bites, are similar to typical hymenopteran stings. However, envenomation by Arizona Bark Scorpions, large *Scolopendra* centipedes, widow spiders, recluses, and reduviids may be systemic or involve much greater and enduring pain. In these cases, one may opt to seek medical attention. Call a poison control center if you are not sure what to do, because their highly trained staff can help you decide if medical attention is warranted. Bites by recluse spiders are not algogenic, so may go unnoticed, and symptoms may not appear for several hours. The outcome of a recluse bite is always uncertain, so should be closely monitored. The bites of some spiders and reduviids may be painful at first, but other symptoms may take some time to develop. The bites of widows and stings of *Scolopendra* centipedes are very painful and symptoms may quickly escalate, so the victim will usually know to seek medical attention—at least for some prescription analgesics. There is antivenom for widow spiders and Arizona Bark Scorpions, but not the other native arthropods. If you think you have been bitten by a recluse spider, try to capture the spider, dead or alive (not too squished), so that its identity can be confirmed or denied by a spider expert. I recommend reading the sidebar on the enigma of spider bites, and one should note that a diagnosis of a spider bite (especially a recluse) without a confirmed culprit usually leads to a misdiagnosis, and more harm can come by assuming it was a spider than the underlying cause.

Tarantulas and most stinging caterpillars have urticating hairs, and envenomation by these irritating setae does not usually require medical attention. The exceptions include the potentially dangerous puss caterpillars, or any species that causes an allergic reaction or systemic involvement. Also, if hairs get into the eyes, you should visit a physician. The puss caterpillar is quite venomous and capable of sending someone to the ER. However, the general consequence of being stung by the hairs is an irritating rash and welt. The hairs can sometimes be removed from the skin by using tape to peel the hairs away. Antihistamines and anti-itch medications may help alleviate the pain, as can analgesics. Cold therapy may also help.

Reptiles

Bites by Gila Monsters, pit vipers, and coralsnakes should always be considered a medical emergency, while bites from venomous colubrid, dipsadid, and natricid snakes *usually* are not. Highly venomous snakes typically bite, inject venom, and release, but Gila Monsters may hang on with tenacity, and should be removed as quickly as possible to reduce the amount of venom entering the wound. There is no absolutely effective method to get a Gila Monster off, but they often let go if lowered to the ground and allowed to escape. If this doesn't work, try squirting rubbing alcohol into the face and mouth or prying it off (not with fingers!). There are instances where a Gila Monster has bitten someone else trying to remove the animal from the first victim. Similarly, if someone is bitten by a colubrid snake or its relatives, don't allow it to hang on.

If bitten by a venomous snake or Gila Monster, follow these general guidelines:

DO:
- Remain calm. Movement and excitation help spread venom.
- Get to a hospital (*not* a clinic). Even if being driven by a friend, call 911 to alert the hospital and poison control centers, so they can find the nearest source of antivenom and start prepping for the patient's arrival.
- Remove rings, watches, bracelets, necklaces, and tight-fitting clothing (e.g., boots if bitten on the foot or ankle).
- Try to keep the affected limb immobile. A sling may help. Elevate above the heart, if possible.
- With a pen having indelible ink, mark the progression of swelling with a line, and write the time at the line, starting with time of the bite near bite mark.
- Carry with you a snakebite treatment protocol from Lavonas et al. (2011) or a protocol of equal or greater utility, in case attending physicians are not familiar with snakebite treatment. Having it allows you to make recommendations to a physician merely by handing them a piece of paper written by snakebite experts, rather than acting like you know more about treatment than a doctor. You can also recommend they call a poison control center.

First Aid   71

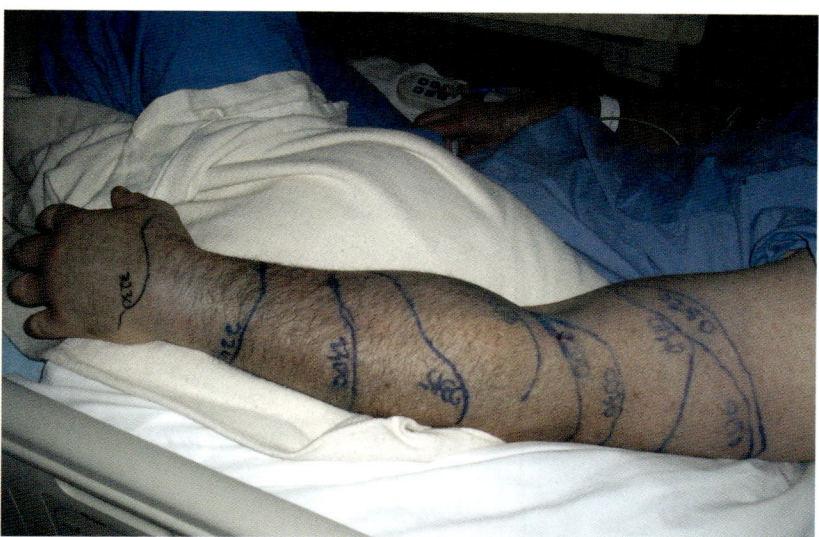

Learn what to do for snakebite—before it happens. *Courtesy of Phil Rakoci*

DO NOT:
- Incise the area of the wound.
- Suck out the venom. Extractors have not been proven to be effective, and using them may waste valuable time. Antivenom is what saves lives and minimizes tissue damage.
- Apply ice, heat, urine, meat tenderizer, or any other substance.
- Use electroshock.
- Use a tourniquet.
- Drink alcohol or take any medications.
- Apply a pressure bandage: Note that the jury is out on this for some species. While this is commonly practiced in Australia to slow the movement of lymph during neurotoxic envenomation by elapids, there is little information on its efficacy for North American species. More research is needed.
- Try to capture or kill the snake. Additional bites can occur and snakes inject more venom with each strike. Physicians may treat symptoms and progression on a case-by-case basis, regardless of the

snake. It should be noted, however, if it was a coralsnake, or a pit viper, and if you know which species with certainty, that information doesn't hurt. If you have a photograph from a safe distance, it *may* help the physician, but is not necessary.

In short, if you are bitten by an elapid or viperid, you will need to get to a hospital and start receiving antivenom in a timely fashion. There are protocols for how physicians should proceed when they receive a patient bitten by a venomous snake. If at all possible, you should know where the nearest hospital is, and which hospitals have snakebite specialists. Physicians who are well versed and experienced in the treatment of snakebite are a great resource. For colubrids and their relatives, the victim should never let the animal hang onto them for more than a few seconds. Little is known about bites of shrews, but they are not expected to cause any more than a local reaction.

Marine Organisms

I'll start with this important announcement: The use of hot water to alleviate pain is used for stings from most marine creatures. Most researchers agree that it is useful for most cnidarians, stingrays, scorpionfishes, leatherjackets, and glaucous nudibranchs. It has not necessarily been proven to be effective (or ineffective) for most other taxa, but I can find no information that it will harm, unless the water is so hot that it will cause tissue damage. It is generally stated that one should immerse the area in water as hot as one can stand it. Prolonged immersion in very hot water is more likely to cause tissue damage, but there are almost no accounts of people scalding their flesh—if it is

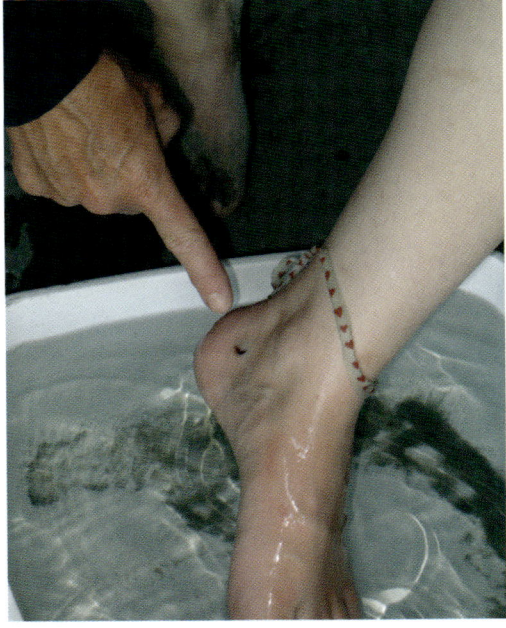

Cold treatment helps with arthropod envenomation, but hot water is excellent first aid for a variety of marine species. *"Symac"/CCA-SA/Wikimedia Commons*

that hot, victims don't want to have their appendage immersed. A common practice is to leave the hand or foot in hot water for a while, then allow it to cool a bit if needed, then re-immerse. One reference showed that in 100% of twenty cases, the intense pain of stingrays was completely relieved by immersion in 46–48°C water for ninety minutes. Hot towels, hot showers, and hot tubs may work on areas of the torso. Victims often say the pain starts to reappear as the affected member cools. Hot water probably denatures some of the venom, which effectively reduces toxicity somewhat. First aid is not mandatory in mild envenomation: many people just wait out the pain.

When someone is stung by a cnidarian, there will be nematocysts that have fired their venomous threads, and others that have not. So, before immersion in hot water, it is recommended that the unfired nematocysts be irrigated in a sting inhibitor before removing tentacles. Sting inhibitors include vinegar and commercially available pre-sting ointments, and post-sting sprays. Sting-no-More was developed to combat Hawaiian cnidarians and has been proven effective at inhibiting 80% of the nematocysts from discharging in *A. alata*. It is also effective for *P. utriculus*, so probably works for certain other cnidarians as well. Although vinegar can keep nematocysts from firing, it may cause the pain to intensify. Do not use cold therapy, apply fresh water or sand, or rub with a towel, as this will actually cause additional nematocysts to discharge. After inhibiting further nematocyst discharge, safely and gently remove tentacles with tweezers, gloves, or other no-contact tools. Then the hot water can be applied. Lidocaine is also an effective pain reliever. There are many so-called remedies in the literature and online, but most are untested, ineffective, or even dangerous to use. These include meat tenderizer, ice, urine, alcohol, baking soda, and even ketchup and mustard. In at least two cases of fatal envenomation, meat tenderizer was applied.

Most stings by cnidarians do not require medical attention, but they do when there are severe, systemic, or allergic reactions. There have been cases of fatal envenomation by jellyfishes (rarely), box jellyfishes, and Portuguese Man-of-War. At least some species of box jellyfishes can cause Irukandji-like symptoms. In all cases of serious envenomation, first aid can be applied (e.g., inhibiting stinging cells from discharge and removing tentacles), but don't delay seeking medical attention. Often, lifeguards are trained in first-aid treatment and know what to look for and do in case of medical emergencies. Over-the-counter analgesics, antihistamines, and anti-itch medications are commonly used in addition to hot water.

Nearly all of the venomous fishes and some invertebrates (sea urchins and Crown-of-Thorns Starfish) have spines that deliver venom, but these spines can also cause physical trauma. The spines in some (lionfishes and some sea urchins) are long and slender, but in most stinging fishes (stingrays, sharks, and most bony fishes) the spines are heavy and stout. Thin or stout, all can break off in the victim, sometimes deeply embedded in the flesh. When this happens, the venom glands and other epidermal tissues may also be introduced into the wound. It is important to get these foreign objects removed, and that is usually best done in a medical facility, after debridement. It is not that unusual for part of an embedded spine to go undetected, even when an X-ray is used. An unusual type of accident is when the long spine of a stingray enters the body cavity. These can be life-threatening. Call 911 immediately, but don't pull out the stinger unless directed to do so. Generally, direct pressure can help suppress excessive bleeding, but in certain circumstances it can cause harm (e.g., applying pressure near a stinger next to an artery) so it is best to get advice from emergency medical specialists.

Most of the other marine invertebrates cause itchy rashes. These taxa include fireworms, sponges, stinging hydroids, and Fire Coral. For sponges, the spicules can be removed with tweezers or tape. A common first-aid practice when rash appears is the use of diphenhydramine and hydrocortisone. The use of heat treatment versus ice packs is poorly known and varies by accounts. If needed, the victim can try them, then discontinue if pain worsens. Analgesics are said to alleviate pain in some instances. There is little information on first aid for cone snail and polychaete bites. Allergic reactions to bloodworm bites are not uncommon, and any of the marine organisms can result in severe allergic reaction.

A final note: many stings and bites by marine organisms are prone to secondary infection from a host of pathogens, including the tetanus bacterium. Many physicians routinely administer prophylactic antibiotics for marine stings.

The Making of Antivenom

Antivenom is also known as antivenin and sometimes antivenene. These are drugs to help slow or stop the progression of the effects of human and pet envenomation. Although there are several types of antivenom produced, the emphasis here is on that for pit vipers. Antivenom is somewhat specific for particular venomous animals. It is produced using the animal's actual venom to build up immunity in horses (equine), sheep (ovine), or sometimes other animals. If venom from a single species is used (and targeting bites from that species), it is termed monovalent, whereas a polyvalent antivenom is made with the venom of several venomous species and is effective for a variety of species with similar venom attributes. For example, pit viper antivenom is polyvalent, made with the venom of several species of pit vipers, so it can be used to counter the effects from any species.

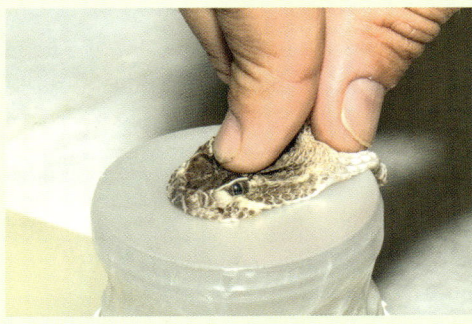

The first step of the process is to obtain raw venom from the target species, using a procedure called milking. For snakes, a typical method is to hold the animal just behind the head, coaxing the animal to open its mouth. The fangs then puncture a membrane covering a collection vial and the technician will massage the jaws and venom glands as needed to entice the snake to release venom. For spiders and scorpions, an electric current will cause the animal to release venom, which may be gathered in a pipette, due to the tiny volume. The collected venom is then stored or injected into the horse or sheep.

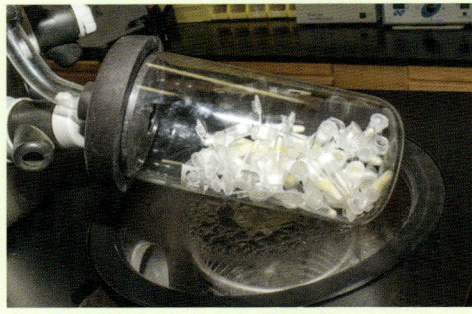

(Top) Milking a Mojave Rattlesnake, NNTRC. (Middle) Snake venom in a beaker. (Bottom) Lyophilized venom is used for both research and making antivenom.
L.L.C. Jones

The first injections use very small doses of venom, so that the animals can start building antibodies without physical harm. Increasing doses are given periodically over time, and as the animal builds more antibodies, it becomes more resistant to the venom. When the level of antibodies is sufficient, the animals are periodically drained of some blood, and the antibodies are then extracted and processed.

Whole antibody IgG molecules were formerly used for pit vipers, but now the IgG is fragmented to decrease the chances of an allergic response to the serum. Two types of fragmented antivenom are Fab and F(ab')$_2$. "Fab" is short for "fragment-antigen binding," and there are two of them on each Y-shaped IgG molecule. Fragments derived from digestion by the enzyme papain yield two Fab fragments from each other and from the Fc fragment, or "tail." The Fab fragments are the antigen binding site. Fragments derived from digestion by the enzyme pepsin separate the Fab fragments from the Fc tail but not from each other, yielding one F(ab')$_2$ fragment containing both binding sites. In both cases, the allergy-causing Fc fragments are discarded. Venom is the antigen, so the antigen-binding fragments of the antivenom are what neutralize the venom. F(ab')$_2$ is generally more effective because the larger molecule is not so readily excreted from the body. The Fc section, especially in whole IgG venom, is the usual cause for allergic reactions to antivenom—both immediate and potentially life-threatening anaphylaxis, as well as delayed serum sickness in some patients.

Antivenom in the United States

In the United States, there are few commercially available antivenoms produced, although some are undergoing clinical trials or poised to become available soon. For scorpions, there is a recently FDA-approved antivenom available called Anascorp. It's full name is *Antivenin Centruroides* (scorpion) F(ab')$_2$ Immune (Equine) Injection, produced by Instituto Bioclon S.A. de C.V., Mexico City. It is a polyvalent antivenom produced with four taxa of *Centruroides* in Mexico (*C. noxius*, *C. l. limpidus*, *C. l. tecomanus*, and *C. s. suffusus*). It has been shown to be effective for eight species of *Centruroides*, including the Arizona Bark Scorpion (*C. sculpturatus*).

The only spider antivenom approved in the United States is Antivenin (*Latrodectus mactans*) (Equine). It is used for envenomation by several species of widow spiders and is produced by Merck and Company. This is the oldest marketed antivenom in the United States. There has been some controversy about its use, which stems from a single instance of a child who died from hypersensitivity, but research has also been published on its effective use. There is another antivenom, Analatro, Antivenin *Latrodectus* (Black Widow) Equine Immune F(ab')$_2$, also being produced by Instituto Bioclon in Mexico City. At the time of this writing, clinical trials by sixteen health-care facilities in the United States have been completed, but it has not yet been approved by the FDA. There are no *Loxosceles* antivenoms in the United States, but there is currently research and development. An inherent problem with developing and using antivenom for recluse spiders is the delay and wide range of symptoms, and misdiagnoses are common.

There is currently one antivenom available for coralsnakes in the United States, Antivenin (*Micrurus fulvius*) (Equine), produced by Wyeth (a subsidiary of Pfizer). It is a monovalent antivenom; however, *M. tener* was until fairly recently considered a subspecies of *M. fulvius*. Wyeth discontinued this antivenom around 2003, with the final lot labeled to expire in 2008. Since then, the FDA has periodically tested the remaining stock and extended the expiration date, most recently through January 2020. This antivenom has been used for both *M. fulvius* and *M. tener*, but it seems ineffective against *Micruroides euryxanthus*. A new antivenom is being studied, Coralsnake (*Micrurus*) North American Immune F(ab')$_2$ Equine. It is being developed for both *M. fulvius* and *M. tener*, but not *M. euryxanthus*, which rarely, if ever, causes severe symptoms (I heard rumors of two somewhat significant envenomations from the Arizona Coralsnake, but could not confirm these with details).

Before about 2000, Wyeth produced the only commercially available product for pit vipers, Antivenin Crotalidae Polyvalent (Equine). It is produced from the venom of Eastern and Western Diamond-backed Rattlesnakes, Tropical Rattlesnakes, and Common Lancehead (*Bothrops atrox*). It was used for Copperhead, Cottonmouth, all rattlesnakes, and lancehead pit vipers. It is still licensed but not available. The widely available product since 2000 is CroFab, Crotalidae Polyvalent Immune Fab (Ovine), produced by BTG. It was produced using venom from Eastern and Western Diamond-backed Rattlesnakes, neurotoxic Mojave Rattlesnake, and Cottonmouth, and is licensed for treatment of all North American pit viper bites. The incidence of allergic reactions to CroFab is small, as it was a great improvement over the older Wyeth antivenom. However, due to the small molecular size of this Fab product, it is often cleared by the kidneys before all the venom is neutralized, requiring more antivenom to be administered for a longer period of time. There is a new antivenom that has been licensed by the FDA and available in the United States since October 2018, Anavip, Crotalidae Immune F(ab')$_2$ (Equine). It is produced by Instituto Bioclon and its larger F(ab')$_2$ molecule is not as easily removed by the kidneys and seems to be efficacious without the additional "maintenance doses" needed with CroFab. There are published studies comparing effectiveness of CroFab and Anavip.

There are many other types of antivenom in use or being developed outside the United States, such as those for wandering spiders, box jellyfishes, stonefish, and exotic snakes. These are sometimes used in the United States, as when a zoo worker is bitten, but there are legality and availability constraints that may need to be overcome. Since the world of antivenoms is dynamic, to see the current status of clinical trials, go to ClinicalTrials.gov, and for other updates, use key words with a computer search engine.

6   How to Use this Book

This book is intended to be useful to a broad audience, from outdoor enthusiasts, gardeners, and field-workers to biologists, poison control center staff, emergency medical technicians, physicians, and researchers. Given the wide range of potential readers, it was a challenge to find a way to make the text readable and interesting to people of diverse backgrounds and education. Since this is primarily a book on the biology of these interesting animals, a certain amount of technical terminology is unavoidable. While I did not avoid most terms one would learn in a general biology course, I did avoid most technical medical terminology, substituting common for medical terms. As an example, I use "swelling" rather than "edema" and "redness" rather than "erythema." The biology terms are all defined in the glossary and may be shown in labeled anatomy figures.

The many color photographs and graphics are not only used to help the reader identify venomous creatures they may encounter, but also to stimulate interest in these fascinating animals. As a group, venomous animals have evolved many interesting attributes that can only be appreciated when seen in life or in photographs. Although I included numerous examples of species and their variation, space constraints did not allow me to include all taxa. For example, there are hundreds of species of stinging insects, but only a few could be represented. Interested readers should consult the many excellent field guides available online or in libraries.

The "meat" of this book is the taxa accounts. A taxon is any level of hierarchy in the classification, or taxonomy, of living organisms. The hierarchy starts out at a broad level and gets subdivided until reaching the species level. To keep things simple, I generally used traditional methods of classification. At the species level, animals are given a scientific name using the genus, which is always italicized and has an initial capital letter, followed by a specific epithet which is also italicized but is lower case. This system of having two words for

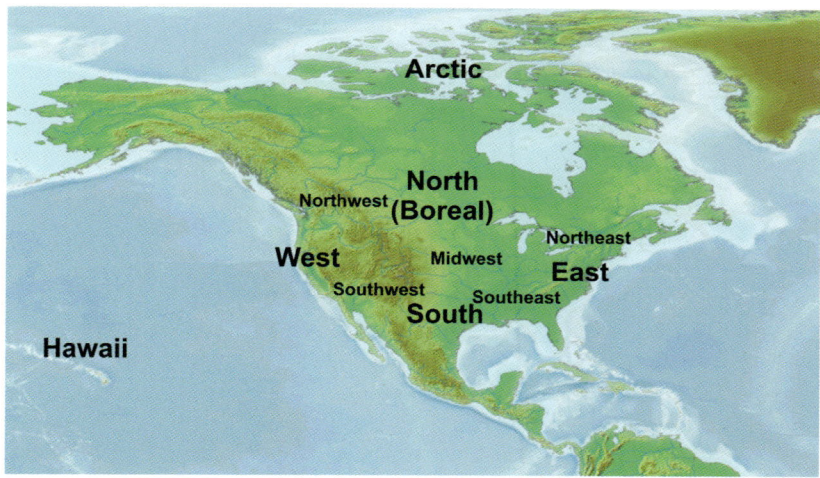

Biogeographic areas of the U.S. and Canada that harbor venomous terrestrial organisms. These regions are not precisely defined (see How to Use this Book) and there are large areas of overlap. Typical taxa for each region is given below, starting with those that are widespread in all cold and warm areas, except for Hawaii. Diversity increases from north to south, and the Southwest has the highest diversity, as it is contiguous with Mexico, with a diverse topography and climate. N.A. refers to North America, north of Mexico.

Widespread taxa: Agelenid and orb-weaving spiders, other spiders (except recluse, widows), polistine and vespine wasps, European Honey Bees, Bumblebees, certain ants, assassin bugs, stinging caterpillars. **Hawaii:** Nonnative Asian Giant Centipede, Lesser Brown Scorpion, some spiders, wasps, bees, fire ants, Stinging Nettle Caterpillar. **Northwest:** Hobo Spiders, yellowjackets, bald-faced hornets, Northern Pacific Rattlesnake. **Southwest:** Many centipedes including Giant Desert, most N.A. scorpions including Arizona Bark, Western and Southern Black Widow, most N.A. recluse spider species (except Brown), most tarantulas, most tarantula hawks, most velvet ants, most harvester ants, Africanized Honey Bee, all N.A. kissing bugs, Gila Monster, most N.A. rattlesnake species, Arizona and Texas Coralsnake, colubrids and their relatives. **Midwest:** Brown Recluse, Grassland Massasauga, Prairie Rattlesnake, colubrids and their relatives. **Southeast:** Florida Keys Giant Centipede, bark scorpions, Southern Black Widow, Red Widow, Florida Harvester Ant, Eastern Velvet Ant, copperheads, cottonmouths, Eastern Diamond-backed Rattlesnake, Timber Rattlesnake, Pygmy Rattlesnake, Harlequin Coralsnake, short-tailed shrews. **Northeast:** Northern Black Widow, European Hornet, Eastern Copperhead, Timber Rattlesnake, Eastern Massasauga, short-tailed shrews.

Map image Public Domain/Wikimedia Commons

a species' scientific name is known as binomial nomenclature. Subspecies represent a next lower level that is sometimes recognized; below the species level, names are referred to as infraspecific taxa. The example below is how an Eastern Copperhead would be classified into the standard taxonomic categories.

Kingdom: Animalia (animals)
　Phylum: Chordata (chordates; animals with notochords and backbones)
　　Class: Reptilia (reptiles)
　　　Order: Squamata (snakes and lizards)
　　　　Family: Viperidae (vipers)
　　　　　Genus: *Agkistrodon* (moccasins)
　　　　　　Species/epithet: *contortrix* (Eastern Copperhead).

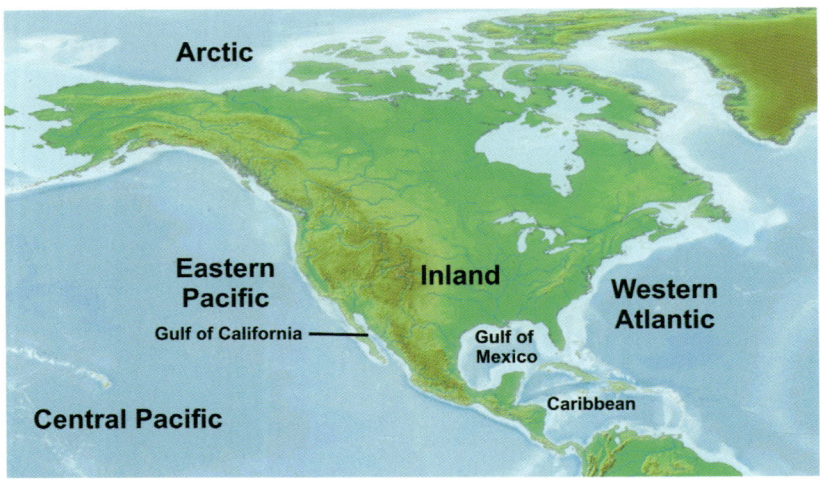

Biogeographic areas of the U.S. and Canada that harbor venomous aquatic organisms. Typical taxa for each region is given below, starting with those that are widespread in all cold and warm oceans and seas. Warmwater areas include the Central Pacific, Gulf of California, Gulf of Mexico, Caribbean, and western Atlantic (mostly south of New England). The eastern Pacific refers primarily to southern California and points south; north of that the marine taxa fall into the widespread category. The Gulf of California is technically out of the U.S., but is frequently discussed due to proximity to California and Arizona. Western Atlantic includes the Gulf of Mexico and Caribbean, due to similar warm water fauna, but also New England and Canada.

Widespread marine taxa: octopuses, squids, biting polychaetes, scyphozoans, sea anemones, chimaeras, squaliform sharks, rockfishes and thornyheads. **Warmwater:** cone snails, glaucous nudibranchs, fireworms, box jellyfishes, physalids, long-spined sea urchins, stingrays, sea catfishes, scorpionfishes, tangs. **Central Pacific:** dangerous cone snails, Pacific Bluebottle, Crown-of-Thorns, Hawaiian Box Jellyfish, native lionfishes, Yellow-bellied Seasnake. **Eastern Pacific:** Humboldt Squid, Purple-striped and Pacific Sea Nettle, Spotted Ratfish, Horn Shark, California Bat Ray, Haller's Round Ray, other stingrays, California Scorpionfish, many rockfishes, Yellow-bellied Seasnake. **Gulf of California:** Fitch's and Diguet's Octopuses, Dall's and Purple Cone Snail, Pacific Bluebottle, many stingrays, Yellow-bellied Seasnake, and other warmwater forms. **Western Atlantic:** Large cone snails, bloodworms, Atlantic and Chesapeake Bay Sea Nettles, Clinging Jellyfish, Portuguese Man-of-War, Atlantic Cownose and Atlantic Stingrays, other stingrays, sea catfishes, typical scorpionfishes, invasive lionfishes. **Inland:** Giant water bugs, naucorids, North American and invasive freshwater catfishes, plus cottonmouths and Eastern Massasauga sometimes in freshwater.

Map image Public Domain/Wikimedia Commons

In the above example, you will notice that the intermediate categories are not included, for simplicity. For example, the Eastern Copperhead is obviously a snake, yet the category that only includes snakes is not shown because they represent a suborder of the Squamata (i.e., suborder Serpentes). Snakes and lizards are very closely related, which is why they are in the same order. Similarly, there is no category only for pit vipers, because they are a subfamily of vipers (subfamily Crotalinae). This is my general approach to the classification section of the book, discussed in greater detail, below.

Given that venomous animals have evolved many times along diverse lineages, I had to make a judgment call on how to categorize them into the most appropriate taxon for each account. For example, most venomous snakes were treated at the species level, while widow and recluse spiders were given genus accounts, and assassin bugs were considered at the family level. These different taxonomic levels were selected at a sort of common denominator. For example, snakes were addressed at the species level because they can have very different toxicity and natural history, while widow spiders are very similar, mostly differing by geographic distribution. In a few instances, I combined taxa, such as "other spiders" or "black-tailed rattlesnakes," because they were difficult to deal with at the basic taxonomic levels.

For nonscientific nomenclature, I used generally accepted "standard" English names whenever possible and mentioned where I strayed from the proposed name. The term "standard" is in quotes because not everyone recognizes these names as standards, and many biologists do not accept a standard English name concept. Per the English standard names I use, each species or subspecies is considered a proper noun, so initial letters of the names are capitalized, such as Eastern Black-tailed Rattlesnake. The moniker "black-tailed rattlesnake" is not capitalized because there are two species. Most authorities that publish standard names also capitalize them because it avoids confusion. For example, a Rock Rattlesnake is always *Crotalus lepidus*, while a rattlesnake found around rocks could be any number or species, such as the black-tailed rattlesnakes, Tiger Rattlesnake, or Twin-spotted Rattlesnake. I followed suit throughout the book for all animals, treating species and subspecies names as proper nouns to avoid confusion, even if not done by all professional societies. For example, the Entomological Society of America does not capitalize arthropod names; it has a list of "common" names, rather than standard English names. In order to maintain text flow among accounts, some lesser-known animals that have a common name for a group of similar taxa are treated as one entity having a single proper name, as that is how they are often perceived and referred to. This mostly applies to aquatic invertebrates, especially Clinging Jellyfish, Fire Coral, Fire Sponge, Upside-down Jellyfish, and Crown-of-Thorns Starfish. For example, any of the dozens of species of *Tedania* that sting are called Fire Coral. Plants are too diverse and known by too many vernacular names, so are not standardized and are not capitalized, which is consistent with plant taxa accounts in the literature.

Taxa accounts are divided into four major sections: terrestrial invertebrates, terrestrial vertebrates, aquatic invertebrates, and aquatic vertebrates. This is a somewhat artificial approach, but it should be useful to the reader. First, one might want to know what the risks are from animals in the water vs. land, as you would never be at risk from a stingray, for example, when walking through a forest. Second, the reader should be able to quickly assess if an animal is a vertebrate (having a backbone) or an invertebrate (without a backbone). In each of these four sections, I have a very brief introduction to the group. The group introduction mentions the types of taxa that follow in the accounts and usually has some information on number of species and basic external anatomy but lacks detail because of space constraints. If you pick up a standard invertebrate zoology text, for example, you will see that the thick treatise itself only includes a very basic synthesis of major taxonomic groups. The introduction to terrestrial vertebrates is a little different because it mostly refers to pit vipers, and there are commonalities to discussion of snakebite.

In some taxa accounts and elsewhere in the book, I discuss exotic pets that may be related to or similar to native taxa, to help increase awareness of the dangers of exotic pets. For example, baboon spiders are types of tarantulas that are sometimes kept in the home as pets, but their disposition and toxicity is much worse than native species. Also, captive, exotic animals are a frequent cause of envenomation in North America. For example, in one study of 188 lionfish stings in Texas, all of them occurred in home aquaria, rather than the Gulf of Mexico, although the study was conducted before they became widespread invasives in the Gulf.

Taxa Account Headings

Classification: This section shows the general classification of taxa in the accounts. All taxa in this book are in the kingdom Animalia, so that level is assumed. Thus, what follows are phylum, class, order, family, genus, species, and sometimes subspecies. One thing to note is that biologists denote the family level with the -idae suffix, such as Vespidae for a group of wasps, and they are referred to as a common name with the -id suffix, which in this case would be "vespids." For some reason, this is not done at other taxonomic levels, except subfamily, which has the -inae suffix. Discussion of other intermediate levels are added only if needed. For example, I used Polistinae for the subfamily of paper wasps and Vespinae for the yellowjackets, because they are distinct and important enough to warrant separate taxa accounts.

Where appropriate I discuss some recent taxonomic changes and alternate classifications. To help me determine which names I would accept, I considered the approach of several sources, such as WoRMS (World Register of Marine Species), the World Spider Catalog, BugGuide, and two "standard" English names lists for reptiles. Ultimately the choices I made were mine. Don't rely on this book as a reference for taxonomic standing; rather, you should research the group of animals you are interested in and come to your own conclusions. Some of the taxonomy as it appears in the text will no doubt be different by the time the book is in print. Taxonomy is a moving but improving target. Also, experts often disagree on taxonomy and nomenclature, so I selected what I felt was most appropriate and acceptable. Biologists often state that taxonomy and nomenclature are only hypotheses to be tested. When there are too many taxa to discuss individually (e.g., the dozens of species of rockfishes), I highlighted some of the most common, researched, and noteworthy taxa.

Identification: When needed, I included information on basic external anatomy as part of the identification section of taxa accounts. Basic anatomy may also be covered in the introduction to the four taxa account groupings. There may also be graphics to help the reader understand the design of an animal group. The identification section is primarily intended to give the reader a general idea about what the species or species groups look like, in terms of gross morphology, including shape, size, and color, and sometimes more specific characteristics used for identification. However, this book is not intended to be a field guide; there are numerous books available to aid in species identification. Invertebrates in particular have many detailed characters that need to be considered to correctly identify an animal to species, or even genus; often, these characters are only observable under a dissecting microscope, and information on identification is only found scattered among scientific papers which may use different taxonomies of the time. If there are few taxa in a group, I tend to have more information on identification. For example, a Gila Monster is easy to identify to species, based on only a few obvious features, whereas a species of agelenid spider would require a microscope, scientific literature, knowledge of specific characters, and probably input from an expert . . . and, perhaps, not all experts would agree on its taxonomy and nomenclature.

Distribution: Descriptions of the geographic range focuses primarily on North America, north of Mexico, and Hawaii, but I often discuss total ranges. The discussion of basic distribution of taxa is commensurate with our knowledge of the range but tempered by the number of taxa in an account and space constraints. For example, a fairly specific range is given for each species of rattlesnake, whereas there are bumblebees of one type or another throughout North America. It is also often difficult to make accurate range assessments of deep-sea fishes. In many cases, it is not possible to determine the geographic distribution of certain taxa from the literature. This is part of the reason I do not include distribution maps, which also take up valuable space. Many field guides and online references have distribution maps, so if the maps exist, they are readily accessible.

You will find a certain amount of U.S. bias discussed in the Distribution section and elsewhere in the text, and even a bias to the southern and southwestern states. This is not done to belittle the northern states and provinces but to reflect the higher diversity to the south. For example, the majority of species of scorpions, recluse spiders, tarantulas, wasps, rattlesnakes, cnidarians, stingrays, and scorpionfishes occur in the South, and many of these are endemic to the American Southwest. Another reason for the Southwest bias is that the original draft of the book only covered the American Southwest, then had to be retrofitted to include the rest of North America north of Mexico. So, my humblest apologies for any distributional bias, and I will gladly take helpful tips on how to better include the rest of the United States and Canada for future revisions. Not all taxa occur in warm climates, of course. Some animals, including bumblebees, jellyfishes, and rockfishes, are adapted to the cold-temperate, boreal climate of the northern United States and Canada.

When I discuss a specific geographic area of North America north of Mexico, I refer to it as a proper noun. When there is an adjective with the -ern suffix, it is not regarded as a proper noun. Some descriptions of the regions I often reference follow. The Southwest is the southwestern United States; the South is the southern tier of the United States; the Southeast is the southeastern United States; the East generally refers to the area east of the Mississippi River, from Florida into Canada; the West generally refers to the area from California into Canada to an ambiguous demarcation extending to middle of the continent; the Midwest includes north-central United States and adjacent Canada; and the North or boreal areas refers to Canada and adjacent United States. Each of these large biogeographic regions have their own similar faunal elements. In the marine environment, I often refer to eastern Pacific, from California (or farther south) to the Arctic Ocean; central Pacific, which is the area around Hawaii; the western Atlantic, which goes along the eastern seaboard; and the Arctic Ocean in the far North.

Because of the connectivity of the Caribbean with the U.S. Atlantic waters, I often mention Caribbean species, even if they are not known from the United States. Similarly, because of the proximity of the Gulf of California to the states of Arizona and California, I often mention species from that inland sea. The Caribbean and Gulf of California are short-distance destinations frequented by divers, anglers, and swimmers. The reader should also note that I discuss venomous animals in U.S. territorial holdings in a chapter that follows the taxa accounts.

Natural History: This section includes basic information on the life and natural history of the animal(s). The habitat affinities of taxa are key to understanding wildlife and their ecosystems. One of the natural history attributes that is most important for safety concerns is the time of year or day when an animal is active. For example, many snakes brumate in the winter and are not usually encountered until the spring, especially in the North. In warm areas, they are often diurnal in the spring and then switch to nocturnal activity in the summer. Many species are also stimulated by rainfall.

Another aspect of an animal's life history is its reproductive biology. For example, there may be courtship rituals, territorial disputes, internal or external fertilization, oviposition, or giving birth to live young. Each of these events coincides with yearly and daily activity patterns. Some species make migratory movements, such as snakes coming out of dens to feeding grounds or stingrays coming into estuaries to breed. Some species become more surface active or migrate to look for mates, as when male tarantulas are on the move.

Feeding and predation are essential components of the natural history of animals. The feeding habits of some species are relatively well known, whereas for others, such information is a little more difficult to ascertain. In the wild, predation is difficult to observe, so that information may be limited. However, we can sometimes make an educated guess and mention likely predators. Just because these taxa are venomous does not mean they are safe from predation. Predation pressure is a key element in how animals evolve, and it is often responsible for a predator-prey arms race that has resulted in animals becoming more venomous or having a more advanced venom apparatus.

Encounters: This section includes information on when and where people are likely to encounter the particular taxa. It is intended as a precaution to increase awareness, so that the reader can proactively know when and where they might expect to see the taxa in question. Of course, nature doesn't always like being pigeonholed, so there are always exceptions. In some instances the information comes from studies about encounters, and in others, it is based on the animals' life history traits. When available, I have included information in this section on statistics of exposures. Two main sources of information come from summaries using the national database for poison control centers, and summaries from hospital emergency room records. It is important to understand the cautions and biases of using these data; in some circumstances the data can be both informational and misleading. In most cases I tried to make sense of the summaries and convey that information to the reader.

Venom Apparatus, Venom, and Symptoms The venom apparatus is specifically that part of the external and internal anatomy that has been modified to deliver venom. In many cases it involves a venom gland, musculature, a venom duct, and a stinger or teeth. A scorpion is a good example of this. It has venom glands with associated musculature, a stinger (aculeus), and anatomical features including the metasoma and telson to deliver venom. The many types of venomous animals have an array of variable designs, so these are each described in this section. To help the reader visualize how venom is introduced, I sometimes provide labeled illustrations of the venom apparatus.

Characteristics of venom have been introduced and generally described in the chapter, "A Venom and Envenomation Primer." The venom section in taxa accounts then briefly describes the type of venom and some, but certainly not all, of the major components. When the information is readily available, I give some LD_{50} estimates and the method of injection if known (e.g., subcutaneous or intravenous); in many cases the method of injection is not reported. In larger animals, I usually give the venom yield, as a range or maximum. I assume the reader knows that tiny animals have correspondingly small venom yields. In many instances certain venom components and allergens have been named after the organism from which they were isolated, such as sistruxin, from the venom in snakes of the genus *Sistrurus*. These specific components are usually named only if they have important functions (e.g., high lethality to vertebrates), or are used in other studies, such as having a potential pharmaceutical applicability or are an indicator of the evolution of related taxa. Most venom fractions, however, are not named, but merely classified into categories or noted for their attributes.

It is often a leap to go from what components are in venom to the symptoms they cause in humans, but of course they are related. Unfortunately, the physiological response of humans to various venom components is often not known. What we do know is which symptoms humans typically exhibit when someone is bitten or stung. In this section I synthesize basic information on symptoms from case reports, studies, and published syntheses. I try to describe the gamut from typical to severe envenomation. For example, on rare occasions, the confirmed bite of a Brown Recluse may cause extensive dermonecrosis, but in typical cases, the effects are localized and self-resolving. Dry bites and stings are not necessarily mentioned because they are common among certain taxa (e.g., spiders, vipers, and elapids). Nevertheless, they are something to be aware of, and there are no dry stings from animals with passive envenomation (e.g., jellyfishes). Some other taxa are incapable of metering their venom because there is no muscular control of a venom gland, as in a stingray or lionfish. If you are stung by one of these fishes and the integument is intact, you cannot receive a dry sting.

Remarks: This section is for snippets of information that may be useful or interesting to the reader.

References and Resources: I try to have ten or more references for each account, but for every reference I show there may be dozens or even hundreds more on the topic. Each of these references cited in this section corresponds with the bibliography in a companion stand-alone publication. I follow a common method of citation I am familiar with for major herpetological journals. I generally selected books, plus journal articles that the reader would have access to, at least as abstracts, online. Many readers may not be able to access obscure journals, which are functionally only available to researchers with subscriptions—or by paying exorbitant amounts of money for reprints. I should also note that I am a fan of open access journals, so that we can all benefit by learning and furthering the advancement of science without being financially gouged. I also tended to refer to recent articles, although there are some classics I could not avoid mentioning. As an example, there are thousands of papers on rattlesnakes, but Laurence M. Klauber, the "Father of Rattlesnake Knowledge," is only cited once—for his two-volume treatise on rattlesnakes. Many of his scientific papers culminated in that book, and since that time, there were hundreds of researchers who have contributed to the knowledge he set the stage for. The bottom line is that you should dive as deeply into the subject matter as you desire, and I hope that these references are a good starting point.

Pain Indices

Two numeric pain scales are used in this book: the Ten-scale Pain Index (0, plus 1–10) and the Hymenopteran Sting Pain Index (0, plus 1–4). In both, zero refers to no pain whatsoever, so you pretty much won't read about it. Even a dry bite or sting has some sort of poke, so that is something, and it could at least be a 1 on both scales. The default for referencing pain in this book is the Ten-scale Pain Index. The Hymenopteran, or Schmidt Sting Pain Index, is specifically reserved for ants, bees, and wasps. Even then, the Ten-scale Pain Index can be used. These two scales have not been cross-referenced. There is a reason for this: hymenopteran stings pretty much vary in intensity of pain, then fade to nothing, while other pain, such as from snakebite, can be very deep and last for months.

The Ten-scale Index is almost a universal measure, and there are countless references to it. Often it is just called "the pain scale." It is frequently used in the medical profession to assess how much agony a person is in. It is not standardized and is subjective. There are different descriptions for each numeric value, but it is usually broken down into minor (usually 1–3), moderate (usually 4–6), and severe (usually 7–10). Sometimes, however, it is just a sliding scale. Pain is a personal thing, as some people are tolerant and some are intolerant. Nevertheless, everyone knows that a 10 is at the very top, being the most excruciating and unbearable pain imaginable. Also, people generally agree a bee sting is a 3–4 on a 1–10 scale (and a 2 on the Hymenopteran Sting Pain Index). Despite what you may read, there is no such thing as a 10+ or 10+++++++, just as there is no A+ grade, despite what teachers may give you. But that does paint a nice picture of what the person may have experienced. One commonality of most charts with descriptions is how your normal activities are affected. At the extremes, for a 1, you just carry on, almost like nothing happened, while at a 10 you are screaming and slamming into walls because you are out of your mind with pain. Medical practitioners often use a "face chart," to show how a person's facial features are reacting to pain, which is especially helpful for children who cannot adequately describe what they are going through. One thing you might note is that the pain experienced by victims in the Envenomation Stories is very frequently in the severe range. It is frequently a 10, and many people say it is the worst pain they have ever experienced.

The Hymenopteran Sting Pain Index was developed by Justin O. Schmidt, and also published by Christopher K. Starr. Justin is affectionately known as the "King of Sting." His reputation came about because he is an avid field entomologist and studies stinging insects. As a result of his passion, stinging insects were an occupational hazard, and he has been stung by more than eighty species (almost always accidentally, despite popular belief). That puts him in a nearly unique position of being able to compare how painful hymenopterans are when they sting. He felt that sting pain fell into four

Justin O. Schmidt excavating ants in Brazil. *Krista Schmidt*

categories. Essentially, these equate to 1 = not bad, perhaps like a pinch or slightly worse; 2 decidedly painful, but nothing to write home about; this is what a European Honey Bee or yellowjacket ranks; 3 = something that really gets your attention because it hurts like Hell, such as a Maricopa Harvester Ant; 4 = a seriously painful sting that may make you lose control of your faculties. Some of the large tarantula hawks are the only ones to rate a 4 in the United States. Even more noteworthy than the four-scale index he developed is Justin's flowery prose when he describes stings, as if they were fine wines to be compared. His descriptions were made famous in an interview with *Outside* magazine in 1984.

TABLE 3 A Ten-scale Pain Index. This particular version is a modification of charts found online. Most online charts are for care of patients, while this relates more to pain from envenomation; nevertheless, pain is pain. A bee sting is generally regarded as a 3–4 by most people, but it depends on personal pain thresholds and other factors.

Category	0–10 value	Description, examples
None	0	No pain whatsoever.
Mild. Does not interfere with most activities. Able to adapt to pain psychologically and with medication.	1 Very mild	Very light pain, hardly noticeable.
	2 Discomforting	Light pain, unpleasant, like being pinched.
	3 Tolerable	Very noticeable, but you can carry on. Like bee sting, especially as it starts to dissipate.
Moderate. Interferes with many activities. Unable to adapt to pain.	4 Distressing	Strong, sometimes deep pain, such as throbbing and burning.
	5 Very distressing	Strong, deep, piercing pain, such as sprained ankle.
	6 Intense	Pain now dominates your thoughts and you cannot think about much of anything else.
Severe. Unable to engage in normal activities and may lack the ability to function properly.	7 Very intense	Like 6, but you can't focus on anything but pain.
	8 Utterly horrible	Pain so intense it grossly alters your behavior. Examples include childbirth and bad migraine.
	9 Excruciatingly unbearable	Pain so intense you cannot tolerate it without medication.
	10 Unimaginably unspeakable	The worst pain you could possibly imagine, and is absolutely unbearable in all regards. Loss of all control because it is so overwhelming.

PART II TAXA ACCOUNTS

7 Terrestrial Invertebrates

The venomous terrestrial invertebrates in the United States and Canada are all arthropods. These include centipedes, scorpions, spiders, and insects. Arthropods are an extremely diverse group of organisms found everywhere on the globe, on land and in sea. No one really knows how many arthropods there are, but they are by far the largest group of all living animals. There are over 1,000,000 described species with many more awaiting discovery. Arthropods were first found in the fossil record some 555–540 million years ago, in the early Cambrian period.

Grasshopper mice are major predators of arthropods, venomous and otherwise. This one is finishing off a Striped Bark Scorpion. *Jillian Cowles/Amazing Arachnids*

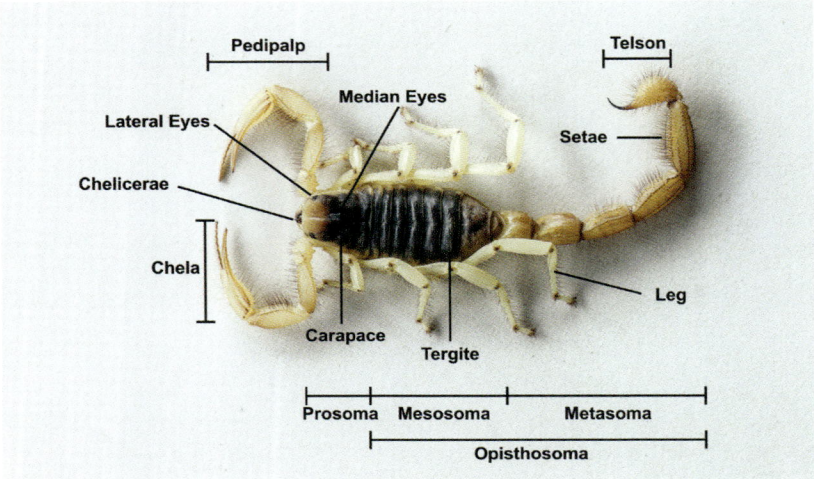

Dorsal aspect of the anatomy of a scorpion (*Hadrurus arizonensis*). *L.L.C. Jones*

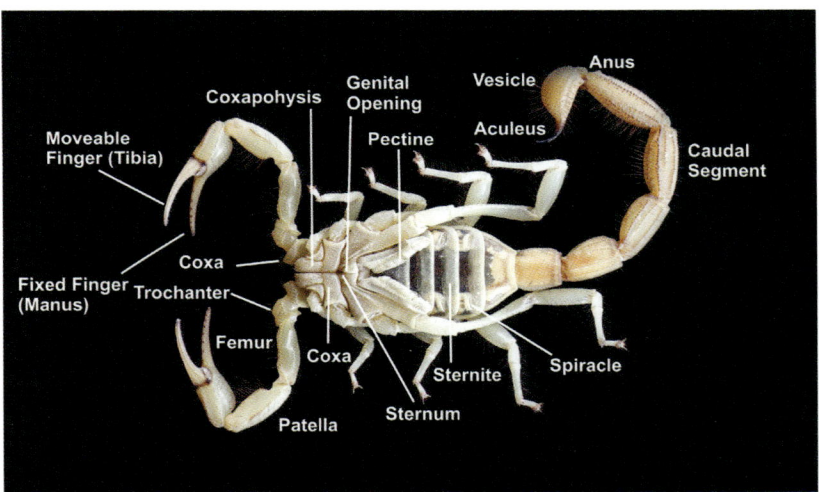

Ventral aspect of the anatomy of a scorpion (*Hadrurus arizonensis*). *L.L.C. Jones*

Arthropods ("joint-footed" in Latin) have segmented bodies and paired, jointed legs, and an exoskeleton. Their legs are segmented into numerous parts to allow mobility and other functions. Athropods must molt periodically to allow growth from juvenile to adult. They are extremely diverse in form and function. While they may have had primitive origins and perhaps even maintained primitive traits, they can also be supremely adapted for their environ-

Scanning electron micrograph of peg sensilla on the pectine of a scorpion.
Jillian Cowles/Amazing Arachnids

ment, which is why they have persisted for hundreds of millions of years.

Centipedes belong to a group of animals known as myriapods (subphylum Myriapoda), along with millipedes. Myriapods are known for having many legs associated with their many body segments. Typically, centipedes have one pair of legs per segment and millipedes have two. Millipedes are not venomous, but they can give off noxious secretions and may be poisonous. Centipedes are probably all venomous. There are several groups of centipedes that range from small soil-dwelling species to the large *Scolopendra* centipedes. These larger centipedes are the ones that have been well-documented as causing envenomation. Over 3,000 species have been described, but there are many more awaiting description.

Chelicerates include the scorpions, spiders, and some other arthropods. They get their name from the first pair of appendages called chelicerae. These appendages have become associated with the mouth of chelicerates, so are considered mouthparts; hence, a spider bites and envenomates with its chelicerae and fangs, but a scorpion does not envenomate with chelicerae, it only feeds with them, so it stings. Scorpions are very primitive but highly adapted arachnids. They have a fused cephalothorax (prosoma) anteriorly, followed

A scorpion (*Chihuahuanus coahuilae*) giving birth in the birth basket.
Jillian Cowles/Amazing Arachnids

The instars then crawl onto the dorsum. If they fall off, they may be eaten.
Jillian Cowles/Amazing Arachnids

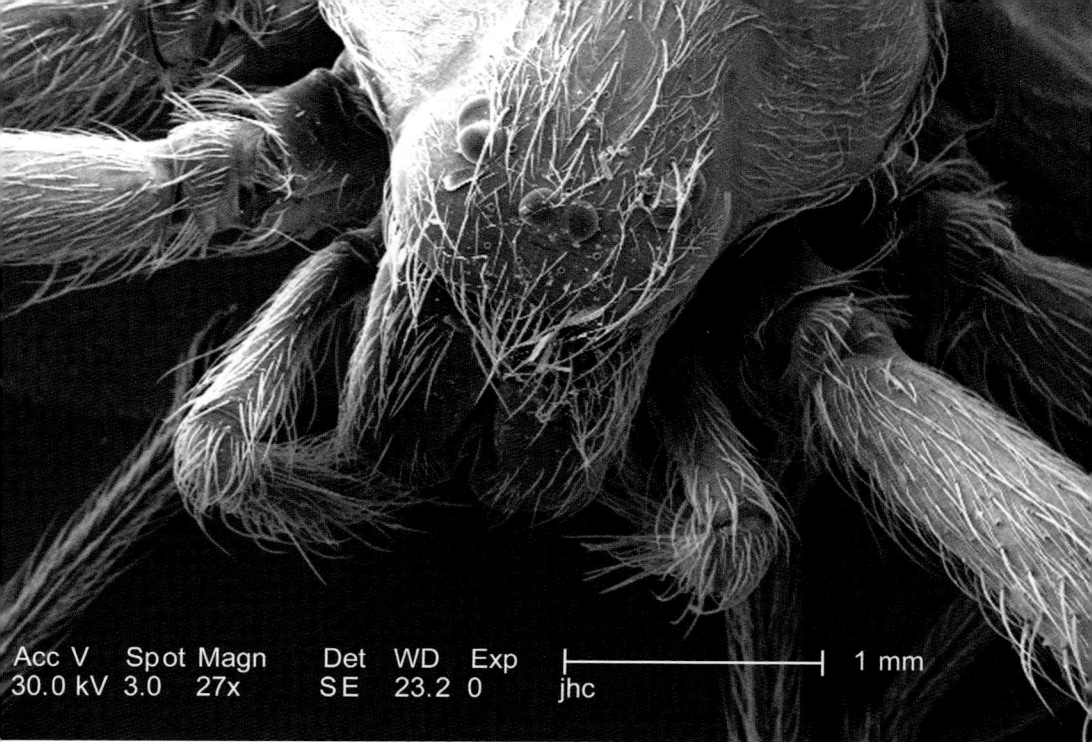

The arrangement of spiders' eyes are good characters to identify them. In this case, the arrangement of the six eyes shows this is a recluse spider.
Leah Mowrey and Michael Smith/Public Domain/CDC

by the middle abdominal section (mesosoma), and the tail-like posterior abdominal section (metasoma). The terminal end of the metasoma has the telson, the stinging apparatus. There are eight segmented walking legs and claw-like pedipalps that can project in front of the prosoma. In the United States, they reach their greatest diversity in the Southwest, with few species found to the north and east. Scorpions are unique in that their cuticle glows under a black light. Taxonomy of scorpions is difficult to follow because of discrepancies between authorities and a general lack of information, as there are frequent revisions and new taxa are described from the United States every year. There are more than 2,300 species worldwide. Tiny scorpion-like animals are pseudoscorpions. They are similar in appearance, but lack a metasoma. Technically, they are venomous. They quickly immobilize their invertebrate prey through their pincers, but they are either not capable of envenomating humans, or I can find no reports of its occurrence.

Spiders are highly successful invertebrates that primarily feed on other invertebrates. They are arachnids, having a cephalothorax, abdomen, and eight legs, but they lack pincers and the abdomen is usually globular. The most

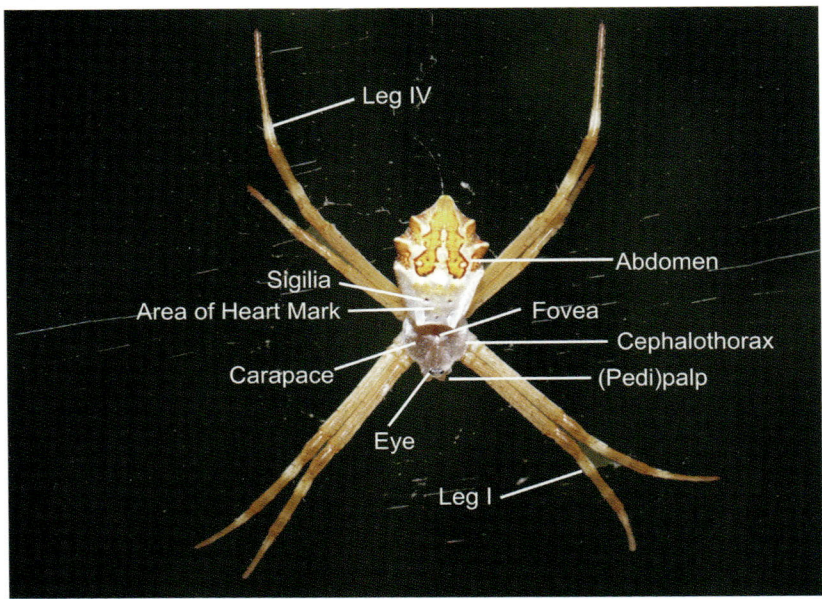

Dorsal aspect of the anatomy of a spider (*Argiope florida*). *L.L.C. Jones*

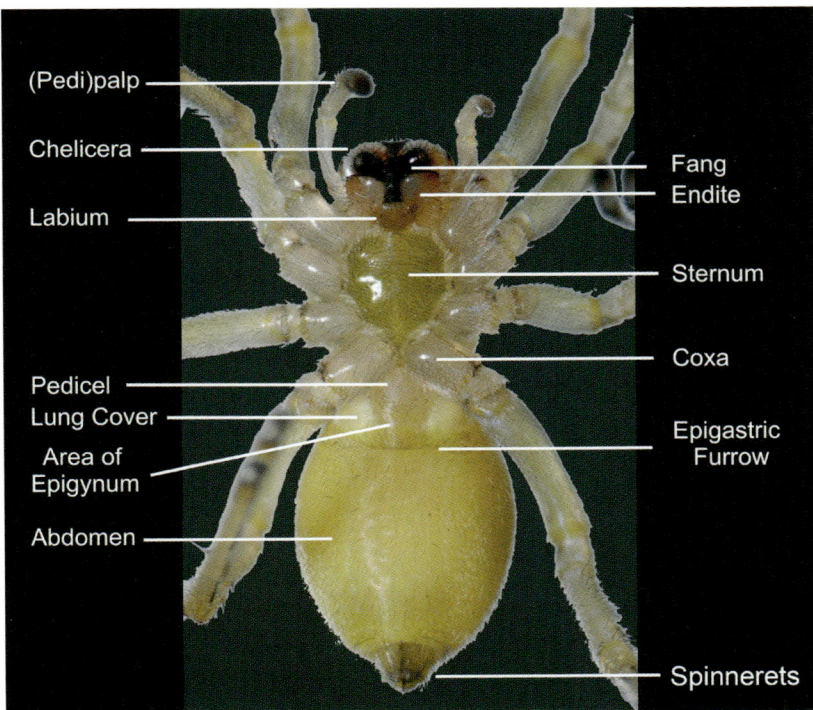

Ventral aspect of the anatomy of a spider (*Cheiracanthium mildei*).
Basal photograph, Don Laorie/CCA/Flickr

Terrestrial Invertebrates   97

The palp of a male spider, such as this one from a tarantula (*Aphonopelma chalcodes*), is often a character needed to identify the species.
Jillian Cowles/Amazing Arachnids

The dense setae on the feet of tarantulas make them quiet walkers.
Jillian Cowles/Amazing Arachnids

unique anatomical features are the spinnerets, which produce silk used in webs. Spiders have interesting behavioral traits. Some spit silk to capture their prey, some can leap great distances, some can "lasso" their prey, and so on. Nearly all species are predatory and venomous, but the bite of most species cannot penetrate human skin or the bite is fairly innocuous. Most spiders use venom to overpower and immobilize prey, but some do bite defensively. There are approximately 113 families worldwide, composed of 46,357 species. In North America north of Mexico, as of 2013, there were more than 3,800 described species. The number of species grows every year as new ones are described.

While most species of animals are arthropods, most of these are insects. Of these, only a small percentage are venomous. Insects typically have three major body parts: a head, thorax, and abdomen. There are six legs (compared to the eight of arachnids), compound eyes, and antennae, plus many have wings. The number of insect species is anybody's guess, but there are probably 850,000 to 1,000,000 described species, although it has been estimated that there are roughly between 2,000,000 and 8,000,000 extant species. It has been suggested that insects comprise 90% of the species on Earth. The lion's share of the venomous taxa belong to the Hymenoptera, the group that contains bees, wasps, and ants. Those that sting are called aculeate hymenopterans. They evolved a stinging apparatus and venom to quickly deter predators from their nests. There are more than 150,000 Hymenoptera described; it is the third-largest insect order. Besides Hymenoptera, there are venomous bugs, moth larvae, aquatic insects (refer to that section), and flies that can sting or bite humans.

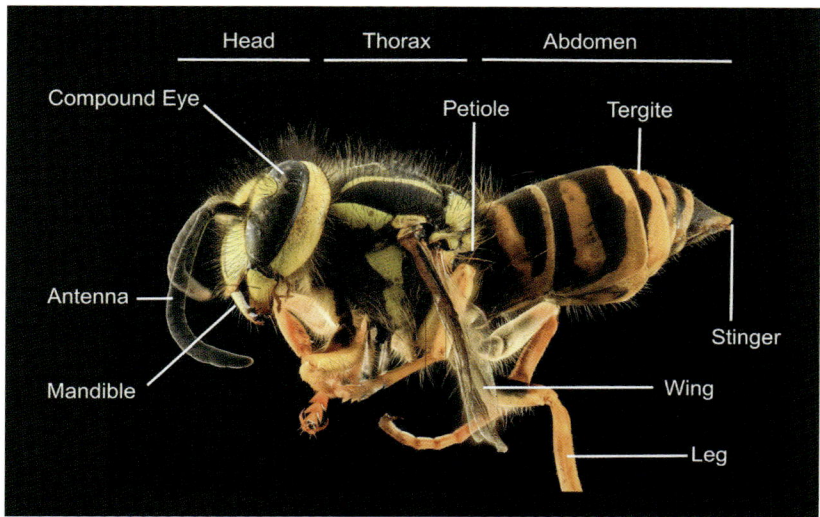

Lateral aspect of the anatomy of a wasp (*Vespula maculata*). *Sam Droege/Public Domain/ USGS, BIM*

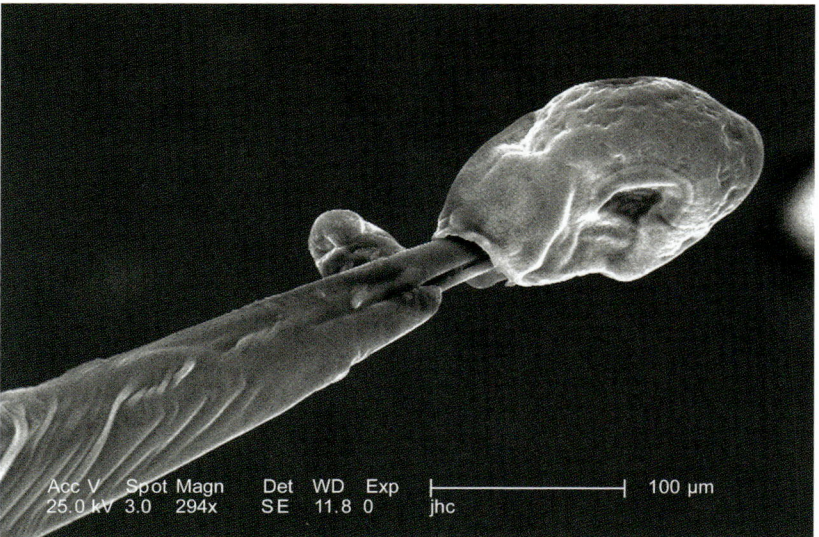

Scanning electron micrograph of a velvet ant's stinger, with sheath still attached. *Janice Haney Carr/Public Domain/CDC*

Centipedes (Class Chilopoda)

"Ultimately, it became the most excruciating pain I ever experienced . . . it was worse than my rattlesnake bite, broken ankle, scorpion [*Centruroides sculpturatus*] sting, vasectomy, tarantula hawk sting, and broken/dislocated finger." *(Envenomation Story, this volume)*

Scolopendra heros, Arizona. L.L.C. Jones

Classification: Phylum Arthropoda (arthropods); Class Chilopoda (centipedes); Order Scolopendromorpha (tropical centipedes); Family Scolopendridae (scolopendrids); Genera *Hemiscolopendra* and *Scolopendra* (giant centipedes).

North American species include *H. marginata* (Eastern Bark Centipede), *S. heros* (Giant Desert or Red-headed Giant Centipede, depending on locality), *S. longipes* (Florida Keys Giant Centipede), *S. polymorpha* (Common Desert Centipede), and *S. viridis* (Green or Florida Blue Centipede). Other scolopendrids have been introduced, including *S. morsitans* and *S. subspinipes* (many vernaculars). While other centipedes are technically venomous, only the scolopendrids seem to be medically important.

There are four orders of centipedes in our area, all of which can sting, but the most notable are the tropical centipedes, because they are large and can deliver the most painful and symptomatic stings. There are nine genera and more than twenty species of scolopendromorphs in the continental United States, with about six introduced species. *Scolopendra heros* is the native species most commonly mentioned in reports of envenomation. There are three subspecies recognized by some authorities: Arizona Giant Desert Centipede (*S. h. arizonensis*), Giant Red-headed Centipede (*S. h. castaneiceps*), and Giant Black-tailed Centipede (*S. h. heros*). Some species of large *Scolopendra* are kept in terraria as exotic pets, and can also inflict dangerous stings, including *S. subspinipes*, *S. gigantea*, and *S. hardwickei*.

Identification: Tropical centipedes are typically large and dorsoventrally flattened. *Scolopendra* are the largest centipedes in the world, with *S. gigantea* of South America reaching at least 27 cm in length. *Scolopendra* spp. have twenty-one or twenty-three pairs of legs. The tergites are variable in size, but the variation is not very pronounced. There is usually one pair of legs per segment. Anterior to the segmented body is the head, which has a protective cephalic shield, three pairs of mouthparts, antennae, and eyes. At the posterior end of the body are the gonopods, which almost look like a false front end.

Perhaps the most distinguishing feature for adult *S. heros*, *S. longipes*, *S. morsitans*, and *S. subspinipes* is the size. They are the largest centipedes in the United States. Adults measure between 15 and 20 mm, but are sometimes reported to reach larger sizes. The color pattern of *S. heros* varies, depending on geographic range, which led to the naming of subspecies. This color polymorphism is being studied by geneticists to determine if it consists of more than one species. In the western part of the species' range, the Arizona Giant Desert Centipede has an orangish body, with a black posterior trunk and black head and anterior trunk. The Giant (or Texas) Red-headed Centipede is somewhat the opposite, having a black trunk and orange or red head and anterior trunk; the posterior trunk and/or gonopods may also be orange or black. The Giant Black-tailed Centipede has a bluish to black posterior trunk. Some color morphs are banded. Some literature states that this species has aposematic coloration, but not all color phases are bright and showy. The wide-ranging species, *S. polymorpha*, is variable in color, but tends to be banded, especially at the junction of tergites. The head and terminal segments may be reddish. A vernacular is sometimes "tiger centipede," but that moniker is used for several other species. The Green Centipede is usually green to greenish blue, especially mid-dorsally. In Hawaii, adult *S. spinipes* has a brown head and dark green body, although the first tergite is light green. It is the only medically important one of at least twelve centipedes there. The Florida Keys Giant Centipede is overall brownish. The Eastern Bark Centipede is very similar in appearance to *S. viridis*, but can be distinguished by behavior. Upon exposure from its cover, *S. viridis* will rapidly dart off, while *H. marginata* is slow to respond. *Scolopendra morsitans* is typically banded and may have a red head.

Young centipedes are hatched white, but gradually take on adult coloration as they develop. Because they are epimorphic, the young have the full complement of legs and segments.

Distribution: There is some discrepancy in the literature about the range of *S. heros*, but it has been recorded (at least) from Arizona, New Mexico, Colorado, Texas, Oklahoma, Kansas, Arkansas, Louisiana, Missouri, and northern Mexico. The Green Centipede ranges from Arizona, Nevada, and Utah to North Carolina and Florida, and south to Panama. The Common Desert Centipede is also wide ranging, mostly found in the western United States, but also as for north as South Dakota and Montana. The Florida Keys Centipede is found in Florida and the Bahamas. *Scolopendra subspinipes* was introduced to Hawaii and many other tropical and subtropical areas. Its actual native range is unclear, although certainly Australasian. *Scolopendra morsitans* has been documented in Florida, Pacific Islands, and elsewhere, where it is considered invasive.

Natural History: There is more information on the Giant Desert Centipede than other native species, so it will be emphasized here as an example of the group. *Scolopendra subspinipes* is also well studied, but is a more tropical species and non-native in the United States. The common name Giant Desert Centipede is accurate with regards to size, but is a bit of a misnomer about the habitat. While it is found in the eastern Sonoran and Chihuahuan Deserts, it is absent from the western Sonoran (including Baja California), Mojave, and Great Basin Deserts. East of the Southwest deserts, the species is generally found in shrublands and woodlands. It is most often encountered beneath cover objects such as rocks and logs, but will emerge on the surface to forage when conditions are right, especially after summer showers and during mild nights, generally avoiding the hot daytime sun. Other species of centipedes are found in many habitats, from deserts to mountains and

Centipedes have maternal care. The young *S. heros* instars are not pigmented yet. Howard Byrne

mild coastal areas. Various species are found throughout the United States, and in most habitats. Because the Scolopendridae is largely a tropical group, they are found throughout many warm, humid areas, including Florida, the Caribbean, and the Pacific Islands

The reproductive traits of Giant Desert Centipedes are probably typical of all scolopendromorphs. There is a courtship display, after which the female will pick up the spermatophore to fertilize her eggs. After oviposition in a secluded nest, this species will brood the eggs and young, wrapping around them with her body. The female will protect the eggs, and may keep them off the ground. The young individuals will remain with the female until they molt, turn brownish, and disperse.

Centipedes are carnivorous. Because of their venomous sting, large tropical scorpions are able to overpower, immobilize, and consume animals as large as bats, mice, small snakes, and lizards, in addition to a host of arthropod prey. Depending on the size of the prey, they can use their entire body and all sharp legs to subdue the animal. Centipedes are preyed upon by small- to medium-sized mammals, reptiles, amphibians, birds, and other arthropods. Pallid Bats are adept at feeding on centipedes and scorpions, as they will land on the ground to capture arthropod prey, but a large *S. heros* is generally too formidable, and may in turn feed on the bat. Grasshopper mice are also effective predators that can feed on relatively large centipedes. Other predators include tarantulas, scorpions, and certain species of snakes, which may include them as a regular part of the diet.

Encounters: Tropical centipedes are uncommonly encountered in the open in the wild. They may venture indoors, but do not seem to seek out domiciles in particular. However, even desert centipedes need moderately mild and humid conditions, so homes, garages, and other structures may provide good environmental conditions. They are mostly observed when looking under a log or rock, but a sleeping bag or clothes on the ground may also provide shelter. There are reports of them getting into homes and even people's beds. The Florida Blue Centipede is said to frequent homes, and stings are not uncommon. If intentionally seeking out centipedes under objects, it is best to use a tool (e.g., potato rake) to lift the object. Centipedes of any type should never be handled without protection. Large species can easily wrap around a short glove to sting the arm. On Oahu, *S. subspinipes* is relatively common in urban and wild areas. During times of heavy rains they are driven from yards and neighboring greenbelts into homes. In one study, Hawaiian Vector Inspection Reports from 1990–1999 documented 221 domicile occurrences on Oahu.

Because large scolopendromorphs are popular as pets among invertebrate aficionados, care should be taken when dealing with them, such as cleaning the terrarium. The animals have potent venom, and bites are extremely painful. Some websites seem to underplay the potential for harm that a tropical centipede is capable of. It is often stated that "the bite [sting] is painful, but they are harmless." However, these animals can deliver a very painful and sometimes dangerous sting; they are *not* harmless. Centipedes are not aggressive animals but will readily defend themselves if they feel threatened. These animals can hang on with tenacity when they sting.

Venom Apparatus, Venom, and Symptoms: I consider centipedes to sting with a modified front leg pinch, while others consider the legs to be mouthparts, so they believe centipedes bite. The reason is that the animal appears to deliver a bite, but venom is injected by the modified front legs on the first tergite, below and to the side of the head. The modified legs are known as forcipules. In larger centipedes (most to all Scolopendromorpha, plus some others, including house scorpions, order

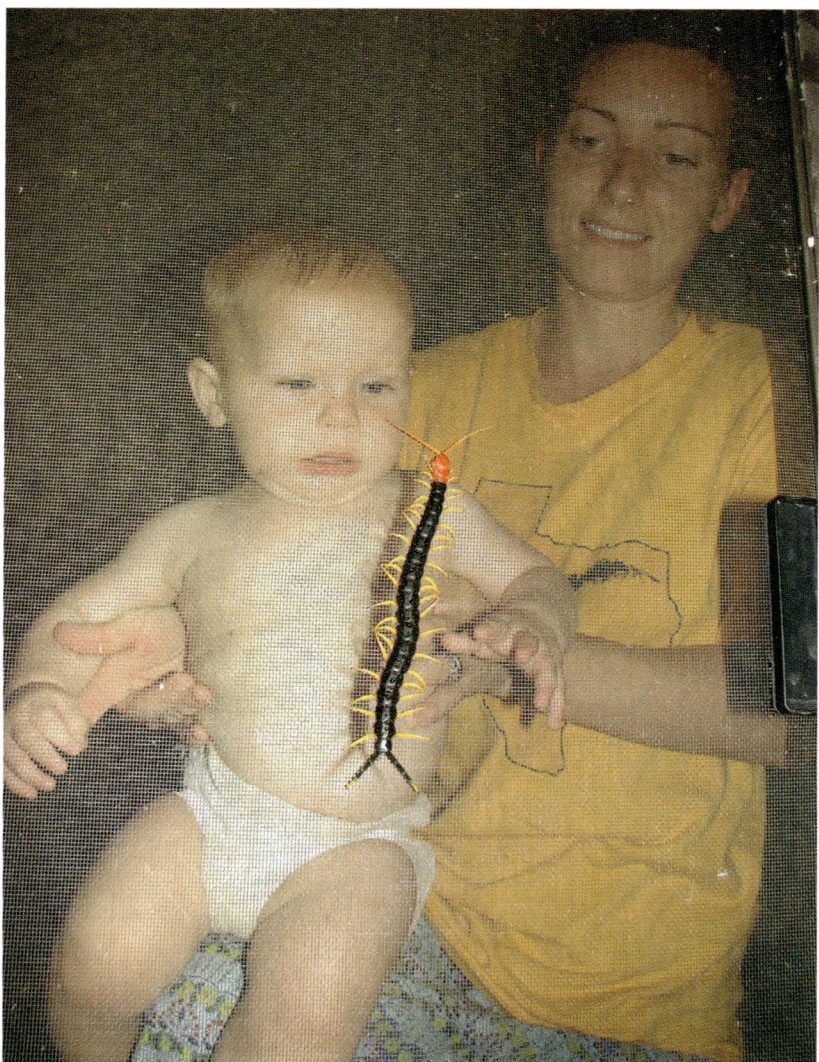

The same species in Texas has a different color pattern and is known as the Texas Red-headed Centipede. *Andy Gluesencamp*

Scutigeromorpha), they can easily puncture human flesh and probably go through some clothing. Small species may have a difficult time envenomating a person. The forcipules are short compared to other legs, but stout and sharp. The movable part is known as the telopodite. The venom duct and glands are normally inside the telopodite, but may extend fur-

Scolopendra polymorpha being preyed on by an Arizona Giant Hairy Scorpion. The scorpion doesn't always win. *L.L.C. Jones*

ther into the leg segments. The venom duct and glands are basically invaginated cuticle and epidermis, with the venom being produced in epithelial cells. The venom is expelled through the meatus in the forcipule into a groove for dispersion.

The venom of large scolopendrids, especially *S. subspinipes*, has been a topic of some research. Myotoxic, cardiotoxic, and neurotoxic properties have been described. There are numerous peptide, protein, and non-protein compounds in the venom, including serotonin, and histamine or histamine releasing agents, which help explains the pain when one is stung. There may be metalloproteinases, esterases, hyaluronidase, phospholipase A_2, lipoproteins, and other proteins in the venom, depending on the species. Several antimicrobial peptides have been isolated from *S. subspinipes*, including Scolopin 1 and 2. A synthetic peptide from *S. subspinipes*, scolopendrasin VII, shows anticancer properties by causing necrosis of leukemia cells. The LD_{50} for crude venom varies greatly between species. The LD_{50} (intravenous) of *S. subspinipes* is reported to be 0.75 mg/kg, and one virulent toxin fraction has an LD_{50} (intravenous) of 41.7 µg/kg. Although venom yield is relatively small, the venom is potent, and stings can be dangerous.

Symptoms include mild to excruciating pain, burning, itching, swelling,

Scolopendra subspinipes in Hawaii, where it is an invasive. Ryan McMinds/CCA/Flickr

numbness, and redness. Severe pain with some swelling and redness are the norm for large species. More severe stings can include systemic symptoms that can last days or weeks. Large *Scolopendra* can also cause changes in blood pressure (drop, then rise, or elevated), respiratory distress, cardiac irregularity, and necrosis. In a case report of a 44-year-old woman in southern Arizona stung on the foot by *S. heros*, she suffered extreme pain, swelling (the entire leg), dizziness, nausea, near-unconsciousness, severe myonecrosis, and acute renal failure. The victim was in a life-threatening situation but fortunately survived, thanks to hospital care. Symptoms lasted at least four weeks. There are several case reports of effects of stings from larger *Scolopendra* in the literature, including those of exotic pet owners in the United States. Human deaths have reputedly been attributable to *Scolopendra*, but there is little information on the details of most of these reports. There is one confirmed report of *S. subspinipes* that caused the death of a child, who had been stung on the head. The amount of venom and toxicity would suggest that even large specimens should not be able to kill an otherwise healthy human adult under normal conditions.

It has been reported that the walking legs of *Scolopendra* are also somewhat venomous, as the tips are sharp and there may be a skin reaction (dermatitis) on contact. However, the venom and apparatus (if present) is not well known, and probably pales in comparison to effects caused by the forcipules.

References and Resources: Bush et al. (2001), Dugon (2012), Fox (2006), Goddard (2012), González-Morales et al. (2009), Jangi (1984), Lee et al. (2015), Lewis et al. (2005), Logan and Ogden (1985), Mercurio (2016), Peng et al. (2010), Shelley (1999, 2002, 2006), Undheim and King (2011).

Bark Scorpions (Genus *Centruroides*)

"I haven't been 'scorpionized' yet, other than a little brush against one where I felt some electricity. We have a karma agreement; they don't sting me, I don't kill them. I catch and carry outdoors to release. My husband squishes them, hence his lack of karma." [He received 14 stings by *C. sculpturatus* in their home]. (D. Drobka, Pima, Arizona, personal communication)

Classification: Phylum Arthropoda (arthropods); Class Arachnida (arachnids) or Scorpionida (extant and extinct scorpions); Order Scorpiones (scorpions); Family Buthidae (buthids); Genus *Centruroides* (bark scorpions).

The family Buthidae is the largest family of scorpions worldwide, having 50–90 genera and 500–1,000+ species, depending on your taxonomic leanings. Although most of the buthids are considered mildly venomous, this family also contains nearly all of the most dangerously venomous scorpions in the world, in the genera *Androctonus*, *Buthus*, *Buthotus*, *Centruroides*, *Leirus*, *Mesobuthus*, *Parabuthus*, and *Tityus*. Bark scorpions are the only native buthids. There are five species (three native) usually encountered in the United States: *C. gracilis* (Florida Bark Scorpion), *C. guanensis* (Cuban Bark Scorpion [= Keys Bark Scorpion, *C. keysi*]), *C. hentzi* (Hentz Striped Bark Scorpion), *C. sculpturatus* (Arizona Bark Scorpion), and *C. vittatus* (Striped Bark Scorpion). Another buthid, the Lesser Brown Scorpion (*Isometrus maculatus*) is non-native. The highly dangerous buthid genera listed above can all be purchased online or in some pet shops. *Centruroides* and *Tityus* are well represented in the Caribbean, as discussed in the chapter on American territories.

The taxonomy of the Arizona Bark Scorpion has fluctuated between the names *C. sculpturatus* and *C. exilicauda* over the years, and both names are commonly seen in the literature for populations in the United States and Mexico. However, research on the venom and DNA of these animals in the southwestern United States and Baja California, Mexico, shows the two names represent two distinct species. *Centruroides sculpturatus* is in the southwestern United States and adjacent mainland Mexico, while *C. exilicauda* is endemic to Baja California, Mexico.

Identification: The anatomy of scorpions is unique. Scorpions are similar in appearance to one another, so one may need to be a bit of an expert to

Arizona Bark Scorpions have a very slender metasoma that usually curls to the side, and very slender pincers. *L.L.C. Jones*

differentiate genera and species. Bark scorpions are somewhat distinctive by their slender proportions. The Arizona Bark Scorpion reaches about 70 mm total length. It is variable in color, but generally yellow or tan, and often referred to as "straw colored." Some animals may be orangish, especially on the limbs. The prosoma and mesosoma may be plain or have

Female Arizona Bark Scorpion with juveniles. *L.L.C. Jones*

some markings, such as thick dark stripes, dorsally. The Striped Bark Scorpion is similar in size and appearance, but tends to have two broad, distinct dark stripes on the dorsum and a dark triangle on the carapace between the lateral and median eyes. To complicate identification, there are striped individuals (e.g., "gertschi" morph) for *C. sculpturatus*, plus unstriped *C. vittatus*. The Hentz Bark Scorpion is striped, much like *C. vittatus*. The Florida Bark Scorpion is variously colored, but usually dark brown. The best way to readily differentiate bark scorpions from other genera is by noting the very long chelae, which are about 6 times as long as wide at the widest point, and the long, slender metasoma. Males have a more slender metasoma than females. Bark scorpions have a tubercle on the telson, and the sternum is roughly triangular, as opposed to roughly pentagonal in other genera. The metasoma is usually off to the side in a coiled position when at rest or waiting for prey.

Isometrus maculatus is slightly smaller at maximum size than native species but is also slender in proportions (including the chelae). It is highly mottled with dark brown on a lighter brown to yellow background.

Distribution: The Arizona Bark Scorpion is found in all counties of Arizona and adjacent areas of southeastern California, southwestern New Mexico, southern Nevada, and southwestern Utah. It is generally believed to have been introduced into California and possibly elsewhere, where it has caused envenomations. It is also found in northern Sonora.

The Striped Bark Scorpion is the most widely distributed species of scorpion in the United States. To the west, its range is east of the Rio Grande in New Mexico, while *C. sculpturatus* is found west of the river. It occurs throughout Texas and is also found in Arkansas, Colorado, Illinois, Oklahoma, Kansas, Louisiana, Mississippi, Missouri, and Nebraska, North Carolina, and South Carolina. Some of these peripheral populations (especially urban populations) are undoubtedly introduced. There are non-native, reproducing populations in Tennessee and South Carolina. It has also been introduced to California and Arizona. The Striped Bark Scorpion naturally crosses the Rio Grande of Mexico, where it occurs in northern Mexico.

Centruroides hentzi is found in Alabama, Florida, and Georgia. The so-called Florida Bark Scorpion was probably introduced from Central or South America, but it is a non-native species across much of the tropics. In North America, it has been reported from Florida and Georgia, but other introduced populations occur in California, Texas, and parts of the Southeast, plus much of Mexico and the Caribbean. *Centruroides guanensis* is known from south Florida, including the keys, Cuba, and the Bahamas.

Striped Bark Scorpion from Texas. *L.L.C. Jones*

Florida Bark Scorpion. *Brent Hendrixson*

Buthids are natural stowaways. They can cling to objects, so are easily transported undetected. Lesser Brown Scorpions are probably native to Sri Lanka, but have been introduced and/or established in southern California, south Texas, Hawaii, Florida, and other parts of the world. Along with some more dangerous buthids, they are available and somewhat popular in the pet trade, so accidental or irresponsible introductions are always possible.

Natural History: The Arizona Bark Scorpion is typically associated with the Arizona upland subdivision of the Sonoran Desert and parts of the Mojave Desert, but it is also in semidesert, shrubland, and oak to pine/oak woodland habitats. This species is abundant in relatively moist areas, such as riparian zones and thickets of shrubs, as well as rock piles and rock walls. Striped Back Scorpions are found in a wide variety of habitats, including Chihuahuan Desert, Tamaulipan thorn scrub, grasslands, shrublands, piñon/juniper woodlands, oak woodlands, pine forests, and other vegetation communities. They have been recorded from sea level to 1,800 m in the Guadalupe and

Chisos Mountains of Texas and 2,340 m in the mountains of Coahuila. By day, bark scorpions hide under and between surface objects, such as rocks, logs, woodpiles, fences, and debris. They are negatively geotaxic, so may cling to the undersides of surface objects. At night, bark scorpions come out to forage and mate. They are known as errant species, meaning they actively move about the surface more than other sympatric scorpions. They are excellent climbers and may ascend anything with a surface that is rough enough for it to gain purchase. In my study of scorpions near Tucson, *C. sculpturatus* is much more abundant in areas with steep or vertical structure, as in deep washes with rocky banks, bedrock walls, rough-bark riparian vegetation, and rocky uplands. They are often seen up in trees, shrubs, cacti, cliffs, and woodrat nests, being virtually absent from flat desert habitat. Bark scorpions are often found in human habitations. Bark scorpions are often among the more common scorpion species and may reach high population densities, even in suburban areas.

In mild climates, bark and Lesser Brown Scorpions may be encountered any time of year, but are most active in the summer months. Striped Bark Scorpions are, in general, more cold-adapted than other species of scorpions, which is why they have such an extensive range. They can actually tolerate freezing and "come back to life" when thawed, even though they lack the antifreeze agents of other cold-adapted arthropods. Different age classes may have different optimal surface activity conditions. During the winter and cold nights, scorpions will brumate, but can be surface active if temperatures are warm. Although bark scorpions are considered solitary, they don't appear to be very territorial. They often aggregate, especially during brumation, when 20–30 animals can be found under a single object.

Bark scorpions may live to about ten years and have up to about thirty young. They feed mostly on insects and other arthropods, including small centipedes. In turn, they are preyed upon by a number of animals, including other arachnids (such as other scorpions), lizards, snakes, rodents, bats, desert shrews, and larger mammals and birds. Grasshopper mice and Round-tailed Ground Squirrels, which are sympatric with *C. sculpturatus*, have a natural resistance to the venom. Grasshopper mice are notorious for feeding on scorpions and other small venomous prey. Pallid bats will also feed on scorpions by landing on the ground, catching and dispatching the prey, then flying off.

Encounters: Bark scorpions are sometimes drawn to human dwellings, probably due to the mild climate, relatively moist indoors, and vertical structure.

They seem most common in newly constructed homes on the outskirts of towns and cities. In one study of the Phoenix metropolitan area, a researcher plotted bark scorpion sting occurrences by area codes to determine where stings were most prevalent. There was a correlation of high levels of reported stings with low-density, residential housing. Bark scorpions are avid climbers, so will get onto walls, closets, furniture, and sometimes beds.

Studies in Arizona estimated that about 10,000–16,000 scorpion stings per year are reported to poison control centers. Most stings probably go unreported. Calls are reported year-round, but are most frequent June through September. Most of these are probably from Arizona Bark Scorpions, since the symptoms can be painful and disconcerting. In one study, it was reported that about 12% of stings involved hospitalization, and mostly these were infants. In another study, however, only about 0.68% of Arizona stings required hospitalization, but these were mostly children, also. About 2.6% of Arizonans stung receive antivenom (n = 3,307 in a 10-year period)—presumably, mostly for toddlers. Although adults are stung more often, toddlers are the most at-risk group for serious consequences. An adult friend of mine has been accidentally stung by scorpions 14 times inside his home (presumably all or most were Arizona Bark Scorpions). He was usually stung while walking barefoot in the house late at night or while he was in bed. This is the basic modus operandi for stings—in one study, 94% of the poison control center scorpion-sting calls were from incidents inside the home. My friend's case is not atypical for infested homes—I read a blog by someone stung thirty-one times by Arizona Bark Scorpions in the home. Stings also occur while putting on clothing, including shoes. Outside, stings often come from people picking up objects, such as wood in woodpiles or debris, where the scorpions may be clinging to the underside of an object. Another friend of mine was stung while picking up a Sonoran Desert Tortoise—the scorpion was clinging to the tortoise's plastron.

Even though species of scorpions are not tracked by poison centers, statistics on stings from 2005–2015 suggest that most serious envenomations in the United States are caused by *Centruroides* species. This is evidenced by the areas where stings occur, and symptoms that led to calls by the victims. For example, there were 182,402 calls reported during this time from twenty-four states, but 62% were from Arizona (11,500 per year, where *C. sculpturatus* occurs). Other states with relatively high numbers of sting calls include Texas (where *C. vittatus* occurs; 1,743 per year), and Florida (where *C. gracilis, C. guanensis, C. hentzi,* and *I. maculatus* occur, 567 per year).

Local effects of pain, redness, and swelling were common in most states except Arizona, while systemic effects that include muscle twitching, involuntary eye movement, respiratory problems, agitation, numbness, excessive secretions, and hospitalization were most prevalent in Arizona and Nevada. Some of the national hotspots for sting reports were Phoenix, Tucson, Las Vegas, Roswell, El Paso, San Antonio/Austin/Temple, and Oklahoma City. *Centruroides sculpturatus* or *C. vittatus* occurs in all these areas.

Another sting report for Texas between 1998 and 2003 showed there were 11,545 calls to poison control centers; presumably most were for *C. vittatus*, which occurs throughout most of the state and has a painful sting. Most calls were from May and June, and west Texas had the highest proportion of calls for the state. Ninety-four percent of the calls were from stings inside or near the home. Only 4% required hospital care.

Venom Apparatus, Venom, and Symptoms: The venom apparatus is the same for all scorpions. There is a telson and aculeus at the end of the metasoma. Inside the telson are the venom gland and duct. Muscles squeeze venom through the duct and out the meatus. Bark scorpions are rather fast and agile, so can make numerous, quick jabs before you are even aware of being stung.

The venom of the Arizona Bark Scorpion has been studied by numerous researchers, often in concert with other, more dangerous species. It has a primarily neurotoxic venom composed of peptides and low-molecular-weight proteins that interfere with sodium and potassium ion channels in excitable membranes of nerve and muscle cells. One such toxin that has been isolated from this species is an α-toxin, known as CsE-V (from <u>C</u>*entruroides* <u>s</u>*culpturatus* <u>E</u>wing <u>V</u>enom). It causes spontaneous rhythmic contraction of muscle. Arizona Bark Scorpions also have cytotoxic venom components and other fractions. The LD_{50} for this species has been reported at 1.12 mg/kg in mice, but there is variability within the species. Striped Bark Scorpion venom is similar, being largely neurotoxic and having peptides and enzyme activity. In one study, the neurotoxic venom acted upon sodium channels and associated pain centers to cause even more pain than *C. sculpturatus* (and *Paravaejovis spinigerus*). Generally, *C. sculpturatus* has a sting that is worse than other U.S. bark scorpions, although I have read enough accounts to believe there are instances in which the sting of *C. vittatus* and *C. gracilis* can also be extremely painful or problematic. Some references flat-out state that *C. sculpturatus* is the only medically important species north of Mexico and that it is the only one that can cause systemic reactions. Some references also state that children and people with allergies can have serious consequences from *C. vittatus*. One reference addressing scorpions worldwide ranks other *Centruroides* north of Mexico as being highly toxic and states that *C. sculpturatus* is among the most dangerous species in the world, although case reports and summaries of typical stings do not really support that assertion.

Symptoms for scorpion stings may be rated to show relative degree of envenomation and reaction on a scale of Grade I–IV. Grade I: Local pain and paresthesia; II: Pain and paresthesia remote to the sting site (in addition to local); III: Cranial nerve *or* somatic skeletal muscle dysfunction; or IV: Cranial nerve *and* somatic skeletal muscle dysfunction. Cranial nerve dysfunction can include blurred vision, wandering eye movements, hypersalivation, trouble swallowing, tongue fasciculations, problems with upper airway, and slurred speech. Skeletal muscle dysfunction can include jerking of extremities, restlessness, and severe involuntary shaking and jerking. The onset of symptoms is generally less than one minute and in serious envenomations may upgrade quickly.

In a normal, healthy adult, there is usually immediate pain, although swelling is not normally excessive. Intense pain usually lasts 2–4 hours or more, then decreases over time. Stings on the extremities are often said to be more painful than those on the body. Sensitivity, numbness, tingling, and the feeling of electrical shock are common at the sting site and often remote from the sting site, as in the lips. Nausea, vomiting, chest tightness, and abdominal cramps may be symptoms of severely envenomated people. There may also be mental anguish. In the literature, there are many references to case studies and symptoms, and there is a range of different individual reactions. Most people I speak to about *C. sculpturatus* stings mention these symptoms and add that they could not sleep the first night. Symptoms may resolve in a couple days but may last for a week or more. I have heard that the sting pain ranges from "not unpleasant" (certainly the exception) to "worse than childbirth—way worse!" The take-home message I usually get from people envenomated by Arizona Bark Scorpions is that it is never a pleasant experience and symptoms may take too long to subside.

The real medical problem, in addition to an allergic reaction or generally severe symptoms, lies with small children, generally less than about 2–5 years of age. The onset of severe systemic symptoms can be rapid. In children, stings by Arizona Bark Scorpions should be considered potentially life-threatening. The same has been said by some authorities for *C. vittatus*.

However, stings are very rarely fatal, and there have been no reports of fatalities by *C. sculpturatus* since 1968. In early literature, death has also been attributed to *C. vittatus*, but details are sketchy. Death by bark scorpions is usually attributable to cardiac or respiratory failure from the effect of the neurotoxins. However, there is debate as to whether the venom of these species is actually capable of taking human life, or if previously reported mortalities may have been caused by other factors. There are numerous blog accounts of stings by *C. gracilis*, which is sometimes kept as an exotic pet. Symptoms range from "not-too-bad" and resolving quickly, to excruciatingly painful with symptoms including redness and extensive swelling. Symptom resolution ranged from minutes to days, and in one case, they escalated over time.

I found few accounts of stings by *I. maculatus*, but they are said to be quite painful, and there may be swelling, redness, tenderness, tingling, itching, formation of a hard welt, numbness, weakness, and feeling of electric shock. One person described it as being like slamming your hand down on a thumbtack. He also described a general feeling as similar to a moderate hangover. One set of thirteen case reports from Singapore found that symptoms were usually local and not severe. However, one of these patients had anaphylaxis. In usual cases, the pain subsides within an hour or so, but there may be lingering sensitivity.

Allergic reactions may be common, and at least nine allergenic compounds have been described for *C. vittatus*. Patients have had allergic reactions upon first sting, but it has been suggested this may be due to cross-reactivity with allergens in Imported Red (or Black) Fire Ants.

Remarks: There is an FDA-approved antivenom, Anascorp. Studies of its development and use show it to be highly effective. It is particularly important in cases where a small child has been envenomated.

There are occasional case reports of people being accidentally stung by exotic species in the wild. Stings by some of the much more venomous exotic species, such as the Deathstalker (*Leiurus quinquestriatus*) should be considered medical emergencies and victims should seek medical attention. Some species have become established from accidental or intentional releases.

In Mexico and other parts of Latin America, there are several species of *Centruroides* that are far more dangerous than our own. About 1,000 human deaths are attributable to those species annually in Mexico, with as many as 58,000 stings per year.

If native *Centruroides* aren't dangerous enough, you can buy a Deathstalker online or in your neighborhood pet shop! *L.L.C. Jones*

References and Resources: Babin et al. (1974), Boyer et al. (2009), Brown and Formanowicz (1995), Brown and O'Connell (2000), Brown and Formanowicz (1995), Brown et al. (2002), Chippaux and Goyffon (2010), Curry et al. (1983), Forrester and Stanley (2004), Jablonsky et al. (1995), Jones (2017, 2018a, b), Kang and Brooks (2017a, b), Langley (2005), Likes et al. (1984), Loret and Hammock (2001), LoVecchio and McBride (2003), Mazurkiewicz and Bertke (2005), McIntyre (1999), McReynolds (2008), Melville (2011), Micks (1960), More et al. (2004), Nugent et al. (2004), Polis (2001), Rowe and Rowe (2008), Rowe et al. (2011), Russell and Madon (1984), Shelley and Sissom (1995), Stahnke and Calos (1977), Stockmann and Ythier (2010), Valdez-Cruz (2004c), Wang and Strichartz (1983), Webber and Graham (2013), Whitmore et al. (1985), Yamashita (2004), Yamashita and Rhoads (2013).

Giant Hairy Scorpions (Genus *Hadrurus*)

Sting on 8 July, 5:18 p.m. "The pain came hot, throbbing, and fast. I knew she gave me a heavy dose, and the pain lasted for over an hour and a half in the index finger. The middle finger pain lasted only 20 minutes. At 7:00 to 10:00 my lips, teeth, and feet became tingly, and the finger was dead numb. The next morning all was well save for the finger still being stiff. Later that night, I experienced a muscle cramp in my left calf. 12 July: the cramp returned. 13 July: my left calf is sore and my index finger is still slightly stiff." [Señor Ocho, Arachnoboards online forum]

Classification: Phylum Arthropoda (arthropods); Class Arachnida (arachnids) or Scorpionida (extinct and extant scorpions); Order Scorpiones (scorpions); Family Carabactonidae (caraboctonids) or Iuridae (iurids); Genus *Hadrurus* (giant hairy scorpions).

Hadrurus spadix of the Mojave and Great Basin Deserts. *L.L.C. Jones*

In the United States, the genus *Hadrurus* is composed of four similar species, collectively known as giant hairy scorpions: *H. arizonensis*, *H. anzaborrego*, *H. obscurus*, and *H. spadix*. There are three additional species from Baja California, Mexico: *H. concolorous*, *H. hirsutus*, and *H. pinteri*. Some authorities include Mexican mainland giant hairy scorpions as belonging to the genus *Hoffmannihadrurus*, while others recommend retention within *Hadrurus*.

Identification: Giant hairy scorpions are the largest scorpions in the United States, reaching about 140 mm in total length. The "hairy" moniker comes from the extensive array of long setae. They have heavy bodies, strong pedipalps, a relatively thick metasoma, and a large, intimidating telson. Because of their impressive size, they are conspicuous animals and they are often kept as pets by scorpion aficionados.

One of the most familiar and well-studied species of scorpion in the American Southwest is the Arizona Giant Hairy Scorpion. The species can be differentiated from other scorpions by size, range, coloration, and anatomical features. There are light and dark color morphs, dependent upon the amount of dark pigmentation on the carapace. The "pallidus" morph is pale in coloration. Dark or variegated color morphs typically have a light-colored area between the lateral eyes that does not reach the median eyes. A recently described species, *H. anzaborrego*, is similar and has both light and dark morphs, as well. However, this species has a light marking that reaches the median eyes, and it also has dark pigmentation on the fingers of the chelae, which is lacking in *H. arizonensis*. As its name implies, the Black-backed Scorpion has an all-dark carapace. *Hadrurus obscurus* is also dark, but it may or may not have a light area between the median eyes.

Distribution: If considered distinct, the caraboctonids are strictly New World species, while the related iurids are Old World species. All species of *Hadrurus* are found in the United States and/or Mexico. There is confusion in the literature as to just which species are found where. They are primarily desert dwellers in the United States and Baja California. *Hadrurus arizonensis* is a broadly distributed species found primarily in the Sonoran Desert and part of the Mojave Desert (although some state *H. obscurus* is the Mojave species). They have been reported from southern and western Arizona, southeastern California, southern Nevada, and extreme southwestern Utah, as well as northwestern Mexico. There are reports of them in New Mexico, but I cannot find specific information to support that claim. *Hadrurus spadix* is primarily a Great Basin Desert form, although the range does overlap somewhat with *H. arizonensis*. It is found in central-eastern California, much of Nevada, northwestern Arizona, southern Utah, and a small area of western Colorado, southwestern Idaho, and southeastern Oregon. As its name suggests, *H. anzaborrego* is limited to the area of Anza-Borrego Desert State Park in southern California. The range of *Hadrurus obscurus* is somewhat unclear. Some authorities state it is primarily a Mojave Desert species, while maps show it in two disjunct semiarid areas of the southern Sierra Nevada and Coast Range Mountains of California. Although the southern Sierra Nevada does transition into the Mojave Desert, the Coast Range is clearly not in the Mojave Desert. This may be an artifact of studies focusing on small areas of southern California (e.g., Kern County, rather than Mojave-wide). Maps showing *H. obscurus* in the Sonoran Desert of southern California are now considered *H. anzaborrego*.

Natural History: Giant hairy scorpions are primarily desert dwellers, although they are absent from the Chihuahuan Desert. They are mostly found in open areas of scattered vegetation, as in creosote bush flats, sagebrush flats, and stabilized sand dunes. In my study area near Tucson, *H. arizonensis* is much more common in creosote flats, especially in sandy areas, than in rockier terrain. It is most common at the edge of sandy washes and dunes, where it can easily burrow, but also in areas with gravel substrates. They are sometimes found farther up into foothills of mountains, or some semiarid (but non-desert) habitats. Their burrows may reach 2.5 m deep. However, they are often found beneath surface objects as well. At night, they emerge on the surface to feed, assuming a basic sit-and-wait position, or may remain at the entrance of their burrow. They are sometimes spotted crossing roads at night. Given their size, they are easy to detect under a black light. The black back of *H. spadix* does not luminesce.

They are most active during the spring through early fall months, and tend to brumate over winter. In my study areas, they seem to prefer warmer temperatures than sympatric species. Like other scorpions, they have a courtship display, the "promenade a deux," which is usually initiated by the male and ultimately results in transfer of the spermatophore, which the male places on the substrate. Gestation is 6–12 months (average ten). They have 25–35 (average thirty) young. The young climb onto the mother's back until the first molt, at about three weeks old. They reach adulthood in 3–4 years.

Hadrurus arizonensis in defensive posture under black light. L.L.C. Jones

They live up to about twenty years. Predators include other scorpions, centipedes, lizards, snakes, small and medium mammals, and birds.

These animals are aggressive feeders and will readily consume other scorpions, including their own species, and other large arthropods. In a one-week period at the same study site, I witnessed two battles between similarly sized *H. arizonensis* and the centipede, *Scolopendra polymorpha*. In the first instance, the centipede won and consumed the scorpion, while in the second battle the scorpion won and consumed the centipede. It appeared as though the animal that got the best grip was the victor. Because of their size, they are also known to feed on vertebrate prey, including small lizards and snakes. They will try to flee predators, often down a burrow, but when confronted in

Arizona Giant Hairy Scorpion eating a Goode's Horned Lizard. *Jim Rorabaugh*

close proximity they may be quick to defend themselves. They are esteemed in the pet trade because they are large, relatively active, can become docile, and do well in captivity. It is often said they are good beginner scorpions for the terrarium, but they can still sting when handled, or when it is time to clean the cage.

Encounters: For people frequenting deserts (except the Chihuahuan), it is not uncommon to encounter these animals at night, especially during the summer. They are more likely to be seen in their natural habitat or backyards than inside a home. If they do get inside a home, they are more conspicuous than other species, so are generally easier to avoid. They are not as good climbers as bark scorpions, so not likely to crawl into bed with you.

Venom Apparatus, Venom, and Symptoms: Giant hairy scorpions do not have virulent venom. *Hadrurus arizonensis* has an LD_{50} of about 168 mg/kg, according to one study. Although the species sometimes feeds on small vertebrates, the venom does not seem to immobilize vertebrate prey, and they rely more on their pedipalps for suppressing their victims. In one instance I saw a large individual consuming a live Western Threadsnake. I do not know if it had attempted to grab or sting the prey, but the small snake was restrained only with the scorpion's mouthparts—the pincers were not even used, despite the fact that the still-living snake was thrashing around. *Hadrurus* are

Detail of *H. arizonensis* subduing a Common Desert Centipede with a well-placed stinger. *L.L.C. Jones*

reputedly able to spray their venom up to 25 cm, but that is probably a rarely used defense mechanism. This behavior may be more useful for dispersing the antimicrobial qualities of the venom.

The venom is used in prey procurement, defense against predators, and antimicrobial defense. *Hadrurus arizonensis* venom is known to have acetylcholinesterase. The venom of some Mexican *Hadrurus* species (especially *H. gertschi* and *H. aztecus*) has been studied recently, and several peptides have been described. One peptide from *H. gertschi* is known as hadrurin. This compound has antimicrobial and cytolytic properties. It is suggested these peptides are used for microbial protection in their subterranean haunts. Like other scorpions, they also have neurotoxic peptides, including hadrucalcin, and some peptides that appear to be precursors of the venom of other scorpions. The antimicrobial properties of the venom have made them important in research.

There is a saying in the Southwest: "Large scorpions are harmless, while small scorpions are dangerous." This undoubtedly arose in parts of the Southwest where most people are aware of only two types of scorpions: the large, conspicuous *Hadrurus,* and the smaller, but infamous, *C. sculpturatus.* Of course, there are numerous scorpion species within this area, and all are smaller than *Hadrurus*. These other small scorpions tend to deliver a less painful sting than *Hadrurus*, so this saying is not accurate. Published case reports of envenomation are few, probably because they are not considered medically important. However, there are numerous sting reports from online forums (usually pet keepers). Giant hairy scorpions are sometimes regarded as being harmless, but they can deliver a painful sting. Envenomation is often likened to less than or equal to a bee sting, but in some instances, it can be decidedly more painful, and one reference gives them a 3 of 4 on a pain and toxicity-to-humans scale among scorpions worldwide. A colleague of mine was stung twice by *H. arizonensis*. The first sting was to the heel and the pain was less than that of a bee, but the second sting to the tip of the finger was "far in excess of a bee sting." He said the pain was intense, ranking an 8 out of 10. The immediate effect of the venom is the peak pain, which may last from 5–30 minutes or longer. There is usually a burning and tingling feeling, as well as numbness and the feeling of electrical shock near the bite area. There is usually some swelling and redness, and the area may become raised and sometimes turn whitish. Muscle cramps are sometimes reported. The symptoms completely disappear after a few minutes to a few days, although minor long-term effects have been reported, such as general soreness of the affected area and tightness in the knuckle.

References and Resources: Fet and Soleglad (2008), Francke and Prendini (2008), Hadley and Williams (1968), Jones (2017, 2018a, b), Polis and Mc-Cormack (1987), Saunders and Johnson (1970), Schwartz et al. (2007, 2009), Soleglad and Fet (2010), Soleglad et al. (2011), Stockmann and Ythier (2010), Tallarovic et al. (2000), Torres-Larios et al. (2000).

Vaejovid Scorpions (Family Vaejovidae)

Paruroctonus borealis sting: "The pain was immediate and was VERY VERY hot. The hot pain lasted about 5 minutes and then after about 10 minutes my finger was completely numb. It was numb for about half an hour. I had no lasting effects except for localized pain of the sting site." [Scolopendra55, Arachnoboards online forum]

Pseudouroctonus williamsi. Chris Brown

Classification: Phylum Arthropoda (arthropods); Class Arachnida (arachnids) or Scorpionida (extant and extinct scorpions); Order Scorpiones (scorpions); Family Vaejovidae (vaejovids).

The Vaejovidae is a large family found in North and Central America, and the largest family in the United States and Canada. There is no universally accepted standard for taxonomy and nomenclature, so the number of genera and species is open for interpretation. As of July 2014, one source cited twenty-three genera and 191 species worldwide, while another cited 10 genera and 153 species. The latter report also stated that there were fifty-nine known but undescribed species and an estimated seventy unknown, undescribed species. The last decade or so has been very exciting in the scorpion world, as there were many new taxonomic revisions and descriptions proposed, especially to Vaejovidae. While it is not possible to have consensus on taxonomy or which species are valid, below is a list of some recent taxonomic changes. Interested parties should consult the primary literature to find the most recent information.

- *Catalania.* A new genus described in 2017 from southern California and Baja California.
- *Chihuahuanus.* This genus was described in 2013. It proposed to subsume some members of *Hoffmannius* and *Vaejovis*.
- *Gertschius.* This genus was described in 2007. The single species in the United States (*G. agilis*) was proposed as a recombination of *Serradigitus agilis*.
- *Graemeloweus.* A genus described in 2016. It is composed of three species formerly in *Pseudouroctonus* in northern California.
- *Hoffmannius.* This genus was described in 2008 and proposed to include several species of *Vaejovis*. In 2013, it was proposed to be abolished, with species transferred to *Paravaejovis*.
- *Kochius.* This genus was described in 2008. It was proposed to include 10 existing species of *Vaejovis* and one new species.
- *Kovarikia.* This genus was described in 2014, transferring three species previously in *Paruroctonus* over to *Kovarikia*, *K. angelena*, *bogerti*, and *williamsi*.
- *Maaykuyak.* This genus was described in 2013 and was proposed as a recombination of *Hoffmannius/Vaejovis waueri*.

- *Paravaejovis*. Although this genus was described in 1980, it was considered monotypic at the time (*Paravaejovis pumilis*, endemic to Baja California, Mexico). In 2013, several species of *Hoffmannius/Vaejovis* were proposed to be subsumed by this genus, including some of our more familiar species in the American Southwest, such as *P. spinigerus* and *P. confusus*.
- *Paruroctonus*. This genus was described in 1934, but in 1983, some species were separated into a separate genus (*Smeringurus*). This is one of the more speciose vaejovid genera.
- *Pseudouroctonus*. This genus was described in 1974. It contains numerous species in the American Southwest.
- *Serradigitus*. This genus was described in 1974. It contains numerous species in the American Southwest.
- *Smeringurus*. This genus was long considered in a subgenus of *Paruroctonus*, but was separated into a distinct species in 1983.
- *Stahnkeus*. This genus was described in 2006, subsuming some *Serradigitus*.
- *Uroctonites*. Described in 1991, with only four species, all from the American Southwest.
- *Uroctonus*. This genus was described in 1876. There are only three species generally recognized, although others in *Uroctonites* and *Pseudouroctonus* were within this genus. One has three subspecies. This genus is sometimes placed in the Chactidae.
- *Vaejovis*. In its heyday, nearly all vaejovids were in its nominal genus, but as can be seen from this list, scorpiologists have been especially active in the last decade. Still, there are probably more than seventy species generally recognized worldwide, making it the third most speciose scorpion genus in the world. It is well-represented by many species in the American Southwest, and new species are being described every year.
- *Wernerius*. This genus was described in 2008. Two species were transferred from *Vaejovis*, and one new species was described in 2012.

Identification: Most species are small to medium scorpions that are yellowish to brownish in coloration. Most lack easily observed characters that will render a positive identification. To identify vaejovids, one generally needs a stereo

Smeringurus mesaensis is one of the most studied scorpions. Inset shows the stiff setae on the legs that help support it on loose sand. *L.L.C. Jones*

dissecting microscope and familiarity with scorpion microanatomy. Even this can prove difficult. Dichotomous keys that do exist may be hidden in older literature that has an outdated taxonomy, possibly using different characters than are typically used now. This leaves the reader with digging through scientific literature to make identifications. Much of our current understanding comes from DNA analysis and a huge suite of morphometric and meristic characters. Suffice to say, a taxonomic understanding of this family is evolving.

Paruroctonus boreus is found in more northern latitudes than most scorpions. Robert Hansen

Distribution: Vaejovids are endemic to North and Central America. They are found across much of the United States and one species is found in Canada, but they are only diverse in the Southwest. Only one species, *Paruroctonus boreus*, has been able to survive in northern latitudes, reaching southern Alberta, British Columbia, and Saskatchewan. It occurs as far south as Arizona. This species is so widespread that genetic research has identified ten clades across its range. *Vaejovis carolinianus* is the only species east of the Mississippi River.

As its name implies, *Chihuahuanus* is associated with the geographic area of that desert, being found in Texas, New Mexico, and Arizona. The two species of *Gertschius* are limited to southern Arizona and southwestern New Mexico. *Kochius* is found in desert habitats in Arizona, California, and Nevada. The three species that constitute *Kovarikia* and *Graemeloweus* are endemic to California. The genus *Maaykuyak* has been recorded from Texas. The large recombined genus, *Paravaejovis*, has representative species in Arizona, California, Nevada, New Mexico, and Utah. *Stahnkeus* occurs in California and Arizona. *Wernerius* occurs within part of the Mojave Desert of California, Arizona, and Nevada.

Although an obligate dune dweller, this *Paruroctonus stahnkei* is hunting in vegetation. *L.L.C. Jones*

Most of these genera are also found in Mexico, in areas south of their U.S. distribution, and Mexico has an even more diverse vaejovid scorpion fauna, including several endemic genera.

Natural History: Various members of the family Vaejovidae cover the gamut of habitat types used by scorpions. They range from the lowest elevations of deserts (below sea level at Death Valley National Park), along the coastlines, to high in the mountains (e.g., like the *V. electrum* that shared a cabin with me at 2,895 m above sea level). Lithophilic species are those typically associated with rock, such as under rocks in deserts and woodlands, in talus slopes, and in limestone outcrops. Many *Vaejovis* belong to this group. Some of these are even found in caves (e.g., *Uroctonus* and *Pseudouroctonus reddellii*), and sometimes termed troglobitic, but they can also be found outside of caves.

Many species are psammophilic, inhabiting loose soils and dunes throughout the Southwest. *Paruroctonus* and *Smeringurus* are well known as dune dwellers. One species in particular, *S. mesaensis* (formerly *P. mesaensis*), has been deemed, "the most researched scorpion on earth." That no doubt

Courtship display in *Chihuahuanus crassimanus*. L.L.C. Jones

relates to its lifestyle and attention paid by a primary researcher, rather than its venom; many papers discuss the life and natural history of the species. It is a large scorpion, reaching about 80 mm in total length, and is dubbed the Giant Dunes Scorpion, Giant Sand Scorpion, Dune Scorpion, or Sand Scorpion. They and their relatives are conspicuous members of sand dunes in much of the Southwest. They are supremely adapted for life in dunes, and their morphology is replete with specialized features (e.g., sand-shoe setae and slit sensillae). Their size, abundance, extremely arid habitat, and spacing patterns on dunes, along with a proclivity for feeding on other scorpions (including their own kind) have made them ideal targets for study.

Encounters: Vaejovids may be encountered in the wild and sometimes in homes. In fact, while writing this very account, I had three *P. spinigerus* in my Tucson outskirts home this week. I even accidentally stepped on one (I always wear shoes). Shortly thereafter, I found *P. confusus* in my bathtub. Sweeping the yard and house with a black light is prudent to reduce encounters—this same week I also found *C. sculpturatus* in my yard.

While reading online forum accounts of sting reactions, I also got insight into methods of exposures. Many accidental stings came from people turning

The Stripe-tailed Scorpion is common and widespread in the Southwest and probably accounts for many stings. *L.L.C. Jones*

over rocks to look for scorpions or other animals. Tailing, or picking up a scorpion by the telson, is a common practice, but based on the number of sting reports, it is not a very safe technique, at least not without puncture-resistant gloves. Most forum reports seem to be from people being "tagged" by their terrarium pets.

Venom Apparatus, Venom, and Symptoms: The venom apparatus is as with other species of scorpions. None of the vaejovids are considered to be of medical importance, but many species are common and certainly add to the thousands of scorpion stings every year. Some species, such as *V. intermedius* are quick to sting and are said to have a very painful sting, while others are generally no worse than a bee sting. There is very little information on vaejovid venom, which is surprising, considering how many species there are and how common they are. The venom is not well studied for the group, but it has been shown that there are two hundred different venom components in *V. mexicanus*. The venom of that species has also yielded a potential new antibacterial drug, Vaejovin, which may be effective for multi-drug resistant bacteria.

Reports of the stings on online forums of *P. spinigerus*, a very common species in Arizona and elsewhere, ranged from "much less than the pain of

Uroctonus mordax, a common forest dweller in California. L.L.C. Jones

a wasp" to "the pain was immense! It felt like someone pulverized my thumb with a hammer!" According to the Arachnoboards sting report forum, stings of *Paruroctonus* spp. (*P. boreus, P. sylvestris, P. gracilior*) sounded fairly painful, from "that of a hornet" and "very, very hot" to "felt like someone stuck an icepick through my foot!" *Uroctonus mordax* and *Serradigitus gertschi* stings are generally reported to be fairly mild, less than a bee sting. Stings of *Smeringurus mesaesis* and *S. vachonis* are reported as being moderately painful and toxic, rating a 2 of 4 among scorpions worldwide. Most other vaejovids rank a 1 of 4. Stings by vaejovids and other scorpions are often reported to poison control centers, but are not separated from *Centruroides* or *Hadrurus* species.

References and Resources: Ayrey (2016), Fet and Soleglad (2005), González-Santillán and Prendini (2013), Graham and Soleglad (2007), Jones (2017, 2018a, b), Prendini and Wheeler (2005), Soleglad and Fet (2003, 2006, 2008), Soleglad et al. (2014, 2016, 2017), Stockmann and Ythier (2010), Stockwell (1989, 1992).

Other Scorpions (Families Chactidae, Scorpionidae, and Superstitioniidae)

Sting from *A. pococki*: "11 November, 6:00 p.m. Stung on left pinky finger. Mild stinging sensation lasted for approximately 1 hour. Moderately mild, tingly "funny bone" feeling began shortly thereafter and lasted approx. 4 hours. Redness and very slight swelling at time of sting lasted approx. 4 hours. All symptoms then disappeared until 17 November, 2:00 a.m. Awakened by intense itching at site of sting. Found it to be red and slightly swollen. Scratching caused swelling and redness of finger and feeling of heat. Finger is now moderately swollen, unbearably itchy, swollen and hard." [MzM, Arachnoboards online forum]

In the pet trade, this handsome California scorpion *(A. pococki)* is called the Mafia Scorpion. L.L.C. Jones

Classification: Phylum Arthropoda (arthropods); Class Arachnida (arachnids) or Scorpionida (extant and extinct scorpions); Order Scorpiones (scorpions). There are three remaining families of scorpions in the United States not covered elsewhere: Chactidae, Scorpionidae, and Superstitioniidae, although there are differing views on family placement.

In the United States, there is one genus represented by the Chactidae, *Anuroctonus*. Some authorities place this genus in the Iuridae, while others have suggested it may constitute its own family. There are two species: *A. pococki* and *A. phaiodactylus*. There are two subspecies of *A. pococki*: *A. p. pococki* and *A. p. bajae*, which intergrade. *Anuroctonus phaiodactylus* is monotypic, and it is geographically separated from *A. pococki*. The genus has vernacular names of black-clawed (or blackclaw) scorpions, swollenstinger scorpions, burrowing scorpions, or mafia scorpions (this latter name is sometimes also used for *H. spadix*, especially in the pet trade).

The genus *Diplocentrus* is the only genus of family Scorpionidae in the United States. Some arachnologists suggest it belongs to the sole family Diplocentridae. *Diplocentrus* are sometimes referred to as toothed scorpions, although few references suggest a common name. I prefer "chestnut scorpions" because of the color and satiny luster of the exoskeleton. Recent literature lists about sixty-two species in the genus. *Diplocentrus* are found mostly in Mexico and Central America, but extend into the American Southwest. All five U.S. species (with my assigned common name) are found in the Southwest: *D. diablo* (Devil's Chestnut Scorpion), *D. lindo* (Trans-Pecos Chestnut Scorpion), *D. peloncillensis* (Peloncillo Chestnut Scorpion), *D. spitzeri* (Madrean Chestnut Scorpion), and *D. whitei* (Big Bend Chestnut Scorpion). *Diplocentrus bigbendensis* is a junior synonym of *D. whitei*.

The last family, Superstitioniidae, is composed of one genus and species, *Superstitionia donensis* (Superstition Mountains Scorpion). By some accounts, the family also includes the genera *Alacran*, *Sotanochactus*, *Typhlochactus*, as well as the nominate genus. Most researchers, however, consider all non-*Superstitionia* to belong to a separate family, Typhlochactidae. The typhlochactids, per se, are troglobitic species.

Identification: *Anuroctonus* are relatively stocky in overall dimensions compared to most scorpions, although the metasoma is not particularly long or thick. They reach about 65 mm in length. The base color is tan to medium brown, but they may have a gray cast dorsally. They have a conspicuous and characteristic vesicle at the base of the aculeus (i.e., the "swollen stinger"). The claws are stout, heavy, powerful, and conspicuously ridged. The chelae are blackish or tan with black fingers. The two species and subspecies of *A. pococki* differ in numerous meristic features that are best viewed under a dissecting microscope, but knowing the range is the easiest way for the casual observer to differentiate species.

This Madrean Chestnut Scorpion was found under a rock with its unpigmented instars. *L.L.C. Jones*

Diplocentrus are medium to large, solidly built scorpions. *Diplocentrus whitei* is the largest U.S. member of the genus, reaching about 75 mm or more in total length, but the other species tend to be around 40–50 mm as adults. All U.S. species are dark brown to nearly black in color and have a glossy appearance. They have very large, heavy, ridged chelae, heavy bodies, and a relatively short metasoma. Members of the genus have a diagnostic subaculear tubercle with whitish setae. The species can usually be differentiated by microscopic characters.

The Superstition Mountains Scorpion is a tiny species, reaching only about 25 mm in total length. It is dark in overall coloration, from brown to almost completely black. However, most individuals have a light to medium base color with a dark brown median stripe on the dorsum that extends onto the metasoma, and a pair of wide dorsolateral dark brown stripes. It is heavy-bodied, with a short metasoma, short pedipalps, and stout (but not large) chelae with short fingers. This scorpion is also unusual in that it only has two lateral eye pairs, rather than the usual three. It has a fairly unique gland on the telson that is used in courtship. The gland appears as a wrinkled elliptical patch under magnification. One specimen was reported to have atypical pedipalp dentition, which could confound identification, as the arrangement of denticles is an important character.

Distribution: *Anuroctonus pococki* is endemic to southern California and Baja California Norte. In the United States, the subspecies, *A. p. pococki*, is found along the California Coast Ranges from the Mexican border to the south end of Monterey Bay, California. It is also found along the west side of the Baja California Peninsular Range in Mexico. *Anuroctonus p. bajae* is limited to the Baja California Range on the desert slope of southern California and northern Baja California Norte. *Anuroctonus p. pococki* × *A. p. bajae* are found in the Laguna Mountains area of San Diego County in southern California. *Anuroctonus phaiodactylus* is found in south-central California, south and southeastern Nevada, western Utah, and just into southern Idaho. It may occur in northwestern Arizona and eastern Oregon.

Diplocentrus diablo is found in the upper Rio Grande Valley of Texas, in Hidalgo, Starr, and Zapata counties, as well as Tamaulipas, Mexico. There are two species in the Trans-Pecos/Big Bend area: *D. lindo* is found over much of the region, while *D. whitei* is limited to Brewster and Presidio counties (i.e., the Big Bend, per se). Both species are also found in adjacent Mexico. *Diplocentrus peloncillensis* is endemic to the Peloncillo Mountains of extreme southeast Arizona (Cochise County) and southwest New Mexico (Hidalgo County), as well as adjacent Mexico. It is similar in appearance to the slightly more widespread species, *D. spitzeri*, which has been recorded from Santa Cruz and Cochise counties, Arizona, and adjacent Sonora.

Superstitionia donensis was described from the Superstition Mountains, near Phoenix, Arizona, but is much more widespread. In the United States, it occurs in Arizona, western New Mexico, extreme southern Nevada, and southern California. In Mexico, it is known from Sonora, Baja California Norte, and Baja California Sur. The specific epithet ("*donensis*") is derived from the type locality, at an area known as "Don's Camp," a place where *dons* (men, usually of high social stature) used to gather.

Natural History: Before *A. pococki* was described as a phenotypically cryptic species in 2004, the literature on *Anuroctonus* refers to only a single species, *A. phaiodactylus*. They live in areas of relatively compact soil, often in grassy, shrubby, or wooded hillsides. *Anuroctonus pococki* is found in relatively arid areas of chaparral in coastal and montane areas, being absent from true deserts. *Anuroctonus phaiodactylus* is a species of the Basin and Range province of the Great Basin Desert and edge of the Mojave Desert. Black-clawed scorpions are obligate burrowers. The burrows are not randomly distributed; rather, they tend to be aggregated in some areas, while absent in others.

The tiny *Superstitionia donensis* rarely ventures to the surface, so is not usually detected with a black light. *Margarethe Brummermann*

The burrows may be recognizable during the day, as they are slightly oval in shape, with discarded soil around the entrance, and dug at about 30–35 degrees into hillsides. They do not tend to be at the base of hills, where water may accumulate. According to one study, burrows averaged about 32 cm deep, with females having deeper burrows than males. There is a terminal chamber. Burrows are maintained for long periods of time, perhaps for the entire lifetime of the scorpion.

They wait at the entrance of the burrow for prey to cross their path. Because they cannot employ the sting from inside the burrow, they rely more on their large, strong pedipalps to subdue prey. Although they are reported to be obligate burrowers, one observer noted that he encountered several out in the open on a single evening when searching for them by black light.

Like some other burrowing scorpions, they use exudate from the telson and mix it with saliva to help clean their body, especially after burrowing. However, *A. phaiodactylus* venom was tested for antimicrobial properties and was found to be negative.

Known predators include Desert Shrews, Jerusalem crickets, darkling beetles, and several species of owls. One study of remains of Barn Owl pellets showed that almost all individual *Anuroctonus* consumed were adult males. They are probably preyed upon by a host of other animals, especially predators that can access their burrow or take them when on the surface.

A fair number of studies have been conducted on the biology of *Diplocentrus*. The Texas species occur from the Chihuahuan Desert up into the mountains, where they can be found in desert scrub, semi-desert grasslands, and oak and pine woodlands. In the Peloncillo Mountains of Arizona and New Mexico, *D. peloncillensis* is primarily in semi-desert grasslands and oak/juniper woodlands. *Diplocentrus spitzeri* is found in the mountains and their foothills near the Sonoran Desert, in some of the mountains of the Madrean Archipelago (i.e., the disjunct mountains allied to the northern Sierra Madre Occidental). The Madrean Chestnut Scorpion also occurs from semi-desert grasslands up to pine/oak woodlands. The Peloncillo Mountains are also part of the Madrean Archipelago, although the Peloncillo Chestnut Scorpion is considered distinct from the other species in the nearby mountains.

Chestnut scorpions are obligate burrowers, usually seen in the entrances of burrows or at burrows under rocks. They are sometimes surface active, as well, but are not as common on the surface as most other scorpions within their range. They may also be abundant under boulders. In one study of *Diplocentrus* in west Texas, they were detected at densities up to about 2 per 100 m^2 while during diurnal searching (e.g., looking under rocks), but generally less than 0.1 per 100 m^2 during nocturnal surveys with black lights. Activity seemed to peak in spring, but they were active throughout the warm season. In the same study, their biomass during some months exceeded that of Striped Bark Scorpions, even though that species accounted for 90% of all scorpion detections. During a recent visit to the Big Bend area, *D. whitei* was seen in each of the several areas I looked. They could be easily seen with a black light in the entrances of burrows. Burrows were generally in mixed rock and soil, but were also seen in areas of relatively loose silt substrates where the soil was stabilized by shrubs. They generally allowed slow approach to within a meter or so, and then scooted back into the burrow. In the Davis Mountains, I found *D. lindo* to be common under boulders; based on observations from my limited effort, they were nearly as common under rocks as *C. vittatus* in the area searched.

In this genus, the range of litter size has been reported as 7–38; the lowest were from *D. spitzeri* and *D. peloncillensis* (the smallest species), and highest in *D. whitei* (the largest species). *Diplocentrus* has few large offspring compared to other sympatric species, such as *Centruroides* and vaejovids. Larger females don't have larger young, but they do have larger litter sizes. There are usually 6–9 instars before maturity. Like other scorpions, young are born live (telson first) into a birth basket. In studies on captives, mothers ate young that failed to climb onto her back.

Predators and prey are probably similar to other burrowing scorpions. In Sonora, American Black Bears were documented preying on *D. peloncillensis*; bears are adept at flipping rocks and logs and digging to search for prey.

There is very little known about the natural history of *S. donensis*. It is not commonly encountered but may be locally abundant. It is generally known from desert-slope hillsides under stones on sandy soils, but can be found into the mountains. It comes out on the surface at night. It is said to be most active after rains. Some suggest this is primarily the monsoonal rains, while others suggest they are more likely to be seen during cooler rains. It is sometimes found in creosote bush litter. Given the wide range of the species, actual research on habitat affinities is warranted.

Encounters: None of these species are commonly encountered in the wild, unless one is specifically looking for them. Two of the three are burrowers, and the other is a tiny, uncommonly encountered species. They do not seek out homes, although like any scorpion, they might occasionally enter a domicile. Males wandering for mates are most likely to end up in a house.

Venom Apparatus, Venom, and Symptoms: The venom apparatus is the same as that of other scorpions. *Anuroctonus* has a swelling at the base of the aculeus, which is diagnostic for the genus. *Superstitionia* has a gland on the telson, but that is not involved in envenomation.

The venom of *Anuroctonus* has been studied somewhat, but the neurotoxic components target invertebrate prey, rather than mammalian predators or humans. Phaiodotoxin is a peptide that was isolated from *Anuroctonus* in Mexico. Another protein, phaiodactylipin, a specific type of phospholipase A_2, has also been isolated from *Anuroctonus* in Mexico. Five types of phospholipase A_2 have been identified from *A. pococki*, (Pha 1–5). Interestingly, the names of these venom compounds refer to the specific epithet "*phaiodactylus*," but the *Anuroctonus* in Mexico is now *A. pococki*, since its taxonomic split from *A. phaiodactylus* in 2004. These toxins are lethal to crickets and isopods, but are considered nontoxic to mice, at the levels tested in studies. In insects, they have a sodium ion-channel toxin. Another neurotoxic peptide is known as anuroctoxin. However, the venom has additional compounds and there are low-level haemotoxic and cytotoxic effects in mammals.

As with other non-medically important species, information on stings to humans is not common in the literature, but there are some sting reports on internet forums. These scorpions are not quick to sting but will defend themselves if they feel threatened. In humans, there can be pain, swelling, redness, and numbness. The sting can be rather painful (6 out of 10 in one report). One account listed "unbearable itching" the day after the sting. Effects can last more than one day, but usually symptoms dissipate in less time.

Chestnut scorpions tend to overpower prey with their strong pedipalps, and their metasoma and telson are fairly small and narrow. The venom has not been described. None of the U.S. Scorpionidae and/or Diplocentridae are considered medically important. The literature states that *Diplocentrus* spp. stings are typically mild and cause local pain with pronounced but temporary localized swelling, although more severe and irradiating pain may be possible. The little information I can find on stings by our native species is that they are pretty innocuous. Most reports are of slight pain, redness, and tingling, self-resolving within minutes. The sting has also been described by one victim as "less painful than a pinch."

I can find no information on the venom or sting of *Superstitionia*, except one report referred to it as mild, ranking a 1 of 4 among scorpions worldwide.

References and Resources: Brown (2004), Brown and Formanowicz (1996), Brown et al. (2002), Bücherl (1971), Crawford and Krehoff (1974), Francke (1974, 1981, 1984), López-González et al. (2009), Soleglad and Fet (2004), Stahnke (1940), Stanley (1966), Stockmann and Ythier (2010), Stockwell and Baldwin (2001), Torres-Larios et al. (2000), Valdez-Cruz et al. (2004a, b, 2007) Vignoli and Predini (2009).

Widow Spiders (Genus *Latrodectus*)

"We report a previously healthy patient bitten by a brown widow, resulting in a serious reaction requiring hospitalization. Symptoms included severe pain, cramps, nausea/vomiting, and fasciculations in the pectoral and quadriceps muscles. This report signals a need to re-evaluate previously held ideas that brown widow bites are of minor consequence." (Goddard et al. 2008)

Female *L. hesperus*, showing the tell-tale red hourglass on the underside. *L.L.C. Jones*

Classification: Phylum Arthropoda (arthropods); Class Arachnida (arachnids); Order Araneae (spiders); Family Theridiidae (cobweb, tangleweb, or comb-footed spiders); Genus *Latrodectus* (widow spiders).

The widow spider genus, *Latrodectus*, is found worldwide. There are more than 30 described species. In our area there are five species: *L. bishopi* (Red Widow), *L. geometricus* (Brown Widow), *L. hesperus* (Western Black Widow), *L. mactans* (Southern Black Widow), and *L. variolus* (Northern

Black Widow). Other members of the family include *Steatoda* (false black widows; *S. grossa* and *S. borealis*) and *Parasteatoda tepidariorum* (Common House Spider).

Identification: Adult female *Latrodectus* are about 13 mm in length (cephalothorax + abdomen) and nearly 60 mm across outstretched legs. The body form of all adult female widow spiders consists of a relatively small cephalothorax and a proportionately large, nearly spherical abdomen. The legs are long and slender. There are two pairs of four eyes. Another characteristic feature is the one that also places it in the family Theridiidae—the "comb." The comb is a distinctive set of serrated setae on the tarsus of the fourth leg. It is used to comb out the silk from its spinnerets to build its web and wrap prey.

Most *Latrodectus* are known as black widows because adult females of most species have a shiny black (or very dark brown) ground color. The Brown Widow Spider, *L. geometricus*, is the only one that has a base color that is distinctly brown, but it resembles black widows that are not fully mature. *Latrodectus* typically have an hourglass-shaped (sometimes triangular) red or orange (sometimes yellow) marking on the ventral side of the abdomen, but in Red Widows, the marking is reduced to a bar or spots.

Male *L. hesperus*. Whitney Cranshaw/Colorado State University/Bugwood

Northern and Southern Black Widows and Red Widows may have additional coloration and markings on the dorsum, sides, and/or legs. The Red Widow usually has red spots with yellow margins, and is unique in having a reddish celphalothorax and legs. The Brown Widow, besides having a brown abdomen, may be variously marked, having a series of spots on the dorsum, or the spots may be fused into stripes. There are a few color characters that can be used to differentiate black from brown widows.

All species are sexually dimorphic. The males are much smaller than the female, and have a proportionately smaller abdomen than adult females. Males only reach about 6 mm in cephalothorax + abdomen length, but with their long legs, are about 43 mm. Their markings are quite variable, but typically some variation of brown or black. Black with red markings or brown and reticulated are common abdominal color patterns; they often have orange and red markings on the legs. Adult males also have enlarged palps. Juvenile widows are also more variously marked than adult females and start out white to yellow, gaining pigmentation with each molt. Young of both sexes may have pigmentation that resembles adult males.

The web itself is distinctive, although it may be similar to that of other theridiids, including Common House Spider and false black widows. The

Some widow spiders have red markings elsewhere on the body, like this *L. mactans*.
L.L.C. Jones

webs appear messy or haphazard in design. They are commonly known as cobwebs, but that term has been misused to represent nearly any spider web. Black widow webs are fairly large and conspicuous but may be hidden among the structural components the web is attached to. Around homes, cobwebs are often among debris piles, structures such as brooms and garden tools, lawn furniture, barbeques, and so on. The egg sacs are also somewhat distinctive, being spherical or pear-shaped. Brown Widow egg sacs are unusual because they are covered in silky protrusions, like little spiky puffballs. The egg sacs of all *Latrodectus* are usually about the same size or larger than the adult female's abdomen, 12–15 mm in diameter.

Distribution: The Western Black Widow ranges from British Columbia and Alberta well into the west coast of Mexico, and from California to Oklahoma, Kansas, and Nebraska. The Southern Black Widow is found from Texas to Florida and north to New Jersey, Ohio, and Illinois. Its range overlaps extensively with the Northern Black Widow, which is found from Texas and Oklahoma to Florida, and north to Michigan, Wisconsin, Massachusetts, and Ontario. The Red Widow is endemic to central Florida. The Brown Widow is a non-native species that is now found throughout much of

Immature black widows (pictured) are similar to Brown Widows. *L.L.C. Jones*

the world, especially in tropical areas. In the United States, it was apparently introduced into Florida, but has now spread into every southern state. For example, it can be very common in southern California and is becoming increasingly widespread in southern Arizona. There is fear it may displace native black widow spiders. The Brown Widow is a supreme example of globalization, which in this case involves the spread of adaptable non-natives across the globe. Widow spiders make excellent stowaways, as in packed-up contents from garages or in lawn furniture. Western, Southern, and Brown Widow Spiders have become established in Hawaii.

Natural History: The biology of most widow spiders is well studied because they are abundant, relatively conspicuous, found around our homes, and medically important. Although widows are found worldwide, they tend to prefer warm areas, such as the deserts, tropics, subtropics, and warm temperate areas. They also occur in shrublands, grasslands, woodlands, and some forest habitats. They are more abundant in the southern United States than most of the more boreal areas. Red Widows are unusual in that they are rare endemics of a small area of Florida, in the central sand pine habitats. They avoid homes and are rarely encountered in palmettos.

Widows are seasonally active during the warm times of the year, when the temperature is at least about 21°C. Black widows and Brown Widows are normally found in dimly lit areas, such as rocky crevices, garages, and under tables, rather than in the open. A few theridiids are social, but widows are solitary, except when mating. Widows mate in the spring and summer. The moniker "black widow" is somewhat a misnomer. Although they are indeed black, *Latrodectus* females in North America rarely eat their mate after copulating. Males may mate several times in a year. Widows may have several egg sacs through the summer, and each sac can contain hundreds of eggs. The female guards the egg sacs and may bite intruders. Within a few days of hatching, the spiderlings send out silk threads, and are carried away in the wind—a phenomenon known as ballooning. The spiderlings will find a place to build a web and feed. They need to find overwintering areas to survive into the next year. Immature stages between molts are known as instars. Most widows live about a year, but some may overwinter and live up to three years.

Although the webs look to be haphazardly made, there is a rhyme and reason to their design. The web is generally attached to several surfaces to cover a fairly large area, and at least part is under an overhang. These

Brown Widows are nearly cosmopolitan, and rapidly populate new areas. *"Incidence Matrix"/CCA/Flickr*

animals are nocturnal hunters. By day, they hide in a dark area, near the top of their web, usually under the overhang. At night, they lay down hunting threads. Ground-dwelling, nocturnal animals, such as crickets, will run into the hunting threads, and be pulled up into the web, only to become more entangled. The widow will then race down to the victim and wrap it in silk and paralyze it. Like scorpions, spiders have a fluid diet, so the soft tissues are liquefied with digestive enzymes for consumption. Black widows often feed on large prey—from large arthropods to frogs, lizards, snakes, and even small mammals. After feeding, the prey is cut off the web and dropped to the ground; this is probably a mechanism to keep ants from entering the web. Red Widows are not as well known as other species, but they are known to build their webs in two species of palmetto trees. The webs are difficult to spot, except when they are covered in morning dew.

Encounters: Because of their proclivity for human milieu (except for the Red Widow), encounters with widow spiders are common and bites are not rare. While widows do occur within homes, they are generally more common in yards, sheds, and garages. Because there can be hundreds of juveniles

The Red Widow of Florida is an uncommon species that does not favor urban environments. *Florida Department of Plant Services/CCA/Bugwood*

produced every year from a single adult female, there can be local infestations. It is possible to just live with them, but there is some risk involved. A walk through the yard, house, garage, and sheds to look for the suspect cobwebs during the day can reveal the presence of widows. Be sure to look for the egg sacs. At night, visit the same areas with a flashlight to confirm the presence of widows. Other types of webs represent other types of spiders. Picnic tables and toilet covers are great widow habitat. It is not uncommon for victims to be envenomated in the buttocks or genitalia when they sit down to do their business in outhouses, so always look under the lid. Red Widows are rarely encountered, and populations may be low within their small geographic range.

It is difficult to assess the actual number of spider bites that occur, but for black widows, the culprit is often known, as opposed to recluse or other spiders. The spider is distinctive and often seen, twin pinprick marks may be visible on the skin, symptoms occur fairly soon after envenomation (not so with recluses), and the effects are often pronounced. Most bites occur in the southern and western United States. In a report on bites treated in U.S. emergency rooms from 2001–2004, 1,016 were attributed to widow spiders.

More than 2,500 cases are reported annually to U.S. poison control centers, but the actual number of bites in the United States is probably higher, since bites are not always reported and may simply be recorded as "spider." The different species of widows are not differentiated in these reports, and it is not known how Brown or Red Widow Spiders contribute.

Venom Apparatus, Venom, and Symptoms: Venom is used to procure prey, but is also used for defense, because a female may bite and envenomate an intruder that appears to threaten the egg sacs. Males are too small to break the skin and have little venom yield. The venom of widow spiders is produced in a pair of glands within the cephalothorax, immediately behind and attached to the chelicerae. The glands are whitish, and about 1.5–2 mm in length and 1 mm in diameter. They are composed of an inner lumen, surrounded by secretory cells, which in turn are surrounded by muscle fibers and other tissues. The secretory cells produce the venom, which is collected in the lumen. When an animal is bitten, the glandular muscles contract and venom is extruded through the needle-like fangs of the chelicerae. Thus, the venom delivery apparatus is much like that of a rattlesnake. The fangs are only about 1–2 mm in length. A spider cannot generally bite through thick clothing, so bare skin is usually the site of a bite. Also like a rattlesnake, a widow spider may "meter" its dosage when envenomating a victim. For example, pinching the leg of the spider, a body part that can be lost without adverse effects, often results in a dry bite or slight envenomation, whereas pinching the abdomen can result in a large dose of venom.

Widows have five neurotoxins that target insects: α-, β-, γ-, σ-, and ε-latroinsectotoxins. Interestingly, there is a neurotoxin that targets crustaceans, α-latrocrustatoxin; although most crustaceans are aquatic, some isopods (pillbugs or woodlice) are terrestrial and part of a widow's normal diet. There is one compound that has been isolated that affects mammals and other vertebrates, α-latrotoxin. It does not affect invertebrates. It appears to be the protein that causes symptoms in humans.

Alpha-latrotoxin affects neurons and neurotransmitters. It allows ions to pass through ion channels (calcium and a host of other ions) and precipitate neurotransmitter release, particularly norepinephrine, acetylcholine, and dopamine. An LD_{50} has been reported in two separate studies as 0.9 to 1.39 mg/kg for white mice. A lethal component isolated from α-latrotoxin has been called fraction B, and in pure form is lethal to mice at an LD_{50} of 0.048 mg/kg intraperitoneally. Although the potency is high, the volume of venom is not. A widow spider only has about 0.7 mg of venom.

It is likely that envenomation by all widow species is similar, except Brown Widows have a slightly less toxic venom, and they have a smaller venom yield. However, there are published case reports of bites by Brown Widows that required hospitalization, due to severe pain and/or systemic reactions. The medical term "latrodectism" is specific to symptoms caused by envenomation from widow spiders, while "arachnidism" is generic to spider bites. People bitten by female widow spiders may feel no pain or just "pin-pricks" at first. At the site of the bite, there may be a small red target. However, within about 5–15 minutes, symptoms appear. People bitten on the upper part of the body and upper extremities tend to have most pain and cramping in the thorax and upper extremities, while those bitten on the lower body or extremities tend to have most symptoms in the abdomen and lower extremities. Diagnostic symptoms of widow envenomation include extreme pain and muscle cramps (e.g., back, torso, neck, leg, hand, arm, foot), but may also include anxiety, difficulty in breathing, headache, high blood pressure, increased salivation, increased sweating, light sensitivity, muscle weakness, nausea, vomiting, numbness, restlessness, and possibly seizures. Contractions and premature labor have been reported in pregnant women.

It is difficult to know, based on literature and reports, just how deadly widow spider bites can be. There is a frequently recurring statistic of 5% mortality without treatment and about six deaths per year in the United States. However, some researchers suggest that deaths from widow bites are essentially nonexistent, including no confirmed cases in the United States. All reports tend to agree that normal, healthy adults are not at risk of fatal injury, but young children and those with suppressed immune systems are at a greater risk of complications.

Other theridiids are venomous, but the bite is less virulent than *Latrodectus*. Most notably are the false black widows and the Common House Spider. The latter can inflict pain and swelling, but it is usually not long-lasting. *Steatoda* bites tend to be a bit worse and are sometimes regarded as medically significant. *Steatoda grossa* can cause pain, both locally and remote from the bite site, nausea, muscle spasms, and malaise. In one study in Australia, symptoms of *Steatoda* spp. envenomation ("steatodism") was similar to that for *Latrodectus*, including some systemic effects; in most cases bites were less severe, but in some cases steatodism was indistinguishable from latrodectism. In one instance, Latrodectus antivenom was administered by accident (from initial false identification of the spider) for a *Steatoda* bite, but it did relieve symptoms.

Remarks: Antivenin Latrodectus mactans is an approved, commercially available antivenom. Its use, however, is the subject of debate among the medical community, spawned by a single fatality case due to hypersensitivity of the victim who received antivenom. However, there are journal articles that have shown excellent responses of victims. In these cases, it was administered because of envenomation pain, rather than because a physician or patient feared a fatal outcome. Typically, reported cases responded well to antivenom treatment.

References and Resources: Almeida et al. (2009), Carrel (2001), Clark et al. (1992), Cowles (2018), Garb et al. (2004), Goddard et al. (2008), Grishin (1998), Isbister and Gray (2003), Isbister et al. (2005), Jones et al. (2011), Kaston (1968, 1970, 1978), Magazanik et al. (1992), McCrone (1964), Meldolesi et al. (1986), Offerman et al. (2011), O'Neil et al. (2007), Reese (1944), Russell (1961, 2001), Tu (1984), Ushkaryov et al. (2004), Vetter (2013).

Why Do Widow Spiders Have Such a Potent Venom?

In his 1974 book, *Venomous Animals of the World*, Roger Caras posed the following question to the reader, regarding the severity of a case study of human envenomation by a black widow spider: "Why and how had it [the spider] been equipped to so seriously impair the health of a 160-pound teenage human being? It ate flies." This basic question of virulence of venoms to humans came up repeatedly in his book.

Since that time, we have learned a great deal about venomous animals and the toxins they sport. Numerous neurotoxic (and other) compounds have since been isolated from the venom of widow spiders. However, I don't think anyone actually answered his question. In 2011, I tried to publish a possible explanation when writing up a note on Western Black Widows feeding on Sonoran Coralsnakes. Although the manuscript was accepted, the editor said to leave out the possible explanation, as "it was too speculative." My view, however, is that speculation can be appropriate and valuable, as long as it is recognized as a theory and a hypothesis to test. So, here is my hypothesis, and maybe someone can test it.

In the course of doing a literature review for the manuscript, I discovered that published reports of vertebrate prey of widow spiders were surprisingly common. The literature helped me piece together a story of how widow spiders

Widows have toxins strong enough to subdue large thrashing vertebrates, such as this Sonoran Coralsnake. *L.L.C. Jones*

feed. In a nutshell, in addition to their messy webs, at night they put down hunting strands on the ground to capture nocturnal terrestrial prey (hence, not just eating flies). The strands tend to target large, ground-moving prey. As the prey is caught in the hunting threads, the strands dislodge from the ground and draw the animal upward into the web, where it becomes entangled in more silk. The spider then wraps its prey in silk and immobilizes it with its venom. After feeding, the prey is cut out of the web and dropped to the ground, where ants often run off with the spoils. This also keeps ants out of the web.

Meanwhile in the lab, biochemists discovered that the neurotoxic venom compounds of widow spiders include five types of toxins that target insects, one that targets crustaceans, and one that targets vertebrates. We know that insects and crustaceans (isopods or woodlice) are normal prey items, so why not small vertebrates, like amphibians, lizards, small snakes, and even mammals? The growing literature suggested that small vertebrates are not so unusual in the diet and are probably underrepresented as prey items because they get cut out of webs . . . and how many people really inspect widow webs to document prey, anyway?

My hypothesis is that the venom is "overkill" in humans primarily because it is designed to quickly immobilize relatively large thrashing prey. If not quickly immobilized, the prey could damage the web or the spider itself. Small vertebrates, along with insects and woodlice, are an abundant food resource, so why not capitalize on it? Although the venom can cause intense pain in humans it is not immediate (i.e., not effective as an algogenic agent). The pain from widow bites in humans comes from severe muscle cramps, which is the same protein that results in immobilization of smaller prey.

This is not to say that widow venom is not used for defense, but humans are probably not the primary target of a female widow's defense; rather, animals such as larger arthropods, frogs, lizards, small snakes, and mammals are known to eat spiders and might make a meal out of a widow and her young, so if she can turn the tides, she can turn a funeral into a banquet.

Recluse Spiders (Genus *Loxosceles*)

I woke up with a stinging sensation on my face. . . . I looked and in my right hand was a dead spider [confirmed Brown Recluse]. Over the course of the last 9 days, I have experienced the most insane symptoms. From excruciating nerve pain in my face, muscle spasms, full body rash, extreme swelling . . . etc. (M. Lindey, Instagram)

Brown Recluse Spiders are docile, but don't try this at home. *L.L.C. Jones*

Classification: Phylum Arthropoda (arthropods); Class Arachnida (arachnids); Order Araneae (spiders); Family Sicariidae (sicariid spiders); Genus *Loxosceles* (recluse spiders).

There are about 120 species of sicariids worldwide belonging to two genera: *Sicarius* and *Loxosceles*. About one hundred species of *Loxosceles* are known, nearly worldwide. In the United States, there are eleven native species: *L. apachea* (Apache Recluse Spider), *L. arizonica* (Arizona Recluse), *L. blanda* (Big Bend Recluse), *L. deserta* (Desert Recluse, formerly *L. unicolor*), *L. devia* (Texas Recluse), *L. kaiba* (Grand Canyon Recluse), *L. martha* (Martha Recluse), *L. palma* (Baja California Recluse), *L. reclusa* (Brown Recluse),

L. russelli (Russell Recluse), and *L. sabina* (Sabino Recluse). There are two non-native species in the United States: *L. laeta* (Chilean Recluse), introduced from South America, and *L. rufescens* (Mediterranean Recluse), from the Mediterranean. There are also native species of the Caribbean mentioned in the chapter on American territories.

Recluse spiders are also known as violin, fiddleback, or brown spiders. The name "Brown Recluse" is misused for any member of the genus, but that should be reserved for one species, *L. reclusa*.

Identification: Recluses are medium-sized spiders (11–17 mm cephalothorax + abdomen) that are overall brown to tan. The abdomen varies in size substantially among sexes and individuals, so range from about the size of the carapace to much larger. The abdomen is usually unicolored, or nearly so, and varies in color. Their long legs are sparsely haired and lack spines or a pattern. Recluse spiders have a characteristic pattern of three sets of two eyes on the anterior cephalothorax, forming a semicircle with one pair anteriorly and two pair posterolaterally; this eye arrangement is diagnostic for the Sicariidae, but also found in the family Scytodidae (spitting spiders), and Diguetidae (desertshrub spiders) have six eyes in a different pattern. Most species have a distinctive violin- or fiddle-shaped marking on the carapace. However, in *L. deserta* and *L. blanda*, the marking may be faint or lacking, so it is best to rely on eye pattern and other characters for positive identification. The "body" of the violin marking is on an elevated portion of the anterior carapace, while the "neck" of the violin is posterior. Even in *L. deserta* and *L. blanda*, which lack the darker pigment, there are rows of hairs that show the vague violin pattern.

To an arachnologist, these spiders are distinctive and easy to recognize, but to a layperson they may not be. Other spiders may have violin-like markings, including Pholcidae (cellar spiders) and male *Kukulcania* spp. (Filistidae). One rather convincing genus of spider is *Titiotus* spp. (Tengellidae), a genus endemic to California and Oregon. While it superficially resembles *Loxosceles* in general shape and coloration, it has eight eyes. Spitting spiders have a similar eye pattern, but otherwise do not resemble recluse spiders. The high degree of layperson and physician/clinician misidentification has been shown in studies where people were asked to send what they believed to be recluse spiders to experts; a wide variety of species was submitted, including those mentioned above, but also species far less similar, and many from outside the range of any *Loxosceles*. It is

This *L. arizonica* was found in the sink of my Tucson home. *L.L.C. Jones*

advised that the reader become familiar with the characters and go online or read pertinent literature to see examples and view images. Binoculars that are used by butterfly watchers can focus as close as 0.5 meters; this can make spider identification much easier, as one can even discern the eye pattern. The various species of *Loxosceles* are difficult to differentiate from one another, but they have largely different geographic ranges. However, identification of species is not necessarily important, because all should be considered to have similar venom qualities.

Distribution: *Loxosceles* is predominately a New World genus, occurring mostly in tropical and subtropical areas and deserts. Some species also occur in Africa, Europe, and Asia. Some have been introduced to other regions, including Australia and other islands.

In the United States, they reach their highest diversity in the "hot" deserts (Sonoran, Mojave, and Chihuhuan Deserts), and absent from the "cold" desert (Great Basin Desert). Five species are found in Arizona: *L. deserta* in the western Sonoran and Mojave Deserts; *L. kaiba*, which is endemic to the Arizona Strip; *L. arizonica* is endemic to central Arizona; *L. sabina* is endemic to the Tucson area (along with perhaps two other species), and *L. apachea* is found in southeastern Arizona, southern New Mexico, and extreme west

Loxosceles deserta lacks the "violin" marking, but has the diagnostic eye pattern.
L.L.C. Jones

Texas. *Loxosceles deserta* also occurs in adjacent Mojave and Sonoran deserts in Nevada, California, and extreme southwestern Utah. *Loxosceles russelli* is endemic to California. *Loxosceles palma*, primarily a Baja California species, and *L. martha* are also known from southern California. In addition to *L. apachea* in the El Paso/Franklin Mountains area, Texas has two endemic or near endemic species: *L. blanda* of the Big Bend and Trans-Pecos areas of the Chihuahuan Desert (just entering southwest New Mexico) and *L. devia* of south Texas. The most wide-ranging and most commonly encountered species is *L. reclusa*, which is found throughout the remainder of Texas, in the Midwest as far north as Nebraska, Illinois, and Iowa, east to the western Carolinas, and in the Southeast as far as the Florida panhandle. However, it is rare in the Gulf Coast.

Despite this seemingly broad range of distribution in the United States, most of the country lacks any species. *Loxosceles rufescens* has appeared in various places in the United States, beyond the range of native species. The recluse spider of Hawaii is the introduced Mediterranean Recluse. The Chilean Recluse (*L. laeta*) has been reported from the Los Angeles area and elsewhere. There have been some infestations and subsequent eradications of non-native populations.

Natural History: The western species are all desert and semi-desert dwellers. In South Texas, *L. devia* occurs in the subtropical Tamaulipan thorn scrub. The Brown Recluse Spider is the most wide ranging, and it occurs in grasslands (i.e., southern Great Plains), woodlands, and parts of the Southeast. Most species do not tolerate cold winter temperatures, which explains the generally southern and tropical distribution.

As the name implies, recluse spiders are secretive. They are largely nocturnal, but may be active during the day in dark areas. In the wild, they occur singly or in aggregations in crevices, rock piles, under bark and vegetation, and in caves. For example, in a study in the Tucson Mountains, recluse spiders were commonly found under fallen saguaro cactuses.

As spiders go, they are somewhat long-lived. They may live two to five years as adults, overwintering outside, or having an extended active season inside homes. They also tend to have relatively few eggs and sacs. The Brown Recluse Spider generally has about fifty eggs per sac, and two or three sacs per year. Only about half the eggs will hatch. Spiderlings do not balloon, so dispersal is limited; populations in dwellings are no doubt based on a very localized, expanding population from individuals inside the home. The Brown and Chilean Recluse Spiders and the South American *L. intermedia* are the most studied—relatively little is known about other species.

Recluse spiders feed on invertebrates. Spiders, crickets, and beetles are important food items. They use silk to hunt prey, but do not build conspicuous webs. If there is a web, it is usually small and misshapen, with the spider's retreat nearby. These sit-and-wait predators will remain motionless for most of their lives. Movement is usually to mate, feed, and manage hunting threads. They put out non-sticky web strands to alert them to arthropods passing by, then will rush out to grab the prey and envenomate it. They may go extended periods without food or water, which is part of the reason infestations can build over time in homes. It is likely that many spider-eating animals probably feed on them, including other arthropods, lizards, some snakes, birds, and small and medium mammals.

Encounters: Except for the Brown and Mediterranean Recluse Spiders, *Loxosceles* do not have an urban or suburban affinity, so they are rarely seen. In some studies, hundreds or even thousands of individual Brown Recluses have been shown to occupy a single home and may be rare immediately outside of the infested human dwelling. As in nature, they tend to hide in dark, tight places, such as bookshelves, cabinets, among cardboard boxes,

Terrestrial Invertebrates   **159**

The Sabino Recluse has a tiny known distribution, mostly in caves. This one has killed a cave-dwelling scorpion. *Jillian Cowles/Amazing Arachnids*

Loxosceles rufescens is a widespread invasive species from the Mediterranean. *Jacinto Lluch-Valero/CCA-SA/Flickr*

under pictures, among clothing, and in drainpipes. Some experts suggest other species do not enter homes, but this is not entirely true. I occasionally have *Loxosceles* spp. visiting in my Tucson home, usually through drainpipes.

Even in homes with high densities of recluse spiders, there are rarely bites. An often-cited example is one Kansas home that had more than 2,000 Brown Recluse Spiders, yet no bites occurred. Neighboring homes averaged nearly one hundred spiders, also without reported bites. The animals are inoffensive, and bites occur when humans accidentally come into contact with them. Most bites in the United States, according to one study, occur on the legs and feet, suggesting socks, shoes, and pants are a common method of transmitting the spider to the skin. They tend to occupy clothing that has not been worn for a while.

Venom Apparatus, Venom, and Symptoms: The venom apparatus is typical of spiders, having venom glands in the chelicerae and fangs to deliver the venom. The fangs of adults are about 1.5–2 mm long, capable of breaking human skin, while those of juveniles are shorter and probably cannot break skin, at least through clothing. The venom yield is miniscule, about 0.70 mg or less.

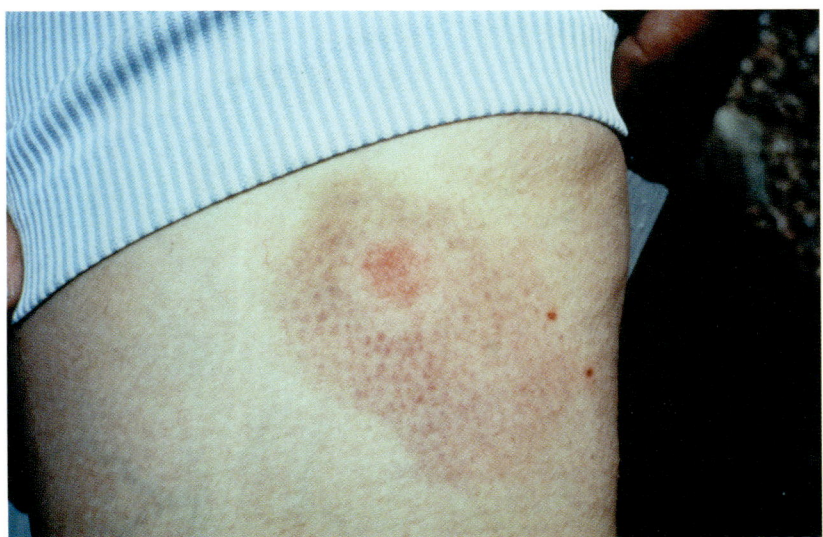

Lesion caused by the bite of a Brown Recluse. *Findlay Russell*

This is the only spider genus in the United States that has proven to cause true necrosis in humans; the phenomenon is often referred to as dermonecrosis (necrosis of the skin), necrotic arachnidism (spider necrosis), or loxoscelism (symptoms from the bite of *Loxosceles*), even if not involving necrosis. It is this symptom that has caused such a spider-bite scare from the literature and media. There has been considerable research on *Loxosceles* venom, especially in recent years. Many venom components have been identified. The venom is cytotoxic and hemotoxic. It is composed of proteins, glycoproteins, peptides, and several destructive enzymes. Enzyme activity (at least for various species studied) includes hyaluronidase, esterase, acetylcholinesterase, lipase, protease, ribonuclease, and phospholipases. The causative agents of necrotic lesions are due to the phospholipase-D family of enzymes. These enzymes were previously termed sphingomyelinase-D enzymes. The venom is also a potent insecticide, so it probably evolved for subduing arthropod prey rather than defense against vertebrates.

There are differences in toxicity among species and among sexes, but most species have not been studied extensively. In one study there was an LD_{50} of 1.45 mg/kg for *L. laeta* and 0.48 for *L. intermedia*, both of which are South American species that may inhabit dwellings. *Loxosceles reclusa* has also been reported to have an LD_{50} of 0.48. It is surmised that at least some South American species are more toxic than North American species, but this has not been well studied. However, there is usually greater severity of bites, in general, between the United States and South America (especially Chile, where *L. laeta* is common). Phospholipase-D enzymes are probably found throughout the genus, even though it was reported as absent in earlier literature. Specific LD_{50} tests are lacking for most species and may be highly variable between and among species in different studies. All species are venomous, however, and probably have similar chemical properties in their venom. In some species studied, females have been shown to be more venomous than males. There can be antigenic cross-reactivity.

It should be recognized that a "bite from a recluse spider" is extremely over-diagnosed in the medical profession and likely over-reported at poison control centers and emergency rooms. There are several studies on this topic, but a good example is of one that demonstrates 80% likely misdiagnoses, and another that showed that 60% of the diagnosed cases were from areas outside the range of any *Loxosceles* spiders. Bites from species other than *L. reclusa* in the United States are rarely confirmed, although there are some reports of human envenomation. The authority on U.S. recluse spiders informed me he only knew of perhaps one bite by a recluse spider west of the range of the Brown Recluse. Part of the problem is that *Loxosceles* bites are not usually felt, pain is not immediate, and the spider is not usually seen—hence, no "smoking gun." Spiders of some sort seem common just about everywhere, including homes and other structures. To laypersons and many physicians and clinicians, there is an erroneous assumption that the presence of spiders in a house indicates the cause of dermonecrotic wounds. This is problematic because misdiagnoses allow other serious maladies to go untreated.

Several studies have demonstrated that there are dozens of other medical causes that can be misdiagnosed as recluse spider bites, including bacterial infection (particularly by methicillin-resistant *Staphylococcus aureus*), herpes, diabetic ulcer, fungal infection, gangrene, squamous cell carcinoma, syphilis, Lyme disease, cowpox, anthrax, and many others.

Recluse spiders can deliver dry bites, because venom glands are controlled by voluntary muscle contraction, but loxoscelism generally refers to local, dermonecrotic, and systemic reactions. There are four categories of effects by recluse spider bites sometimes referenced: (1) unremarkable (very little damage, self-healing); (2) mild redness, itching, slight lesion, but self-healing; (3) dermonecrotic (necrotic skin-lesion forming); and (4) systemic or viscerocutaneous (affects vascular system, very rare, potentially fatal). In reports from the United States, most confirmed cases are self-healing and do not require medical intervention. About 90% of envenomations by native species do not cause serious reactions and less than 1% are systemic. For those bites that cause necrosis, two-thirds heal normally. Case reports tend to contain information on extreme cases, rather than typical or mild cases. Nevertheless, the consequences of a bite can be serious, especially for small children and people with compromised immune systems. In cases of necrosis, the lesion may be 40 cm in diameter, cause scarring, and take months to heal. Some may require skin grafting.

Symptoms usually include some pain and swelling that does not appear for 2–8 hours. A blister may form at the site. A scab can form, which may heal or drop off after 2–5 weeks. In place of the scab, a necrotic lesion sometimes forms. In a small percentage of confirmed U.S. envenomations, symptoms became systemic, particularly in children. When they occur, these reactions include malaise, muscle pain, nausea, vomiting, and fever. In extreme cases, there can also be blood clotting, hemolytic anemia, rhabdomyolysis (breakdown of muscle tissue that releases harmful products into the bloodstream), renal failure, coma, and death. Fatal outcomes are very rare for native species, and usually attributable to renal failure.

The Chilean Recluse reputedly has the most virulent toxin and/or high venom yield of those in the United States, and necrotic lesions are not uncommon in bites from that species. In one study in Chile, *L. laeta* bites were often serious, with fatal outcomes in 3–4% of the cases. No bites at *L. laeta* infestations are known to have occurred in the United States. The few infested areas were temporal in nature, being eradicated before they could spread, but spider introductions are a moving target.

Remarks: There are many published treatment options in the literature. There are no approved antivenoms in the United States but there are in Latin America. Results of antivenom treatment on *L. reclusa* using sources from Latin America are encouraging, but it needs to be administered early to be effective, usually before symptoms appear.

If someone knows for certain that a particular spider bit them, it can be killed (preferably not completely smooshed), or coaxed into a jar, where rubbing alcohol can be added to kill and preserve the specimen. The identity needs to be confirmed by a spider expert, which is usually not the victim or medical personnel. If confirmed by an arachnologist, it will help the physician to determine treatment options. There is enzyme-linked immunosorbent assay (ELISA) designed to detect *Loxosceles* venom that has shown promise.

References and Resources: Barbaro et al. (1996, 2005), Bey et al. (1997), Binford et al. (2005, 2008), Blackman (1995), Cowles (2018), Cramer (2015), Davidovici and Wolf (2007), de Oliveira et al. (2005), de Roodt et al. (2007), Forrester et al. (1978), Geren and Odell (1984), Gertsch and Ennik (1983), Gomez et al. (2002), Hogan et al. (2004), Ori (1984), Richards (2008), Richman (1973), Sandidge and Hopwood (2005), Swanson and Vetter (2006), Vetter (2005, 2008, 2015), Vetter and Barger (2002), Vetter and Bush (2002), Vetter et al. (2003a), Waldron (1969), Zobel-Thropp et al. (2012).

Spider Bites: Tracking a Mystery Wrapped in an Enigma

Here are some facts and estimates. With a few exceptions, all spiders are venomous. As of May 19, 2018, the World Spider Catalog listed 47,518 valid species of spiders on Earth, with some 4,000 species in North America, north of Mexico. A study in 2017 estimated about 25 million tons of spiders on Earth, which eat 400–800 million tons of prey—mostly insects—exceeding the total biomass of humans. One estimate of grassland habitats showed an average 152 spiders/m². One house in Kansas had more than 2,000 Brown Recluse Spiders, and a study in urban Arizona showed a density of 0.15 Western Black Widows/m² in good habitat. There are a lot of spiders out there.

Here are some bite statistics. A 25-year synthesis of telephone calls to poison control centers (1983–2009) reported 362,617 bites by spiders in the United States, of which 58,304 were from black widows and 42,544 were from recluse spiders. This comes out to about 14,000 spider bites reported to poison control centers per year. The same publication reported that 13% of the fatal envenomations were due to spider bites. Another study of envenomation from 2001–2004 treated in U.S. emergency rooms reported 123,000

Spider bites, especially those alleged from recluse spiders, are sensationalistic media fodder. This story tells of symptoms that cannot possibly be due to a spider. False news of spider bites is still rampant, despite the attempts by Rick Vetter and other professionals to separate fact from fantasy.

cases of spider bites, *annually*. Another study lists sixty-six fatalities from spiders in one decade. It sounds pretty ominous. These statistics come from peer-reviewed scientific journals, and given how many spiders there are out there, they should reflect actual numbers, right? Not so fast.

Richard Vetter and several other researchers have pointed out that it is much more likely that the majority of these reports were from uninformed victims and misdiagnoses by medical staff. There is a sort of spider panic, fueled by disdain, misinformation, and sensational media blitzes. Without going into details, Vetter and his colleagues have shown how easily spiders are misidentified by laypersons and how medical personnel are a little too quick to point the accusing finger at these arachnids. What is usually lacking is the smoking gun; a diagnosis is often made in the absence of a culprit at the scene of the crime. A spider in the house or even the room means very little, as spiders are reluctant to bite. In some instances, a victim does see a black widow that bites them, but a confirmed recluse spider bite is a rare thing. Recluses and Hobo Spiders are often blamed simply because uninformed medical personnel immediately think "spider" when dermonecrosis appears. This malady is actually an uncommon outcome of recluse bites, anyway, and never confirmed in Hobo Spiders, but dermonecrosis is a common symptom of dozens of other diseases. In the United States, spider bites are almost never fatal, so nearly all reports of fatalities are of misdiagnoses.

What we need is to put things in perspective. A small subset of U.S. spiders is capable of human envenomation, as covered in this book. In general, a spider with large enough fangs can bite a person if it is pressed against the skin. Many spiders are mildly venomous to humans, and only a few can cause significant local or systemic symptoms. The real problem with bite statistics lies in the current inability of poison control centers and emergency rooms to track meaningful information. Until we have a reporting system that includes only confirmed cases, tracking bite statistics in spiders will be fraught with error. Thus, this is a call to readers to report known, not surmised, bites to a reliable data source or publication. One reliable method is through published case studies with the culprit confirmed by someone versed in spider identification, and vouchers are always recommended. Also, it is important to report asymptomatic reactions, as well as those of little consequence; most case studies of envenomation tend to report only severe cases. Except for widows and recluses, the bites of most species are poorly known. Interestingly, due to the paucity of information for most species, one of my best sources of first-hand accounts of arachnid bites came from online forums, where knowledgeable arachnid aficionados report their envenomations, including dry or mild bites.

References: Bradley (2013), Langley (2005), Nyffeler and Birkhofer (2017), O'Neil et al. (2007), Trubi et al. (2011), Ubick et al. (2005), Vetter (2015).

Funnel Weaver and Grass Spiders (Family Agelenidae)

[Following the confirmed bite of *Agelenopsis aperta* in Southern California] "A 9-year old boy bitten on the neck experienced neck rigidity and pharyngeal edema [swelling] within 1 h of envenomation, quickly progressing to headache, nausea and disorientation. The bite site was painful upon palpation with arthralgia and myalgia [joint and muscle pain] in the neck and shoulders without edema or erythema [redness]. Symptoms later progressed to pallor, malaise, throbbing headache, unsteadiness with leg heaviness, and a sensation of throat tightness, regressing by morning and dissipating over the next 3 days." (Vetter 2012; a summary of his 1998 case report)

Classification: Phylum Arthropoda (arthropods); Class Arachnida (arachnids); Order Araneae (spiders); Family Agelenidae (funnel weaver spiders).

Hololena nedra. Ron Wolf

Genera and species known to cause human envenomation include: *Agelenopsis aperta* (Desert Grass Spider), *A. oregonensis* (Oregon Grass Spider), *A. pennsylvanica* (a grass spider), *Eratigena* (=*Tegenaria*) *agrestis* (Hobo Spider), *E. duellica* (Giant House Spider), and *Hololena* spp. (no common name). The *Hololena* species implicated in human envenomation include *H. pacifica*, *H. sula*, and *H. nedra* (but possibly *H. adnexa* or *H. turba*; taxonomy is problematic at the species level). *Tegenaria* is a large genus and some species were transferred to *Eratigena*. See Distribution section below for other genera.

There are more than seventy genera and about 1,200 species of Agelenidae worldwide. The family Agelenidae is known as funnel weaver or funnel-web weaver spiders, but should not be confused with the funnel-web spiders (families Hexathelidae and Dipluridae). The Hexathelidae has some seriously venomous species, but none of these are found in North America. The Dipluridae occur in the United States, but are not known to be particularly venomous to humans. Some members of the Agelenidae are also known as grass spiders, particularly those in the genus *Agelenopsis*.

The Hobo Spider is sometimes referred to by a more sinister common name, "aggressive house spider," but that is an artifact of a misinterpretation of its Latin epithet, *agrestis*, which means "from the fields," rather than "aggressive." Media sensationalism probably encouraged that moniker, but "Hobo Spider" is the generally accepted English name.

Identification: Superficially, agelenids resemble wolf spiders (family Lycosidae). As with lycosids, they can move rapidly to procure prey. Wolf spiders (except one genus) do not make sheet webs, a signature identifier for agelenids. Lycosids and agelenids can also be differentiated by eye pattern; the former has a distinctive rectangular shape. The eight eyes of agelenids are in two strongly procurved rows, except for *Eratigena* and *Tegenaria*, which are slightly procurved to nearly straight. Agelenids have feathery setae on the body.

Adult female agelenids range from about 7–19 mm (cephalothorax + abdomen) and about 7–10 mm for males. The dorsal cephalothorax (carapace) of most species is tan to grayish with a pair of darker longitudinal stripes. The abdomen tends to be striped in two parallel rows, or is spotted along those rows. Some have a pattern between the parallel rows. The Hobo Spider usually has a "herringbone" pattern on the abdomen. The legs are of medium thickness and length. They are often banded, but some species are unbanded to weakly banded. There are conspicuous setae and trichobothria on the legs. *Agelenopsis* are among the largest of the genera, with females

Agelenopsis sp. Identification to genus and species is difficult for agelenids, but they should all be considered venomous to humans until proven otherwise.
Joseph Berger/Bugwood

reaching about 19 mm. They have a narrow abdomen and long spinnerets. *Melpomene* also has long spinnerets. *Hololena* and *Calilena* have shorter spinnerets, typically a two-line spotted abdomen and are yellowish (*Hololena*) to grayish (*Calilena*), with distinctly banded legs. *Hololena* has the shortest spinnerets, which are barely discernible when viewed from above. The many genera and species of agelenids are similar, and identification may require an expert armed with a microscope. Even then, some species are poorly defined or may be difficult to differentiate. For our purposes, it seems prudent to avoid contact with all species of agelenid.

Agelenids are also identified by their web, which they rarely leave. The web is sheet-like, having a funnel at the rear. The spider is often seen just inside the entrance of the funnel.

Distribution: Agelenids are common throughout most of North America, but are particularly diverse in the American Southwest. North of Mexico there are thirteen species of *Agelenopsis*, two species in *Eratigena* and thirty species of *Hololena*. Other U.S. genera include *Barronopsis* (5 species), *Calilena* (19 species), *Melpomene* (1 species), *Novalena* (6 species), *Rualena* (9 species), *Tegenaria* (3 species), and *Tortolena* (1 species). In addition, *Coras* (15 species) and

This *M. rita* kept me company while I was writing the first draft of the agelenid account. Its web was in my windowsill. *L.L.C. Jones*

Wadotes (11 species) are now usually considered to be agelenids, Canadian species include *A. actuosa*, *A. naevia*, *A. oregonensis*, *A. potteri* (introduced), *A. utahana*, *E. agrestis* (introduced), *E. atrica* (introduced), *T. domestica*, *W. calcaratus*, and *W. hybridus*. These and the rest also occur in the United States.

The Hobo Spider and its kin are mostly Old World species. *Eratigena agrestis* is a native of Europe. It was likely introduced into the port of Seattle in the early 1900s. In the United States, it is predominately found in the Pacific Northwest, and adjacent southwestern Canada. It has become established in Washington, Oregon, British Columbia, Alberta, Wyoming, Idaho, and Montana, but also northern Utah and isolated areas of Colorado, and probably occurs in northern California and perhaps Nevada. The species will likely expand its range but has a preference for boreal climates, so is not expected to become widespread in the southern states.

A larger species of the genus is the Giant House Spider, *E. duellica*. It is more likely to invade homes than the Hobo Spider. It is mostly a boreal species in the United States, but has at least been recorded in Colorado. The Barn Funnel Weaver (*T. domestica*) is also a common spider of human domiciles, and is essentially cosmopolitan. It can be extremely common in homes in the West. The other *Eratigena* are also non-native species from

Europe. The only native *Tegenaria* is *T. chiricahuae*, from southeastern Arizona, although there are some species found in Mexico.

Agelenopsis are the typical grass spiders that are found throughout North America. The greatest diversity is in Texas, where northern, western, eastern, and endemic forms converge. The most broadly distributed species is the Desert Grass Spider, found from California to Texas and north to Wyoming. Other species of grass spiders are distributed across most of the United States. *Hololena* is the largest genus of agelenids in the United States, where it is endemic, or nearly so. It is most diverse and distributed in the West. Twenty-two of the thirty species are found in California—mostly in the southern and central parts of the state. Similarly, *Novalena* and *Calilena* are mostly western, also with high diversity in California. *Rualena* is endemic to southern and central California. Several of the species are island endemics. The single species of *Melpomene*, *M. rita*, is found in Arizona.

Natural History: Agelenids are most often found in open areas, such as grasslands, shrublands, and sometimes among human milieu, including yards, parks, vacant lots, and old fields. Some species may be found in homes, especially the Hobo Spider and its Old World relatives. They are active when temperatures are warm enough, and adults peak in size in the summer and early fall. They tend to be nocturnal, but may sometimes be seen inside the entrance of the funnel during the day. They may also race out to grab prey during the day.

Agelenids build distinctive sheet webs. There may be accessory webbing that can knock down flying prey. The webs are not sticky. In a corner of the sheet is a silk funnel, where the resident takes refuge, mates, and guards egg sacs. The webs may be more intricate than they appear on the surface. *Agelenopsis* have among the more conspicuous webs in occupied habitat. The largest species, *A. aperta*, has a large web, which may be half a meter across. In optimal habitat, there may be dozens or hundreds of the webs visible, especially when they collect dew. Hobo Spiders and their relatives build smaller sheet webs, primarily on vegetation or on and under debris, or sometimes in human structures, including the home. The other genera tend to build webs similar to *Eratigena*—small sheet webs with funnels that may lead to a protected structure such as a crevice or under a rock. The webs are somewhat similar to those of unrelated mygalomorph spiders: funnel web spiders (family Dipluridae, with two genera and five species in the United States),

A large unidentified agelenid on an outhouse, at the entrance of its funnel at night.
L.L.C. Jones

and southern house spiders, *Kukulkania* (family Filistatidae). Both mygalomorph families are not considered to have species of medical importance.

Agelenids often wait near the entrance of the funnel for prey and will quickly take refuge if danger approaches. When insects or other arthropods get onto the web, the spider will race out, grab its prey, and then haul it back to its lair inside the funnel. Invertebrate prey is not caught in the web; rather, the web provides a tactile platform to detect prey that is perusing the habitat. During mating season, August through October, males may wander to search for mates.

Agelenids usually mature in the late summer and fall. Males usually die in the fall, but females may overwinter. Egg sacs are normally produced in the fall or following spring. Fall egg sacs may also overwinter. The Hobo Spider females lay eggs in up to four or more egg sacs, which are spherical, about 8–13 mm in diameter. Egg sacs are typically deposited near the web under a surface object, such as yard debris. Each sac can have between fifty and 150 eggs. The female guards the egg sac, so it is possible she could deliver a defensive bite to a potential threat when nesting. After the young mature, they leave the area to overwinter. By the next summer, they may be grown enough to mate, but will usually die within the next year or two.

Encounters: Based on the sensationalist media reports and circumstantial journal articles about alleged bites from Hobo Spiders, one might think encounters with this species are fairly common. Actually, it is difficult to know how common encounters are. Certainly, this species, along with the other *Eratigena* and *Tegenaria*, can be common around dwellings in some areas, but one would normally have to be in contact with the web to be in contact with the spider, except during mating season. This species is known to enter homes, where they are usually at ground level, in tubs, or in basements. Other spider species, including those that frequent buildings, have been shown to prey on Hobo Spiders, and may compete with them. Other species of agelenids can be found in homes on occasion. Although human envenomation has been confirmed for *Agelenopsis* and *Hololena*, there are few case reports, suggesting contact and bites are rare. Case reports only happen in those extremely rare cases when a spider bite specialist gets a positive identification of the culprit, and publishes the reports. Most species of agelenids are much more likely to be encountered out of doors. Exposure is rare because these animals are usually inside their funnel and can be easily avoided. These animals are natural insecticides, so are beneficial residents of yards.

Venom Apparatus, Venom, and Symptoms: The venom apparatus is as described for other species of spiders. Agelenid spiders have a largely neurotoxic venom. Neurotoxic compounds have been isolated from the venom of some agelenids, including *Agelena opulenta*, *A. aperta*, and *H. curta*. Sulfated nucleosides have been isolated from Hobo Spider venom, as well as from *A. aperta*, *H. curta*, and three species of *Loxosceles* tested (*L. arizonica*, *L. deserta*, and *L. reclusa*). These compounds are probably present in most agelenids and sicariids. The effects of sulfated nucleosides in venom are being explored. These chemicals have an ability to stimulate the immune system in mammals. In addition to the sulfated nucleosides, *A. aperta* venom consists of acylpolyamines and peptides. Some isolated compounds include numerous fractions of AG and α-, μ-, and ω-agatoxins (named for the genus). These all affect insects, but α- and ω-agatoxins also affect vertebrates. The venom affects pre- and post-synaptic sodium and calcium channels. The venom of *H. curta* has been investigated for its insecticidal qualities, as it quickly paralyzes insect prey. The compounds have been dubbed curtatoxins.

The Hobo Spider has been the subject of debate. In 1987, a researcher induced Hobo Spiders to bite four shaved rabbits. In the scientific literature, he described skin lesions from the bite of the male spiders. It was suggested that cases of necrotic arachnidism in humans might be caused by Hobo Spiders (known as *T. agrestis* at the time) in the Pacific Northwest, as recluse spiders do not occur there. Shortly thereafter, there were numerous reports of "tegenarism," symptoms of necrotic arachnidism allegedly caused by Hobo Spiders, from back when they were in the genus *Tegeneira*. Then there were numerous published articles implicating the species as the cause of necrotic lesions. What followed was a sensational media frenzy about the dread Hobo Spider. Not too long after this, however, some researchers began to question if Hobo Spiders could actually cause necrosis in humans. Researchers could not induce necrotic lesions in study animals, nor did any of the case reports in humans provide the confirmation of a Hobo Spider as the culprit. In addition, in Europe, Hobo Spiders and their relatives are common house spiders that have never been implicated in significant envenomation of humans. In the U.S. case reports from confirmed bites are rare and fail to show necrosis as a symptom, although they do show other symptoms indicating envenomation.

For Hobo Spiders, one of the problems in determining the effects of the bite is due to relatively little information on the venom itself, or the effects of envenomation on vertebrates. In one study in 2011, the venom failed to cause hemolytic action on human red blood cells. In the same study, the bacterial fauna was assayed, and the mechanical transfer (via fangs) of methicillin-resistant *Staphylococcus aureus* (MRSA), the bacterium whose infection has led to misdiagnosis of spider bites, was investigated. The spider was found to have ten ubiquitous, nonpathogenic types of bacteria, but the MRSA-exposed spider did not transfer the bacterium. The only record I could find of actual symptoms of known Hobo Spider envenomation occurred in Oregon. The victim had bite-site pain, redness, and muscle twitching; the symptoms self-resolved in eight hours.

There are two published case reports of envenomation in southern California by *A. aperta*. The symptoms caused by the bite in one of these cases are featured above in the opening quote. The other victim was a 58-year-old man in good health that was bitten on the thumb while removing a web from a pipe. He killed the spider, which was later identified by a spider expert. The bite was painful and within a minute started to swell and turn red. Within fifteen minutes, a 15-cm blue-black lesion with a white center developed. Symptoms disappeared in about a week. There are three case reports of bites in Oregon by *A. pennsylvanica*, all exhibiting pain, redness, itching, and swelling. One case also involved back pain. These bites resolved in 1–10 days. Another bite was by *A. oregonensis*, which showed symptoms similar to *A. pennsylvanica*, except no itching, no back pain, and resolving in four days. There are also three case reports of confirmed bites by members of the genus *Hololena*. One was misidentified as a Hobo Spider in Washington State. Three of the reports were well-documented and the culprits were positively identified. All bites were accidental, with the spider being pressed against the skin. The bites were heralded by a burning, pinprick-like pain. The pain extended to mid-calf on the ankle-bitten victim. In the cases with significant pain, it was described as having a very hot to burning sensation. The burning pain lasted fifteen minutes to two days. In one case, the victim's pain was described as "a deep burning on his stomach as if he were being touched with a hot coal." None of these cases involved swelling, although numbness was present in at least one case. All cases included redness, extending up to 20 cm from bite site. Other symptoms in the more severe cases were headache, neck pain, and 4- to 6-hour bouts of vomiting (5–24 hours post-exposure, in two cases). In the one case of mild envenomation, the victim said her worst symptom was anxiety for the first thirty minutes following the bite. In all cases, the symptoms resolved in three days or less, except for one case when resolution was at least ten days post-bite.

Remarks: The genus *Eratigena* is an anagram of *Tegenaria*.

References and Resources: Adams (2014), Adams et al. (1990), Ayoub et al. (2005), Binford (2001), Bradley (2013), Cowles (2018), Gaver-Wainwright et al. (2011), McKeown et al. (2014), Ori (1984), Quistad et al. (1991), Schroeder et al. (2008), Skinner et al. (1989), Tzouros et al. (2005), Ubick et al. (2005), Vest (1987a, b), Vetter (1998, 2012), Vetter and Isbister (2004), Vetter et al. (2003b).

Tarantulas (Genus *Aphonopelma*) and Similar Taxa

"I have more than one hundred Orange Baboon Tarantulas [a common exotic pet species] . . . But BEWARE!! . . . During the last 2 months, I was bitten on 3 occasions . . . The fireworks came in about 20 minutes . . . intense burning . . . Imagine a hot electric iron placed on your hand . . . for about 20 hours. Painkillers have no effect. Yes, it happened again [third bite], but this time it was worse. The intense burning pain can make a grown man cry. The muscular cramp affected the calf muscle, the ankle and the toes. I suffered for a week." (Magician Ang, *Tarantulas and Others* blog)

Native tarantulas are generally docile, but they can bite if they feel threatened.
L.L.C. Jones

Classification: Phylum Arthropoda (arthropods); Class Arachnida (arachnids); Order Araneae (spiders); Family Theraphosidae (tarantulas); Genus *Aphonopelma* (North American tarantulas).

There are more than one hundred described genera and nine hundred "species" of tarantulas worldwide. In the United States, there is only a single genus, *Aphonopelma*. According to a taxonomic revision in 2016, there appears to be twenty-nine valid taxa, down from fifty-five species before the study. There were many synonymies that accounted for the reduction, although fourteen new species were described in the monograph.

There are six species groups in the United States that share anatomic similarities; they also show geographical patterns of distribution. Refer to the section on Distribution for more information. Some of the more studied and/or widespread species include: *A. anax* (Texas Tan Tarantula), *A. chalcodes* (Arizona Blonde Tarantula), *A. eutylenum* complex, *A. hentzi* (Texas Brown Tarantula) complex, *A. iodius* (Desert Tarantula), *A. marxi*, *A. prenticei*, *A. reversum* complex, and *A. steindachneri*. Four dwarf desert species include *A. joshua*, *A. mojave*, *A. paloma*, and *A. saguaro*. Some of the Madrean Sky Island species include the aptly named (corresponding with the area) *A. chiricahua*, *A. madera*, and *A. peloncillo*.

Other species of theraphosids or tarantula-like spiders are popular as pets in the exotic pet trade or may be in shipments of international cargo. Some of these are much more venomous than our native species.

This Velveteen or False Tarantula is in a different family but is a megalomorph related to tarantulas. These animals are quicker to deliver a painful bite than true tarantulas native to the United States. *Marshal Hedin/CCA/Flickr*

A similar group of mygalomorph spiders of the family Nemesiidae are commonly called velveteen tarantulas or aggressive false tarantulas. There is one genus (*Calisoga*) and five questionable species, all in northern and central California, plus just into Nevada. Nemesiids are not as docile as theraphosids and are quick to give a defensive display. It is not a bluff and they will deliver a painful bite if harassed. They are large and have a velvety covering of silvery to brown setae. They are 16–30 mm in length.

Identification: Adult tarantulas are the largest spiders in the world. They are heavy-bodied terrestrial spiders with relatively thick, heavy legs. They are conspicuously "hairy," having a large number of setae on the legs and urticating bristles on the abdomen. Their movements seem slow and deliberate, but they can dart surprisingly fast when startled or procuring prey. The species in the United States are similar, except in size, as indicated by the fact that they are all in the same genus. They are typically tan or gray to brown or black. There is sexual dimorphism. Adult males range from about 15–55 mm cephalothorax + abdomen length, depending on species. They possess modified pedipalps and have tibial spurs. The male's abdomen is smaller than that of a female, and legs relatively longer and more spindly. Females are larger and range from 16–70 mm.

The business end of a tarantula, showing the very long fangs. *L.L.C. Jones*

Some species, such as *A. paloma*, are actually quite small as tarantulas go, representing the lower end of the size scale. These are commonly known as dwarf tarantulas, and are often mistaken for juveniles of larger species.

Nemesiids are similar, but have very long posterior spinnerets and lack tufts on the feet.

Distribution: Tarantulas of the genus *Aphonopelma* are found in the New World, from the American Southwest to Central America. Their greatest diversity is the southwestern United States and central Mexico. In the United States, they occur primarily from California to Texas, but into Oklahoma, Arkansas, Kansas, Louisiana, and Missouri. These latter three states have only one species, *A. hentzi*. By state, Arizona has the highest diversity, with 16 species recognized at the time of this writing, followed by California with ten species, and Texas and New Mexico, each with six species. Two diversity hotspots in the Southwest are the California Floristic Province and the Madrean Archipelago (southeastern Arizona and adjacent New Mexico). The species groups are generally distributed as follows:

- ***A. iodius* group.** This group is also referred to as the western group, but the West contains three other species groups. It comprises four species found west of about Cochise County, Arizona (i.e., the Cochise Filter-Barrier), into California.
- ***A. paloma* group.** This is a group of about a dozen dwarf tarantulas from the West, especially the California and Arizona deserts.
- ***A. marxi* group.** This is one of the four western groups, and is primarily distributed in the Madrean Archipelago area of southeastern Arizona, adjacent New Mexico, and adjacent Mexico. There are six U.S. species described, and there are undoubtedly other species yet to be described from southeastern Arizona and Mexico. Several species are endemic of one or more of the sky islands. *Aphonopelma marxi* itself is a major exception, being found in Arizona, Colorado, New Mexico, and Utah.
- ***A. steindachneri* group.** This is currently considered monotypic in the United States, being found only in California, but more Mexican species will likely be added as research continues.
- ***A. hentzi* group.** This group is one of the eastern groups (i.e., east of the Cochise Filter-Barrier). It is related to the moderatum species group, also an eastern group.

- ***A. moderatum* group.** This is the other eastern group, more closely related to *A. hentzi* group than the other western groups. There are three species.

Natural History: Based on the ranges of most species in the United States, one might think tarantulas are desert obligates, but this is not the case. Certainly, each of the four deserts is home to a variety of species, but tarantulas are also present in coastal sage scrub, coastal and interior chaparral, shrublands, grasslands, juniper and oak woodlands, pine/oak woodlands, and even some mixed and conifer forests. In the United States, their diversity and habitat use is fairly similar to that of scorpions—they are predominately in the Southwest, and they range from the coasts of California to Texas, and from valleys to mountains. As one goes north or east of the Southwest, the numbers decline precipitously.

There is a lot of variation in some of the natural history traits of tarantulas, but many of the nuances probably have yet to be teased out, so generalized information presented here is based on some of the work done on some of the larger and more common species in the Southwest.

Tarantulas are ground-dwelling, burrowing spiders. A tarantula will usually excavate the burrow itself, but it may also use the burrows of other

One of the dwarf tarantulas (*A. paloma*). *Jillian Cowles/Amazing Arachnids*

animals, or be found under debris. A typical burrow might be in the order of 200 cm deep, ending in a 100 mm nest chamber. There may be side chambers. The entrance to the burrow may be lined and enforced with silk, but not necessarily. The entrance is often flush with the surface, but it may be elevated into a vertical turret, crescent-shaped, or under an overhang or surface object. The spiders often stay inside or near the opening of the burrow during the active season, but males venture outside during mating forays. This is why most tarantulas collected are males. Females and burrows of many species may be difficult to locate.

In general, surface activity coincides with the warm temperatures of spring through fall, but tarantulas may remain at their burrows during much of this time. Timing of surface activity has been the topic of study by numerous biologists. These above-ground movements often coincide with the onset of summer rains or just the reproductive cycle. Summer thunderstorms can include extremely heavy downpours, which threaten tarantulas and their burrows. During heavy rains, many tarantulas may block their burrows with soil, soil and silk blocks, or even with their bodies, to avoid the flooding of their tunnels.

Before breeding, males will construct a sheet web and deposit semen onto it, which they then absorb in the pedipalps. The sperm stays viable for an extended period. Males may wander up to a mile to seek a mate. To entice a female at her burrow, he taps the ground to lure her out. If receptive, he will hold her up by her fangs using the tibial spines on his front legs, then insert the modified pedipalps into the female's genital opening on the ventral side of the abdomen. Females house sperm in spermathecae. After mating, the male will leave. In rare instances, females have been known to cannibalize males. The following summer, the female will lay 200–300 eggs in silken egg sacs in the nest chamber of the burrow. The female guards the eggs and young spiderlings, at least as long as it takes them to use their yolk reserves. Spiderlings do not balloon, but disperse overland. *Aphonopelma* are long-lived, taking about 7–10 years to reach maturity. When young, they can molt several times per year, but as they reach adult size, molting becomes much less frequent (e.g., once a year). Males live only 1–1.5 years after reaching maturity, while females can live another 10–15 years.

Tarantulas feed on just about any small animal they can overpower with their legs and chelicerae. Prey usually consists of insects and other invertebrates (crickets, grasshoppers, cicadas, and other spiders), but lizards, small snakes, and even small mammals are potential meals. Tarantulas are ambush predators. They wait near the entrance to their burrow and detect prey with their array of sensory setae, which gather movement, chemical, and tactile clues. When prey is detected, they rush out and grab the victim, which is crushed and/or envenomated. As with other chelicerates, they can only feed on a fluid diet, so tarantulas may crush their prey with their strong chelicerae, then liquefy the contents with digestive juices. The liquescent meal will then be sucked into the mouth and gut. Indigestible parts, such as exoskeletons, are left behind.

Predators include a number of small- and medium-sized animals, but the most notorious are the tarantula hawks. These are large wasps that seek out and paralyze tarantulas in order to deposit a single egg per tarantula. The paralyzed tarantula becomes a source of nutrition for the larval wasp. Tarantulas rarely win the battle against tarantula hawks. There is a chemical cue used by the wasp, so it is not detected as a threat. Refer to the section on tarantula hawks, the largest of the spider wasps, for more information, as they are also venomous to humans. Tarantulas can also be parasitized by certain types of flies.

Encounters: Tarantulas rarely enter homes even where they are common, partly due to their large size, but wandering males may enter through open doors and large gaps while in search of a mate. This is usually during the summer into fall. Most encounters are in the yard, in the wild, or in captivity.

North American tarantulas are not inclined to bite, and many people who are fond of them do not hesitate to handle them gently. However, they can bite, especially when encountered in the wild and handled roughly, or if they are pinched against the skin or otherwise feel threatened. They can move quickly when threatened. To relocate them, they can usually be captured by coaxing into a container, but beware of their urticating bristles.

In captivity, there is always the potential for a bite and receiving a dose of bristles, so caution is advised. Many people handle their pets without problems, because our native *Aphonopelma* are mild tempered and do not seem to be particularly venomous. The bigger issue comes from those pesky bristles. Some species of exotic tarantulas lack urticating setae, some have more dangerous bristle types, and some species have a much more venomous bite than our native species. Baboon spiders (subfamily Harpactirinae) are popular as pets, but they tend to be more aggressive than our native species and can deliver a nasty bite. Common species include the King Baboon and Orange Baboon spiders.

Brazilian Wandering Spiders, *Phoneutria* spp. (in a related family, Ctenidae) rarely show up in shipments from South America, but they are often confused with (and outnumbered by) a more common genus of ctenid wandering spider (*Cupiennius*) that is not nearly as toxic. Although very rarely encountered, ctenid spiders do occasionally show up in shipments, such as bananas, so they could constitute an occupational hazard for some. One study showed that only seven of 135 spiders in banana shipments were *Phoneutria*, but if you are bitten by one, percentages of occurrence in shipments don't matter much. Training in awareness and identification are in order for those with such an occupation. I highly recommend people do not keep Brazilian Wandering Spiders as pets in their homes because they are probably the most venomous spiders on Earth and very short tempered. If accidentally released into the wild, they could become established in some southern areas.

Venom Apparatus, Venom, and Symptoms: As noted, North American tarantulas have two mechanisms for envenomation. The first comes from the bite and the second comes from the urticating bristles.

The venom apparatus for a bite is similar to other spiders, but the venom glands are completely inside the chelicerae. The secretory cells of the glands

An urticating hair of a tarantula. *Jillian Cowles/Amazing Arachnids*

excrete venom into the lumen, and the muscles of the glands force venom through the fangs. The tarantula is capable of delivering a dry bite, as are wandering spiders. At rest, the curved fangs are folded underneath the chelicerae. When annoyed, a tarantula can put on an impressive threat display, by rearing up on its two pairs of hind legs, elevating the front part of the body and legs, and having its fangs at the ready. This is enough to keep most humans at bay. The display may be a bluff, but occasionally they will bite, depending on species and circumstances.

There have been relatively few studies on venom of *Aphonopelma*, presumably because none of the native species are considered to be medically important, although there are very few case reports. More is known about venoms of other species that have more potent venom. Regardless, there are numerous chemical components in the venom of any tarantula, to liquefy tissue of their prey. Compounds include the polyamines spermine, spermidine, cadavarine, and putrescine. There are some acylpolyamines that are neurotoxic and venomous to insects. There are reports of proteolytic toxins, but this may be due to digestive fluids used in assays, rather than venom glands. Adenosine 5′ triphosphate and adenosine 5′ diphosphate is present in some *Aphonopelma* venom. Some *Aphonopelma* venoms also contain chitinase, protease, and phospholipase A, among other enzymes. An LD_{50} of *Aphonopelma* sp. has been reported to be 14.1 mg/kg in mice.

Our native *Aphonopelma* are often considered docile, although one avid spider aficionado said they are not really docile, just tolerant. Some exotic species have a reputation for being much less tolerant, and are sometimes regarded as aggressive. Native tarantulas are often handled by pet owners without being bitten, but bites happen occasionally. There are few actual case reports, presumably because reactions tend to be localized, mild, and self-resolving within minutes to hours. The bite of native *Aphonopelma* is generally likened to a bee or wasp sting, with local redness, swelling, and some pain, or even less symptomatic, though itching can be considerable.

There are a few bite reports for native *Aphonopelma*. They include apparently dry bites to mild to moderate pain, some swelling, burning, and itching, which may be severe. Most bites are probably not reported. The medical community could benefit from reports of bites, even if nonsensational. No one has ever died from the bite of a tarantula, of any genus or species.

Another problem may come from the urticarial setae or bristles ("hairs"). There are at least six types of urticarial bristles, termed Type I to VI. The bristles are usually barbed, so that they can remain lodged in the victim.

The bristles are clearly used as defense, but tarantulas may also use them to line their burrows or mark their territories. When used in defense, they are rubbed off the abdomen with the rear legs, and flicked out to become airborne. The bristles generally affect the skin and mucous membranes of the predator or victim. There are case reports of effects from urticating bristles in humans. In the eyes, they can become lodged in the cornea, causing pain, redness, and vision problems, such as "floaters." In the lungs, they will cause coughing and shortness of breath, especially if allergic. In the nose, they will cause rhinitis. On the skin, they usually cause itching and a rash, which is typically worse in areas of thin skin, such as under the armpit. Effects will subside, but in some instances, may linger for weeks. Acute allergic reactions requiring hospitalization have been documented from exposure to urticarial bristles of tarantulas. Hobbyists who own tarantulas may have repeated exposure and should be aware of this. *Aphonopelma* in the United States have the Type I bristles, which are not as problematic as some other types found in exotic species.

Exotic pet owners should be aware that some of the large tropical tarantulas, such as *Poecilotheria* and baboon spiders, are decidedly more venomous than our local species, and symptoms can be dramatic. The effects can include extreme pain that lasts for days, and the affected area can become locked in spasm due to the neurotoxins. Other systemic reactions are possible, as well.

Remarks: The name "tarantula" originated in Europe. A European species of wolf spider (family Lycosidae) is *Lycosa tarantula*, named from the Italian village of Taranto, where the species is common. Its bite was believed to be severe and cause "tarantism." It was also believed that the victim would have to engage in a dance, the "tarantella," to avoid severe symptoms. Now it is known that the venom of *L. tarantula* is not medically important and bites are unlikely to occur. The term tarantula came to be used for New World theraphosids because of their general resemblance to *L. tarantula*. Common usage has become so widespread that "tarantula" is now used for the theraphosids, rather than *L. tarantula*, which is now usually called the Tarantula Wolf Spider to avoid confusion.

The bite of *Phoneutria* (not actually a tarantula) should always constitute a medical emergency, as human fatalities have occurred following a bite. There is no antivenom for native theraphosids, as there is no real need, but there is an experimental antivenom for *Phoneutria*, equine $F(ab')_2$ that is raised against the venom of *P. nigriventer*.

Also not tarantulas, but tarantula-like members of the genus *Phoneutria* are dangerously venomous animals. They arrive rarely in shipments of bananas—and some are even kept as exotic pets. They are not only very venomous but also very aggressive. *John Murphy*

References and Resources: Baerg (1958), Blackman (1995), Cabbiness et al. (1980), Cowles (2018), Hamilton et al. (2011, 2014, 2016), Hendrixson et al. (2013), Isbister (2002), Janowski-Bell and Horner (1999), McCormick and Meinwald (1993), Murray (2006), Ori (1984), Prentice (1992), Punzo and Henderson (1999), Rash and Hodgson (2002), Russell et al. (1973), Smith (1994), Stahnke and Johnson (1967), Vetter and Hillebrecht (2008).

Yellow Sac Spiders (Genus *Cheiracanthium*)

Bite at 3:45 a.m. "awakened by a sharp pain near the right eye . . . a small spider [later positively identified as *C. inclusum*] . . . chelicerae firmly embedded in the skin . . . cold compresses/2.0 gm of acetylsalicylic acid [aspirin] did not alleviate the pain. Within three hours after the bite, the throbbing pain had spread over the cheek to the right mandibular area. By 10:30 a.m. the pain was still intense but seemed to be abating, and by 4 p.m. it was gone. The venom appeared to have only local neurotoxic effects." (Furman and Reeves 1957)

Classification: Phylum Arthropoda (arthropods); Subphylum Chelicerata (chelicerates); Class Arachnida (arachnids); Order Araneae (spiders); Family Eutichuridae (long-legged sac spiders); Genus *Cheiracanthium* (yellow sac spiders). *Cheiracanthium* is sometimes incorrectly spelled *Chiracanthium*. Most are called yellow sac spiders.

Yellow sac spiders are sometimes thought of as aggressive, and this display of *C. inclusum* is impressive. *Joseph Berger/CCA/Bugwood*

Yellow sac spiders were considered to be in the family Clubionidae or Miturgidae, but the *World Spider Catalog* (version 14.5) lists them as belonging to Eutichuridae. The eutichurids are represented by 12 genera and 308 species found nearly worldwide. Yellow sac spiders are predominately from the Old World. There are two species north of Mexico: *C. inclusum*, often called the Black-Footed or American Yellow Sac Spider, and *C. mildei*, known as the Northern Yellow Sac Spider or Long-Legged Sac Spider. *Cheiracanthium mordax* is established in Hawaii. In its native Australia, it is known as the Biting Garden Sac Spider, among other names. *Cheiracanthium* are also known as wandering spiders, but this is also the term for a guild and not to be confused with Ctenidae, and a host of other vernaculars.

Identification: Yellow sac spiders are small, reaching only about 4–10 mm total length and 25 mm leg span. The male has longer legs than the female. The cephalothorax is about the same length as the abdomen. They are usually yellow to yellowish green, especially on the abdomen, but can be straw-colored, tan, or orangish. They lack an obvious pattern, but the abdomen often has a middorsal stripe that tapers posteriorly, known as the heart mark. They have two rows of four eyes, with the upper row procurved toward the lower row, which is nearly straight. Their legs are relatively long and slender, and the first pair is longer than the second and third pairs. The

Northern Yellow Sac Spider (*C. mildei*). *Don Laorie/CCA/Flickr*

legs are generally monocolored, with dense setae on the "feet" that makes them appear dark. The spinnerets are visible from above.

Another identifying character is the sac. *Cheiracanthium* are sometimes called "sleeping bag spiders." The sac is a tube-like cocoon of silk where the spider resides during the day and where the female deposits her eggs. This sac is often inside of a rolled leaf or other vegetation, amongst debris, or in the home. Some related taxa also have a silk sac. In this regard and general external anatomy, the Eutichuridae are similar to Clubionidae and Miturgidae. *Cheiracanthium* lacks conspicuous thoracic grooves, and the posterolateral spinnerets have short, rounded tips. They have longer legs than clubionids.

Distribution: *Cheiracanthium inclusum* is often regarded as a native in the New World, from the United States and Canada south to South America and the West Indies, but is sometimes considered to have a non-native origin. It is found throughout most of the United States, and has presumably been introduced into Africa and the island of Réunion. *Cheiracanthium mildei* came from Europe. It was thought to have been introduced to New York in 1947, and has spread to every state in the country, except Hawaii, and all provinces in Canada. Hawaii has its own resident species, *C. mordax*, which has also been implicated in envenomation there. It is apparently native to lands of the western Pacific in both hemispheres, where it is fairly widely distributed.

Natural History: Yellow sac spiders occur in many vegetation communities. Those in the continental United States and Canada are found widely in temperate areas, while the Hawaiian species is tropical to subtropical. The American Yellow Sac Spider is more commonly observed outdoors in natural habitats, where it can be extremely abundant, but can be in disturbed habitats, including agricultural lands. The Northern Sac Spider is more commonly encountered indoors or in agricultural lands, where it can also be extremely abundant. In its native home of Europe, *C. mildei* is often found in shrublands, but it has taken up a partially urban to agricultural lifestyle in North America. The Northern Sac Spider is well studied as a common spider in agricultural settings. In California, it has been shown to be a dominant spider in vineyards, habitats that tend to have low diversity and low density of spiders. *Cheiracanthium inclusum* also occurs in agricultural areas but is typically more abundant in the surrounding natural environs. There are numerous studies on species of *Cheiracanthium* acting as biological control of insect pests in agricultural lands.

Cheiracanthium spp. are related to clubionids (pictured), which also sleep in a silken sac. There is little information on the effects of clubionid envenomation in humans.
Marshal Hedin/CCA/Flickr

By day, yellow sac spiders reside in their sacs. They do not construct webs, other than their protective sacs, but can have a trail of silk to suspend them if falling. At night, yellow sac spiders become surface active. Behaviorally, they are termed hunting, wandering, or prowling spiders, because they move about rapidly, searching for small arthropods on the ground, vegetation, or other structures, such as walls and ceilings. When hunting, they hold their front legs elevated in front of them. Prey includes fruit flies, leafhoppers, fleahoppers, small true bugs, and butterflies and moths. After each night of foraging, they construct a new sleeping sac. They also feed on nectar, which helps them maintain their high level of energy and ensure reproductive success.

During mating, the male crawls underneath the female, upside down, and inserts his pedipalps into her genital opening. She will lay one to five sets of eggs covered in silk sacs, about a month apart. There are generally 30–40 eggs per sac, but there may be as many as seventy. Females defend the eggs. They are short-lived spiders. Males will live about a year and females about 1.5 years. In homes and other mild areas, they can be encountered year-round.

Encounters: Depending on where you live, yellow sac spiders range from uncommon to extremely abundant. They may be common in nature, yards, agricultural fields, and in some areas often enter homes, especially when searching for prey, seeking a mate, or coming in from the cold. Look for their telltale sleeping sacs in corners, as with the junction of walls and ceiling. They often select two surfaces in which to build their sacs, so may be between items, such as folded sheets, towels, blankets, or clothing. They are commonly encountered in drapes and windowsills. They are nocturnal, so be extra wary at night.

Most bites occur when the victim is in bed or putting on clothing. The female may also bite when she is guarding her egg sac. Encounters also occur when the spider is brushed up against during yard work or while going about business around the house. Inside, one can seek out the spiders during daylight hours by searching for their sacs, but some may remain unseen. They are quite small, so may be hard to locate.

Yellow sac spiders have been characterized as being highly aggressive by some accounts, yet nonaggressive by other accounts. These spiders mostly defend themselves when being pressed against the human body, rather than attacking someone with little provocation. However, they do not usually need to be prodded to bite like some other species. Some literature suggests that most spider bites in the United States are due to *Cheiracanthium* spp., but other reports say this is not the case. It is difficult to track that type of information, but a report in Australia (which is useful for comparison, as it has a similar spider fauna) demonstrated that *Cheiracanthium* and other sac spiders accounted for only about 2% of the confirmed spider-bite cases, out of about eighteen families of spiders. In one study of bites in the continental United States, all bites were indoors. In a study of confirmed bites in Oregon, *C. mildei* was the most commonly implicated species (n = 10 of 33), tied with agelenids, and closely followed by orb-weaving spiders (n = 7). There are at least four records of bites by *C. mordax* in Hawaii.

Venom Apparatus, Venom, and Symptoms: For their small size, these spiders possess an impressive biting apparatus. The chelicerae and fangs are tucked under the cephalothorax, but when extended to capture prey or defend themselves, they are surprisingly long; they can easily bite through thin clothing. When they bite, they leave a 6–8 mm fang gap, and may bite repeatedly.

Experiments on the venom of *C. mildei* demonstrated it causes hemolysis and cytolysis in sheep blood cells, but the venom did not cause skin lesions

Distinctive eye pattern of *Cheiracanthium (C. mildei)*. Don Laorie/CCA/Flickr

in rabbits. It was the only genus of spider except *Loxosceles* to lyse sheep red blood cells, among the dozens of species from numerous families tested. The confirmed dermonecrotic agent found in *Loxosceles*, phospholipase (sphingomyelinase) D, is absent in *C. mildei*. The hemolytic compound in yellow sac spider venom is believed to be a phospholipase A_2.

This genus has been extensively cited in medical journals, popular articles, and online as producing dermonecrotic skin lesions. Recent studies have shown this is almost never the case. The only studies that reported confirmed bites have shown a single instance of skin lesions due to a yellow sac spider bite, and that was very minor. That example did not include species in the New World.

Although the bite does not usually cause dermonecrotic lesions, it does cause considerable discomfort. A common symptom is immediate and sometimes-severe pain, which may radiate from the bite site. The pain may last for less than one hour to three days, but in rare cases, symptoms may last up to three weeks. There is usually redness, swelling, and wheal formation. Intense itching may occur. Based on the few case reports, systemic effects are uncommon. These include nausea, tremors, fever, vision changes, and sore throat.

Remarks: Yellow sac spider trivia: There was a recall of Mazda cars because yellow sac spiders were found to be attracted by fumes and holed up in car engines. There were no reported envenomations resulting from this phenomenon, but the media capitalized on the idea of venomous critters in cars.

References and Resources: Cowles (2018), Foradori et al. (2005), Furman and Reeves (1957), Hogg et al. (2010), Isbister and Gray (2002), Isbister and White (2004), Lehtinen (1967), McKeown et al. (2014), Ramírez (2014), Vetter et al. (2006).

Other Spiders (Order Araneae)

"I had venom sprayed in my eye while viewing a Green Lynx Spider up close. The eye turned red, there was general discomfort, and my vision was fuzzy. I rinsed out the eye with fresh water. The symptoms were gone in about an hour." (A. Kristensen, Spider Pharm, personal correspondence)

Classification: Phylum Arthropoda (arthropods); Subphylum Chelicerata (chelicerates); Class Arachnida (arachnids); Order Araneae (spiders).

In addition to the other spiders covered in this book, members of several other families have been implicated in human envenomation, although case reports of bites are scant. It is likely there are other spiders in our region that can cause human envenomation that I am unaware of. We could benefit from more case reports of confirmed bites, even if asymptomatic or minor. Spider families and genera usually implicated in human envenomation in North America, north of Mexico, not otherwise included in this book are:

- Hacklemesh spiders (Amaurobiidae: *Callobius*)
- Folding-door trapdoor spiders and turret spiders (Antrodiaetidae: *Antrodiaetus*)
- Cork-lid trapdoor spiders (Ctenizidae: *Bothriocyrtum, Ummidia*)
- Wafer-lid trapdoor spiders (Euctenizidae)
- Orb weavers and garden spiders (Araneidae: *Araneus, Argiope, Argiopes, Neoscona*)
- Woodlouse spiders (Dysderidae: *Dysdera crocata*)
- Ground spiders (Gnaphosidae: *Herpyllus, Drassodes*)
- Wolf spiders (Lycosidae: *Lycosa* and other genera)
- Crab spiders (Thomisidae: *Misumenoides*)
- Lynx spiders (Oxyopidae: *Peucetia*)
- Jumping spiders (Salticidae: *Phidippus*)
- Huntsman or giant crab spiders (Sparassidae)
- Thick-jawed orb weavers (Tetragnathidae, includes *Tetragnatha* spp. and *Nephila clavipes*)

Those in or related to the above spider groups are likely venomous to humans to some degree. Other spider families are known to occasionally bite humans, and may cause minor discomfort. These include Clubionidae, Coriniidae, Dipluridae, Filistatidae, Pholcidae, Tengellidae, and Zodariidae—but documentation of bite symptoms is poorly documented. Any bite that breaks the skin by any species has the potential to cause discomfort.

Identification: It is beyond the scope of this book to adequately describe all of these spiders. There are a few good guides out there to help with identification and natural history information. All of the spiders in our list have one thing in common—they are not tiny. They range from the size of a recluse spider on up to that of a dwarf tarantula. Because of their size, they have the ability to penetrate human flesh if they bite, and because they are spiders, they have venom.

The tarantula-like spiders include the cork-lid, wafer-lid, and folding-door trapdoor spiders, as well as the velveteen tarantulas (see also tarantula account), which sometimes have trapdoors. They are smaller than most *Aphonopelma*, but belong to the same primitive group, the infraorder Mygalomorphae. They are large and heavy bodied, but may or may not be hairy. Trapdoor spiders are famous for their trapdoors that are inconspicuous when closed. Female California Trapdoor Spiders (*B. californicum*) are dark brown, very stout, and have short, thick legs. They reach about 33 mm in length. The other genus in the family implicated in human envenomation is represented by *U. audouini*. It is similar in appearance and habits to *B. californicum*, and females reach about 28 mm. The antrodiaetids includes several species in our region. The genus *Antrodiaetus* has nineteen species in the United States, based on the *World Spider Catalog* at the time of this writing. One is *A. riversi*, the California Turret Spider. Others include *A. apachecus*, *A. gertschi*, *A. montanus*, and *A. pacificus*. The euctenizids include numerous species in several genera in California. They range from about 15–28 mm long.

Another group of spiders are those that are shaped more-or-less like crabs. Their legs are relatively long and spread out laterally, so that they remain prostrate, rather than other spiders that are more elevated above the substrate. Their front legs are arched at the front, with legs to the side and pointing forward. This group includes the crab spiders and giant crab or huntsman spiders. There are other families that have a crab-like posture. Crab spiders are smaller than huntsmen and are usually found

A trapdoor spider's burrow (*A. californicus*). There are different families of mygalomorph spiders that construct trapdoors. *Marshal Hedin/CCA/Flickr*

while hunting on vegetation, often flowers. They have a wide abdomen and carapace, further adding to the appearance of a crab. The largest true crab spider is *Misumenoides formosipes*, one that is sometimes implicated in human envenomation. It reaches a moderate 11 mm in length. It is highly variable in coloration, usually being white, yellow, or pink. Individuals can match the flowers they hunt on and have the ability to change to the color of the flower. The huntsman spiders are larger than crab spiders and have a narrower abdomen and cephalothorax. Also, they are not quite as crab-like as the thomisids. Two common species are *Heteropoda venatoria* (Pantropical Huntsman Spider) and *Olios giganteus* (Golden Huntsman Spider). Female *H. venatoria* reach about 10 cm in leg span, while *O. giganteus* reaches about 7.5 cm leg span. They are similar in appearance, being mostly tan, although the Golden Huntsman looks like it is wearing dark socks.

The orb weavers are familiar to many of us because they are common and their web design is the quintessence of what people think of as spider webs. This is the third largest family of spiders with about 170 genera and more than 3,000 species worldwide. There are more than thirty genera and 160

A Golden Huntsman is disinclined to bite, and should be a welcome pest-eater inside homes. *L.L.C. Jones*

Large crab spiders are known to envenomate humans. Some, like this one, are cryptic on the flowers on which they hunt. *"Ilyetssuti"/Public Domain/Pixnio*

species in the United States. The species that tend to be noted when they occasionally bite humans are the larger species. The genera most often reported and confirmed as biting humans are *Araneus, Argiope*, and *Neoscona*. Large *Araneus* include *A. bicentenarius, A. cavaticus* (Barn Orbweaver), *A. diadematus* (Cross Orbweaver), *A. gemmoides* (Plains Orbweaver), and *A. trifolium* (Shamrock Orbweaver), among others. These range in size from about 20–28 mm in length. These species all have large abdomens, small cephalothorax, and all except *A. trifolium* have what is termed an angulate abdomen, having two humps on the dorsoanterior end. This is especially pronounced in large adult females. The large *Argiope* are the familiar garden spiders, including *A. aurantia* (Black and Yellow Garden Spider) and *A. trifasciata* (Banded Garden Spider). Females reach 28 and 25 mm, respectively, while males are only 4–8 mm. These spiders have large orb webs and females may have a zigzag web structure known as a stabilimentum. Garden spiders hold their long legs in pairs, with the anterior legs pointing forward and the rear pointing back. The Black and Yellow Garden Spider has distinctive banding of those colors. Both species have a silvery cephalothorax. Three of

(Left) At least some of the angled orb weavers like this *Araneus diadematus*, have caused disconcerting symptoms in humans. (Right) A large orb weaver with a huge web that may be face-high in vegetation is *Nephila clavipes*. Note the tiny male in the web above the much larger female. *L.L.C. Jones*

Little is known about bites from tetragnathids, like this *Tetragnatha montana*. The long chelicerae are impressive up close, but the spiders are inconspicuous in wetland vegetation. *Janet Graham/CCA/Flickr*

the larger *Neoscona* are *N. crucifera* (Arboreal Orbweaver), *N. domiciliorum*, and *N. oaxacensis* (Western Spotted Orbweaver). Females reach about 18 mm and have a boldly marked, contrasting abdominal pattern. There are other families of spiders that also build orb webs, but these are not normally implicated in human envenomation. Another common family of spiders that have orb webs is the Tetragnathidae. There are eleven genera and about forty species in the United States. They often build their orb webs in vegetation over water. A giant among tetragnathids is *Nephila clavipes*, the Golden Silk Orbweaver of Texas to Florida and South Carolina. It can reach 34 mm (abdomen + cephalothorax), and the leg span can be over 10 cm. Members of the common genus *Tetragnatha* are also long and slender, but only measure about 7–14 mm (abdomen + cephalothorax) in length.

Another group of spiders that many people are familiar with are jumping spiders, family Salticidae. They are not only common and speciose, but their interesting habit of making long, quick hops is legendary. Overall, they are stocky, with a large cephalothorax and relatively small abdomen. They have short legs and the anterior legs are strong and muscular, allowing them to make remarkable leaps. Some are vividly colored, and they have elaborate courtship rituals that are often captured on film in wildlife documentaries.

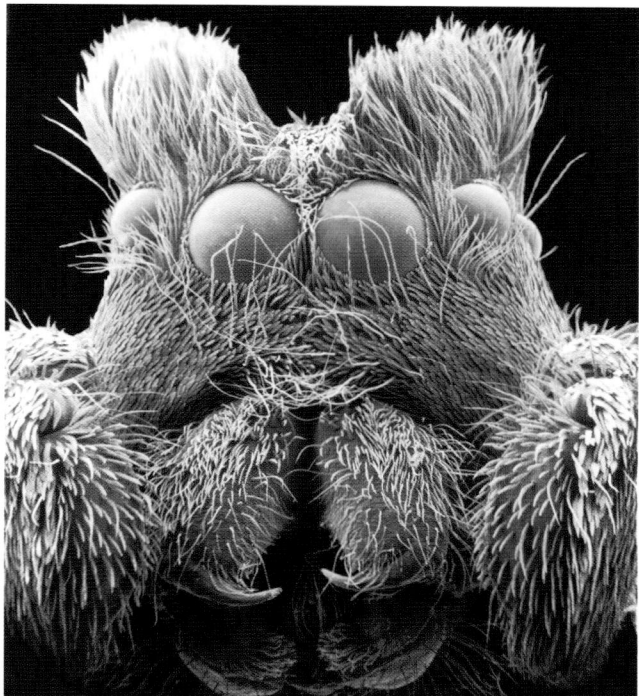

Scanning electron micrograph of the face of a jumping spider (*Habronattus ophrys*). Salticids have a very distinctive eye pattern. Even some people who dislike spiders think jumpers are cute.
David Hill/CCA/Flickr

This is the largest family of spiders, with more than 5,700 species worldwide and more than three hundred species in the United States. *Phidippus* is the largest genus, and the one that is usually implicated in human envenomation. The largest species is *P. regius* (Regal Jumping Spider), at 23 mm. It may be red in color, although there is a gray color phase. *Phidippus audax* (Bold Jumping Spider) is also variously colored, but has metallic green chelicerae, as do some other species. There are several large species reaching 14–17 mm that have aposematic red coloration, at least on the abdomen, including *P. apacheanus*, *P. ardens*, *P. cardinalis* (Cardinal Jumping Spider), *P. carneus*, and *P. johnsoni* (Johnson's Jumping Spider). Salticids have excellent eyesight, and when viewed up close their eye pattern is distinctive; it includes a large pair of eyes that look like goggles in the middle front of the cephalothorax. There are numerous other species of *Phidippus*, as well as many other genera, but most are too small to bite humans.

Large jumping spiders with aposematic coloration, like this *Phidippus carneus*, are probably capable of delivering a painful bite. L.L.C. Jones

The Gnaphosidae and Lycosidae are diverse and common ground-dwelling spiders that are somewhat similar in appearance to agelenids. However, they have distinctive features, especially the eye patterns. These are also among the largest families. There are more than 2,100 species of gnaphosids worldwide and 255 species in the United States. There are more than 2,300 species of lycosids worldwide and more than 230 species in the United States. The gnaphosid genera being reported as venomous to humans include *Drassodes* and *Herpyllus*. There are seven species of *Drassodes* across most of the United States. These are small to moderate "little brown jobs," that have few distinctive markings. There are thirteen species of *Herpyllus* in the United States that are widespread across the country. These are commonly known as parson spiders. Most are dark with a light dorsal stripe on the abdomen—the marking is supposedly similar to a cravat (neckwear) warn by parsons. *Herpyllus ecclesiasticus* (Eastern Parson Spider) is responsible for most reports of envenomation, but the nearly identical *H. propinquus* (Western Parson Spider) has also been implicated. The Lycosidae are a little more mysterious as potential culprits for envenomation in humans, especially in North America. The genus name *Lycosa* was formerly in widespread use for North American

Western Parson Spider, *Herpyllus propinquus*. Jillian Cowles/Amazing Arachnids

species, but is now restricted to the Old World. There are some large lycosids including those in the genus *Hogna*. The Carolina Wolf Spider (*H. carolinensis*) is the largest; it may reach 35 mm and have an impressive leg span. The Rabid Wolf Spider (*Rabidosa rabida*) and its congeners are common, large species (to over 20 mm), but is usually dubbed harmless.

The hacklemesh spiders, represented by *Callobius*, among other genera, are widespread across the United States. There are about twenty-five species of *Callobius*, most of which are found in the West. These spiders look similar to gnaphosids and other ground spiders, but have a distinctive eye pattern of parallel rows, among other features. Unlike ground spiders, they do make webs. Some *Callobius* grow to more than 20 mm. They are brown to dark gray and/or orange to reddish and may have an abdominal pattern.

The lynx spiders are exemplified by the Green Lynx Spider (*Peucetia viridans*), a beautiful little spider that often (but not always) has a bright green cephalothorax and abdomen with a herringbone pattern. They may also be tan and variously colored when viewed up close. Their color depends on their diet. The legs are long and spindly, and noticeably spiky with dark, thick setae. They reach about 16 mm in length.

Distribution: Given the breadth of taxa covered in this section, suffice to say all of these animals are found somewhere in the United States and Canada, and most areas will have multiple species.

Another common foliage dweller is the Green Lynx Spider, a species that may spray venom in the eyes. *L.L.C. Jones*

As a genus, *Callobius* is found in much of the United States but is most speciose in southern California. The similar and related *Amaurobius* is endemic to California, except for an introduced species.

The two trapdoor spider families are most speciose in the West. Of the Ctenizidae, *B. californicum* and *Hebestatis theveneti* are monotypic California endemics. *Ummidia* is more diverse. Of the presumed fifty species or so found in both New and Old Worlds, only eleven are described from the United States. Most species occur in the Southwest. *Ummidia audouini* is found in Oklahoma, Texas, and the other Gulf States. The other trapdoor spider family (Antrodiaetidae) is widespread, with 12+ species of *Antrodiaetus* in the United States. *Antrodiaetus pacificus* is found in northern California; *A. montanus* is found from northern California to northern Utah; *A. apachecus* is found in Arizona and New Mexico; *A. riversi* is found from California to the Pacific Northwest; and *A. gertschi* is found in northern California. Curiously, there are two species of *Antrodiaetus* in Japan.

Of the crab-like spiders, the Sparassidae are found in the southern states. *Heteropoda venatoria* is a tropical species introduced into Hawaii, California, Texas, Florida, and many island territories worldwide. In Hawaii, *H. venatoria* is known as the Cane Spider. *Olios* spp. occurs from California to Texas and northern Mexico. The crab spiders (Thomisidae) are found everywhere. The large *Misumenoides* is found from California to Florida and north to the Great Lakes. Two other genera of flower dwellers, *Misumena* (2 species) and *Misumenops* (25 species) are also widespread in the United States and Canada.

The remaining families (and most genera) of orb weavers, long-jawed orb weavers, ground spiders, wolf spiders, lynx spiders, jumping spiders, and cobweb weavers are widespread in the United States and many are found in Canada.

Natural History: These families and genera occupy valleys to mountains of most habitats and niches of North America. However, a given species may be somewhat specialized for habitats where they exist. The reader is advised to learn the local denizens of the area they live in by perusing a good field guide.

Each spider family tends to have a distinctive set of general patterns of habitat use and habits. Their webs (or lack thereof) are often indicators of their lifestyle. The orb-weaving spiders often create large, intricate, flat webs between trees and shrubs and overhangs. Long-jawed orb weavers usually build their webs over or near water. They catch their prey in the webs. The host spider is sometimes present in the web, but may also be hidden. They may be diurnal or nocturnal, depending on species and habitats.

The ground hunters (Lycosidae, Gnaphosidae, and Salticidae) do not build obvious webs, but actively hunt for prey. They may be surface active day or night, depending on species and environmental factors. Some hide under rocks and other cover, coming out primarily at night to prowl. Some species remain in leaf litter or vegetation where they hunt for insects. The crab-like species (Thomisidae, Sparassidae) are sit-and-wait predators. Some of the crab spiders remain in or near flowers. They are targeting insects looking for nectar, such as honey bees. Huntsmen spiders are found in cracks, crevices, prickly pear pads, homes, and in other situations.

The loosely built webs of *Callobius* are often under or among rocks and other objects, with a funnel entrance. The Antrodiaetidae line burrows with silk and have a collapsible collar, hinged trapdoor, or turret webs. The Ctenizidae have a thick, cork-like trapdoor. During the day, they remain inside their webs and burrows, and by night, they wait by the entrance to rush out and seize prey. Both kinds of trapdoor spiders may be hard to locate, since the lid of the trapdoor is made with silk and the local substrate.

Spiders are most active in the spring through fall, depending on the species and latitude. In boreal climes, spiders tend to be most active during the warmer months, but to the south, they may be active for an extended part of the year. Hawaii, the California coast, Florida, and south Texas are particularly mild, allowing a long active season.

Males are generally smaller than females. The male builds a small web, deposits semen into the web, and then draws it up into his pedipalps. He then leaves the web to search for females. When he finds a female, he will court her. This can be a simple or sometimes complex display—that of jumping spiders is the stuff of legend, as seen in many nature shows or online videos, including the well-publicized Peacock Spider. Depending on species and circumstances, they lay eggs in the spring through fall. If late in the year, the eggs and/or spiderlings overwinter. The spiderlings grow and develop in the spring through summer and usually reach maturity in the late summer to early fall. There are usually 5–10 molts to reach maturity. After maturity, most spiders no longer molt, although mygalomorphs may continue to molt periodically. Depending on the species, spiders live one to several years. Overwintering sites are usually in burrows (e.g., trapdoor spiders) or under rocks and logs. Most species deposit eggs in egg sacs, although some lycosids are well known for transporting eggs, then young, on their back. Egg sacs may be protected by the female, although in some species, the female may

A large wolf spider, *Rabidosa santrita*, with young on her abdomen. These large spiders are disinclined to bite, and it is usually reported to be mild. *L.L.C. Jones*

die before the eggs hatch. After hatching, many species disperse immediately. Some species climb to high branches, then let out silk lines and let the wind carry them off, a process known as ballooning.

All of these spiders feed on insects and other invertebrates commensurate with their habits. For example, orb weavers target flying insects, while ground hunters feed on terrestrial arthropods. Flower-dwelling crab spiders feed on pollinating insects. Jumping spiders find prey where they hunt, such as caterpillars in shrubs. Trapdoor spiders hunt ground dwellers that venture near their lairs. In turn, spiders are preyed upon by a variety of other animals, including birds, amphibians, lizards, grasshopper mice, and a host of larger invertebrates, including other spiders. Many fall prey to predatory wasps and other parasites.

Encounters: Spiders are commonly encountered by humans, primarily in the spring through fall, and they usually peak in abundance and size in the fall. Encounters are usually casual; we see them and may walk through their webs, but we do not come into physical contact with them, or if so, they are not likely to bite. Some of the encounters are outdoors, but they can venture indoors, and a few species seem to capitalize on human habitations. The orb weavers tend to stay outside, although this may include human structures. They are also drawn to gardens because of the moist, lush environment and ample supply of prey, including the garden spiders. One species of orb weaver is known as the Barn Spider (*A. cavaticus*) because of its habit of nesting around barns. It is primarily a species of the Northeast, but there are several species of similar *Araneus* throughout North America. Tetragnathids are very common near water, including over streams and pond edges, so are often found among vegetation, such as cattails.

Jumping, lynx, and ground-dwelling spiders are mostly encountered outdoors, but they may wander indoors. These families of spiders can be found in nearly every habitat and area of North America. Males looking for mates are more likely to be found indoors than the more sedentary females. Other frequently indoor species include the sparassids, *Olios* spp. and *H. venatoria*. These animals are commonly found indoors where they hide by day under curtains, pictures on walls, under pillows, and virtually anywhere there is a narrow hidey-hole for them to spend the daylight hours. Clothes, bedclothes, and drapes are excellent places to find them. At night they come out to feed and are often seen on walls.

Few of these spiders are prone to defend themselves with little provocation. However, *Calisoga* and *Drassodes* are said to have a bit of an attitude, so are best left alone. Although the spiders in this section are known to bite humans, most are reluctant to do so. Bites generally occur when they are accidentally pressed against the flesh, or handled.

Venom Apparatus, Venom, and Symptoms: The venom apparatus is the same as for other spiders. The hollow fangs deliver toxins from the venom glands in the chelicerae and anterior cephalothorax. In addition to biting, the Green Lynx Spider has a novel way of introducing venom into predators: spraying into the eyes. The spitting spiders (Scytodidae) are the only other spiders that spit, but they do so for prey procurement and spit silk, not venom.

Little is known about the venom of most of these spiders, but some are being investigated for potential pharmaceutical uses. Across the breadth of spider families and genera, there are low molecular weight compounds (e.g., organic acids, nucleosides, nucleotides, amino acids, amines, and polyamines), acylpolyamines, peptides, and proteins. Most are small proteins. Some foreign wolf spiders, ctenids (see chapter on tarantulas), araneids, sparassids, and others have various enzymes including (as a group) hyaluronidases, proteases, phospholipases, and metalloproteinases, but native species are poorly known. Some wolf spiders in the genera *Lycosa*, *Geolycosa*, and *Hogna* have cytolytic properties, as does *P. viridans* and *Phidippus* spp. Venom of some araneids (*Araneus*, *Neoscona*, *Argiope* [*aurantia*], and *Nephila clavata*) has been shown to interfere with synaptic membranes of the brain in both vertebrates and invertebrates. Some spiders have neurotoxic venoms that may affect vertebrates. However, the amount of venom is very small in spiders. *Heteropoda venatoria* has a potassium ion channel blocking protein, HpTx2.

General symptoms for envenomation by most of these species include some degree of pain (slight to severe), local swelling, itching, and redness, but there is variation among genera, species, and human reactions. There may be dry or superficial bites. Most symptomless

A large orbweaver, the common Black and Yellow Garden Spider. *L.L.C. Jones*

bites undoubtedly go unreported, and even bites with symptoms are lost to a lack of information about the species identification. On occasion, there may be some systemic reactions (usually slight) and possibly slight necrosis. There are few published case reports where the offending spider was positively identified.

In one of the few studies of known bites in Oregon, symptoms of a number of species were reported. Several were from species discussed in other sections, while some were from a few of the genera/species in this section. There were five confirmed bites by *A. diadematus* and one of *A. saevus*. The bites mostly included the general symptoms listed above, and resolution ranged from one to several weeks. Symptoms of the one case of *A. saevus* lasted twelve weeks, while the others completely resolved in 1–3 weeks. In four of the six incidences, there were systemic reactions that included nausea (2 cases), muscle cramps (1 case), anxiety (1 case), and fever and numbness (the *A. saevus* envenomation). One bite by *A. occultus* resulted in pain, swelling, and redness. It resolved in four days. There were two bites by *Phidippus* sp. These resulted in pain and itching, and one of them also had swelling and redness. They resolved in 1–2 days, without systemic reactions. A single reported bite by *Callobius* spp. had pain, itching, swelling, redness, and nausea; time to resolution could not be recorded.

There are a few case reports of envenomation by *P. viridans*. The species is easily recognizable, even when squashed, which may help with positive identification. These bites generally included local pain, swelling, induration (area becomes hard), redness, burning, itching (sometimes severe), and sensitivity to touch, but usually with no systemic effects. There are also a very few case reports of the effects of their spraying people in the eyes. In the first such report, a man claimed that he was sprayed in the eye at a distance of about 30 cm. He complained of great pain in the affected eye, experienced severe conjunctivitis, and was visually impaired for two days. Based on the few other reports (including the opening quote in this section), his reaction was somewhat worse than that of others. Parson spider bites can be painful, with swelling, itching, redness, and nausea. Allergic reactions have been reported for the Eastern Parson Spider. Bites by the Pantropical Huntsman are said to cause pain, redness, swelling, and itching. Bites by *O. giganteus* are said to range from asymptomatic to about that of a bee sting.

At the time of this writing there are a few accounts in online forums (Arachnoboards) for some of these species. A bite by the cellar spider, *Pholcus phalangoides*, resulted in some pain, warmth, redness, and some numbness.

One correspondent of mine has been bitten several times by these, and while they cause a reaction, it is always slight. One bite by a captive California Trapdoor Spider only resulted in a slight ache. The circumstances were interesting, though. As the pet owner was forcing the spider to abandon its burrow, it sunk its fangs deeply into his finger, and he had to pry it off with a paintbrush. A spider researcher said in a post on Flickr that a bite by the euctinizid spider *Aliatypus torridus* was not too bad and there was probably a day's worth of lingering pain, but, as he pointed out, he deserved it. A pet owner posting on Arachnoboards poked his Carolina Wolf Spider in the leg and it turned around and bit him. The bite "stung" at first and there was some redness and itching. The bite area became dark red and very sore for two days. He found it hard to concentrate on the effects of the spider bite because he was also attacked by yellowjackets while mowing his lawn. A bite by a Woodlouse Spider to a spider collector caused no reaction beyond the usual poke. The Woodlouse Spider, by the way, was a subject of a recent scare, being dubbed the most venomous animal on Earth; this is pure poppycock. They do have an impressive gape and fangs, which probably led to the hoax, but the few bites that have been reported are asymptomatic, to less than that of a bee.

A trapdoor spider (*Aliatypus torridus*) biting the hand that feeds it.
Marshal Hedin/CCA/Flickr

The maligned Woodlouse Spider has some wicked mouthparts, and they may be quick to defend themselves, but the bite is usually mild. They use the long fangs to pry open pillbugs. *Jillian Cowles/Amazing Arachnids*

Remarks: The spider in *Charlotte's Web*, a popular children's book, was *A. cavaticus*, the Barn Spider. In the book, the female lead's name was "Charlotte A. Cavaticus." That book is one of the few that refers to a spider in a positive manner.

Peptides from spider venom, including those of taxa in this section, show great promise as insecticidal agents for their ability to interrupt the neural junctions of insects.

References and Resources: Adams (2014), Anderson (1997), Bernard et al. (2000), Boyer et al. (2001), Bradley (2013), Campbell et al. (1987), Cowles (2018), Diaz and Leblanc (2007), Fink (1984), Isbister and Gray (2002), King and Hardy (2013), Kuhn-Nentwig et al. (2011), McKeown et al. (2014), Michaelis et al. (1984), Nieuwenhuys (2008), Russell and Waldron (1967), Ubick et al. (2005), Vetter and Isbister (2006), Wong et al. (1987).

Yellowjackets and Hornets (Subfamily Vespinae)

Allergic reaction: "While in a very confused state, I waved down a car and garbled to the driver to get me an ambulance. For the next half hour I slumped over, unable to move, drooling on my vest, and fighting to maintain consciousness. I had an overwhelming urge to sleep, but I thought that if I fell asleep I would never wake up again. I experienced that infamous feeling of impending doom associated with anaphylaxis." (Author's testimonial, this volume)

Western Yellowjacket, *V. pensylvanica*. L.L.C. Jones

Classification: Phylum Arthropoda (arthropods); Class Insecta (insects); Order Hymenoptera (wasps, bees, ants, and sawflies); Family Vespidae (vespid wasps); Subfamily Vespinae (yellowjackets and hornets).

The Asian Giant Hornet (*Vespa mandarinia*) has recently been found in British Columbia and Washington State. If these become established in North America, they could be a serious invasive species that not only has a powerful sting, but also feeds on European Honey Bees. These large, true hornets have captured the imagination of sensationalistic reporters who refer to them as "Murder Hornets." *Li Cheng Shih, CCA, Flickr*

The Vespidae includes eusocial species of aculeate wasps, as well as solitary species. The eusocial species include the yellowjackets and hornets (subfamily Vespinae) and paper wasps (Polistinae). There are more than 5,000 species of vespid wasps worldwide, but just over sixty species are within the Vespinae. There are four genera of Vespinae, two of which are found naturally in North America. These are yellowjackets (genus *Vespula*), with about twenty-two species, and bald-faced hornets or aerial yellowjackets (genus *Dolichovespula*), with about fifteen species.

There are thirteen *Vespula* found north of Mexico. These include *V. acadia* (Forest Yellowjacket), *V. alascensis*, *V. atropilosa* (Prairie Yellowjacket), *V. consobrina* (Blackjacket), *V. flavopilosa* (Downy Yellowjacket), *V. infernalis* (Cuckoo Yellowjacket), *V. intermedia* (Northern Red-banded Yellowjacket), *V. maculifrons* (Eastern Yellowjacket), *V. pensylvanica* (Western Yellowjacket), *V. squamosa* (Southern Yellowjacket), *V. sulphurea* (California Yellowjacket), and *V. vidua*. *Vespula germanica* (German Yellowjacket) is introduced. *Vespula rufa intermedia* is now recognized as *V. intermedia*.

There are six *Dolichovespula* in the United States and Canada, which are commonly referred to as aerial yellowjackets. These include: *D. albida*

(Arctic Yellowjacket), *D. alpicola* (Rocky Mountains Aerial Yellowjacket), *D. arctica* (Parasitic Yellowjacket), *D. arenaria* (Common Aerial Yellowjacket), *D. maculata* (Bald-faced Hornet), and *D. norvegicoides* (Northern Aerial Yellowjacket).

The genus *Vespa* represents the true hornets, which are native to the Old World. *Vespa affinis* (Lesser Banded Hornet), *V. crabro* (European Hornet), and *V. simillima* (Japanese Hornet), have been reported or become established in parts of mainland United States and Canada.

There are other subfamilies, genera, and species of wasps within the Vespidae such as potter wasps (subfamily Eumeninae), which make mud nests and rarely sting, and Mexican Honey Wasps (*Brachygastra mellifica*), which produce honey but can readily sting. Some of these are covered in the Other Terrestrial Invertebrates section.

Identification: Vespids have three distinct body sections: the head, thorax, and abdomen. The Vespinae and Polistinae (paper wasps; next chapter) are similar, but yellowjackets and hornets appear stouter than Polistinae. The petiole is slender in Vespinae, but not as apparent or elongate as in Polistinae. The head has large compound eyes and ocelli above and between the compound eyes. The antennae are short and stout and the powerful mandibles are conspicuous. The paired wings join the dorsal thorax, just below the scutum. The six medium-sized legs are joined to the thorax ventrally.

These are medium-sized insects. Queens are larger than workers and males. Most species range from about 10–15 mm in length. Bald-faced Hornets are larger (to about 20 mm) and European Hornets larger still, reaching about 25 mm in males and 35 mm in overwintering queens. Yellowjackets are generally marked with vivid black and yellow aposematic coloration, although some species may have white, rather than yellow, markings, and some have reddish coloration. Bald-faced Hornets, which are not true hornets, have ivory and black markings. Other aerial yellowjackets are yellow and black. The European Hornet is brown to black and yellow. Vespid genera and species can usually be identified by the pattern on the face and dorsal abdomen.

Distribution: While the family is distributed worldwide, the subfamilies and genera are less cosmopolitan. The Vespinae are primarily boreal and/or montane, or in areas with mild climates. They are essentially distributed throughout the United States and Canada, except in arid areas. This is one of the few groups of venomous animals that is widespread and diverse in

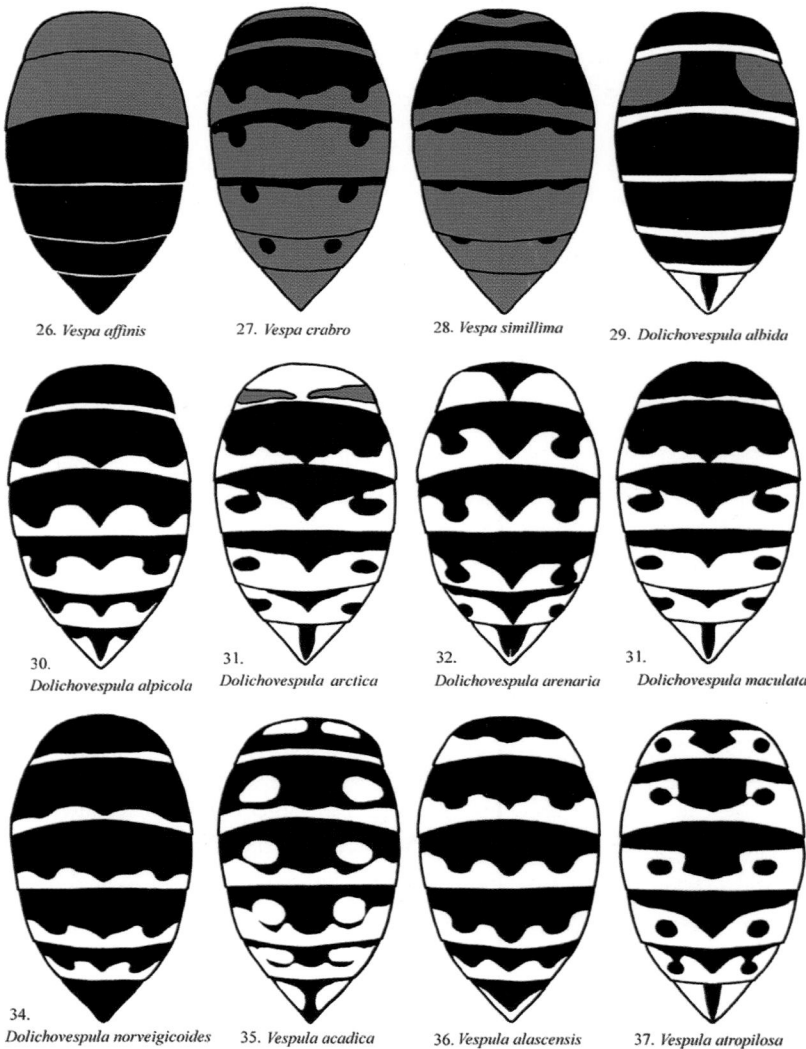

26. *Vespa affinis*   27. *Vespa crabro*   28. *Vespa simillima*   29. *Dolichovespula albida*
30. *Dolichovespula alpicola*   31. *Dolichovespula arctica*   32. *Dolichovespula arenaria*   31. *Dolichovespula maculata*
34. *Dolichovespula norveigicoides*   35. *Vespula acadica*   36. *Vespula alascensis*   37. *Vespula atropilosa*

Color patterns of abdomens of yellowjackets are diagnostic.
Kimsey and Carpenter/CCA/Journal of Hymenopteran Research

Canada. The number of species dwindles going south into Mexico, although European Hornets have been introduced as far south as Guatemala. Most of the species occur primarily in the northern United States and in much of Canada, with some exceptions. *Dolichovespida arenaria* and *D. maculata* have a broader distribution in the United States, as well as Canada. *Vespula*

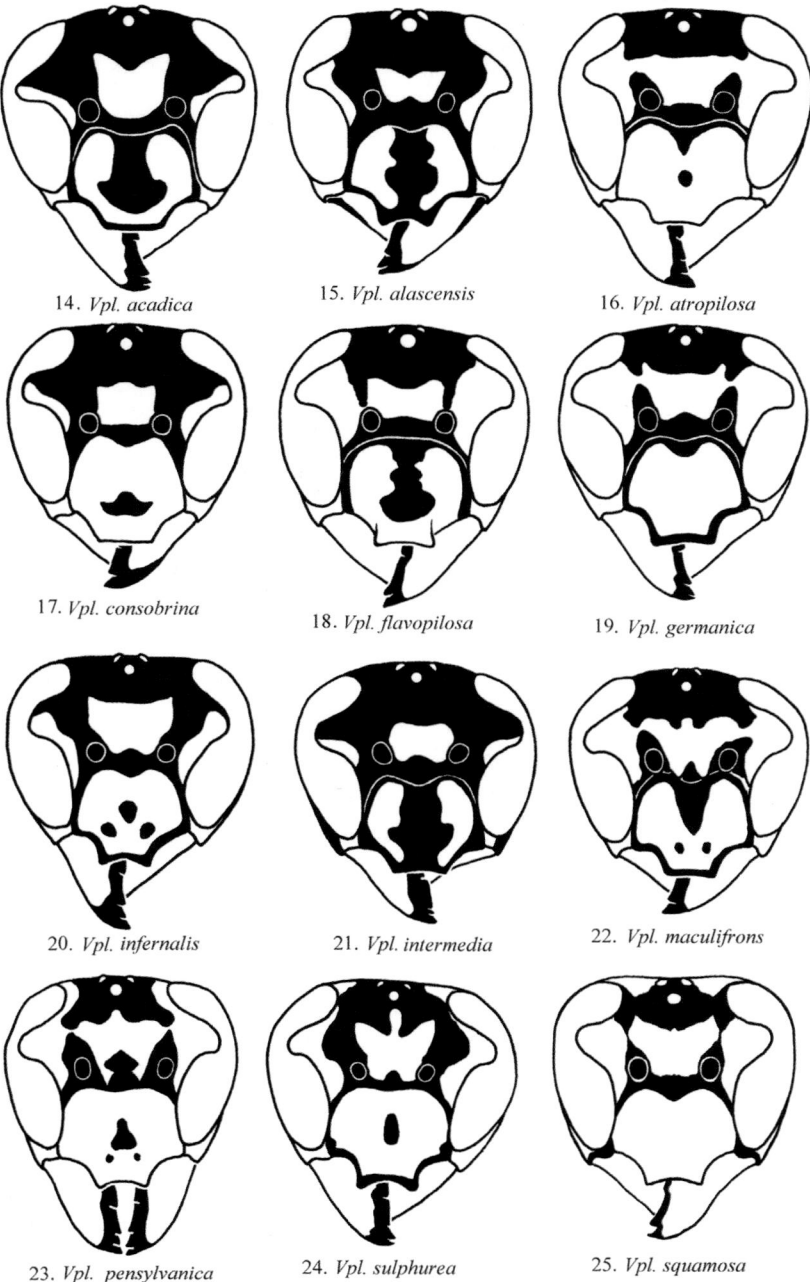

Facial patterns of yellowjackets are also diagnostic.
Kimsey and Carpenter/CCA/Journal of Hymenopteran Research

Bald-faced Hornet, *D. maculata*. Gary Alpert/Public Domain/CDC

maculifrons is primarily an eastern U.S. species, while *V. pensylvanica* is the western U.S. counterpart. Both species also occur in Canada. *Vespa crabro* and *Vespula squamosa* are primarily in the eastern United States and Ontario. *Vespula sulphurea* is the only species that does not occur in Canada; it has been reported from California and Oregon.

The introduced *V. germanica* is found throughout North America. It was probably introduced to Montreal in the 1960s and by the 1970s had expanded to much of the East. By 1989 it had extended its range to California. All of the *Vespa* are introduced to North America, but only *V. crabro* is relatively widespread. It was introduced in New York in the mid-1800s, but is now widespread east of the Mississippi River. A colony of *V. affinis* was reported in Los Angeles County, California, but it is unclear if the species has become established. *Vespa simillima* was introduced into British Columbia, but has apparently not become established (yet). Vespids, in general, have a tendency to fare well in some exotic areas, and many islands have non-native populations. There are excellent field guides and online resources that have specific distributions for all vespids.

European Hornet, *Vespa crabro*, an invasive species that is the only true hornet in North America. *"Smudge 9000"/CCA/Flickr*

Natural History: Vespines are generally thought of as woodland and forest dwellers. That is somewhat true, but some taxa are found in a variety of non-desert and suburban habitats.

They are eusocial animals, the highest level of sociality among animals. Eusocial behavior includes the caring of young and division of duties for the benefit of the colony. The duties are performed by different castes; each has different morphology and behavioral roles in the colony. It starts with the queen, which is the reproductive female. She will overwinter alone in a hibernaculum, such as in a log or under bark. She will then emerge in the spring to start a nest. Most species have a new nest every year and a single queen builds the population, but there are exceptions. In areas with mild climates, like coastal California, the Southeast, and Hawaii, nests of some species can be multiyear and have multiple queens. These nests can be become quite large—up to several meters long. Upon emergence from brumation, the queen feeds on nectar and gathers weathered wood to mix with saliva, to start building the papery "queen nest." She lays several eggs in the cells of the nest. After hatching, the new female workers begin the task of nest building, while the queen focuses on laying eggs. The eggs hatch into more workers to feed the larvae, build the nest, and protect the colony.

(Left) exterior and (Right) interior view of an aerial yellowjacket nest. *Terry Poutry*

The nest is composed of a series of downward-opening cells in vertical layers surrounded by a papery case. Most species of yellowjackets build nests underground, often in rodent burrows, but they may be in other natural or artificial cavities. Rotting logs are frequently used. The aerial yellowjackets build the same sort of nest, except that it is usually external and hung from a tree or other similar structure. As summer progresses and the nest grows, the workers become more protective of the colony. Reproductive males and females are produced in special cells; they participate during mating flights later in the season. By fall, there will be hundreds to thousands of individuals in the nest. The nest will have grown into an impressive structure that may be basketball-sized (a typical size) to a few meters in diameter, depending on the species and whether or not the nest is multiyear. Most nests have 1,000 to 3,000 individuals and can have 10,000 to 15,000 cells by fall, but some massive multiyear nests can have more than 100,000 individuals. *Vespula squamosa* is one species that can survive the winter in mild areas and build these huge nests. In general, though, the reproductive females will overwinter, and then start new nests the next spring.

Yellowjackets and hornets are primarily predators of insects, spiders, and other small animals, but also supplement their diet with energy-rich nectars of plants and herbivorous insects, such as aphids. In Washington State, I often marveled at how yellowjackets systematically hovered around the base of my house, picking off harvestmen that rested on the walls a few centimeters above ground. I had no doubt they were targeting these animals specifically with this specialized hunting tactic. Some species of yellowjackets are also well known as picnic and campground pests. They are sometimes (erroneously) called "meat bees" by campers because of their habit of scavenging on picnic foods, such as fried chicken and tuna sandwiches. They will also forage among beverages for moisture and sugary delights. When not at a picnic table, they may also be seen scavenging on carrion. Some species have more of a proclivity for scavenging than others.

Because of their infamy as stinging insects, most people don't realize how important they are at keeping pest insect populations in check. However, some of the larger species (*Vespa* spp. and Bald-faced Hornets) may feed on honey bees and may impact local populations. Also, vespines can become serious pests in fruit-growing regions, where they feed on the nectar and may sting the workers. In turn, vespines are fed on by other animals. A large nest full of fat, juicy larvae is a pretty good source of nutrition. This is what facilitated the predator-prey arms race in colonial-nesting insects. Their defenses are pretty effective, as nest predation is not common, although skunks have been shown to be pretty good at consuming combs. Other predators on nests include Coyotes, Northern Raccoons, and Black Bears. Individuals may be preyed upon by a variety of arthropod, amphibian, reptilian, mammal, and bird predators. Vespines are also parasitized by various insects.

Encounters: Depending on where you live, your potential for encounters can range anywhere from low (e.g., deserts) to extremely high (mountain forests, woodlands, and boreal areas). Vespids are among the most likely venomous animals to be encountered by the general populace. Most encounters are outdoors, such as natural areas, parks, and zoos, although nests can be in yards, homes (e.g., in walls), garages, storage units, and public places, including school grounds. Encounters at picnic tables, trash cans, and water sources are common in some areas. It is important to note that encounters with individuals outside of the nest area are less likely to end badly than those where workers are protecting the nest.

The biggest problem is when one unwittingly stumbles upon a nest or must work in areas of high densities of wasps. For example, a main cause of sick leave in field-going personnel in the U.S. Forest Service is due to stings by Vespinae. Timber harvesters and those operating logging equipment are at repeated risk of disturbing nests, especially when operating loud, vibrating equipment. There is considerable economic loss in recreation due to outbreaks of yellowjackets. Economic losses can also be significant for agriculture workers, such as in fruit orchards and meat processing plants.

Subterranean nests are difficult to see and avoid. Aerial nests may seem much more obvious, but people tend to look down at their feet when walking through the woods. For me, the best precaution is recognizing a "buzzing threshold." There is an ambient background buzzing always present in the woods due to yellowjackets and flies, but when the volume increases, it means a nest has been disturbed, and you are being warned by aposematic buzzing. I became adept at running full throttle before I knew why I was running. It is difficult for people to hear buzzing during rain or when they are using chainsaws, operating equipment, have earplugs in, or are near a stream.

It is also possible to see yellowjackets coming and going into their nest entrances if you are really paying attention. Moving slowly and attentively through the woods is much more prudent than charging through the woods.

Yellowjacket nests can also be found in areas where you had not encountered them in the previous year. Along with paper wasps, yellowjackets and aerial nesters can take up residence in recreational vehicles, logging equipment, heavy machinery, and so on. For example, a queen may find a vent in an RV and can start a nest. By the time you decide to go camping, there may be a thriving colony inside. Always inspect your equipment before dewinterizing it. RV owners can also purchase yellowjacket filters before winterizing. Another option is simply to use hardware cloth or screen over openings.

One study reports that from 2000–2004, there was an estimated annual average of more than 230,000 people admitted to emergency rooms due to hymenopteran stings in the United States. Most were identified as "bees," with only 44,000 being vespids, but many laypersons think yellowjackets are bees. Fewer than 1% of these stings required hospitalization. In another U.S. study on calls to poison control centers, from 2001–2005, insects accounted for 57% of the more than 94,000 calls per year, and bees, wasps, and hornets combined accounted for more than 12,000 calls per year. Since most stings undoubtedly go unreported, it does suggest the much greater magnitude of the problem. In reports on animal-related fatalities in the United States, there seem to be about 50–60 deaths by stinging hymenopterans per year, which is more than any other group of animals, but most are due to allergy, rather than the venom per se. It has been stated that allergic reactions are

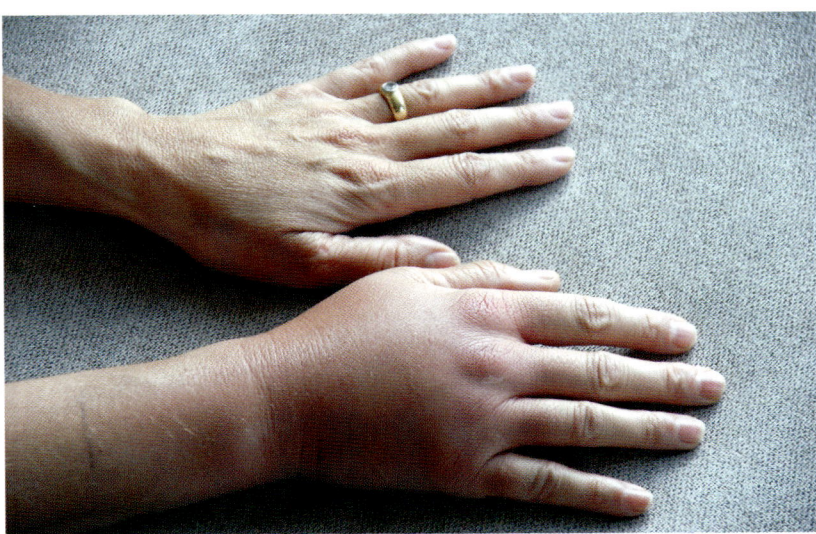

Stings by vespids usually result in inflammation. If inflammation is abnormal, one should consider allergy testing, and possibly immunotherapy. *Stephanie Young Merzel/CCA/Flickr*

more frequently reported for vespids than any other group of animals. Again, these numbers are difficult to track, but stinging hymenopterans are arguably the most medically important group of venomous animals, despite generally delivering a localized envenomation that completely resolves without medical intervention within hours to a few days. Compounded by hundreds of thousands of envenomations in the populace, the <1% of severe systemic reactions add up. One paper states that while only forty people die per year due to wasp stings, 4,000,000 individuals in the United States are potentially at risk of hypersensitivity and anaphylaxis for their *next* sting.

Venom Apparatus, Venom, and Symptoms: The venom apparatus is a modified ovipositor of the female, so the workers are the primary cause of human envenomation. The queen is focused on egg-laying in the nest most of the year, but early spring queens can also be a hazard. Males cannot sting. The venom apparatus is a fairly complex structure with several component parts. When at rest, the entire stinging apparatus is inside a sheath in the abdomen. Terga 7 and 8 enclose the stinger. The stinger itself is composed of a shaft of the stylet and two lancets—holdovers from an earlier ovipositor function of sawing, drilling, and egg-laying. When the animal is at rest, the stinger is not ready to sting yet. When the wasp becomes agitated, the component parts are

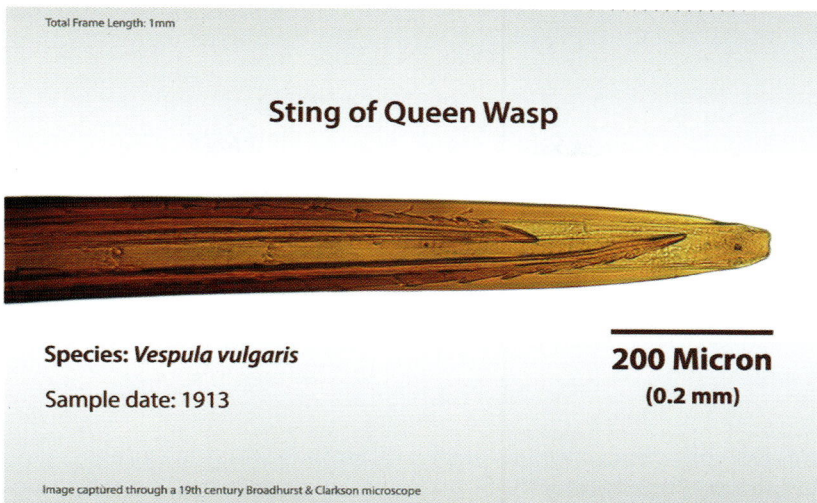

Stinger of European Wasp, *V. vulgaris*, another invasive species.
Tom Blackwell/CCA-SA/Flickr

positioned to be at the ready. The stylet and lancets slide together and form a channel for venom delivery. As with an ovipositor, the stinger saws into the victim's flesh to gain penetration, although the barbs are not long enough for the stinger to get lodged in the flesh. This way it can be inserted and withdrawn several times to deploy venom at multiple sting sites.

There are two glands associated with the venom apparatus, but the venom gland is what gives rise to the algogenic compounds. The other gland is Dufour's gland, which excretes other types of chemicals; the function is more closely related to reproduction and release of pheromones than defense. Muscular contractions force venom from the venom gland into the stinger. The yellowjacket will often gain purchase on the victim with legs and mandibles, and often under clothing for support, since the semi-flexible stinger may take some maneuvering to penetrate flesh. Unlike honey bees, which are capable of delivering only a single sting, wasps do not sacrifice their lives for the good of the colony.

The stings of yellowjackets are completely defensive. The algogenic compounds are very effective at targeting mammals, presumably because mammals are nest robbers. The LD_{50} of vespine venom is moderate (e.g., *V. germanica* at 2.8 mg/kg and *D. maculata* at 6.1 mg/kg), with some reports of up to 50 mg/kg. Venom yield is low, so it is estimated that healthy humans would require 1,000 stings or more to be fatally stung, although it varies by species. Pharmacological and biologically active compounds include histamine, serotonin, dopamine, and noradrenaline. Enzymes present include phospholipases A and B, as well as hyaluronidase, esterase, lipase, protease, kallikrein-like peptidase, acid phosphatase, alkaline phosphatase, and phosphodiesterase. Vespid venom also contains the antigen 5s protein, a human allergen. Phospholipase and hyaluronidase are also human allergens in wasps.

Typical local effects of the venom include immediate pain, swelling, redness, itching, and burning. The pain is fairly intense for several minutes, and then subsides. On the Hymenopteran Sting Pain Index, on a scale of 0–4, yellowjackets and hornets rate a 2, with a duration of about 4–10 minutes for *Vespula* and 3–4 minutes for *D. maculata*. Although the pain intensity subsides, there can be lingering effects that may take days to completely dissipate. The immediate, but short-lived pain is enough to dissuade would-be predators. There are a few records of mass envenomations, but these represent unusual circumstances. My personal experience is that there are usually 1–15 stings per encounter, with more than one sting almost always associated with nest protection. With large, multiyear nests, it is always possible to receive dozens to hundreds of stings if the victim cannot reach safety. People who cannot run away, such as children and those with limited mobility, are at a much higher risk of mass envenomation. There is a case report of a 3-year-old who was stung about 120 times while hiking with his family in Oregon.

Allergic reactions are relatively common for stings by vespids (1–4%) due to the potential for repeated encounters and the presence of at least fourteen named allergens among *Vespula* and four among *Dolichovespula*. Systemic reactions of somewhat sensitive people can include swelling of the entire limb, fever, headache, malaise, and hives. Moderate systemic reactions can include constriction of the throat and chest, abdominal pain, nausea and vomiting, dizziness, wheezing, and generalized edema. In severely sensitive people, the following symptoms can occur: labored breathing, difficulty in swallowing, hoarseness and labored speech, confusion, weakness, fainting (sudden drop in blood pressure), and a feeling of impending doom. In severely sensitive people, death is possible due to effects of anaphylaxis, such as from respiratory failure. There are various estimates of the number of people stung each year and the number of fatalities attributable to aculeate hymenopterans, but these are difficult to gauge for a number of reasons. First, most stings never get reported to those collecting such data. Second, there is variation in sting sensitivity and numbers of stings received. Third,

Vespula maculifrons with an earthworm. Eastern Yellowjackets, like other vespids, are fierce predators. *Katja Schultz/CCA/Flickr*

there is no perfect tracking system to record pertinent data. And lastly, cause of death is not necessarily well understood or documented. Cause of death determination may be difficult due to multiple symptoms that may or may not have been attributable to the sting itself.

Cross-sensitivity has been the topic of numerous studies. However, it is a difficult to accurately predict how sensitive to hymenopteran venoms people will be when confronted with numerous stinging insects of various taxa. Cross-sensitivity is expected to be high among vespids. Some studies suggest there is little to moderate cross-sensitivity between yellowjackets and other hymenoptera, or even among genera (but genera can change, and that is not based on allergy assessments). There is considerable variability, but many of the same or similar types of allergens are found among species. Skin tests vary, and it is not really possible to have the wide array of hymenoptera venom available for testing. Nevertheless, skin tests for hymenoptera are recommended for anyone showing more than a local reaction.

References and Resources: Akre et al. (1980), Goddard (2012), Greene and Breisch (2005), Habermann (1972), Jakob et al. (2017), Kimsey and Carpenter (2012), King et al. (1978, 1983, 1996), Krombein et al. (1979), Langley (2005, 2008), Leluk et al. (1989), Lopez-Osorio et al. (2014), McDonald et al. (1976), Müller et al. (1992), O'Neil et al. (2007), Pickett et al. (2001), Ratnieks et al. (1996), Schmidt (1983, 2008, 2009a, b, c, 2014, 2016), Schmidt et al. (1986), Vetter et al. (1999), Visscher and Vetter (2003), West et al. (2011).

Paper Wasps (Subfamily Polistinae)

Description of the sting of *Polistes canadensis* (Red Paper Wasp): "Caustic and burning, with a distinctly bitter aftertaste. Like spilling a beaker of hydrochloric acid on a paper cut." (Schmidt 2016)

Classification: Phylum Arthropoda (arthropods); Class Insecta (insects); Order Hymenoptera (wasps, bees, ants, and sawflies); Family Vespidae (vespid wasps); Subfamily Polistinae (paper wasps).

Being in the family Vespidae, paper wasps are related to yellowjackets and hornets, so are similar in many regards. The Polistinae are more speciose, however, with more than 1,100 species worldwide, belonging to four tribes, according to some authorities. Most of the species in North America belong to the tribe Polistini, in the genus *Polistes*. One species, *P. dominula* is considered an invasive, non-native species and has quickly colonized most of the United

European Paper Wasp, Colorado. *"Louis"/CCA/Flickr*

States and Canada. There are nearly thirty species and subspecies, depending on taxonomic leanings. *Polistes* are too diverse to be detailed, but common species are discussed in the Distribution section. There are two other genera in the United States and/or Canada that are similar: *Brachygastra* (one species, *mellifica*; Mexican Honey Wasp) and *Mischocythrus* (three species).

Identification: Paper wasps are similar to yellowjackets, except they are more gracile and there is a distinct petiole between the abdomen and thorax—hence, they appear to have a slender waist, unlike bees and yellowjackets, but similar to some other wasps. They are typically a little larger than most yellowjackets, about 20–25 mm total length. Terminology for their anatomy is essentially the same as for yellowjackets and hornets. Also like yellowjackets, they tend to be aposematically colored, often with black and yellow, orange, or red, which aids in species identification. Unlike yellowjackets, the queen is not significantly larger than the workers.

They are often identified as a group by their papery, upside-down umbrella-shaped nests, which is why they are sometimes called "umbrella wasps," although Mexican Honey Wasps have a paper nest similar to aerial yellowjackets. The nest is suspended by a stalk, known as a petiole (the same term as a wasp's "waist," from the same Latin root, *petiolus*, which means "stalk, stem, or little foot"). *Polistes* is Latin for "founder of a city," referring to the colonial nest.

Distribution: *Polistes* wasps are the most cosmopolitan group of wasps, but are better represented in the southern United States than the more boreal yellowjackets. Relatively widespread species in the Southwest include *P. apachus*, *P. arizonensis*, *P. carolina*, *P. comanchus*, *P. dorsalis*, *P. exclamans*, *P. flavus*, *P. fuscatus*, and *P. major*. *Brachygastra mellifica* is found in Texas and Arizona, south to Central America. Southeastern species include *P. annularis*, *P. bellicosus*, *P. carolina*, *P. d. dorsalis*, *P. exclamans*, *P. metricus*, and *P. rubiginosus*. Boreal species include *P. aurifer* and *P. fuscatus*. Texas has a particularly diverse *Polistes* fauna, as there are Southwest, Southeast, and Mexican species represented. *Mischocytharus* are also found along the southern United States, but one (*M. flavitarsus*) ranges into British Columbia.

Natural History: Paper wasps are a successful group of animals and can be found in virtually every terrestrial habitat, from coastal to inland, xeric to mesic, and desert to montane.

Polistes spp. normally have an umbrella-like paper nest, as opposed to the vespids that have large, oval paper nests, or nests in the ground. *L.L.C. Jones*

Paper wasps are related to yellowjackets and hornets, so are also eusocial. They also build paper nests from wood and saliva, but their nests are smaller than those of vespines. The bottom of the nest is open, with the cells facing down. An overwintering queen will first build a "queen nest." Eggs are laid in the cells by the queen, and she feeds early-season larvae. The cells are closed off when the larvae are ready to pupate. When workers mature, they take over nest-building chores and nurture the young. Adults are frequently seen milling about the nest. As with typical yellowjackets, nests are annual. There can be up to a few hundred hexagonal cells in a mature nest, but most encountered have a few to a few dozen cells.

Paper wasps are predatory insects, but do not tend to scavenge as do certain *Vespula*. Like yellowjackets they dismember prey with their powerful mandibles, and feed chunks of prey to the young. In return, the young give off a high amino acid liquid that the adults will consume. While they may eat any number of insects, many paper wasps feed almost exclusively on caterpillars, so are important economically. They are sometimes used as a natural biological control to keep caterpillar pests in check. For example, paper wasps are used to control tobacco hornworms and cotton bollworms.

Yellow Paper Wasp on a prickly pear fruit. *L.L.C. Jones*

They are also a natural biological control for home gardens and ornamental landscaping. *Polistes* are often seen hunting at tent caterpillar pavilions. Tent caterpillars are not only destructive defoliators, they may be mildly venomous to humans.

Encounters: Paper wasps are frequently encountered in most areas of North America, but are not usually found indoors. They sometimes occur in urban and suburban settings and may be common around homes and other structures. Nests typically are attached under an overhang, such as a branch or an awning. Public restrooms and similar buildings can have multiple nests under the eaves. These nests are often exposed, but they may also be hidden in vegetation and other situations. These wasps are sometimes considered aggressive, but are not usually as quick to antagonize as are yellowjackets and hornets. Fortunately, they do not have massive nests with potentially thousands of workers. Mexican Honey Wasps do have massive nests, but are not inclined to sting—but they can and will if they feel threatened. Because paper wasps do not tend to scavenge, they are not usually pests at picnics, but they can be attracted to beverages. Workers will sting under certain circumstances, and

The stinger of a Carolina Paper Wasp. *"Insects Unlocked"/Public Domain/Flickr*

stings are not uncommon, but some species seem to be more antagonistic than others. Stings usually occur if one bumps into their nest, accidentally brushes against them, or disturbs a nest near human activity, as when a door is slammed. *Polistes* have an aposematic posture. The wings are held nearly perpendicular to the body as an early warning sign. If the intruder does not retreat, they may initiate an attack and each individual can sting multiple times.

Venom Apparatus, Venom, and Symptoms: The venom apparatus is similar to their close relatives, the Vespinae, so refer to that section.

Paper wasp venom contains histamine, serotonin, a hemolytic enzyme appropriately called Polisitin, and kinin-like compounds referred to as polisteskinins, as well as phospholipases A and B, acid phosphatase, esterase, hyaluronidase, and lipase. The LD_{50} for *P. comanchus navajoe* is 5.0 mg/kg for mice, but the venom yield is small and it is estimated to take about 1,250 stings to kill a healthy human. Of course, the number of animals in the nest never approaches that size.

Symptoms of human envenomation are similar to that of yellowjackets, including pain, redness, and swelling. Some paper wasps have a more painful sting, however. The Hymenopteran Sting Pain Index for yellowjackets and

Polistes comanchus grabbing a drink in a desert tinaja. L.L.C. Jones

hornets is 2 on the 0–4 scale, but native species of *Polistes* generally score a 2–3, with a rating as high as 4 in the neotropics. Some of my correspondents claim the sting of some species is more painful than that of a honey bee or yellowjacket, and at least for *P. rubiginosus*, I concur. Unlike vespines, the venom of some species may also cause hemorrhaging. The sting can be highly allergenic, and people may become sensitized after a single sting. At least eight specific allergens have been identified and named from native U.S. *Polistes*, which includes four enzymes and four antigen 5 family antigens. Symptoms of hypersensitivity include hives, pruritis, and extensive swelling beyond the sting site, but there may be more serious symptoms, such as anaphylaxis and pharyngeal swelling. People with allergies are sometimes at risk of asphyxiation without medical treatment. Paper wasp venom may be cross-sensitive with yellowjacket venom.

References and Resources: Findlay et al. (1977), Goddard (2012), Grant et al. (1983), Greene and Breisch (2005), Habermann (1972), Jakob et al. (2017), Langley (2005, 2008), Leluk et al. (1989), O'Neil et al. (2007), Pickett and Wenzel (2003), Reisman et al. (1982), Richards (1971), Richter (2000), Schmidt (1983, 2008, 2009a, b, c, 2014, 2016), Schmidt et al. (1986), Turillazzi and West-Eberhard (1996).

Tarantula Hawks
(Genera *Pepsis* and *Hemipepsis*)

"It felt like it laid an atomic bomb in my hand!" (P. Rakoci, personal communication)

A red-winged tarantula hawk, genus *Pepsis*. Members of the genus may be difficult to differentiate in the field, and some species can have red or black wings. *L.L.C. Jones*

Classification: Phylum Arthropoda (arthropods); Class Insecta (insects); Order Hymenoptera (wasps, bees, ants, and sawflies); Family Pompilidae (spider wasps); Genera *Pepsis* and *Hemipepsis* (tarantula hawks).

Worldwide, there are about 5,000 species of Pompilidae, with six subfamilies commonly recognized: Ceropalinae, Ctenocerinae, Epipompilinae, Notocyphinae, Pepsinae, and Pompilinae. Those in the Pepsinae include the large species we know as tarantula hawks, but other genera and species are also venomous to varying degrees. Pompilinae is also a large subfamily of spider wasps, but do not reach the physical size of tarantula hawks.

There are about 18 species of *Pepsis* north of Mexico, which include many previous synonymies. Some of the more referenced species include *P. P. chrysothemis*, *P. grossa* (=*formosa*), *P. mexicana*, *P. mildei*, *P. pallidolimbata*, and *P. thisbe*. Species of *Hemipepsis* include *H. mexicana*, *H. toussainti*, and *H. ustulata*.

Identification: Tarantula hawks are large wasps, ranging from about 12 to 60+ mm in length. Our larger species are generally about 40 mm in length. Both sexes have wings. Females may be larger than males, but sexual dimorphism may not be pronounced. Females tend to have curly antennae, which are straighter in males, but this is often considered a poor indicator, and it would be risky to handle a straight-antenna tarantula hawk without precautions. These are slender wasps with long legs. Most are black or blue-black, with black to blue or yellow-orange to red wings. Some species have variation in the wing color, such as *P. grossa*, which usually has reddish-orange wings, but sometimes has black wings. A reflective metallic hue to the body and wings is often apparent. Besides the aposematic coloration, there is also an aposematic odor. They are generally considered to be in the same superfamily with vespids (Vespoidea) and have similar body proportions of Polistinae, due to elongation and narrowing of the first two abdominal segments. Tarantula hawks differ from vespids because the wings are laid back against the dorsal abdomen at rest, rather than being elevated or off to the sides. When moving about, tarantula hawks periodically flick their wings. Also, the middle sclerite along the sides of the thorax is divided. Their long, spiny legs are adapted for burrowing, feeding, and host capture. There is a conspicuous spine on the tarsus. There are numerous insect taxa that are Batesian mimics of tarantula hawks.

Distribution: The family Pompilidae is cosmopolitan. The genus *Pepsis* is found throughout most of the New World, and generally follows the distribution of tarantulas, so in Latin America, where tarantulas are most diverse, so are *Pepsis*. The literature states that there are 130 to 250+ species in the genus, mostly in Latin America. There are about 15 species of *Pepsis* in the United States, all but two of which (*P. marginata* and *P. saphirus*) are found in the Southwest—the same basic distribution of tarantulas. Two of the more commonly encountered species are the desert-dwelling *P. grossa* and *P. thisbe*, but most other species are not uncommon and occur from California to Texas and into Mexico. The genus *Hemipepsis* is found in both the Old and New Worlds. *Hemipepsis ustulata* is found in all Southwest states, *H. toussainti* occurs in Arizona and New Mexico, and *H. mexicana* occurs from Texas to Panama.

Natural History: Pompilids can be found in a variety of habitats throughout the country—anywhere there are spiders. Tarantula hawks can be found wherever there are tarantulas—primarily the arid parts of the Southwest. They are found in all four deserts, as well as chaparral, coastal sage scrub, grasslands, thorn scrub, shrub-steppe, and lower mountain slopes of piñon, juniper, oak, and pine woodlands.

While the related vespids are social wasps, pompilids are solitary, so do not form or defend colonies. Spider wasps are relatively well studied because of their requirement for having a spider as a host to feed their larva. They are a type of parasitoid, which is essentially an intermediate between parasite and predator, because their larvae parasitize the host. The developing larva does not kill the host until it has matured. Tarantula hawks are the largest of the spider wasps because they must tackle the largest of the spiders. Adults actually feed on nectar of certain plants, including milkweeds, soapberries, mesquites, and other species.

Males search for females by hilltopping—watching for female passersby from hilltops and the tallest vegetation in the area. After mating, female tarantula hawks fly overhead searching for burrows of tarantulas or the free-ranging males (e.g., those on mating forays). When the wasp finds a burrow, she will tap around the entrance with her antennae to try and coax the spider out. If the tarantula does not emerge, she will cut the web blocking the entrance with her mandibles and force the spider out. During the confrontation, the wasp contacts the spider with her antennae. The spider may rear up in a defensive posture, but it does not strike, presumably because of chemosensory cues on the wasp's exoskeleton. The wasp will move away and groom its antennae with its mandibles before engaging an attack. As the tarantula rears up, the wasp will grab its legs and sting it on the underside between sternal plates at the junction of the cephalothorax and abdomen. This will paralyze the spider within seconds. The wasp will groom again and then take the tarantula back to a burrow. The burrow may be pre-dug by the wasp or may be the tarantula's own burrow. They will sometimes use another burrow of similar size. The female will lay a single egg on the abdomen of the still-living spider, which will be fed on by the developing wasp larva. The larva will eat non-vital organs first to keep the spider alive as long as possible,

Terrestrial Invertebrates 233

(Left) A black-winged tarantula hawk engaging a tarantula (*Aphonopelma chalcodes*). (Right) The *Pepsis* wasp nearly always wins, and the prey is immobilized in seconds. *Susan Sferra*

A red-winged tarantula hawk with orange antenna tips has just immobilized the host for its larva. *Jim Rorabaugh*

then build a cocoon and pupate, emerging as an adult several weeks later in the spring. Eggs laid on large tarantulas develop into female wasps, while those laid on smaller tarantulas become males.

Different species of tarantula hawks have differences in host species, which may dictate regional availability, rather than having single parasitoid-host specificity. For example, *P. grossa*, *P. thisbe*, and *P. mildei* will use *Aphonopelma steindachneri*, *A. hentzi*, *A. echina*, *A. harlingenum*, and *A. heterops* in various areas of Texas, while in Arizona and adjacent California, *P. grossa* will hunt *A. chalcodes*. In southern California, *P. thisbe* hunts *A. reversum* and *A. eutylum*. *Pepsis mildei* in southern California reputedly preys on trapdoor spiders, but this is the exception to the rule.

Not many animals will even attempt to prey upon tarantula hawks, male or otherwise, but there are records of predation by the Greater Roadrunner. These crafty birds do not hesitate to feed on rattlesnakes and other venomous animals, aided by their armored feet and legs and icepick-like beak. However, tackling a tarantula hawk is a rarely reported event, even for a roadrunner. I have heard eyewitness accounts of large toads eating tarantula hawks, and while the toads showed some discomfort, they ingested the prey and survived. After eating one, they sometimes ate more, even though they were apparently stung during feeding. A Tiger Whiptail (*Aspidoscelis tigris*) was documented feeding on a tarantula hawk. *Pepsis* sometimes overwinters and lives into the following spring, but they apparently do not live more than about a year. *Hemipepsis* are more likely to be active in mild areas in the winter than *Pepsis*.

Encounters: If one spends time in the outdoors, particularly in deserts and other arid parts of the Southwest, tarantula hawks are frequently seen in the spring through fall. These wasps may be very common when feeding on nectar—there may be dozens on plants at a single time, during optimal conditions. They are also frequently seen flying over the ground, when females are in search of tarantulas and their burrows. Unless there is an open door or window, they are not likely to enter buildings. They are not attracted to people or their food. Males cannot sting, so there are no threats from half the tarantula hawk population.

Problematic encounters primarily come from not seeing a wasp and brushing up against it. This could happen when walking through flowering vegetation or thick brush. Another potential encounter could be to motorcyclists who are not wearing helmets with face guards. Convertible cars with the tops down also pose a hazard. Getting hit by a large *Pepsis* wasp at 60 miles per hour not only makes me cringe, but is potentially a life-ending experience. People stung by tarantula hawks often lose control of their faculties because of the immediate, intense pain. Researchers and invertebrate aficionados intentionally handle them, but that is a risky proposition. Tarantula hawks are not quick to sting, unless provoked, but collecting them is sufficient provocation.

Venom Apparatus, Venom, and Symptoms: The venom apparatus is similar to vespids, as females have a modified ovipositor with a venom gland, Dufour's gland, and stinger apparatus. The venom gland is surrounded by muscle. The gland is convoluted, as with other wasps, except Mutillidae. The stinger itself can be huge; that of *P. grossa* can be 7 mm long. That is about twice the length of a fang of an adult Rock Rattlesnake, a creature that is nearly ten times the length of the wasp!

The main components of *Pepsis* wasp venoms studied (e.g., *P. chrysothemis*, *P. grossa*, *P. pallidolimbata*, and *P. thisbe*) include various proteins, including glycoproteins, and biologically active amines. Compared to the aculeate social wasps and bees, there are relatively low levels of histamine and no serotonin, but relatively high levels of dopamine, norepinephrine, and acetylcholine. Although the venom is an extremely powerful algogenic agent, it does not translate into a very toxic substance, with an LD_{50} of about 60 mg/kg to greater than 120 mg/kg, depending on species tested. Hence, the active venom components for defense (which target vertebrates) are not

Pepsis mildei at a streamside with a Two-striped Gartersnake. *L.L.C. Jones*

the same as those used to immobilize hosts (which target invertebrates). The venom of tarantula hawks is highly effective for both defense and host capture. The results of a defensive sting are intense and immediate. When delivered to a host spider, the victim is completely paralyzed within seconds. Immobilization usually lasts for the duration of the host's life, meaning that while the host is being fed upon by the larva, it will not awaken (although this has happened). The host generally dies only after the larva has wreaked considerable bodily harm before it pupates. In many other spider wasps, paralysis is only temporary and the spider resumes bodily function while hosting the larva.

The hallmark symptom of a sting is pain. The tarantula hawk, particularly *P. grossa*, is said to have the most immediately painful sting of any hymenopteran in North America, and perhaps, the world. It ranks a 4 (highest possible) on the Hymenopteran Sting Pain Index. The South and Central American Bullet Ant, so named because it feels like you have been shot when you have been stung, is also extremely painful; the pain is not as intense and immediate but it lasts much, much longer. There are numerous colorful quotes among those describing the pain of tarantula hawks, including:

- "instantaneous, electrifying, excruciating, and totally debilitating"
- "blinding, fierce, and shockingly electric"
- "immediate, excruciating pain that simply shuts down one's ability to do anything, except, perhaps, scream"
- "a running hair dryer has just been dropped into your bubble bath"
- "pouring boiling oil on you"
- "It felt like it laid an atomic bomb in my hand!"
- "felt like razor blades being jammed into every joint"
- "If you eat a Carolina Reaper [world's hottest pepper] immediately after being stung, you will forget all about being stung."

While the pain can be extremely intense, it is usually not long-lasting, and may be completely gone within a few minutes—but that is long enough to get the message across. In some instances, however, the pain may linger with less intensity for up to twelve hours. The pain may actually traumatize someone more than the bite or sting of a more toxic animal, such as a rattlesnake. Having painted this ghastly picture of a tarantula hawk sting, some people who have been stung say it was no worse than a bee sting, so no doubt there

Close up of the face and orange antennae of *Pepsis ruficornis*, a Caribbean species that occurs in some U.S. territories. *Public Domain/USGS,BIM*

is variation among species and other factors may be at play. Also, there are many more species of spider wasps besides *Pepsis* and *Hemipepsis*, and we know very little about the effects of their stings on humans.

Remarks: The tarantula hawk is the state insect of New Mexico, although technically it is a group of two genera and several species. Since it is difficult to get accidentally stung by one, they are of more value as "watchable wildlife," and New Mexico recognizes that.

References and Resources: Leluk et al. (1989), Pitts et al. (2005), Punzo (1994, 2005, 2007), Punzo and Ludwig (2005), Schmidt (1983, 1990, 2004, 2009a, b, c, 2016), Schmidt et al. (1986), Schoeters et al. (1997), Starr (1985), Vardy (2000, 2002, 2005).

Velvet Ants (Family Mutillidae)

[Note: "cow killer" is a vernacular for some velvet ants] "'COW KILLER?' he bellowed out . . . Damn, boy . . . I thought you said 'Caterpillar!' From then on, every time we'd visit, he'd always ask me if I'd been picking up any more of them 'atomic caterpillars,' and we laughed about that till the day he died." (Anonymous, Walter-Reeves.com [a gardening website])

Thistledown Velvet Ant (*Dasymutilla gloriosa*), shown on a creosote bush. It is thought to be a mimic of the fruit of that shrub. When it rapidly travels along the ground, it appears to be a creosote bush fruit blowing in the wind. *L.L.C. Jones*

Classification: Phylum Arthropoda (arthropods); Class Insecta (insects); Order Hymenoptera (wasps, bees, ants, and sawflies); Family Mutillidae (velvet ants).

These hymenopterans are quite different than true ants (Formicidae), so despite their name they are more akin to wasps and are often placed in the superfamily Vespoidea. There are more than 8,000 species with numerous subfamilies and dozens of genera worldwide, and North America has about 480 species. They are well represented by the large and conspicuous genus, *Dasymutilla*, which has more than 150 described species in North America. Other speciose genera include *Pseudomethoca*, *Sphaerophthalma*, and *Timulla*.

Some of the commonly referenced *Dasymutilla* species include the whitish velvet ants (*D. gloriosa* and/or *D. sackenii*), the red or red-haired velvet ants *D. occidentalis*, *D. magnifica*, and *D. klugii*, *D. quadriguttata*, *D. nigripes*, and Satan's Velvet Ant, *D. satanus*. *Dasymutilla occidentalis* is commonly known as the Eastern Velvet Ant (ironic, as "occidentalis" refers to "western."); it is also the species most commonly referred to as the "cow killer."

Identification: The common name "velvet ant" describes the general external appearance of these wasps, which have a dense covering of what looks like velvety hairs. The terrestrial, wingless females are superficially similar to large hairy ants. The "hairs" are actually densely packed setae. Velvet ants range in size from a few mm to about 25 mm in length. Males are winged; females are not. In addition to the presence or lack of wings, there is marked sexual dimorphism. Males and females may be so different in appearance that they have sometimes been described as separate species, and it has proven difficult to match them up for a proper taxonomic description. There are differences in size, coloration, setae, and gross morphology. In fact, many species have not been matched up, so numerous synonyms probably exist.

The females tend to be robust, having a thick, stout abdomen and thorax. There is a waist, but in females, especially, it tends to be short in length. Unlike ants, there is no petiole. The cuticle of velvet ants is extremely tough, as noted by entomologists when they try to pin specimens. There has also been research on how much pressure is required to crush a velvet ant. The legs are shorter than typical wasps, adding to the ant-like appearance. Males often look like a cross between a female velvet ant and a typical wasp or winged ant. Females may be conspicuous as they move about the substrate. They move very rapidly and rarely stop for any length of time. Coloration, pattern, density of setae, and microscopic traits help identify them to species.

Scanning electron micrograph shows how densely packed the setae are.
Janice Haney Carr/Public Domain/CDC

The head of a velvet ant, showing why these wasps have that common name.
Wayne Boo/Public Domain/USGS, BIM

One of the *Dasymutilla* with red aposematic coloration. *L.L.C. Jones*

The pattern of orange to yellow and black aposematic coloration is commonplace among velvet ants. The Thistledown Velvet Ant and similar species have whitish setae, while the Satan's Velvet Ant is yellowish. The reddish species may have alternating patterns of orange or red to black, or a black thorax and reddish abdomen. The coloration of the whitish velvet ants is similar to that of the fruits of the creosote bush, so these animals are cryptic, rather than aposematic. Their movement on the ground may look like a creosote bush fruit rolling in the wind. The white also reflects intense sunlight. Müllerian mimicry is found between and among velvet ants. Batesian mimicry is found among other arthropods, including beetles, ant lion larvae, and other hymenoptera (e.g., scoliid wasps and ants). Studies of Müllerian mimicry in *Dasymutilla* in the United States has shown that there are seven mimic complexes, or rings, based on pattern similarities within specific geographic areas: western, eastern, desert, Madrean, Texan, and tropical. These complexes are composed of several species each that share similar aposematic or cryptic coloration. Evolving convergent patterns is believed to benefit all species in those areas, by having shared warning signs.

Distribution: The family Mutillidae is found worldwide, often in deserts and dry tropical areas. There are several hundred Nearctic species among

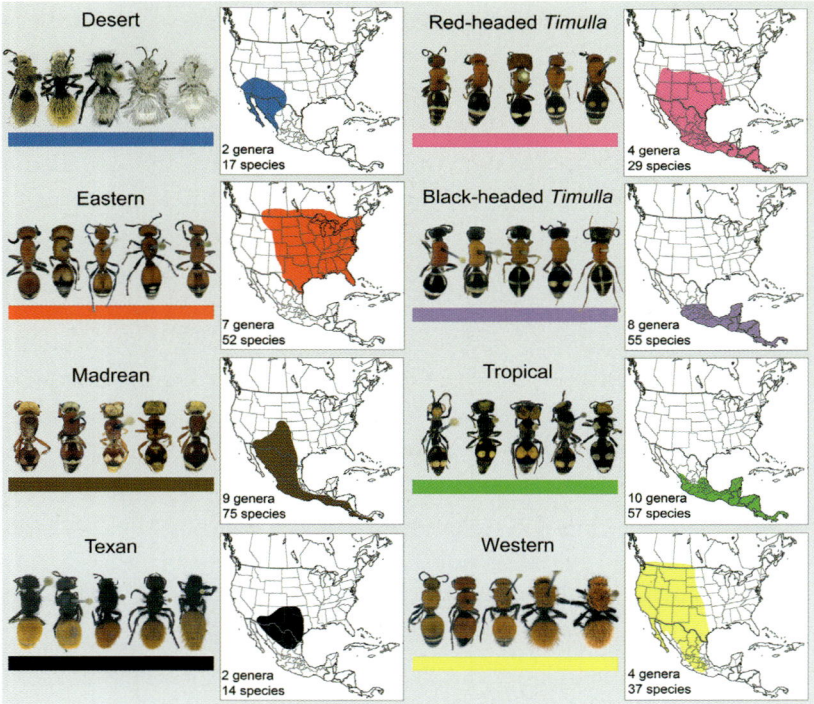

Velvet ants are a textbook example of a Müllerian mimicry ring. *Joe Wilson*

numerous genera, but they reach their peak U.S. distribution in the American Southwest. Most of the studies have been done in California and other southern Border States. The Eastern Velvet Ant is the commonly encountered *Dasymutilla* species from east Texas to the Atlantic.

Natural History: Velvet ants are solitary wasps. They are mostly arid-land inhabitants of the American Southwest, but are not limited to those environs, as they are found in a variety of other habitats throughout the United States.

Females are quick and are usually seen as they run rapidly over the ground. The Eastern Velvet Ant has been clocked at an average of 0.5 km/hr. They will also move about in shrubs in search of nectar for food and prey to serve as larval hosts. Their active period tends to be spring to fall. Surface activity is generally dictated by temperature. They are mostly active outside of the heat of the day in hot weather. Most species are regarded as crepuscular to nocturnal, while some have more diurnal tendencies. The short-lived winged males are often seen feeding on nectar. Males seek mates by flying about 30 cm or so over the ground. When females are located, the male may land, and if both parties are willing, they quickly mate, then go about their business. Mating may occur on vegetation. Mating has rarely been documented in the wild, but likely involves the use of pheromones and stridulation. The sounds made by both sexes of mating pairs have been described as a "honk" or "squeak." One way that researchers have learned about mating—and how to link male and female species for taxonomic associations—is by caging a female in the wild and waiting for males to approach.

Mutillids nest in a variety of microhabitats, but are often found in subterranean burrows, within wood, or within the nests of their hosts. They do not have strong nest-site fidelity, but they do have good cognitive abilities to recognize structures in their activity area. These animals are ectoparasitoids that oviposit on hymenopteran hosts. Their specific hosts are mostly not known, but New World species mostly target wasps and bees. They generally lay a single egg on a single host. The hymenopteran host is usually a prepupa or pupa.

Velvet ants are supremely adapted at avoiding predation. Two things in their favor are their algogenic venom delivered through a long stinger and their very hard cuticle. Animals that typically feed on small invertebrates find them unavailable as food. Animals with good visual acuity, such as mammals, lizards, and birds, will likely remember the aposematic coloration on repeat encounters. In laboratory trials, lizards will attempt to feed on species that do not sport aposematic colors, but will avoid those with warning colors—or at least only try them once. In addition, mutillids secrete exocrine substances, especially a ketone, 4-methyl-3 heptanone, that acts as an aposematic odor to repel potential predators. Velvet ants also have aposematic sound. Potential predation events tend to end at the onset of stridulation, as when humans encounter a rattling rattlesnake or hive of buzzing bees. By using this arsenal of defenses, velvet ants are rarely eaten by potential predators.

Encounters: Human encounters often occur when female mutillids are scurrying about the ground. Such is the case with bathers that are barefoot on beaches or near pools. Another chance encounter is if the velvet ant gets into clothing or is otherwise brushed up against skin. Entomology constitutes an occupational hazard when these wasps are targeted. While some species are very attractive, they are not popular in the exotic pet trade, probably because they are small and hyperactive. There is no evidence that

The stinger of a velvet ant is impressive, and so is the pain it can deliver.
Wayne Boo/Public Domain/USGS, BIM

Dasymutilla is drawn to humans, their refuse, clothes, aromas, food, or other situations where people are involved in their daily chores. Mutillids do not tend to enter homes or structures, but that is always possible. They are not particularly prone to stinging humans, preferring to run away, but will defend themselves if they feel threatened.

Venom Apparatus, Venom, and Symptoms: The venom apparatus is a modified ovipositor, as with other female wasps, although they have particularly long stingers for their size. The stinger is also maneuverable.

The venom is not particularly well studied compared to many other wasps, presumably because stings are uncommon, so velvet ants do not constitute a significant medical hazard. The venom is clearly algogenic in nature. It is composed of a number of compounds similar to other wasps and bees, including amines, peptides, proteins, and enzymes, but they do not seem to have kinin-like substances. Serotonin, histamine, and acetylcholine contribute to the pain factor. In species studied, enzymes include small amounts of phospholipase, hyaluronidase, esterase, and cholinesterase. The LD_{50} of *D. klugii* and *D. lepeleteirii* was reported to be 65 and 75 mg/kg, respectively. Hemolytic activity is very low in *D. lepeleteirii* and not well researched in other species.

The pain is often reported as immediate and intense to excruciating, although there are many species and there is no doubt much variation. The Hymenopteran Sting Pain Index is 2 or 3 depending on species. Some species are regarded as delivering the second most painful sting among domestic hymenopterans, after tarantula hawks. In addition to pain, redness and swelling are the usual local symptoms. Stings from velvet ants are presumably uncommon, and tracking of information on these animals is poor. In sting reports from emergency rooms and poison control center calls, velvet ants are lumped with wasps, ants, Hymenoptera, or just "insects." Systemic reactions are poorly documented. I can find no reports in the literature of anaphylactic reactions to mutillid stings, but those are possible, given the many same venom qualities as other hymenopterans, a group that is well-known for its allergens. Cross-reactivity with other hymenopteran venoms is virtually unstudied.

References and Resources: Hurd (1951), Linsley et al. (1955), Manley (1977), Manley and Pitts (2002), Manley and Radke (2006), Pilgrim et al. (2008, 2009), Pitts et al. (2008, 2009), Schmidt (2009a, b, c, 2016), Schmidt and Blum (1977), Schmidt et al. (1986), Spangler and Manley (1978), VanderSal (2008a, b), Vitt and Cooper (1988), Wilson et al. (2012).

Fire Ants (*Solenopsis geminata* Group; Primarily *S. invicta* and *S. richteri*)

"From the 29,300 physicians surveyed, reports of 83 fatal and two near-fatal fire ant-sting reactions were received. Most anaphylactic deaths were reported from Florida (22) and Texas (19)." (Rhoades et al. 1989)

Red Imported Fire Ant, arguably the most important venomous animal in the United States. *Eli Sarnat/Public Domain/Animal and Plant Health Inspection Service (APHIS)*

Classification: Phylum Arthropoda (arthropods); Class Insecta (insects); Order Hymenoptera (wasps, bees, ants, and sawflies); Family Formicidae (ants); Genus *Solenopsis* (fire and thief ants). The ants we typically know as fire ants are in the *S. geminata* group.

Worldwide, there are more than 200–280 species of *Solenopsis*. In the *S. geminata* group, there are four native and two non-native species in the

United States. The native species include the Southern Fire Ant (*S. xylori*), Golden Fire Ant (*S. aurea*), Tropical Fire Ant (*S. geminata*), and Desert Fire Ant (*S. ambychila*). The non-native species are the Red Imported Fire Ant (*S. invicta*) and Black Imported Fire Ant (*S. richteri*).

The Little Fire Ant, *Wasmannia auropunctata*, is not related to *Solenopsis* but has a similar sting, and is emerging as a serious non-native pest. In the wild it is known from Hawaii and Florida within the United States, but also northern Mexico. As a greenhouse pest, it occurs in California. It is invasive on many islands.

Among our species of fire ants, the Red Imported Fire Ant is by far the most medically and socioeconomically important species. Its acronym, RIFA, has come into common usage. With the arrival of the Black Imported Fire Ant (BIFA), these invaders are sometimes called IFA because the red and black species are not always discernible and may hybridize. In this account I refer to traits as those of RIFA, which are essentially the same as BIFA, unless indicated otherwise.

Identification: Fire ants are small animals with a polymorphic worker caste. Workers can vary in size from about 2–4 mm in total length, but the queen is always larger than the largest worker, up to about 6 mm. Virgin queens and males have wings, but after mating, the queens lose their wings. The female workers are wingless. As a group, fire ants vary in color from red to black, with some yellows to browns, and individuals are not necessarily monocolored. When seen en masse at the nest, RIFA tend to look red, but up close their coloration is a little more complex.

Members of the genus *Solenopsis* are easy to recognize when viewed with a stereo microscope. Their antennae have ten segments with a two-segmented club at the end. There are some other diagnostic features that can be seen with a microscope that can be used to distinguish species. However, one does not necessarily need a microscope to identify RIFA, as they can usually be recognized based on their large, conspicuous nest mounds with the massive colony of small, reddish ants. In some areas these nests are extremely abundant.

Distribution: The Southern Fire Ant is the most widespread species, occurring in southern states, from California to Florida, ranging as far north as Utah and Colorado. In the Southeast, it is being replaced by the invasive RIFA. The Golden Fire Ant is found in the southern parts of California, Nevada, Utah,

The face of a native fire ant, *S. geminata*. *Eli Sarnat/Public Domain/APHIS*

Although not closely related to *Solenopsis*, the Little Fire Ant is an invasive species. *Eli Sarnat/Public Domain/APHIS*

New Mexico, and most of Arizona. In Texas, it is found in the west and central parts of the state. The Desert Fire Ant has a similar range, but does not occur in northern Arizona, Nevada, or Utah. The Tropical Fire Ant is found from Texas to Florida and North Carolina, as well as being introduced into Hawaii, where it is now widespread. The Little Fire Ant is found on several, but not all of the Hawaiian Islands, and is established in Florida. Both of these and the RIFA are considered invasive species in Hawaii.

The RIFA occurs primarily in the Southeast, but most common in Texas and Florida; however, it is spreading to the west and there are populations in Arizona, California, New Mexico, and Hawaii. There are even scattered populations in parts of New England. It was first noted in the United States in 1945, in Alabama, but quickly spread throughout the southeastern United States, where it was well suited to the warm, humid environment. In the desert Southwest, it is only found in humid pockets of habitat; deserts can be rendered humid in parks, agricultural lands, and greenhouses. According to some models, most of the Southwest is predicted to be invaded. Many of the introduced populations west of central Texas have been extirpated, or have died out naturally, but other introductions are likely to occur. The Black Imported Fire Ant (BIFA) is found in Mississippi, Alabama, and Tennessee, at least, but its range is spreading and they are hybridizing with RIFA.

Natural History: Native fire ants tolerate arid conditions better than IFA. However, they are rather adaptable and can tolerate drought and flood. During drought, they maintain moisture under the surface. During floods, they form floating rafts or balls, with workers on the outside and queens on the inside. When they reach a structure, they will climb it until the water recedes. This was well publicized following the flooding from recent hurricanes. There are volumes written about IFA natural history, but relatively little about the native species. They are eusocial insects that can have complex and variable natural history traits. They build large nests in the form of soil mounds. Large colonies can have up to 300,000 individuals.

When breeding, spring through fall, winged males and sexual females (virgin queens) emerge from the mound and mate high in the air. After mating, the queens come to the ground, lose their wings, dig simple burrows, and begin egg-laying to establish a colony. A queen that establishes a colony is termed the foundress. The first workers are generally small, but help the queen to survive, build the nest, and feed the larvae. Subsequent workers tend to be larger. Most colonies of fire ants are monogynous.

However, some colonies can be polygynous. These colonies can grow quite large, but are not necessarily stable, as workers in some colonies will even attack their own queens. The density of colonies in infested areas can be quite high, and the rate of spread is rapid. In Texas, studies have shown that polygynous colonies are decimating native ant populations at a much greater rate than monogynous colonies. The larvae molt four times, pupate, then molt into adulthood. Some colonies produce more males than others.

The mounds are large and conspicuous, but they do not have entrances on the top (except when swarming), as seen in most other mound-building ants. Workers enter the mound from subterranean foraging tunnels. Inside the mound there is an extensive network of tunnels. By keeping the mound sealed, they can have climate-controlled housing and they are somewhat resistant to damage during storms.

Fire ants are omnivorous, being foragers, scavengers, and predators. They will eat a wide variety of plant material, including seeds, fruits, and insect exudates but will gladly accept live or dead arthropods in their path. By attending to aphids and feeding on their honeydew, fire ants protect the insects and the host plant from predators, in a classroom example of mutualism. Fire ants will even feed on amphibians, reptiles, birds, and mammals. Normally, these are small, helpless individuals, as in nestlings, or injured animals. Disabled humans have even been known to be fed upon by IFA. When a large food source is located, workers will communicate to others about their bounty. Large food items can be significantly reduced in size. Adult fire ants do not consume solid food but rather take it back to the colony where larger larvae convert it to liquid for the workers.

Encounters: All boreal states and provinces and most of the Southwest are currently free of IFA, and although some native species, such as the Southern Fire Ant, can sting, native species are not as aggressive and the sting is not as intense. For those living in the Southeast, encounters with RIFA can be an everyday fact of life. It is uncertain how and when (or if) RIFA will spread through the rest of the country, but colonies are starting to appear in all states along the southern international border. In the heart of IFA country (central Texas to Florida) millions of stings occur every year. If one walks barefoot across an infested lawn, being stung is almost a certainty. For those living in Hawaii, Tropical and Little Fire Ants can also be a painful menace.

Dozens of distinctive RIFA mounds may exist in a patch of lawn. *L.L.C. Jones*

The "sneaker test" reveals this is a RIFA mound, as hundreds of ants rapidly emerge to protect the nest. *L.L.C. Jones*

IFAs seem to be the ideal human pest. They are small and inconspicuous, yet form massive colonies in high densities. A density of about fifty mounds per hectare is considered an infestation. They are quick to defend themselves and their queen with a series of venomous stings, and multiple stings can come from large numbers of individuals in rapid succession. In some areas they are so common that they are a continual health menace. They affect our yards and farmlands, costing an estimated $750 million dollars in agricultural losses. They displace or destroy native ants and eat native wildlife, and our control methods also displace or destroy native ant and wildlife species.

Venom Apparatus, Venom, and Symptoms: The venom apparatus of fire ants consists of a modified ovipositor, so it is the female workers that sting when a nest is threatened. There is a sheath, venom sac, stinger, and the associated Dufour's gland. The sheath is at sclerites 8 and 9. The stinger has a fused dorsal shaft and ventral two lancets. When extended, a syringe is formed to deliver venom from the sac into the victim. Fire ants also bite, but the bite is nonvenomous and is done to anchor itself to the victim in order to sting effectively. These animals are small, and the stinger and venom gland mechanism is only about 1.5 mm, but that is enough to break the skin. The venom is used both for defense and procuring prey.

The venom is composed of 95% alkaloids and 5% or less protein. The alkaloids are termed solenopsins. There have been at least five solenopsins isolated. They function to penetrate the exoskeleton of invertebrates, but are hemolytic in mammals. The venom also has herbicide, fungicide, and bactericide qualities, which helps to reduce infection in people stung. Hyaluronidase is present and phospholipases A and B activity has been reported. Although histamine is not present, there are several allergens

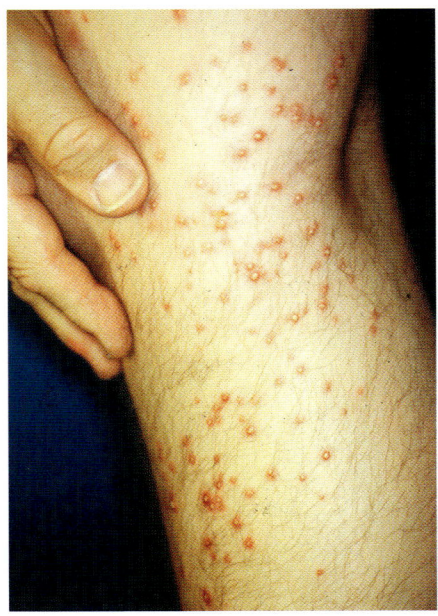

A researcher with 250 RIFA stings on the leg. Most victims form pustules after being stung. *Daniel Wojcik/CCA/Bugwood*

identified in the venom. Sol i (for *Solenopsis invicta*) I-IV and Sol r (for *S. richteri*) II-III are allergens. Sol i I is a phospholipase A_1 and some others are in the antigen 5 family.

The sting from a single individual is only mildly painful and generally described as a burning sensation. Along with the red color, this accounts for the moniker "fire ant." The fire feeling is usually the result of multiple stings as they swarm up a leg. The relatively mild pain per sting does not reach the "bee sting" threshold of most animals in this book—but few people are only stung by one ant—and people are sometimes stung by hundreds. Within 1.5 hours, a red mark will form at each sting site, and a wheal up to 5 cm across can form. The trademark symptom of IFA is a characteristic pustule that forms within twenty-four hours at each of the sting sites. The pustules last from days to weeks, then slowly diminish over time. In some people, the marks may still be visible after a month. This causes a rash and itching that can last quite some time after being stung. Not all individuals (like me) develop pustules. Stings from other species of native fire ants are similar in pain and symptoms, except they do not generally form pustules, so burning and itching subsides within an hour.

Because each ant can sting repeatedly and most individuals in a colony can sting, large numbers of stings can be accumulated in short order. In some instances, mass envenomations have occurred. These are generally to a person with compromised mobility. Two deaths from mass envenomation (i.e., due to the venom itself, not the allergens) occurred to a child and an elderly person. However, it is possible to survive mass envenomations. People have been stung thousands of times and survived.

Allergic reactions are not uncommon because millions of people are stung annually, victims are often stung repeatedly over time, and there are several allergens present in the venom. Certainly the biggest medical problems are due to hypersensitivity and anaphylactic shock. Symptoms of severe systemic reaction include hives, fever, headache, dizziness, nausea, vomiting, perspiration, hoarseness, swollen larynx, respiratory distress, reduced heart rate, low blood pressure, and loss of consciousness. Without rapid emergency medical treatment, severe cases can quickly become fatal.

There can actually be cross-sensitivity to Striped Bark Scorpion venom, and in Texas, the ants and scorpions have a similar range. This is probably why Striped Bark Scorpion stings sometimes cause allergic reactions to people never stung by scorpions before.

The magnitude of IFA as a medical problem is immense. According to a 1985 study, of the 40 million people living within the range of IFA, it was estimated that 14 million are stung at least once each year. It is estimated that 80,000 people per year require medical attention. Six in 1,000 people are estimated to be at risk of anaphylaxis. Even with these staggering statistics, only about five deaths per year are reported from fire ant stings. Nearly all of these are from allergic responses. Human mortality due to RIFA is increasing in Texas and Florida. There is a fatal case of a 3-month-old baby from native *S. xyloni*. She was found covered in ants in her crib and flown by helicopter to the hospital. She was in respiratory distress when she reached the hospital, where she died. Although she was stung hundreds of times, tests confirmed she died from anaphylaxis. It is estimated that five billion dollars is spent on medical treatments and control activities.

Remarks: There is a sort of fire ant panic within areas that do not even have IFA. Just seeing ant mounds or clearings in yards is enough to send people scurrying to hardware stores, where they can find entire rows and shelves loaded with insecticides to control ants. This large-scale, widespread killing of ants across the landscape is irresponsible and undoubtedly causes environmental havoc in some areas. For example, the decline in horned lizards is linked in part not only to IFA effects on native ants, but also RIFA control measures. In areas outside the range of IFA, indiscriminate use of ant poisons will kill the hundreds of native species of ants and other animals.

References and Resources: Allen et al. (1994), Ascunce et al. (2011), Baer et al. (1979), Eubanks (2001), Fisher and Cover (2007), Greenberg et al. (1985), Hoffman (1993), Jakob et al. (2017), Kemp et al. (2000), MacConnell et al. (1971), Macom and Porter (1996), Markin et al. (1972), Porter and Savignano (1990), Rhoades (1989), Schmidt (1986, 2004, 2009a, b, c, 2016), Stafford (1996), Stafford et al. (1989), Summerlin (1976), Taber (2000), Tschinkel (2006), Vinson (1997), Wetterer and Porter (2003), Wojcik et al. (2001).

Harvester Ants (Genus *Pogonomyrmex*)

"... the New World harvester ants in the genus *Pogonomyrmex* ... are famous for their piercing, throbbing, deep pain that lasts for hours and has been described as feeling like the ripping of muscles and tendons. These stinging ants were also used in manhood rites by indigenous peoples in California because their painful and toxic stings facilitated life-guiding visions." (Schmidt 2008)

Maricopa Harvester Ant, the most venomous insect on Earth (along with a South American species). *L.L.C. Jones*

Classification: Phylum Arthropoda (arthropods); Class Insecta (insects); Order Hymenoptera (wasps, bees, ants, and sawflies); Family Formicidae (ants); Genus *Pogonomyrmex* (harvester ants). The genus *Ephebomyrmex* is generally considered a junior synonym.

The vernacular "harvester ant" is a descriptive title that can be used for any number of seed or plant gathering species of ants, but generally this name specifically refers to *Pogonomyrmex*. There are nearly seventy species, of which about twenty-five occur in North America north of Mexico. Some of the more widespread or commonly referenced species include: *P. apache* (Apache Harvester Ant), *P. badius* (Florida Harvester Ant), *P. barbatus* (Red Harvester Ant), *P. californicus* (California Harvester Ant), *P. desertorum* (Desert Harvester Ant), *P. maricopa* (Maricopa Harvester Ant), *P. occidentalis* (Western Harvester Ant), and *P. rugosus* (Rough Harvester Ant). The Maricopa Harvester Ant likely includes several cryptic species.

Identification: Harvester ants are affectionately called "pogos" by entomologists. These ants are mostly known as being large and conspicuous, but there are smaller, less conspicuous species. Workers are generally between about 4–10 mm total length and queens of larger species can reach about 17 mm. Like other eusocial insects, pogos have castes that include a queen (or queens), workers, and males. The vast majority of ants in a colony are the female workers. They lack wings and have the ability to sting. Polymorphic castes of queens include both winged and wingless individuals. Winged individuals (alates) are breeders. Harvester ants can be recognized by their pectinate (comb-like) middle and hind tibial spurs. The petiole has a long anterior peduncle. *Pogonomyrmex* means "bearded ant," which alludes to a good identifier—a beard-like psammophore of elongated setae.

Pogos range from orange and red to brown or black in overall coloration. Some species may vary in color or have more than one color per individual. Harvester ants can sometimes be identified by their coloration or nest types, but it may take an expert to identify them to species, using several microscopic characters. Besides size and color, characters include rugosity of the various parts of the ant's anatomy, the arrangement of setae, the spines before the peduncle, the structure of the mandibles, presence of a circumorbital ring, and so on. Knowing the geographic range and habitats is also useful in identifying species. If the reader really wants to identify a particular species, they need a dissecting microscope and published keys to species.

Distribution: Harvester ants are only found in the New World, from Canada to the tip of Tierra del Fuego in South America. There are also species in Haiti. Nearly all of the U.S. species are found in the Southwest. There are two boreal species that range as far north as North Dakota (*P. occidentalis*) and into southwestern Canada (*P. salinus*). Only the Florida Harvester Ant is found east of the Mississippi River, in Florida, Louisiana, and the Carolinas. Distribution maps of most species have been published in a couple books on harvester ants.

Head of *P. comanche*. Note the rugose surface and beard. *Alex Wild/Public Domain/Flickr*

Pogonomyrmex barbatus. Note the club-shaped antennae and beard. *Antweb/CCA*

Natural History: Harvester ants are mostly insects of arid lands, being best represented in deserts and other arid localities, including coastal habitats, interior and coastal chaparral, shrub communities, grasslands, oak woodlands, and the Colorado Plateau semi-desert. They are usually found in open areas and avoid dense forest cover of the high mountains, although there are montane species. All pogos are ground nesters, although two species are social parasites, lacking their own nests. Harvester ants are common in disturbed situations such as clearings for power lines, campgrounds, vacant lots, and dirt roads. They are also commonly used in commercial ant viewing boxes. While these are an artificial and potentially controversial habitat, viewing boxes have sparked a sense of awe and wonder in many a youth.

These are perhaps the most conspicuous ants in arid regions, not only because of their size and abundance, but also because of their large nests. Pogos are generally regarded as mound builders, but some species have no mounds or the mounds are not large, especially when colonies are young. In some instances, they can even be seen nesting beneath cracks in concrete or asphalt. To someone familiar with the ants in a particular area, different species of harvester ants can often be recognized by their nests. In general, the nests are characterized by a zone of cleared vegetation in a large perimeter around the entrance. In some cases, the clearing may be more than 5 m across. There are many explanations for this nest design, including its utility as a firebreak, humidity control, thermoregulation, predation control, and so on. The clearing is associated with a central mound or entrance that may have sand, gravel, and vegetative debris. In sandy areas, Maricopa Harvester Ants can have very large mounds, more than a meter in height. Nests of harvesters are particularly evident in grasslands, since the clearing is starkly differentiated from tall grass. Unlike fire ant mounds, there is an entrance at the top or center of the mound where harvester ants actively come and go. There may be an auxiliary entrance. The main entrance is often inside a cone that helps provide sunlight, or shade, depending on time of day. Some species clear trails radiating from the cone out into the surrounding environments, where they travel in search of food. They are active at various times of the day, depending on environmental conditions, but often have a daily pattern of foraging by day, often in the morning and afternoon, then remaining in the nest at night. During hot weather, they may be nocturnally active. The activities around the outside of the nest include foraging, trail building and maintenance, pebble collection, patrolling, defense, and communication at the entrance. Internally, nests can be several meters deep, composed of an intricate network of tunnels and rooms with various functions.

Mound of Maricopa Harvester Ant. *L.L.C. Jones*

Harvester ants are eusocial. The queen is the egg-laying caste. Males are winged, specifically for the purpose of mating. The majority of adult animals in the colony are sterile female workers. During swarming events, the alates simultaneously take flight to mate. Males often land on tall vegetation to wait for females. A female can mate with several males. In some areas, swarming is triggered by the arrival of the North American monsoon in summer, especially after heavy thunderstorms. After mating, females come to the ground and lose their wings, but males perish. Females become the foundresses for new colonies. Their colonies can be quite large, having up to about 20,000 individuals. However, some species, such as *P. imberbiculus*, may have small colonies, numbering in the tens. Some colonies can also be perennial, lasting for decades. The foundresses themselves can be up to thirty years old.

The food of harvester ants is generally vegetation, but arthropods or other organic matter are also consumed. Workers head out daily along the paths radiating from the mound in search of vegetation. They eat a wide variety of seeds, mostly small seeds on the surface, because they are highly nutritious. Seeds of some plant species are preferred over others. Generally, harvester ants do not remove more than about 10% of the seeds in an area. Harvester ants compete with other granivores, including several genera of ants, rodents, and birds. Sympatric harvester ant species may compete over seed resources, or may avoid competition by resource partitioning. Because the ants are so

Experts assured me that these *P. rugosus* were not rafting but rather climbing on one another to avoid drowning, following heavy summer rains. L.L.C. Jones

abundant and colonies can be large and old, these ants can have a profound impact on their ecosystems. They can modify the vegetation community in the area of colonies, cause increases or decreases in microfauna, and cause physical changes in the microtopography. The nests themselves can be problematic for humans, as in airport runways, parking areas, and residential areas.

The introduced Red Imported Fire Ant and Argentine Ant (*Linepithema humile*) are considered to be among the hundred worst invasive species. Both of these species affect populations of harvester ants and may completely displace them. They compete for food, and Argentine Ants have been shown to raid nests of *P. subnitidus* in southern California. On their raids, they take food, capture the brood, and attack workers. When there is a raid, harvester ants must defend the nest, so workers will not forage. One of the ecological consequences of this invasion from South America is the decline in Blainville's Horned Lizard from California, a species that prefers to feed on harvester ants. Argentine Ants have formed a supercolony that stretches 900 km along the California coastal areas and into Baja California, Mexico. In Europe, one supercolony is 6,000 km long!

A wide variety of animals prey on harvester ants, but the best known are the horned lizards. These lizards may be completely dependent on harvester ants, or pogos are at least their primary diet. They evolved mechanisms to feed on harvester ants without being stung and have resistance to the venom. It's possible that the evolution of harvester ant venom is the result of an arms race with horned lizards and other predators. Horned lizards have a range similar to harvester ants in North America. During swarming events, it is common to see lizards of various species feeding on the grounded alates.

Encounters: The most problematic species are the large, "aggressive" pogos. According to one researcher, the following commonly encountered species are aggressive: *P. barbatus*, *P. californicus*, *P. maricopa*, *P. occidentalis*, *P. rugosus*, *P. subnitidus*, and some populations of *P. salinus*. Some uncommon species are also aggressive—just not as likely to be encountered. However, some species are considered docile, since they can hardly be encouraged to defend their nest or even sting. These include *P. apache*, *P. comanche*, *P. desertorum*, *P. subdentatus*, and some less common species.

Harvester ants can be common in nearly any open area, especially in arid portions of the Southwest. They are not attracted to the indoors. Around homes, they are mostly a concern if they become established in lawns. They may also be common along roadways and driveways, vacant lots, golf courses, city parks, and other places frequented by people. In many areas they seem to be everywhere on the ground. In campgrounds, for example, there can be thousands of them on the surface in the vicinity of campsites. It may be difficult to even find a spot to pitch a tent that is not too close to a nest or a major travel corridor. They can also be common in urban and suburban settings.

Stings are not uncommon, but not commonly reported. For example, during a two-year period (March 2002–2004) 237 ant stings were reported to poison control centers for Arizona, exclusive of Maricopa County (the greatest population center of Arizona). Undoubtedly far more stings went unreported during that time period. It is believed most reported stings were caused by harvester ants, since the pain and symptoms can be troubling. During the same time, 623 bee and wasp stings, 4,655 scorpion stings, and 346 snakebites were also reported.

Venom Apparatus, Venom, and Symptoms: While the mechanism of envenomation is stinging, harvester ants also bite with their strong mandibles, but this minor pain is rarely noticed when one is being stung. The stings are completely defensive. The harvester ant sting apparatus is similar to other ants. From the proximal end of the venom reservoir to the distal end of the sting shaft is about

1.5 mm in length. Harvester ants also use the abdominal glands as part of their chemical communication, including trail marking and aggregating behavior.

The venom of some species of harvester ants is commonly cited as being the most toxic insect venom in the world, having the lowest LD_{50} of any insect studied to date. In mice, the LD_{50} ranges from about 0.09–0.7 mg/kg. The most venomous species in our area is the Maricopa Harvester Ant, a wide-ranging species found from California to Texas, north to Utah and Colorado, and south into Mexico. The LD_{50} is often cited as 0.125 mg/kg. Only *P. cunicularius* (a South American species) has a higher toxicity, having been reported to have an LD_{50} of 0.09 mg/kg, but statistically that is probably not significantly different from the Maricopa Harvester Ant. All harvester ants studied have very potent venom. The good news is that the venom sacs of a harvester ant have a volume of about 20 µg, so it would take the venom of about 50 million ants to fill a liter bottle—but that would be a seriously intimidating batch of liquid! The venom is not particularly virulent to insects, as it is primarily defensive to target vertebrate predators. Other LD_{50} values for different species are 0.42, 0.47, and 0.62 for *P. badius*, *P. rugosus*, and *P. barbatus*, respectively.

The venom has a variety of compounds, including those with a kinin-like activity. Harvester ant venom seems particularly rich in enzymes. These include phospholipases A_2 and B, hyaluronidase, lipase, esterase, acid phosphatase, and alkaline phosphatase. The venom has a hemolytic quality that lyses red blood

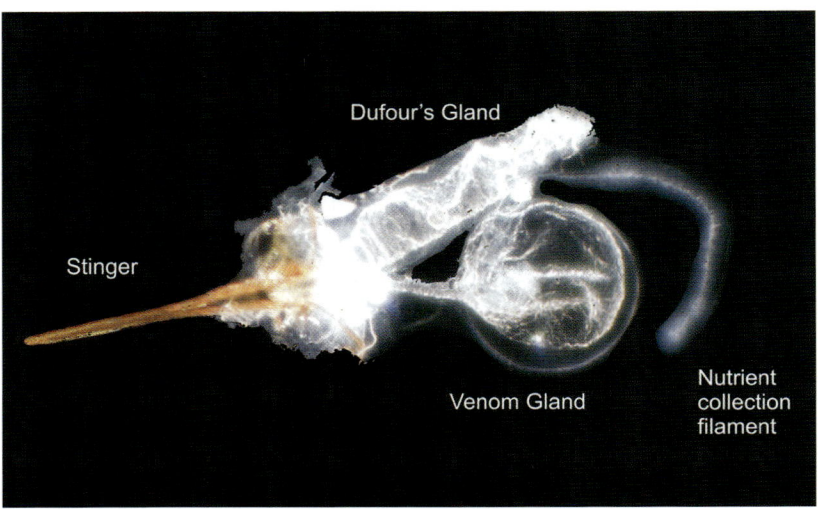

Venom apparatus of a harvester ant. *Justin Schmidt*

cells and mast cells, and in doing so, releases serotonin, histamine, and an SRS-A (the acronym for "slow-reacting substance of anaphylaxis," now known as leukotrienes), which results in smooth muscle contraction and vasodilation.

The usual initial symptom of envenomation is pain. Harvester ants rank a 2–3 on the Hymenopteran Sting Pain Index, depending on species. Species that are ranked a 2 are about as painful as a wasp or honeybee sting, but usually not as intense initially. Pain may increase over the first few hours of being stung. Those that rate a 3 are decidedly more painful; these include the Florida and Maricopa Harvester Ants. In some cases, the pain has been described as "ripping muscles or tendons," or "turning a screw in the flesh." The usual symptoms that follow are redness, swelling, burning, and itching. Symptoms of severe stings can include vomiting, aching for twenty-four hours, localized piloerection, sweating, and long-lasting pain and tenderness in nearby lymph nodes. The pain can last several hours and some symptoms may linger for days. Some victims can become hypersensitive to harvester ant stings. Systemic reactions include wheezing, coughing, labored breathing, hives, welts, and light-headedness. Systemic reactions have been reported from *P. barbatus*, *P. maricopa*, and *P. rugosus*. There are at least two human deaths reported from immunological reactions to *P. barbatus*.

I am not aware of any reports of mass envenomations. When only a single to a few stings are expected, the amount of venom would never be enough to kill a healthy human, so all serious reactions are due to immunological responses. However, it is possible that if someone were incapacitated at a nest site, there would be a potential for mass envenomation and even death due to the effects of the venom.

People who are sensitive to particular harvester ants are usually cross-sensitive to other species of harvester ants, or fire ants, but tend to not be sensitive to venom of wasps or bees. It has been shown that hymenopteran stings in which severe local reactions are noted can result in 5% chance of anaphylaxis during the next sting. However, people who have a positive skin test or a previous anaphylactic reaction are 60% more likely to have an anaphylactic reaction.

References and Resources: Cole, Jr. (1968), Cole (1994), Gordon (1986), Heinze et al. (1992), Hölldobler (1976), Johnson (1992), Klotz et al. (2005), Lockey (1974), MacMahon et al. (2000), Pinnas et al. (1977), Schmidt (1986, 1990, 2008, 2016), Schmidt and Blum (1978), Schmidt et al. (1986, 1989, 2009a, b), Taber (1998), Wagner et al. (2004), Whitford and DiMarco (1995), Zee and Holway (2006).

European Honey Bee (*Apis mellifera*)

"One man died and another was injured during a bee attack Wednesday in Douglas [Arizona] while the men were cleaning a yard. . . . While the workers turned on lawn mowers, the crew was attacked by a swarm of bees. . . . Firefighters were dispatched . . . and worked on one man who had collapsed. . . . The 32-year-old man went into cardiac arrest, and paramedics treated him and took him to Cochise Regional Hospital where he was pronounced dead. . . . A witness said his face and neck were covered with bees. . . . It was estimated that the hive had between 300,000 to 800,000 bees . . . the hive was 4 feet wide and 6 feet long . . ." (*Arizona Daily Star*, Carmen Duarte reporting, 8 October 2014)

Although non-native, European Honey Bees are not generally considered invasive. *Judy Gallagher/CCA/Flickr*

Classification: Phylum Arthropoda (arthropods); Class Insecta (insects); Order Hymenoptera (wasps, bees, ants, and sawflies); Family Apidae (bees); Genus *Apis* (honey bees).

In the United States, the European (or Western) Honey Bee, *A. mellifera*, is the most commonly domesticated species, and usually the only one encountered in the United States and Canada. There are about seven species of *Apis* generally recognized, but several more have been recognized by some authorities. There are about thirty subspecies of *A. mellifera*, and they may intergrade. A few subspecies in the United States are used for agriculture, including *A. m. carnica*, *A. m. caucasia*, *A. m. mellifera*, and *A. m. linguica*. The African Honey Bee is also a subspecies, *A. m. scutullata*. It is well known to intergrade with the other European Honey Bee subspecies, and progeny are termed "Africanized Honey Bees." The African subspecies and intergrades are commonly referred to as the infamous "Killer Bees" by the media and general public.

Identification: European Honey Bees are about 5–17 mm in total length. The general color is banded black and dull yellow-orange on the abdominal segments. They are relatively broad; the abdomen almost appears to be

Detail of *A. mellifera*. In the southern United States, virtually all are considered Africanized "Killer Bees." *Public Domain/USGS, BIM*

contiguous with the thorax, but there is a narrow constriction between these sections, so they superficially resemble yellowjackets. The head has three ocelli and a pair of compound eyes. The tongue is very long for nectar collection. There are two pairs of wings attached to the thorax. The forewings are longer than the hindwings. The thorax itself, and even the head, is noticeably hairy, having a dense covering of setae. The legs are medium in length, and the hind pair is modified to collect pollen, having a structure known as a pollen basket.

Distribution: Honey bees are indigenous to the Old World, originating in Europe, Africa, and Asia, depending on species and subspecies. Since then, *A. mellifera* have been introduced all over the world for their honey, beeswax, and pollination. They were brought from Europe to the United States in the early 1600s. Honey bees are found throughout the United States, from deserts to mountains and coastal habitats to inland areas.

African Honey Bees were imported to Brazil from Africa in the 1950s to increase honey production. In 1956, twenty-six swarms escaped quarantine and began a northward spread. In 1985 they were documented in California's San Joaquin Valley. In 1990, the first colonies became established in south Texas, and in about a decade most of the Southwest had become Africanized. They are cold-limited, so are not expected to thrive and persist in more northerly latitudes. Currently, most to all of the honey bees from southern California to south Florida are Africanized.

Natural History: Honey bees are eusocial. Much of what we know about social insects comes from research on these animals, because they are important both socioeconomically and medically. A single foundress queen starts perennial colonies by laying eggs that hatch into workers (sterile females) and drones (males). All castes are winged. Eggs are deposited singly into the hexagonal cells of the nest, and the larvae are tended by the "nurse" workers. They are fed nectar, pollen, diluted honey, and royal jelly. Royal jelly is a secretion fed to all castes but is especially important in developing queens. After about a week, they are then sealed in their nest chamber by the nurse to pupate. After another week, more bees emerge, especially workers. The new workers clean the nest and build more cells; then as they become older, they concentrate on foraging. The drones are larger than workers, but do not forage—their primary purpose is to mate with new reproductive females in flight to found new colonies. Drones are haploid, having only a single set of chromosomes derived from the foundress queen.

Honey bees collect nectar from a wide variety of flowers. Since they travel many miles, flying from flower to flower, they are outstanding pollinators. They collect pollen on their rear hind leg pollen baskets and setae on other parts of the head and body and then spread it to other flowers. This is a typical flowering plant–pollinator symbiotic relationship, and flowers have adopted ornate floral parts to attract such pollinators. Nectar is processed within the foraging workers, then stored in honeycombs and sealed as a source of nutrition.

Hives can be formed in a variety of places, but are usually inside of cavities, such as a hollowed tree trunk, caliche pocket, karst features, or even in homes and other structures. Occasionally, they are more in the open, as in trees. The most familiar nesting site for a colony is an artificial, layered box, maintained by beekeepers and farmers. Bee boxes are also used for property protection. There may be thousands to tens of thousands of bees in a hive. Nests are perennial, but drones are banished from the hive before overwintering. When hives grow too large or resources become limited, swarming occurs. When the founding queen leaves, up to 60% may leave with her in a "primal swarm." Smaller, subsequent swarms can also occur, and virgin queens will accompany the swarm to found new colonies. Scouts will look for a suitable area to start a new colony.

The well-known honeycomb of the hive of this eusocial insect yields honey, royal jelly, and beeswax. *Florida Department of Plant Services/CCA*

Honey bees are preyed upon by a wide variety of animals. Other invertebrates include mantids, spiders, wasps, assassin bugs (e.g., the Bee Assassin), and many other species. Amphibians, lizards, birds, and other vertebrates will feed on European Honey Bees. Some mammals (including humans) are fond of honey and risk digging into nests to get honey and eat larvae. This activity is the reason honey bees evolved a defensive strategy and venom that targets mammals.

Encounters: Honey bees are commonly encountered by humans. African and Africanized Honey Bees have a number of behavioral traits that differ from the European Honey Bee, making them more likely to be encountered. Compared to European Honey Bees, they tend to swarm more frequently, are more aggressive during swarming, and establish colonies in natural settings. They have more "guard" bees and are quicker to protect the hive. When a threat is perceived, they deploy more workers to defend the hive.

Beekeeping is a popular commercial and recreational pastime.
Mary Hightower/CCA/University of Alaska/Flickr

The workers will pursue perceived threats longer. They are more often involved in mass envenomations than European Honey Bees. However, even non-Africanized Honey Bees sting, and mass envenomations are possible.

European Honey Bees can enter homes and sometimes nest within walls. In general, though, they are more of a risk to humans outdoors. People can be attacked when they are within about fifteen meters of a nest. The bees can be stimulated into a mass defense strategy when power equipment is used within about thirty meters of the hive; the vibrations may trigger an attack, so be mindful when using power equipment, such as lawn mowers. Africanized Honey Bees are relentless pursuers and can follow for a long distance—much more so than yellowjackets and hornets.

Honey bees are often encountered around garbage bins, water sources, and other similar circumstances. Be careful in these situations and don't antagonize them. You may not know how close the hive is, and you don't want the sentries to send out signals of a threat.

Not surprisingly, bee stings account for a large percentage of envenomations. In one study of stings and bites treated in emergency rooms from 2001–2004, bee stings averaged about 163,000 per year, or about 18% of all bite and sting treatments. In another study around the same time period, from calls to poison control centers, bees and wasps collectively were responsible for 12,407 annual calls. Nearly 10% of these resulted in treatment in a health-care facility. Undoubtedly, most stings go unreported. While other species of bees can sting, the majority of bee stings are due to European Honey Bees.

Venom Apparatus, Venom, and Symptoms: The venom apparatus is similar to other hymenopterans; it is a modified ovipositor, so drones do not sting. Inside the sheath there is a venom gland, Dufour's gland, acid gland, and alkaline gland. The stinger is composed of the lancets and stylus, with the venom duct between these structures. Venom is expelled into the venom duct by muscular contraction. The stinger of the worker bee is barbed and stays inside the victim's flesh, unlike the stingers of wasps and ants. The stinger of the queen is unbarbed, and it may actually sting repeatedly over time. The worker bee will usually die after stinging a victim, whereas the queen will not. When a European Honey Bee stings a victim, the entire mechanism is ripped out of its abdomen, taking some abdominal contents with it, so the bee rarely survives such a physical assault.

The venom is used only for defense. It contains simple organic compounds, peptides, and proteins. The best-known peptide is melittin. It is the primary algogenic compound, and accounts for 40–50% dry weight of the venom. It has fungicidal and antibiotic properties. Another well-studied peptide is apamin, which is neurotoxic, affecting sodium and potassium channels. Similar to apamin is the MCD peptide (mastocyte degranulating peptide), which is probably responsible for a massive release of histamine. The venom also has dopamine, acetylcholine, serotonin, low levels of histamine, and several enzymes. These include phospholipase A_2, acid phosphatase, hyaluronidase, and collagenase. There are numerous allergens in honey bee venom. As of 2017, *Apis mellifera* had twelve named allergens, Api m (for *Apis mellifera*) I-XII.

A single sting is rather painful, a 2 on a scale of 0–4 of the Hymenopteran Sting Pain Index. Typical local reactions include swelling, redness, and itching around the sting site. The pain decreases after several minutes, but effects can linger for a day or two or possibly longer. In the case of mass envenomations, the melittin can cause the victim to lose consciousness. Stings from mass envenomations can be life-threatening. A single victim may receive hundreds to thousands of stings, in part due to the nest size, and in part due to the tenacity of pursuit. In at least one instance, a victim received about 10,000 stings. Throughout North and South America, about 1,000 deaths from mass envenomation have been recorded, mostly from Africanized Honey Bees. These incidences often lead to systemic reactions. Symptoms include swelling, fatigue, dizziness, nausea, vomiting, fever, and unconsciousness. Renal failure is often reported as the cause of death. The LD_{50} for European Honey Bees (Africanized or otherwise) is about 3 mg/kg for mammals, which is about the same for true wasps (*Vespa*), but more than yellowjackets and Bald-faced Hornets. It is often stated that a healthy adult can survive up to about 1,000 stings, but there is variability, and there are cases of fatal attacks from far fewer stings. Elderly people, children, and those who are otherwise of limited mobility are most at risk. An estimated five hundred stings is often cited for the number of stings a child can survive. Severe systemic reactions usually develop within twenty-four hours, but may be offset by 2–6 days. However, some people who have had large numbers of stings have survived, thanks to medical care.

Of course, even a single sting can be fatal if the victim is allergic to the venom. Allergic reactions can include anxiety, confusion, respiratory distress, swollen tongue and airway, and cardiac distress. This is a concern for much of the general populace, because many people throughout the country have previously been stung by European Honey Bees. It is estimated that 60–100

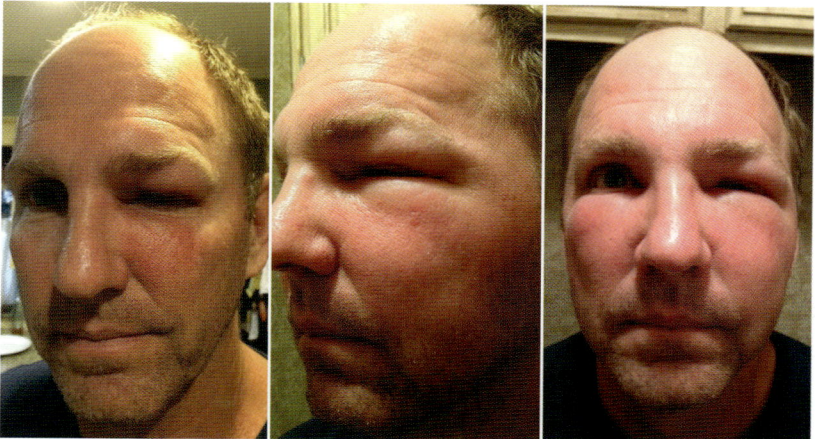

This beekeeper was stung on the face. Stings to the face often look worse than those elsewhere on the body, but a call to a poison control center may be in order.
"Oakley Originals"/CCA/Flickr

people die in the United States each year from bee stings. While some are from mass envenomations, most are from anaphylactic shock. Cross-reactivity among hymenopterans is complicated and varies among individual humans and species of insect. Cross-reactivity with wasp or ant venom should always be considered a possibility, and there are numerous studies on this topic in the literature.

Remarks: Honey bees have helped humans over the centuries by pollinating crops and producing honey and beeswax. There has been recent concern about declines in European Honey Bees as pollinators and the trend toward Africanized intergrades. However, there were many native pollinators around before *A. mellifera* appeared in the United States.

References and Resources: Caron and Connor (2013), Collins et al. (1982), de Lima and Brochetto-Braga (2003), Engel (1999), Gould (1998), Jakob et al. (2017), Langley (2005, 2008), O'Neil et al. (2007), Ruttner (1988), Schmidt (2008, 2009a, b, 2016), Schmidt et al. (1986), Schumacher and Egen (1995), Schumacher et al. (1990), Seely (1985), Vetter et al. (1999).

Bumblebees (Genus *Bombus*)

"I had this unfortunate experience [stepping on a bumblebee nest] when I was a boy of 10, and my recollection of the event is still vivid some 30 years later. I was pursued by bees and stung repeatedly while running more than 100 meters to my house. It was estimated that I had been stung more than 100 times, and I was extremely ill for several days, vomiting frequently for the first 24 hours." (Smith 1982)

Classification: Phylum Arthropoda (arthropods); Class Insecta (insects); Order Hymenoptera (wasps, bees, ants, and sawflies); Family Apidae (bees); Genus *Bombus* (bumblebees).

Western Bumblebee, an important native pollinator. *L.L.C. Jones*

There are about 250 known species of bumblebees, with forty-six species belonging to eight subgenera occurring in North America, north of Mexico. Cuckoo bumblebees were previously placed in their own genus, *Psithyrus*, but that is now considered a subgenus. There are eight other subgenera. Refer to the section on Distribution for a list of some common species by region.

Identification: Although bumblebees have the same general body plan of other bees, they are rather distinctive animals, having large, wide bodies that are covered by a dense covering of branched setae (called pile), giving them a fuzzy appearance and feel. They are about 20–40 mm in total length. They are aposematically colored, with patterns consisting of (depending on species) black, yellow, white, and/or orange to reddish bands. They are related to honey bees, and have similar characters, such as central ocelli and compound eyes, a pollen basket on the hind legs (except in cuckoo bumblebees), and long tongues for feeding on flower nectar. There is a great deal of variation in color and pattern in some species and among sexes, so they are sometimes difficult to identify to species. Fortunately, there are numerous published and online keys and guides to help identify species based on geographic range, color patterns, and other characters. DNA barcoding has proven useful in determining variation among taxa.

Western Bumblebee hovering to drink nectar. *L.L.C. Jones*

Distribution: Bumblebees are found in both the Old and New Worlds, including North and South America, Europe, Asia, and a small area of northern Africa. They are primarily in temperate zones, especially in northern latitudes, but they do occur in the tropics. Bumblebees are important commercial pollinators, so have been introduced into many parts of the world. In many cases, imported species have become naturalized and have outcompeted native species. At least one species is found in nearly every part of the United States and Canada, although diversity is highest in coastal California and Oregon, the Rocky Mountains, and the northeastern United States and adjacent Canada. In some areas there may be more than twenty species present. Specific distributions by species is sometimes complex and many species are declining, but here are some of the more common and/or widespread species of *Bombus* by region, with common name in parentheses (and "Bumblebee" is understood):

- **Cosmopolitan (United States and/or Canada):** *B. fervidus* (Yellow), *B. griseocollis* (Brown-belted), *B. pensylvanicus* (American; it pretty much stops at the border)
- **Boreal (mostly Canada and Alaska):** *B. borealis* (Boreal), *B. frigidus* (Frigid), *B. hyperboreus* (High Arctic), and *B. polaris* (Polar)
- **West (includes the Western United States and/or adjacent Canada, usually including Rocky Mountains):** *B. bifarius* (Two-form), *B. flavifrons* (Yellow-headed), *B. huntii* (Hunt), *B. mixtus* (Fuzzy-headed), *B. nevadensis* (Nevada). *B. occidentalis* (Western), and *B. sylivcola* (Forest)
- **East (especially northeast United States and/or adjacent Canada):** *B. bimaculatus* (Twin-spotted), *B. impatiens* (Common Eastern), *B. perplexus* (Confusing), *B. sandersoni* (Sanderson), *B. ternarius* (Tri-colored)

One species endemic to extreme northwestern California and adjacent Oregon, *B. franklini*, is thought to have gone extinct in recent years.

Natural History: Bumblebees are found in essentially all terrestrial habitats of the United States and Canada, although they are uncommon in deserts and other arid habitats. They prefer mild to cold temperate climates of northern latitudes, maritime climates, and mountains. They are found in coastal areas, grasslands, chaparral, shrub and shrub-steppe, woodlands, forests, and tundra. Bumblebees are one of the few animals in this book that

Hive of an Eastern Bumblebee. *L.L.C. Jones*

are widespread across all latitudes of Canada. *Bombus polaris* and *B. hyperboreus* have been found as far north as the northern tip of Ellesmere Island, Nunavut Territory; that puts them well north of the Arctic Circle and makes them the northernmost eusocial insect species.

Unike their cousins the honey bees, bumblebees do not form particularly large colonies nor produce honey. The number of individuals in a mature colony may be about twenty to a few hundred, although large colonies can have upwards of 1,000 bees. Colony numbers peak in late summer and fall. There is a single reproductive queen for the colony. Queens are larger than workers and males, and are most often encountered in the spring and early summer. Foundress queens die before winter, but new queens overwinter and start a colony in the following spring; the queen lives up to about a year. Nests and colonies are typically annual, although some tropical species have longer-lived queens and perennial nests. Nests are typically in the ground, as in old rodent burrows, especially near rocks and logs, or may be in vegetation or on the surface of the ground. Nests may have a waxy protective covering. Eggs are laid singly in waxy pots. The larvae go through four instars, then pupate, and emerge as gray-pigmented adults in around five weeks. They will get normal adult coloration in a few days. The sterile female workers help with the chores, such as feeding and caring for larvae, nest maintenance, and foraging. Males (haploid, from the queen's DNA) and fertile diploid females (new queens) appear later in the season. They are evicted from the colony and live singly, but will mate for future colonies. The males die, but the now-fertile queens will then overwinter and start a new colony the following year.

Bumblebees are arguably the most important pollinators of native plants throughout much of their ranges. Bumblebees feed on a wide variety of flowering plants, so are extremely important pollinators for native vegetation as well as agricultural crops. They are the most important commercial pollinators in greenhouses. They are used for pollinating tomatoes, eggplants, chiles, berries, fruits, and seed crops, among others. They typically use their long tongue to lap nectar from flower tubes, but they may actually bite a hole in the base of the flower with their mandibles to drink the nectar if their tongue cannot reach. They also feed on the pollen itself. Pollen is attracted to the pile, so is not limited to the pollen basket. When a bee lands on another flower, the pollen is then transferred to the next flower.

Cuckoo bumblebees are so named because they, like cuckoo birds, are social parasites, and will invade a nest and displace the queen—then let the bumblebee workers raise her larvae.

Encounters: Bumblebees are common in some areas, but generally less frequently seen than honey bees. They are most commonly encountered in the wild in fields of wildflowers, or among flowers in greenbelts, gardens, and yards. They are not particularly attracted to the indoors, water, garbage bins, or picnic areas. They do not tend to pester people, so the biggest threat comes from stepping on a nest or accidentally brushing up against one. While not as aggressive as many eusocial hymenopterans, they will defend the nest. Bumblebees may constitute an occupational hazard to farm workers who use them as pollinators. While they may be used to pollinate open fields, they are more often used in greenhouses.

Venom Apparatus, Venom, and Symptoms: The sting apparatus is similar to other hymenopterans, being a modified ovipositor. Only queens and workers can sting. Although bumblebees are most closely related to honey bees, bumblebees have less-barbed stingers, so the sting mechanism stays in place in the bee's abdomen. Like wasps, bumblebees do not die after stinging and can sting repeatedly.

The structure of bumblebee venom is similar in character to wasps, hornets, and honey bees, having biologically active amines, peptides, and enzymes. Unique peptides called bombolitins are present in bumblebees. Pure venom extracts from *B. pensylvanicus* shows high levels of hyaluronidase, phospholipase A_2, phosphatase, and protease activity. The LD_{50} of two species (*B. sonorus* (= *pensylvanicus*) and *B. impatiens*) is about 12 mg/kg. The relatively high lethal dose, coupled with the bee's generally unaggressive nature and usually small colony size, means that fatal stings due to action of the venom are highly unlikely, even if stepping on a nest. However, bumblebees do have a larger dose of venom than honey bees and wasps.

The sting ranks a 2 on the Hymenopteran Sting Pain Index, although it is often stated that the sting is less painful than a honey bee. While the highest intensity of pain subsides within a few minutes, effects can linger for days in some individuals. In addition to pain, symptoms include redness, swelling, and itching in the area of the sting. Stings from multiple individuals can result in a more serious and symptomatic envenomation.

The medical profession is becoming more cognizant of allergic reactions caused by bumblebees, rather than just focusing on European Honey Bees and wasps. Part of the reason is because stings by bumblebees, though uncommon in nature, are an occupational hazard for agricultural workers. Two human allergens have been named from *B. pensylvanicus*, Bom p 1, a phospholipase A_2, and Bom p 4, a protease. Cross-sensitivity is high between bumblebees and honey bees. Allergic reactions are less commonly reported for bumblebees, but there have been several case reports of anaphylaxis.

Remarks: Overall, bumblebees are only of minor medical importance, although for someone who is allergic to bees, they can be of paramount importance. As pollinators of native plants, they are indisputably important for ecosystem health. A large decline in many bumblebee populations in recent years has been a cause for concern. Declines are due to competition or displacement from non-native bumblebee species (e.g., greenhouse escapees that become invasive), habitat degradation, introduction of exotic parasites and diseases, and climate change. There is a pollinator awareness campaign among public agencies and private organizations to increase awareness and make recommendations on how to think globally and act locally to conserve bumblebees and other pollinators.

References and Resources: Alford (1974), Argiolas and Pisano (1985), Bucher et al. (2011), de Groot (2006), de Groot et al. (1995), De Jong et al. (1999), Hatfield et al. (2012), Heinrich (2004), Hoffman and Jacobson (1996), Jakob et al. (2017), Koch et al. (2008), Schmidt (2008, 2009a, b), Stern et al. (2000), Williams (1998, 2007), Williams and Osborne (2009), Williams et al. (2009, 2012, 2014), Van der Zwan et al. (1992).

Large Carpenter Bees (Genus *Xylocopa*)

"The male bee is unable to sting. It is the male carpenter bee, which is most often noticed. They hover in the vicinity of the nest and will dart after any other flying insect that ventures into their territory. A common behavior of the males is to approach people if they move quickly or wave a hand in the air. The males may even hover a short distance from people causing unnecessary panic." (Penn State Entomology Department online)

Eastern Carpenter Bee. *Judy Gallagher/CCA/Flickr*

Classification: Phylum Arthropoda (arthropods); Class Insecta (insects); Order Hymenoptera (wasps, bees, ants, and sawflies); Family Apidae (bees); Genus *Xylocopa* (large carpenter bees). There are other genera of carpenter bees, but the genus *Xylocopa* is most noteworthy because of its size.

There are 500 to 700+ species of large carpenter bees worldwide. Taxonomy of the *Xylocopa* is problematic due to similarity between species and a lack of distinctive anatomical characters. They are largely tropical animals but at least nine species belonging to six subgenera occur in the United States. Some widespread and commonly encountered species include *X. californica*

(Western), *X. micans* (Southern), *X. tabaniformis* (Horsefly-like), *X. varipunctata* (Valley), and *X. virginica* (Eastern).

Identification: Large carpenter bees are stout and about 15–30 mm in length. Most are difficult to differentiate from one another. They have a body design similar to bumblebees, but are not covered in pile. While they do have dense, long setae on the thorax and legs, the abdomen appears shiny and is usually black or blue-black. Male Valley Carpenter Bees are an exception; they are light brown in color, as opposed to the shiny black female. There may be some lighter-pigmented setae on the thorax, but not to the extent of bumblebees. They have very large compound eyes, which accounts for the common name of *X. tabaniformis*, which superficially resembles a giant horsefly. They lack the pollen basket of bumblebees, but they do collect pollen on the hind legs and thorax. The Eastern Carpenter Bee has yellowish setae on its thorax.

Distribution: *Xylocopa* are found on every continent except Antarctica. About 90% of the species are in the tropics and subtropics, with the remainder in temperate zones. Large carpenter bees are found throughout most of the United States, but are most diverse in the American Southwest. Only

Face of an Eastern Carpenter Bee. *Bee Inventory and Monitoring/USGS, BIM/Public Domain*

Female Valley Carpenter Bee, the usual black color for the genus. *Xylocopa* are important native pollinators. *L.L.C. Jones*

The male Valley Carpenter Bee, which does not sting, is brown. *L.L.C. Jones*

Eastern and Southern Carpenter Bees are found in the American Southeast, although Valley Carpenter Bees have been reported from Florida. The Eastern Carpenter Bee is the only one that ranges north into New England, and in Canada, into Ontario.

Natural History: Large carpenter bees are generally regarded as solitary species, although they exhibit some social behavior, with multiple related females living in or near the natal nest. They are not eusocial because they do not have distinct castes. Their simple social structure is studied because it is a somewhat intermediate system between completely solitary and eusocial Hymenoptera. The lek mating system has been studied in *X. varipuncta*. Males leave their natal nests and hover above nonflowering plants away from nests to try to attract females. They secrete sex pheromones to attract females, but sometimes other males will visit the bee. They are territorial, so will try to repel other males. If females are willing, they will land on foliage to mate.

The entrance of the nest of large carpenter bees is a perfectly round hole in wood. The hole looks as if it were created with a power drill—hence the name, carpenter bee. The bee "drills" the hole by biting out dead chunks of wood with its powerful mandibles and excavating galleries in which to store food and lay eggs. Each egg compartment has a single egg up to 15 mm. The compartments are separated from one another by a mix of saliva and wood fiber, having the consistency of particleboard. The chambers contain "bee bread," a mixture of pollen and nectar, for the larvae to feed on. The cells are sealed off for pupation. As the larvae grow, young females may stay in the nest with the female parent for some time before dispersing, and social interactions can occur in the nest. However, there are no true castes of queens and workers. Usually the nest is formed in naturally occurring soft wood such as pine and redwood. Nests are often in a standing dead tree, but the bees (especially *X. virginica*) may also drill holes into fencing, beams, or supports of buildings and other structures.

Xylocopa are good pollinators of native plants and some agricultural crops and ornamental flowers. As they go from flower to flower, their buzzing action can attract pollen, which can then be dispersed onto other flowers. Sometimes carpenter bees, like some bumblebees, are nectar thieves, reaching nectar by biting a slit in the flower's corolla, rather than using their tongue to access nectar from the corolla's opening. This is often the case when the corolla is deep. Carpenter bees can be used as home garden pollinators. Strategically placed blocks of wood can be used to lure them and keep them away from more important structures.

Eastern Carpenter Bee entering its round hole in the beam of a porch. *L.L.C. Jones*

Carpenter bees are may live 2–3 years. They are often dedicated to particular sites over the years and may use the same sites for the duration of their life, where they can nest, lay eggs, and overwinter. They have been studied for their thermoregulation capabilities, such as coping with desert heat or warming up to become active in cold areas.

Encounters: While it is not uncommon to see carpenter bees, it is pretty difficult to get stung. Stings are not well-documented. Most stings probably involve people with occupational hazards, most likely entomologists attempting to capture them, and perhaps people who use them as pollinators. People trying to evict bees from their nest may also be at a little higher risk, as are those who accidentally rub against them. Carpenter bees may appear aggressive when they buzz up to people, but they are announcing their territory and do not intend to sting. These are usually males, which cannot sting, and females have little inclination to sting, but will defend themselves if they feel threatened. If you get close to a nest they may make warning flights toward you but rarely sting during these encounters. For example, while trying to photograph *X. virginica* for this book, several individuals aggregated in the beam of a cabin would dive-bomb me, but never made contact. Nevertheless, it is probably best to avoid their nests.

Venom Apparatus, Venom, and Symptoms: Large carpenter bees have a venom apparatus similar to other hymenopterans, having a venom gland, Dufour's gland, free filament, and sting mechanism, encased in sclerites. The stinger is a modified ovipositor, so only females sting. Stings to humans are uncommon, and the venom itself is not particularly potent, so the venom has not gotten as much attention from researchers as the more aggressive species. The venom is composed of pharmacologically active compounds, proteins, and enzymes. The venom shows histamine, phospholipase A, hyaluronidase, and small amounts of acid phosphatase activity. It has a relatively low hemolytic capability. The LD_{50} for *X. virginica* in two studies was measured at 22 and 75 mg/kg.

The pain of the sting has been reported as 1–3 on the Hymenopteran Sting Pain Index for North American species. It is generally regarded as being about as painful as a honey bee sting, but some species are reported to be more painful. The Eastern Carpenter Bee is said to deliver an extremely painful sting. There is not much in the literature to compare the various species of large carpenter bees, or even those occurring in the United States.

Other than pain, symptoms include redness around the sting site, swelling, burning, and itching. The venom can also cause local paralysis. Systemic reactions likely occur in hypersensitive people. Little is known about allergens specifically for carpenter bees.

Remarks: Carpenter bees can cause damage to wooden structures, but not usually to the extent that the structures become unsafe and require replacement. Often there is but a single external hole in the wood, although several bees may nest near one another. For exposed structural lumber, it can be painted to dissuade the bees from nesting, as they prefer a natural wood surface.

References and Resources: Balduf (1962), Chappell (1982), Frankie and Vinson (1977), Gerling et al. (1989), Heinrich and Buchmann (1986), Hurd (1978), Leys et al. (2000, 2002), Marshall and Alcock (1981), Minckley (1998), Perveen et al. (2012), Piek (2013), Schmidt (1986, 2008, 2016).

Assassin Bugs (Family Reduviidae)

"Unfortunately, I am one of the few that can provide a first-hand testimonial to the venom [Wheel Bugs] dispense; I was bitten on the finger by one in Chino Valley [Arizona] and the pain was excruciating. The intense pain lasted for several hours, followed by a week of dull throbbing. Skin surrounding the wound became inflamed, and in a few days the tissue surrounding the bite sloughed off, leaving a depression. The bitten finger felt much warmer than the uninjured fingers for months." (Smith 1982)

Classification: Phylum Arthropoda (arthropods); Class Insecta (insects); Order Hemiptera (true bugs); Family Reduviidae (assassin, ambush, and kissing bugs).

Hemiptera is a huge cosmopolitan order of 50,000–80,000 species worldwide with many different families that are quite variable in design, living in terrestrial, aquatic, and even marine habitats. The family includes insects as diverse as scale insects, mealy bugs, aphids, spittlebugs, stinkbugs,

Adult Wheel Bug. *Mike Keeling/CCA-EQ/Flickr.* Inset shows an instar. *L.L.C. Jones*

whiteflies, cicadas, water scorpions, and giant water bugs. The suborder Heteroptera (sometimes considered its own order) includes typical true bugs. Although the term "bug" refers to this order of insects, it is commonly misused to represent practically everything that is vaguely insect-like. Even lobsters (which are crustaceans) are called bugs by divers because of their superficially similar appearance to insects.

Reduviidae is one of the largest families of true bugs, with more than 7,000 species worldwide. There are approximately 180 species in fifty genera in the United States, belonging to a dozen subfamilies. Genera include *Apiomerus, Arilus, Melanolestes, Microtomus, Narvesus, Pselliopus, Rasahus, Rhiginia, Sirthenea,* and *Zelus,* among others. The kissing bugs (*Triatoma*) are in the subfamily Triatominae; while technically they are parasites and should not be included in this book, they bite, are venomous, are medically important, and are physically similar to other reduviids. Smaller taxa are too small to be of concern to humans, but any reduviid large enough to puncture the skin should be considered venomous to humans.

Identification: The Hemiptera have several distinct features. They possess piercing and sucking mouthparts. In those that have wings, there are two pairs, with the upper wings being partially chitinous and the posterior parts being membranous. The lower wings are entirely membranous. Because of the configuration, many Heteroptera (including most Reduviidae), appear to have a sort of "X" pattern dorsally that can be easily recognized at a glance. Most hemipterans are nonvenomous and feed on plants.

What makes an assassin bug distinct from other Hemiptera are the elongated head and neck with a transverse groove, a stout three-segmented proboscis, four-segmented antennae, ocelli (in addition to compound eyes), and two-segmented tarsi. The proboscis fits in a groove under the prosternum. At a glance, leaf-footed bugs (Coreidae) and some other true bugs are superficially similar and often misidentified by laypersons as assassin bugs (especially kissing bugs), but one can differentiate them by the neck, which is narrower than the head in assassin bugs, and broad in other true bugs. Leaf-footed bugs also have broad rear appendages. The various genera and species of reduviids have their own sets of distinctive characters. Assassin bugs may stridulate.

To try and characterize all species would be too ambitious, especially considering the variation of different stages of nymphs, so the reader is invited to learn more about them in the references and resources section, but here are some highlights of a few common taxa. The genus *Arilus* has one species, *A. cristatus,* the Wheel Bug. It is distinctive with its cog-like "wheel" on the thorax. It is also the largest assassin bug at 28–36 mm total length; females are larger than males. The nymphs are quite different in appearance from the gray adults, having black legs, head, and thorax, and a round, bright red abdomen. They are almost spider-like. *Zelus* are elongate species. There are seven species of *Zelus* in the United States. Other slender genera include *Narvesus* and *Rasahus*. The corsairs include the Black Corsair (*Melanolestes picipes*), *Sirthenea carinata*, Western Corsair (*Rasahus thoracicus*), and other species of *Rasahus*, which are about 15–23 mm. The name comes from the French word "corsair," which refers to pirates, perhaps alluding to the scimitar-like mouthparts or their tendency to attack other insects. The Black Corsair is jet black as an adult, but instars have a red abdomen. The Masked Hunter is similar in appearance to a Black Corsair. It derives its name from the nymphs, which are masked by household dust and lint, which adheres to them. Another similar genus is *Microtomus*, which is also large, to 30 mm. They have a bold wide or red band across the anterior dorsal abdominal area.

The several species of bee assassins (*Apiomerus* spp.) may be difficult to differentiate, and reach about 15–20 mm. Adult *A. flaviventris* has a yellow belly and dorsally a pattern of red, black, and yellow. While the coloration is aposematic, it is also cryptic on red and yellow plants.

The eleven U.S. species of kissing (or conenose, cone-nosed, conehead, or blood-sucking conenose) bugs belong to the genera *Triatoma* and *Paratriatoma*. They are similar in appearance to other reduviids, but tend to be broader than most species. They are fairly large, about 13–33 mm in length.

Distribution: Assassin bugs are found worldwide and are well distributed throughout the United States and into Canada. The Wheel Bug is the best-known species, particularly in the eastern United States; it is found across the United States from California to Florida, north to New England and Canada, and south to Central America. Corsairs are mainly found in the southern United States. The Black Corsair is broadly distributed from California to Florida, north to Wisconsin, and south to Brazil. Bee assassins are found throughout most of the United States. Of the 140 or so species of kissing bugs, nearly all are in the New World, but there are only eleven north of Mexico, as most members are tropical and subtropical. They are found primarily in the southern United States, from coast to coast, but the highest diversity is in the Southwest, especially in the southern Border States. Some of the more wide-ranging species include: *T.* [=*Paratriatoma*] *hirsuta*, *T. lecticularia*, *T. protracta*, *T. recurva*, *T. rubida*, *T. rubrofasciata*, and *T. sanguisuga*.

The Yellow-bellied Bee Assassin sports aposematic coloration that also renders it cryptic on flowers, where it hunts bees and other insects. *L.L.C. Jones*

Natural History: The various genera and species are found in a wide variety of habitats from coastal areas to inland, and valleys into the mountains. They occur across North America, but are most abundant and diverse in the southern United States. Assassin bugs are all terrestrial.

The majority of hemipterans use their piercing and sucking mouthparts to drink plant juices—hence, they can be major agricultural pests—but the reduviids (and some other families) are major exceptions. Assassin bugs are aptly named because they prey on ("assassinate") other insects. The exception is that kissing bugs actually drink blood from vertebrates—usually mammals. Many reduviids ambush their prey, sometimes waiting in flowers to do so. Other assassins are more active and fly between hunting grounds to prowl for invertebrate prey. Vegetation is the usual hunting ground. Some species, such as the male corsairs and *Rhiginia* are strong fliers. The various species of reduviids can be diurnal to nocturnal, depending on species and time of year. Some species come to lights at night, including *Melanolestes* and *Microtomus*. Kissing bugs are nocturnal.

As a group, reduviids may be active year-round, especially in the South, but are active mostly during spring through fall. Summer to fall is peak time for adults of most species. They live from several months to over a year. Those that survive over the winter will either have subdued activity or remain in

Triatoma recurva, a large kissing bug. L.L.C. Jones

hibernacula until emergence the following spring. Eggs may be laid singly or in clusters. The eggs of the Wheel Bug appear to be a cluster of dozens of little bottles on vegetation, while those of the Black Corsairs are laid singly under stones. Kissing bugs lay eggs singly in areas near their food source (e.g., woodrat dens or rarely inside of houses). Nymphs of all taxa may look quite different from adults. After emergence, nymphs go through several instars before becoming adults. Even young nymphs are efficient predators, feeding on appropriately sized prey. Assassin bugs use their forelegs, which may be sticky with highly specialized setae, to grab prey. Their proboscis is used to deliver the fatal blow of venom and digestive juices.

Assassin bugs are important predators of insect pests. The Masked Hunter (*R. personatus*) is important because it can reside in human dwellings and feed on bedbugs, which are human parasites. The nymphs of that species are sticky and gather debris, such as dust and lint, for concealment. Black Corsairs feed on May beetles (June "bugs"). *Zelus* often eat caterpillars, leafhoppers, and weevils that damage agricultural crops. The large Wheel Bug feeds on a wide variety of insect types, from pest caterpillars to beneficial pollinators, but caterpillars seem preferred. Some species feed on spiders. They will attract spiders on their own webs, and then attack them when they venture too close. Assassin bugs are used in agriculture as a biological control. Their venom is extremely insecticidal, so their venom is of interest to researchers for control of crop pests.

Kissing bugs differ from the other reduviids because they are parasitic, rather than predatory; they feed on vertebrate blood. Woodrats are the most common prey, although they will also feed on the blood of other animals, including Nine-banded Armadillo, Virginia Opossum, Raccoon, and other species. They may also feed on humans and their pets.

Encounters: Human encounters are generally accidental, as when one is walking through vegetation or picking flowers, but kissing bugs will seek out a meal, which may be human. Typical assassin bugs may be inconspicuous in vegetation, especially when they are remaining immobile. If we brush against them, or get them in our hair or clothing, they can defend themselves. They may also bite when handled. Some species such as *Microtomus*, Wheel Bugs, Black Corsairs, and bee assassins are large attractive animals, so may be an inviting target to handle.

Kissing bugs are so named because they may bite humans on or near the lips or other thin spots, especially while they are sleeping. The host may also be a household pet, usually a dog. These bugs move slowly and deliberately, and then select a spot on the skin to insert their stylets to feed on the victim. Kissing bugs often hole up just outside of homes, in nests of woodrats and other animals, or other dark places near their prey, and may venture into houses. Once in the home, they may set up shop in the vicinity of their newly found, reliable food source.

Bites from assassin bugs are not well tracked in emergency rooms or 911 calls. One report from 2000–2004 lists about 1,000 emergency room visits per year due to kissing bug bites, but I do not know if this includes other reduviids.

Venom Apparatus, Venom, and Symptoms: The venom apparatus of all reduviids includes a salivary gland complex, with the proboscis being the delivery system. The proboscis has chemosensory sensillae and trichobothria. Inside the proboscis, attached to the head by muscles are maxillary and mandibular stylets. The stylet bundles are barbed to penetrate and tear flesh, and, together, they form a tube to pump saliva into the prey and then suck out the partially digested food. In one Australian species examined in detail (*Pristhesancus plagipennis*), it was found to have two distinct venom glands, the anterior main gland and the posterior main gland, plus an accessory gland. Researchers have shown that each of the main glands has a distinct proteome, each with different complex venom. Secretions from the accessory gland are relatively innocuous. The posterior main gland is highly effective

Adult Western Corsair. L.L.C. Jones

for predation, while the anterior gland is probably used for defense, and possibly other functions. Depending on the stimulus, they can meter the venom discharge from each of the glands. It is not known if our native species also have this capacity, or how widespread it is in the venomous animal world because few people have looked.

Much of what we know about assassin bug venom is from research on several genera and species from outside the United States. The venom of assassin bugs probably evolved primarily for predation, with defense being a secondary feature. The venom of species studied includes peptides, proteins, and enzymes. There are similarities to venom components of elapid snakes and cone snails. At least three specific peptides have been isolated from reduviid saliva. Although it varies by species, the saliva can have proteases, hyaluronidase, lipase, esterase, phospholipase, amylase, invertase, acid phosphatase, trypsin, pepsin, and adenosine triphosphate. The venom is both neurotoxic and cytotoxic to insects. Insects are paralyzed almost immediately, even if the prey is hundreds of times larger. Apparently they are resistant to their own venom, although there are reports of intraspecific

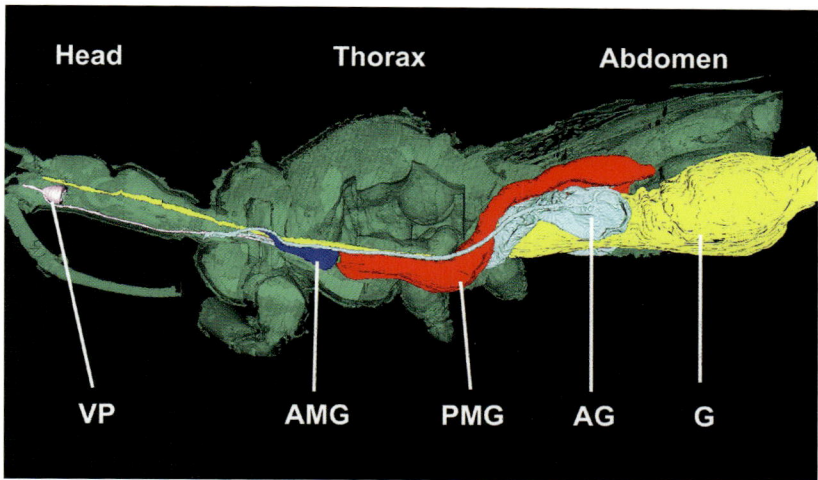

Dual venom system of a reduviid, from Walker et al. (2018). AG = accessory gland, AMG = anterior main gland, G = gut, PMG = posterior main gland, and VP = venom pump. *Andrew Walker/Nature Communications (2018)*

predation among hatchlings. After the insect is paralyzed, it is digested with the cytotoxic elements of the venom, and then the contents of the prey are sucked up through the straw-like proboscis. Some of the venom compounds can affect both invertebrates and vertebrates. For example, hyaluronidase with proteases can break down tissues—in an insect, that is digestion, while in a vertebrate, it is a painful local reaction.

In humans, Wheel Bugs and other large species, such as the Western and Black Corsairs, can deliver an extremely painful bite that is said to be immediate and excruciating. The area of the bite becomes red, hot to the touch, and swollen. Numbness can last for several days. The wound may then become hard and white due to tissue growth, and a lesion may form, with tissue sloughing off, leaving a small hole. It often heals completely in about two weeks but may take longer. Symptoms sometimes persist for many months to over a year. Most long-lasting effects come from secondary infection, despite antimicrobial properties to the saliva.

Kissing bugs are a different story. The bite itself is usually painless. The saliva has an anesthetic quality to it, which makes sense for an animal that needs to feed on blood unmolested for about ten minutes. Symptoms come later. The main symptoms are cutaneous. The bite area often becomes red and swollen and may have welts, but there may also be redness on the extremities. Intense itching is a common complaint. Often the venom causes gastrointestinal difficulties that include nausea, vomiting, and diarrhea. People bitten by kissing bugs have a high frequency of allergic responses, including respiratory difficulties and anaphylactic shock. Sensitized people have higher levels of immunoglobulin G (IgG).

Remarks: Although pathogens are outside the scope of this book, kissing bugs are also infamous vectors (carriers) of Chagas disease, which is caused by the protozoan parasite, *Trypanosoma cruzi*. The parasite is spread by the bug when it defecates while feeding. If the feces enter the victim's bloodstream through the bite or subsequent scratching, the person can become infected. Early symptoms of the disease include fever, headache, swollen lymph nodes, and local swelling of the bite. These symptoms may be delayed and they are also similar to those of many other disorders. In 8–12 weeks, victims progress to a chronic phase of the disease, but two-thirds of the victims do not develop further symptoms. However, the remainder may have serious health problems, which may result in death, especially from heart failure.

This devastating disease affects seven to eight million people from Mexico to South America. About 12,500 people die from the disease each year. However, Chagas disease is extremely rare in the United States. Only six cases have been documented, in California and Texas, even though at least seven species of *Triatoma* in the Southwest have been shown to be carriers. Studies have shown the protozoan to be present in twenty species of wildlife in at least thirteen states. Dogs can catch the disease. It can be fatal in 0–60 weeks. Chagas in dogs is most common in southeast Texas and Louisiana, but does range as far north as Minnesota, and the vectors are found throughout the southern United States.

References and Resources: Berniker et al. (2011), Corzo et al. (2001), Eaton and Kaufman (2007), Edwards (1961), Evangelin et al. (2014), Galvão et al. (2004), Hagerty and McPherson (2000), Hall (1924), Henry and Froeschner (1988), Kirchhoff (1993), Kumar and Sahayaraj (2012), Lent and Wygodzinsky (1979), Marshall et al. (1986a, b), McPherson et al. (1992), Readio (1927), Schmidt et al. (2011), Smith et al. (1958), Snodgrass (1944), Stevens et al. (2011), Walker et al. (2018), Weirauch et al. (2014), Wignall and Taylor (2008).

Stinging Caterpillars (Order Lepidoptera)

"Survey of an 'epidemic' [of *Megalopyge opercularis*] in southern Texas showed a high percentage of severe symptoms, including constitutional reactions that required hospitalization in 3 cases. In fact, severity of reactions observed in some patients suggests the possibility that death could result from these stings. . . . The symptoms most frequently observed were marked local pain, local swelling, lymphadenopathy, and headache; shock-like symptoms and convulsions were also noted." (McGovern et al. 1961)

Eastern Puss Caterpillar, juvenile of the Southern Flannel Moth. *L.L.C. Jones*

Classification: Phylum Arthropoda (arthropods); Class Insecta (insects); Order Lepidoptera (moths and butterflies).

The order Lepidoptera is the second largest insect order, with an estimated 175,000 species in 126 families and forty-six superfamilies. Even with those staggering numbers, there are very few venomous species, and nearly all of those are larvae. According to one source, there are nearly 13,000 species of lepidopterans in the United States. Moths are much more speciose than butterflies, accounting for 94% of the lepidopterans. The vast majority of venomous species are moth larvae.

Adult Southern Flannel Moth. *L.L.C. Jones*

It is often reported that about 50 species in the United States are known to be venomous. This is probably an underestimate, as many moth larvae are poorly known. Many others bear spines but have not been reported as venomous. Some caterpillars may have urticating hairs, but they may lack a venom apparatus. Seven genera of three families consistently rise to the top as being of medical importance. All are moths in the families Limacodidae (slug caterpillar moths, about fifty U.S. species), Megalopygidae (flannel moths, twelve U.S. species), and Saturniidae (giant silkworm moths, about seventy-five U.S. species). They include:

- Puss caterpillars or Asps (*Megalopyge* spp.; family Megalopygidae). There are four genera in the Megalopygidae (flannel moths). *Megalopyge* is well known as a venomous genus. There are seven U.S. species. The genus *Norape* is also known to be venomous, with three U.S. species.
- Saddleback (*Acharia* (=*Sibene*) *stimulea*; family Limacodidae).
- Silk moths, Io Moth group (*Automeris* spp.; family Saturniidae). There are seven species in the United States. All are presumed venomous. There are many other genera of silk moths, some of which are venomous. Another silk moth larva, *Coloradia pandora*, is venomous.
- Spiny oak slugs (*Euclea* spp.; family Limacodidae). There are five species.

- Hag Moth (*Phobetron pithecium*; family Limacodidae). At least some related genera are probably venomous.
- Stinging Rose (*Parasa indetermina*; and Smaller Parasa, *P. chloris*; family Limacodidae). Taxonomy is unclear, but other species have been named.
- Buck moths and sheep moths (*Hemileuca* spp.; family Saturniidae). There are more than twenty species in the United States.
- Stinging Nettle Caterpillar (*Darna pallavitta*; family Limacodidae).

Other families and genera of moths that are reported to be venomous or cause dermatitis or allergies from hairs and skin secretions in the United States include:

- **Arctiinae** (tiger moths and wooly bears), *Arctia* spp., *Hyphantria* spp., *Spilosoma* spp., but perhaps other genera.
- **Lymantriidae** (tussock moths), *Euproctis* spp., *Lymantria dispar* (Gypsy Moth, introduced), *Orgyia* spp., *Lophocampa caryae* (Hickory Tussock Moth).
- **Lasiocampidae** (tent caterpillars), *Malacosoma* spp., and other genera.
- **Notodontidae** (prominent moths), *Lochmaeus* spp., *Thaumetopoea wilkinsoni*.
- **Noctuidae** (owlet moths), *Acronicta* spp.
- **Zygaenidae** (leaf skeletonizer moths), *Neoprocris* spp., *Zygaena* spp.
- **Nymphalidae** (brush-footed butterflies). The only group of butterflies with venomous larvae and only in the genus *Nymphalis*. The Mourning Cloak (*N. antiopa*) caterpillar is mildly venomous.

Identification: After hatching, lepidopterans undergo a metamorphosis from a larval stage (caterpillar) to pupa (in a cocoon [moth] or chrysalis [butterfly]) to a winged adult. Caterpillars are soft-bodied, generally slender, worm-like organisms. From front to back, they are divided into a head, thorax, and abdomen. The head has small ocelli rather than compound eyes. The mandibles are well developed for chewing leaves. The thorax has three segments, each with a pair of true legs on the venter. There are seven abdominal segments, with prolegs on abdominal segments III to VI. Abdominal segment X (which is really two fused segments, X and XI) has an anal proleg.

Prolegs may be absent, as in slug caterpillars. Often the last abdominal segment may have some sort of adornment, such as a "horn." The number of setae is highly variable, from sparsely "haired" to being completely covered in dense "fur." Clusters of spicules and spines are often called bristles or tufts.

Most species of venomous caterpillars can be recognized by their wicked-looking spines that are often showy, intricate, and branched. The Saddleback, Stinging Nettle Caterpillar, spiny oak slugs, Stinging Rose and Smaller Parasa, buck moths, and Io Moth group all fit into this category. Some of these species are aposematically colored, but some are cryptically colored. The buck moths are highly variable, but tend to be more cryptically colored. The Saddleback is very distinctive. It is brown with a green saddle-like marking dorsally. There are four appendages armed with venomous spines. The spiny oak slugs are variously colored, but all have numerous spine-bearing appendages. Many tend to be greenish, but there may be other colors and some are more subdued in coloration. The genus *Parasa* is morphologically similar in appearance, having numerous spiny appendages. The Stinging Rose is brightly colored with yellow, red, and/or orange, with black markings, including stripes in some individuals. The Smaller Parasa is subdued yellowish to gray. Caterpillars of the Io Moth group are mostly bright green, usually with a white or white and red stripe along the side (*A. io, A. iris,* and *A. randa*). *Automeris cecrops* and *A. zephyria* are green with black and white markings on the side, while *A. patagoniensis* is more of a brown color. All are morphologically similar, covered in branched bristle clusters dorsally. Buck moths are mostly brown, but there are other colors and patterns. They have armament similar to the Io Moth group, but the spine clusters are not usually as pronounced. *Coloradia* species are similar to buck moths.

The species that don't show wicked-looking spines tend to be "hairy." The puss caterpillars do not look wicked at all—in fact, they get their name because they look like a fluff of hair from a pussy cat. They are also called Asp, Tree Asp, and Wooly Slug. In Mexico, they are known as El Perrito, Spanish for "little dog," also alluding to the ball-of-fur appearance. Young instars look more like typical caterpillars, although they do have some long hairs. As they grow, the hairs completely surround the body and its venomous spines. The adult moth is also fluffy, but nonvenomous. The Hag Moth caterpillar, or monkey slug, is a very odd looking animal that can easily be identified by its bizarre shape alone, which doesn't much resemble a caterpillar. It has six pairs of hairy appendages dorsally, making it look like some sort of hairy, contorted spider. Other

Larva of an Io Moth. *Courtesy of Clemson University/USDA extension*

Adult Io Moth. *Ronald Billings/Texas A&M/Bugwood*

related genera are similar in appearance. Most of the other genera and species that cause urticaria are also hairy, such as the wooly bears, tussock moths, and tent moths.

Adults of some of these caterpillars are among the most beautiful lepidopterans, rivaling many of the butterflies. This is especially true of the

Saddleback Caterpillar. *Judy Gallagher/CCA/Flickr*

Hag Moth larva. *Jerry Payne/CCA/USDA, Agriculture Research Service (ARS)*

saturniids, some of which are fairly large and variously marked with eye spots, stripes, and bright colors. *Automeris* species, for example, have large eyespots on the upper hind wings, while the upper forewings look like leaves. Some of the *Hemileuca* are also dazzling. *Euclea* and *Parasa* have beautiful green bands on the forewings.

Distribution: There are venomous caterpillars throughout the world. The most common species of puss caterpillar in the United States is *M. opercularis*, found from Texas to Florida and north to New Jersey, and it is also recorded from Arizona. Other eastern species include *M. crispata*, *M. lacyi*, and *M. pyxidifera*. *Megalopyge bissesa* is the common species from southeastern Arizona to west Texas, while *M. lapena* is generally regarded as a stray there; *M. immaculata* is also known from west Texas.

The Io Moth proper is primarily an eastern species that reaches Colorado and Utah to the west and Ontario to the north. One subspecies is found in New Mexico and adjacent northwest Texas. New Mexico also has *A. cecrops*, *A. randa*, and *A. zephyria*. Texas also has *A. louisiana*, a primarily Louisiana species, and *A. zephyria*. Arizona has four species of the Io group: *A. cecrops*, *A. iris*, *A. patagoniensis*, and *A. randa*.

Buck moths and sheep moths are found in nearly all U.S. states, plus British Columbia and Alberta, although the majority of species are found in the Southwest. The number of species by state, according to an online database, in descending order, is: Arizona (thirteen species); New Mexico (ten species); California and Nevada (nine species each); Colorado and Utah (eight species each); and Texas (seven species). Compare this to one species in North Dakota, Florida, and Louisiana. The most wide-ranging species in the Southwest are: *H. hera*, *H. eglanterina*, and *H. nevadensis*. The most commonly referenced species with regards to stinging is *H. maia*, an eastern species (or complex) that is found from Texas to Florida, and north to New York. It is simply called the Buck Moth Caterpillar in most of the literature. All *Hemileuca* are presumed venomous; some species are confirmed to be so. Their close relatives in the genus *Coloradia* have a similar distribution and are only found in the western United States or Mexico. *Coloradia pandora* is the most widespread, occurring from Texas to Washington.

The Saddleback (or "Packsaddle") is an eastern species, found from Texas to Florida, and as far north as the Great Lakes. The Hag Moth is found from Texas to Florida, and north to Ontario and Quebec. Some of the relatives of the Hag Moth are found farther west, but they are not discussed much in the literature as being venomous. The Stinging Nettle Caterpillar is native to Southeast Asia, probably arriving in Hawaii from Taiwan. It is now established on all islands except Molokai and Lanai. It is not only a pest to humans, but also native, ornamental, and commercial crops. One primarily eastern Canada species is the Hickory Tussock Moth, which ranges south to Tennessee, but it has also been reported from as far south as Texas. A field guide should be consulted for the many forms.

Buck Moth larva. *L.L.C. Jones*

Hickory Tussock Moth larva. *Kevin Ripka/CCA/Flickr*

Natural History: Moths are essentially the nocturnal counterpart of butterflies, although nearly all buck moths fly by day. Most of a lepidopteran's limited time on Earth (generally less than a year) is in the larval form. For larvae, there is not such a distinction between diurnal and nocturnal, as they typically remain on the plants they feed on. This means lepidopteran larvae may be tasty morsels to any manner of animal that needs protein, day or

Spiny Oak Slug. *Jerry Payne/CCA/USDA, ARS*

night. The larvae of most species have preferences for larval food plants. The eastern species are typically found on hardwoods. Most are not host-plant specific and many species of trees are used, but there are exceptions, and most have their favorites. The species of *Hemileuca* have different host plants within their ranges, so have different preferences. For example, *H. maia* generally prefers oaks, *H. nevadensis* prefers willows, *H. oliviae* (the Range Caterpillar) prefers grasses, and *H. hera* prefers composites. The so-called Spiny Oak Slug actually will feed on many plant species.

The life cycle begins when the female moth lays eggs on a tree or shrub that will serve as a host plant for her larvae. After hatching, the larvae feed on the leaves, and as they grow, go through several instars. Each successive instar usually becomes more adept at envenomation than the previous one. Those with aposematic coloration also develop their late instar pattern with successive molts. After the animal reaches its final larval instar, it spins a silk cocoon and pupates inside. In the cocoon, the caterpillar goes through metamorphosis, and emerges as an adult moth. Butterflies are similar, but instead of a cocoon, they pupate into a chrysalis. The adults are not venomous, but many species are poisonous, such as the famous Monarch Butterfly and its kin.

The timing of events from egg-laying and hatching, then emergence and mating, varies by species. There is sometimes just one brood per year, although the puss caterpillar may have up to three (two is the norm) and the Io Moth ranges between one and four. With multiple broods, adults and caterpillars may be active over an extended season. In some species, overwintering occurs as prepupae or pupae, whereas others overwinter as eggs. Variations may be species-specific or depend on local climate. After pupation adults emerge for the primary purpose of mating and oviposition. Hatching (at least first brood) is usually in the spring and the caterpillars spend weeks to months consuming leaves. Some species may be very abundant in a particular area where they can be pests, capable of defoliating trees. For example, Io Moth caterpillars feed on agricultural crops like citrus, berries, and ornamentals, including *Hibiscus*, as well as dozens of native species.

Caterpillar predators include invertebrates, such as salticids, mantids, and paper wasps, as well as vertebrates, such as amphibians, lizards, birds, and mammals. Birds, in particular, probably helped drive their evolution, as many species feed on insects within shrubs and trees. Birds have good color vision, so aposematic coloration is a logical deterrent.

Encounters: These animals do not usually venture into homes or seek out people, but for those living, hiking, or working in the woods or parks, they can be encountered fairly regularly. Some people, such as forest workers and recreational area employees have an occupational hazard. Some species do seem to favor ornamental trees that are often planted in yards, public parks, school grounds, and developments, so we have inadvertently brought them closer to us. Caterpillars eat and live on vegetation, typically trees and shrubs, but also grasses and annuals, so anyone can come into contact with them if they walk through vegetation or brush up against trees. Caterpillars can also get dislodged and be knocked out of trees, then fall to the ground. For example, puss caterpillars can be knocked out of trees during high winds, and end up under passersby collars. Some other species may descend to the ground or disperse by using silk. *Hemileuca* may be so abundant and cryptic that they are difficult to avoid.

Not all encounters are accidental. Some people pick up the caterpillars, out of curiosity, or as an occupational hazard. Children often like to handle caterpillars. One problem is that small children tend to put things in their mouths . . . bad idea! One published account discusses the problem of ingestion of

Hemileuca tricolor larva. L.L.C. Jones

venomous and poisonous caterpillars by children. Although few species are venomous, many more species are poisonous, even in the adult stage.

The highest incidence of stings from puss caterpillars in the United States is in Texas, where outbreaks are not uncommon. There have been cases of school closures in Texas when puss caterpillars were so numerous that stings to children constituted a public health hazard. In one report that actually tracked puss caterpillars, they caused an average of 147 visits per year to emergency rooms during 2001–2004. In the same report, caterpillars (all species) accounted for 460 visits to emergency rooms. Around the same time period, an average of 2,094 calls per year was made to poison control centers due to caterpillar stings. The majority of stings undoubtedly go unreported. During seasonal infestations of common species such as *Hemileuca*, it may be a part of life that you are going to get urticarial dermatitis, depending upon where you live and how much you get into the outdoors.

Venom Apparatus, Venom, and Symptoms: Caterpillar toxins are only used in defense. Many species of caterpillars are poisonous, a ploy to keep

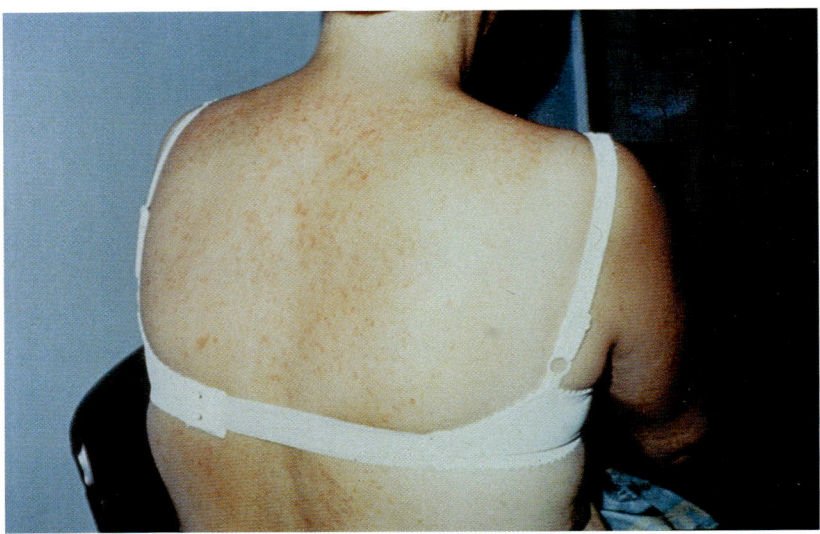

Lesions caused by a puss caterpillar. *Findlay Russell*

from being eaten, but they fall outside the scope of this book. Poisonous species often sport aposematic coloration.

The venom apparatus of caterpillars is associated with specialized setae. Classification of caterpillar setae is complex. Not all setae are venomous. Some have chemosensory or other functions. There are many published accounts of the anatomy of caterpillar setae, particularly those that cause irritation from mechanical trauma and envenomation. Venomous setae fall into two general types: spicules and spines. Spicules are easily dislodged where they can cause irritation in the skin and mucous membranes, as do the spicules of sponges. Saturniids often have these breakaway spicules in their bristles. Venomous spines are not easily dislodged. The venom production and delivery system remains intact in the epidermis where they are innervated. Puss caterpillars have good examples of venomous spines. The spines are hidden under the cover of nonvenomous setae that resemble mammal fur. Spines and spicules are formed from an evagination of the body wall. The venom is produced from epithelial cells at the spine's origin in the skin and stored in the spine. Venomous and irritating setae are sometimes referred to as urticating hairs, as in tarantulas. Some species generally regarded as venomous may not be, because they exude caustic secretions that irritate human skin, but they lack the mechanism to introduce the toxins into the flesh. Conversely, some caterpillars may have urticating hairs, but they do not possess venom. In both cases, and in the case of some truly venomous species, the end result is often the same: dermatitis.

Venom of stinging caterpillars has not been studied to the degree of some other taxa. Much of the venom research has been conducted on species outside our borders, where some more seriously venomous species occur, particularly the South American species of *Lonomia*, which can deliver a fatal sting. However, there are undoubtedly similarities in exotic and domestic species, as many of the same families and genera are involved. For example, there are several studies on South American *Megalopyge*. As a group, stinging caterpillars may have histamine, acetylcholine, formic acid, serotonin, and other chemicals. Kinins may be present. Proteins and enzymes may be present. These may include hyaluronidase, proteases, phospholipase A_2, esterase, and high molecular weight proteins.

Effects of envenomation in humans is called "lepidopterism," but like similar medical terms for envenomation by animals, may not be very useful because of the often dissimilar array of lepidopteran venoms and symptoms—and of course, most Lepidoptera are not venomous, and some may be poisonous—so the term confuses envenomation and poisoning. Most species of stinging caterpillars (Stinging Rose, Smaller Parasa, stinging oak slugs, Saddleback, Hag Moth, Io Moth complex, Stinging Nettle Caterpillar, buck moths) cause moderate pain, swelling, and redness. Saddleback and buck moth caterpillars can also affect nearby lymph glands. The venomous abilities of the Hag Moth caterpillar are not well known. One authority could not seem to get stung or envenomated when he tried. However, other literature says their stings are instantaneously painful and cause dermatitis, though quickly self-resolving. Dermatitis is a frequent symptom of envenomation by many species, usually in the form of a burning, itchy rash. It is not only caused by the venom, but also by physical trauma from the spicules and spines.

The breakaway spicules of some species (e.g., Io Moth complex and buck moths) can also cause health problems when they become airborne. This includes respiratory distress, sneezing, coughing, and irritation in the mouth, throat, esophagus, lungs, and nostrils. If the urticating hairs get lodged in the victim's eyes, they can cause *ophthalmia nodosa*, a potentially serious problem requiring medical attention. People who are frequently exposed often have increasingly severe symptoms over time, and effects seem to be cumulative if the victim stays in contact with airborne spicules.

The most seriously venomous species in the United States are the puss caterpillars. The sting site is usually evidenced by a series of red marks from

Stinging Nettle Caterpillar of Hawaii. *Kim Starr/CCA/Flickr*

the spines. Blisters and welts may form. The wound site may become hemorrhagic and develop into pustules and large, bulbous lesions. Envenomation is instantaneously painful, with a stinging and burning sensation in most victims. Many victims also have swelling, redness, and radiating pain. Lymph glands are often involved, and redness may radiate. In more serious envenomations there may be numbness, fever, headache, nausea, vomiting, increased heart rate, hypotension, chest pain, respiratory distress, acute abdominal pain, muscle spasms, vision problems, shock-like symptoms, and convulsions. No deaths have been attributable to native species, but some authorities believe that may be a possibility, albeit rarely.

Allergic responses or cumulative sensitivity to caterpillar venom is common, especially for those with occupational exposures. Anaphylaxis is not common but has been reported.

References and Resources: Diaz (2005), Eagleman (2007), Elston (2007), Foot (1922), Henwood and McDonald (1983), Hossler (2009, 2010a, b), Kawamoto and Kumada (1984), Langley (2008), Lemaire et al. (1992), McGovern et al. (1961), O'Neil et al. (2007), Pinson and Morgan (1991), Pitetti et al. (1999), Powell and Opler (2009), Stipetic et al. (1999), Tuskes (1996), Wirtz (1984).

Other Terrestrial Invertebrates (Phylum Arthropoda)

"Yes they [Blue Mud Daubers] can sting and will!!! I am deadly allergic to this species. I live in Oklahoma. I have been stung 4 times now over a period of about 10 years. This last time it has taken me over a week of returned visits to my doctor to recuperate, after a 6 hour E.R. visit. Please don't let people think these wasps can't be dangerous or deadly!!!!!" (D. Pruitt, BugGuide online forum, 24 June 2011)

A robber fly with its prey, a large sphinx moth. *L.L.C. Jones*

Classification: Phylum Arthropoda (arthropods).

There are various taxa that can be put into this category of additional stingers and biters, as many other Hymenoptera are capable of stinging, and there are some parasites or other species that can bite. Parasites (except kissing bugs) are not included in this book, except in a sidebar discussion, but all have toxins of some sort. There are also species that may cause irritation on the skin from poisonous compounds, but these are not venoms. This would include blister beetles (order Coleoptera, family Meloidae).

The following bulleted list shows some nonparasitic species that are somewhat noteworthy, although the list is not comprehensive (e.g., there are other ants, bees, and wasps that can sting, but they are generally mild). Examples include the Elongate Twig Ant (*Pseudomyrmex gracilis*), Cactus Bee (*Diadasia rinconis*), a Cuckoo Bee (*Ericrocis lata*), Club-horned Wasp (*Sapyga pumila*), Iridescent Cockroach Hunter (*Chlorion cyaneum*), Scarab Hunter Wasp (*Triscolia ardens*), and Water-walking Wasp (*Euodynerus crypticus*). Some others I feel are worth some mention include:

- **Sweat bees** (order Hymenoptera, family Halictidae). A large family including those of the genera *Dieunomia* and *Lasioglossum*.
- **Leafcutter and mason bees** (order Hymenoptera, family Megachilidae).
- **Digger bees** (order Hymenoptera, family Apidae, tribe Anthophorini and others).
- **Mud daubers** (order Hymenoptera, family Sphecidae and/or Crabronidae). The Blue Mud Dauber (*Chalybion californicum*) is noteworthy among this group (see opening quote and Envenomation Story).
- **Other sphecid wasps** (order Hymenoptera, family Sphecidae and/or Crabronidae). This includes the impressive cicada killers and other species. Species include the Eastern Cicada Killer (*S. speciosus*), Western Cicada Killer (*S. grandis*), and Pacific Cicada Killer (*S. convallis*).
- **Potter wasps** (order Hymenoptera, family Vespidae, subfamily Eumeninae).
- **Short-tailed ichneumons** (order Hymenoptera, family Ichneumonidae, genus *Ophion*). There are eleven U.S. species. Other ichneumons tend to have very long ovipositors and essentially do not or cannot sting.
- **Wood, mound, and field ants** (order Hymenoptera, family Formicidae, genus *Formica*). A large genus with more than eighty U.S. species. Perhaps the best known is the *F. rufa* group, including *F. obscuripes* (Western Thatching Ant).
- **Trapjaw ants** (family Formicidae, genus *Odontomachus*). Six U.S. species include *O. brunneus*, *O. clarus*, *O. desertorum*, *O. haematodus*, *O. ruginodus*, and *O. relictus*. These ants have a rather painful sting

(and bite), but stings are rarely reported and details of symptoms are scarce. *Odontomachus haematodus* is a South American species that is invading the U.S. Gulf of Mexico states, at least from Louisiana to Florida. It is a very aggressive species compared to others of the genus.

- **Asian Needle Ant** (family Formicidae, *Brachyponera* (= *Pachycondyla*) *chinensis*). An introduced species from Asia that has been established in the United States since the 1930s, but has only recently become a pest and a noteworthy health risk.
- **Miscellaneous ants.** Ants of the genera *Hyponera*, *Pseudomyrmex*, and *Tetramorium* have been reported to cause severe allergic reactions in the United States.
- **Robber flies** (order Diptera, family Asilidae). A very large and diverse family in the United States and worldwide. Robber flies may have a bite that is comparable to a bee sting, but little has been reported on the effect of their bites on humans.
- **Other flies.** Larvae of some dipterans have a venomous bite, perhaps most notably horse and deer flies, family Tabanidae.
- **Others?** There are a lot of ants, wasps, bees, and perhaps other invertebrates out there that may sting and cause some degree of harm.

Cicada killer wasps, like this *Pepsis*-mimicking *Triscolia ardens*, are disinclined to sting, and its bark is worse than its bite (sting, actually). *L.L.C. Jones*

These carpenter ants are tending to honeydew-producing scale insects. Carpenter ants bite, and the pain is increased when they spray formic acid into the wound, so technically, they are venomous. *Bob Peterson/CCA/Flickr*

Identification: As in other chapters of miscellaneous biters and stingers, this is a very diverse array of species, some of which are not closely related, so identification should be done with one of the many field guides. Of these groups, all are Hymenoptera of some sort, except robber flies. Reading the sections on other hymenopterans can give a general overview on what these animals look like; for example, some members of the Apidae (true bees), Formicidae (ants), and Vespidae (social wasps) are discussed in the chapters on their more notorious members. To generalize, sweat bees are small bees often of metallic colors that are drawn to human sweat. The megachilids are about the size of honey bees, but are mostly solitary. They don't have pollen baskets but collect pollen on ventral abdominal setae. They are well known for their ability to cut out circles from leaves. Digger bees are similar, but

A bumblebee-mimicking robber fly in Georgia. *Gail Hampshire/CCA/Flickr*

tend to have a hairy thorax and head. Also, they dig characteristic burrows in soil, often in aggregations. Mud daubers are typical wasps, except that they have a very long petiole. They build characteristic mud nests that are usually side-by-side cylinders. The potter wasps also build mud nests, but the chambers look like little urns. Cicada killers are ominous-looking beasts, but their bark is worse than their bite (well, sting actually). They superficially look like a yellowjacket, but may be up to about 5 cm long! Short-tailed ichneumons, on the other hand, do not look terribly threatening, with a slender body and long legs somewhat reminiscent of a crane fly. They tend to be drab brown. The huge family Formicidae has many stinging and biting members, some of which are described in this book. Those in the genus *Formica* are superficially similar to harvester ants, but often build huge mounds, especially in woodland areas. Trapjaw ants are easily recognized upon close inspection by their massive, elongate jaws, which can snap shut. Asian Needle Ants are medium-sized at about 5–6.5 mm long, and relatively slender. They differ from other ants by their inability to climb smooth surfaces, like glass.

Robber flies differ from all of the winged Hymenoptera by having only a single pair of flying wings, as they are true flies. These animals are fairly distinctive as a group, with their characteristic pose on rocks and vegetation. The pose consists of their heavy, long legs holding the head and body well

Although their large, snapping mandibles are impressive, members of the genus *Odontomachus* can also give a nasty sting. *Eli Sarnat/Public Domain/APHIS*

off the substrate, and their long abdomen is pointed back or up. They may be quite large, reaching about 5 cm, although there are many smaller species in this huge family. The mydas flies are similar, but lack the conspicuous mouthparts. Many species of robber flies are mimics of other venomous animals, including bumblebees, yellowjackets, and tarantula hawks.

Distribution: Most of these species are distributed throughout the United States. The diversity seems particularly high in the arid Southwest for most of these groups. Trapjaw ants are more diverse and prevalent in the Southeast. Asian Needle Ants have now spread from the Carolinas and Georgia to

Florida, Ohio, New York, and Maine. Please refer to specific references to find out distributions, although these may not be very precise for the hundreds of species making up this list.

Natural History: Again, for hundreds of species for a large and disparate group of taxa, more in-depth references are in order for anyone wanting to learn about these animals, but here are some highlights. Many species are found in virtually all terrestrial habitats, especially in the southern states. The nesting and feeding habits are rather interesting for many of these taxa. The two types of wasps that use mud to build nests are both solitary hunters, although their mud nests may be aggregated. Mud daubers are spider hunters and Blue Mud Daubers feed on black widow spiders. They will line the inside of their cylinders with their arachnid booty. One colleague mentioned how he likes to demonstrate their hunting prowess by slicing open their mud cylinders and watching large numbers of dead or paralyzed spiders fall out. Potter wasps are similar, except they parasitize caterpillars. Ichneumons are also caterpillar predators. Not surprisingly, the cicada killers kill cicadas—because these are large insects, the wasps are also large to handle such prey. They live in burrows that may be as large as a small rodent's. Digger bees are also burrowers that line their holes with a waxy substance to help stabilize it. Aggregations of digger bee burrows can be fairly dense in some areas. Leaf-cutter bees usually nest in gaps and natural cavities, such as those in stems of vegetation, and mason bees may also use mud to help make their nests. Some sweat bees are eusocial insects that nest in woody materials. All are extremely valuable pollinators. The *Formica* ants have very large mounds, which are often covered by workers. They are eusocial, and hundreds of thousands of ants may make up the nest colony, depending on species. They typically feed on honeydew from aphids, but will feed on a variety of other insects. Trapjaw ants nest under rocks, logs, or in grasses, and the sites may be mesic (e.g., mountains or along streams) to xeric (e.g., in desert areas). They are effective predators with their highly specialized mandibles that can snap shut with significant force—in fact, it has been estimated they have the strongest and fastest bite force in the animal kingdom, although the bite force of a solpugid is also said to hold the title. But it is their sting that is venomous. Asian Needle Ants prefer mesic areas for nesting, but this ranges from logs in the forest to homes and schools. Robber flies are voracious predators that use a venomous bite to subdue insects, even in flight.

Mud nest of the Organ Pipe Mud Dauber Wasp. At least some mud daubers can deliver a painful sting, but most are mild and rarely sting. *Bart Everson/CCA/Flickr*

Encounters: While most of these taxa are fairly common in our environments, most are not inclined to bite or sting. Many species of ants are common in and around homes and yards and some species will guard the nest. Asian Needle Ants may be locally abundant, but do not swarm to protect their nest. They usually sting when accidentally pressed against the skin. Trapjaw ants are not likely to be encountered by most people and are generally considered uncommon. Sweat bees may be locally abundant and often sting people, especially those who try to wipe them off their skin, but the sting is minor, only rating a 1 on the Hymenopteran Sting Pain Index, although the giant sweat bees may reach 1.5 of 4 on the scale. Digger bees may also be locally abundant. Their stings are more painful, but they are generally not inclined to sting. People walking on their burrows or brushing against them, however, have a chance of being stung.

Most species of mud daubers are uninclined to sting, but some researchers have shown that Blue Mud Daubers are not so docile when their nest is disturbed. You pretty much have to encourage most mud daubers, sphecids, potter wasps, ichneumons, and robber flies into envenomation. Entomologists are the most at-risk group due to handling these animals, but the casual person can also be stung when the creatures feel threatened, as when one sits on them. Robber flies are extremely common and one can easily go a lifetime without being bitten, but if you are unaware of their venomous nature, you may be at risk. For example, robber flies commonly wind up in outdoor pools. You should not remove them from the pool with your bare hands—use a scoop or gloves.

Venom Apparatus, Venom, and Symptoms: The stinging apparatus has been covered in the chapters of the various Hymenoptera. Only the females sting, because the stinger is a modified ovipositor. Males of some Hymenoptera have a false stinger. It is a sharply pointed terminal sclerite that can be jabbed into the assailant. Field ants are a bit different than other ants because rather than injecting venom with their stingers, they bite the assailant and spray venom into the wound. This novel apparatus for getting venom into predators is said to be quite effective. The robber flies are predators and different from hymenopterans because they bite, as do many parasitic flies, such as horseflies and mosquitoes. They possess venom glands (likely in both the thoracic and labial glands) and have biting mouthparts that can inject venom into predator or prey. The short, stout proboscis is easily seen and can be used to differentiate robber flies from the similar mydas flies, which don't bite humans. Robber flies have extraoral digestion, so saliva and stomach contents are injected into prey primarily for the purpose of paralyzing and digesting insects. Parasitic flies, on the other hand, may take bits of flesh from us or drink our blood, as they actually feed on us.

The venom of bees is completely defensive. Venom of some wasps and ants, and all robber flies is primarily for subduing prey, but is also used for defense. The non-native trapjaw ant, *O. haematodus*, is notorious for defending its nest and will aggressively sting people repeatedly. Humans can be stung numerous times before they realize they've been stung the first time. Among some trapjaw ants studied (mostly in the Neotropics) the venom contains organic proteins, lipids, and vasoactive amines. Among these proteins are the enzymes phospholipase A_2 and hyaluronidase. Asian Needle Ants have an antigen 5, known as Pac c 3, that is highly allergenic, and it has been reported that they are about four times more likely to cause anaphylaxis than other species. The chemical composition of the venom of most of the other species in the list are poorly known, although one could surmise they are similar to their relatives. Robber fly venom is not well researched, but at least in some species, it has neurotoxic and proteolytic components that may affect both invertebrates and vertebrates. The species that seem to affect humans the most are larger species, like the giant robber flies (genus *Promachus*). In one study, the venom from the salivary glands of one or two robber flies was able to kill a 20 g lab mouse and 1/126 of one-half gland was enough to kill a 1 g cicada. Lab mice injected with the venom *Promachus* spp. exhibited lethargy, bristling hairs, labored breathing, and body contractions.

Of the species in our list, the digger bees, Blue Mud Daubers, field ants, trapjaw ants, and robber flies can actually produce a sting or bite about the same as a honey bee. The sting of trapjaw ants are said to be even more painful than these hymenopteran gold standards, but are short-lived. As best I can tell, the primary symptom in humans from all species is pain on a 1–2.5 level on the 0–4 Hymenopteran Sting Pain Index, often with some redness, itchiness, and swelling. The sting of an Asian Needle Ant is reported to be a little different, heralded by some slight burning that gradually expands to the size of a quarter, with needle-like sensations being felt. Sensitivity can last a couple weeks, bringing on flashes of pain when the area is touched. Sweat bees are a solid 1, while some trapjaw ants are 2.5. As mentioned for Asian Needle Ants, there are cases of sensitivity and allergy

The invasive Asian Needle Ant is increasing its North American distribution.
Animal Diversity Web/CCA

Digger bees, like this *Anthophora curta*, are unlikely to sting but will if pressed against the skin. However, this is a male, so it can't sting. *Alexis Roberts/Public Domain/ Univ Texas, Austin*

that can sometimes be severe, including anaphylaxis. The bite of horsefly larvae is often an occupational hazard, and it has been reported to produce a very painful bite. Its venom has neurotoxic and lytic properties. In reports from Japan, the pain lasts ten minutes to two days and causes intense itching, sometimes including swelling and lymph gland involvement. There are other fly larvae known or thought of as being venomous, but there is little information on the effects to humans.

Remarks: The Asian Needle Ant has gone from being a relatively innocuous, locally distributed non-native species to a notorious invasive species in the past decade or so. Not only are the stings highly allergic, but they are transforming ecosystems by displacing native ants, fueled by climate change and other factors.

References and Resources: Arnett (2000), Artigas and Papayero (1997), Borror and White (1998), Eaton and Kaufman (2007), Irving and Hinman (1935), Johnson and Triplehorn (2004), Kahan (1964), Pauchard et al. (2016), Rice and Silverman (2013), Rodriguez-Acosta and Reyes-Lugo (2002), Schmidt (1982, 1986), Schmidt et al. (1986), Silva et al. (2015), Smith (1982), von Reumont et al. (2014a).

8 Terrestrial Vertebrates

When we speak of venomous terrestrial animals, one group immediately pops into our minds: reptiles. There are two to several types of venomous lizards, depending on your definition of venom, but the vast majority of venomous reptiles are their legless relatives. Snakes are the stuff of nightmares for some people, and a source of fascination for others. Snakes

Mexican Beaded Lizards occur naturally within a few hours south of Arizona, but they are also esteemed exotic pets. Although docile in captivity, accidents happen and owners sometimes get bitten. *L.L.C. Jones*

and lizards are very closely related; both are placed in the same order, Squamata. In fact, some lizards are legless, while some snakes have limb vestiges. The presence of limbs is the primitive condition, so early squamates had limbs, and as some evolved to exploit the underground environment, or other nooks and crannies, they evolved a loss of legs.

There are more than 6,000 species of lizards worldwide. The five species of Helodermatidae (Gila Monster and beaded lizards) are venomous. Some researchers also consider the Varanidae (monitor lizards) to be venomous. The bite from a helodermatid will leave no one in doubt that they are venomous. The Gila Monster is endemic to the American Southwest and adjacent Mexico, where it is considered an icon of the eastern Sonoran and Mojave Deserts. The other species occur from northern Mexico to Guatemala. Helodermatids are easily recognized by their large size and osteoderms.

There are nearly 3,500 recognized species of snakes. Of these, about 600 are considered venomous, and it is sometimes stated that only two hundred of these are medically important. However, these numbers are difficult to agree upon, because of various definitions of what constitutes a venomous snake and what constitutes medical importance or significance. In North America, venomous snakes belong to the families Elapidae (coralsnakes, cobras, and their relatives) and Viperidae (vipers), plus a few members of the family formerly known as Colubridae (typical snakes). Sea snakes are sometimes placed in their own family, Hydrophiidae, but most researchers consider them to belong to a subfamily of the Elapidae, Hydrophiinae. Vipers also have two subfamilies, the true vipers (Viperinae) and the pit vipers (Crotalinae). The former family Colubridae is polyphyletic and the group is used as a depository for some species of uncertain taxonomic placement. However, the main groups of venomous species have been placed into three families (Colubridae, Dipsadidae, and Natricidae), sometimes still regarded as colubrid subfamilies.

In the United States, there are only three terrestrial elapids, and all are coralsnakes. There are two genera, *Micruroides* and *Micrurus*. These are slender, cylindrical, active snakes that often sport vivid aposematic colors that form rings around the body. All of the U.S. species have the pattern of red-yellow-black rings encircling the body, but the rhyme, "red touch yellow, kill a fellow," isn't entirely accurate. There are numerous Batesian mimics of coralsnakes. Elapids have a proteroglyphous skull morphology, with fixed, front fangs. Coralsnakes account for a small percentage of bites by venomous snakes in the United States.

Pit vipers are highly evolved at venom delivery, having a solenoglyphous venom apparatus. *Paul Condon*

Venom apparatus of a pit viper. *L.L.C. Jones*

Most venomous snakebites in the United States are caused by the Crotalinae. There are two types of pit vipers—those with rattles and those without. All are heavy-bodied snakes that are generally slow moving, although their strike is rapid. Those without rattles are called moccasins, and belong to the genus *Agkistrodon*. There is disagreement about the number of species and subspecies in the United States, so to simplify things, I discuss taxonomy in the copperhead and cottonmouth accounts. These are well-known vipers of the eastern United States. There are other species of *Agkistrodon* in Mexico and Central America.

Like the Gila Monster, rattlesnakes are also an icon of the Southwest, where they reach their highest diversity, although at least one species occurs in most areas of the United States and some range into southern Canada. There are about thirty species and subspecies in the United States and more in Mexico, and some reach the Neotropics. They are a taxonomically challenging group, so there is some disagreement about species designations. Those of the former "Western/Prairie" Rattlesnake species complex, in particular, are confusing in terms of taxonomy, and scientific and common names. Rattlesnakes are the only serpents in the world that have rattles. They add a rattle segment every time they shed their skin, and segments may break off, so the length of the rattle is not an indicator of age. There are two genera, *Sistrurus* and *Crotalus*, which can be differentiated by their head scales and other features. Species of *Sistrurus* have large shield-like scales on the top of the head, and tend to be small with small rattles, while *Crotalus* has small scales on top of the head, and most species are larger with larger rattles (there are exceptions). Rattlesnakes and mocassins have heat-sensing pits; a solenoglyphous skull morphology, with large folding fangs in the front of the upper jaw; and vertical pupils, a feature that a few other taxa also have.

The colubrids and their relatives are a huge group of nonvenomous to mildly venomous species, although there are a few species outside North America that are highly venomous. The venomous U.S. species are discussed in the single chapter. Most of the mildly venomous colubrid-like snakes are rear-fanged, having an opisthoglyphous skull morphology. Others lack fangs, having an aglyphous skull morphology.

In addition to reptiles, a handful of mammals are venomous, including a few species of North American shrews. There are very few venomous amphibians (none in North America), and no venomous birds, by the way.

Snakebite is a major health concern, as all bites by elapids and pit vipers should be considered a medical emergency. Elapids and some crotalids

Pit vipers and Gila Monsters have a male-male combat ritual, as shown here with Panamint Rattlesnakes. *Rich Gassaway*

Mating Western Diamond-backed Rattlesnakes. *L.L.C. Jones*

Snakes are often maligned and killed; many people are envenomated by snakes they are trying to dispatch, or by those presumed dead. *L.L.C. Jones*

Not only are pit vipers just trying to survive, they are excellent rodenticides and help in disease control. *John Murphy*

Pit vipers can open their mouths widely and strike quickly at a perceived threat. *Bill Love*

have a predominantly neurotoxic bite, whereas most pit vipers have a predominantly tissue-destructive bite. The latter is more commonly known as a "hemotoxic" bite, but this is a problematic term, as these destructive toxins affect more than blood. However, the term is often legitimately used and much of the literature refers to hemotoxins, and I follow suit with much of what is reported in the literature. Also, many snakes have a combination of two or more of the following components in their venom: neurotoxic, hemotoxic, cytotoxic, myotoxic, and cardiotoxic. It is important to know if one is bitten by a coralsnake or a pit viper, to know which polyvalent antivenom is needed. There is no antivenom available for the Sonoran Coralsnake, but this species does not have a large venom yield and serious cases are unreported. While it is somewhat important to know the type of pit viper one is bitten by, it is more important for a physician to use polyvalent antivenom, and monitor all body systems to counter any symptoms that arise, including allergy, serum sickness, and secondary infection.

There are many references to aid physicians in snakebite treatment options and how to administer antivenom, as well as general information on venomous snakes throughout North America. Some of these references are given below. Many states and regions have one to several excellent field guides with general information on North America's venomous and nonvenomous species.

Red Diamond Rattlesnake. *L.L.C. Jones*

Snakebite and Antivenom References: Boyer et al. (2015), Bush et al. (2002), Cannon et al. (2008), Carroll et al. (1992), Clark et al. (1997), Dart and McNally (2001), Dart et al. (1997, 2001), Degenhardt et al. (1996), Fry (2015), Gold et al. (2002), Heard et al. (1999), Hill et al. (2001), Jones (2016a), Lavonas et al. (2011), Mackessy and Baxter (2006), Minton (1987), Rubio and Keyler (2013), Offerman et al. (2002), Russell (1980), Seifert et al. (2009), Walter et al. (2009), White (1991).

General Pit Viper and Select Regional Reptile References: Aldridgeand Duvall (2010), Behler and King (1979), Campbell and Lamar (2004a, b), Conant and Collins (1991), Degenhardt et al. (1996), Ernst (1992), Ernst and Ernst (2011a, b), Greene (1997), Hammerson (1999), Hayes et al. (2008), Hubbs and O'Connor (2009), Jones et al. (2016a, b), Klauber (1972), Lemm (2006), Lowe et al. (1986), Mackessy (2010a, b, c), Minton and Minton (1969), Powell et al. (2016), Rubio (1998, 2010, 2013), Schuett et al. (2002a, 2016a, b, 2018), Stebbins (2003), Stebbins and McGinnis (2012), Werler and Dixon (2000), Wright and Wright (1957).

Self-Immunization

Self-immunization for snake venom (SISV) is a rarely used practice that involves injecting oneself with snake venom, either through a syringe or sometimes via the snake itself. Practitioners say it is done to help build up antibodies against snake venom and possibly boost the body's immune system. It is essentially the same principal as making antibodies in horse serum for antivenom or becoming desensitized to insect venom by immunotherapy. People who practice this often cite the Bill Haast example. Bill milked thousands of venomous snakes for the Miami Serpentarium Laboratories, and was bitten 172 times, but he lived to be one hundred years of age and rarely got sick. To some, this suggested he had not only built up resistance to snakebite, but he may have boosted his own immune system in general. There are some practitioners out there that are outspoken on the benefits of SISV. However, without going into detail, most authorities consider SISV a dangerous practice.

There are several potential issues with SISV, which are frequently discussed on blogs, forums, and in videos. First, SISV has never been scientifically tested. Opinions, observations, and anecdotes do not constitute the scientific method, such as the "gold standard" placebo-controlled randomized clinical trial. Because there have been no controlled and repeatable experiments, SISV is done rather haphazardly by snake fanciers and thrill seekers in their homes. There are too many variables to control to make these observations useful. As you see repeatedly in this book, venom virulence, composition, and yield vary greatly by population, species, behavior, and age of the snake. While practitioners may build up antibodies, they could become sensitized to venom through prolonged exposure. There is a fine line between sensitization and immunological therapy, especially outside medically controlled situations. Although I found no fatal cases of overdose from SISV, that possibility remains. However, there was a recent case of fatal envenomation when a man tried to immunize a woman's toddler with rattlesnake venom.

Although SISV should never be attempted at home, it might be prudent to test snake immunization in a medical facility, using proper clinical trials, to see if the predicted benefits occur. For example, immunization might be beneficial for someone with an occupational hazard to snake-venom exposure.

Gila Monster (*Heloderma suspectum*)

Case report of circumstances and results of a bite: "Placed Gila Monster on his shoulder, it bit him on the neck. The Gila Monster fell to the ground, he picked it up, placed it in a hat. It bit him through the hat." Results: "abdominal cramps, edema, edema of airway, tongue swollen on its sides. Admitted to non-critical care unit." (French et al. 2015)

Classification: Phylum Chordata (chordates); Class Reptilia (reptiles); Order Squamata (snakes and lizards); Family Helodermatidae (beaded lizards and Gila Monster); Genus *Heloderma*; Species *suspectum* (Gila Monster).

There are five species currently recognized in the family, all in the genus *Heloderma*. There are four species called beaded lizards, and the sole representative in the United States is the Gila Monster. Helodermatid fossils date back at least 23 million years. They belong to an ancient suborder of lizards (suborder Varanoidea) that is more closely related to snakes than most other lizards, except the Old World monitor lizards (family Varanidae).

Identification: Gila Monsters are typical of lizards in that they have a head, neck, torso, four legs, and a tail, but they are easy to identify and differentiate from all other lizards. The head is robust and the legs and tail are thick

Banded Gila Monster on red sand dunes in Utah's Mojave Desert. *BLM/Public Domain*

and short. As adults, they are the largest native lizard in the United States, in both length and mass. They reach a length of about 570 mm total length. Gila Monsters are also very robust, weighing between 450–900 g. They are readily recognizable by the aposematic coloration of yellow to orange or pink markings alternating with black. Two subspecies are sometimes recognized (Banded, *H. s. cinctum*, and Reticulated, *H. s. suspectum*), although variation makes recognition of these as subspecies questionable. The generic name *Heloderma* is Latin for "studded skin," referring to the small, granular scales, which are actually skin containing osteoderms. Unlike other lizards, Gila Monsters have forked tongues like their relatives, the monitor lizards and snakes. Gila Monsters are not sexually dimorphic. Hatchlings always have the banded pattern of *H. s. cinctum* and as they mature, they may develop the reticulated pattern of *H. s. suspectum*. Neonates are about 16 cm long.

Distribution: In the United States, Gila Monsters are endemic to the Southwest. Their distribution centers on southern and western Arizona, but they are peripherally found in adjacent California, Nevada, New Mexico, and Utah. They are also found in the Mexican state of Sonora, and have not been reported from Baja California. Their close relative, the Mexican Beaded Lizard

The skull of a helodermatid is covered with bony osteoderms. *L.L.C. Jones*

Reticulated Gila Monster in the Sonoran Desert. *L.L.C. Jones*

(*H. horridum*), is also found in Sonora, and the two species co-occur in the area of Alamos. The other species of *Heloderma* are all found in Mexico, and one species, *H. charlesbogerti* (Guatemalan Beaded Lizard), occurs in Guatemala. Assuming subspecies are recognized, the Banded Gila Monster is found in western Arizona and adjacent California, Nevada, and Utah, while the Reticulated Gila Monster is found in southeast Arizona and adjacent New Mexico.

Natural History: Gila Monsters are primarily found in the Sonoran Desert, especially in an area called the Sonoran Desert Upland, characterized by giant saguaro cacti and palo verde trees. They are also found in parts of the Lower Colorado Desert subdivision of the Sonoran Desert and parts of the Mojave Desert. They also occur in semi-desert grasslands at the transition to the Chihuahuan Desert. They seem most frequently encountered in foothill areas of desert mountains, and often in or near washes. In Sonora, they are also found in foothill thorn scrub.

Gila Monsters are often considered rare because they are infrequently encountered in the wild. Some people living in Phoenix and Tucson, the two large metropolitan areas within its range, have never seen one—perhaps because urbanites may not spend much time in the wilds, but also because *H. suspectum* tends to spend up to 95% of its life underground. Because of the presumed rarity, and the potential for poaching animals for the black market

Juveniles of both subspecies have a banded pattern. Even hatchlings will defend themselves. *L.L.C. Jones*

exotic pet trade, Gila Monsters are protected by law in every state they occur in, as well as in Mexico.

Surface activity is usually in the spring, when they are diurnal, or in the summer (the monsoon period), when they are largely nocturnal. However, during the monsoon, they may also be surface active on warm, overcast, humid days, or following a thunderstorm. This activity pattern is largely driven by temperature. They do not tolerate heat as well as most other desert lizards and do not spend a significant amount of time basking, as most other lizards do. They maintain a body temperature below about 30°C. They have various adaptations for staying cool and retaining water, both physiologically and behaviorally. The reason for such a reduced period of surface activity is because they have an incredibly low metabolism and are active only during peak breeding and foraging times. In one study in Utah, they were active during about ten days per month during the active season (primarily March through September). They time their activity to intersect with the breeding of ground-nesting birds and small mammals, upon which they feed. Their movements are slow and precise; they have been clocked at a maximum speed of 1.7 km/h. They are simply not able to take on active prey and are not sit-and-wait predators, so they specialize on eggs and nestling mammals. Their venomous nature, body armor, and aposematic coloration fit well with this lifestyle; otherwise, they would be easy targets for potential predators.

Gila Monsters mate in the spring, after winter emergence. Males seek out females, often at their burrows. Males have a highly ritualized combat display when competing for mates. This is similar to their relatives the varanids and pit vipers. Females lay 2–12 (average 5–6) eggs per clutch in an underground nest. Oviposition is usually 6–8 weeks after mating, between June and August. Hatchlings don't appear on the surface until April at the earliest, but May to August is more typical. Hatchling activity is largely nocturnal and loosely coincides with the North American Monsoon, when the climate is moderate and food resources plentiful. The eggs hatch in the fall, but hatchlings overwinter in the nest, emerging in the spring or summer when they will have a bounty of prey to sustain them. Nests are rarely found in nature. They are long-lived animals. Large adults can be at least twenty years old. In captivity, they have been known to exceed thirty years of age.

Encounters: To the general public living in Gila Monster country, encounters are few and far between. Some southern Arizonans have gone their whole lives without seeing one in the wild, but these are not usually outdoorsy people. Gila Monster researchers may have to work hard to find a sample of study animals. In my lizard study area near Safford, Arizona, a good area for Gila Monsters, of nearly 15,000 lizards recorded, only five were Gila Monsters. This translates to about one Gila Monster for every 3,000 other lizards or one Gila Monster seen for every hundred hours of driving in good habitat. Another study near Tucson has about the same encounter rate. However, in one particular hotspot near Tucson, it only takes about five hours of driving during good conditions, in what may be optimal habitat to detect a Gila Monster. Gila Monsters are not attracted to homes, garbage, or other human interfaces, although they may be attracted to moist and lush areas, as in large yards at the fringes of urban domiciles or golf courses. They sometimes show up in swimming pools. They are most often encountered in remote areas. Encounters are usually accidental, as when one is hiking or mountain biking. When these animals are seen, they are usually quite obvious.

In a recent study of 105 bites (70 unique bites were investigated) in Arizona between 2000 and 2011, based on calls to poison control centers, most were from intentional handling. Most of the victims were males (83%) and most bites occurred on the extremities (77%). This is the same stereotypical victim and placement of snakebite, though that demographic is changing, but pure accidents do rarely happen. For example, a landscaper was bitten while moving a rock, a man was bitten while working on his furnace, and one person stepped on an unseen Gila Monster. There were also other interesting examples of circumstances that led to envenomation, including people "saving" someone else from a potential bite, trying to remove the lizard as it was biting someone else, handling by researchers and zoo personnel, removing the lizard from a hazard (e.g., pool, road, rodent trap, and humans attempting to kill it), photographing the animal, and handling an animal believed to be dead.

Many herpetologists and hobbyists are not hesitant to handle Gila Monsters. Gila Monsters are generally inoffensive and usually have slow movements. When caught in the wild, they are less tolerant of handling. In captivity (which is illegal without special permits), they can become quite docile. However, people get caught off-guard and make mistakes, or an external stimulus can arouse the animal. Some people are bitten when a captive Gila Monster thinks it is feeding time. These lizards can move surprisingly fast to twist and bite someone viewed as a threat. The best precaution is to not handle the animal.

If bitten, the $64,000 question (which could be a hospital bill) is how to remove them. Sometimes they will have a slashing bite or bite and release, but they are notorious for their tenacity and hanging on for an extended period with their powerful jaws. The longer they hold on, the more venom is introduced into the victim. A common method includes putting the animal down on the ground, so it feels it can escape if it lets go. If the animal still does not let go, squirting alcohol or other irritants into the mouth and face may be effective. The more physical approach is to try and pry it off, or in desperation, rip it off. There are examples of lethal removal, but I cannot support this if other methods can be successful. There is actually little information on what to do, but I would suggest putting the animal on the ground and allow it an opportunity to escape. If the lizard has no intention of letting go, then prying the jaws apart is probably the next best option, but if not done correctly, this may stimulate the animal to bite down harder. To pry, two strong objects should be used to push the jaws apart, away from the victim, so in most cases, level-headed people need to assist. Never use hands to pry the jaws apart. People attempting to help remove a Gila Monster from someone else have also been bitten in the process. The best bet, of course, is to not find oneself in a situation to make that sort of decision; instead, treat these magnificent animals with respect.

Venom Apparatus, Venom, and Symptoms: The venom apparatus includes the teeth and venom glands, which are modified salivary glands. The venom glands are composed of three or four independent lobes on each side of the lower jaw. Externally, the lower jaw in this area appears swollen due to the presence of the glands. The venom is produced in columnar cells in alveoli of the venom glands, and then excreted to the surface, inside the mouth, via a lumen that empties into a venom duct between the lower teeth and lips. There is one venom duct per gland lobe and these are concentrated in the midsection of the mouth. The dentary teeth are the most specialized for introducing venom, but venom is in the animal's mouth, so any of the teeth can introduce it. The dentaries are especially sharp, recurved, and grooved. During or after a bite, the animal may lose teeth, but they will be replaced. When the animal bites a victim, venom is released into the area between the teeth and lips, and then flows into the grooves via capillary action. Venom is not controlled by constricting musculature around the glands; rather, as the jaws are moved, venom is expelled into the area of the dentaries.

Venom is used for defense rather than overpowering prey. However, it is possible that some of the venom components (e.g., exendin-4) are used to aid or regulate digestion during predation. The venom has been well studied. It contains numerous components, including serotonin, hyaluronidase, protease, phospholipase A_2 (one study examined five variants), and novel proteins and peptides named after the animal itself, including helodermin (exendin-2), helodermatine, helospectin I and II (exendin-1), gilatoxin (a kallikrein-like protease), and gilatide. Exendin-4 is a peptide that has been extensively researched for its utility in the control of Type II diabetes in humans. Exenatide (a synthetic compound) is a name for a diabetic control compound (Byetta—Amylin; Lilly). Exendin-3 is similar, found in the Mexican Beaded Lizard. There is another peptide that is lethal in mice, and is presumably the most virulent component of the venom in Gila Monsters, with an LD_{50} of 0.135. The venom targets endotherms (birds and mammals), being largely inactive in ectotherms. Gila Monsters are resistant to their own venom. The LD_{50} of whole venom has been studied by several researchers and ranges from 0.4 to about 4 mg/kg, with the average about 2.7 mg/kg.

The venom is highly algogenic. A yarn is that a Gila Monster bite is lethal—but only if the victim has a gun nearby! Beyond the excruciating pain, bites from Gila Monsters can have serious systemic consequences, although most do not require hospitalization. Symptoms are invariably local, but can be systemic, especially if large amounts of venom are introduced. The usual symptoms are excruciating pain, local swelling, redness, weakness, dizziness, and nausea. There can also be increased or irregular respiration, sweating, marked and rapid fall in blood pressure, rapid heartbeat, hypotension, vomiting, hypothermia, and hemorrhaging in the eyes and internal organs (intestines, kidneys, lungs, and liver). Some victims experience airway obstruction, a possibly lethal situation. Studies in other mammals include a host of other systemic reactions, including paralysis and convulsions.

In nonhuman mammals studied, death usually comes from respiratory distress, but cardiac complications are common. Myocardial infarction has been recorded in a case study of a human that was bitten while handling a Gila Monster. Whether or not Gila Monster bites can cause death has been the subject of debate. There are published records of fatal injuries, but these reports are old and may be questionable (e.g., death may have been due to causes other than direct effects of the venom). Some have suggested that bites could not be fatal in a healthy human adult, but recent evidence of airway blockage in some individuals, plus the potential for myocardial or respiratory complications, suggest a bite could cause a potentially life-threatening situation, albeit rarely.

Remarks: There is no antivenom for Gila Monster envenomation. There is little information on envenomation by beaded lizards, but it happens occasionally, at least to zoo personnel and hobbyists. One case report of envenomation by this species showed signs of severe envenomation, including systemic reactions and anaphylaxis.

References and Resources: Beck (2005, 2009a, b), Bou-Abboud and Kardassakis (1988), Brown and Cantrell (2003), Carmony (1999), Christel et al. (2007), DeNardo et al. (2004, 2018), Eng et al. (1992), Ernst (1992), French et al. (2015), Goldberg and Lowe (1997), Gomez et al. (1989), Hooker and Caravati (1994), Lovich and Beaman (2007), Malhotra et al. (1992), Mebs (1970), Patterson (1967), Piacentine et al. (1986), Raufman (1996), Russell and Bogert (1981), Strimple et al. (1997), Triplitt and Chiquette (2006), Tu and Murdock (1967), Vandermeers et al. (1991).

Copperheads
(*Agkistrodon contortrix* and *A. laticinctus*)

"The generalization of copperhead snakebites as mild or of benign clinical significance should be reconsidered." (Scharman and Noffsinger 2001)

Western Copperhead, Trans-Pecos form, from western Texas. *L.L.C. Jones*

Classification: Phylum Chordata (chordates); Class Reptilia (reptiles); Order Squamata (snakes and lizards); Family Viperidae (vipers); Genus *Agkistrodon* (in part; moccasins); Species *contortrix* (Eastern Copperhead) and *laticinctus* (Western Copperhead).

There are two groups of moccasins in the United States, copperheads and cottonmouths. Although congeneric, cottonmouths are quite different in appearance from copperheads, and differ in ecology and venom attributes.

There was one species and five subspecies of copperheads recognized until recently: *A. c. contortrix* (Southern Copperhead), *A. c. laticinctus* (Broadbanded Copperhead), *A. c. pictigaster* (Trans-Pecos Copperhead), *A. c. mokasen* (Northern Copperhead), and *A. c. phaeogaster* (Osage Copperhead). Recent genetic research, however, suggests there are two species: a western clade (*A. laticinctus*) and an eastern clade (*A. contortrix*), with no subspecies recognized. The western clade would include the former *A. c. pictigaster* and *A. c. laticinctus*, and the others would be in the eastern clade, except that the Osage Copperhead would largely be a hybrid form. Using "Broad-banded Copperhead" as a standard English name for *A. laticinctus* (as proposed by Crother 2017 and the Center for North American Herpetology), is problematic, due to the same name being used for the earlier taxon with the same "standard" name—but it is a different geographic entity. Hence, I refer to the new *A. laticinctus* as the "Western Copperhead," a novel name to avoid confusion. I concur with *A. contortrix* being called the Eastern Copperhead, as that name was not previously used for any copperhead taxa. While discussing the taxonomy of copperheads with some herpetologists, they are not all on board with the proposed "standard lists" mentioned above for *Agkistrodon*, so the overlap of names among different geographic entities leads to taxonomic and nomenclatural confusion. Without getting too wrapped up in this controversy, especially given broad hybridization (or intergradation) zones, for our purposes, we can consider all copperheads to be essentially the same type of animal, albeit with subtle differences.

Identification: Copperheads are small to medium snakes. Adults are generally 45–60 cm, although the longest recorded individual was 132 cm. They are heavyset and have a wedge-shaped head, distinctly larger than the neck. Like other moccasins, copperheads lack rattles, but are otherwise similar to rattlesnakes. Neonates are about 16–25 cm. The name "copperhead" comes from the coloration on the top of the head. They are beautifully marked animals with broad bands of tan to pink earth tones with darker bands of various hues of brown (e.g., reddish, chestnut, or cinnamon). The former subspecies were distinguished by their color patterns and ranges, but the Western and Eastern Copperheads also differ phenotypically. *Agkistrodon laticinctus* has relatively uniform dark and light bands, while *A. contortrix* has dark and light alternating saddles that are somewhat hourglass-shaped. Young copperheads have a yellow-tipped tail, as do young cottonmouths and some young rattlesnakes.

Distribution: Copperheads are primarily denizens of the eastern and central United States. The Eastern Copperhead is found from east Texas to Iowa and east to Massachusetts and the eastern seaboard, although most of Florida is unoccupied. The Western Copperhead is found from the Trans-Pecos of Texas and adjacent Mexico east to east Texas and north to Kansas, and possibly Nebraska. The broad hybrid zone is from east Texas north to east Nebraska; the Osage pattern class predominates in eastern Kansas and Nebraska.

Natural History: In general, copperheads are primarily found in deciduous woodlands with downed logs, branches, and leaf litter, or in longleaf pine habitats, often in mesic habitat, especially near streams and other wetlands. In the Trans-Pecos area, they can also be found in open creosote bush habitat. Those in the Midwest are often found in prairies, but are also most abundant in moist canyons and riparian areas.

Copperheads are cryptically colored, blending in with leaves and dappled sunlight of their woodland surroundings. They are often found in a resting-hunting position, curled up near a log, where they hope to intercept prey using the log as a travel corridor. When not surface active, they may be found under any number of objects, such as logs, slabs, rocks, and debris piles. Their immobility saves energy and allows them to go unseen, both by predator and prey. Copperheads can be on the move when seeking mates, hibernacula, and hunting grounds.

Western Copperhead, Broad-banded form from central Texas. *L.L.C. Jones*

Daily surface activity is dictated by the weather, season, and latitude. Northern populations are adapted to cooler environments and a shorter active season than those in the South. The preferred body temperature of copperheads is between about 20°C and 30°C, so they are most active during similar temperatures. They may be active in cooler temperatures when warming up. They emerge from brumation in the spring and may bask in the sun to gather warmth to gain energy for their daily activities. In general, they are often most active at dusk and dawn. During warm spring rains, there is both warmth and moisture, so they tend to be diurnally surface active. They can be nocturnal when night temperatures are high enough to allow activity. In October or November, they seek out hibernacula, which are often in rocky escarpments on hilltops or ledges. In areas that lack rocky habitat, copperheads usually overwinter in or under logs, woodpiles, tree stumps, and mammal burrows. They may come out to bask during the winter during exceptionally warm, sunny days. Although they are not social animals, they may aggregate in areas that have good conditions for brumation.

Males may engage in combat, particularly in the spring. Copperheads give birth to live young. Mating occurs in the spring and/or fall. Those at the northern extremes only mate in late summer to early fall. Pregnant females may congregate in areas that provide good shelter from predators. They give birth in August and September in the South. Litters can range from a few to a dozen. At birth, they are ready to defend themselves and feed on small prey. Neonates use their bright yellow tail tip to lure in small animals on which they feed. They wriggle their tail tips to mimic a caterpillar to attract small frogs, toads, lizards, and shrews. Larger individuals have a wider variety of choices: they mostly feed on mammals, but will also eat amphibians, reptiles, and even insects. When in season, cicadas seem to be a preferred prey.

Encounters: Copperheads can be extremely abundant. In many areas they are the most common venomous snake, and in some of these areas they are the most common snake species. Non-brumating populations as high as 49.4 individuals per hectare have been documented in Kansas, but typical densities are considerably less. Stories of ten to thirty-five individuals per night encountered by snake-watchers are not uncommon, but not typical. Even when copperheads are not on the prowl, they can be surprisingly abundant under surface objects. A single debris pile can provide shelter for several individuals. They can be similarly abundant in areas of hibernacula. Even the "rare" Trans-Pecos pattern class can be numerous under certain circumstances. In one well-known case, a herpetologist collected fifteen individuals from an area of cane grass along the Rio Grande in one night, and many more were seen.

Copperheads do not shun large recreation areas such as forested parks, vacant lots, or even suburbs. For example, some suburbs of Houston provide good habitat. Unfortunately, animals living in populated areas tend to be wantonly killed by humans. Copperheads seek refuge in any manner of surface objects, natural or otherwise, including logs, rocks, old dilapidated buildings, wood piles, fallen fence posts, railroad ties, and debris piles such as carpeting, roofing, and lumber. Abandoned sawmills and woodlots provide good sources of shelter. While copperheads may seek out human debris and disturbed areas, they tend to be most common in natural habitats, like forests and woodlands. They are commonly encountered in rocky areas and those with wood debris, such as fallen logs, especially near streams, rivers, and lakes.

Copperhead bites are second only to rattlesnakes in the United States among venomous snake species. In one study there were nearly 3,000 copperhead bites annually, of which 44% were in Texas. In another study from

Eastern Copperhead, Southern form, from eastern Texas. *L.L.C. Jones*

2001–2005, there were 5,156 copperhead bites nationally, of which 925 were from Texas, far more than any other state. It is likely that many bites go unreported, especially for dry bites or when financial considerations are involved. Plus, many people probably just do not report bites by copperheads, if they think they can ride out the symptoms.

Venom Apparatus, Venom, and Symptoms: The venom apparatus is the same as for other pit vipers. The fangs of copperheads are small, reaching only about 5 mm in length, and usually less. Venom yield is about 40–75 mg.

The venom is tissue-destructive, having hemotoxic and myotoxic fractions. It has been estimated that 100 mg would be needed to kill a human, but bites rarely approach full capacity of the venom glands. The subcutaneous LD_{50} for the Southern pattern class is about 8–25 mg/kg, making it one of the least venomous pit vipers. Human deaths have been reported variously in the literature, but most are older reports and there are always additional factors to consider, including allergy to antivenom. At best, death by copperhead bite is extremely rare, approaching nonexistent in a healthy adult. It has been published that a rate of mortality is about 0.01 to 0.3%. Nevertheless, these animals are not to be trifled with, as they have some potent tissue-destructive agents in the venom

Eastern Copperhead, Northern form, from Kentucky. *Bill Love*

that can cause significant physical distress and tissue damage. The fibrinolytic enzymes that cause anticoagulation are well studied.

Symptoms include local pain, swelling, hemorrhage, weakness, giddiness, fever, headache, nausea, vomiting, sweating, and thirst. Swelling and hemorrhaging is higher than for cottonmouths, coralsnakes, and even rattlesnakes (as a whole). Serum-filled blisters may form in the affected area. Rarely are there more serious systemic symptoms involved, such as low blood pressure, rapid heart rate, and shock. In more severe bites, swelling can extend well beyond the area of the bite and the affected area can be grossly discolored by hemorrhage. Gangrene or other secondary infection can be a serious side effect. About 20–30% of bites by copperheads are estimated to be dry, but only about a quarter of bites are treated in emergency care units.

Remarks: There is some literature on effectiveness on the use of antivenoms for copperhead bites. Crotalidae polyvalent immune Fab (ovine) (CroFab; FabAV) is not indicated for copperhead bites, but being of ovine origin (i.e., not as allergic as equine), it has been clinically tested, usually with good results. Physicians often treat symptoms without antivenom, especially in milder cases, presumably because bites are not generally considered life-threatening. However, some physicians think antivenom is valuable for halting symptoms, even if not life-threatening.

References and Resources: Refer also to the introduction for terrestrial vertebrates. Burbrink and Guiher (2015), Fitch (1960), Guan et al. (1991), Johnson and Ownby (1993), Kudo and Tu (2001), Langley (2008), Lavonas et al. (2004), Lomonte et al. (2014b), O'Neil et al. (2007), Parrish and Carr (1967), Porras et al. (2013), Price (2009), Retzios and Markland (1988), Sanders and Jacob (1981), Scharman and Noffsinger (2001), Seifert et al. (2009), Smith et al. (2009), White and Weber (1991).

Cottonmouths (*Agkistrodon conanti* and *A. piscivorus*)

"Venomous snakes are often perceived as aggressive antagonists, with the North American Cottonmouth having a particularly notorious reputation for such villainy. Our findings challenge conventional wisdom about aggressive behavior in an animal perceived as more dangerous than it is." (Gibbons and Dorcas 2002)

Adult Northern Cottonmouth, Texas. *Troy Hibbitts*

Classification: Phylum Chordata (chordates); Class Reptilia (reptiles); Order Squamata (snakes and lizards); Family Viperidae (vipers); Genus *Agkistrodon* (in part; moccasins); Species *conanti* (Florida Cottonmouth) and *piscivorus* (Northern Cottonmouth).

Until recently, one species and three subspecies generally recognized were *A. p. leucostoma* (Western Cottonmouth), *A. p. piscivorus* (Eastern Cottonmouth), and *A. p. conanti* (Florida Cottonmouth). Based on genetic traits, the species was split into *A. conanti* (Florida Cottonmouth) and *A. piscivorus*

(Northern Cottonmouth). The Florida Cottonmouth is essentially the same entity, and the other former subspecies are now lumped as the Northern Cottonmouth. Cottonmouths are often called water moccasins. They are treated together as they have similar traits, both in natural history and venom attributes. Although they are in the same genus as copperheads, they are phenotypically and ecologically different.

Identification: Cottonmouths are typical of pit vipers, having heavy bodies with a triangular head and vertical pupils; like other *Agkistrodon* have no rattles. They reach about 190 cm, but most adults generally are between about 60 and 90 cm. Adult cottonmouths are generally olive to brown to dark gray-brown, or even nearly black, depending on regional variation, but there is usually an underlying pattern visible. Cottonmouths usually have distinct facial markings. Young individuals may be similar in appearance to sympatric copperheads, with alternating dark and light bars or saddles. Young ones invariably have a bolder color pattern than older individuals, and their contrasting facial markings are more distinct than adults. Copperheads also tend to be more reddish in overall coloration. Neonates have a yellow-tipped tail, used for caudal luring. Cottonmouths can often be identified by their open-mouth threat gesture, lacking in copperheads.

Juvenile Northern Cottonmouth, Texas. *L.L.C. Jones*

Cottonmouths are also superficially similar to certain watersnakes, which often share their habitat. However, watersnakes do not have the same patterns and lack the vertical pupils and loreal pits. Many watersnakes are killed because they are erroneously thought to be cottonmouths. The reader should take the time to differentiate the venomous and harmless species.

Distribution: The Northern Cottonmouth is found from the eastern half of Texas to extreme western Florida, and north to Illinois and Kentucky. The Florida Cottonmouth is found throughout Florida, and in adjacent Alabama and Georgia. There is a zone of hybridization from Mississippi to Florida and Georgia.

Natural History: Cottonmouths are usually regarded as a semiaquatic species of Southeast deciduous woodlands. They occur in and near swamps, ponds, reservoirs, sluggish streams, rapidly flowing streams, bogs, wetlands, and marshes. They are adept swimmers and take to the water readily.

Cottonmouths are active when the temperatures are most conducive to foraging and mating. In the summer, they may be nocturnal, while in spring and fall they tend to be diurnal. They can be active at temperatures below that preferred by many sympatric species. They are generally inactive during

Adult Florida Cottonmouth. *Bill Love*

This neonate Florida Cottonmouth shows the yellow tail, presumably used for caudal luring. *L.L.C. Jones*

the winter, especially in the northern part of their range. In northern areas, cottonmouths may seek deeper hibernacula, sometimes brumating communally, whereas southern animals may seek winter shelter in less permanent spots. During brumation and when inactive, they are generally found under surface objects, burrows, in debris piles, or under overhangs near streams and ponds. Young have been found inside crayfish burrows. During periods of activity, they are often found resting beside riparian logs, in branches of trees, or in similar streamside situations.

Cottonmouths give birth to live young. There is a courtship ritual preceding mating, and the pair may copulate in the water. Courtship generally occurs in the spring, but mating may occur any time of the active season under favorable conditions. Males may engage in combat rituals. Cottonmouths have between three and twelve offspring (usually 4–8), which usually appear on the surface in summer to fall. They live up to at least twenty-one years in captivity.

The specific epithet "*piscivorus*" alludes to the alleged propensity of the animal's feeding on fishes. While they do feed on fishes, many fish species are too fast to be captured. However, *A. piscivorus* feeds on a wide variety of animals in their aquatic and near-aquatic realm. Prey includes fishes, am-

Cottonmouths are often found in or near water, an area to exercise caution.
Troy Hibbitts

phibians, waterbirds and their eggs, small mammals, and reptiles, including watersnakes, copperheads, other cottonmouths, small rattlesnakes, and even juvenile turtles and alligators. Invertebrates are sometimes consumed, including insects, crayfish, and snails. The choice of prey depends on where the animal lives, and what potential prey is abundant, available, and catchable. In turn, they are preyed upon by a variety of animals, including American Bullfrogs (which have resistance to cottonmouth and copperhead venom), snakes (e.g., their own ilk and kingsnakes), American Alligators, wading birds, raptors, Northern Raccoons, and Virginia Opossums.

Encounters: Cottonmouths can be extremely abundant and may be the dominant species of snake in certain habitats. They are often encountered in a wide variety of situations, but usually near water. They can be active in the water or on the shore. Watersnakes are also semiaquatic and are often encountered in the same situations, so it is good to know how to differentiate harmless watersnakes from cottonmouths.

Cottonmouths are cryptically colored in their realm, so may be difficult to spot among riparian debris. When approached, they may remain motionless or flee into water. They sometimes vibrate their rattle-less tail (a behavioral precursor to the evolution of the rattle) or emit a foul-smelling liquid to ward off potential predators. If these ploys fail, they may display their cottony white mouths in a threat posture. They are not particularly inclined to bite, but when they do it is often with short, quick jabs. They can bite while floating on water, or even when completely submerged. Bites usually occur May–September, their most active period. Juveniles may be more prone to bite than adults. Bites are said to often be dry, but I can find little information on this.

A cottonmouth handled or stepped on is likely to bite. In Texas, only about 7% of the venomous snakebites are due to Northern Cottonmouths. From 2001–2005, nationally there were 173 calls to poison control centers annually for cottonmouth envenomation, compared to 961 calls for copperheads and 1,193 calls for rattlesnakes. Florida, Texas, and Louisiana had the highest rates of cottonmouth calls. Considering how abundant these animals are in the Southeast, this is a low figure.

Venom Apparatus, Venom, and Symptoms: Their fangs are about 11 mm long. The venom glands contain about 75–240 mg of venom. The LD_{50} of a cottonmouth is about 5 mg/kg. The estimated lethal dose for a human is about 100–150 mg, but pit vipers very rarely deplete their venom supply on a defensive bite. The venom of cottonmouths is highly hemorrhagic and tissue-damaging, less so than some larger *Crotalus* rattlesnakes, but more so than *Sistrurus* rattlesnakes. Enzymes found in the venom include hyaluronidase, phospholipase A_2, L-amino acid oxidase, disintegrins, hemorrhagins, kallikrein-like enzymes, and others. Specific fibrolytic enzymes, piscivorase I and II, have been isolated from the Northern Cottonmouth.

There is pain at the site of the bite, and the venom generally causes hemorrhaging and prevents coagulation of the blood. The affected area tends to darken from internal bleeding, and tissue may even liquefy. Swelling may be significant, starting with the affected area (e.g., a toe), then radiating farther along the appendage. Tissue damage may include loss of digits, significant loss of muscle mass, and loss of appendage function. Gangrene is a serious consideration for cottonmouth envenomation; in one study it was reported that 50% of cottonmouth bites resulted in gangrene-crippling of digits. Other symptoms may include weakness, giddiness, cardiac problems, muscle twitching, nausea, vomiting, and rarely paralysis. It is estimated that about 1% of cottonmouth bites could be fatal. With antivenom and medical care, fatalities are rare but have been reported. Cottonmouths are used in the manufacture of CroFab/FabAV, so that antivenom should be effective for serious envenomations.

The Cantil (A. bilineatus) is also known as the Mexican Moccasin. It is a dangerously venomous relative of cottonmouths and copperheads, ranging from Sonora to Guatemala. *L.L.C. Jones*

References and Resources: Refer also to the introduction for terrestrial vertebrates. Burbrink and Guiher (2015), Burkett (1966), Eskew et al. (2009), Ford (2002), Gibbons and Dorcas (2002), Glaudas and Winne (2007), Hahn et al. (1995), Heatwole et al. (1999), Kardong (1975), Langley (2008), Lomonte et al. (2014b), Roth (2005), Savitzky (1992), White and Weber (1991).

Grand Canyon Rattlesnake (*Crotalus abyssus*)

"We called her the 'tough-as-nails' snakebite victim. She was bitten twice in the knee, which had swollen greatly, but she insisted that since the swelling wasn't getting any worse, she was going to endure it while she finished her Grand Canyon trip." (P. Kolar, National Park Ranger, personal communication)

A Grand Canyon Rattlesnake at the National Natural Toxins Research Center that gives its venom to science. This color phase gives it the name, "Grand Canyon Pink." *L.L.C. Jones*

Classification: Phylum Chordata (chordates); Class Reptilia (reptiles); Order Squamata (snakes and lizards); Family Viperidae (vipers); Genus *Crotalus* (typical rattlesnakes); Species *abyssus* (Grand Canyon Rattlesnake).

This animal is often considered a subspecies of *C. oreganus*. In earlier literature, it was considered a subspecies of the wide-ranging Western Rattlesnake (*C. viridis abyssus*). This is one of several species of Western/Prairie Rattlesnake species complex in this book, for which information may be muddled by changing taxonomic assignments. There is recent evidence that

C. abyssus is not a valid species, or even a subspecies, based on biochemical research, suggesting it is a variant of the *C. lutosus*. It is retained here as a species because there is a long history of reference to it. Also, it occupies a finite geographic area (part of the Grand Canyon), so is sometimes encountered by visitors, hikers, park staff, and tour guides.

There are various reports of *C. abyssus* hybridizing with other species of Western/Prairie rattlesnakes in contact areas, but these seem to be based more on appearance than genetic confirmations.

Identification: This is a medium-small species of rattlesnake, measuring between about 45 and 80 cm. The record is 101.6 cm. Neonates are about 24 cm. The general color is sometimes a pinkish hue, earning it the vernacular of "Grand Canyon Pink Rattlesnake," but individuals can be tan, cream, sandy brown, yellowish, greenish, salmon, orangish, or brick red. The ground color usually matches the substrate. There are 36–48 darker dorsal blotches. *Crotalus abyssus* is not usually strongly patterned, and the anterior blotches contrast with the background more than the posterior blotches. However, some adults retain strong markings more typical of juveniles. The tail is banded, generally with dark ground color to possibly black and white; just before the rattle there can be black. There is a dark eye stripe extending

A yellowish individual. *William Wells*

to the rear of the mouth, bordered by lighter coloration above and below, but in faded animals, there is little contrast. The juveniles have more pattern that fades with age. One of the best ways to identify this species is by knowing where it came from.

This species can be similar in appearance to *C. concolor*, *C. lutosus*, *C. viridis*, and *C. pyrrhus*, all of which occur near the range of *C. abyssus*. Hybridization in the area is not well understood. Specimens from the west end of the range, south of the Grand Canyon may be dark, similar to *C. cerberus*.

Distribution: This species is essentially endemic to the greater Grand Canyon area. It ranges from around Tuckup Canyon to the west, in Arizona, up canyon into the Little Colorado River, and up the Colorado River into adjacent Utah. In southern Utah, it has been reported from original boundaries of the Grand Staircase-Escalante National Monument, the Kaiparowits Plateau, and Paria River. As it gets into this northern area, specimens have been regarded as being someone different than those in the heart of the range. Several other similar species make their home in the greater Grand Canyon area, but this is the one that is found in the heart of the canyon. In the lower elevations of the canyon are Mojave Desert dwellers such as the Mojave Rattlesnake and Sidewinder, while up on the Colorado Plateau are high-elevation species, such as the Great Basin, Prairie, and Midget Faded Rattlesnakes.

Natural History: Information on this species is limited, primarily due to the constraints of where it lives and access by researchers. It occurs as a pure form from desert scrub and mesquite scrublands into piñon/juniper woodlands within the Grand Canyon, at an elevation of 550–2,435 m. It is often found in rocky and riparian areas, such as the canyon bottom and tributaries, but does occur in uplands. In the river bottoms, it is not usually found in the floodplain; rather, it is encountered in the terraced or hillside colluvium. It is not usually found in steep canyon walls, preferring areas where it can be more effectively mobile.

Grand Canyon Rattlesnakes have somewhat linear home ranges, defined by the linear habitat of navigating between canyon walls. Activity is generally diurnal, but that depends on time of year and elevation. In the lower elevations, they have been reported on the surface during the winter, although they are primarily active spring through fall.

Reproduction is poorly known. Male combat has been reported. Courtship has also been reported, and both times it occurred at night. Typically,

A greenish individual. *Bill Love*

this species breeds every other year. It has between two and thirteen young, usually six to eight.

Feeding habits are poorly known. They seem to feed predominately on small mammals, including chipmunks, woodrats, deer mice, and pocket mice. Desert Spiny Lizards and possibly Tiger Whiptails have also been consumed. Amphibians are also reported in the diet. One was seen feeding on Spotted Sandpiper young in the canyon bottom. Neonates include insects in their diet.

Encounters: Unless you are hiking, rafting, or camping in the Grand Canyon, or are employed by the Grand Canyon National Park or other entity within its range, this animal will not be encountered in the wild. It is not particularly common, or at least commonly seen, but is not rare either. This species is generally reported to have a mild disposition. It is cryptic and often does not rattle when approached. However, some accounts state than some individuals are more irascible. Certainly, any individual trod upon or picked up may deploy a defensive strike. Bites by this species rarely occur.

Venom Apparatus, Venom, and Symptoms: The venom apparatus is the same as for other pit vipers. The fang length is 5.3–8.5 (average 6.5) mm. It has a moderate venom yield of 60–97 mg, maximum of 137 mg. The Mean LD_{50} (intraperitoneal) is 4.6 mg/kg and intravenous 2.05 mg/kg. It contains venom components common to other rattlesnakes of the Western/Prairie species complex. Although snakebites by this species are rare, there has been a fair amount of research on the venom. It has high phosphodiesterase and kallikrein-like activity; plus metalloproteinases are relatively high compared to others of the western rattlesnake complex. A novel phospholipase A_2 (D49) has been isolated from its venom. A natriuretic peptide (which causes sodium excretion in the urine) has been studied for medicinal uses, as it reduces blood pressure and increases nitrite levels. Neurotoxin has been found in one sample.

There is a general dearth of published information on bites by the *C. abyssus*. One correspondent, who was a back-country ranger in the Grand Canyon at the Phantom Ranch from 1995–2005, said she and the other staff knew of four bites during that period. One was the "tougher-than-nails" victim mentioned in the opening quote. In another instance, a young male was walking barefoot on the beach of the Colorado River when he stepped on a stick—but the stick turned out to be a snake—which promptly bit him. In another instance, a male river guide was bitten on the finger. He planned to ride out the envenomation, but the swelling extended to his arm and he couldn't feel his fingers, so four days later he was medevaced out of the canyon. The last bite was by another river guide who was pointing out (or possibly handling) a "Grand Canyon Pink Rattlesnake" to the group she was leading when it struck her on the hand. She was medevaced out ASAP. None of these cases were followed up on, so the symptoms beyond swelling, medical treatment, and outcomes aren't known.

The fairly high toxicity with high metalloproteinase activity and moderate venom yield probably causes significant problems for a victim, so swift medical attention is always warranted. There are no reports of death by envenomation, but the species should be considered to have that potential. There is one case of a death related to a hiker who encountered a Grand Canyon Rattlesnake, however, in 1933. He had an irrational fear of snakes (ophidiophobia) and saw a rattlesnake while hiking on—get this—the Rattlesnake Trail. He was extremely frightened when the snake struck at him (and missed), but the encounter elicited a fatal heart attack. This was reputedly verified by a physician. People who can be scared to death by snakes should probably not hike or live in Arizona.

References and Resources: Refer also to the introduction for terrestrial vertebrates. Da Silva et al. (2011, 2012), Davis (2016), Douglas et al. (2002), Ernst and Ernst (2011b), Feldner et al. (2016a), Ghiglieri and Meyers (2001), Glenn and Straight (1982), Mackessy (2010), Martins et al. (2014), Pook et al. (2000), Reed and Douglas (2002), Schuett et al. (2018), Stahlecker (2004), Young et al. (1980).

Eastern Diamond-backed Rattlesnake (*Crotalus adamanteus*)

"In 2000, a 2-year-old boy from Florida was bitten on the knee by a rattlesnake and received 90 vials of polyvalent crotalid antivenom. He had only 1 fang mark and bled from a cutdown site, oral orifices, and the gastrointestinal tract. He received vasopressor agents and transfusions but had had no response to pain within 24 hours and was declared brain dead." (Schulte et al. 2016)

A large individual with a long rattle, Big Cypress Swamp National Preserve. *L.L.C. Jones*

Classification: Phylum Chordata (chordates); Class Reptilia (reptiles); Order Squamata (snakes and lizards); Family Viperidae (vipers); Genus *Crotalus* (typical rattlesnakes); Species *adamanteus* (Eastern Diamond-backed Rattlesnake).

Identification: This is the largest venomous snake in the United States. Most adults encountered are between 90–130 cm, but occasionally individuals reach about 180 cm; large animals are rarely encountered these days, due to persecution and habitat destruction. A record length of 251 cm is generally accepted, although there are some claims of slightly larger animals. Neonates are about 41 cm, larger than some adult rattlesnake species.

The ground color tends to be brown to yellowish or grayish, with dark brown to black light-edged markings. There is a series of 25–35 diamonds along the dorsum. Anteriorly, the diamonds become blotches and posteriorly they become bands. The end of the tail is usually black. The diamonds contrast much more than that of the Western Diamond-backed Rattlesnake, and *C. adamanteus* lacks the distinct "coon-tail" of *C. atrox* and some other similar western species. It has a distinct black eye stripe edged by a white to yellow margin.

Distribution: *Crotalus adamanteus* is endemic to the Southeast. It is found throughout Florida, including some of the keys and barrier islands, and historically ranged west to eastern Louisiana and north to southern North Carolina. Its peripheral distribution has been shrinking, and it is nearly extirpated from Louisiana. It is primarily found on the Coastal Plain, so is absent from upland areas of Alabama, Georgia, Mississippi, and the Carolinas. Populations are often disjunct; northern Florida seems to be its stronghold.

Natural History: Many aspects of the life and natural history of the Eastern Diamond-backed Rattlesnake have been well studied. The species is a lowland form, found from sea level to about 500 m. They inhabit most dry, lowland habitats within their range, including longleaf pine forests, palmetto thickets, hardwood hammocks, and wiregrass flatwoods. Although they are not usually encountered in swamps and marshes, they can be found in adjacent uplands. They are sometimes found in vegetation right up to beaches and may swim to nearby islands. There are extant populations on some of these islands.

This species is surface active spring through fall, and will brumate during winter. They are primarily diurnal, being mostly active in the morning. During both hot and cold weather they seek cover in burned-out root cavities, logs, woodpiles, and the burrows of Gopher Tortoises, Nine-banded

Armadillos, and other mammals. They are more inclined to brumate for longer periods in the northern part of their range than in the south. On warm days in the winter, they will often emerge from brumation. In a study of a southern population, they were recorded being surface active on about half the winter days.

They breed in August and September in the northern part of the range, while in the south the breeding period may extend into December. They give birth from mid-July to early October, generally every other year. Parturition occurs in natal dens that include Gopher Tortoise burrows and hollow logs, where the female remains with her young until their first shed. The number of offspring ranges from 4–32, but is usually about a dozen.

Eastern Diamond-backs are mammal eaters. They employ both ambush and active hunting modes, depending on circumstances. An individual may remain in a hunting coil for at least a week before moving on. The prey is commensurate with the size of the snake. Adults prefer larger prey like Swamp Rabbits, Eastern Cottontails, squirrels, and even Domestic Cats, while young feed primarily on small rodents and lagomorphs. Birds and their eggs are sometimes taken, but ectotherms are not usually on the menu. After a mammal is bitten and released, the snake can track down its prey, able to distinguish the chemical cues of envenomated prey from non-envenomated mammals.

Large *C. adamanteus* are top-level predators with few predators themselves, save humans and American Crocodiles. The young have a variety of predators, including American Bullfrog, mammals (e.g., skunks, Northern Raccoon, Wild Boar, Coyote, and Domestic Cat), reptiles (e.g., Eastern Indigo Snake, Coachwhip, and Harlequin Coralsnake), and large birds (e.g., Great Horned Owl, Red-tailed Hawk, and Wood Stork). Humans and their pets are probably the major predators, and with an expanding population, these snakes hardly have a chance to persist in suburban areas. During rattlesnake roundups, wild *C. adamanteus* are collected and killed for public "entertainment." This practice has exacted a heavy toll, resulting in extirpation and fragmentation of existing populations, reduced density, and smaller adult size. The good news is that the many rattlesnake roundups have changed to Rattlesnake and Wildlife Festivals, where conservation and education have replaced carnage. In the few shows remaining that collect wild specimens and kill them, Eastern and Western Diamond-backed Rattlesnakes are often collected by pouring gasoline into burrows, a practice that has been outlawed in some, but not all states where *C. adamanteus* occurs. This is dangerous for the snakes and any inhabitant of the burrows, including the federally endangered Gopher Tortoise. The Eastern Diamond-backed Rattlesnake is also targeted by illegal collectors for specimens and hides. Although not currently listed under the Endangered Species Act, it has been petitioned for listing.

Encounters: *Crotalus adamanteus* is infrequently encountered in most of its range. In some areas they have been extirpated. However, they are still relatively common in some areas, particularly those with tracts of longleaf pine habitat. As they are primarily active during the morning hours of the spring through fall, that is when they are most likely to be seen. Adults are large, heavy animals with a large rattle that is probably the loudest of any rattlesnake. They have a relatively mild disposition and may remain coiled or motionless without rattling when humans are near, which is one reason people are accidentally bitten while hiking or working in the woods. They have the longest striking distance of any rattlesnake; a large adult can strike more than a meter away.

Bite statistics for this species are difficult to assess. Reports of *C. adamanteus* bites are sometimes included with those of *A. conanti*, *A. contortrix*, and *M. fulvius*. In Florida, there were 3,143 venomous snakebites reported to poison control centers from 2000–2005, of which 792 were rattlesnakes

Eastern Diamond-backed Rattlesnakes are the largest species and one of the most dangerous of rattlesnakes, but their numbers are declining due to persecution. *Bill Love*

(*C. adamanteus, C. horridus,* and *S. miliarius*), compared to 255 bites by Harlequin Coralsnakes, 218 Florida Cottonmouths, and 31 Eastern Copperheads, without including the 1,847 unidentified venomous snakes. It is reported that there is about one snakebite fatality per year in Florida, and this is usually due to *C. adamanteus.*

Venom Apparatus, Venom, and Symptoms: The Eastern Diamond-backed Rattlesnake has the longest fangs of any rattlesnake, averaging about 15 mm, but they may be as long as 27 mm. They also have the largest venom yield, with adults averaging about 400–700 mg, and reaching about 850 mg. Neonates average about 14 mg. The venom of *C. adamanteus* is potent. LD_{50} values differ by study and method of delivery, but averages seem to be about 1.7 mg/kg (intravenous), 1.9 mg/kg (intraperitoneal), 7.7 mg/kg (subcutaneous), and 28 mg/kg (intramuscular). An estimated lethal dose for a human is about 100 mg. It has been estimated that without medical intervention, lethality rates could be as high as 40%.

The venom is among the best studied of rattlesnakes. It generally has a strong hemotoxic activity and may contain procoagulants or anticoagulants. The venom can cause a great deal of tissue damage. Some individuals or populations also have neurotoxins, similar to Mojave toxin. The common components of rattlesnake venom are also present, such as phospholypase A_2, hyaluronidase, and L-amino oxidase.

Symptoms of bites are often serious. They invariably include pain, swelling, hemorrhaging, weakness, and giddiness, but can also include bloody diarrhea, weak pulse, fainting, cardiac distress, respiratory distress, convulsions, and sometimes death. Tissue damage is often severe, and the tissues may become necrotic. Bites often leave victims crippled.

References and Resources: Refer also to the introduction for terrestrial vertebrates. Fischer et al. (1961), Heinrikson et al. (1977), Hoss et al. (2010), Kurecki et al. (1978), Margres et al. (2014, 2015, 2016), Means (2009), Rokyta et al. (2011, 2012).

Western Diamond-backed Rattlesnake (*Crotalus atrox*)

"A man in Texas was nearly killed when a rattlesnake bit him. The scary part: The snake's head wasn't attached to its body when it attacked. The scarier part: It's not uncommon. . . . He picked up a shovel and swung, splitting the snake's head from the rest of its body. . . . But when he reached to grab the severed head, it latched both its fangs into his fingers. . . . Rattlesnake venom takes effect almost immediately, and the bite was brutal. . . . Doctors induced him into a coma where he stayed for five days. . . . His kidneys failed because of shock. Now, he's on dialysis and receiving antibiotics." (CNN, A. Archie and S. Ahmed, reporting, 8 June 2018)

Classification: Phylum Chordata (chordates); Class Reptilia (reptiles); Order Squamata (snakes and lizards); Family Viperidae (vipers); Genus *Crotalus* (typical rattlesnakes); Species *atrox* (Western Diamond-backed Rattlesnake).

Western Diamond-backed Rattlesnakes are sometimes called coon-tails vernacularly, although some other species have banded black and white tails. *L.L.C. Jones*

Identification: Most adults are between 1–1.3 m in length, but some individuals can acquire a much larger length. The most generally accepted length record is 2.26 m. The largest individuals are found in South Texas and Mexico, where the occasional animal is nearly 2 m long. They are generally gray to light brown in overall coloration, although the pattern may be suffused with reddish or yellowish pigments. They have 24–45 dark markings on the back, edged to various degrees with a lighter border. This species sometimes has the local vernacular of "coon-tail rattlesnake" because the tail is vividly patterned with alternating white and black bands.

Crotalus atrox can be very similar in appearance to the Mojave Rattlesnake, which overlaps much of its range. However, *C. scutulatus* usually has much broader white tail bands compared to the black ones. Also, *C. scutulatus* tends to be more distinctively patterned than *C. atrox*, having crisper dorsal markings with cleaner light-scale borders. Some populations of *C. scutulatus* are noticeably greenish, a color not seen in *C. atrox*. There are also differences in the head scales and eye stripe. The eye stripe angles backward to a greater extent in *C. scutulatus*. Hybridization may occur between the two species, but is uncommon at best. The Red Diamond Rattlesnake also looks like *C. atrox*, but occurs farther west and is usually of a reddish hue. Natural hybrids have been documented with *C. horridus* from Texas.

Juveniles are similar in appearance to adults but have a more distinct pattern. It is even more difficult to differentiate juvenile *C. atrox* and *C. scutulatus* in the field. Juveniles start life with a single button, and then add more with successive molts.

Distribution: This is the most wide-ranging rattlesnake in the Southwest. It occurs in southeastern California and extreme southern Nevada, throughout southern Arizona and much of New Mexico, and occupies the western two-thirds of Texas. It also occurs in parts of Oklahoma, Arkansas, and seems to be expanding into southern Kansas. Its range extends south into Mexico.

Natural History: Western Diamond-backed Rattlesnakes are found in a variety of habitats, from deserts into open coniferous forests. However, they seem most abundant in sparsely vegetated arid and semiarid habitats. Occupied deserts include the eastern Sonoran Desert (although it is absent from southwestern California and Baja California), Chihuahuan Desert, and a small part of the southern Mojave Desert. It is absent from the Great Basin Desert. In many open desert areas, such as creosote bush flats, they may be

A yawning C. atrox shows off its formidable dentistry. The fangs are within a sheath when not in use. *R. C. Clark*

absent or outnumbered by lower desert species, including the Sidewinder and Mojave Rattlesnake. It is also found in semi-desert grasslands, subtropical thorn scrub, and prairie grasslands of northern Texas and Oklahoma. In parts of the range, Western Diamond-backs are sometimes found in open oak woodlands or occasionally open coniferous forest. In South Texas, their habitat extends onto the barrier islands in the Gulf of Mexico, where they may be quite abundant. They are a completely terrestrial species, but can be found in riparian areas, especially along seasonal washes. They are not rock dwellers per se, but may be abundant in generally rocky areas.

Western Diamond-backs are most active March through October. During the spring and fall, they tend to be diurnal, but during the summer are most active at night. During cloudy, mild weather in the summer, they may be diurnal, particularly after a summer thunderstorm, but are also active at night during these times, provided the temperature is warm enough. They typically brumate during the winter, but it is possible to see them any time of the year, particularly near dens, when warm periods prevail. They often den communally, numbering from about thirty in Arizona to 100–200 in Texas and Oklahoma. They may den with other species of snakes.

Male combat, courtship, and mating are well-documented. *Crotalus atrox* typically mate in the spring and give birth to live young in the late summer to early fall. Litter sizes range from about 3–12, but 5–8 young is a more typical number. The female stays with and guards the young for 5–10 days after parturition. When the young disperse, they may be much more abundantly encountered than adults, when *atrox* "pencils" are well distributed on little-traveled roadways at night. The young are ready to fend for themselves at birth and will feed on a variety of small animals.

Western Diamond-backs are generally regarded as sit-and-wait predators, but may actively forage by following a scent trail. They repeatedly flick the tongue to pick up scents. Often, though, they coil up near a rodent scent trail to ambush a passing mammal. In typical fashion of a sit-and-wait pit viper, the prey will be struck, injected with venom, and then tracked to its final resting place, where it will be consumed whole. *Crotalus atrox* feeds primarily on mammals. Prey includes mice, rats, rabbits, hares, and squirrels. They have also been known to feed on birds, lizards, and even (rarely) Sonoran Desert Tortoises. Unlike many rattlesnake species, juveniles do not selectively target lizards. They may scavenge, and while feeding on a roadkill may become one themselves. Occasionally insects or amphibians may be preyed upon.

Although it would seem like this large predator might be at the top of the food chain, it is not, although actual observations of animals preying upon *C. atrox* are uncommon. The several species of animals that have been documented feeding on *C. atrox* include Coyotes, Collared Peccaries, skunks, raptors, and Greater Roadrunners. Kingsnakes and some other species of snakes are quite adept at handling a rattlesnake nearly their length or smaller. Contrary to popular belief, kingsnakes do not preferentially seek out rattlesnakes as prey, but they are resistant to the venom. Of course, the most notorious predator of Western Diamond-backed Rattlesnakes is the Human Being.

This species can live to about twenty-five years in captivity, and it is likely some individuals can be similarly aged in the wild. However, the massive size attained by some animals does not correlate completely to age.

Encounters: *Crotalus atrox* is often the most abundant species of rattlesnake (or snake) where it occurs. This is difficult to qualify based on density studies, but anecdotal information suggests very high numbers in some areas. For example, a rancher killed 1,200 while clearing a 4,000 ha plot of land in Texas. Another anecdote is one where five tons of Western Diamond-backs were caught in one season by a collector in South Texas. They are also very abundant in Arizona, New Mexico, and Oklahoma, where it is often the most common species of snake. The Lower Colorado River area and other desert flats have a lower density than upland areas to the east.

Crotalus atrox can get very large, especially in South Texas, although most are considerably less than maximum size. *Noah Fields*

Western Diamond-backed Rattlesnakes do not shun human habitation, particularly in the suburbs. They are all-too-frequently encountered in people's yards, porches, and sometimes their garages, storage areas, and homes. They may actually be drawn into the vicinity of humans because we inadvertently supply shelter (e.g., buildings and debris piles); mesic conditions, in an otherwise arid region (e.g., gardens, landscaping, parks, golf courses); and food (e.g., by using bird-feeders or attracting woodrats, ground squirrels, and hares to our domiciles). In some desert communities, such as Tucson and Phoenix, there are dozens of calls per day to fire departments or removal services to extricate rattlesnakes, and hundreds or even thousands of individuals are removed every year. This is such a burden on fire departments that many now will not respond to a "snake in the yard" call. Fortunately, for people not wanting to deal with uninvited snake guests, there are many removal services or willing neighbors. Some of these will do it as a community service and others are more professionally inclined. There are also organizations that will train people how to safely remove and relocate venomous snakes.

More people are bitten by Western Diamond-backed Rattlesnakes than any other venomous snakes, except perhaps copperheads. *Crotalus atrox* is generally regarded as causing the highest number of fatalities, although some sources suggest the Eastern Diamond-backed Rattlesnake is at the top of the list. According to one source, using data from poison centers from 2001–2005, California and Arizona top the list for rattlesnake bites (1,226 and 1,143, respectively). Certainly those in Arizona (not California) were largely due to the Western Diamond-backed Rattlesnake. Texas had 668 rattlesnake bites during this time (Western Diamond-backs are the main culprit), compared to 925 bites from copperheads, 155 from cottonmouths, and 89 from coralsnakes. Arizona ranks highest in per-capita rattlesnake bites nationally, followed by New Mexico, due to frequent encounters with Western Diamond-backs.

Venom Apparatus, Venom, and Symptoms: The Western Diamond-backed Rattlesnake is sometimes regarded as the most dangerous snake in the United States. While it does not have the most virulent venom, the animal is widespread and common, frequently coming into contact with humans; its body length, striking distance, and fangs are long; it has a high venom yield as an adult; and it is often quick to defend itself. This species is well known for its ability to rear up well above the ground to deliver a strike. The fangs are an impressive 10–13 mm long, the second longest among U.S. rattlesnakes. Venom yield is also high—it can be between 600 and 1,145 mg and possibly more. However, an average venom yield for an adult snake is probably closer to 200–300 mg. There are many studies reporting an LD_{50}, as this species is considered the gold standard for comparison of venom potency. However, this "standard" is variable, depending on the study and injection route. This ranges from 1–6.3 mg/kg (intravenous), 3.7–20 mg/kg (intraperitoneal), and 11.7–19.3 mg/kg (subcutaneous). The lethal dose for a healthy human has been suggested to be about 100 mg. This would imply that this species is more deadly than it actually is. Even serious bites are rarely fatal, partly because of the amount of venom actually injected is considerably less than full capacity, but also due to medical care. However, bites by medium to large adults can be fatal. Estimates of fatalities vary widely, but 5–15 deaths are generally reported annually in the United States, from all pit viper species combined.

The venom of the Western Diamond-backed Rattlesnake varies but is composed of about 53% hemotoxic or tissue-destroying enzymes, 30% proteases, and 17% neurotoxins. While the bites are rarely fatal, they can be quite serious, given the hemorrhagic and myotoxic nature of the venom. The venom also contains enzymes that promote blood clotting. In some individuals or populations, there may be neurotoxic elements that can ultimately be more dangerous than the tissue-digesting enzymes.

Symptoms include immediate and intense burning pain, swelling, discoloration of the affected area, hemorrhaging, necrosis, decreased or increased heart rate, low blood pressure, fever, sweating, weakness, giddiness, nausea, vomiting, and respiratory distress. Secondary infection and gangrene are not uncommon and the loss of a digit or limb is not unusual.

Remarks: Like other rattlesnakes, Western Diamond-backs are often wantonly persecuted by humans, despite being important ecologically to control rodent populations and the parasites and diseases they transmit to humans (e.g., Chagas disease, bubonic plague, hantavirus). They are still killed in rattlesnake roundups that have not switched to rattlesnake and wildlife festivals. Fortunately, the species is adaptable, so they have not been hunted to extinction, but the methods used for roundups (e.g., putting gasoline into burrows or setting fire to habitat) causes significant damage to rare animal populations. The good news is that times are changing. Round Rock, Texas, now sponsors the Texas Rattlesnake Festival, with the slogan "Education, Not Eradication." This popular event demonstrates how thousands of rattlesnakes and other species need not be harmed to entertain and enlighten us.

References and Resources: Refer also to the introduction for terrestrial vertebrates. Beaupre et al. (1998), Beck (1995), Bjarnason and Tu (1978), Bjarnason et al. (1988), Castoe et al. (2007), Curry et al. (1989), Landreth (1973), McCue (2007), Minton and Weinstein (1986), Ownby et al. (1978), Schuett et al. (2016c), Seifert et al. (2009), Suchard and LoVecchio (1999), Tanen et al. (2001), Taylor and DeNardo (2005).

Sidewinder (*Crotalus cerastes*)

"That's not how you catch a Sidewinder. You do it like this." (Author's recollection of his friend's last words before he was bitten, as he tried to pick it up behind the head during a YMCA outing in Junior High School.)

Colorado Desert Sidewinder. L.L.C. Jones

Classification: Phylum Chordata (chordates); Class Reptilia (reptiles); Order Squamata (snakes and lizards); Family Viperidae (vipers); Genus *Crotalus* (typical rattlesnakes); Species *cerastes* (Sidewinder).

There are three commonly recognized subspecies: *C. c. cerastes* (Mojave Desert Sidewinder), *C. c. cercobombus* (Sonoran Sidewinder), and *C. c. laterorepens* (Colorado Desert Sidewinder).

Identification: The Sidewinder is a small species of rattlesnake, usually between about 55 cm and 70 cm. Various maximum lengths have been reported in the literature, ranging from about 82–98 cm. They are generally tan, but may be gray or reddish, with somewhat darker ochre to brownish blotches on the back. They blend in with the sandy areas in which they inhabit. For example, those in reddish sands have a pink tint. They are easily identified by the horn-like projections over the eyes—actually, these are modified supraocular scales that are conspicuously raised more than in other rattlesnakes. The function of the "horns" is not clear, but they probably shade the eyes and protect against sand. There are species of horned vipers in North Africa that are not only similar in appearance, but also independently evolved a similar design and sidewinding abilities for the same reason, efficient locomotion in sandy areas. The rattle of the sidewinder is small and difficult to hear. The Sidewinder can also be detected by the telltale sign of its track, which looks like a series of elongate J's in the sand.

A pink-hued Mojave Desert Sidewinder on pink dunes, doing what it does best—sidewinding. *L.L.C. Jones*

Sonoran Desert Sidewinder, showing the raised supraocular scales. *L.L.C. Jones*

Distribution: Sidewinders are endemic to the Southwest and adjacent Mexico. They are only found in parts of the Mojave and the Sonoran Deserts. They occur in southern California, southwestern Arizona, and the extreme southern part of Nevada, plus extreme southwestern Utah. The subspecies align with biotic communities within this area: *C. c. cerastes* is found in the southern Mojave Desert, *C. c. cercobombus* is found in the eastern Sonoran Desert, and *C. c. laterorepens* is found in the western Sonoran (Colorado) Desert. The Colorado Desert is in California, and named for the river, not the state.

Natural History: The Sidewinder is a denizen of the most hostile environments of the desert. It is usually found in sparsely vegetated areas of sand or gravel. It is exquisitely adapted to move across these loose surfaces by its characteristic sidewinding motion. They do not frequent open dunes without vegetation for cover. In dunes and other sandy areas, they may be the only species of rattlesnake. Few other snakes are adapted to sand dunes, although shovel-nosed snakes and other species that "swim" through sand are notable exceptions. The Sidewinder is not a sand-swimmer; instead, it moves across the surface by "throwing" loops—difficult to explain on paper, but interesting to watch. Sidewinding can be seen in many snakes on loose surfaces, but Sidewinders have refined the motion to an art. It is an efficient way to move on sand and they can move with remarkable speed. By sidewinding, contact with the hot surface is reduced—only two small parts of the snake touch at any given time, and those areas shift as the animal moves. They are not obligate dune dwellers and may be quite common in areas of gravel or even desert pavement. However, they are almost never seen in rocky uplands or above 1,800 m elevation.

Sidewinders are generally nocturnal during their activity period, although they will come out in the day during cooler times. They are usually active from about March or April to about October or November. They are sometimes seen coiled beneath shrubs, especially creosote bushes and mesquites, partially concealed by sand. While this appears to be a resting position, they are poised to ambush small prey that gets within striking range. Shrubs in open, sandy areas are heavily laden with burrows used by rodents and lizards where Sidewinders may take refuge from the heat on hot days.

Sidewinders often bury themselves in sand to rest, hide, cool down, or ambush prey, but they are always alert. *L.L.C. Jones*

Prey consists mostly of lizards and small mammals, but occasionally they will eat other snakes and birds. Neonates are only 15–20 cm, so can only eat small lizards and small mice. Young 'winders have been seen to use caudal luring to attract lizard prey.

There is little direct information on population size, but two studies from one area indicate they reach about 1 individual per hectare. There are anecdotes of herpetologists seeing between thirty and fifty on stretches of roads during a single night.

Sidewinders brumate in burrows, especially those of kangaroo rats. They generally emerge from brumation in the late winter to early spring and are active through the fall. Male combat, courtship, and mating has been documented. They typically mate in April and May, but may also mate in the fall. They usually give birth to 7–12 young in August through November, but most often in October. The female will stay with young after parturition, until their first shed. *Crotalus cerastes* does not live as long as most rattlesnakes. A record of more than twenty-eight years in captivity has been reported, but in the wild, they typically live 3–7 years.

Encounters: Sidewinders are rarely encountered by the general populace because they occur only in lowland desert areas and are usually on the move at night. However, they can be incredibly abundant in certain areas. There are no bite statistics, but it is assumed few rattlesnake bites are due to Sidewinders. Most bites probably occur from people handling them.

My first introduction to snakebite is alluded to in the opening quote. When I was in grade school at a campout in Valley of Fire State Park, Nevada, a counselor captured a Sidewinder in a butterfly net, and then another counselor showed us all the "proper way" to catch a rattlesnake. He picked it up behind the head, and the snake twisted around and jabbed him in the finger with a single fang. He was rushed to the hospital in Las Vegas, where he was told he was allergic to both antivenom and venom. His arm became darkly discolored and swelled to the shoulder. It was months before he regained full functionality—but this was better than the prognosis of the other uninformed counselors and kids in the camp ("He's going to die, for sure!").

Venom Apparatus, Venom, and Symptoms: For their size, the fangs are actually rather long, being about 5–8 mm long in an adult snake. The venom yield is rather low, ranging from about 20–68 mg, although it is probably 20–40 in average adults. The subcutaneous LD_{50} is about 5.5 mg/kg, based on two studies, rendering it slightly more virulent than the Western Diamond-backed Rattlesnake. Average bites have been reported to be 18 mg. It has been estimated that the lethal dose in a human would be about 40 mg, which seems low, compared to the relatively well-studied Western Diamond-back. However, fatalities have rarely been reported. The general consensus is that bites from Sidewinders are milder than most rattlesnakes, partly because of the low venom yield and moderate toxicity.

Sidewinder venom is high in protease activity, although bites suggest hemolytic components are present in sufficient quantity to cause considerable hemorrhage in serious bites. Sidewinder venom is not usually the topic of research, although it is often used in studies involving multiple species. Until recently, Sidewinders were not considered to have any neurotoxic components to their venom. However, one recent report documented that a patient bitten by an alleged Sidewinder exhibited classic neurotoxic symptoms. It is interesting that the victim was bitten through the toe of a leather boot, attesting to the sharpness and length of the fangs—and how leather boots are not necessarily bite-proof.

More typical symptoms include those of a typical rattlesnake bite with tissue-destructive venom. The bite may or may not include immediate pain, but subsequent pain seems inevitable. In one often-cited quote, one victim in a case report likened it to "having the arm soaking in a bucket of boiling oil." Other symptoms include swelling at the bite site or beyond, discoloration and hemorrhaging, weakness, giddiness, dizziness, increased body temperature, nausea, and even paralysis (which suggests neurotoxic elements may be present). Necrosis may develop. Anaphylaxis has been reported for a Sidewinder bite in a victim without prior exposure, but details are not well known.

Remarks: Sidewinders are among the more interesting species of watchable wildlife. I never tire of watching them sidewind. They are commonly seen on little-traveled, paved roads amid sandy areas in the Mojave and western Sonoran Desert during the early evening hours, especially in the spring.

References and Resources: Refer also to the introduction for terrestrial vertebrates. Bosak et al. (2014), Funk (1965), Jayne (1988), Moore (1978), Reiserer (2016), Russell (1960), Secor (1994, 1995), Secor and Nagy (1994), Secor et al. (1992), Washington and Ruha (2007).

Arizona Black Rattlesnake (*Crotalus cerberus*)

"Watchdog of the god Hades, Cerberus has multiple depictions in mythology, including a three-headed dog with a snake-like tail, and a ruff or mane of snakes, and lion claws. Posted at the gates of the underworld, Cerberus controlled entry of souls of the dead and prohibited them from leaving." (Davis et al. 2016)

Adult Arizona Black Rattlesnake. Some forms are not so dark, but this is the normal condition. *L.L.C. Jones*

Classification: Phylum Chordata (chordates); Class Reptilia (reptiles); Order Squamata (snakes and lizards); Family Viperidae (vipers); Genus *Crotalus* (typical rattlesnakes); Species *cerberus* (Arizona Black Rattlesnake).

The Arizona Black Rattlesnake was considered a subspecies of the much wider-ranging and highly variable, former Western/Prairie Rattlesnake; complex (*C. viridis cerberus*).

Identification: This is a medium-sized rattlesnake, reaching about 60–90 cm as an adult, although they have been recorded to reach about 1.2 m in length. Neonates are generally 200–260 mm in length. Despite its name, the Arizona Black Rattlesnake has a variable color pattern, although primarily black overall is typical of large adults in much of the range. Adults may also be a lighter brownish, olive, yellowish, or grayish color. Dorsal blotches are present, but may be effectively lost in the dark pigmentation. Some black individuals have white to bright yellow or gold bars along the back between the inconspicuous dorsal blotches. Juveniles always have a light gray to tan background color, but also have distinct, darker brownish dorsal blotches. As juveniles mature, they will normally take on the darker adult pigmentation and the blotches will become less distinct.

If this isn't confusing enough, in addition to color pattern differences between adults and juveniles, there is regional variation, individual variation, and physiological variation. The latter is a curious phenomenon—much more unusual for snakes than lizards. Some dark animals may lighten up and their blotches may become more evident with various stimuli, such as being put in captivity, changes in temperature, or after being envenomated by another snake. As with other species of animals that exhibit

Juvenile Arizona Blacks have a distinct pattern of dorsal blotches. *L.L.C. Jones*

color change, the physiological response is due to dispersion or contraction of pigments within chromatophores. In this case, the pigment is melanin, so the chromatophores are melanophores. Arizona Blacks do not change color as rapidly and frequently as do many other taxa capable of color change. The function of color change in Arizona Black Rattlesnakes is not entirely clear.

Because there is tremendous variation in color pattern among individuals, populations, and geographic areas, there may be confusion of species identity, especially when near the contact zone with other species of snakes, particularly those of the Western/Prairie rattlesnake species complex. Hybridization has been reported and has occurred in captivity with *C. atrox*, but hybrids in the wild with other taxa of the complex are generally lacking in evidence. Juveniles have a typical Western/Prairie rattlesnake pattern. Large, dark Arizona Black Rattlesnakes are superficially similar to Southern Pacific Rattlesnakes, although their ranges are widely separated.

Distribution: Arizona Black Rattlesnakes are found almost exclusively in central Arizona, along the Mogollon Rim and associated mountains, and south into the Santa Catalina, Rincon, Galiuro, and Pinaleño Mountains. Putative *C. cerberus* with the phenotypic appearance of *C. oreganus* are found in the Little Dragoon Mountains. Along the Rim, they just enter extreme western New Mexico.

Natural History: This species can be found in some locales in the Arizona upland subdivision of the Sonoran Desert, but they are more typically mountain dwellers. They range from about 850 m to over 2,800 m in elevation. They are mostly encountered in interior chaparral and evergreen woodlands, and sometimes into higher-elevation coniferous forests. They are usually associated with rocky areas, often near streams. Other rattlesnake species in the area where they occur are ecologically separated to various degrees, although there is quite a bit of overlap with Black-tailed Rattlesnakes.

Arizona Blacks may be diurnal or nocturnal, depending on elevation and weather. Being primarily mountain dwellers, they prefer temperatures lower than desert species, so are often diurnal. They are usually active during late spring through early fall. The North American Monsoon seems to influence surface activity of this species, and it has helped shape speciation in the Western/Prairie Rattlesnake species complex. As with many snakes that occur within monsoon region, they tend to become most active after showers, day

Neonate *C. cerberus* at their natal den. They will disperse after shedding. *Rich Gassaway*

or night, when temperatures are conducive. Arizona Blacks have their peak breeding period during the monsoon (July and August).

Many rattlesnakes could loosely be termed social species simply because of aggregations at hibernacula and the fact that the female stays with the young until the first shed. After leaving communal hibernacula, rattlesnakes often go their own way. However, Arizona Blacks seem to be more social than most—or at least more studied than most. They aggregate during the spring at communal basking sites and during the summer. These aggregations appear to be intentional and based on kin, suggesting the species is social. During these gatherings, females recognize and interact with their kin; males may even interact with females and young. We are just scratching the surface of this interesting topic, but hopefully it will help the public to regard snakes as social vertebrates, rather than primitive, unintelligent, biting machines.

Arizona Black Rattlesnakes feed primarily on mammals, including mice, rats, pocket gophers, and even hares. However, birds (including nestlings and their eggs), lizards, snakes, and amphibians have been reported as prey. In turn, these animals are likely food for a variety of predators, particularly larger carnivorous mammals, raptors, and humans.

Encounters: This is an animal of sparsely populated areas, although there are some mountain cities near populations, as well as some large urban areas in the nearby valleys. Throughout most of its range, it is infrequently encountered, although it may be locally abundant. Hikers, backpackers, bird-watchers, snake-watchers, and herpetologists are among the most likely people to encounter this species. People who live in mountains are at a higher risk than most. Arizona Blacks are cryptically colored, so difficult to spot. Fortunately, they tend to have a largish rattle and can be heard by most at 5–6 m away.

Venom Apparatus, Venom, and Symptoms: Specific information for this species has been somewhat concealed in research and assays for either *C. viridis* or *C. oreganus*, because it was considered a subspecies of one of those until recently. The fangs are about 6–10 mm. The venom yield of the Arizona Black Rattlesnake has been reported to be about 90–180 mg.

The LD_{50} averages 6.0 mg/kg (intramuscular) and 5.4 mg/kg (intravenous). Major components of the venom of Arizona Black Rattlesnakes were compared to other species and subspecies in the Western/Prairie rattlesnake species complex. It possesses Mojave-like toxin, but at a lower level than either *C. oreganus* or *C. concolor* (0.7 mg/ml for *C. cerberus* vs. >2.0 mg/ml for the others); *C. concolor* is often regarded as primarily neurotoxic, while *C. cerberus* is not. Among the Western/Prairie Rattlesnake species complex, the Arizona Black had the highest level of metalloproteinase activity, but the lowest toxicity of any of the species or subspecies, consistent with the inverse relationship between metalloproteinase activity and LD_{50}.

Reports of bites by Arizona Black Rattlesnakes are practically nonexistent. One case report of a 45-year-old man bitten on the finger indicated he had local pain, redness, and swelling that rapidly extended beyond the bite site. The swelling reached his upper arm within an hour. He had muscle twitching in his arm, leg, and chest. He was treated with a regimen of CroFab and responded well. Within a week, the swelling was reduced and there were no long-term hematological or systemic problems. Given the high level of metalloproteinases, it would be expected that a serious bite could lead to severe tissue destruction.

Remarks: This is a good example of the value of specifying subspecies and geographic areas in publications. Much of the information that could have been gleaned for this species was lost due to the fact that it was (at times)

These Arizona Blacks were basking outside a communal den in the mountains of central Arizona. *L.L.C. Jones*

thrown in the mix of the Western, Prairie, and Northern Pacific Rattlesnake groups. This makes it difficult to extract valuable information on its natural history, ecology, and nature of the venom. Fortunately, some researchers are well aware of this issue.

References and Resources: Refer also to the introduction for terrestrial vertebrates. Amarello and Smith (2011, 2012), Davis (2016), Davis et al. (2016), Douglas et al. (2002), Flesch et al. (2010), Mackessy (2010a), Pook et al. (2000), Whitlow et al. (2008).

Midget Faded Rattlesnake (*Crotalus concolor*)

"Within 30 min, this numbness was more widespread on the left side of the face and appeared to travel down the left arm. Shortly thereafter, fasciculations occurred in both lips, and a parent commented that it looked 'like worms crawling under the skin.'" (Mackessy et al. 2003)

Adult Midget Faded Rattlesnake. *Stephen Mackessy*

Classification: Phylum Chordata (chordates); Class Reptilia (reptiles); Order Squamata (snakes and lizards); Family Viperidae (vipers); Genus *Crotalus* (typical rattlesnakes); Species *concolor* (Midget Faded Rattlesnake).

This animal is sometimes considered a subspecies of *C. oreganus*. In earlier literature, it was considered a subspecies of the wide-ranging Western/Prairie Rattlesnake species complex (*C. viridis concolor*).

Identification: Among the Western/Prairie Rattlesnake species complex, it is the smallest species, reaching only 41–66 cm. Neonates are about 19 cm. The "faded" part of its moniker alludes to the fact that it usually has poorly defined blotches, almost to the point of being uniform in color, especially as an adult. When present, the edges of the dorsal blotches are darker than the ground color, and the interior of the blotches less so. Overall coloration matches the local substrate well, and across its range it can be any number of earth tones, including tan, gray, brown, yellow, pink, red, orange, and salmon. Young animals tend to have more conspicuous markings than adults. There are 37–47 blotches. There may be faint banding on the tail and the base of the rattle may be black. There is a dark and light eye stripe, but these may also be faded.

Crotalus concolor may be similar to *C. abyssus*, which may also have faded markings, especially as adults. However, *C. concolor* is generally found north of the Colorado River confluence of the Paria River, while *C. abyssus* is to the west and south, although there is overlap near the Paria River junction. Grand Canyon Rattlesnakes attain a larger size. There is also overlap or near-contact between ranges of *C. viridis* (including "*nuntius*") and *C. lutosus*. The *nuntius* morph of *C. viridis* (the Hopi Rattlesnake) is also similar in size and may have a similar pattern.

Distribution: This animal is primarily found in the Colorado Plateau region from extreme northern Arizona near Lake Powell (where most of the local historic range is now underwater), throughout much of eastern Utah, and in adjacent western Colorado. It is associated with canyons carved by the Colorado and Green Rivers. It ranges north into southwestern Wyoming in the Green River Basin.

Natural History: This species primarily occupies sparsely vegetated sagebrush and other scrublands. It is often found in rocky to gravelly and sandy areas. It ranges from about 550 to 2,100 m or more. Important habitat features include the rocky areas on south-facing slopes used for denning. Neonates and post-partum females will stay in the den area, whereas males and other females tend to wander away. Den areas are critical habitat features that are considered a limiting resource, so are important for conservation. Away from dens, they seek out good areas for hunting in shrubby habitat and riparian areas. They often use draws and riparian corridors for movements, so their ranges may be fairly linear. Migrating animals tend to return to the same den to overwinter every year and occupy the same foraging area.

They usually emerge from brumation in May. There may be up to a hundred individuals in hibernacula. Mating occurs in July and August during

Juvenile Midget Faded Rattlesnake. The dorsal blotches are usually more distinct than the adults. *Stephen Mackessy*

the migration away from dens. Females usually breed every three years. At any given time about 25% of the female population is pregnant. Females will store sperm over the winter and give birth to young during the following August. They average four or five young, which are born at the dens. This low reproductive output and reliance upon specific denning sites, plus their generally low densities on the landscape helps explain why these animals are of conservation concern across much of their range. Longevity is probably around 10–20 years.

Neonates and juveniles tend to eat several types of lizards within their range, including Sagebrush Lizard, Plateau Fence Lizard, Tiger Whiptail, and Plateau Striped Whiptail. The diet shifts to predominately small mammals as individuals reach adulthood. The principle mammalian prey is the Deer Mouse. Other small mammals include Least Chipmunk, Ord's Kangaroo Rat, Western Harvest Mouse, and Bushy-tailed Woodrat. They have sometimes been found associated with woodrat nests. Adults may also eat large lizards, including spiny lizards and Common Chuckwallas.

Encounters: These animals are generally considered uncommon or even rare in parts of their range, but are locally abundant in some areas. They are typically found in fairly remote locations, but for people frequenting these areas, encounters can happen. They are a somewhat nervous species that will rattle and defend themselves, but they can also be cryptic and quiet. They are most likely to be encountered in rocky outcrops or hillsides, or in riparian or canyon areas. Bites by this species are presumably rare and/or rarely reported.

Venom Apparatus, Venom, and Symptoms: The fang length is small for a rattlesnake, 4.1–5.2 (average 4.7) mm. Venom yield is also small at 6–34 mg (dry weight), averaging about 9–22 mg.

The venom of this animal is fairly well studied. Although human envenomation is presumably rare, researchers have done quite a bit of work on its venom because it is primarily neurotoxic and possesses one of the most lethal venoms in North America. The LD_{50} ranges from 0.13–0.45 mg/kg. Fortunately, the venom yield is small for a rattlesnake, but toxicity is so high that it makes this snake dangerous. The neurotoxic Mojave-like toxin present in this species is generally known as concolor toxin. Two types of myotoxins have been isolated from the venom. This species has some other venom components typical of the Western/Prairie Rattlesnake species complex, although it is the only species of the group that shows no metalloproteinase activity. It does show L-amino oxidase, phosphodiesterase, arginine ester hydrolase, phospholipase A_2, kallikrein-like, thrombin-like, and plasmin-like activity.

There are few examples of case reports of envenomations by this species. One published report has two case studies of human envenomation; both bites were "legitimate." In one case, the victim was bitten on the toe. Symptoms began in about ten minutes and included numbness of the face and the mouth was hard to open. Numbness spread in about thirty minutes to the arm and fasciculations began in the lips. The leg became swollen and the victim had a difficult time maintaining balance. The victim was treated with Wyeth polyvalent antivenom and the symptoms began to resolve, although the leg remained swollen for about three days. Consistent with the lack of metalloproteinases in the venom, there was no hemorrhaging or necrosis. In the second report, a person near Moab was bitten on the finger. Within an hour there was numbness around the mouth and in the back of the throat. The finger was swollen, and this spread to include the hand. Numbness spread down both sides of the body to include the ribs. There was throbbing pain in the hand and mild pain in the ribs. There was slight hemorrhage at the bite site. After eight hours, the victim was administered ten vials of the Wyeth polyvalent, and more (unknown quantity) was given later, but treatment was complicated by an allergic reaction. The symptoms resolved rather

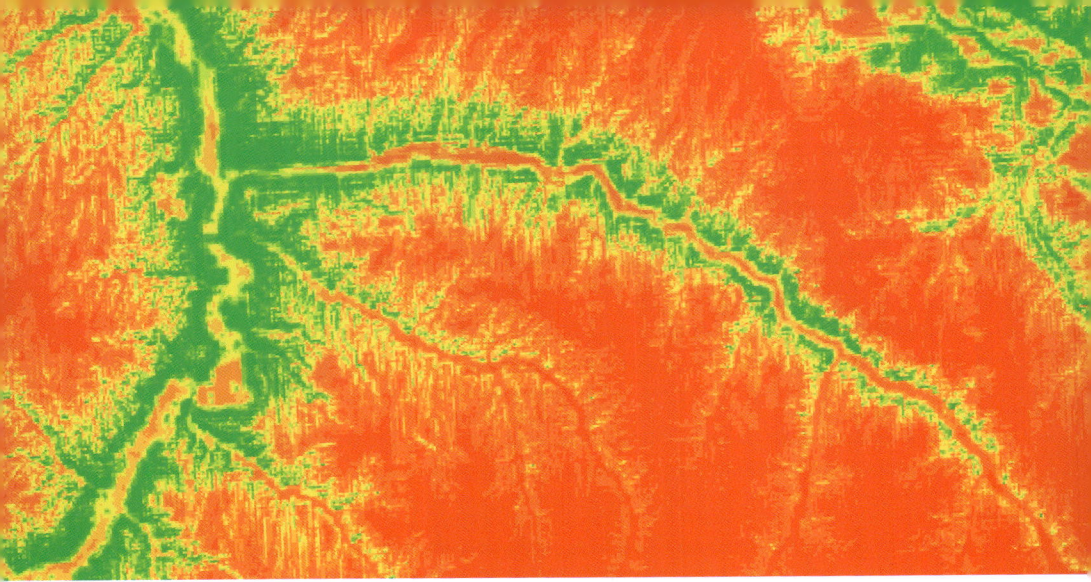

Because populations of *C. concolor* are declining due to habitat loss and persecution at their dens, scientists are making predictive models to find potential denning sites to protect. In this model of a Colorado management area, the green is where dens are likely to occur, while in the red area, they are unlikely. This allows biologists to concentrate conservation guidelines in the green areas. *Stephen Spears.*

suddenly after about thirty-eight hours, with no long-term effects. Given the high virulence and presence of concolor toxin, this species should be considered capable of delivering a fatal bite.

References and Resources: Refer also to the introduction for terrestrial vertebrates. Aird (1985), Aird and Kaiser (1985), Ashton and de Queiroz (2001), Ashton and Guyer (2003), Ashton and Patton (2001), Bieber et al. (1987), Douglas et al. (2002), Engle et al. (1983), Feldner et al. (2016b), Glenn and Straight (1977), Kaiser et al. (1986), Mackessy (2010a), Mackessy et al. (2003), Ownby et al. (1988), Parker and Anderson (2007), Pook et al. (2000), Pool and Bieber (1981), Schuett et al. (2018), Weinstein et al. (1985).

Southern Pacific Rattlesnake (*Crotalus helleri*)

"If you're walking through the flat desert of Phelan, California, and you're bitten by a Southern Pacific rattlesnake, you will start to bleed badly. The snake's venom is loaded with proteins that break down the walls of your blood vessels and that prevent the now-leaking blood from clotting. Let's say you survive. You drive up some twisting mountain roads to the town of Idyllwild, . . . the Southern Pacific rattlesnake lives here too, and you get bitten again. And this time, the venom doesn't go for your blood. The toxins of these snakes include proteins that stop nerves from sending signals into muscles. They start to paralyse you." (*National Geographic,* 27 January 2014, E. Yong reporting)

A large, dark *C. helleri. L.L.C. Jones*

Classification: Phylum Chordata (chordates); Class Reptilia (reptiles); Order Squamata (snakes and lizards); Family Viperidae (vipers); Genus *Crotalus* (typical rattlesnakes); Species *helleri* (Southern Pacific Rattlesnake).

Crotalus helleri was considered a subspecies of the former Western/Prairie Rattlesnake species complex (*C. viridis helleri*). Some consider *C. helleri* to be a distinct species, while others consider it a subspecies of *C. oreganus*. Two subspecies are sometimes recognized, *C. h. helleri* and *C. h. caliginis* (Coronado Island Rattlesnake), but it is usually considered monotypic. Only the nominate subspecies is found north of Mexico.

Identification: This is a relatively large, heavy-bodied snake. Adults are generally between about 56 and 112 cm in length, although the record is 163 cm. Large adults are fairly dark overall, and some can be almost completely black. Young animals generally have a much more pronounced pattern, consisting of 27–43 (usually 35–36) dark blotches with light edges. Between these are lateral dark blotches and below those are ventrolateral dark blotches. The background color is generally lighter than the blotches, being brown to somewhat gray, olive, or yellowish. One of the most distinctive features is the pattern on the head. In smaller animals and young adults, there is a distinctive, boldly marked mask across the eyes, continuous laterally with a pronounced eye stripe. The eye stripe is bordered by white and extends to the corner of the mouth. The top of the head is dark brown. As animals age, these features become less distinct. The tail, just before the rattle, is usually black.

This individual is even darker, actually black, so it is considered melanistic. *Bob Hansen*

Large, dark adults are similar in appearance to the Arizona Black Rattlesnake, but the ranges of these two animals are widely separated. They are also similar to the Northern Pacific Rattlesnake, with which it may hybridize at the contact zone (e.g., San Luis Obispo County, California). Some individuals are similar to Mojave Rattlesnakes, which may be found nearby, but the head and tail markings differ. These two species have hybridized in nature. *Crotalus helleri* has also been reported to hybridize with *C. ruber* in nature.

Distribution: *Crotalus helleri* is found in coastal southern California, from San Luis Obispo and Kern counties south to the Mexican border. In Mexico, it is found in Baja California, just to the extreme northwest of Baja California Sur. It is also found on Santa Catalina and Santa Cruz Islands off southern California. At the north end of its range, it segues into the range of the Northern Pacific Rattlesnake. The Coronado Island Rattlesnake is endemic to South Coronado Island, just south of the U.S. border, off extreme northwestern Baja California.

Natural History: This is an animal of arid and semiarid lands, but within the generally maritime western climate zone (i.e., the Mediterranean climate). It does not occur in the deserts, but may follow stringers of riparian areas and canyons into more inhospitable deserts. It is found from sea level to over 3,350 m in elevation. In southern California, it occurs up to the highest mountains. It can be encountered right up to the edge of the Pacific Ocean in coastal sage scrub and chaparral habitats. Going east, it occurs into the mountains, including the west slope of the Baja California Peninsular and Coast Ranges. Here it occupies grasslands, chaparral, oak woodlands, piñon/juniper woodlands, and montane coniferous forests.

Crotalus helleri is most active on the surface from March through October or November. It is occasionally seen during the winter on warm days. During the milder times, it is usually diurnal, but may be nocturnal during hot weather, particularly at lower elevations. It is typically often most active in early morning and late afternoon.

This species mates in both spring and late summer/fall. Combat and courtship displays have been documented. Parturition usually occurs in late September through late October. Litters of 1–16 (average 7–8) have been reported. Individuals are normally solitary, although small aggregations of two or three individuals have been reported. Hibernacula are not well-documented, but this species is not known to use communal hibernacula. Also, in mild areas,

A large adult *C. helleri* with exceptional markings, similar to *C. oreganus*, with which they may hybridize. Bill Love

they probably brumate near the surface, and may emerge on warm days. In the higher mountains, they probably have more substantial hibernacula.

They use olfaction to find sites from which to ambush potential prey. They are often seen in rocky areas, vegetation, near logs, along rodent trails, and even near debris piles waiting to ambush prey. Southern Pacifics have been documented eating a wide variety of prey, including reptiles, amphibians, birds, and insects. There are several accounts of cannibalism. Young snakes feed predominately on lizards, while adults feed almost entirely on mammals. Mammal prey includes mice, kangaroo rats, chipmunks, ground squirrels, gophers, and lagomorphs. Serum in some ground squirrels (but not all), Gray Woodrat, Cotton Rat, and California Kingsnake effectively neutralize its venom.

Encounters: Southern Pacific Rattlesnakes are native to an area that has become densely populated by humans—coastal southern California. They are found in a wide variety of terrestrial habitats. While they avoid or have been extirpated from most urban areas, they can be surprisingly common in suburban areas, including vacant lots, greenbelts, and parks. They are most commonly encountered in campgrounds, mountain resorts, on hiking trails, and in rural areas. They are mostly encountered in the spring, with about one-half of the annual bites occurring in April to June. This species accounts for about 80% of venomous snakebites in southern California, where it is either the most common or only species of rattlesnake. It is often the most commonly encountered snake of any sort within its range. Poison control centers surveyed between 2001 and 2005 reported 1,226 rattlesnake bites statewide; the other common species responsible for many snakebites farther north in California is the Northern Pacific Rattlesnake. Annually, two hundred or more people are reportedly bitten by *C. helleri*.

Venom Apparatus, Venom, and Symptoms: The fangs average about 9 mm in length. Venom yield ranges from about 75 to 390 mg, averaging about 110 mg. Venom injected per bite has been reported to range between 0 and 235 mg. The LD_{50} has been well studied. It is reported to average 3.6 mg/kg, but there is high variability among populations, individuals, and age classes. Reports of LD_{50} range from 0.84–2.13 (intravenous), 1.4–2.4 (intraperitoneal), average 3.6 (subcutaneous), and average 5.1 (protein weight basis, intramuscular). Two colors of venom have been described: yellow and white. The white venom has been reported to have an LD_{50} of 2.95 mg/kg and the yellow venom at 1.6–1.8 mg/kg. The lethal dose in a human has been estimated at 50–70 mg. The Southern Pacific Rattlesnake is often considered one of the more dangerous rattlesnakes in the United States.

The venom itself is complex and may contain both tissue-destroying and neurotoxic fractions, depending on locality and other factors. Those in southern California do have a relatively higher likelihood of having neurotoxic components than animals from Mexico. *Crotalus helleri* from Mt. San Jacinto, California, were shown to have a neurotoxic Mojave-like toxin;

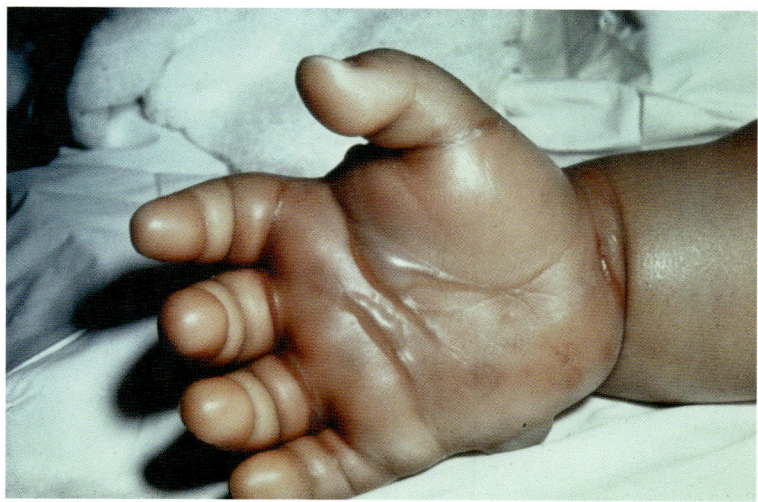

Southern Pacific Rattlesnakes cause most rattlesnake bites in southern California, many of which can be serious. This is one treated by the late, great Findlay Russell. *Findlay Russell*

this population is isolated from *C. scutulatus*, so recent hybrid origin seems an unlikely cause. The venom includes arginine esterase, catrin, catrocollastatin, catroxase I, chymotrypsin-like activity, defibrizyme, dipeptyl peptidase IV, deoxyribonuclease, esterase, hemorrhagic toxin II, kinin-releasing activity, plasmin-like activity, kallikrein-like activity, L-amino acid oxidase, Mojave toxin, NAD-nucleotidase, 5'-nucleotidase, peptide C, phosphodiesterase, phospholipase A_2, phosphomonoesterase, protease, metalloproteinase, ribonuclease 1, thrombin-like activity, and trypsin-like activity. Many of these same or similar components are also in other rattlesnake venoms. Curiously, hyaluronidase was not found in this species or those of the western rattlesnake species complex.

It has been shown that juvenile snakes have high toxicity (but low venom yield) with high levels of phospholipase A_2, while larger snakes have less toxicity (but high venom yield) with more protease activity. This makes sense from a feeding standpoint—small snakes need to quickly immobilize small ectothermic prey (lizards) and small mammals, while adults need to digest larger endothermic prey, which is easier to track. The higher volume of large prey requires more digestion, although it is unclear if predigestion speeds the digestive process.

There has relatively recently been a flurry of sensationalistic journalism claiming that snakes are becoming more toxic—*C. helleri* and *C. scutulatus* are at the center of this topic. However, researchers cannot find scientific validity for this assertion.

Symptoms are well-documented and quite variable, so health-care workers need to be vigilant of symptoms in determining treatment options. There is generally immediate pain associated with a bite, but not always. There is usually swelling and often discoloration of the bite site and affected limb. Hemorrhaging may occur and blebs may form. There may be hypotension, changes in heart rate, dizziness, giddiness, fever, difficulty in speaking, nausea, and weakness. Numbness and/or tingling sensations may be present. In severe envenomations, respiratory distress, cardiac distress, partial paralysis, and shock may occur. Fatalities have occurred from *C. helleri*, but with modern medical care, this is a rare outcome.

References and Resources: Refer also to the introduction for terrestrial vertebrates. Ashton and de Queiroz (2001), Bush and Siedenburg (1999), Bush et al. (2002), Douglas et al. (2002), Dugan et al. (2008), Figueroa et al. (2008), French et al. (2004), Hayes and Mackessy (2010), Lemm (2006), Mackessy (1988, 2010a), Pook et al. (2000), Seifert et al. (2009), Sunagar et al. (2014), Wasserberger et al. (2006), Wingert and Chan (1988).

Timber Rattlesnake (*Crotalus horridus*)

"Few, if any of the other-than-human kinds of people that populate the mythical realities of the North American Indians are held in greater esteem than the rattlesnake man-being . . . all rattlesnake man-beings are revered as grandfathers, as sorcerers with their medicine rattles, and as warriors with their deadly bite." (Hamell and Fox 2005)

Classification: Phylum Chordata (chordates); Class Reptilia (reptiles); Order Squamata (snakes and lizards); Family Viperidae (vipers); Genus *Crotalus* (typical rattlesnakes); Species *horridus* (Timber Rattlesnake).

The Canebrake form of the Timber Rattlesnake. These are large, dangerous snakes.
L.L.C. Jones

Crotalus horridus is commonly known as the Timber Rattlesnake in the northern part of its range, while it is commonly known as the Canebrake Rattlesnake in the southern part of its range. These have been considered different subspecies, *C. h. horridus* for the northern group and *C. h. atricaudatus* for the southern group, but most taxonomists no longer recognize subspecies because observed variation is not consistent with distinct geographical ranges. The standard English name Timber Rattlesnake prevails for the species.

Identification: This is the third largest rattlesnake in the United States; only Eastern and Western Diamondbacks can reach greater lengths. Adults are typically 1 to 1.5 m long, but the species has been recorded to 1.89 m in length. Neonates are about 20–40 cm. Timber Rattlesnakes are sexually dimorphic, as males are larger and less boldly marked than females.

The color pattern is highly variable across the range with regional variation between populations that corresponds to named color morphs. The 15–34 crossbands are distinctive. They are often referred to as chevron-shaped,

The northern form of the Timber Rattlesnake. This large snake lost its rattle.
L.L.C. Jones

but are more typically zigzagged, across the back and often down the sides. The crossbands may become dorsal blotches anteriorly. The ground color varies from brown, tan, gray, yellowish, reddish-orange, to black. The darkest individuals are found in the Northeast, although lighter color morphs occur there as well. The anterior part of the snake is often brighter, becoming dark posteriorly. The typical coloration of the canebrake form consists of a gray or tan background color, often with a yellow or pink hue. There is a cinnamon middorsal stripe, which is usually more defined in the canebrake form. The posterior part of the body is banded and tail is usually black, or at least darker. This species has a facial stripe.

Distribution: The Timber Rattlesnake is the widest ranging eastern U.S. species. It is the only large rattlesnake in most of the East, save the Eastern Diamond-backed Rattlesnake. It occurs as far west as eastern Texas to southeast Nebraska and extends to the Atlantic. Along the eastern seaboard, it ranges from northern Florida to New Hampshire, New York, and Vermont in the Northeast, and southern Minnesota and Wisconsin to the west. The species has apparently been extirpated from Maine and Ontario. It tends to be spottily distributed and generally uncommon, but may be locally abundant.

Natural History: This was the first species of rattlesnake to be encountered by European settlers and the first to be described scientifically. It was described by Linnaeus in 1758. It was quite the biological novelty of its day, as no other snakes on Earth are known to possess a rattle. Throughout much of its range, it is the only large-bodied rattlesnake, so researchers wishing to study large rattlesnakes in the East often have only one choice of subject animal. To this, add the issues of conservation concern (see below) and its medical importance and you have a well-studied animal.

In the North, Timber Rattlesnakes favor rocky areas of woodlands and forests, whereas in the South, they generally occur along river bottoms, swamplands, and other low-lying mesic habitats, or in adjacent uplands. The canebrake form was named because of its proclivity for occupying dense cane (a giant grass, similar to bamboo) habitat ("brakes"). Before European settlement, canebrakes were a common and ecologically important feature of the Southeast. Overgrazing, land-clearing, urbanization, hydrological alteration, fire-regime alteration, and agriculture have reduced this plant community to about 2% of its former range. While canebrake habitats are few and far between these days, the snake still occupies other types of dense vegetation in mesic low-lying

Some Timber Rattlesnakes are very darkly hued, such as this individual. *Bill Love*

areas. The northern populations were also more widespread throughout most of the eastern United States, but they are also generally uncommon.

In the Southeast, these animals are most often found in areas that have large amounts of woody debris and litter. Logs, root masses, stumps, and debris piles are important microhabitats. Throughout their range, rocky areas provide excellent habitat for this species. These surface features are used for hunting and hibernacula. Timber Rattlesnakes in Texas also have arboreal tendencies, being found in shrubs and trees almost one-fifth of the time they are encountered, according to one radio-telemetry study.

Timber Rattlesnakes in the Southeast are active almost any time of the day when weather conditions are appropriate. Various references characterize them as either being nocturnal or diurnal, but it is more complicated. The thermal, moisture, and ecological requirements dictate diel activity patterns. In northern populations, surface activity is restricted by cold weather, so they may be limited to 4–5 months above ground. In southern populations, there is an extended yearly activity period of 9–10 months, from about March to November.

Northern populations often use communal hibernacula. Historically, as many as 50–200 snakes inhabited the same hibernaculum, usually rocky outcrops. Recently, numbers have been reduced by wanton killing. They may brumate with a host of other snake species. In the South, however,

Timber Rattlesnakes tend to brumate singly or in small numbers. Hibernacula include large logs, debris piles, burrows, and burned-out stumps or subterranean refugia, as in burned-out root holes following wildfires. Silvicultural practices of removing large trees and exclusion of a natural fire regime has reduced these types of critical overwintering habitat. They can be loosely social animals; neonates often aggregate with adults, particularly related adults, and can follow scent trails of family members to hibernacula or other aggregation areas.

The Timber Rattlesnake is generally a sit-and-wait predator. It will typically seek out a feeding area by using its keen olfactory senses, then coil up in an ambush pose to await prey. It is often seen lying by or on a log, because these structural features constitute a runway route for small mammals. A typical ambush pose is a partial coil, with the head and fore body on the log. Small mammals form the bulk of the prey. Unlike many rattlesnake species to the west that switch from lizards to mammals as they mature, lizards are not usual fare for young *C. horridus*. However, some lizards, snakes, and amphibians are included in the diet. Juveniles tend to eat White-footed Mice, while adults take larger prey including rats, squirrels, pocket gophers, and lagomorphs.

Males have been observed in combat during the spring and again during the late summer and fall breeding season. Mating occurs as early as the spring, but is typically mid-July to October (usually August). Sperm is stored until ovulation during the following spring. Timber Rattlesnakes only breed every two to three years. Usually 6–10 young are born. They generally disperse from the mother after about a week.

Encounters: Timber Rattlesnakes are not commonly encountered, as they have been reduced in numbers from historic levels and may be absent in densely populated areas. They are more likely to be found in the woods than in a yard, although that depends on the yard. They can be locally abundant, especially in sparsely populated areas. They do not often crawl across particularly long distances, as they are more likely to be hunkered down in an ambush coil than on the move. It has been shown that they avoid crossing roads, so pockets of habitat that are encompassed by roads can isolate populations. The exception to these generalizations is when the animal is migrating to or from its hibernaculum. At this time it may cross roads, clearings, agricultural lands, and other generally avoided habitats, to stay on its learned path. Some populations may have been reduced up to 50% due to road mortality. The best way to encounter these animals is by going into their habitat, such as hiking or working in the woods. They are often in dense vegetation that may be generally avoided by people, although clearing vegetation has a risk of encounter. They are most often encountered in the summer and fall, although they do emerge from hibernacula in spring. However, in the Southeast, one should consider them to be potentially present year-round, especially on mild days.

Although I have set the stage to suggest snakebite by this species is rare, the fact is that there are millions of people within the range of this large and dangerous snake. Most of its range is populated by humans, unlike some of the vast expanses of the West. Snakebites are not as common in the Midwest and Northeast as they are in the West and South, although frequency is lower than it was historically, when Timber Rattlesnakes were more abundant and widespread.

Bite statistics for Timber Rattlesnakes are difficult to quantify, as reporting centers do not separate out rattlesnake species. However, in states where *C. horridus* predominately occurs, we can glean information from a paper on bites reported to poison control centers from 2001–2005. The states that only have *C. horridus* are West Virginia, with thirty-five bites, and Virginia, with fifty-five bites. Similarly, Timber Rattlers are widespread in Kentucky, Missouri, and Pennsylvania, but also have scattered populations of *Sistrurus* spp.; they reported 52–57 bites. The number of "unknown crotalid or venomous" is much higher, ranging from 199 (West Virginia) to 539 (Kentucky), but *Agkistrodon* spp. and *Sistrurus* spp. add to this total.

The best way to avoid encountering these animals is to stay out of their habitat, such as thickets, wooded areas, and rocky areas. They are cryptically colored, blending in well with leaves and dappled sunlight, although they are easier to spot on trails than when bushwhacking. They typically have a mild temperament and are reluctant to rattle, so could easily be trod upon (remember, "Don't tread on me!").

Venom Apparatus, Venom, and Symptoms: They have long fangs and a large venom yield. The fangs are usually 9–10 mm in length. Neonate fangs are about 3 mm in length. Venom yield is about 60–240 mg, depending on size of the animal, but has been reported to exceed 300 mg. The LD_{50} ranges from about 1.6–2.6 (intravenous), 0.26–7.8 (intraperitoneal), and 3–24.9 (subcutaneous) mg/kg, although some reported averages include 0.26 and 0.8 mg/kg. Toxicity changes with age class: neonates are less toxic, but toxicity increases and peaks in juveniles, then drops in adults. One report sums it all up by simply stating that the venom is particularly lethal to mammals. The lethal dose for an adult human is estimated to be about 75–100 mg, and numerous fatalities have been recorded.

The venom of *C. horridus* is well studied. It is primarily hemotoxic and myotoxic, although some neurotoxin fractions may be present. The canebrake form is more neurotoxic than northern forms. Some of the venom components include hyaluronidase, phospholipase A_2, canebrake toxin, crotoxin, crotalocytin, metalloproteinases, L-amino acid oxidase, cholinesterase, bradykinin-releasing enzymes, Mojave-like toxin, and numerous other compounds. Some potential prey mammals show resistance to the venom, including Virginia Opossum, Southern Plains Woodrat, and Hispid Cotton Rat. This is interesting because the Hispid Cotton Rat may form a significant part of the diet.

Symptoms of the bite include pain with redness, discoloration, and swelling that progresses from the bite site. Blood blisters often form on the skin and hemorrhaging and necrosis can be extensive. There can also be general weakness, giddiness, nausea, tingling, breathing difficulty, weak pulse, lowered blood pressure, and gastrointestinal problems. Severe envenomations may also include paralysis, shock, cardiac failure, and renal failure. Symptoms can include disseminated intravascular coagulation, a potentially fatal medical condition in which the normal balance of bleeding and clotting are affected. Secondary infection can come from tetanus and bacteria that have been documented in the mouth of Timber Rattlesnakes. There are many case reports of envenomation by Timber Rattlesnakes, including those of religious groups in the Appalachians who handle venomous serpents as an integral part of their ceremonies.

Remarks: The snake on the Gadsden Flag ("Don't Tread on Me") appears to be a cross between an Eastern Diamond-backed Rattlesnake and Timber Rattlesnake. It not only has diamonds, but also a dorsal stripe. Both of these species were known to early colonists.

References and Resources: Refer also to the introduction for terrestrial vertebrates. Civello et al. (1983), Clark, A. M. et al. (2003), Clark, R. W. (2002, 2004a, b), Clark, R. W. et al. (2010, 2012), Hamell and Fox (2005), Hasiba et al. (1975), Kitchens et al. (1987), Levin (2016), Minton (1953), Reinert and Rupert (1999), Reinert et al. (1984), Rokyta et al. (2013), Rudolph et al. (1999, 2004), Schmaier et al. (1980).

Rock Rattlesnake (*Crotalus lepidus*)

Describing the bite from a 60-cm *C. l. klauberi*: "He described the pain as being deep-seated down to his bones. He said it felt like having his hand 'repeatedly being hit hard by a hammer.' The pain continued to grow over time and ultimately reached a 10 on a 0–10 scale. He said it was pure agony when the hand was touched." (Envenomation Story, this volume)

Banded Rock Rattlesnakes from West Texas are cleanly marked with narrow dark bands and little interband mottling. *Troy Hibbitts*

Classification: Phylum Chordata (chordates); Class Reptilia (reptiles); Order Squamata (snakes and lizards); Family Viperidae (vipers); Genus *Crotalus* (typical rattlesnakes); Species *lepidus* (Rock Rattlesnake).

There are two subspecies in the United States: *C. l. lepidus* (Mottled Rock Rattlesnake) and *C. l. klauberi* (Banded Rock Rattlesnake). There are two additional subspecies endemic to Mexico: *C. l. morulus* (Tamaulipan Rock Rattlesnake) and *C. l. maculosus* (Durango Rock Rattlesnake). It has been suggested *C. l. morulus* is not a valid subspecies. There are differences between northern subspecies (those in the United States) and the Mexican forms, and there have been published accounts suggesting *C. l. klauberi* and *C. l. lepidus* are not distinctive enough, anatomically and genetically, to warrant subspecific recognition. Other researchers have suggested there are additional taxa (species or subspecies) among some of the populations, particularly in Mexico, yet to be described.

Identification: Adults range between about 35 and 75 cm. In a study in the Chiricahuas, 105 individuals averaged 38.8 cm, but large adults are generally between 45 and 60 cm. The record is 82.8 cm. The smallest neonate measured 166 mm. They are highly variable in color and pattern. The Banded Rock Rattlesnake is sexually dichromatic in many populations. In southeastern Arizona and adjacent New Mexico, the male usually has a greenish background color, while the female is gray to bluish. The ground color of the females matches the rockslides they inhabit, and the green of the males matches rocks covered in lichens. These colors reflect the microclimate of warm, dry sites (without lichens) favored by females, and mesic areas (with lichens) favored by males. In other populations they may be tan, pinkish,

Adult Mottled Rock Rattlesnake from the Big Bend of Texas. *L.L.C. Jones*

or even whitish. Both sexes of all color phases have distinctive jagged crossbands darker than the ground color. The crossbands are thinly edged with white. Between the crossbands, there may be markings. In the classic *klauberi*, there are dark, highly contrasting crossbands, with no markings between the bands. The tail is often yellowish and may have faint banding. The tail of juveniles is yellow in both subspecies. Juveniles tend to be more heavily marked between crossbands than adults.

The Mottled Rock Rattlesnake has an even more variable pattern, but is not sexually dichromatic. There are two general color phases, however. Those in the western part of the range tend to be darker than those in the eastern part of the range. This corresponds to the color-matching of predominate rock types. Those in the western part are darker like the volcanic rocks of the area, while those in the Pecos River and Edwards Plateau area are pale, to match the limestone substrates. In addition to these general color morphs, there is generally a great diversity in ground color, ranging from nearly white to pink, bluish, grayish, olive, or tan. The Mottled Rock Rattlesnake earns its name from the tendency to have more mottling between the crossbands than *klauberi*, and the crossbands do not have as much contrast. However, the two subspecies often resemble one another in some areas.

Putative natural hybrids of *C. l. klauberi* × *C. willardi obscurus* and *C. l. klauberi* × *C. pricei* have been reported.

Distribution: The species ranges from southern Arizona, New Mexico, and Texas south to Aguascalientes and northern Jalisco. In the United States, the Banded Rock Rattlesnake occurs in southeastern Arizona, southwestern New Mexico, and extreme western Texas. In Arizona, it is known to occur only in the Canelo Hills and Chiricahua, Dos Cabezas, Dragoon, Huachuca, Santa Rita, Peloncillo, and Whetstone Mountains. In the Peloncillo Mountains, the first record of a Ridge-nosed Rattlesnake from that range was actually a hybrid with a Banded Rock Rattlesnake. In New Mexico, the subspecies occurs from the southern border and farther north than Arizona populations, into Socorro County. In Texas, its range is limited to the Franklin Mountains near El Paso. In the United States, the Mottled Rock Rattlesnake only occurs in the Guadalupe Mountains of southeastern New Mexico, and Texas. In Texas, it is found east of El Paso to the Edwards Plateau, and questionably as far east as about Eagle Pass, where the allegedly, but possibly erroneous, type locality is. Banded Rocks occur south to northern Durango, and Mottled Rocks occur south to northern San Luis Potosí.

Natural History: The Banded Rock Rattlesnake is a mountain form, found from about 1,220 m to about 2,930 m, although they are most often found between about 1,525 m to 2,130 m. It is primarily associated with pine/oak woodlands of the Madrean Archipelago (sky islands allied to Sierra Madre Occidental). The Mottled Rock Rattlesnake occurs at lower elevations, overall, than its Madrean conspecific. At its lower elevation, it can be found in Chihuahuan Desert habitat, characterized by creosote bushes and lechuguilla agaves. It ranges higher into the Chisos, Davis, and Guadalupe Mountains, into oak and pine woodlands. In the eastern part of its range in the Edwards Plateau, its habitat is dominated by shrubs and small hardwood trees, including persimmon and shin oak.

Being primarily a mountain dweller, it is largely a diurnal snake, although its surface activity period is influenced by weather. On particularly warm nights and at lower elevations it can be encountered at night. Most of the range of this species is within the influence of the North American Monsoon, where it is most active during or after summer thunderstorms or when there is cloud cover to moderate temperatures and keep humidity high. Most specimens are found on the surface between March and October, but they are not commonly encountered until late spring and summer. Most surface activity is in the morning or late afternoon. They seek subterranean refugia during the heat of the day, although summer rains and cloudy periods will draw them to the surface. Peak nocturnal activity is in August. Brumation probably occurs in their rocky domains.

This species is aptly named, as it is rarely found far from its rocky habitat. In fact, it is usually found amid cobble and boulder talus fields or rocky outcrops and ledges, but will occasionally wander into the surrounding habitat. When surface active they are often sedentary, in an ambush coil among rocks, although Chihuahuan Desert animals have been shown to move an average of 20 m per day. In the ambush coil, they can wait for hours for prey to cross their paths. Despite their apparent inactivity, they are always alert to the possibility of predators and prey. They are habitat specialists that feed primarily on lizards, although some populations also feed on small mammals. The Banded Rock Rattlesnake's preferred prey throughout most of its range in Arizona and southeast New Mexico is the Yarrow's Spiny Lizard, which is by far the most abundant lizard species in its typical rockslide habitat. The particular prey species of a population probably depends on what is available. Rock rattlesnakes will also consume snakes (including their own species), mammals, amphibians, and invertebrates. Banded Rock Rattlesnakes

This juvenile *C. l. klauberi* is about as big as this Arizona Sister Butterfly, but can still cause significant harm to humans. Juveniles tend to be more mottled. *L.L.C. Jones*

also prey on centipedes of the genus *Scolopendra*, which are quite venomous themselves. In the Big Bend of Texas, Mottled Rock Rattlesnakes feed on a variety of small mammals and lizards, but not centipedes (at least in studied populations). Juvenile Rock Rattlesnakes have yellow tails used for caudal luring. When lizards are captured, they are not released, as would be done with mammals; rather, the venom is allowed to take effect while the snake has hold of the prey. Studies have shown that this species is not as good a mammal-tracker as is the Prairie Rattlesnake, which specializes on mammals. However, when Rock Rattlesnakes bite and release mammal prey, they are usually successful in finding their meal.

Relatively little is known about the life history of this animal. Males have been found in combat displays. There is limited information on mating, and much of what we know is from Mexican subspecies or captive animals. Mating takes place sometime between spring and fall. Sperm are stored over the winter and young are born in late summer, which coincides with the monsoon, and its bounty of young lizards and mammals. Relatively few young are born per litter—the average is about four.

Predators are not well-documented, but presumably these are the usual array of large carnivorous animals that share their habitat, such as kingsnakes, raptors, and some mammals. In Arizona, and presumably elsewhere, White-nosed Coatis have been documented feeding on *C. lepidus*.

Encounters: The casual observer will probably never see a Rock Rattlesnake in the wild. Outdoor enthusiasts are more likely to encounter them occasionally. In general, in order to run into one of these pretty little animals, you would have to seek them out. If detected by humans, Rock Rattlesnakes may remain immobile, hoping to go undetected, but more often they quickly seek shelter in their extensive network of crevices. They often glide, rattling, to just under the surface rocks, but will venture back out when they think the danger has passed. When under these surface rocks, they may give away their presence by continuing to rattle. They are small and the rattle is difficult to hear. Although there is little information on the subject, it seems likely that most bites happen to hobbyists that keep venomous animals as pets and researchers who handle their test subjects.

If they feel threatened and cannot escape, they can be quite nippy. In one of the few "legitimate" bites on record, a herpetologist was bitten while walking past one that was hidden in the hollow of a tree. He did not even brush against the tree and did not even see the snake until after he was bitten. There are stories of people who are bitten while tailing a Rock Rattlesnake.

Venom Apparatus, Venom, and Symptoms: The fangs are among the shortest of rattlesnakes, about 3.5 mm or slightly longer. The venom yield is relatively small, but variable. Venom yield has been reported to be between about 10 and 35 mg, but may be as low as 6 mg and as high as 129 mg. The reports of LD_{50} are also extremely variable, and Rock Rattlesnakes in some areas have a relatively virulent venom. The Mottled Rock Rattlesnake is reported to have an LD_{50} that ranges from 0.72–2.2 mg/kg, 0.15–0.64 mg/kg, or average 1.55 mg/kg, based on different studies. The Banded Rock Rattlesnake includes LD_{50} values of 2.8–13.5 (average 8.8) mg/kg, and 0.9–23.4 mg/kg, based on different studies. More recent studies have been conducted on lethal toxicities (intraperitoneal) of all four subspecies in mice, lizards, and crickets. Raw venom of *C. l. klauberi* is extremely toxic to lizards, having an LD_{50} of 0.17 mg/kg (0.11–0.24 95% confidence intervals), but only 1.36 mg/kg (1.04–1.65 C.I.) in mice, and 4.7 mg/kg (2.6–6.87 C.I.) in crickets. This makes sense for their penchant for feeding primarily on lizards, but not so for centipedes—although crickets are not centipedes. The raw venom of *C. l. lepidus* is slightly more toxic to lizards than mice (1.16 [0.92–1.41 C.I.]

vs. 1.59 mg/kg [0.85–2.3 C.I.], respectively) and less toxic to crickets (3.41 mg/kg [2.58–4.23 C.I]). It seems likely that toxicity may reflect the relative feeding habits and prey selection of any given population, in concert with ontogenetic and individual variation.

All populations seem to have hemorrhagic, myotoxic, and fibrinolytic toxins, but some populations also have Mojave-like toxins. Components include phospholipase A_2, metalloproteinases, serine proteases, esterase, lectins, phosphoesterase, phosphodiesterase, and L-amino acid oxidase. Populations with neurotoxic venom have been found in Cochise County, Arizona, Hidalgo County, New Mexico, and Chihuahua; however, other populations in Cochise and Hidalgo counties, physically close to the others, showed no neurotoxic activity. All studied populations in west Texas lacked neurotoxins. The LD_{50} of two populations with neurotoxins were 0.15 and 0.64 mg/kg, while those without were 2–15.5 mg/kg. Interestingly, all populations, neurotoxic or not, had strongly hemolytic activity.

Symptoms of bites include pain (in one account dubbed "unbearable and excruciating"), burning, numbness, swelling (reaching well beyond the bite site), hemorrhaging, blistering, necrosis, nausea, and vomiting. There are no records of fatalities, but it should be considered capable of delivering a fatal bite, especially if one is bitten by a large individual from a population with high virulence. The venom may be highly algogenic.

Remarks: The Banded Rock Rattlesnake is protected by law in Arizona. It is one of the three sky island rattlesnakes highly sought after in the exotic pet trade, both domestically and internationally. While collecting in Arizona is illegal without a state permit, and state permits are not given for commercial collecting or for the pet trade, their omnipresence for sale on the Internet is curious. Some individuals may have been bred in captivity, but poaching remains a problem, particularly for the Banded Rock Rattlesnake and the two other sky island species—*C. willardi* and *C. pricei*. The Mottled Rock Rattlesnake is protected in New Mexico.

References and Resources: Refer also to the introduction for terrestrial vertebrates. Beaupre (1995), Campbell et al. (1989), Chiszar et al. (1986), Farallo and Forstner (2012), Forstner et al. (1997), Holycross and Jones (in press), Holycross et al. (2002), Jacob and Altenbach (1977), Martínez-Romero et al. (2013), Norris (2005), Price, M. S. (2009), Prival (2008), Prival and Porter (2016), Rael et al. (1992), Swann and Bell (1999), Vincent (1982).

Great Basin Rattlesnake (*Crotalus lutosus*)

"A third unusual death [in Utah from *C. lutosus*] was a tragic fatality (1987), recorded as a homicide, which resulted when a large rattlesnake (*G. v. lutosus*) bit a 22-month-old girl after the snake had been placed around her neck (Washington County). The child died in approximately 5 hours." (Straight and Glenn 1993)

Yellowish variant of *C. lutosus*. Jackson D. Shedd

Classification: Phylum Chordata (chordates); Class Reptilia (reptiles); Order Squamata (snakes and lizards); Family Viperidae (vipers); Genus *Crotalus* (typical rattlesnakes); Species *lutosus* (Great Basin Rattlesnake).

This animal is often considered a subspecies of *C. oreganus*. In earlier literature, it was considered a subspecies of the wide-ranging Western/Prairie Rattlesnake species complex (*C. viridis lutosus*). As with others of the former species complex, information may be muddled by previous taxonomic assignments.

Identification: This is a moderately large rattlesnake, ranging from about 75–135 cm as adults, but most are about 90–120 cm. The pattern and color

Gray variant of *C. lutosus*. Bryan Hamilton

are variable. The ground color of this snake is generally similar to the areas it inhabits, including ashy gray, tan, buff, yellowish, or olive brown. The eye stripe ranges from distinct (especially in young animals) to virtually nonexistent. It generally has 32–49 distinctive, well-separated dorsal markings. The dark markings of this handsome snake contrast with the ground color. The blotches are brown to black, often with light centers, usually edged with a light border. A common pattern of dorsal markings includes narrow, jagged-edged blotches, but there may be blotches more reminiscent of a typical member of the Western/Prairie Rattlesnake species complex. Anteriorly, the dorsal markings are wider and typically more distinct than the posterior markings, which transition into bands. The tail is barred, but not usually black and white as with some other rattlesnakes. Neonates are about 25 cm. The juveniles are even more distinctly marked. They have an intricate pattern on the top of the head that diffuses with age.

This is the only rattlesnake within most of its range (i.e., the Great Basin), except in the south end. On the perimeter of its range, it may hybridize with neighboring rattlesnakes of the Western/Prairie Rattlesnake species complex (e.g., *C. abyssus* and *C. concolor*), although this is not well understood and individuals of all species may be quite variable.

Dark variant of *C. lutosus*. Bryan Hamilton

Distribution: This is a well-named rattlesnake, as it occurs primarily within the Great Basin. It occurs throughout Nevada, except the Mojave Desert portion in the south end of the state, plus adjacent areas of extreme eastern and northeastern California, southeastern Oregon, southern Idaho, western Utah, and the Arizona strip (extreme northwest corner of the state, north of the Grand Canyon).

Natural History: This animal is primarily a denizen of sagebrush desert and other shrubland communities, from around 910–1,220 m in the lowest valleys of the Great Basin Desert, to about 2,130 m in the adjacent mountains. It also occupies grasslands, interior chaparral, piñon/juniper, and open forests. Densely forested areas are generally not occupied. It can be found in a variety of microhabitats, including rocky, sandy, and riparian areas, and even alkali flats and agricultural areas.

These snakes live in places having a harsh climate for a rattlesnake (northern latitudes, relatively high elevations), so have a limited period of activity. Precipitation is predominately in the form of snow. Great Basin Rattlesnakes have long periods of brumation, spending from about one-third to two-thirds of their life in hibernacula. They are most active from May through September, but may move out of their dens in early spring. As with other northern species, they tend to hibernate in large colonies. There are often more than 100 individuals of mixed ages and sexes, as well as other species of snakes and lizards found within hibernacula. Those inhabiting relatively mild climates (lower elevations to the south) tend to brumate in smaller numbers and are surface active for longer periods. They breed every 2–3 years. Mating occurs in July and August. Parturition usually occurs in September. They have 3–13 (usually 5–8) young.

Great Basin Rattlesnakes search for ambush spots by following the scent of mammal activity, and then assume an ambush position to wait for animals passing by. They feed primarily on lizards when young, but as they mature switch to mammalian prey. Lizard prey includes Common Side-blotched Lizards, Sagebrush Lizards, and whiptails. Mammalian prey is mostly a variety of rodents, especially pocket mice and kangaroo rats. Woodrats, pocket gophers, squirrels, and lagomorphs are also consumed. They occasionally feed on ground-nesting birds and their eggs, lizard and snake eggs, amphibians, and carrion.

Encounters: While the Great Basin is the largest of the U.S. deserts, human population density is very low. In fact, U.S. Route 50, which crosses Nevada's Great Basin, has been dubbed "The Loneliest Road in America" because signs of civilization and other cars are few and far between. Thus, these animals generally inhabit remote areas. For those living, working, or recreating in areas occupied by Great Basin Rattlesnakes, this is perhaps the only venomous reptile, and encounters may be relatively common.

These animals are cryptically colored, having a ground color that matches the environment, although the dorsal pattern may be relatively bold. Against a background of dappled light or uneven surfaces such as rock and vegetation, the blotches are inconspicuous. *Crotalus lutosus* can be found in a wide variety of habitats where they can be encountered by humans, such as open areas and greenbelts, parks, rocky areas, abandoned buildings, garbage dumps, and near homes and yards. They can be extremely abundant near hibernacula when they have not dispersed.

This species is often considered one of the calmest, so is not often easily aroused from a feeding coil. Thus, it may not be detected, even when surface active, and it may not rattle, but will defend itself if it feels threatened. Also, some individuals or populations are decidedly more nervous than others. If they do rattle, the rattle is large and easily heard by most people.

Great Basin Rattlesnakes in Great Basin National Park are basking near the entrance of their den. *Bryan Hamilton*

According to one study of venomous snake exposures reported to poison control centers from 2001–2005, there were 119 venomous snake bites reported in Nevada, including 78 from rattlesnakes, 4 from unknown viperids, and 36 from other unknown venomous species. While the species of rattlesnakes were not tracked, one would think a fair number of these were due to bites by Great Basin Rattlesnakes, which is the most widespread species in Nevada.

Venom Apparatus, Venom, and Symptoms: In the few specimens measured, fang length was 6–8.8 mm. Venom yield is large, at 65–240 mg (65–110 mg average range). The LD_{50} has been reported as 4–4.6 mg/kg (intramuscular) and 1.0–2.2 (mg/kg intraperitoneal).

The venom has relatively high levels (among the Western/Prairie Rattlesnake species complex) of metalloproteinases, phospholipase A_2, and phosphodiesterase, and shows kallikrein-like activity. Hemorrhagic ability varies among populations: those in the north end of the range have a higher capacity than those to the south end of the range. The Great Basin Rattlesnake is not resistant to its own venom.

There are few case reports of bites by confirmed *C. lutosus*, but at least one fatality was recorded (see introductory quote); two other cases of fatality by *C. lutosus* in Utah were of older, questionable reports. Two case reports of nonfatal envenomation include a bite on the finger of one person and one on the back of another. However, it should be mentioned these reports from 1945 may include symptoms of the bite as well as the treatments. Symptoms from the bite to the finger included profuse bleeding at the bite site, swelling, discoloration, severe pain in the stomach, thirst, pain-induced insomnia, necrosis, sloughing of tissue at the bite site, and depression. The instance of the bite to the back included stinging and burning pain, generalized then excruciating pain, swelling, discoloration, nausea, vomiting, chills, rapid pulse, diarrhea, heavy perspiration, paralysis of the arm, and a rash. In this latter case, it was apparent that the victim was seriously allergic to the horse-serum antivenom, which undoubtedly added to the laundry list of symptoms. Two herpetologists I know were bitten by *C. lutosus* while handling them, and exhibited the usual symptoms of pain, swelling, and discoloration, among others. One of them required extensive hospital care.

Remarks: Historically, hundreds of Great Basin Rattlesnakes could be found at dens, but over the years, the average densities have decreased as dens were dynamited and closed. This species is protected in Utah.

References and Resources: Refer also to the introduction for terrestrial vertebrates. Adame et al. (1990), Aird (1985), Aird et al. (1988), Cobb and Peterson (2008), Davis (2016), Diller (1990), Douglas et al. (2002), Feldner et al. (2016c), Glaudas et al. (2008, 2009), Hamilton and Nowak (2009), Mackessy (2010a), Minton (1987), Parker and Brown (1974), Pook et al. (2000), Seifert et al. (2009), Straight and Glenn (1993).

Black-tailed Rattlesnakes (*Crotalus molossus* and *C. ornatus*)

". . . if [compartment] pressure increases had been recognized earlier, this would have allowed for earlier surgical intervention. Surgical decompression of compartment syndrome is most effective and results in fewer complications if it occurs within a 'golden period' of 8 to 12 hours post-onset. Delay of diagnosis and treatment in this case likely contributed to the neuropathy and muscle necrosis sustained by the patient." (Hardy and Zamudio 2006)

The honey-yellow variant of *C. molossus* is highly esteemed by pit viper aficionados.
L.L.C. Jones

Classification: Phylum Chordata (chordates); Class Reptilia (reptiles); Order Squamata (snakes and lizards); Family Viperidae (vipers); Genus *Crotalus* (typical rattlesnakes); Species *molossus* (Western Black-tailed Rattlesnake) and *ornatus* (Eastern Black-tailed Rattlesnake).

Until recently only one species was recognized in the United States, *C. molossus*, the Black-tailed Rattlesnake, with one subspecies, *C. m. molossus*, the Northern Black-tailed Rattlesnake. In 2012, the species occurring in the

United States was taxonomically reevaluated and a formerly described taxon was resurrected as a valid species, *C. ornatus*, the Eastern Black-tailed Rattlesnake. Unfortunately, the literature does not distinguish between these taxa before 2012, so one needs to refer to locality data to determine which of the two species are being referenced.

There are other taxa in the *C. molossus* group in Mexico, including *C. m. nigrescens*, *C. m. oaxacus*, and *C. m. estebanensis*; the latter species is endemic to an island in the Gulf of California and is sometimes regarded as a full species. *Crotalus molossus* hybridizes (or intergrades) with *C. basiliscus* in Sonora, Mexico, and these two forms may be difficult to distinguish.

Identification: Although *C. molossus* in the United States was considered monotypic for a long time, herpetologists were aware of common phenotypic differences between *C. molossus* and *C. ornatus*. The difference was noted in unofficial names reflecting general color morphs: yellow morph (*C. molossus*) and dark, gray, or Texas morph (*C. ornatus*). However, there is also variation among both species, so it is best to rely on geographic range and scale characters to distinguish species.

Crotalus molossus is a medium-large rattlesnake, with the largest recorded specimen being 133.1 cm, but most large adults are about 90–120 cm. This is one of the most beautiful of rattlesnakes, being even more ornate than *C. ornatus*. They often have an overall lemon-honey to greenish-yellow coloration, with dark contrasting dorsal and lateral markings. Anteriorly, the dorsal markings are blotches, which give way to saddles posteriorly. The saddles then become crossbands and fade on the posterior part of the body. Some individuals or populations are more boldly colored than others. As the name suggests, the tail is black. The head has distinctive black shading from the eyes to the rostrum that extends posterolaterally as an eye stripe past the mouth. There is an average of 32 dorsal blotches, which are not joined dorsally in the front.

Not surprisingly, it is similar to *C. ornatus*, but that species tends to have more of a gray to brown or olive background color, with blotches that may not have high contrast. In many individuals, the crossbands fade posteriorly, to the point where they are barely noticeable. There is an average of 29 dorsal blotches, which are joined dorsally in the front, sometimes into a dorsal stripe-like pattern that is somewhat similar to the Trans-Pecos Ratsnake (*Bogertophis subocularis*), which often shares its range. *Crotalus ornatus* also has more white in and between the anterior dorsal blotches. There are also scalation differences.

A more common hue of *C. molossus*. L.L.C. Jones

Crotalus molossus and/or *ornatus* have been reported to hybridize with other species, including *C. atrox*, *C. scutulatus*, and Mexican species of the *C. molossus* complex, but hybridization is not well known.

Distribution: The ranges of the two species are not precisely known, in part due to limited genetic sampling in the western part of the range. For example, the Chihuahuan Desert (typified by *C. ornatus*) extends into extreme southeastern Arizona, but these western forms are presumed to be *C. molossus*, so more research is indicated.

Crotalus molossus occurs in Arizona, east of the Colorado River, and in southwestern New Mexico. It occurs primarily in the upper Sonoran Desert, Madrean Archipelago, and Mogollon Mountains, and possibly western Chihuahuan Desert. In New Mexico, its range is also in the Madrean Archipelago (e.g., Peloncillo and Animas Mountains) and the Mogollon Mountains, as well as the Pinos Altos Mountains. *Crotalus ornatus* is essentially found east of this area in New Mexico and Texas. These two species apparently do not hybridize and their ranges are not known to overlap, but more sampling is probably in order. Their ranges are mostly separated by valley grasslands, but there is some contiguous mountain habitat. *Crotalus molossus* also ranges

The gray tones of *C. ornatus* are less ornate than most *C. molossus*. L.L.C. Jones

into Sonora, including Isla Tiburón, and Chihuahua. In Chihuahua, its range meets that of *C. m. nigrescens*, with which it may hybridize. *Crotalus ornatus* also occurs in Chihuahua.

Natural History: The Western Black-tailed Rattlesnake is generally considered a mountain species although it can be found at sea level in Sonora. It occurs in desert mountains to the east of the Colorado River, near the valley edge, to nearly 3,000 m in alpine forests. Although it can be common in some desert mountains, where it shares its habitat with *C. tigris* and *C. pyrrhus*, it is most often encountered in oak and pine/oak woodlands. Where ranges and elevations overlap, it shares its habitat with other montane species. It is seemingly less common in higher-elevation conifer forests. Within these vegetation communities, it is usually found in the vicinity of rocks, often in drainages and canyon sides. The Eastern Black-tailed Rattlesnake is found in the same type of habitat, from Chihuahuan Desert foothills, up into mountains, including the Guadalupe, Chisos, and Davis Mountains, and the hills of the Edwards Plateau of Texas.

They are most active from April through October. The range is primarily within the region of the North American Monsoon, and they tend to be

A browner variant of *C. ornatus* resembles *C. molossus,* and the overlap in phenotypes is why herpetologists were not quick to recognize subspecies of the former *C. molossus*—before advancements in genetic technologies. *L.L.C. Jones*

relatively inactive until the monsoon arrives, when they search for mates and prey. They brumate in rocky retreats, caves, burrows, and possibly mine tunnels, where they also seek shelter from summer heat, although shrubs, cacti, and surface objects may also be used. In some areas they tend to overwinter alone, but are also found in communal hibernacula, sometimes with other snakes, in more northerly areas and higher elevations. These rattlesnakes are generally diurnal during the spring and fall and nocturnal during the summer, depending on weather conditions. As with others snakes within the monsoon zone, they may be surface active in the day during the summer after thunderstorms and when cloudy conditions prevail, as well as at night.

Black-tailed rattlesnakes have been reported to engage in male combat rituals. They breed from mid-July to early September. Males and females may remain together for several days during courting, and male combat may occur if new suitors arrive. Females will store sperm over the winter until needed for fertilization. Parturition occurs during late July to September, when the monsoon's bounty of small mammals is high, providing food for neonates. There are 3–16 young born live (average about 6, usually less than 10). Females bask and select parturition sites that may be near their hibernacula or dens. The neonates remain with the female for about a week, until the first molt. The male may also stay with the female during the breeding period and for a short time after parturition. Researchers who have been studying a population of black-tails for many years have suggested these animals are much more engaged socially than we normally think of for snakes (but see accounts for Arizona Black and Timber Rattlesnakes).

This is primarily a mammal-eating species. They may hunt actively or passively. When ambushing prey, they will select sites at which to position themselves that have odors of small mammal activity. A number of mammal taxa are included as prey, including mice, woodrats, kangaroo rats, squirrels, and lagomorphs. However, they have also been known to feed on birds, lizards, snakes, amphibians, and insects. Unlike many species of Southwest rattlesnakes, the neonates do not seem to favor lizards (which may be common in their habitat), but generally feed on small mammals.

Encounters: While black-tails mostly occur in remote locations, such as mountains in sparsely populated areas, they are common, so in communities or recreational lands within the species' range, they are likely to be encountered. For example, herpetologists doing a long-term study on this species in the Chiricahua Mountains of Arizona are sometimes called on to remove Western Black-tails from residences in the nearby towns of Portal and Paradise. Hikers in mountains sometimes encounter this species more than any other sympatric species. In Big Bend National Park, *C. ornatus* is said to be the most commonly encountered rattlesnake.

Among rattlesnakes, these two species are particularly mild-mannered. But they are rattlesnakes, so if one picks one up or steps on it, they have a good chance of being bitten. Also there is variation in temperament among all rattlesnakes. For example, I recently found a *C. molossus* on a dirt road, so I got out of my car to move it out of harm's way. It struck vigorously at me and bit the tongs several times while unloading venom. After I moved it off the road, I placed it near a tree trunk for a photo opportunity. It then struck the tree several times in rapid succession before it turned around to note where I was, and then it resumed striking at me repeatedly. Because these snakes are common, it behooves the backcountry adventurer to stay on the path and pay attention to their surroundings. They are not as likely to rattle as most other rattlesnake species, but the rattle is large and loud.

People living in black-tail country should be especially wary of this large and potentially dangerous snake. Even though they can be vibrantly colored against a plain background, they are cryptic against a background of shrubs, leaves, or other debris. They are also popular among hobbyists who may keep them as pets. Their placid demeanor may lead the hobbyist into a false sense of security.

Venom Apparatus, Venom, and Symptoms: The fang length, venom yield, and LD_{50} are clouded by the fact that *C. ornatus* was lumped within *C. m. molossus* for a very long time, and subspecific variation (i.e., the Mexican forms), was not always noted in publications. Between these species, the fangs are long, measuring between 9.6 and 13.5 mm, and the strike range is presumably high, given the length of this species. These are important metrics for those who think that boots alone will keep the observer from being bitten. The venom yield for an adult (for the former *C. m. molossus*) averages about 286 mg, but has been measured as high as 540 mg.

The LD_{50} estimates for *C. molossus* and/or *C. ornatus* ranged from 8.4–17.4 mg/kg in previous studies. The venom is generally regarded as being of moderate toxicity compared to other species of rattlesnakes, but fang length is long and venom yield high, so these are potentially dangerous snakes. No fatalities have been reported but there are case studies of serious envenomation.

Venom components for black-tailed rattlesnakes (species combined) consist of phospholipase A_2, phosphodiesterases, esterases, and fibrinolytic/hemorrhagic proteins (M4 and M5). Mojave-like neurotoxins have not been found in the venom. The venom is strongly hemorrhagic, even compared to other rattlesnakes.

Symptoms of envenomation include pain, swelling, leakage of blood into surrounding tissues, some blood coagulation, blebs at the bitten area, elevated blood pressure, reduced platelet count, and fibrinogen deficiency. In one case study, compartment syndrome was noted, but pressures had not been monitored. CroFab had been used initially, but was discontinued. Due to a lack of monitoring and discontinuation of the antivenom, compartment pressures continued to increase and tissue damage was extensive, requiring fasciotomy. It took two years for the patient to regain nearly complete function of the affected limb.

References and Resources: Refer also to the introduction for terrestrial vertebrates. Anderson and Greenbaum (2012), Chen and Rael (1997), Greene et al. (2002), Hardy and Greene (1999, 2005), Hardy and Zamudio (2006), Hardy et al. (1982), Persons et al. (2016), Rael et al. (1992, 1997), Sánchez et al. (2001).

Northern Pacific Rattlesnake (*Crotalus oreganus*)

". . . just after my 13th birthday, I was bitten by a Northern Pacific rattlesnake . . . near Yosemite National Park, California . . . . I had my arms dangling at my side, and a 5-foot-long rattlesnake bit me in the middle of my left palm. . . . 13 surgeries, $700,000 worth of helicopter flights, surgeries, and hospital stays (paid by my insurance, of course), and 20 months later, I am very happy with the outcome of this experience and my good fortune of getting through all this without any significant loss." (RattlesnakeBite.org, accessed 12 October 2016)

Individuals from Carrizo Plains, California, may include hybrids, as this is a contact zone with *C. helleri*. L.L.C. Jones

Classification: Phylum Chordata (chordates); Class Reptilia (reptiles); Order Squamata (snakes and lizards); Family Viperidae (vipers); Genus *Crotalus* (typical rattlesnakes); Species *oreganus* (Northern Pacific Rattlesnake). This taxon is often considered a subspecies of the "new" Western Rattlesnake, as *C. o. oreganus*.

In earlier literature, it was considered a subspecies of the "old," wide-ranging "Western Rattlesnake" (*C. viridis oreganus*). As with others of the former Western/Prairie Rattlesnake species complex, information may be muddled by taxonomic changes. To avoid confusion with so-called standard names that vary, I recommend that "Northern Pacific Rattlesnake" should only be used for *C. oreganus* if considered monotypic or *C. o. oreganus* if considered polytypic.

Identification: This is a medium-large rattlesnake, usually between about 75 and 120 cm, with a record of 152.4 cm. The ground color ranges from gray to tan, brown, olive, pinkish, salmon, yellowish, or orangish, and even to essentially black. Anteriorly, the dorsal blotches are broad and fairly close together. They are typically medium to dark brown in the centers, or a dark version of the ground color, edged by darker pigment, and separated by contrasting, light interspaces. Posteriorly the blotches merge into bands that may be essentially black and white near the tail. They typically have a very distinct eye stripe of dark edged by white. The top of the head is usually dark. The pattern may diffuse and darken with age, although some large animals still retain distinct markings. *Crotalus oreganus* is similar in appearance to *C. helleri*, but is usually more vividly marked. These two species hybridize at the contact zone in San Luis Obispo County, California. *Crotalus oreganus* may also hybridize with *C. lutosus* at their contact zones, in California, Oregon, and Idaho.

Distribution: This species' distribution essentially picks up to the north where the *C. helleri* leaves off to the south in San Luis Obispo and Kern counties. It ranges north along the Pacific Coast Ranges and Sierras, along the Cascades foothills, and in the Central and Willamette Valleys. It is generally uncommon to rare in the valleys. Proceeding north, its range becomes disjunct, as the species apparently cannot exist in the cool climates of the northern Oregon and Washington coastal ranges or Puget Sound. The species' range at this latitude is limited to the more arid inland areas of the Columbia River Plateau, as far north as south-central British Columbia.

Natural History: The natural history of *C. oreganus* is fairly well studied. The Northern Pacific Rattlesnake occurs across a huge elevational breadth, from sea level to about 3,350 m. It does not occur in true deserts, but can be found in semiarid shrublands, including coastal sage scrub, coastal chaparral, valley shrublands, mountain foothills, and Great Basin steppe. This snake also occurs in grasslands, oak woodlands, and even into conifer forests. Fairly high elevations of the Sierras are occupied, but not those in the more northerly Cascade Range. Wherever they occur, they are often in the vicinity of rocks, where they spend the cold months in hibernacula. After brumation, they will venture far out into neighboring habitats.

Typical of northern-latitude snakes, they have a reduced activity period compared to southern species. The activity period varies by latitude and elevation. In the Northwest, they may head into hibernacula as early as mid-September and not emerge until May. In the Southwest, they can emerge as early as late March and not overwinter until October to December. The more northerly populations also tend to have the highest site fidelity for hibernacula and brumate in large numbers (up to 500). Also depending on latitude and elevation, surface activity can be diurnal to nocturnal, as they seek optimal temperatures.

Females give birth in August and September, and gestation may be as short as ninety days (at least for an Idaho population). They give birth to 1–15 young, and possibly as high as twenty-five.

This species feeds on a variety of prey, including small mammals, lizards, birds, and amphibians. Typically, they are mammal feeders, foraging on mice, rats, gophers, squirrels, and lagomorphs. Young will either feed on lizards or small mammals, depending on what is available in their vicinity. There is some interesting research on the relationship of ground squirrels and *C. oreganus*.

A pair of Northern Pacific Rattlesnakes mating. *"Buddha Dog"/CCA-SA /Flickr*

Neonate *C. oreganus.* Ron Wolf

California Ground Squirrels have resistance to the venom, and they are very aware of rattlesnakes, taking every opportunity to harass them.

Encounters: Northern Pacific Rattlesnakes may be vividly marked, but their contrasting patterns often make them difficult to see. This snake may be abundant locally, but is often found in sparsely populated areas in much of its range. For example, there are only about 25 bites/year reported to poison control centers in Washington, where it is the only species of rattlesnake—and this figure includes captive animals of any species. In populated areas of central and northern California, bites are much more likely to happen.

Venom Apparatus, Venom, and Symptoms: The fang length is 6.2–7.5 mm (average 6.6 mm). Information on venom yield is a little cryptic in the literature for this species, but a fairly large animal probably yields about 90–200 mg of venom. An average defensive bite volume between 41 and 93 (average 64) mg has been reported. This species has one of the lower LD_{50} estimates among the Western/Prairie Rattlesnake species complex, with an intravenous LD_{50} of 2.8 to 4.8 mg/kg.

The venom is relatively well studied and there are numerous case histories of bites by this species. There are several metalloproteinases present in the venom. The venom is high in metalloproteinase activity, kallikrein-like activity, plasmin-like activity, and thrombin-like activity, but low in phospholipase A_2.

One study of two hospitals in a single county (Santa Clara County, California) reported on 57 case studies of bites by this species, from 1964 to 1981. No deaths occurred. Symptoms common to all bites included local pain,

Humans are a bigger threat to snakes than vice versa. *Sam Beebee/CCA/Ecotrust/Flickr*

progressive swelling from the bite site, numbness, regional lymph node pain, and agitation. In more severe envenomations, patients showed discoloration of the skin, blebs, giddiness, and nausea. In the most severe cases, blood-tinged secretions, renal failure, shock, and coma were suggested. However, in this study, most envenomations were relatively mild and none were severe. While there were no compartment-syndrome emergencies or fasciotomies reported in this series, there are in other studies. This species has been known to cause fatal envenomation.

In one online first-person account of a 13-year-old boy accidentally bitten on the hand while sitting on a rock in Yosemite, the envenomation was obviously severe. The victim claimed to have been unconscious for about twenty-four hours. While general symptoms were not included in the report, there were fasciotomy (palm to bicep) and skin graft surgeries. The victim claims that there were thirteen surgeries and thirty-five hospital days, at a total cost of $700,000 (helicopter flights, surgeries, hospital time). After twenty months he recovered without significant loss to physical abilities.

References and Resources: Refer also to the introduction for terrestrial vertebrates. Butner (1983), Diller and Wallace (1984), Douglas et al. (2002), Gier et al. (1989), Hayes et al. (1995), Mackessy (1988; 1991; 1993a, b; 1996; 2010a), Pook et al. (2000), Poran et al. (1987), Swaisgood et al. (1999, 2003), Wallace and Diller (1990).

Twin-spotted Rattlesnake (*Crotalus pricei*)

"In addition to climate change, populations of *C. pricei* in Arizona are impacted by poaching to various degrees. Laws in Arizona and New Mexico explicitly prohibit collection of *C. pricei*. Nonetheless, poaching does occur and snakes end up as pets in the United States and Europe." (Prival 2016)

Gray variant of *C. pricei*. Some have a more blue to purple hue. *Bill Love*

Classification: Phylum Chordata (chordates); Class Reptilia (reptiles); Order Squamata (snakes and lizards); Family Viperidae (vipers); Genus *Crotalus* (typical rattlesnakes); Species *pricei* (Twin-spotted Rattlesnake).

There are two subspecies, *C. p. pricei* (Western Twin-spotted Rattlesnake), found in the United States and Mexico, and *C. p. miquihuanus* (Miquihuanan Rattlesnake), which is confined to Mexico.

Identification: This is a small, slender species of rattlesnake, ranging from about 30–60 cm as adults, but most encountered are between 30–45 cm. Neonates are only 15–20 cm. The ground color is gray to bluish or purplish gray, or even with a hint of pink. Some individuals or populations are more brownish or reddish-brown. They generally match the background rocks where they are found. As the name implies, they have paired spots that can fuse into blotches or bands, especially on the posterior part of the body and tail. They have between 39 and 64 brown to dark-gray pairs of spots. The head is not as triangular as most pit vipers, the body is slender, and the rattles are tiny, giving this rattlesnake a superficial appearance of a nonvenomous snake. The proximal rattle is usually orange to reddish. Neonates have yellow tails that are used for caudal luring.

Distribution: In the United States, this species is endemic to Arizona, where it occupies Madrean sky islands, the isolated mountain ranges to the north of the Sierra Madre Occidental. They have been documented from the Chiricahua, Dos Cabezas, Huachuca, Pinaleño, and Santa Rita Mountains, all in the extreme southeastern part of the state. In Mexico, it ranges south into Sonora, Chihuahua, and Durango, with an isolated population in Aguascalientes.

Natural History: This is a mountain dweller that inhabits some of the harshest environments endured by rattlesnakes. Its elevational range in the United States is about 1,980–3,250 m, to nearly the top of High Peak (Mt. Graham)

Brown variant of *C. pricei*. Bill Love

in the Pinaleño Mountains, the highest point of southern Arizona. Vegetation ranges from pine/oak woodlands at the lower elevations to conifer forests at the higher elevations, including pines, Douglas-fir, and even subalpine fir and Engelmann spruce at the highest elevations.

These animals are typically associated with rock, especially talus slopes and outcrops, but can also be found venturing into the nearby forest. At the higher elevations, they can be found in the vicinity of mountain meadows. The Banded Rock Rattlesnake is found in some of the same mountains, but occupies oak and pine/oak woodlands generally below the elevational range of *C. pricei*. The other "sky island" rattlesnake, *C. willardi*, sometimes shares the range with these other two species, but only reaches about 2,400–2,750 m. The Black-tailed and Arizona Black Rattlesnakes sometimes share their montane habitat, as well. The Arizona Black is only sympatric in the Pinaleño Mountains.

Crotalus pricei becomes surface active in April or May to October. They can occasionally be seen November through January when temperatures are especially warm and sunny. Their peak activity season seems to be during the summer rainy period of the North American Monsoon, in July and August. They are diurnal, often found at the surface in the morning, afternoon, and following thunderstorms. They avoid basking and hunting on the surface during the heat of the day during the summer, unless storms moderate the temperature.

Courtship and mating is believed to occur in August and September, but information is scant. Parturition occurs in July and August. Females seem to reproduce biennially. Litters range from about 3–9 young and average 5–6.

Twin-spotted Rattlesnakes feed primarily on lizards, and in the United States, this is almost exclusively Yarrow's Spiny Lizards, which share the high mountain habitats. Other lizards, small mammals, and birds are also preyed upon. When lizards are struck, they are not released, but small mammals are bitten, released, and tracked. These snakes ambush their prey by waiting for them to cross their paths. Young twin-spots are born at a time when young Yarrow's Spiny Lizards appear on the scene.

Encounters: Twin-spotted Rattlesnakes are found in sparsely populated, isolated, high-elevation mountains. They are most often found in rocky areas that are not generally used by humans, such as steep talus slopes. They are virtually never seen by the general public, but may be encountered by hikers, campers, researchers, and poachers.

These small rattlesnakes are wary and when detected, they will quickly disappear into their rocky domain, rattling all the way. Their tiny rattle is difficult for many people to hear, so they can easily go unnoticed while on the surface. They are not too inclined to bite—they would rather flee—but will bite if they feel threatened. If one tries to grab them as they are diving under rocks, they can be quick to strike. Thus, virtually everyone bitten has been handling them. Certainly, bites occasionally happen with illegally kept terrarium pets.

Venom Apparatus, Venom, and Symptoms: The fangs are very short, being only about 3 mm, and the fangs of neonates are only about 1 mm. The venom yield is very low for a rattlesnake, about 4–8 mg. The toxicity is not well known and some references have stated that they have "low to moderate toxicity," while others suggest the venom is highly toxic. The intramuscular LD_{50} has been reported as 11.5 and intravenous at 0.95 mg/kg. Either way, with such a low venom yield, this is probably one of the least dangerous species of rattlesnakes.

Bites are rarely reported, but are said to cause both local and systemic reactions. Symptoms were said to be "more serious than expected from such a small rattlesnake." One of the four reported bites resulted in a secondary infection.

Remarks: Twin-spots are one of the three species of small Madrean sky island rattlesnakes of Arizona that are booty for poachers who sell them in the black market reptile trade. The best-known area for *C. pricei* is frequently visited by poachers, and a long-term study to determine the effects of poaching has been conducted in the Chiricahua Mountains. Poaching has not caused significant population declines yet, but has caused a decrease in average size, so larger females are scarce; they give birth to more young than smaller females. It has been suggested that this species could be at risk of extirpation from climate change, as it occupies the highest elevations already, so could not migrate upslope to reach cooler preferred temperatures. Because of these threats, a famous snake-watching (and poaching) site in the Chiricahuas has been closed to the public.

References and Resources: Refer also to the introduction for terrestrial vertebrates. Bryson et al. (2002, 2011), Cruz et al. (1987), Goldberg (2000), Mahaney (1997), Prival (2016), Prival and Schroff (2012), Prival et al. (2002).

Southwestern Speckled Rattlesnake (*Crotalus pyrrhus*)

"The pain was very severe about an hour after the bite. Subsequently, there was much discoloration; the fingers were like red bananas. There were blebs all over the fingers and wrist, and the palm became one great blister." (Klauber 1972)

A typical *C. pyrrhus* is very speckled, matching the background in color and pattern.
Marshal Hedin/CCA/Flickr

Classification: Phylum Chordata (chordates); Class Reptilia (reptiles); Order Squamata (snakes and lizards); Family Viperidae (vipers); Genus *Crotalus* (typical rattlesnakes); Species *pyrrhus* (Southwestern Speckled Rattlesnake).

Until recently, *C. pyrrhus* was considered a subspecies of the Speckled Rattlesnake, *C. mitchellii*, which contained several other subspecies. It is now considered a species complex and former subspecies have either been elevated to species, or subsumed within one of the other species. The Panamint Rattlesnake (*C. stephensi*) and Tiger Rattlesnake (*C. tigris*) are two other taxa of the complex in the United States. Other species in Mexico include *C. mitchellii* and *C. angelensis*. *Crotalus mitchellii muertensis*, an island form in the Gulf of California, has been subsumed into synonymy with *C. pyrrhus*. The common name of *C. pyrrhus* has long been recognized as Southwestern Speckled Rattlesnake.

Identification: This is a medium rattlesnake, usually 60–100 cm as an adult, although males from coastal chaparral habitat in southern California may exceed 120 cm occasionally. The record for *C. pyrrhus* is at least 132 cm. Neonates are born at a length of about 20–30 cm. The color pattern is highly variable, but is usually speckled with light and dark flecks. Ground color ranges from nearly white to tan or brown with strong hues of yellow, orange, red, pink, blue, or black. There are always numerous crossbands, usually ill-defined or obscure. The bands number from about 26–41, with a mode of 34. They may have a beautifully intricate pattern, composed of light and dark mottling, with light and dark patches between the bands. Toward the tail, the bands tend to become more distinct, numerous, and closely spaced. The result is that near the end of the tail, most specimens have dark and light banding that in extreme cases can nearly be black and white. The background color, banding pattern, and speckling match the background substrate extremely well. This is a location and population-level phenomenon, as individuals do not change color or pattern.

A more banded individual, which resembles it relative, *C. tigris*, found in the same area; however, Tiger Rattlesnakes have a proportionately smaller head. *L.L.C. Jones*

The white morph of wild-caught *C. pyrrhus* is protected by law but is still highly sought after in the black market pet trade. *L.L.C. Jones*

The most similar species are the closely related Panamint and Tiger Rattlesnakes. The Panamint typically has more distinct banding and less speckling. It also has less dark and light banding on the tail with a smaller area of black pigmentation. The tail pattern is more similar to color of the body. The Southwestern Speckled Rattlesnake has internasal (nasorostral) scales, which are lacking in *C. stephensi,* and lacks the ridged or creased supraocular scales of the Panamint Rattlesnake. The easiest way, however, to differentiate these species is by their geographic ranges. The ranges essentially abut one another latitudinally, but do not overlap. Hybridization is not known to occur in nature. This species has hybridized in captivity with *C. helleri* and *C. ruber* and such hybrids are also thought to occur in nature. Reported hybridization with *C. tigris* is considered erroneous. However, *C. pyrrhus* has historically hybridized with other species of the *C. mitchellii* complex (*C. stephensi* and *C. angelensis*). There have been several studies addressing biogeography in Panamint and Southwestern Speckled Rattlesnakes.

Tiger Rattlesnakes are also rock dwellers and may have a similar pattern of speckling, banding, and ground-color matching, but they can be differentiated by the relative head and rattle width. The Tiger Rattlesnake has a proportionately much smaller head than other similar rattlesnakes in the United States, with the head as wide as or slightly wider than the rattle.

Distribution: The range includes southern California, western Arizona, extreme southern Nevada, and a tiny area of southwest New Mexico. It also occurs in Baja California del Norte and extreme northwest Sonora. While it is mostly found in the Lower Colorado River subdivision of the Sonoran Desert, it ranges north into the southern Mojave Desert. Panamint Rattlesnakes range north of this species, in the Mojave Desert, but the geographic split between the species does not align completely with the desert delineations. Similarly, the ranges of *C. tigris* and *C. pyrrhus* also abut one another in western Arizona, and there is some overlap.

Natural History: This species occurs on both sides of the Baja California Peninsula Range in southern California and adjacent Mexico. On the west side, it inhabits coastal sage scrub, chaparral, and piñon/juniper habitat, but is not found all the way west to the Pacific Ocean, avoiding the coastal fog belt. On the east side of the mountains, it usually inhabits chaparral and desert scrub. The rest of its range inland is usually in desert habitat. It ranges in elevation from near sea level to 2,213 m. This is a saxicolous species that rarely ventures far from its rocky lair. It is most often seen amid boulder-strewn foothill areas and low desert mountains, generally avoiding open desert flats. However, it may venture from rocks into surrounding desert floor to forage or seek mates. It is occasionally encountered higher in the mountains.

They are surface active around March or April through September or October. They are active in the day during mild conditions of spring and fall, as well as summer days with heat tempered by rain and clouds, but are otherwise primarily nocturnal. During the occasional summer rains that occur within its range, it may take advantage by drinking rainwater that collects in its coils. This is known as rain harvesting, a feature that is shared by some other rattlesnakes. During hot weather, it is usually active from dusk until just after midnight, although when nights are cool, nocturnal activity is limited to an hour or two after dusk. During the heat of the day, the animal seeks shelter under rock cover, in crevices, under bushes, and in mammal burrows. Brumation occurs during the cooler fall and winter months, when up to twenty animals have been recorded in hibernacula. Combat behavior between males has been observed. Mating could occur during any part of the active season, based on gametic cycles, but is presumed to typically peak mid-April to mid-June. Gestation has been reported as 140 days and parturition usually takes place from July to September.

Young *C. pyrrhus* feed primarily on a variety of lizard species, while adults tend to feed on small mammals, but they will also feed on lizards, including the Common Chuckwalla, a large rock-dwelling species that occurs across most of its range. Common prey mammals include kangaroo rats, woodrats, ground squirrels, mice, and lagomorphs. Occasionally birds, and possibly insects, are taken. This species has been recorded as both an ambush predator and active predator. When these snakes strike lizard prey, they hold onto it until the lizard is immobilized and can be consumed. When mammals are struck, they are released and tracked, to be consumed after they have become immobile or dead.

Encounters: There is little published information on encounters and envenomation by this species. It is generally regarded as uncommon, but may be locally abundant. Bites are also presumably uncommon. Some published reports of bites by venomous species in southern California fail to mention this species as being one of the potential culprits. Because they are primarily desert- and rock dwellers, they are not likely to be encountered by the general populace. However, there are geographic areas where this species occurs in the vicinity of suburban settings in southwestern California, and portions of desert communities such as Palm Springs, Borrego Springs, Las Vegas, and Kingman.

They are extremely cryptic on their speckly rocks and gravel substrates, resembling granite boulders. If your activity takes you on rocky slopes, as when bouldering for recreational purposes, keen eyes and ears are necessary. When startled, these animals are usually quick to rattle and defend themselves. They are sometimes seen on roads at night, particularly paved roads that closely parallel rocky slopes.

Due to its great variability and attractive patterns, this species is popular among exotic pet owners who keep venomous serpents. One Internet site I viewed had a "rare white phase" Speckled Rattlesnake that was for sale at over $1,000. These white phase animals are essentially protected by law because they live on Department of Defense lands that require a permit for access. This population is being studied by the Arizona Game and Fish Department and Department of Defense to assess its status and effects of poaching. Given the generally nervous disposition of this animal, having one in captivity is a potential hazard to its owner (and to others if the animal escapes).

Venom Apparatus, Venom, and Symptoms: Some of this information may be clouded because of its being lumped with other members of the speckled rattlesnake complex. Fang length is between about 7 and 11 mm. The venom toxicity is about average among rattlesnakes. The LD_{50} ranges from 4.1–23.5 mg/kg, with an average of about 9.6 mg/kg (intramuscular). The venom yield is relatively high, ranging from 60 (possibly juveniles) to 350 mg. A typical value for an adult probably ranges between 100 and 250 mg. It is the high-venom yield of *C. pyrrhus* that renders this animal so dangerous.

The venom of *C. pyrrhus* is of the tissue-destructive type, being relatively high in metalloproteinases, although other species in the *C. mitchellii* complex may have some Mojave-like neurotoxins. The venom also contains phospholipase A_2, esterase, phosphodiesterase, proteases, and peptide inhibitors. There are few case reports of bites by this species. An old report from the 1930s stated that a bite on the finger of an individual caused "very severe pain" about an hour after the bite, and after antivenom was administered. There was extensive swelling, discoloration, and blistering. A serious bite is documented in a case report in this volume, as well as one of an accidental ocular envenomation. I could find no reports of fatalities due to this species, but it should be considered capable of delivering a fatal bite.

Remarks: This species is protected by law in Utah, where it only occurs in the extreme southeast portion of the state, along with Gila Monsters and Sidewinders.

References and Resources: Refer also to the introduction for terrestrial vertebrates. Douglas et al. (2006, 2007), Glaudas (2009), Glaudas and Rodríquez-Robles (2011), Glenn and Straight (1985), Lawing et al. (2012), Lemm (2006), Meik (2008, 2016), Meik et al. (2015).

Red Diamond Rattlesnake (*Crotalus ruber*)

"If you want to find a Red Diamond Rattlesnake, you need to go to where 'the *ruber* meets the road.'" (one of the author's horrible puns)

The coastal color morph of the Red Diamond Rattlesnake. *Bill Love*

Classification: Phylum Chordata (chordates); Class Reptilia (reptiles); Order Squamata (snakes and lizards); Family Viperidae (vipers); Genus *Crotalus* (typical rattlesnakes); Species *ruber* (Red Diamond Rattlesnake).

The taxonomy has been a topic of research and debate—it ranges from being monotypic to having four subspecies. *Crotalus exsul* is generally considered the same species, and that name technically should have priority, but *ruber* is the officially accepted epithet due to common usage. Subspecies are sometimes not recognized, but if so, our U.S. subspecies would be *C. r. ruber*, which is by far the most studied form.

Identification: This is one of the largest species of rattlesnake in the United States. It attains a size up to 165 cm, and 183 cm specimens have been reported, but unconfirmed. Most adults range between about 60 and 140 cm. Neonates are about 30 cm. It is usually reddish, often brick-red in color,

The desert morph of *C. ruber*, showing the distinct coon-tail pattern it shares with a few other rattlesnakes. *Bill Love*

although some specimens are more grayish. Specimens from Baja California may be darker and the island forms may be slightly different, including different average and maximum sizes. In the United States, those on the desert slopes of the Baja California Peninsular Range tend to be lighter and more pastel in appearance, while those on the coastal slopes are a little darker and may not be as reddish. This is a "coon-tail" species, having distinct white and black bands on the tail. Its pattern of 22–42 (usually 29–42) dorsal blotches is usually more distinct than *C. atrox*. Juveniles tend to be grayer in coloration than adults. This species has hybridized with *C. helleri* in the wild and in captivity.

Distribution: In the United States, it is endemic to southern California. It is found from near the coast to the east slope of the Baja California Peninsular Range, and just into the far western Sonoran Desert. Its range does not overlap with the Western Diamond-backed Rattlesnake, which is found just to the east. In Mexico, it is found in parts of Baja California and several adjacent islands.

Natural History: On the west slope of the Baja California Peninsular and Transverse Ranges, this snake is found in coastal sage scrub, chaparral, and sometimes grasslands. It may co-occur with *C. helleri*, but *C. ruber* tends to be found in rockier areas. It is found from near sea level to 1,525 m at Palomar Mountain, but they generally occur below about 1,200 m. In the upper elevations, it is found in oak and pine woodlands. On the east side of the peninsular range, it occurs into rocky foothill desert scrub, but not far into the open desert.

Red Diamonds may be surface active any month of the year, but during the winter tend to be tucked away in hibernacula. They are mostly active spring through fall, with April to June being the peak activity period. They are diurnal to nocturnal, depending on temperatures. In March and April, they are diurnal, and then switch to nighttime activity. They become diurnal again in the fall. They may also be stimulated into activity by spring or summer rains. Hibernacula may have from one to a few individuals, and occasionally up to about 25. After coming out of brumation, males may engage in combat rituals. Mating occurs March to May. Young are born in August and September. They have 3–20 young per litter.

Red Diamond Rattlesnakes usually hunt in an ambush position, but may actively follow scent trails. Like many other rattlesnakes, the young tend to feed on lizards and small mammals. Adults tend to prey on mammals, and given their large adult size, they can handle fairly large prey animals such as woodrats, squirrels, lagomorphs, and even spotted skunks. California Ground Squirrels are resistant to their venom. There are ontogenetic changes in venom components to accommodate the shift in prey preference. Birds, snakes, and carrion have also been documented in the diet.

Encounters: Within its limited range in the United States, it is fairly common in good habitat, often comprising up to 10% or more of the surface-active, local snake population. Southern California has a large human population, so it is not surprising that encounters are common. While these animals are not generally found in urban settings, they are near countless rural communities from the coast to the desert, and are often present at new housing construction. Red Diamond Rattlesnakes are generally rather passive animals, but they will defend themselves if provoked. They can go from a seemingly docile state to rearing up and striking with surprising rapidity. These snakes are somewhat cryptically colored, but more important, they are very difficult to spot in thick chaparral, sage scrub, or amid rocks. They may also be hesitant to rattle, although the large rattle is easily heard when deployed.

Tracking bite statistics for this species is difficult, but there are numerous bites by this species every year; the Southern Pacific Rattlesnake is more likely to be the culprit of a bite over most of the Red Diamond's westernmost range in California.

Venom Apparatus, Venom, and Symptoms: They have very long fangs, about 9.9–13.2 mm (average 11.6). The venom yield is very high. There are numerous studies on venom yield, with a wide range of reported values, but the average is probably around 350 mg for an adult, while the maximum reported is 670 mg. *Crotalus ruber* has one of the least toxic venoms among rattlesnakes at 21.3 mg/kg, with an intravenous LD_{50} of about 3.5–4 mg/kg. The venom of neonates is more virulent, but the venom yield is much lower. The lethal dose for a human has been estimated at about 100 mg.

The venom of *C. ruber* includes a typical arsenal of rattlesnake venom components, including three metalloproteinases, L-amino acid oxidase, kallikrein-like enzymes, ester hydrolases, esterases, lectin, phosphodiesterase, and phospholipase A_2. This latter enzyme type varies by latitude across the range of the species. In humans and other mammals, the metalloproteinases break down fibrinogen, causing muscle necrosis. Neonates have almost no metalloproteinase activity. Neurotoxic Mojave-like toxins have not been isolated in this species in the United States.

There are numerous accounts of envenomation by this species. Symptoms include severe pain, swelling, discoloration, hemorrhagic blebs, necrosis, bloody diarrhea, and an increase then decrease of blood pressure. This species is capable of delivering a fatal bite.

References and Resources: Refer also to the introduction for terrestrial vertebrates. Brown et al. (2008), Douglas et al. (2006), Dugan and Hayes (2012), Dugan et al. (2008), Halama et al. (2008), Hamako et al. (2007), Lemm (2006), Lyons (1971), Mackessy (1985), Mori and Sugihara (1988), Mori et al. (1987), Murphy et al. (1995), Straight et al. (1992), Takeya et al. (1990).

Mojave Rattlesnake (*Crotalus scutulatus*)

"By the time they reached the ranger, the [6-year-old] boy was already vomiting and foaming at the mouth.... He lost control of his muscles, and his limbs. He had a rash all over his face. He couldn't breathe.... He was brought to Mission Viejo Hospital... where doctors recognized the symptoms as a bite from the especially toxic rattlesnake, the Mojave Green.... It took 42 vials of antivenom just to stabilize him." (R. Jaslow, CBS News, July 2012)

Classification: Phylum Chordata (chordates); Class Reptilia (reptiles); Order Squamata (snakes and lizards); Family Viperidae (vipers); Genus *Crotalus* (typical rattlesnakes); Species *scutulatus* (Mojave Rattlesnake).

The brown color morph of *C. scutulatus* resembles *C. atrox* more closely, but there are subtle differences. *L.L.C. Jones*

There are generally two subspecies recognized, *C. s. scutulatus*, the Northern Mojave Rattlesnake and *C. s. salvini*, the Huamantlan Rattlesnake. The former occurs in the United States and Mexico, while the latter is endemic to Mexico. I use "Mojave" as the preferred English name, rather than "Mohave," for a variety of reasons, such as its use as the standard for seventy years.

Identification: This is a medium-sized rattlesnake. Large adults tend to be about 1 m in length, although they have been documented as reaching at least 1.2–1.4 m. Neonates are reported to be about 265–280 mm snout-vent length. The color varies by locality and even among individuals in a given area. The overall coloration may be a hue of green, yellow, tan, brown, or dark gray. There are 27–44 dorsal markings. Anteriorly, these are usually rhombs or diamonds, and those on the neck area may be misshapen, often into a linear pattern. Posteriorly, the dorsolateral markings become bands. Ultimately, the bands are black and white on the tail, making this one of the "coon-tail" species that includes the Western Diamond-backed and Red Diamond Rattlesnakes.

The frequently used name "Mojave Green" refers to a common color variant found in many areas. As the name implies, it has a decidedly green to yellowish green cast. This color phase makes it distinctive from the similar Western Diamond-backed Rattlesnake, although not all Mojaves are green. A *fairly* good diagnostic character from a distance is that the white bands are usually quite a bit wider than the black bands in *C. scutulatus*, while those of *C. atrox* are about the same size or just a bit wider. Another useful character is the eye stripe. In Western Diamond-backs, the posterior white portion of the eye stripe goes to about the corner of the mouth, while in Mojave Rattlesnakes the stripe extends back beyond the mouth, at a greater angle. Mojave Rattlesnakes tend to have more precisely edged light borders along the perimeter of the dorsal blotches or bands. Markings of juveniles of both species are similar, so it is even harder to distinguish juveniles of these species apart.

Many laypersons confuse the Western Black-tailed Rattlesnake, Prairie Rattlesnake, and Northern Pacific Rattlesnake with the Mojave Rattlesnake. This is probably because the common name of the infamous "Mojave Green" is pervasive, so other species with a greenish hue may be misidentified.

Mojave Rattlesnakes are reported to occasionally hybridize with Prairie Rattlesnakes and Western Diamond-backed Rattlesnakes in nature, but the degree of natural hybridization is not well understood. In captivity, they are known to have hybridized with *C. atrox, C. cerastes laterorepens, C. oreganus,* and *C. uni-*

This color morph of *C. scutulatus* earns it the vernacular, Mojave green. This species is similar to *C. atrox*, but the latter is never greenish. *L.L.C. Jones*

color. One apparent hybrid rattlesnake I encountered in the Tucson Mountains was not retained because it was on National Park lands, but it looked to me like a *C. tigris* × *C. scutulatus*, while a companion thought it looked more like a *C. tigris* × *C. atrox*. Only genetics could answer that definitively.

Distribution: This is a fairly wide-ranging animal. It is found in the Mojave Desert in southern California and adjacent southern Nevada, northwestern Arizona, and extreme southwestern Utah. It is also found in the Sonoran Desert east of the Colorado River in Arizona, and parts of the Chihuahuan Desert of New Mexico, Texas, and Arizona. In Mexico, it is distributed in the Chihuahuan Desert and adjacent areas. *Crotalus s. scutulatus* is absent from Baja California, but on the mainland occurs to Jalisco. *Crotalus s. salvini* occurs just to the south of its range in Mexico.

Natural History: Although this species ranges from about sea level to about 2,600 m, they are most frequently observed below about 1,500–1,800 m. They are usually found in relatively flat or rolling terrain. I have mostly seen individuals in creosote bush flats and in semi-desert grasslands and plains,

An unusual field observation: a Coachwhip consuming a healthy-looking Mojave Rattlesnake. *Phil Rakoci*

such as the Sonoita Plains. It can occasionally be encountered in open oak and piñon/juniper woodlands or even the lower edge of pine woodlands. Not all seemingly typical habitat is occupied. For example, it is absent from the creosote flats of the western Sonoran Desert, parts of the northern Chihuahuan Desert, and Great Plains. Where Mojaves are sympatric with other rattlesnakes, there is usually habitat partitioning. For example, Western Diamond-backs favor slightly higher and more rugged terrain than Mojaves.

Mojave Rattlesnakes are active March through October. They may be diurnal, crepuscular, or nocturnal, depending on conditions. In the Mojave Desert, they often take refuge in rodent, fox, and tortoise burrows. During the winter, they brumate singly or in pairs. Male combat has been noted in fall, although mating occurs in both spring (usually March through May) and again in the summer and early fall. The gestation period is about 170 days. Parturition occurs in August and September. There are 1–17 (average 8) per litter.

Mojave Rattlesnakes feed on a variety of animals, including mammals, reptiles, birds, amphibians, and arthropods. Bird and reptile eggs are also eaten. Juveniles probably feed more on lizards than adults. Adults seem to mostly feed on small mammals, including mice, rats, squirrels, and lagomorphs. It has been suggested that the strong neurotoxic venom ensures that hopping rodents, such as pocket mice and kangaroo rats, can be dispatched and tracked before they get too far, because animals that jump do not leave the extensive scent trail of animals that run along the ground.

Encounters: Although these are very dangerous animals that are relatively widespread, they tend to (not without exception) occupy sparsely populated areas. Mojave Rattlesnakes can be abundant in some areas, including outskirts of the large cities of Tucson and Phoenix, and smaller desert and grassland cities such as Barstow, Mojave, Sierra Vista, and California City. They can be the most commonly encountered species of snake in gravelly creosote flats and some grasslands. Bite statistics are difficult to track, especially in areas where they co-occur with *C. atrox*. Some publications suggest there are relatively few bites that are attributable to *C. scutulatus*, while others suggest up to 30% of the *atrox*/*scutulatus* bites are likely due to the latter. If that is the case, then *C. scutulatus* would be responsible for hundreds of annual bites in the Tucson and Phoenix area. Similarly, it has been suggested that a significant proportion of the 519 bites in Pima and Cochise Counties, Arizona, from 2002–2009, were due to *C. scutulatus*.

Venom Apparatus, Venom, and Symptoms: The fangs are not particularly long, between about 6–9 mm, but long enough to penetrate jeans and legs, and sometimes boots. The venom yield is 50–90 mg.

Based on LD_{50} measures, most populations of the Mojave Rattlesnake are considered among the most venomous snakes on Earth. Because of this, *C. scutulatus* possesses what may be the most studied venom of U.S. snakes. Two general types of whole venoms are recognized: neurotoxic and tissue-destructive. The neurotoxic type is often called Mojave toxin or Mojave-like toxin, even in other snake species. Of course, whole venoms are "cocktails" with many components. The LD_{50} is variable, depending on the references and populations. The venom of neurotoxic populations can be extremely virulent with an LD_{50} of 0.13–0.54 mg/kg (average around 0.34), and for those with the more hemorrhagic venom an LD_{50} of 2.29–3.8 mg/kg has been reported. A human lethal dose has been estimated to be about 10–15 mg in animals with Mojave neurotoxin.

After Mojave toxin was recognized and isolated, other species or populations of rattlesnakes were shown to have neurotoxic Mojave or Mojave-like toxins, including *C. concolor*, *C. helleri*, and *C. lepidus*, among others. Most populations of Mojave Rattlesnakes have the neurotoxic venom. For those bitten by Mojave Rattlesnakes with tissue-destructive venom, the symptoms would be similar to most other rattlesnakes. Symptoms of envenomation (both venom types combined) may include pain, swelling, discoloration, blebs, necrosis, elevated heart rate, lowered blood pressure, double vision,

I almost trod upon this Mojave Rattlesnake during my scorpion study. In the dim glow of a black light, I thought it was a cow pie! One reason I always wear snake gaiters. L.L.C. Jones

difficulty in speaking and swallowing, trouble breathing, drooping eyelids, depression, diarrhea, shock, renal failure, cardiac arrest, and respiratory failure. Snakebite fatalities in Arizona and California have been attributed to this species. Untreated bites are severe and have a higher mortality rate than most other species of rattlesnakes. It was this species that killed renowned herpetologist Dr. Frederick Shannon. However, there is some dispute as to just how dangerous this snake really is, as there are few confirmed fatalities. Some of those fatalities are likely due to anaphylaxis or anaphylactoid syndrome. Fortunately, most cases are likely treated in hospitals, and Mojave venom is used in the production of CroFab. Neurotoxic tropical rattlesnakes are used in the production of Anavip, and are effective against neurotoxic Mojave Rattlesnake bites.

There have been numerous recent articles in the popular media about how rattlesnakes, particularly *C. scutulatus* and *C. helleri*, are becoming more toxic, but there is no scientific evidence to support this claim.

Remarks: This species is protected by law in Utah.

References and Resources: Refer also to the introduction for terrestrial vertebrates. The number of references about Mojave Rattlesnake venom is much greater than suggested here. Bieber et al. (1975), Cardwell (2016), Cate and Bieber (1978), Chippaux et al. (1991), Glenn and Straight (1978, 1989), Glenn et al. (1983), Goldberg and Rosen (2000), Hardy (1983), Jansen et al. (1992), Jones (2016), Massey et al. (2012), Rael et al. (1984, 1986, 1993), Sánchez et al. (2005), Schuett et al. (2002b), Wilkinson et al. (1991), Wooldridge et al. (2001), Zepeda et al. (1985).

Panamint Rattlesnake (*Crotalus stephensi*)

"A California Kingsnake was wrapped tightly around a Panamint Rattlesnake and the snakes were barely moving. Disturbed by the onlookers, the kingsnake retreated under a nearby bush. The rattlesnake was dead by that time, and appears to be biting itself, but was described as biting onto the kingsnake before it died." (CaliforniaHerps.com, accessed 12 October 2016)

A beautifully patterned Panamint Rattlesnake with distinct dorsal blotches. *L.L.C. Jones*

Classification: Phylum Chordata (chordates); Class Reptilia (reptiles); Order Squamata (snakes and lizards); Family Viperidae (vipers); Genus *Crotalus* (typical rattlesnakes); Species *stephensi* (Panamint Rattlesnake).

Until recently, this was considered a subspecies of *C. mitchellii*, but this species complex has since been split into four distinct species. Much of the published information on this species has been muddled by being lumped with *C. mitchellii*, and most emphasis on the former *C. mitchellii* was on *C. pyrrhus*, rather than *C. stephensi*.

Identification: This is a medium-sized rattlesnake, with adults measuring between about 60 and 80 cm. A maximum adult size of 94.3 cm has been reported. Neonates are about 24 cm. *Crotalus stephensi* can be very similar to *C. mitchellii* in general appearance, a species that is also highly variable in color pattern. Like the Southwestern Speckled Rattlesnake, the color matches to the background. The ground color can be pinkish, yellowish, gray, bluish gray, tan, brown, or reddish brown. There are 27–43 (usually 36–37) dorsal bands or blotches. Anteriorly, the dorsal markings are blotches, giving way to crossbands posteriorly. The head and face often have a grayish to bluish cast. The tail has 2–9 dark brown or black bands that often fuse to form a black end before the rattle. Fortunately, we don't need to rely on color pattern to differentiate *C. stephensi* from *C. mitchellii*.

Compared to the Southwestern Speckled Rattlesnake, the Panamint: (1) lacks internasal scales, (2) has ridged or creased supraocular scales, (3) possesses ground coloration in the tail bands, and (4) the black tail bands are limited to the distal 15% of the body. However, the easiest way to differentiate these two species is by knowing their geographic distribution, as they are not known to overlap or hybridize in nature. This species, and *C. pyrrhus*, are superficially similar to *C. tigris*, but the ranges do not overlap. Also, *C. tigris* has a noticeably smaller and narrower head.

Distribution: The range is north of that of the Southwestestern Speckled Rattlesnake. While *C. stephensi* is generally considered a Mojave Desert species, it does not precisely follow the biogeographic boundaries of the desert. It occurs from the Mojave River northward, into the southern Great Basin Desert of western Nevada. At Las Vegas, it occurs north and west of the city, while *C. pyrrhus* is found south and east of the city. Panamint Rattlesnakes are also found in the Sierra Nevada Range of central California and adjacent Nevada.

Natural History: These animals are associated with rocky areas, such as hillslopes of mountain ranges, but after brumation they may venture into the nearby valleys. They are found from about sea level in the low desert ranges (i.e., Death Valley) to about 2,400 m in the Sierra Nevadas. In lower elevations they can be found in creosote and sagebrush deserts, while in the mountains they can be in oak and pine woodlands.

This species is active between about March and September, depending upon elevation. During the winter they may occupy dens with over 100 individuals. During the heat of the summer, they are most active at night. In the cooler months, they may be diurnally active, or may be crepuscular,

Grayish individual near Las Vegas, Nevada. *L.L.C. Jones*

depending on ambient temperatures and the temperature of the substrate. Males have been observed in the combat ritual, especially after emerging from brumation. There is little information on reproduction in this species, but females are reported to have 1–10 ovarian eggs. A wild-caught adult lived over twenty years in captivity at the Los Angeles Zoo.

Panamint Rattlesnakes feed on lizards, birds, and rodents. Rodent prey includes kangaroo rats, pocket mice, woodrats, and antelope ground squirrels. Cannibalism has been documented in captivity, although the prey was regurgitated.

Encounters: These animals occupy remote or lightly populated areas in their range. An exception is probably western Las Vegas, where housing developments are encroaching into the hills and canyons of the mountains to the west. There are other desert communities in the Mojave Desert where human habitation encroaches upon the habitat. They are not uncommon, but not commonly observed by most people. Most are probably encountered in rocky areas, such as Red Rock Canyon, a popular recreation area at the west edge of Las Vegas. They are cryptically colored, often banded with a diffuse speckled pattern, so are not easy to see when hiking. Fortunately, this species tends to be mild mannered, but this means they may not rattle to announce their whereabouts.

Male Panamint Rattlesnakes in combat, after emergence from winter dens. *Rich Gassaway*

Venom Apparatus, Venom, and Symptoms: The fang length averages 7.6 mm. Based on 13 milked specimens, 73 mg was the average venom yield and 129 mg was the maximum. An LD_{50} of 5.5 mg/kg has been reported.

The venom of this species is not well known. Bites are rarely reported. In one study of snakebite in southern California where the culprit was known, only 3/204 (<1%) of the bites were from Panamint Rattlesnakes. I know of no case reports, but they are presumed to be a dangerous snake. Because they were formerly placed within the *C. mitchellii* complex (albeit a distinct subspecies), reported bites could represent a mix of *C. pyrrhus* and *C. stephensi*.

References and Resources: Refer also to the introduction for terrestrial vertebrates. Douglas et al. (2006, 2007), Lawing et al. (2012), Meik (2008, 2016), Meik et al. (2015).

Tiger Rattlesnake (*Crotalus tigris*)

"We report the proteomic and antivenomic characterization of *Crotalus tigris* venom. This venom exhibits the highest lethality for mice among rattlesnakes and the simplest toxin proteome reported to date." (Calvete et al. 2012)

Gray color morph of the Tiger Rattlesnake. Note the painted rattle, which indicates it is a study animal. *L.L.C. Jones*

Classification: Phylum Chordata (chordates); Class Reptilia (reptiles); Order Squamata (snakes and lizards); Family Viperidae (vipers); Genus *Crotalus* (typical rattlesnakes); Species *tigris* (Tiger Rattlesnake).

Although three clades have been recognized, based on mitochondrial DNA, these are not consistent with any anatomical features or geographic patterns in coloration, so no subspecies have been described. Tiger Rattlesnakes are related to Panamint and Southwestern Speckled Rattlesnakes, which they resemble.

Identification: This is a smallish medium-sized rattlesnake, measuring between about 45 and 90 cm, but most adults are about 60–75 cm. Neonates are

27–29 cm. Tiger Rattlesnakes have the smallest head relative to the body size among all rattlesnake species. In adults, the head may not be much broader than the basal rattle. Neonates have a larger head proportionately than adults. Tigers have 37–52 (usually 42–43) crossbands. Although the dorsal markings may be more blotch-like anteriorly, they are essentially banded on the entire body and tail. The bands tend to be diffuse, rather than having distinct edges; they may or may not have much contrast with the ground color. The dorsal to lateral markings are usually darker than the ground color and fuse into the ground color. The ground color is variable, usually matching the substrate, so may be gray, bluish gray, tan, brown, orangish, pink, or lavender. While the banding has more contrast at the tail, it is never black and white, as with the "coon-tail" species. The pattern of *C. tigris* may resemble that of its relatives, *C. pyrrhus* and *C. stephensi*, but can be easily distinguished by the small head.

I saw an apparent natural hybrid with *C. atrox* or *C. scutulatus* in the Tucson Mountains, but the animal was released before photographs could be taken.

Distribution: The Tiger Rattlesnake is found in Arizona and Sonora, including Tiburon Island. In the United States, it is found approximately from the vicinity of Ajo to Phoenix and Tucson. There is an apparently disjunct population in the Peloncillo Mountains, which spans extreme southeastern Arizona and adjacent New Mexico. The full distribution may not be completely known.

Natural History: Rattlesnakes occur primarily in the Arizona upland subdivision of the Sonoran Desert, but they are also found in foothill thorn scrub habitat in Mexico, and in areas of thorn scrub remnants in Arizona. They are sometimes encountered in semidesert grasslands and interior chaparral. In its extreme eastern range, it enters Chihuahuan desert scrub. It is generally confined to lower elevation desert mountain ranges and the foothills of higher mountains, rarely venturing far into the creosote or mesquite flats. Its upper elevational limit has been reported to be 2,440 m, but most individuals are encountered below 1,650 m. It is almost always associated with rocky areas, or immediately adjacent uplands, bajadas, and washes. These animals have strong site fidelity for overwintering rocky areas, but move into the surrounding habitats, particularly along dry washes. Although their mountainous habitats are isolated by valleys and urban encroachment, there is a surprising amount of gene flow between populations.

Neonate Tiger Rattlesnakes are rarely encountered in the wild. *L.L.C. Jones*

They can be surface active spring through fall. They overwinter during November to March in rocky uplands and by late spring and summer will move down washes and adjacent bajadas to forage and mate. However, after emergence from hibernacula, but before migrations begin, they may select a "staging area" where they will remain in relative inactivity until the late summer. They are by far most commonly encountered after the arrival of the North American Monsoon, July through September. In fact, this species is endemic to a region that receives significant monsoon rainfall. Males are more likely to be found on the move than females during the monsoon. They can be found day or night, depending on local conditions. During the peak active period, they tend to be nocturnal, but may be active during mild mornings on the day following thunderstorms.

Male combat has been observed. Mating was reported to occur in the spring in several publications, but this was based on limited information. They typically mate during the monsoon. Copulation, which can last several hours, has been observed. Parturition occurs in July and August, and the young are usually born under rocks or deep in crevices, so maternal care is poorly known. On two occasions, females were seen with a single neonate. Litter size ranges between 2–6, with an average of 4. Neonates are not well known because they are rarely encountered in the wild. A wild-caught adult lived over fifteen years in captivity.

The most toxic of rattlesnake venoms may have evolved in *C. tigris* and *C. scutulatus* to quickly immobilize prey, such as the many types of saltatorial desert rodents in their habitats. *L.L.C. Jones*

Tiger Rattlesnakes use both an ambush strategy and active searching for prey. Several researchers have noted them poking into crevices, burrows, and nests. When they find a likely productive spot, they may settle down into an ambush position. They mostly eat lizards and small mammals. Lizards are probably eaten by young animals. Their diet may then shift to larger (but still small) mammalian prey. The small and narrow head of this animal would suggest it preys on animals in crevices, small burrows, and other tight spots. The relatively short fangs would also imply small prey is normally targeted. It has been shown that a wide rattle is an indication of a well-fed individual.

Encounters: These animals live in areas that are sparsely populated, although as humans encroach into their habitat, interactions could become more frequent. Tiger Rattlesnakes are found near the two biggest urban and suburban centers of Arizona, Phoenix and Tucson. A study of reptiles and amphibians at a golf course in the Tortolita Mountains near Tucson shows that *C. tigris* is common there. This species can also show up in yards of people who live in the suburbs, particularly if their homes are nestled up against the foothills. This is often regarded as among the more inoffensive species, but some reports say they can be cantankerous. If pushed too far, they will surely bite. In one episode of *The Crocodile Hunter*, these animals were free handled, but this is a dangerous practice, even for a so-called "professional." Make no mistake, these are seriously venomous animals that should be left alone. They are sometimes encountered by hikers. For example, these snakes seem rather abundant in Sabino Canyon, a popular recreation area on the east side of Tucson, where 750,000 people visit annually. During the day, the snakes are usually hidden away, but at night they come out to feed. Normally this might not be a problem, but there are popular nighttime moonlight walks for the public at night during the animals' peak activity time. During those walks, hundreds of people are walking the roads and trails. Flashlights are often not used because the visitors want to experience the desert in the light of the moon. Although it is a pleasant experience, cryptically colored snakes would be hard to spot in moonlight and moonshadows. And there are Western Diamond-backed, Western Black-tailed, and possibly Mojave and Arizona Black Rattlesnakes in the area, along with Gila Monsters and Arizona Coralsnakes. It is not too unusual for me to encounter *C. tigris* and other rattlesnakes while doing my scorpion surveys in Saguaro National Park, which is why I have a snake-protection rule for myself and anyone that joins me.

There are no bite statistics for this species and no detailed case reports I know of, although there are about 3 anecdotal accounts of bites.

Venom Apparatus, Venom, and Symptoms: The fangs are relatively small, 4–4.6 in those measured. The venom yield is also small, being 6 to 17 mg. One report claims a maximum of 60 mg. This is the most venomous rattlesnake of all, with an LD_{50} reported as 0.056 mg/kg (intravenous), 0.07 mg/kg (intraperitoneal), and 0.21 mg/kg (subcutaneous). This animal shows neurotoxic and myotoxic elements in its venom, including neurotoxic Mojave-like toxin, phospholipase A_2, two proteases, and several other proteins. The venom lacks metalloproteinases.

The very few recorded bites by this species have resulted in local reactions and no systemic reactions, although one bite from a presumed Tiger Rattlesnake resulted in pain, swelling, and some vascular problems, but ultimately the patient was released and recovered in a couple of weeks. However, this is an extremely venomous animal that should be avoided. The lack of case reports should remain as is. I expect a severe envenomation would result in more than a few local and no systemic reactions.

References and Resources: Refer also to the introduction for terrestrial vertebrates. Beck (1995), Calvete et al. (2012), Goldberg et al. (2003), Goode et al. (2008, 2016), Powell et al. (2004), Weinstein and Smith (1990).

Prairie Rattlesnake (*Crotalus viridis*)

"Dynamite has been used to destroy some dens . . . most of those slain by humans pose no significant threat; many are slaughtered by sadistic individuals in the name of 'fun.' In northern Utah, Parker and Brown (1973, 1974) found that a den population decreased from more than 300 individuals in the 1940s to . . . 12 in 1972, evidently due to predation by humans." (Hammerson 1999)

Classification: Phylum Chordata (chordates); Class Reptilia (reptiles); Order Squamata (snakes and lizards); Family Viperidae (vipers); Genus *Crotalus* (typical rattlesnakes); Species *viridis* (Prairie Rattlesnake). This includes the former subspecies, *C. v. viridis* (Prairie Rattlesnake) and *C. v. nuntius* (Hopi Rattlesnake).

Prairie Rattlesnakes often have a greenish hue, so are sometimes misidentified as "Mojave green" rattlesnakes. *Steve Mackessy*

Until about 2000, *C. viridis* was called the "Western Rattlesnake," and it included eight subspecies, one of which was *C. v. viridis* (Prairie Rattlesnake). After that, most authorities split the species complex into the Prairie Rattlesnake (*C. viridis*) and the newer "Western Rattlesnake" (*C. oreganus*). Most of the subspecies were transferred to *C. oreganus*, or were elevated to species. Some respected publications on venomous snakes recognize the split as valid, but still refer to the common name of *C. viridis* as "Western Rattlesnake," which causes confusion. At the time of this split, the only subspecies retained within the newly recognized *C. viridis* were *C. v. viridis* (Green Prairie Rattlesnake) and *C. v. nuntius* (Hopi Rattlesnake). More recently, it has been suggested that these two subspecies are not genetically or phenotypically distinct enough to be considered taxonomically valid, and there is a broad zone of intergradation, so the Hopi form is now regarded a smaller regional variant.

Identification: The Hopi form is decidedly smaller than the typical Prairie Rattlesnake, being 38–70 cm in length, while the typical form of the Great Plains reaches 80–120 cm, with a record of 121.9 cm. There are 33–57 well-defined dorsal blotches that posteriorly become bands. The ground color of the typical form is often greenish, which accounts for the misidentification of Mojave "Green" Rattlesnake, although the range of these species only partially overlap. It may be brown, reddish, yellowish, or gray. The Hopi form is often pink, tan, or salmon brown. The dorsal markings of the two forms are similar, being distinct, darker than the ground color, and wider than long. There may be banding on the tail, and the distal part of the tail may be black, but there is never a "coon-tail" marking. There is a distinct, dark eye stripe bordered by white/light coloration.

There are many reports of hybridization with other species, including *C. abyssus*, *C. cerberus*, or *C. oreganus*, but one researcher could find no confirmation of such events. However, there is some empirical evidence that *C. viridis* has a hybrid zone with *C. concolor*. It is also believed *C. viridis* occasionally hybridizes with *C. scutulatus*.

Distribution: The Prairie Rattlesnake is a wide-ranging species. It is found from Arizona to Idaho and Alberta in the west, and Coahulia, Texas, Oklahoma, Iowa, and North Dakota to the east.

Natural History: This species is frequently studied. Prairie Rattlesnakes range in elevation from about 100 m along the Rio Grande in Texas to over

This image went viral on the Internet, with all sorts of wild claims about its origin, but in real life, dens are destroyed by unethical poachers if their true localities are revealed. *Steve Mackessy*

2,770 m in Wyoming. In general, their preferred habitat is grassland, but they can also be found in the Chihuahuan Desert and some semiarid areas, such as the Colorado Plateau. They occur in juniper woodlands and sometimes conifer woodlands. Across the range they are typically found in the high plains, but especially in or near rocky areas that include sandstone, shale, and limestone habitats. They may be found in open areas, but tend to be most abundant in canyons and near rocky areas. In the Chihuahuan Desert, they are most commonly encountered in mesquite grassland-desert transition, but do occupy creosote desert in some areas. They can also be found in a wider variety of habitats including lava fields, buttes, dunes, shrublands, and mesas, as well as in riparian areas.

This species is well known for congregating in large numbers—up to hundreds of individuals—in hibernacula in suitable sites, often amid rocky hillsides. They may also be common in prairie dog towns. In northern areas they may brumate for up to nine months, but in southern areas, brumation time is less. They overwinter in subsurface retreats, which can include deep, rocky hibernacula, or in the absence of rocky terrain, mammal burrows. Such burrows include those of foxes, skunks, and prairie dogs, especially

Female Prairie Rattlesnakes remain at the den to protect their neonates (inside the coils) until they shed. *Hunter Johnson*

when the burrows have been further excavated by other animals. They may share their den with other Prairie Rattlesnakes and other snake species, and even potential predators and prey; they are not a threat when torpid. After emergence from brumation, they move into surrounding areas to mate and forage. The surface-active period varies by latitude. In the south (e.g., western Texas), they may emerge from brumation from late February to early April, while in the north they may emerge from late March or mid-April to early May. These snakes may be active day or night, depending on the temperature. Activity is reduced during nights with 75–100% moonlight.

Prairie Rattlesnakes can recognize odors of conspecifics. Males may exhibit combat displays after emergence from the den. Mating generally occurs in the summer, often late summer, so is often at night. Males search for females along scent trails, but may be interrupted by foraging opportunities. When a female is found, there is a courtship, and copulation may last for several hours. The gestation period is 100–120 days. Parturition has been reported as early as spring, but this was in captivity, and more often occurs August through October. Females give birth underground in maternity dens. Based on a large sample, the average number of young is 7.8, ranging between three and twenty-one. This species is long-lived, and captive animals have been known to approach thirty years of age.

Much of what we know about rattlesnake feeding behavior is from studies on *C. viridis*. They can actively search for prey, or may sit and wait. They use their full repertoire of senses to locate feeding hotspots, target prey animals, strike, track, and locate the dead or immobilized meal. The list of prey items is large, but somewhat unclear, due to the taxonomic lumping with other species of "Western" Rattlesnake complex. However, most studies specific to *C. viridis* proper suggests it predominantly preys on a variety of mammals commensurate with availability. Smaller individuals feed on smaller mammals, such as mice, voles, and pocket mice. Neonates in some populations feed extensively on lizards. Other small prey includes frogs and insects. Larger animals typically feed on larger prey, such as kangaroo rats, pocket gophers, prairie dogs and other ground squirrels, woodrats, and lagomorphs. In some populations, however, birds have been reported as the main quarry. A variety of species is consumed, especially ground-nesters such as quail.

Encounters: These animals have frequently been eradicated from areas near dense human habitation. Unfortunately, they have also been wantonly slaughtered in and near dens, in the name of doing society a favor. However, Prairie Rattlesnakes are still common in many areas and there are numerous encounters and reported bites. It has a reputation as being a nervous species

The "nuntius" form (i.e., the Hopi Rattlesnake) is now regarded by most authorities as a small, light-colored variant of *C. viridis*. *Bill Love*

that is quick to defend itself. When humans are at a distance, it tends to remain immobile and does not rattle. At close quarters, however, it may rattle and assume one of two striking poses, including an "S-shaped" striking coil or an elevated "high-S" display, with the upper third of the body at the ready.

Tracking specific species responsible for bites is difficult, but in some states where the Prairie Rattlesnake is the sole rattlesnake or occupies a large portion of the state, bites are not rare—or particularly common. Bites reported to poison control centers between 2001 and 2005 include, Kansas (52), Colorado (161), Nebraska (24), South Dakota (27), Montana (33), and Wyoming (21). Many of these states are sparsely populated, but per capita rates (0.2–2.0 per 100,000 persons) are in keeping with most other regions.

Venom Apparatus, Venom, and Symptoms: Fang length in adults averages about 7.4 mm, while fangs of the Hopi form are smaller (average 4.1). The LD_{50} has been reported as about 5.5 to 7.1 mg/kg in juveniles to 14.3 to 14.8 mg/kg in adults. In intravenous injections, it is below the mean of the Western Rattlesnake complex. There are numerous estimates of venom yield that range from averages of about 45–90 mg/kg and 162 mg/kg. The smaller Hopi pattern class yields 38–72 mg/kg. Due to its large size and venom yield of most populations, it is a dangerous rattlesnake. Deaths have been attributed to its bite.

The venom has been well studied in both neonates and adults; in fact, all age classes have many similarities in venom composition. It is primarily myotoxic and hemotoxic. The venom includes L-amino acid oxidase, metalloproteinases, phospholipase A_2, kallikrein, serine proteases, and other peptides and proteins. There are reports of neurotoxic elements (Mojave and concolor toxin) from some populations. It has a high phospholipase A_2 component, compared to other members of the Western Rattlesnake complex. It also ranks highly in levels of hemotoxic activity. Northern populations have high levels of myotoxin. There are differences between venom composition between age classes and sexes in some populations studied.

Symptoms of envenomated patients include pain, discoloration, swelling, tingling and numbness of the bite area and beyond (e.g., tongue, mouth, and scalp), giddiness, excitability, nervousness, hemorrhagic blebs, swelling, heart problems, weakness, low blood pressure, and difficulty in breathing. Death, when it does occur, is usually attributed to cardiac failure. However, death in this day and age is rare, as some previous case studies suggested suboptimal treatment may have been the cause; also, antivenom is readily

Prairie Rattlesnakes sometimes have a reputation for being irascible; they are quick to rattle and bite in defense. *L.L.C. Jones*

available these days in most areas. However, in 2017, a well-publicized fatality in Colorado lead to area closures for hikers until the snakes went into brumation.

Remarks: It is protected by law in Utah, Iowa, and Colorado.

References and Resources: Refer also to the introduction for terrestrial vertebrates. Cameron and Tu (1977), Chiszar et al. (1986), Clarke et al. (1996), Davis (2016), Duvall and Schuett (1997), Duvall et al. (1990), Goode and Duvall (1989), King and Duvall (1990), Komori et al. (1988), Mackessy (2010a), Murphy and Crabtree (1985), Ownby and Colberg (1987), Ownby et al. (1976, 1979, 1982, 1983), Saviola et al. (2015).

Ridge-nosed Rattlesnake (*Crotalus willardi*)

"I can tell you, through personal lunatic experience, that there is no significant difference between the bites of *C. w. silus*, *amabilis*, or *obscurus*." (K. Peterson, personal communication)

Adult Arizona Ridge-nosed Rattlesnake, the State Reptile of Arizona. *L.L.C. Jones*

Classification: Phylum Chordata (chordates); Class Reptilia (reptiles); Order Squamata (snakes and lizards); Family Viperidae (vipers); Genus *Crotalus* (typical rattlesnakes); Species *willardi* (Ridge-nosed Rattlesnake).

There are two subspecies in the United States, the Arizona Ridge-nosed Rattlesnake (*C. w. willardi*) and the New Mexico Ridge-nosed Rattlesnake (*C. w. obscurus*). There are three other subspecies usually recognized in Mexico, *C. w. amabilis*, *C. w. meridionalis*, and *C. w. silus*. However, it has been suggested each of the subspecies is a full species.

Identification: This is a small rattlesnake, between about 35 and 55 cm with the record of a wild *C. w. obscurus* at 68.8 cm. The Mexican subspecies *silus* is

Close-up of the beautiful facial markings, ridged nose, and heat-sensing pit of *C. w. willardi*. L.L.C. Jones

slightly larger. A distinctive feature is the ridge on the snout, formed by scales along the dorsolateral margin in front of the eyes. The Arizona subspecies is typically brick-red, while juveniles may be gray in overall coloration. Similarly, the New Mexico subspecies is typically reddish, but may be gray as adults and young individuals are gray. However, there is some local variation and these snakes may have a brown, gray, reddish, orangish, or yellowish ground color. Ridge-nosed Rattlesnakes have 18–45 darker blotches separated by whitish crossbands. The Arizona subspecies has characteristic bold white markings below the ridge. These markings are easily viewed at a distance, from both the front and side. The New Mexico subspecies typically lacks distinctive facial markings. In *C. w. willardi*, juveniles are less boldly patterned than adults, but do have some muted white facial markings. Juveniles have yellowish to black tails.

Unlike other rattlesnakes, juveniles lack the typical rattle button. Some reports of this phenomenon suggested the button must have been knocked off, but that is not the case. Neonates are born with a pre-button that does not allow rattles to be added. After the first shed, the pre-button becomes a button, as seen in all other species of rattlesnakes. Rattles are not added until after about the third shed at 9–13 months of age. By thirteen months, most *C. willardi* have shed 4–6 times.

This species has been known to hybridize with *C. lepidus* in the Peloncillo Mountains.

Distribution: In the United States, this species is only found in southeastern Arizona and adjacent New Mexico. The Arizona subspecies is found in the Santa Rita, Huachuca, Whetstone, and Patagonia mountains, as well as the Canelo Hills. The New Mexico subspecies is found in the Animas and Peloncillo Mountains of extreme southwestern New Mexico. The species and subspecies are found in adjacent Mexico. The other subspecies are Mexican endemics.

Natural History: The natural history of this species is not well known. In some of the literature, it has been deemed a generalist species, but given its small range and relatively narrow ecological attributes, this does not seem accurate. It is a montane species that occurs primarily in pine/oak woodlands between about 1,525 and 2,750 m. At the lower elevational range, as in the Canelo Hills, it is found in bunchgrasses with oaks, sycamores, and ashes, but can also be found in more upland situations. At higher elevations, it may be found in conifer forests. They sometimes frequent rocky areas, but less so than other mountain dwellers. *Crotalus w. obscurus* has been reported to frequent rock outcrops devoid of sediments and other materials.

It is most active during the day, but can be nocturnal during hot weather. It is surface active for a surprisingly long period of the year, but is most often active during March through November. It seems to have a proclivity for being active during relatively cool times, rather than in the heat of the early summer, although the monsoon moderates daytime temperatures when storms arrive. They can be active during the monsoon. There is little information on hibernacula of this species, but they do not seem to use large communal hibernacula. They are most often documented singly in situations that are just below the ground, as in rotting stumps, logs, rock crevices, and talus slopes.

Male combat has been observed. While courtship has been noted in captive animals throughout the year, the few observations in nature suggest summer is the norm. The gestation period is probably 4–5 months in the wild, but was recorded as being as long as thirteen months in captivity. Females have been observed with their neonates until the first shed.

Ridge-nosed Rattlesnakes hunt in both ambush and active foraging modes. As ambush predators, they are often coiled near the base of a natural

Crotalus w. willardi female at its den with neonates. *Bryan Hughes*

Neonate *C. w. willardi*. This species is born without a button to its developing rattle. *L.L.C. Jones*

structure, such as a bunchgrass. They have been reported to actively forage, presumably for sleeping lizards, among rocks at night. Juveniles have a bright yellowish tail, and caudal luring has been reported. It has also been suggested they employ facial luring—presumably the facial markings may resemble invertebrates or vegetation worth exploring.

Prey includes invertebrates, lizards, and mammals. It is likely young animals are more dependent on lizards as the primary prey, while adults target small mammals. One study has shown that Yarrow's Spiny Lizards and centipedes are probably the most common prey of young animals, while small mammals are the predominant prey of adults. Other consumed lizard species include whiptails and Madrean Alligator Lizards. In addition to centipedes, which are eaten by several species of pit vipers, they also consume scorpions and probably other invertebrates. Mammals in the diet include shrews, pocket mice, grasshopper mice, Deer Mice, Brush Mice, Cliff Chipmunks, and pocket gophers. Birds have also been reported in the diet. Lizards are held during envenomation while mammals are bitten, released, and tracked.

Encounters: This species is sometimes considered rare, although all known historic localities are still believed to be occupied. Much of its range is so isolated in remote montane habitats that it is essentially never encountered by the public. It is much more likely to be encountered when specifically being targeted. Thus, the only people likely to be bitten by *C. willardi* in the wild are those seeking it out for research, nature-viewing, collecting, photography, or poaching. Most or all bites come from the animals being handled. Envenomations also occur to pet owners who keep them in captivity. It should be noted they are protected by state law in Arizona and New Mexico. *Crotalus w. obscurus* is federally protected as a threatened species.

Venom Apparatus, Venom, and Symptoms: The fang length of this serpent is between 5.3 and 6.0 mm, with an average of 5.7 mm. This is enough to go through a boot in some circumstances. In early literature LD_{50} estimates were highly variable, but recent estimates have been made for all four subspecies on mice, lizards, and crickets. The mean LD_{50} (intraperitoneal, with 95% confidence intervals) estimates for *C. w. obscurus* is 1.58 mg/kg (0.94–2.22 C.I.) and for *C. w. willardi* is 2.45 mg/kg (1.93–2.99 C.I). The venom yield is small. There is not much information on this, but the extraction amounts have been reported as 3.1–3.7 mg, based on few samples,

Adult New Mexico Ridge-nosed Rattlesnake. Even though the pattern is much more bland than its Arizona cousin, it is highly sought after by poachers in the black market exotic pet trade due to its rarity. *Bill Love*

but 37 mg has been reported for large adults. Although the venom is fairly potent, with the small venom yield, a bite would not be expected to be able to kill a healthy adult.

The venom of all four subspecies has been studied and there are both similarities and differences between all, but the number of populations sampled is not great. All subspecies had varying levels of metalloproteinases, L-amino acid oxidase, phospholipase A_2, thrombin-like and kallikrein-like serine proteases, and phosphodiesterase. The venom of *C. w. obscurus* was more toxic toward mice than lizards or crickets, while *C. w. willardi* was more toxic to lizards than mice or crickets. This generally agrees with the studies of feeding habits of these taxa.

There are few case reports; two known bites were poorly documented in the literature, and there are others in this volume. One correspondent (see opening quote) has apparently been bitten multiple times. Symptoms include pain, swelling, and discoloration of the affected area, with possible bleb formation. In the brief accounts of two envenomations by one individual in this volume, the pain from *C. w. willardi* was "less than a bee sting," with the neonate being less painful than the 30 cm specimen. In another story in which one person was bitten on two occasions, one was

Neonate *C. w. willardi* from Sonora. Neonates can have either black or yellow tails.
L.L.C. Jones

a fairly significant reaction. A bite to the finger from a large *C. w. willardi* caused extreme pain, with swelling extended up the arm. There were also liquid-filled blisters on the finger, hand, and arm. Systemic effects included light-headedness and nausea, and the victim felt as if he were going to lose consciousness, perhaps suggesting a drop in blood pressure. There were also later complications from serum sickness. The bite took 6 weeks to resolve, and there was permanent damage resulting in arthritis, joint pain, and limited mobility of the affected area.

Remarks: This is perhaps the most sought-after species of sky island rattlesnake for the black market reptile trade. It is highly coveted for its fabulous markings (the Arizona subspecies) and small terrarium requirements. The New Mexico species is not so desirable for its facial markings, but is because of its rarity.

References and Resources: Refer also to the introduction for terrestrial vertebrates. Barker (1991, 2016), Holycross and Goldberg (2001), Holycross et al. (2002a, b), Saviola et al. (2017), Setser et al. (2011), Smith et al. (2001).

Eastern Massasauga (*Sistrurus catenatus*)

"The 13-year-old Madison Heights girl says she didn't even know Michigan had rattlesnakes. She was simply out for a walk with family at Stony Creek Metropark on Tuesday and picked up the snake in an effort to keep her younger brother, 10-year-old [name withheld] . . . from stomping on it." (*USA Today*, 5 September 2013, Eric D. Lawrence reporting)

Light, but well-patterned individual of *S. catenatus*. Mark Spangler/CCA/Flickr

Classification: Phylum Chordata (chordates); Class Reptilia (reptiles); Order Squamata (snakes and lizards); Family Viperidae (vipers); Genus *Sistrurus* (massasauga and pygmy rattlesnakes); Species *catenatus* (Eastern Massasauga).

Until recently, there was only one species of massasauga recognized, but the Eastern Massasauga was considered a distinct subspecies. In 2011–2013, the scientific names of massasaugas were uncertain until the International Commission on Zoological Nomenclature ruled on retaining *S. catenatus* and *S. tergeminus* for the northeastern and western forms, respectively.

Dark, well-patterned individual. *Bill Love*

Identification: Most adults are 45–75 cm, with a record of 100 cm. *Sistrurus* have large plates on the top of the head, rather than small scales like *Crotalus*. They are generally gray to brown with 30–50 dark-brown to black dorsal blotches and distinctive lateral blotches. The epithet *catenatus* is Latin for "chain," alluding to the light chain-mail-like pattern between the blotches. Some individuals are virtually black with little trace of a pattern. The blotches near the tail may fuse to form bands. The side of the head is distinctly marked with light and dark facial stripes. The rattle is small.

Distribution: The Eastern Massasauga occurs farther east than other populations of massasaugas, but it is more of a Great Lakes/Midwest species than an eastern one. They are found in much of Michigan, Indiana, Ohio, and Illinois, plus parts of New York, Pennsylvania, Missouri, Iowa, Minnesota, and Ontario. It is the only species of rattlesnake throughout much of its range.

Natural History: By rattlesnake standards, this is an ecologically unusual species. It is not only associated with wetland habitats but is also found in northern latitudes. It is generally found near marshes, bogs, ponds, swamps, and wet meadows, although it can also be found in adjacent upland areas. The several telemetry studies of habitat and microhabitat use across its range indicates considerable area-specific variation and the life-history needs of individuals and the population (e.g., males vs. gravid females and differences in available habitat). This is also one of the few rattlesnake species to occur in dense coniferous forests, a relatively cool environment for a snake. Other upland situations include agricultural areas where they may be accidentally killed by farm equipment during harvest time in late summer. Hardwood forests are sometimes used, depending on conditions and available habitats. Non-native habitats, such as residential areas, golf courses, and areas of invasive vegetation tend to be avoided.

Eastern Massasaugas are most active on warm days from late spring to early fall, and will brumate in late fall to early spring. They are primarily diurnal, but in the summer may be active during late afternoon to early evening. Unlike most other northern rattlesnakes, they do not den communally in hibernacula. Rather, they seek out rodent burrows, crayfish burrows, stump holes, rocky areas, downed wood, and other retreats, often near groundwater level. When they emerge, they may sit atop brush piles, grasses, and even beaver lodges to bask in the sun.

They are well known for preying on amphibians, and a number of frog and toad species have been recorded in the diet, but they seem to feed primarily on endotherms, including several species of mice (especially voles) and shrews. In one study, neonates fed primarily on shrews. They have also been reported to feed on lizards, snakes, carrion, and invertebrates. Young individuals have yellow-tipped tails, which they use for luring in prey.

Encounters: Although once common, this species is now considered rare throughout most of its range due to urbanization, wetland loss, persecution, and other factors. It is federally listed as endangered and is also protected in all states where it occurs. Random encounters with wild *S. catenatus* are unusual and bites are uncommon to rare. Some people living in Eastern Massasauga country are unaware they have a native rattlesnake in their area, although bites do make headlines.

The complex pattern of blotches renders them difficult to see in dappled light and amongst vegetation and leaves. They often remain still and do not rattle, but if they do, the small rattle is difficult to hear. They may flee or enter water if disturbed, but may be quick to bite, repeatedly, if they feel threatened.

Venom Apparatus, Venom, and Symptoms: The fangs are small, only reaching about 5 mm. The venom yield is about 15–45 mg, but it is thought a single bite would deliver about 5–6 mg of venom.

The venom is potent, having neurotoxic, myotoxic, and hemotoxic fractions. There is a presynaptic Mojave-like neurotoxin known as sistruxin in *S. catenatus*. There are also several proteolytic enzymes present, including phosphomonoesterase, phosphodiesterase, L-amino oxidase, and PLA_2. There are several estimates of LD_{50}, but all of them suggest a high toxicity, ranging from

Black or melanistic phase of *S. catenatus*. *James Tuich/CCA/Flickr*

0.22–0.90 from most studies, but as low as 6.8 mg/kg, depending on the study and method of injection.

Although human envenomation is rare, it is well-documented. Many bites are apparently dry, but others can be serious. The most common symptoms of a mild bite are immediate pain, swelling, and discoloration. More serious bites include fever and chills, cold sweats, numbness, lymph gland involvement, tremors, and rarely, death. Bites from *S. catenatus* may cause anticoagulation, extensive bleeding, and tissue damage. Despite their small size, a bite should always be considered a medical emergency. In a case report of a women bitten by *S. catenatus* in Ontario in 1957, in the late stages of her symptoms, she complained of chest pain and difficulty in breathing, and died shortly thereafter in a hospital. There was extensive tissue damage, a lack of clotting, and renal failure.

Remarks: Adding to the number of threats to conservation of populations of Eastern Massasauga is an emerging disease, a fungus, often simply known as snake fungal disease, caused by *Ophidiomyces ophiodiicola*. This fungus can disfigure and debilitate snakes to the point where they can no longer feed.

References and Resources: Refer also to the introduction for terrestrial vertebrates. Allender et al. (2011, 2013, 2016), Harvey and Weatherhead (2006a, b), Jaffe (1957), Jellen and Kowalski (2007), Jones et al. (2012), Kubatko et al. (2011), Moore and Gillingham (2006), Shepard et al. (2004, 2008), Weatherhead and Prior (1992).

Pygmy Rattlesnake (*Sistrurus miliarius*)

"The attack on the child's coagulation system, which can lead to internal bleeding, is a relatively new development that has threatened four children in Oklahoma. . . . The pygmy rattlesnake is interesting; it has a different type of antigen, one we can't describe yet, in its venom. . . . Snakes evolve and venom can change and adapt, and we're seeing it becoming quite resistant to the CroFab now. We think the pygmy rattlesnake venom is morphing." (AP October 2, 2017, originating with Tulsa World; NOTE: I cannot find biochemical or research evidence to substantiate this physician's claim.)

Classification: Phylum Chordata (chordates); Class Reptilia (reptiles); Order Squamata (snakes and lizards); Family Viperidae (vipers); Genus *Sistrurus* (massasaugas and Pygmy Rattlesnake); Species *miliarius* (Pygmy Rattlesnake).

There are three subspecies: *S. m. barbouri* (Dusky Pygmy Rattlesnake), *S. m. miliarius* (Carolina Pygmy Rattlesnake), and *S. m. streckeri* (Western Pygmy Rattlesnake). It has been suggested that subspecies are not well

Dusky Pygmy Rattlesnake, *S. m. barbouri*. Bill Love

differentiated and may not be valid, but this is based on limited sampling for a genetic analysis. "Pygmy" is sometimes spelled "Pigmy" in the literature.

Identification: The Pygmy Rattlesnake is among the smallest species of rattlesnake, although the Desert Massasauga is often regarded as having a smaller average adult size. Adults are usually between about 40 and 55 cm, with a record of 83.2 cm. Neonates are diminutive, being only 14–18 cm. As with massasaugas, it has a relatively narrow head with large scales on the top. The head is generally well marked, usually with a pattern on top and a light/dark eye stripe on the side. The rattle is small to virtually nonexistent, sometimes being little more than a button. About one-third to one-half lack a rattle. There are subtle differences between the subspecies, and all are variously patterned. They are generally gray to brown in ground color. The Dusky and Western subspecies are usually gray, ranging from light gray to almost black. There may be a reddish middorsal stripe between the blotches. There are beautiful reddish morphs of the Carolina Pygmy in North Carolina that are coveted by reptile aficionados. They are also protected by law to prevent overcollecting. The 21–50 dorsal markings are irregular blotches or crossbands that are darker shades of the ground color or gray to black. The tip of the tail may be dark to inconspicuously banded. Young animals have a yellow-tipped tail, used for caudal luring.

Carolina Pygmy Rattlesnake, *S. m. miliarius*. *Troy Hibbitts*

The beautiful red phase of the Carolina Pygmy Rattlesnake. *Bill Love*

Because these snakes have an absent to inconspicuous rattle, they are sometimes misidentified by laypersons as hog-nosed snakes, but rattlesnakes have vertical pupils, a heat-sensing pit, and lack the upturned rostral scale of hog-nosed snakes. There is little geographic overlap in range with the Western Massasauga, and massasaugas have larger, more distinct dorsal blotches.

Distribution: *Sistrurus miliarius* is a Southeast species. The Western Pygmy Rattlesnake is the most wide-ranging subspecies. It is found in eastern Texas; eastern Oklahoma; most of Arkansas, Louisiana, and Mississippi; southern Missouri; western Tennessee; west-central Alabama; and extreme southwestern Kentucky. It is absent from the bottomlands of most of the Mississippi River, except the delta. The Carolina Pygmy is found from North Carolina west to extreme northeastern Mississippi. The Dusky Pygmy is found throughout Florida and into parts of the adjacent states.

Natural History: Pygmy Rattlesnakes occur in both uplands and valley bottoms, often in the vicinity of streams, lakes, ponds, marshes, swamps, and even roadside ditches and agricultural areas. The Western Pygmy is most common inland in areas with sandy soils. They also occur in the northern coastal plain, but usually in areas with less permeable substrates. It is often associated with pine and deciduous woodlands or grassy areas. The Dusky

Western Pygmy Rattlesnake, *S. m. streckeri*. Bill Love

Pygmy is common in swamps and wetlands, especially the Everglades, but also in upland pine forest. The Carolina Pygmy is found in pine and scrub flatwoods on the Atlantic coastal plain and in pine habitats to the west.

This animal has been recorded year-round in southern parts of the range, but is mostly abundant spring through fall. During moderate weather, they are active during the day, but become increasingly nocturnal during hot summer weather. Male combat has been reported. Parturition usually occurs in July and August; sperm can remain viable in the female's oviduct until they fertilize eggs the following spring. The Carolina Pygmy has small litters, usually of 2–8 (maximum 11). Duskies also have 2–11 young, while the Western Pygmy has 2–19 (usually 4–10) young, although a record of 32 young has been documented in Oklahoma. The diet of *S. miliarius* is primarily frogs, lizards, small snakes, and insects. Occasionally, nestling birds and fish are consumed. Young animals wave their yellowish tail-tips to attract small animal prey.

Encounters: These little snakes range from being locally uncommon to abundant, but in most areas are infrequently encountered by the public, even where common. The species is small, inconspicuous, and often hidden under litter and debris, which can include palmetto fronds, logs, and woodpiles. Abandoned mills may harbor large numbers in the sawdust and under slabs and abandoned equipment. The very small to vestigial rattle can hardly be heard by people with good hearing from more than a couple meters away. The Pygmy Rattlesnake normally tries to hasten a retreat upon discovery, but can be pugnacious if threatened.

In Florida, bites by Pygmy Rattlesnakes are said to be not uncommon, but in Texas and Oklahoma, accidental bites are rare. The strike range is only about 10–13 cm. Many to most reported bites are from people who misidentify them as hog-nosed snakes and intentionally pick them up. Bites also occur to people walking barefoot. These are popular snakes among exotic pet owners. Their small size and attractive color suits them well for home terraria. There are undoubtedly bites that occur with captive animals.

Venom Apparatus, Venom, and Symptoms: Most research of the Pygmy Rattlesnake has been conducted on *S. m. barbouri*. The fangs of this subspecies are 5–6 mm in length. Although the venom yield has been reported as high as 125 mg, the typical volume is usually between about 12–35 mg. The LD_{50} for the species as a whole has been reported as 24.3 mg/kg subcutaneously, but other reports reported intravenous toxicity as 0.28–6.84 mg/kg intravenously. As such, it is difficult to gauge just how virulent the venom is, and it has been variously reported as being highly toxic to not-so-much, compared to other rattlesnakes.

Venom composition of the two western subspecies has been compared. It is similar across the subspecies, but there are differences in relative amounts of venom components that likely reflect differences in prey selection. The venom is largely tissue-destructive. It contains serotonin, L-amino oxidase, ATPase, metalloproteinases, kallikrein-like proteases, phospholipase A_2, phosphomonoesterase, phosphodiesterase, myotoxin α-like proteins, trypsin-like compounds, and other components.

Symptoms for the species include immediate pain, swelling, hemorrhaging, discoloring of the skin, weakness, giddiness, nausea, breathing difficulty, passing of bloody urine, renal failure (rarely), and prolonged numbness of the bite site. There are no known fatalities from the bite of this species. Nevertheless, even the smallest rattlesnakes have a relatively powerful venom, so bites should be taken seriously.

References and Resources: Refer also to the introduction for terrestrial vertebrates. Clark (1963), Gibbs et al. (2013), Kubatko et al. (2011), Rowe et al. (2002), Wooten and Gibbs (2012).

Grassland Massasauga (*Sistrurus tergeminus*)

"At 8 hours after the bite [by *S. t. tergeminus*], I briefly awoke, sweating, shaking uncontrollably with chills, with a fever of 38.3°C., mild nausea, and was very "light-headed." (Baldwin 1999)

Grassland Massasauga, the Desert variety (now *S. t. edwardsii*). L.L.C. Jones

Classification: Phylum Chordata (chordates); Class Reptilia (reptiles); Order Squamata (snakes and lizards); Family Viperidae (vipers); Genus *Sistrurus* (massasaugas and Pygmy Rattlesnake); Species *tergeminus* (Grassland Massasauga).

Until recently, a single species was recognized, the Massasauga (*S. catenatus*), with three subspecies. The northeastern subspecies (*S. c. catenatus*) was found to be a strongly divergent clade and was elevated to species (*S. catenatus*, Eastern Massasauga); it is treated in a separate account. The Grassland Massasauga has two subspecies, the Desert Massasauga (*S. t. edwardsii*) and the Western Massasauga (*S. t. tergeminus*). I do not follow suit with the two "standards" (Crother 2017; Center for North American Herpetology, accessed March 2019) for English names for this species and one of its subspecies because of the confusion it introduces by overlapping names of different geographic entities. The novel name "Grassland Massasauga" is used for the species rather than the proposed "standard" name, Western Massasauga, because that name was previously used for *S. t. tergeminus*, which did not include *S. t. edwardsii*. The "standards" proposed the name "Prairie Massasauga" for *S. t. tergeminus*, but I retain the original name for the same geographic entity, the Western Massasauga. Ironically, the name "Prairie Massasauga" would have been a good choice for a novel name for the western species, as the western clades are dwellers of prairies, but since that name has now been introduced for *S. t. tergeminus*, it would only add more confusion at this point. The same problem of nomenclature arose when they assigned "standard" English names to copperheads, and it occurs in other taxa as well, including the nomenclaturally chaotic "western rattlesnake" complex.

Identification: This is a small serpent, with adults generally between 30 and 65 cm in length. The Western Massasauga has a record size of about 90 cm. The Desert Massasauga is even smaller, only reaching about 53 cm maximum size. The genus *Sistrurus* differs from typical rattlesnakes by the large scales on the top of the narrow head. The background color is gray to tan or brown. They have a distinct, dark eye stripe, bordered by white to cream lines. There are 21–50 distinct dorsal blotches, having a color similar to, but darker than, the ground color. The blotches usually have a thick, dark border edged with lighter coloration. The tail is typically banded, but less so than many other rattlesnakes. The banding is usually shades of the ground color and darker dorsal markings. The terminus of the tail may be black. The rattle is small.

Distribution: The Desert Massasauga is found in southeastern Arizona, New Mexico, and Texas. The Western Massasauga is found in parts of Texas, Oklahoma, Kansas, Nebraska, and Iowa, with a disjunct population in Missouri. A disjunct population in eastern Colorado has been reported to be an intergrade, but is usually considered the Desert Massasauga now. Intergrades occur in Kansas and the panhandle of Texas. This species occurs from about sea level in Texas to about 1,600 m in Colorado.

Natural History: The Grassland Massasauga is primarily an inhabitant of grasslands and prairies. The Desert Massasauga is not really a desert species, although it does occur in part of the northern Chihuahuan Desert of New Mexico and small areas of Mexico. It too is usually associated with grasslands

Western Massasauga, *S. t. tergeminus*. Refer to the text to better understand nomenclatural confusion of massasaugas. *Terry Hibbitts*

habitat, but also occurs in woodlands of the Edwards Plateau, Tamaulipan thorn scrub, and Coastal Plain of Texas. Similarly, the Western subspecies is most often found in semiarid grasslands in central and northern Texas and adjacent Oklahoma. It is often in the vicinity of riparian or rocky areas, but is not limited to those situations. Some populations of *S. t. tergeminus* occur in boggy areas, a habitat more typical of *S. catenatus*.

This species is most active in low elevations and southerly areas from as early as March to as late as November, while those in more northerly latitudes and higher elevations have a shorter duration of surface activity. The Desert Massasauga is predominantly nocturnal and most often encountered May to mid-September, when daytime temperatures are high and nighttime temperatures mild.

Males engage in combat rituals. The mating period of *S. tergeminus* is unclear due to literature including *S. catenatus*, plus observations during captivity may not be typical for the wild. Mating seems to occur throughout much of the surface-active season. There is a gestation (all massasaugas combined) of 71–115 days. Parturition peaks in August and September. There are generally only 5–6 neonates. The young may remain with the mother for a short time after parturition.

Sistrurus t. edwardsii in Arizona and New Mexico feeds primarily on lizards (whiptails, Southwest Fence Lizards, earless lizards, Ornate Tree Lizards, etc.), followed by mammals and centipedes. In Texas, mammals were reported more often in the diet than lizards, for both subspecies. Mammals eaten include shrews and small mice. Massasaugas sometimes feed on snakes (including conspecifics), birds, invertebrates, and, rarely, frogs and toads. Early published literature on Grassland Massasaugas stated they primarily eat amphibians, but there was no basis for that claim. This misinformation was likely from extrapolating dietary traits of the Eastern Massasauga. Based on the few studies of the western subspecies, amphibians are extremely rare in the diets of both the Desert and Western Massasauga.

Encounters: Grassland Massasaugas range from being rare to abundant, depending upon locality. In southeastern Arizona, it has apparently been extirpated from two of its three range clusters due to overgrazing, and was only known from the San Bernardino Valley, where it is uncommon. Recently, however, a new population was discovered in Arizona. In Colorado, Texas, and presumably parts of New Mexico and Oklahoma, it may be locally abundant. For example, 43 were collected in a single evening in May along roads in Texas, west of Fort Worth. In another cited example, a farmer reported 50–60 Massasaugas during one growing season. In two seasons, in Colorado, 254 individuals were picked up for study during road cruising in 12 counties. Despite some high local densities, this animal is of conservation concern throughout much of its range due to development, vegetation conversion, overgrazing, habitat destruction, prescribed fire during the active season, road kill, and human persecution.

Massasaugas have a variable temperament. *Sistrurus t. edwardsii* can be rather nippy and *S. t. tergeminus* ranges between mild to sometimes irascible. The rattle is small, so it is difficult for many people to hear. Given their proclivity for grasslands—a notoriously difficult environment in which to see snakes—one should be cautious in these habitats. They are mostly nocturnal and may become active at dusk, when it is hard to see, anyway.

There is little information on encounters and how this species figures into snakebite statistics, but it is a small species that is often in fairly remote areas, so most encounters are likely from those living in areas where densities are high near human habitation. They are kept by some exotic pet owners who maintain venomous species, and being small, can be kept in smaller confines.

Venom Apparatus, Venom, and Symptoms: Information provided here mostly includes *S. catenatus*, as it was lumped with *S. tergeminus* in many studies. The fangs are short, averaging about 5 mm. The venom yield is about 14–45 mg (averaging 25–35 mg), but it has been suggested that the usual yield may be closer to 5–6 mg. There is a potent venom. The LD_{50} of all massasaugas combined varies according to the particular study, but ranges from about 2.9–6.8 mg/kg, with higher toxicity for intraperitoneal injection (0.22–0.9 mg/kg), in one study. In other studies, intraperitoneal LD_{50} estimates include 2.9 mg/kg, plus 0.76 and 0.9 mg/kg. The human lethal dose has been estimated at 30–40 mg. Human death has been recorded for *S. catenatus* and/or *S. tergeminus*.

Massasaugas have a tissue-destructive venom. Myotoxin α-like proteins have been isolated from the eastern species, but not the western. However, *S. t. tergeminus* has been shown to have a presynaptic Mojave-like neurotoxin, sistruxin, which is also found in *S. catenatus*. *Sistrurus t. edwardsii* has venom that is high in metalloproteinases, serine proteases, phospholipase A_2, and other compounds, including some novel components. It has five three-fingered toxins, which are similar to those found in some elapids and

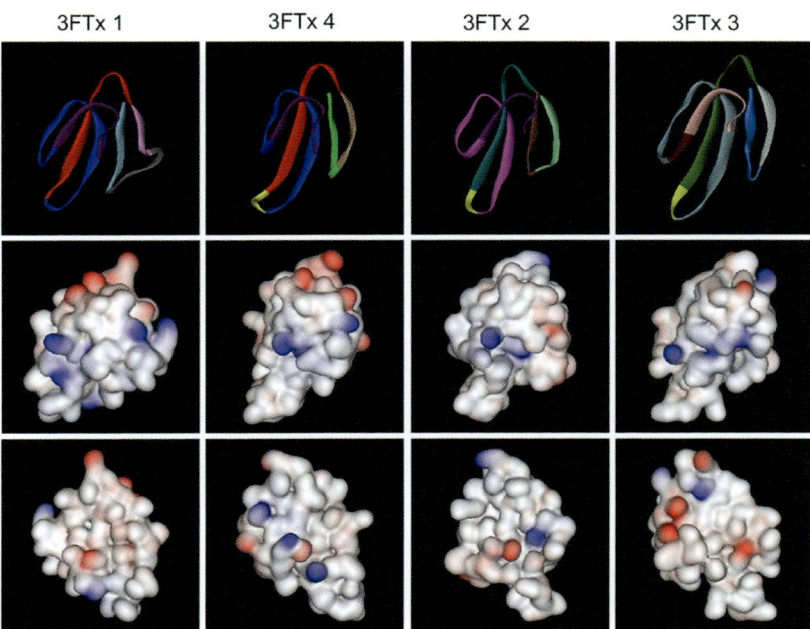

Three-fingered toxin found in at least some massasaugas. *Doley et al. (2008)/CCA*

seasnakes. Typical rattlesnake proteases also include L-amino oxidase, phosphodiesterase, and phosphomonoesterase.

While some populations may have neurotoxic elements in the venom, most published accounts of bites by massasaugas indicate a highly hemorrhagic toxin. Despite their small size, the venom is more potent than that of most rattlesnakes and a bite should be taken seriously. The bite of the Western Massasauga can be very painful and tissue discoloration from leaking capillaries is a common symptom. It has been reported that swelling is usually limited to the vicinity of the bite site, and while there may be some faintness and nausea, an uneventful recovery follows in a few days. However, details of these bites are sketchy. In one well-documented case report by the victim, without medical intervention, he described a bite on the finger that became swollen to the elbow, and then into the chest and torso. There were a variety of local and systemic reactions including considerable pain, numbness, tingling, a metallic taste in the mouth, light-headedness, nausea, uncontrollable shaking, fever, sweating, sensitivity to touch, and lymph node pain in the axilla and groin. The muscle and joint pain was significant. At fourteen days post-bite, the entire layer of the skin sloughed off the arm. At seventeen months, he still had problems with the bitten hand and believed there to be permanent damage to the hand muscles and functionality. I am aware of no case studies of bites by the Desert Massasauga.

Remarks: *Sistrurus t. edwardsii* is protected in Arizona and Colorado, and the nominal form is protected in Iowa, Missouri, and Nebraska.

References and Resources: Refer also to the introduction for terrestrial vertebrates. Baldwin (1999), Doley et al. (2008), Feldner et al. (2016d), Hobert et al. (2004), Holycross (2003), Holycross and Mackessy (2002), Kubatko et al. (2011), Pahari et al. (2007), Sanz et al. (2006), Wooten and Gibbs (2012).

Sonoran Coralsnake (*Micruroides euryxanthus*)

"Although venomous, the Arizona coralsnake (*Micruroides e. euryxanthus*) is not medically important based on past clinical experience [ten cases]. . . . However coral snakes should not be handled casually and anyone bitten should be observed in a hospital for at least 24 hours." (Hardy 1991)

Sonoran Coralsnakes in Arizona and New Mexico often have no dark pigments on their red and yellow bands. *Bill Love*

Classification: Phylum Chordata (chordates); Class Reptilia (reptiles); Order Squamata (snakes and lizards); Family Elapidae (elapids); Genus *Micruroides* (Sonoran coralsnakes); Species *euryxanthus* (Sonoran Coralsnake).

This is a monotypic genus. There are three subspecies: *M. e. euryxanthus* (Arizona Coralsnake), *M. e. australis* (Coralillo de Sonora), and *M. e. neglectus* (Coralillo de Mazatlán). Only the Arizona Coralsnake occurs in the United States. This species is not closely related to other coralsnakes.

This *M. euryxanthus* has black specks on the bands, more closely resembling U.S. species of *Micrurus*. L.L.C. Jones

Identification: This is a small, slender snake, usually 30–45 cm, with a record of about 66 cm. This is arguably the most beautiful U.S. serpent. The body and tail is completely encircled by black, yellow, and red bands. Unlike other U.S. coralsnakes, they often lack any black pigment within the red (and sometimes yellow) bands, giving them a precisely demarcated ringed pattern. It does not overlap in range with the other coralsnakes in the United States, but can also be differentiated by the anterior banding pattern. The band after the first yellow head-and-neck band is red, rather than black in the other species. Also, *M. euryxanthus* has broader yellow bands than U.S. *Micrurus*.

There are many species of coralsnake mimics that are banded with black, yellow, and red. They include Western Shovel-nosed Snake, Sonoran Shovel-nosed Snake, Milk Snake, mountain kingsnakes, and Long-nosed Snake. The Arizona Coralsnake differs from all mimics by having broad bands that completely encircle the body. There is a rhyme that goes something like this: "Red touch yellow, kill a fellow; red touch black, friend of Jack." There are other similar iterations. This might be useful to a layperson, except that it is easy to mix up some of the words and get the recipe wrong. Also, it isn't actually true. For example, when viewed from above, the Sonoran Shovel-nosed

Red touch yellow, kill a fellow? Sorry, don't use that rhyme. This is the Sonoran Shovel-nosed Snake, which co-occurs with *M. euryxanthus* in part of its Arizona range. Coralsnakes elsewhere may have a completely different color pattern. *L.L.C. Jones*

Snake (*Chionactis palarostris*) is quite a convincing mimic that has red touching yellow, although it differs by having a yellowish snout (*palarostris* means "pale snout"), and the bands do not encircle the body. If you visit another country in Latin America, you may encounter far more dangerous species of coralsnakes that do not have the red-touch-yellow pattern of our U.S. species.

Distribution: The species occurs from Arizona and New Mexico in the north to Sinaloa in the south. In Mexico, *M. e. euryxanthus* is found in much of Sonora, including Isla Tiburón, and just into northern Chihuahua. *Micruroides e. australis* occurs in southern Sonora and western Chihuahua. *Micruroides e. neglectus* is known from the vicinity of Mazatlán, where only three specimens have been collected. The Arizona Coralsnake occurs from sea level to about 1,800 m.

In less than a day's drive from southern Arizona, in southern Sonora, its range overlaps with the northern end of the range of a much larger and more dangerous species of coralsnake, the West Mexican Coralsnake (Coralillo Bandas Claras, *Micrurus distans*). Being in the genus *Micrurus*, it is more closely related to other coralsnakes.

A few hours south of Arizona in the State of Sonora, *M. euryxanthus* co-occurs with a much larger and dangerous coralsnake, *Micrurus distans*. L.L.C. Jones

Natural History: The Arizona Coralsnake is found primarily in the Arizona upland subdivision of the Sonoran Desert and semi-desert grasslands of desert mountain foothills. It barely enters the far western Chihuahuan Desert in the eastern extent of its range. It is rarely encountered in the Lower Colorado subdivision of the Sonoran Desert and is apparently absent from most of the lower desert; it has not been recorded in California, for example. It seems to occupy areas with significant moisture from the North American Monsoon. It does not occur farther east than extreme western New Mexico, however, even though this area is within the normal monsoonal range. This species can enter the lower elevations of oak savannas, but does not venture higher into the mountains. In Sonora, it is well distributed in the foothill thorn scrub vegetation type.

Little is known about this secretive little animal, as it seems to live a predominately subterranean lifestyle. It is generally active April through November, but most are encountered during the monsoon, late June through September. It is primarily nocturnal, but may be crepuscular and are occasionally encountered during the day, particularly after rains or on warm, cloudy days. The only known hibernaculum is of an individual found about 0.3 m under rocky substrate. It is usually seen in rocky or gravelly areas, but often in the vicinity of washes.

Virtually nothing is known of its reproduction, including information from captive animals, as they do not usually fare well in captivity. They probably have a spring peak in mating, but may be sexually active for much of the year, based on anatomical characters of reproductive organs. There may be a second reproductive peak in fall. Based on the few records, there are 2–6 eggs produced, usually 2. Females probably do not breed every year.

One aspect of this snake's biology that is fairly well known is its proclivity for feeding on small, smooth-scaled snakes and lizards. The most common prey items are threadsnakes, and coralsnakes have been found in their burrow systems. Other small snakes in the diet include Western Groundsnakes, black-headed snakes, nightsnakes, shovel-nosed snakes, Ring-necked Snakes, hook-nosed snakes, and Variable Sandsnakes. Lizards consumed include whiptails, Madrean Alligator Lizards, skinks, and night lizards. However, laboratory studies show that captives may refuse to eat a large number of suitably sized species, especially rough-scaled species, but also some smooth-scaled species.

Despite being venomous, these animals are preyed upon by a variety of species. This includes carnivorous mammals, Greater Roadrunners, and even Western Black Widows. Predation events are poorly known because they are so secretive.

Encounters: Unfortunately for those of us who really love to see this animal in the wild, it is uncommonly encountered. It is probably not uncommon, just secretive, spending most of its life underground. When encountered, these nervous little snakes tend to crawl away rapidly. If cornered, they are quite twitchy, and usually hide their head in their coils and may elevate their tail to take attention away from their head. They may also engage in cloacal popping, wherein they discharge air and feces from their cloaca, a ploy used by numerous species of snakes. They are well known for being timid, so are not likely to bite unless they feel threatened.

Unless you are walking barefoot and step on one (and people rarely walk barefoot in areas replete with rattlesnakes and cholla cactus), you have to handle them to be envenomated. All known human envenomations have occurred on the hands. Bites may be from captive animals, but may also occur from people who consider them harmless. I heard a tale of a scoutmaster that caught one and let his scouts handle it, knowing it was venomous, but considered the species harmless because he believed they don't bite.

Venom Apparatus, Venom, and Symptoms: The venom apparatus of elapids is unique among snakes and is known as the proteroglyphous (front-fanged) condition. There are fixed front fangs that encircle a groove, to form a tube. The paired fangs are on the anterior margin of the maxilla bone. A venom gland produces the venom and by contraction of muscles, is injected into the victim through a duct in the fang. *Micruroides* (and seasnakes) differs from *Micrurus* in having teeth on the maxilla bone, behind the fangs.

This animal has a very small head and mouth with small fangs, only measuring 0.1 to 1 mm in length. The fangs are so short that it would seem to make successful bites on a human unlikely. This has been a topic of some discussion. There are numerous erroneous beliefs suggesting this species is unable to envenomate humans: (1) they are so mellow, they will not bite; (2) the fangs are too short to penetrate skin, except the webbing between fingers of the hand; (3) they are rear-fanged (hence cannot envenomate unless the skin gets to the back of the mouth); and (4) they need to hang on and chew to introduce venom. The truth is they do have very short fangs, but they are in the front, the snakes can open their mouths widely, they do bite, and venom is injected, not chewed in.

The venom yield is small, about 6 mg. I could not find a figure for LD_{50} for this species, but the protein content is higher than for other U.S. coralsnakes. A different measure used for neurotoxicity is t_{90}, which is a measure of the time it takes to inhibit 90% of the muscle twitching in lab animals with a given dosage. *Micruroides euryxanthus* is forty-six minutes, between that of *M. tener* (thirty-one minutes) and *M. fulvius* (sixty-seven minutes). This measure suggests Arizona Coralsnakes are more neurotoxic (but not necessarily more lethal) than Harlequein Coralsnakes, a decidedly dangerous species, The human lethal dose has been estimated at 6–8 mg, so this is a potentially dangerous snake.

Victims show signs of neurotoxic envenomation, but the venom composition has not been deciphered. It is sometimes written that this species is not medically important, as are the other two U.S. species, and bites are rare. While bites by this species are not well tracked, there were four case studies of bites reported from 1955–1965, an additional six by 1991, and twelve bites reported from 2001 through 2005. Undoubtedly, bites are not always reported.

Just how potentially dangerous these snakes are is a matter of debate. Bites may exhibit little pain or numbness at the time of the bite and more significant symptoms are typically offset by a matter of hours. Thus, the

Sonoran Coralsnakes may raise their curled tail as a false head and emit a cloacal popping sound. *L.L.C. Jones*

victim or even a physician unfamiliar with coralsnake envenomation may believe the bite did not penetrate the skin, the bite was dry, or envenomation was slight. Of course, this can be the case. Symptoms of the few case studies of victims include numbness of the bite site or elsewhere (e.g., face), paresthesia, reduced muscular control, headache, nausea, difficulty focusing the eyes, and drooping eyelids. To my knowledge, there have been no severe bites reported and no fatalities.

There is currently no antivenom for the Arizona Coralsnake, and it is generally believed to be unnecessary. Antivenom for the Texas and Harlequin Coralsnakes and Latin American species of the genus *Micrurus* are not effective for *M. euryxanthus*. However, there are clinical trials with an experimental antivenom, Snake (Micrurus) North American Immune $F(ab')_2$ Equine. The Wyeth *Micrurus fulvius* antivenom has been shown to be effective for envenomation by *M. distans*, at least in lab mice.

Remarks: Batesian mimicry has been studied using coralsnakes and similarly patterned snakes (or models) as test subjects. Research has shown that indeed nonvenomous species are avoided in areas where coralsnakes occur—at least among some predatory species. Where they do not occur, predators will often attempt to prey on banded snakes. However, this may be muddled by migratory patterns of birds that seasonally encounter venomous species. Greater Roadrunners, which are notorious snake eaters, do not pass up many opportunities for a meal and do not avoid coralsnakes or their mimics.

References and Resources: Refer also to the introduction for terrestrial vertebrates. Goldberg (1997), Greene and McDiarmid (1981), Hardy (1991), Jones et al. (2011), Meik et al. (2007), Pfennig et al. (1991, 2007), Roze (1974), Russell (1967), Savage and Slowinski (1992), Seifert (2007), Shaw (1971), Sherbrooke and Westphal (2006), Slowinski (1995), Vitt and Hulse (1973), Yang et al. (2017), Young et al. (1999).

Harlequin Coralsnake (*Micrurus fulvius*)

"A 10-year-old girl from Riverview thought she spotted a necklace on her doorstep, but it turned out to be a venomous snake. . . . Her mom [said] she was very lucky, considering it was a [Harlequin] Coralsnake." (Fox 13 News, Tampa Bay, December 6, 2017)

The Harlequin Coralsnake is the most dangerous coralsnake in the United States, and has taken human life recently—from a person who tried to kill it and did not seek medical attention. *Troy Hibbitts*

Classification: Phylum Chordata (chordates); Class Reptilia (reptiles); Order Squamata (snakes and lizards); Family Elapidae (elapids); Genus *Micrurus* (American coralsnakes); Species *fulvius* (Harlequin Coralsnake).

This species is sometimes referred to as the Florida Coralsnake. Until recently, it was considered the nominate subspecies of a larger ranging form, the Eastern Coralsnake, which included the Texas Coralsnake; both have been elevated to species.

Identification: Coralsnakes are long, slender, and cylindrical, with smooth scales. Harlequins are the largest of the U.S. coralsnakes. Most adults are

60–90 cm, with larger individuals being females. The record is about 1.3 m. This is a tricolored snake with relatively equal-width red and black rings separated by narrower yellow rings. The rings completely encircle the body. The adage, "red touch yellow, kill a fellow" applies to this snake, because no others within its range have that pattern. However, that rhyme does not apply to many Latin American coralsnakes or some nonvenomous snakes in the western United States. There is black speckling, spots, or blotches on the red rings.

Distribution: The Harlequin Coralsnake occurs throughout Florida and ranges along the coastal plains of the Gulf of Mexico to extreme eastern Louisiana, and north along the Atlantic to southeastern North Carolina. It also occurs in central Alabama.

Natural History: *Micrurus fulvius* is found in low-lying forested habitats of the coastal plains, but also ranges into interior forests. They avoid frequently flooded wetland habitats, preferring oak and hardwood hammocks and groves of longleaf pine. They are usually associated with habitats that include dense leaf litter, logs, palmettos, stumps, Gopher Tortoise burrows, and other hiding places. Harlequin Coralsnakes are diurnal to crepuscular, and seem especially active on sunny mornings. They can be surface active any time of the year, but mostly during the spring and late summer to fall.

They breed primarily in the spring, but there may be a second, smaller period of mating in the fall. Oviposition occurs from late May into July. They lay 1–13 (usually 4–7) very elongate eggs underground or in leaf litter. The eggs hatch in August or September, and neonates are only about 20 cm long. Harlequin Coralsnakes feed on many smooth-scaled snakes and lizards they are able to overpower. They are active hunters that poke, prod, and smell their way through their habitat until they come upon suitable prey. When prey is detected through visual or chemical clues, the snake rapidly strikes. If they connect with the reptilian prey, they will hold onto it until the venom takes effect. Although they don't need to chew their prey to envenomate it, they often do, to help work in the venom.

The coloration is aposematic, and it is usually believed the color pattern will keep them from being preyed upon by diurnal predators with color vision, especially birds.

Encounters: Harlequin Coralsnakes may be very common, but are not frequently encountered by most people. However, it is not unusual to find them in suburban yards and public areas, such as parks. Despite their bright warning colors, they may blend in with dappled sunlight and debris, so may be hard to see. Very few bites to humans are truly accidental. If they are trod upon, they cannot bite through a boot and probably not denim jeans. Nearly all bites occur from people handling them. They are often picked up because the captor does not recognize their venomous nature—it is a case of mistaken identity, where a captor believes the snake to be a Scarlet Kingsnake or Scarlet Snake. This is often the case when children are bitten. The other way humans get bitten is when they handle them, knowing they are venomous. This is often the case for young adults. They may think the snake is incapable of successfully biting, except in finger webbing, but this is not true for any coralsnake. Others just want to exhibit bravado to their fellow humans, or be thrill-seeking. Yet others are bitten while trying to dispatch the snake—the situation of the 2006 fatal envenomation noted in the next section. Had it been left alone, the man would still be alive.

Nationally, coralsnake bites only account for 2–3% of envenomations, although most of these are in Florida. According to poison control center statistics from 2000–2005, there were 255 coralsnake exposures reported in Florida. Another Florida poison control center study from 1998 to 2010 reported 553 bites, of which 387 were retained for further analysis; so, while national coralsnake exposure numbers pale in comparison to copperheads and rattlesnakes, in Florida and adjoining areas, they are an important health concern.

Venom Apparatus, Venom, and Symptoms: All coralsnakes have a well-developed venom delivery system (proteroglyphous) with hollow, fixed fangs in the front of the upper jaw, and a venom gland and ducts. Unlike popular belief, they inject venom and do not need to hang on and chew, although this may help expedite venom delivery. The fangs are only about 1.5–2.5 mm, but that is long enough to bite through skin just about anywhere on the human body. The venom yield is generally about 4–6 mg, although large individuals can be in excess of 12 mg, and there is a record of 38 mg. Large females are decidedly more dangerous than smaller snakes, due to the larger venom yield. It has been estimated that a lethal dose for humans is probably around 4–10 mg. The LD_{50} in mice varies, depending on the study and method of injection, and ranges from about 0.3 mg/kg to nearly 1 mg/kg (average probably around 0.6 mg/kg), but this demonstrates that no matter which study is referenced, this is a dangerously venomous snake.

The venom of *M. fulvius* is primarily neurotoxic, but it also has cardiotoxic, myotoxic, and hemotoxic properties. The percentage of toxic fractions within the venom is high. There are numerous PLA_2 and three-fingered toxins, plus hyaluronidase, C-type lectins, metalloproteinases, Kunitz-type protease inhibitors, phosphodiesterase, and L-amino acid oxidase, among other fractions. The PLA_2s form about a third of the venom volume, and are the most lethal components in lab mice.

The bite may or may not be painful, and symptoms may be delayed up to twelve hours. About 25% seem to be either dry bites, or contact was not made. Based on one study, about half of the bites had systemic symptoms, most of which suggest potent neurotoxins at work. There may be contusion and swelling at the bite site, plus abdominal pain and cramping, fasciculation, paralysis, convulsions, weakness, nausea, vomiting, paresthesia, vision problems (e.g., droopy eyelids and double or blurred vision), respiratory distress (including paralysis), and a host of other specific neurotoxic disorders and mental maladies. It has been estimated that about 10–20% of the bites could be fatal without medical care. However, thanks to modern medicine, fatalities are rare. In 2006, a Florida man died within hours from the bite of *M. fulvius*. The 29-year-old man tried to kill the snake, and did not seek medical attention after being bitten. This was the first fatality from *M. fulvius* since 1967.

Remarks: There is an antivenom that is no longer produced by Pfizer/Wyeth (Antivenin Micrurus fulvius). At the time of this writing, it is still available as stocks have been tested and expiration dates extended, but supplies are running out and it is unknown how long dates can be extended. There is currently another antivenom in clinical trials in Florida, and researchers and collaborators at the VIPER Institute are pursuing FDA approval. Also, there have been some studies on efficacy of other coralsnake antivenoms from Latin America, with mixed results for *M. fulvius* and *M. tener*. The potential efficacy of using pressure bandages for *Micrurus* bites needs further research, as it is an effective first aid in areas with large numbers of elapids (e.g., Australia), because it slows lymph flow and allows more time for antivenom injection.

References and Resources: Refer also to the introduction for terrestrial vertebrates. Castoe et al. (2012), German et al. (2005), Jackson and Franz (1981), Kitchens and Van Mierop (1987), Margres et al. (2013), Norris et al. (2009), Tennant (1997), Vergara et al. (2014), Wood et al. (2013).

Texas Coralsnake (*Micrurus tener*)

". . . we carried out an unbiased in vitro screen to identify snake venoms capable of activating somatosensory neurons. Venom from the Texas coral snake (*Micrurus t. tener*), whose bite produces intense and unremitting pain, excited a large cohort of sensory neurons." (Bohlen et al. 2011)

Texas Coralsnakes invariably cause intense pain, and old records claim human fatality. *Troy Hibbitts*

Classification: Phylum Chordata (chordates); Class Reptilia (reptiles); Order Squamata (snakes and lizards); Family Elapidae (elapids); Genus *Micrurus* (American coralsnakes); Species *tener* (Texas Coralsnake).

There are four subspecies of *M. tener* generally recognized. The Texas Gulf-Coast Coralsnake (*M. t. tener*) is the only one found in the United States. The other three are endemic to Mexico: Coral de Fitzinger (*M. t. fitzingeri*), Coral de Tampico (*M. t. maculatus*), and Coral Potosíno (*M. t. microgalbinius*).

Identification: A Texas Coralsnake is a slender snake that is generally about 60 cm as an adult, but they occasionally reach about 90 cm. The record is 121.3 cm. As with other U.S. coralsnakes, they have red, yellow, and black bands. The nose is black and the first band after the yellow neck band is black, as in the Harlequin Coralsnake, but unlike the red band of the Arizona Coralsnake. *Micrurus* lacks maxillary teeth, other than the fangs. It can be differentiated from the Harlequin Coralsnake by its range, but also has more black pigmentation in the red bands, and the black neckband touches the parietal scales.

As with all U.S. coralsnakes, there are numerous nonvenomous mimics. In the range of *M. tener*, this includes the Scarlet Snake, Milksnake (a species complex), and Long-nosed Snake. The Scarlet Snake and Milksnake have black bands that touch the red bands, so the old adage "red touch yellow, kill a fellow; red touch black, venom lack" works, except bitten people almost never die from bites of *M. tener*. The Long-nosed Snake technically has yellow-white touching red, but is quite different in general appearance (long snout and very speckly pattern). Only coralsnakes have a pattern with rings that evenly encircle the body.

Distribution: The Texas Coralsnake is found in south Texas, west to the vicinity of the Pecos River, and east to the Gulf Coast. It is also found in western Louisiana and extreme southwestern Arkansas. Its range does not overlap with *M. fulvius*. The species ranges into Mexico, to about the state of Hidalgo.

Natural History: This snake is primarily a subtropical species. It occurs east of the Chihuahuan Desert in the Tamaulipan thorn scrub of south Texas and the deciduous and mixed woodlands of east Texas to Arkansas. The vegetation in this area includes hardwoods, pine forests, and thorn scrub, but the snake can also be found in grasslands and even around human debris in populated areas. It is a largely fossorial species, so it needs organic ground litter present in its habitat. It can be surface active throughout the year in mild climates, especially in Mexico. In the United States, surface activity seems to peak March through May, and then again in the fall. It tends to avoid surface activity during the heat of summer. While it can be active day or night, it seems most active in mornings, especially following rainstorms. In the United States, this species brumates during most of the cooler months.

Courtship has been documented. Mating usually takes place in April and May and then again in the fall. If mating in the spring, eggs are laid in the late summer to fall, suggesting a gestation period of only thirty-seven days, based on captive animals. If mating occurs in the fall, they probably store sperm until the following spring. Either way, oviposition usually occurs in June and July. The eggs are about 38 mm long, laid in clutches of 2–12. They hatch about two months later, at about 16–20 cm.

They feed primarily on lizards and small snakes. Lizards include whiptails, skinks, glass lizards, and Eastern Fence Lizards. Snakes seem to be a generally preferred prey and include Ring-necked, Glossy, hook-nosed, black-headed, rat, garter, ground, patch-nosed, brown, lined, and thread snakes. They are known to be cannibalistic. Occasionally small mammals, frogs and toads, and possibly invertebrates are taken.

Encounters: Encounters are usually regarded as uncommon, but they are sometimes seen by those living and hiking in areas where they occur. They are sometimes encountered in yards, parks, and other human milieu. They are initially nervous and try to escape upon detection, but if pressed to defend themselves will writhe and "pop" their cloaca, and may be quick to bite. Most of the bites occur when people handle them intentionally, either

Dr. Elda Sánchez taking a Texas Coralsnake out to be milked for research. *L.L.C. Jones*

knowing they are venomous, or believing they are nonvenomous. Both the Texas and Harlequin Coralsnakes figure into a small but important part of snakebite statistics. A common figure is about 2% of the bites in the Southeast are attributed to coralsnakes.

Venom Apparatus, Venom, and Symptoms: The venom apparatus is as described for the Arizona Coralsnake, but the fixed front fangs are longer, measuring about 3 mm. It has been reported as having a venom yield of 10–12 mg, with a human lethal dose of 4–7 mg. Some of the information on venom of Texas Coralsnakes has been muddled by their earlier synonymy with the Harlequin Coralsnake, although they have usually been recognized as separate subspecies. Both species are primarily neurotoxic, but there are differences in the venom, and the Harlequin Coralsnake is believed to be a more dangerous species. The LD_{50} has been reported as 0.53 mg/kg for *M. fulvius* + *M. tener*, and 1.53 mg/kg for *M. tener*. In another study, *M. fulvius* + *M. tener* averaged at 0.279 and 0.779 mg/kg, respectively. In both studies, the virulence of *M. tener* is about one-third that of *M. fulvius*. Another study reports an LD_{50} of crude venom at 1.19 mg/kg, with the most virulent fraction at an LD_{50} of 0.06 mg/kg.

The venom of *M. tener* has recently been studied in specimens from Tamaulipas, Mexico. Using HPLC analysis, some 34 fractions were isolated. The venom had nucleosides and peptides, and was well represented by phospholipase A_2 and three-finger peptides. Some of the fractions caused flaccid paralysis in laboratory mice, while others seemingly targeted snakes, and still others had no observable effect on either.

The hallmark symptom of a Texas Coralsnake bite is excruciating pain. Other symptoms may include swelling (usually slight), redness, giddiness, difficulty speaking, feeling of swelling in the throat, difficulty swallowing, aching in the face region, swelling of lymph nodes, increased salivation, tongue tremors, cold skin, tingling, stiff joints, blurred vision, droopy eyelids, conjunctivitis, weakness, drowsiness, dizziness, hypertension, fever, vomiting, nausea, weak and erratic pulse, paralysis, and cardiac and respiratory distress. Some of these symptoms may have been due to sensitivity or treatment. In one study from 2000–2004 at poison control centers in Texas, 82 patients were studied for bites by *M. tener*. In this study, almost 90% reported local swelling, pain, redness, or paresthesias. Only 7.3% had systemic effects, none of which was severe. Texas Coralsnakes have been implicated in human death, but those records from Texas (two individuals) were reported in 1883, and no deaths have since been attributed to that species in recent times. A bite by *M. tener* should always be regarded as a medical emergency.

One FDA-approved antivenom is available, Wyeth (now Pfizer) Antivenin Micrurus Fulvius, which has been in short supply. This is a monovalent equine antivenom produced from venom of its namesake, but is not effective for *M. tener*. Another antivenom is produced by Instituto Bioclon, in Mexico, Coralmyn. It is a (Fab)$_2$ antivenom of an equine origin, using venom from *Micrurus n. nigrocinctus* (Black-banded Coralsnake). It has been proven effective against the venom of both *M. tener* and *M. fulvius* in lab mice. An experimental antivenom, Snake [Micrurus] North American Immune F(ab')$_2$ (Equine), is in clinical trials. This antivenom is made from both *M. tener* and *M. fulvius*.

References and Resources: Refer also to the introduction for terrestrial vertebrates. Bénard-Valle et al. (2014), Bohlen et al. (2011), Greene and McDiarmid (1981), Morgan et al. (2007), Norris et al. (2009), Sánchez et al. (2008), Savage and Slowinski (1992), Seifert (2007), Shaw (1971), Slowinski (1995), Pfennig et al. (1991, 2007), Young et al. (1999).

Other Snakes (Family Colubridae and their Relatives)

"After treating an 11-year-old boy for a poisonous snakebite . . . when he picked up a "harmless" garter snake . . . Dr. Findlay Russell said: 'We have suspected for a long time that all snakes could be called venomous. . . . clinically this is the first case reported in which a so-called non-venomous snake has been involved in poisoning. . . . We are going to have to go back and reexamine our concepts of what constitutes venomous and nonvenemous.'" (*Los Angeles Times*, 1975, D. Townsend reporting)

Of the so-called "harmless" snakes in North America, the hog-nosed snakes of the genus *Heterodon* are gaining notoriety as being venomous when the victim receives a prolonged bite. Pictured is *H. kennerlyi*, the Mexican Hog-nosed Snake. *L.L.C. Jones*

Classification: Phylum Chordata (chordates); Class Reptilia (reptiles); Order Squamata (snakes and lizards); Families Colubridae (colubrids), Dipsadidae (rear-fanged snakes), and Natricidae (watersnakes, gartersnakes, and their relatives). Collectively in this book, I refer to the group of three families as "Colubridae or colubrids and their relatives."

Historically, the family Colubridae, which included these other families, was known to be polyphyletic, so attempts to make taxonomic sense of this huge group were undertaken. However, the various studies are not always in agreement, so I am generally following the Center for North American Herpetology (CNAH, accessed March 2019), which shows familial assignment for the former Colubridae, if for no other reason than it seems to make bookkeeping sense for groups of similar taxa. Because this website is updated regularly, a revision of small snakes in 2018 placed *Chilmeniscus* and *Chionactis* into the genus *Sonora*, collectively called North American Groundsnakes, but most English names of familiar species remain the same. Colubridae (in the current sense), Dipsadidae, and Natricidae are often treated as subfamilies of Colubridae elsewhere in the literature, and common names of families can be misleading. Colubridae, as currently recognized, are referred to as "harmless egg-laying snakes." They do lay eggs, but because some can cause harm and they may even have enlarged, grooved rear teeth, I refer to them by the familial name as colubrids. The Dipsadidae are known as "rear-fanged snakes," but under the taxonomy of CNAH, some currently recognized colubrids are midly venomous, rear-fanged species; nevertheless, I refer to them also as rear-fanged snakes, or simply, dipsadids. The Natricidae are known as "harmless live-bearing snakes." They do bear live young, but some of these can cause envenomation in humans, so I refer to them as "watersnakes, gartersnakes, and their relatives," or simply, natricids. A few species of colubrids and their relatives outside the United States are seriously venomous animals, such as the Twig Snake, Boomslang, and Tiger Keelback Snake. Of the hundreds of genera worldwide, only five are known to have delivered fatal bites to humans, but none in the United States or Canada. There are dozens of colubrid, dipsadid, and natricid species in the United States. Most cause little to no harm, and bites are usually insignificant. However, humans sometimes experience classical venomous reactions to some species, at least in some circumstances. Those taxa that are considered rear-fanged, mildly venomous, or implicated or suspect in causing some degree of human envenomation include:

Colubridae

- Hook-nosed snakes (*Ficimia streckeri* and *Gyalopion*, 2 species)
- Brown Vinesnake (*Oxybelis* c.f. *aeneus*)

- Leaf-nosed snakes (*Phyllorhychus*, 2 species)
- North American Groundsnakes (*Sonora*, 7 species)
- Black-headed snakes (*Tantilla*, 11 species)
- Lyresnakes (*Trimorphodon*, 3 species)
- Bullsnakes and their relatives (*Pituophis* spp.) and patch-nosed snakes (*Salvadora* spp.) have also been discussed as having toxic saliva, and I can confirm this for *P. catenifer* (Gophersnake) based on an Envenomation Story and a photograph of a separate incident.

Dipsadidae

- Regal Black-striped Snake (*Coniophanes imperialis*)
- Ring-necked Snake (*Diadophis punctatus*)
- Hog-nosed snakes (*Heterodon*, 5 species)
- Nightsnakes (*Hypsiglena*, 3 species)
- Northern Cat-eyed Snake (*Leptodeira septentrionalis*)
- Other dipsadid genera include *Carphophis* (2 species), *Contia* (2 species), *Farancia* (2 species), and *Rhadinaea* (1 species). These genera are undoubtedly venomous to their prey.

Coniophanes imperialis is a small dipsadid, but it is capable of delivering a very painful bite. *Terry Hibbitts*

Natricidae

- North American watersnakes (*Nerodia*, 10 species). At least some species have been discussed in regards to toxic saliva.
- Gartersnakes (*Thamnophis*, 16 species; at least *T. elegans* and *T. sirtalis* have caused human envenomation). Also, I recently became aware of a confirmed human envenomation by *T. eques*, the Mexican Gartersnake.
- Other natricid genera include *Clonophis* (1 species), *Haldea* (1 species), *Liodytes* (3 species), *Regina* (2 species), *Storeria* (3 species), *Tropidoclonion* (1 species), and *Virginia* (1 species), but I am unaware of human envenomation reports.

In the future, we may discover that more species can deliver a toxic bite to humans. Case reports of bites from colubrids and their relatives are rarely reported in the literature but always enlightening. I encourage readers to submit reports of bites by these snakes to published outlets to further our knowledge. For example, in this book I mention novel Envenomation Stories that I became aware of from Gophersnakes, lyresnakes, Wandering Gartersnake (*T. elegans vagrans*), Northern Cat-eyed Snake, and Ring-necked Snake, as well as "western" hog-nosed snakes, which have been fairly well reported in the literature.

Identification: The vast majority of snakes in the United States are colubrids and their relatives. As a group, they are typical snakes, generally being elongate and lacking triangular heads, rattles, and heat-sensing pits, as in pit vipers. Some do resemble coralsnakes. The stout-bodied hog-nosed snakes are superficially similar to pit vipers in body form and pattern, but lack the rattle and pit, and have an upturned snout. Massasauga and Pygmy Rattlesnake bites sometimes occur because people pick them up thinking they are hog-nosed snakes. Nightsnakes, Northern Cat-eyed Snakes, and lyresnakes also somewhat resemble pit vipers with their colors, patterns, and vertical pupils (and the latter two of these have heads much larger than the neck). Lyresnakes also have saddle-like bands, similar to Black-tailed Rattlesnakes and some other species. Some of the North American Groundsnakes are among the coralsnake mimics; the Organ Pipe Shovel-nosed Snake (*S. palarostris organica*) is a rather convincing mimic with the same pattern of colors as the Arizona Coralsnake, with which it shares its range. Among the coralsnake mimics, none have the complete, even-width rings of yellow, red,

The beautiful and rare Texas Lyresnake. *Troy Hibbitts*

and black. Tri-colored kingsnakes and Long-nosed Snakes are among the coralsnake mimics that are not considered venomous; they all kill prey with constriction rather than envenomation.

Dipsadids and some colubrids are often termed "rear-fanged, mildly venomous." Some species have distinctive features, such as the modified rostral scales of the hog-nosed, leaf-nosed, and hook-nosed snakes; the black heads of most black-headed snakes; and the extremely slender body form of Brown Vinesnakes and Northern Cat-eyed Snakes. Ring-necked Snakes and some other species have bright aposematic coloration on their venter. Rather than trying to describe the many species in more detail, I refer the reader to the accompanying photographs, or better yet, one of the many excellent field guides.

Distribution: Gartersnakes, Ring-necked Snakes, Bullsnakes and their congeners, black-headed snakes, and hog-nosed snakes are found throughout much of the United States, especially in the southern U.S. states. Southwest species include shovel-nosed snakes, North American groundsnakes, leaf-nosed snakes, *Gyalopion* spp., Brown Vinesnake, lyresnakes, and nightsnakes. The humid East (especially Southeast) has more species of natricines than the arid West. Some East/Southeast taxa include wormsnakes, *Farancia* spp., Pine Woods Littersnake, *Clonophis*, *Haldea*, *Liodytes* spp., *Regina* spp., *Storeria* spp., *Tropidoclonion*, and *Virginia*. South Texas has three endemic (north

The Brown Vinesnake is considered a colubrid, although it has rear fangs. *R. C. Clark*

of Mexico) species, in addition to those found in the Southwest and Southeast: Regal Black-striped Snake, Northern Cat-eyed Snake, and Tamaulipan Hook-nosed Snake (*Ficimia streckeri*). The strange little sharp-tailed snakes are the only group endemic to the Northwest, an area not particularly diverse in reptiles.

Natural History: The natural history of these dozens of species is variable, so the reader should consult the many field guides for more information. They are mostly warm-temperate to subtropical animals, except for some of the more widely ranging taxa that can tolerate cooler temperatures (e.g., some gartersnakes). As a whole, they range from coast to coast in the lowest valleys to the highest mountains, although snakes are much more diverse in lowlands. Snakes are usually active from spring through most of the fall, until temperatures are too cool to allow surface activity. Colubrids and their relatives range from diurnal to nocturnal, depending on species and weather. They lay eggs, except for natricids, which give birth to live young. Natricids are also the most aquatic. While some live on dry land, most prefer mesic and wet areas. The watersnakes are aquatic to semiaquatic, and the gartersnakes are usually semiaquatic, or live in

The Northern Cat-eyed snake is a dipsadid. *Greg Green*

humid uplands. All snakes are carnivores. The dipsadids are rear-fanged, so they hold on to their prey while it becomes immobilized by venom. Mostly colubrids are constrictors—or just swallow their prey—but some are also envenomators with modified teeth. Many of the small species feed on invertebrates, lizards, salamanders, frogs and toads, small snakes, and small rodents, while the larger species feed on larger prey, including mammals and birds. Snakes, even venomous ones, are frequently preyed upon by a variety of animals, including amphibians (especially American Bullfrogs), raptors, Greater Roadrunners, a host of medium to large mammals, other snakes (especially kingsnakes), large lizards, and even some venomous invertebrates (e.g., large spiders, black widows, large scorpions, and giant water bugs).

Encounters: Not without exception, in their daily activities, most people do not frequently encounter snakes. Many species may even be difficult for herpetologists to find. The exception is probably gartersnakes, which can be extremely abundant in certain situations. Gophersnakes, Bullsnakes, and watersnakes are also commonly encountered, and undoubtedly thousands of people are bitten by them (usually when picking them up), and almost never with ill effects. Most species are secretive by nature and are largely nocturnal. Most to all of the recorded noteworthy envenomations from colubrids and their relatives were from people handling them in captivity, or while collecting

Ring-necked Snakes are dipsadids that behaviorally display their aposematic ventral coloration, but they are very inoffensive. *L.L.C. Jones*

them. Children are sometimes victims because they may pick up a snake out of curiosity. To date, only a few Brown Treesnakes (*Boiga irregularis*) have been discovered in Hawaii, and a population has not become established. If these snakes do get a foothold, it could be a significant problem, as it is on Guam. Refer to the chapter on American Territories to read more about that non-native species, and the venomous snake found in the American holdings in the Caribbean.

Venom Apparatus, Venom, and Symptoms: These species, even hog-nosed snakes, are usually considered nonvenomous or harmless to humans in most literature accounts, partly because they lack a distinct venom gland and a well-developed venom delivery system. Also, a superficial bite is usually asymptomatic, save the minor pain due to mechanical injury of the sharp teeth. Dipsadids are known as opisthoglyphs, meaning they have fangs in the rear of the mouth, at the posterior of the maxilla bone. Most of these species also have grooves in their teeth to expedite the flow of venom. Most colubrids and natricids are aglyphous, which means they lack fangs or grooved teeth. This however, is an oversimplification. For example, the lyresnakes belong to the colubrid genus *Trimorphodon*, which means "three tooth shapes" in Latin, as they have different-sized teeth, including enlarged front teeth, a mouthful

Saddled Leaf-nosed Snake *(Phyllorhynchus browni)* is extremely inoffensive like many small colubrids and dipsadids, so the effects of bites to humans are unknown for many species. *L.L.C. Jones*

of smaller teeth, and enlarged, grooved rear teeth (fangs) on the upper jaw. The toxic secretions of colubrids and their relatives usually come from the Duvernoy's glands above the jaw. Although Duvernoy's gland is often thought of as simply a salivary gland, it should be considered a true, albeit primitive, type of venom gland. Snake venom is, afterall, highly evolved saliva.

Some of these snakes are very small and disinclined or unable to bite, so many references have no problem considering them harmless to humans. This includes wormsnakes (*Carphophis* spp.), earthsnakes (*Haldea striatula*, Rough Earthsnake, and *Virginia valeriae*, Smooth Earthsnake), North American groundsnakes, Pine Woods Littersnake (*Rhadinaea flavilata*), brownsnakes (*Storeria* spp.), sharp-tailed snakes (*Contia* spp.), Lined Snake (*Tropidoclonion lineatum*), black-headed snakes, and hook-nosed snakes. Having said that, the Regal Back-striped Snake is also rather diminutive, but the bite may pack a wallop. Also, species such as the hook-nosed snakes have a well-developed venom delivery system—but I can find no records of anyone ever having been bitten by them. The other species should be assumed to be capable of producing a mildly to moderately venomous bite, particularly if the snake is allowed to hang on for several minutes. Lyresnakes can get rather large,

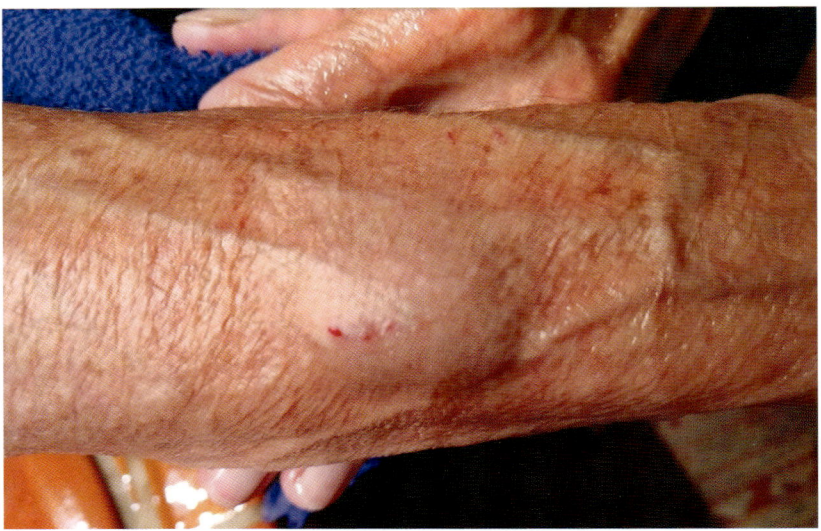

Thousands of people handle Gophersnakes without having ill effects from bites, but occasionally they cause symptoms. *Randy Gray*

so should be handled with care. Of these snakes in the bulleted lists, hognosed snakes and gartersnakes have received a certain amount of attention because of some well-documented cases of envenomation resembling that of a mild to perhaps moderate rattlesnake bite. At least one publication labels these genera as being "medically significant," although most publications label them as harmless.

Historically, the venom of colubrids and their allies has not been the topic of much study, but given several well-documented case studies of venomous reactions caused by so-called nonvenomous species, this is changing. While some of these species may not have venom that is potent to humans, the venom can be quite virulent to their normal prey. Some LD_{50}s in mice (usually measured in μg/g) include 13.9 μg/g (intraperitoneal) for the Wandering Gartersnake, 33.0 μg/g (intraperitoneal) for *Thamnophis sirtalis parietalis*, the Red-sided Gartersnake, and 26.0 μg/g (subcutaneous) for *H. jani texana*, the Texas Nightsnake. The LD_{50} for the Brown Treesnake is 31.0 μg/g (intraperitoneal) in mice, but 1.75 μg/g in chickens and 2.5 μg/g in House Geckos, demonstrating the specificity of toxins targeting birds and lizards, their normal prey, rather than mammals.

Among some of the snakes on our lists, some venom components have been found that are similar to viperids and elapids, including proteases,

The Red-sided Gartersnake belongs to the Natricidae, but this species is not usually thought of as being venomous, although the victim of this bite said it caused profuse bleeding, a symptom of a saliva with anticoagulant properties. *Garrett Craft*

phosphodiesterase, arginine esterase, acetylcholinesterase, CRiSPs (cysteine-rich secretory proteins), and phospholipase A_2. Missing from all of the species of those studied were hyaluronidase and enzymes with thrombin- or kallikrein-like activity. The phospholipase A_2 isolated from *Trimorphodon lambda* has been dubbed "trimorphin." Nightsnakes and Wandering Gartersnakes have been shown to have venom that can induce hemorrhaging. Bites to other snakes by these species cause intense hemorrhage and necrosis. The Bullsnake actually lacks a Duvernoy's gland, but does have phosphodiesterase in its saliva. Brown Treesnakes have three-fingered toxins ("irditoxin"), as well as two myotoxic proteins, acetylcholinesterase, and other compounds.

There are case studies for bites from several of these genera. Hog-nosed snakes have long been recognized as producing envenomation-like symptoms in humans on occasion—so yes, they unequivocally should be considered venomous. *Heterodon* are commonly kept as pets, and bites are not rare. In one case, a "western" hog-nosed snake (formerly *H. nasicus*, before being taxonomically split into three species) bit the victim on the arm and hung on for 3–5 minutes. The entire arm and hand swelled considerably. There was no initial pain or symptoms, but as the swelling increased, so did the pain. The victim sought medical attention. Over the next several days, bruising appeared and large blisters formed. Healing was not complete until after five months. Curiously, the conclusion of the report was that the bite was medically significant and resulted in envenomation, yet it went on to say that hog-nosed snakes should not be considered dangerous or venomous. I include two novel reports of envenomation by "western" hog-nosed snakes in this volume; one of these was fairly serious and resembled the symptoms of the bite mentioned above. However, in this case, there were also systemic effects for about a week, which included headache, dizziness, nausea, and vomiting. It also took five months to heal. In the other story, there were no significant systemic effects, but the bite itself was extremely painful, and the snake was unwilling to disengage its hold on the victim. Quick bite-and-release events by hog-nosed snakes do not usually show signs of envenomation. Hog-nosed snakes are primarily toad-eaters that can quickly paralyze toads with their rear fangs, so venom is used in prey procurement and/or digestion. Hog-nosed snakes deploy an arsenal of interesting defensive tactics, including playing dead and flattening the neck, but biting is not commonly employed unless they feel overly threatened (like when being handled roughly).

Bites from Brown Vinesnakes can result in swelling and slight pain for at least up to two days. The Brown Treesnake could be a danger if established in Hawaii. On Guam, where the snakes are extremely abundant, bites are not uncommon and symptoms usually consist of pain, swelling, and redness. They are not considered a significant medical threat to a healthy adult, but infants are a different story. Bites usually occur when children are sleeping, and in some instances, it was believed the snakes were actually trying to feed on them (e.g., starting with a finger); this would not be possible, of course. In one case report of two infants, they developed respiratory problems within a few hours and required medical treatment for asphyxiation. Swelling, discoloration, blisters, lethargy, diminished sensory perceptions, and drooping eyelids were also recorded in some patients.

Gartersnakes have occasionally been implicated in human envenomation since the 1970s, as in a case I remember well while in grad school—the opening quote is from that case, as it appeared in the *Los Angeles Times*. Another case report involved a 13-year-old boy who was bitten for ten minutes by an Eastern Gartersnake (*T. s. sirtalis*). The bite resulted in rapid swelling, bruising, and lymph node involvement, which lasted at least five days. While preparing this book, I became aware of more gartersnake envenomations, including one from a Wandering Gartersnake, which is featured as an Envenomation Story. There was significant swelling, skin-sloughing, and discoloration that lasted for months. A recent record of envenomation by a Mexican Gartersnake was posted on social media. Photographs showed that within a few hours of the bite there was significant swelling to the hand and wrist, with some hemorrhaging, but there were few other details and the extent of symptoms and resolution is not known, except that the victim got much better over time. One record of a bite by a 65 cm Northern Cat-eyed Snake to a zoo worker resulted in a small amount of swelling, some pain, and some redness—not unlike the sting of a bee. Another bite from the same species is also featured in an Envenomation Story. It was similar, with perhaps more swelling, and it took two days for the swelling to subside and a few extra days for the pain to disappear. Bites from lyresnakes are said to result in local redness, swelling, and itching. One Envenomation Story in this book features the bite of a California Lyresnake (*T. lyrophanes*). The symptoms were fairly significant, with pain, tingling, swelling, bruising, cyanosis, tenderness, and joint stiffness; the symptoms did not completely resolve for about thirty days. Envenomation by the small Regal Black-striped Snake (30–45 cm as adults) is poorly known, but bites result in persistent bleeding at the bite site, suggesting anticoagulant properties. There may be immediate pain, redness, itching, numbness, and swelling that may persist for several days. In one of the two case reports, the pain was "sharp and immediate," and swelling extended to the elbow. The bite site (finger) exuded fluid and the hand became red and swollen. The pain dissipated in about a day, but the swelling and numbness persisted, and the hand was sensitive to the touch. I could find little information on bites by *D. punctatus*, but one correspondent was bitten on the arm by a 30 cm specimen he picked up in the wild. It hung on briefly, but he said pain was "immediate and just like a wasp." There were no other effects and the pain dissipated after a while.

Most herpetologists do not hesitate to handle any of these species because they are not considered seriously venomous to humans and the handlers do not usually allow the snake to hang on for a prolonged period. For example, I experienced a quick bite-and-release by a very large Sonoran Lyresnake (*T. lambda*) without any ill effects (although the bite site had a strangely attractive aroma), and a quick bite on my finger by a Brown Vinesnake caused some redness and persistent itching that was gone by the next day. In some instances, teeth can break off in the skin and cause itching. I have also been bitten by gartersnakes and Gophersnakes, and they were always uneventful.

Lyresnakes are also rear-fanged colubrids. This is the Sonoran Lyresnake. *L.L.C. Jones*

In another Envenomation Story in this volume, a herpetologist reported his most surprising venomous reaction coming from a Gophersnake, a species that is probably responsible for thousands of bites per year, with very few people exhibiting ill effects.

The take-home message is that if one is bitten by any snake in the United States thought to be harmless, significant medical consequences are not expected, unless the animal is allowed to hang on for a prolonged period. To my knowledge, none of our species has caused severe systemic reactions, and no fatalities have been reported. I am aware of no confirmed cases of allergies to venom, but that is always possible with repeated exposure.

References and Resources: Refer also to the introduction for terrestrial vertebrates. Cox et al. (2018), Fritts et al. (1990, 1994), Fry and Wüster (2004), Gutiérrez and Sasa (2002), Hayes and Hayes (1985), Hill and Mackessy (2000), Kardong (2002), Mackessy (2002), Mackessy et al. (2006), McKinstry (1978), Minton (1990), Morris (1985), Pyron et al. (2011, 2013), Saviola et al. (2014), Vest (1981), Weinstein (2013), Weinstein and Kardong (1994), Weinstein and Keyler (2009), Weinstein et al. (2011, 2013).

American Short-tailed Shrews (*Genus Blarina*)

[Regarding the "shrewmouse"] " . . . In Latine it is called *Mus araneus*, because it containeth in it poison or venime, like a spider, and if at any time it bite either man or beast the truth of this will be apparent. . . ." (John Swan, Speculum Mundi, 1671)

Northern Short-tailed Shrew, showing the namesake tail. *Gilles Gonthier/CCA/Flickr*

Classification: Phylum Chordata (chordates); Class Mammalia (mammals); Order Soricimorpha (shrews and moles; previously Insectivora, the insectivores); Family Soricidae (shrews); Genus *Blarina* (American short-tailed shrews)

Taxonomy of shrews has always been problematic, so there are various iterations of species and subspecies assignments and distribution in the literature over time. Until relatively recently, there was a single species recognized, the American Short-tailed Shrew (*B. brevicauda*), but that species was later divided into several species: Northern Short-tailed Shrew (*B. brevicauda*), Southern Short-tailed Shrew (*B. carolinensis*), Elliot's Short-tailed Shrew (*B. hylophaga*), and Everglades Short-tailed Shrew (*B. peninsulae*). The latter is usually considered a subspecies of *B. carolinensis*. It has been suggested that *B. shermani* is a distinct species, rather than a subspecies of *B. carolinensis*. Other shrews may be venomous.

Identification: Shrews are the most diminutive species of mammals in the United States. Besides their small size, they can be identified by their oval body covered in dense fur, small legs, tapered snout, inconspicuous ears, reddish pigmented teeth, and tiny eyes without functional eyelids. American short-tailed shrews are differentiated from all other shrews by their blunt snouts, skull characters, and short tail. Among the *Blarina*, differences among species are slight. Skull characters, morphometrics, karyotype, range, and genome are most often used to separate taxa.

Distribution: The genus *Blarina* is endemic to North America. North of Mexico it is only found in the eastern United States. The Southern Short-tailed Shrew occurs in the southeastern part of the country, south of the range of the Northern Short-tailed Shrew. It has been recorded from the eastern third of Texas to Florida, and north to southern Virginia and extreme southern Illinois. Elliot's Short-tailed Shrew is primarily a Midwest species. It occurs in extreme northeast Texas, most of Oklahoma, east to the Mississippi River, and north to southern Nebraska and southern Illinois, and west to extreme northwestern Colorado. Two isolated populations of the endemic Texas subspecies, *B. h. plumbea*, have only been recorded from Aransas and Bastrop counties, Texas. *Blarina shermani* is limited to Lee County, Florida, in the vicinity of Ft. Meyers. *Blarina peninsulae* is also found in peninsular Florida, although it is more widespread.

Natural History: *Blarina hylophaga* is primarily associated with the Great Plains, where it occupies mesic grassland habitats, usually in pine/oak savannas and in association with floodplain forests. It may occupy brushy, sandy habitats, rocky grasslands, and coastal vegetation. It is generally associated with leaf litter and downed wood or rock, but is sometimes in relatively open areas. *Blarina carolinensis* is often found in pine forests, ranging from early to late successional stages, and is often more abundant in the denser young stands. In addition to pine forests, it has been found in brushy bottomlands and hardwood forests, cane breaks, and even swampy areas, but is probably uncommon in saturated soils. It also occupies disturbed sites, including reclaimed strip mines, old fields, and roadsides. Where the two species co-occur, Elliot's Short-tailed Shrew tends to select more xeric sites

Southern Short-tailed Shrew. *Greg Schechter/CCA/Flickr*

than the Southern Short-tailed Shrew. *Blarina peninsulae* is poorly known, but occurs in the swamplands and hammocks of Florida.

American short-tailed shrews are most abundant in the spring and fall, which coincides with their reproductive cycle. Their populations fluctuate greatly over the course of a year and between years. They nest in areas of dense vegetation, under logs, or in debris piles. They are short-lived animals, with averages ranging from about eight months to two years. In Texas, *B. hylophaga* average 4–7 young per litter and have up to three litters per year. They are born hairless and are weaned at one month. *Blarina carolinensis* is similar in reproductive traits.

Determining population densities of shrews has always been a challenge for mammalogists due to their cryptic nature, high metabolism, population cycles, mass, and general avoidance of live traps. Nevertheless, when shrews are targeted and successfully sampled, they often turn out to be very common. In some studies, *B. carolinensis* and *B. hygrophagus* were considered the most abundant small mammal. In other population studies, other species of shrews were often more abundant. Population densities of Southern Short-tailed Shrews ranged from 2.2 to 13.2 shrews per hectare in the few studies reported. This translates to a lot of short-tailed shrews in good habitat, yet these cryptic animals are rarely seen by humans if they are not in a trap. When I sampled mammals and amphibians in pitfall traps in northern California (outside the range of *Blarina*), thousands were captured; they were the most commonly captured mammals. With their need to feed to sustain their metabolism, they often ate all other small animals in the traps. *Blarina shermani* is an exception to the high-density generalization. It has been suggested they may be extinct in their tiny range, but I can find little recent information.

Like all shrews, these animals have a very high metabolism. They eat about one-half of their body weight per day. They feed on a wide variety of animals, including insects (beetles and their larvae often figure highly into prey percentage and volume), spiders, millipedes, snails, slugs, and earthworms. American short-tailed shrews are also known to eat vertebrates including small mammals (e.g., deer mice), salamanders, young birds, and eggs; it has been suggested that their venom may have evolved (in part) to include feeding on mammals, but most studies suggest invertebrates are primary prey. Some plant material is consumed, including fungi. *Blarina* spp. are known to horde food. They use their venom to paralyze small prey, which can then be kept alive and immobile in their larders for later consumption.

Despite possessing a musky smell that deters some animals, predators of *Blarina* spp. are many, including owls, hawks, copperheads, cottonmouths, ratsnakes, foxes, coyotes, bobcats, skunks, raccoons, opossums, and, rarely, fishes.

Encounters: Although shrews may be among the most common mammals in preferred habitat, they are very small and scurry about under the leaf litter and grasses, so easily avoid detection. Most people rarely if ever see them.

There are a number of bites reported in the literature, dating back to 1889, and some online. Most reports come from the occupational hazard of mammalogists studying the animals in the field or in captivity. *Blarina* are completely nonaggressive unless cornered and threatened, and an accidental bite is highly improbable.

Venom Apparatus, Venom, and Symptoms: The venom is produced in the submaxillary gland at the base of the lower incisors. Other salivary glands do not produce venom. When the animal bites, venom presumably flows along a groove between the teeth and enters the victim at the bite site. *Blarina brevicauda* has a reported LD_{50} of 3.4 mg/kg.

Short-tailed Shrews, and probably other species of shrews, have a venomous bite. *USGS, BIM/Public Domain*

Most of what we know about *Blarina* venom comes from the relatively few studies on *B. brevicauda*, the Northern Short-tailed Shrew, or from *Blarina* spp. during the time when the genus was considered monotypic or when the geographic range of the study animals was not reported. Thus, there are no studies comparing the different species, although it is often stated that all *Blarina* are venomous. The venom is reported to be both neurotoxic and tissue-destructive. The venom not only paralyzes invertebrates, it can be fatal to small mammals.

Despite its potency, the venom is not well researched. However, three proteins have been isolated: blarina toxin, blarinasin-1, and blarinasin-2. These proteins have a kallikrein-like proteolytic activity that is very similar to Mexican Beaded Lizard venom and seems to have originated through convergent evolution. A peptide that causes paralysis, soricidin, has also been isolated. In laboratory mice the venom causes irregular respiration, convulsions, paralysis, and death.

In humans, some bites fail to produce symptoms. These are assumed to be from instances when the skin was not broken during the bite, or at least toxic saliva was not introduced. Some researchers have suggested that differences in reactions to the bite may be due to individual sensitivity, but this seems

Although not a shrew, it is a venomous mammal. Common Vampire Bats *(Desmodus rotundus)* have recently been moving north and now barely make it into South Texas. Of course, they are actually parasites, so not given a taxon account in this book.
A. Catenazzi/CCA-SA/Wikimedia Commons

unlikely for asymptomatic bites, since it involves virulent venom that affects mammals. There are sometimes marked effects in humans from short-tailed shrew envenomation. The venom is algogenic, as some victims have complained of extreme pain. The pain can radiate from the bite site and last for several days. A bite is also accompanied by pronounced swelling and redness. The affected limbs may not be completely functional for several days.

Remarks: The venom of *Blarina* shows promise in medical use for the treatment of cancer, wrinkles, and neuromuscular therapy in humans and as an insecticide. Shrews themselves are great insecticides. *Blarina* (at least *brevicauda*) can be a host for ticks and the bacterium that causes Lyme disease. The Masked Shrew (*Sorex cinereus*) is possibly venomous, but if so is not believed to have particularly potent venom.

References and Resources: Aminetzach et al. (2009), Baumgardner et al. (1992), Benedict et al. (2006), Carson and Rose (1992), Ellis and Kraver, (1955), Genoways and Choate (1998), George et al. (1981, 1986), Halsey (2005), Kita et al. (2004, 2005), Martin (1981), Maynard (1889), McCay (2001), Pearson (1942, 1950), Reilly et al. (2009), Robinson and Brodie (1982), Stewart (2013), Stewart et al. (2011, 2014), Thompson et al. (2011), Tomasi (1978).

9 Aquatic Invertebrates

The aquatic invertebrates are a diverse array of organisms, most of which are found in the marine environment, but there are also some freshwater species. The accounts in this section include those on seven phyla: Annelida, Arthropoda, Cnidaria, Echinodermata, Mollusca, Nemertea, and Porifera.

By far the most important aquatic invertebrates, in terms of diversity of venomous species and number of people stung in North America, are the

Purple-striped Sea Nettle, showing the long, trailing tentacles of many jellyfishes. The tentacles harbor the venom and may break up in the surf, causing envenomation even when detached. *Oliver Dodd/CCA/Flickr.*

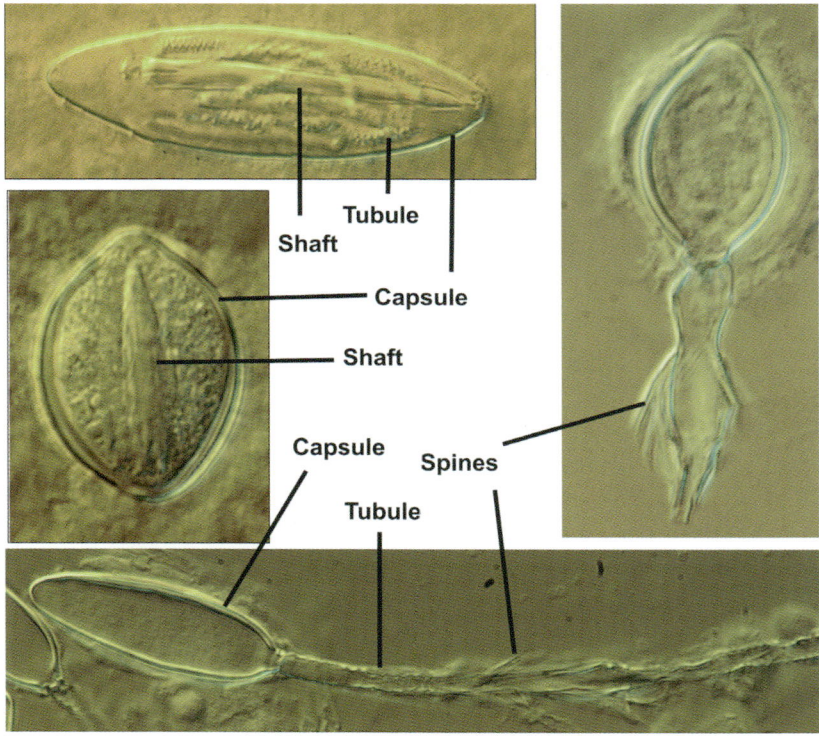

Stinging organelles of cnidarians. The left and upper images from a light microscope show undischarged nematocysts, while the right and lower images are of discharged nematocysts. Venom is introduced through the tubule, which is held in place by the spines. *Allen G. Collins*

cnidarians, the jellyfish-like sea creatures. There are several opinions on how to subdivide the phylum, but one thing they all have in common is that they have a venomous sting. The class Anthozoa includes the true corals and sea anemones. The subphylum Medusozoa includes jellyfish-like organisms, including true jellyfishes (class Scyphozoa), box jellyfishes (class Cubozoa), men-of-war (class Hydrozoa, order Siphonophorae), and Fire Corals and other hydroids (class Hydrozoa). Hydroids are a variety of animals with a polyp life stage, rather than a monophyletic taxon. In addition to polyps, many cnidarians have a different life stage known as a medusa. A polyp has a physical form like a sea anemone and a medusa has a form like a jellyfish. Cnidarians are primarily marine, although there are some freshwater taxa. They occur in all oceans and latitudes. Cnidarians are primitive, as shown by fossils of hard-bodied animals, going back to the Cambrian or Precambrian

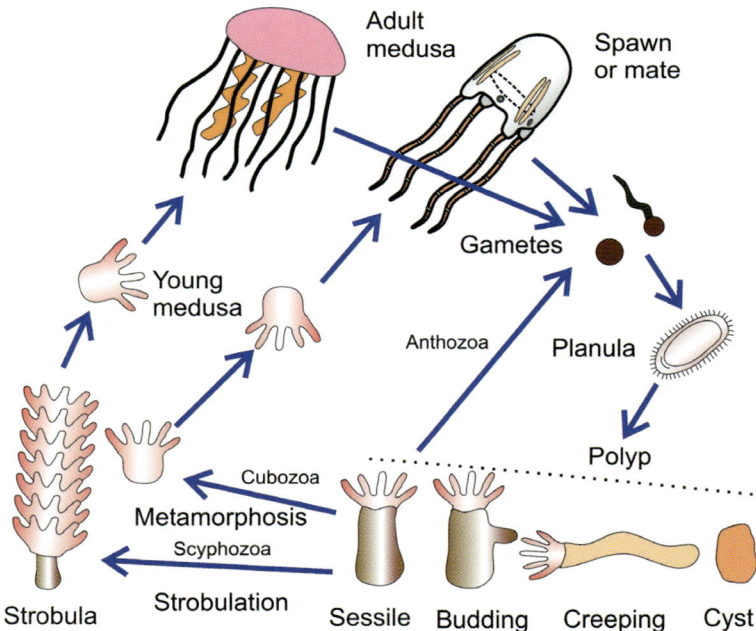

Cnidarians have a complex and interesting life cycle. There are variations on this theme for various taxa. *L.L.C. Jones*

Eras. Currently there are at least 10,000 described species, but these strange, soft-bodied animals are difficult to classify. For the most part, they are radially symmetrical and have tentacles that house the stinging organelles. There are many species of cnidarians that can cause significant envenomation of humans, and potentially death, but the most dangerously venomous species are in the western Pacific.

An even more diverse group is the mollusks, which account for nearly one-quarter of all species of marine animals. They have also exploited freshwater and terrestrial habitats, but the marine forms are the ones known for being venomous. Among the 85,000 or so described species (WoRMS (World Register of Marine Species) has accepted 46,500 marine species), only a relative handful are venomous to people. Mollusks are soft-bodied animals that sometimes have internal or external shells. Those that are venomous belong to the class Cephalopoda, which includes octopuses, squids, and cuttlefishes, and class Gastropoda, which includes snails and nudibranchs. Octopuses are globose with a mantle (the head and body) and eight arms,

Cephalopods envenomate by biting with a horny beak, like this Humboldt Squid, which could break a wrist. *Michelle Christman*

and squids are more streamlined, with eight arms and two tentacles (a term used for a variety of unrelated structures among divergent biota). There are only about 800 species of cephalopods worldwide, including about 300 species of octopuses. Octopuses are usually considered the most intelligent invertebrates. Gastropods that have shells are known as snails or clams, while those that do not are known as nudibranchs or sea slugs, a term also used for other slug-like animals in the sea. Gastropods are the most diverse class of mollusks, having some 65,000 described species but estimates suggest there may be 80,000–150,000 species worldwide. A few species of cone snails are the only gastropods that are dangerously venomous to humans.

None of the members of the other phyla are considered dangerous, but they can be an unpleasant nuisance and occasionally cause worrisome symptoms. The most primitive are the sponges (phylum Porifera). WoRMS recognizes over 8,500 species worldwide, almost all of which are marine. The venomous ones are called Fire Sponges or red sponges, and there are probably some others that may cause dermatitis. Annelids, the segmented worms, are diverse in both terrestrial and aquatic systems, but only a few are venomous to humans. WoRMS recognizes over 22,000 marine species, among them

Ribbon worms are venomous, but there is little information on effects to humans.
Ron Wolf

the polychaete worms (class Polychaeta). Echinoderms include starfishes, sea urchins, and sand dollars. The only starfish that is venomous to humans is the Crown-of-Thorns Starfish, infamous for its coral reef polyp-eating habits. Some sea urchins are venomous, but they are mostly a problem due to the physical damage caused by their spines.

Nemertea are the ribbon worms. While they are possibly venomous to humans, there is little information on their bites, so are not covered in detail, although they occur in marine waters off our coastlines. This is one of those cases where it would be beneficial to hear from someone who has actually been bitten, even if the envenomation was only asymptomatic or mild. The other invertebrates covered in this book are freshwater aquatic insects. These include some beetles and true bugs. Insects may rule the land, but they are almost entirely absent from the marine environment, a realm ruled by other arthropods, most notably the crustaceans. Only one type of crustacean is known to be venomous, but it is not found north of Mexico.

Octopuses and Squids (Class Cephalopoda)

"A Washington state woman landed in the emergency room after she posed with a venomous octopus on her face to win a photo competition. . . . during the photo shoot, she was bitten twice. . . . 'It was a really intense pain . . . and it just bled, dripping blood for a long time.' For two days, she tried to push through the pain and swelling, but ultimately went to the hospital." (*New York Post*, 7 August 2019, Jackie Salo reporting)

Although generally docile, Giant Pacific Octopuses are a potential hazard for divers and aquarium workers. *Michael Bentley/CCA/Flickr*

Classification: Phylum Mollusca (mollusks); Class Cephalopoda (cephalopods); Orders Octopoda (octopuses) and Teuthoidea (squids).

Cephalopods include octopuses, squids, and cuttlefishes (order Sepiida), in addition to several other more obscure forms. Cuttlefishes do not occur in North American waters, although some squids are misidentified by the casual observer as cuttlefishes. Cuttlefishes do occur in central west-Pacific American territories, so refer to that chapter. Although commonly used, "octopi" is an improper plural of octopus; it is octopuses or octopods.

From the giant to the small—size does not necessarily matter, as some of the more venomous octopuses are very small, although I can find no information about envenomation from the California Lilliput Octopus (pictured). *Douglas Klug/CCA/Flickr*

Octopuses in our area include Giant Pacific Octopus (*Enteroctopus dofleini*), two-spotted octopuses (*O. bimaculatus* and *O. bimaculoides*), Day Octopus (*O. cyanea*), Crescent Octopus (*O. hawaiiensis*), Atlantic Pygmy Octopus (*O. joubini*), Night Octopus (*O. ornatus*), Eastern Pacific Red Octopus (*O. rubescens*), and Common Octopus (*O. vulgaris*), among others. Fitch's or Lilliput Octopus (*O. fitchi*) is a peripheral species, but has been recorded in our area. Less common species include Brown-striped Octopus (*Amphioctopus burryi*) and Atlantic White-spotted Octopus (*Callistoctopus macropus*). The blanket octopuses (*Tremoctopus violaceus, T. gracilis,* and *T. gelatus*) are rarely encountered inshore.

The blue-ringed octopuses (*Hapalochlaena* spp.) are worthy of discussion because they are among the deadliest invertebrates on Earth. They are non-native, but occasionally show up in pet shops or online for sale to marine aquarium hobbyists.

Most nearshore squids were primarily in the genus *Loligo*, before being split into different genera, although there is still reference to the common species being within *Loligo*. Species of *Doryteuthis* are commonly known as "inshore squids." In our area, squids include Opalescent Inshore Squid (*D. opalescens*), Longfin Inshore Squid (*D. pealeii*), Arrow or Slender Inshore

Squid (*D. pleii*). Roper Inshore Squid (*D. roperi*), Humboldt Squid (*Dosidicus gigas*), Hawaiian Bobtail Squid (*Euprymna scolopes)*, and Atlantic Brief Squid (*Lolliguncula brevis*). There are numerous deepwater and pelagic species, especially in the tropical waters of the central Pacific and Gulf of Mexico, including Flying Squid (*Sthenoteuthis oualaniensis*), Neon Flying Squid (*Ommastrephes bartramii*), and Hawaiian Flying Squid (*Nototodarus hawaiiensis*), plus giant squids (*Architeuthis* spp. and *Mesonychoteuthis hamiltoni*).

Identification: As a group, mollusks are diverse in body form, but all have soft bodies, with a characteristic body wall known as the mantle. Inside the mantle is usually a mantle cavity. In some species there is also a protective external shell, and in others there may be an internal shell for support. The cephalopods (meaning "head foot") are camouflage experts and have a remarkable capacity for color and pattern change. Some cephalopods can even change their texture. Octopuses are familiar animals to most people. They have eight arms, no tentacles, and a globular, generally spherical body, with a vestigial shell. Like squids and cuttlefishes, the arms have suckers. They move through the water via jet propulsion using their siphon, but are highly agile and adept at moving across the ocean floor and in tight spaces. Cephalopods often discharge ink as a protective device. Octopuses are the most intelligent invertebrates known. The most impressively proportioned octopus is the Giant Pacific Octopus. The maximum size is difficult to pinpoint because of discrepancies in various reports, and octopods are stretchy. They have been reported to average about 15 kg, with an arm span of about 4 m. Larger individuals have been recorded at nearly 50 kg and have a radial span of about 6 m. In the center of the pack are moderately sized octopuses such as *O. bimaculatus, O. bimaculoides, O. cyanea, O. ornatus, O. rufescens,* and *O. vulgaris*. On the other end of the spectrum are several small species of octopus, having a mantle the size of a ping-pong ball, including *O. joubini* and *O. fitchi*; despite its small size, the latter is possibly a dangerously venomous species. The blanket octopuses (*Tremoctopus* spp.*)* are strange pelagic species of tropical and subtropical waters. They have a huge membrane between four arms, resembling a blanket or cape. These animals are said to have the greatest degree of sexual dimorphism of any large animal, as adult females may be 2+ m in length, and males are 1–2 cm. The female weighs about 10,000 times as much as the tiny, rarely observed male!

Squids are elongate animals with eight arms and two tentacles. The anterior part of the body has the arms and tentacles, followed by the large eyes and the posterior part (generally, the mantle), with the triangle-shaped fins. The tentacles are used for catching prey, while the arms handle and subdue prey. The arms and tentacles surround the mouth. Internally, they have a flattened pen (internal shell) for support. They generally remain in the water column and often travel in large schools. Most are of moderate size, usually less than 0.3 m in length, not including tentacles. However, just as there is a large octopus in the eastern Pacific, so too is there a large squid, the Humboldt Squid. These animals reach nearly 2 m in mantle length, and may weigh up to 45 kg. There are some very deep ocean giant squids that dwarf *D. gigas*, but these are so rarely encountered and almost never found alive; they are a mystery. Their maximum size is estimated at about 14 m with tentacles and nearly 5 m in mantle length.

Distribution: The Giant Pacific Octopus is distributed from northern California northward. Recent studies indicate a second cryptic species, known as the Frilled Giant Pacific Octopus, occurs in Prince William Sound. It is anatomically and genetically distinct. The California Two-spotted Octopus is a familiar species of southern California, where it is sometimes found in tide pools or shallow waters. The Eastern Pacific Red Octopus is another common species of shallow waters, found from Alaska to Baja California. The tiny species, *O. fitchi*, is known from the northern Gulf of California, along with a few other species, although there is a single record from the Salton Sea of California. There have been no subsequent records of this rather venomous species from the Salton Sea, but after speaking to an expert, the original record appears legitimate. In the central Pacific there are three nearshore octopuses: the Day Octopus, the Night Octopus, and the Crescent Octopus, a recently described Hawaiian endemic.

In the Atlantic and Gulf of Mexico, there are also several species of octopuses, including the widely distributed Common Octopus complex and the Atlantic Pygmy Octopus. The Brown-striped Octopus is a Caribbean species that was described from the Florida Keys. The White-spotted Octopus is another Caribbean species. The strange blanket octopuses are sometimes seen offshore in the Gulf of Mexico, the Caribbean, and in Hawaiian waters.

Two of the better-known squids from the eastern Pacific include the Opalescent Inshore Squid and the Humboldt Squid. Longfin Inshore Squid and Slender Inshore Squid are common species in the Atlantic, including the Gulf of Mexico. Atlantic Brief and Roper Inshore Squids are also species of the Atlantic and Gulf of Mexico. Some species of squid are economically important, being harvested for calamari. Reports of cuttlefishes in the Florida Keys are probably Caribbean Reef Squid. Similarly, the curious Hawaiian Bobtail Squid is erroneously believed to be a cuttlefish by some laypersons. Many of the pelagic species are known as "flying squids," a moniker sometimes used for the Humboldt Squid. Flying squids are usually offshore in pelagic waters, or in the depths of the open ocean.

Natural History: All cephalopods inhabit the marine environment. While the various species can range from the intertidal zone to the deep sea and open ocean, most discussed here can occur in shallow coastal waters. They can be found on or over a variety of substrates. Octopuses are usually found in rocky or coral reefs, often in crevices by day, or amid shells and coral. Octopuses do not like being out in the open and over sediment bottoms, where they are easy prey for a number of animals. Octopuses are color- and texture-changing experts. Within a second, they can match the color and shape of the background. Their body is so flexible, they can squeeze into tiny nooks and crannies to avoid detection and predation. Squids may be near the bottom, between the bottom and the surface, or near the surface. Many species have a diel pattern of migration, generally being deeper during the day, and ascending toward the surface at night. Even when near the bottom, squids are not benthic like many of the octopods. The blanket octopuses and flying squids, and many other species of cephalopods, are pelagic. Some species are deepwater denizens.

Cephalopods are bisexual and often sexually dimorphic. They have a courtship ritual that culminates in direct sperm transfer, which may be external (in the mantle cavity) or internal. The male has a specialized arm known as the hectocotylus, which transfers a spermatophore into the female's mantle cavity or genital opening. Sperm may be used immediately or stored. The female may lay thousands of fertilized eggs. Octopus eggs are usually in grape-like clusters, and the female will guard and ventilate them. Squids lay eggs in strings on the bottom, although some species have pelagic eggs. There is no larval stage, but young of even benthic species may be planktonic until they attain some size.

Octopuses are well known as crustacean eaters, but will also eat other animals including gastropods, marine worms, fishes, and other cephalopods. Pelagic species often eat shrimps, fishes, and other cephalopods. Whereas octopuses are usually benthic, so feed on benthic prey, squids are suspended in the water column, or on the surface, so feed on animals they can overpower in those areas, such as fishes and shrimps.

In turn, cephalopods are often fed upon by a variety of animals. The archenemies of octopuses are moray eels, but they are also fed upon by other animals, including other types of fishes, marine mammals, and fellow cephalopods. Many people are aware of the epic battle between oceanic behemoths—giant and colossal squids and Sperm Whales. Sperm Whales eat giant squids, and often have the battle scars to prove it. Some of the scars may be dinner plate-sized from the suckers or slash marks from the hooks on the tentacles.

Encounters: Octopuses are mostly encountered by divers and people exploring tide pools, but unless they are being collected for food are not likely to be handled, although some people delight in handling octopuses when they can. A SCUBA diver can hardly pass up an octopus without wanting to play with it, although the Giant Pacific Octopus is more than a handful. It is a very strong animal with powerful arms and suction cups, and with its large horny beak, is capable of inflicting harm. However, so-called video-documented "attacks" by these animals seem to be sensationalized accounts by the media where the animals were not aggressive and the divers were in no real danger. Nevertheless, these are large, strong animals, and they do have a venomous bite, so they are best enjoyed through a mask or aquarium.

Most of our native octopuses tend to be docile, but there are numerous records of people handling them who were bitten, in nature and in aquaria. The opening quote was from someone who did not know that octopuses are venomous, and just thought it would be a memorable experience to hold one. Fishermen may also hook or net octopuses to eat, or use as fishing bait. If caught on a fishing line, anglers should just cut the line, rather than trying to remove the hook from the vicinity of the horny beak. Fishermen often handle octopods more roughly than divers do, so are more likely to get bitten.

Squids are an important commercial seafood and the huge Humboldt Squid is highly sought after by fishermen. Humboldt Squid are normally found in deeper waters south of California, but have been showing up periodically in temperate waters as far north as Sitka, Alaska, and sometimes in relatively shallow waters. These animals have a reputation for being aggressive, strong, and dangerous. They have been known to attack divers as far north as southern California, and are known as *diablos rojos* (red devils) in Mexico, where they are commercially fished. I do not know of stories of envenomation by this species, but they can certainly cause physical injury and potentially death from their strong beaks, toothed suckers, and propensity for dragging prey into the depths. It has been suggested the aggressiveness is overrated, but they have been confirmed to attack humans. In one well-publicized (and videotaped) account, a professional diver had his shoulder dislocated, wrist broken in several spots, and his eardrum ruptured as he was pulled down into deeper water. Another professional biologist had his head munched by one at night. Both divers survived. Humboldt Squid seem aggressive toward humans, and are cannibalistic, when in a feeding frenzy, and a school can reputedly (unconfirmed by first-hand published account) be a thousand strong. For divers, it is best to steer clear of Humboldt Squid in their domain. From a boat, however, they can make for an exciting catch and yield a lot of calamari. Just don't fall overboard or get near the beak.

Another way to encounter cephalopods is via marine aquaria. The Giant Pacific Octopus is commonly kept on display in large public aquaria, and several envenomations have occurred to staff handling the animals or cleaning the aquarium. Other species of octopus often kept in public aquaria include *O. bimaculoides*, *O. vulgaris*, *O. cyanea*, and some exotic species. Of the exotic species, blue-ringed octopuses (*Hapalochlaena* spp.) are by far the most dangerous. They have claimed numerous lives. They do not occur in North America, being from the Indo-West Pacific, but sometimes show up for sale at aquarium stores or online. For example, when I was working my way through college at a retail marine aquarium outlet, one of the wholesalers had a large holding tank housing dozens of blue-ringed octopuses. When I pointed out how seriously venomous these animals were, the owner replied, "only the Australian forms are venomous, and these are from the Philippines." That was not true, of course. He "proved" his claim to me by handling one. He wasn't bitten, so he was not envenomated, but he was satisfied. During my next visit a couple weeks later, there was a sign on the aquarium flatly stating, "Philippine Blue-ringed Octopus: Nonvenomous." Nowadays, online forum sites suggest these animals are difficult (but not impossible) to acquire these days, and nearly all aquarists recognize their extremely venomous nature. Bloggers vehemently try to dissuade others from keeping them as pets. Besides being potentially deadly, octopuses are pre-eminent escape artists. Also, they could potentially invade some U.S. waters if released intentionally or not. Although they are beautiful and interesting animals, they make lousy

Humboldt Squids are large and dangerous, but little is known about their venom. *Michelle Christman*

aquarium pets anyway, generally lasting only a few weeks. Similarly, cone snails, cuttlefishes, and nudibranchs are sometimes kept in aquaria, but are also generally destined to expire in short order because of their specialized feeding habits.

Venom Apparatus, Venom, and Symptoms: Venom is produced in the posterior salivary glands, which are located in the abdomen. Venom is secreted through a tubule into the mouth area. It is introduced into prey when the horny, chitinous beak bites into the prey or predator. The beak is similar to a parrot's beak, and is located between the arms. In the Humboldt Squid, the beak is an impressive 8 cm in length, about the width of a human wrist, which it can snap. The radula of a cephalopod is essentially a rasping tongue. The blanket octopus is a tool-using mollusk: it will bite off pieces of *Physalia* tentacles and use them for defense.

At the top of the list for cephalopod toxicity would be the blue-ringed octopuses, which would only be encountered in the exotic pet trade. They have highly potent neurotoxic venom originally dubbed as maculotoxin, which turned out to be identical (or nearly so) to tetrodotoxin, an extremely virulent poison known from pufferfishes, newts, and poison arrow frogs. Tetrodotoxin is produced by *Vibrio* bacteria in the octopus's salivary (venom) glands, but is also found throughout the body as a somatic poison. Other venom components include histamine, tryptamine, octopamine, dopamine, and acetylcholine. The venom causes numbness of mouth and tongue, blurring of vision or blindness, nausea, paralysis, and possibly death through respiratory or cardiac failure. Death can occur within minutes. It has an LD_{50} of 0.36 mg/kg.

The venom of native species is less well known, and reported bites are rare. None are believed to be deadly. Some venom components among an array of cephalopods include cephalotoxin, hyaluronidase, tachykinins, metalloproteinases, chitinase, and phospholipase A_2. The venom mostly targets crustaceans, mollusks, and other invertebrates, more so than vertebrates. Reported bites by *O. bimaculoides, O. cyanea, O. ornatus, O. rubescens, O. vulgaris*, and *E. dofleini* cause a range of symptoms. In mild cases, there may just be a pinch that bleeds. Other cases are likened to a bee sting. In more severe cases, the pain can be intense. Usual symptoms of moderate envenomation include immediate pain, bleeding, burning, swelling, tingling, and possibly nausea. In two cases of envenomation by *E. dofleini*, there was tingling, itching, swelling, redness, and burning, subsiding over a period of ten days in one case and four weeks in the other. In some cases, swelling has been extensive, involving the entire limb. The swelling may take weeks to completely subside. There have also been reports of necrosis and/or cases of secondary bacterial infection. The Eastern Pacific Red Octopus has been reported to cause necrosis and take a long time to heal. The most serious bites

The Eastern Pacific Red Octopus can give a painful bite. *Ron Wolf*

likely come from prolonged exposure as when a captive animal is feeding and mistakes a hand for food—and will not let go. They are strong animals that may hang on with tenacity, so may be difficult to remove.

In the case of the Fitch's Octopus bite mentioned in the Envenomation Story and opening quote, the victim complained of extreme pain and short-term blindness. Bites by this species are poorly known, but it is said to be "dangerously venomous" in the literature. The other Envenomation Story shows that *O. digueti* is capable of producing a painful or possibly dangerous envenomation. One colleague received a superficial bite by *O. bimaculatus* but reported no obvious signs of envenomation, except perhaps extensive bleeding. Case reports of envenomation by octopuses are rare.

While squids are known to be venomous, there is little information on their bite. To my knowledge, none of our native squids have been implicated in significant human envenomation. They seem generally disinclined to bite, but it does happen, especially to fishermen catching them or using them as bait. One blog for squid fishermen asked if anyone had been bitten and 14/33 respondents said they had. Of these, those who had been nipped were asked to rate the pain on a scale of 1–10, with these results: 1–3 = 14%; 4–7 = 57%, 8–10 = 27%. Some of the comments were "it felt like the jaws

I am unaware of any serious envenomations from either of the two-spotted octopuses from southern California, although one woman I spoke to developed an MRSA infection after being bitten. *Douglas King*

of life" and "I'm never doing that again," suggesting the bites can be quite painful. Unfortunately, we do not know which species were involved (although one respondent said their species was only 5 cm long), where the bites occurred, and how much of the pain was likely from venom, rather than the beak. Another blog centered on the northwestern Atlantic posed the same question. There were several responses in the affirmative, and the take-home message was that it "hurt like Hell," but it sounded more like physical trauma of the beak rather than venom. Even with the lack of more precise information, it suggests one does not want to be bitten by a squid of any size.

References and Resources: Boletzky (2003), Halstead (1978), Hillyard et al. (1989), Holford et al. (2009), Hwang et al. (1989), Jones (1963), Keen (1971), Lane (1962), Morris et al. (1980), Norman (2000), Roper et al. (1984), Ruder et al. (2013), Ruppert et al. (2004), Schwartz and Meinking (1997), Simidu et al. (1987), Suzuki et al. (1986), Thomas and Scott (1997), Voss (1956).

Cone Snails (Family Conidae)

"But when it comes to lethality rates, it's the deadliest cone snail in the world, *Conus geographus*, which takes the prize for invertebrates, with a case-fatality rate of 70 percent. The incredibly high rate reflects the speed with which it kills—those who die succumb in a matter of minutes from sweeping paralysis." (Wilcox 2016) [NOTE: this species does not occur in North America, but is discussed in the chapter on American Territories.]

Living Textile Cone from Hawaii. No humans have died from cone snail envenomation in Hawaii, but several cone snails there have potent toxins. *Harry Rose/CCA/Flickr*

Classification: Phylum Mollusca (mollusks); Class Gastropoda (gastropods); Order Neogastropoda; Family Conidae (cone snails).

Venomous species of marine snails are in the suborder Toxoglossa (meaning "toxic tongue or radula") and superfamily Conoidea, which include cone snails, turrid snails (family Turridae), and auger snails (family Terebridae).

All taxonomic levels are a topic of discussion and interpretation. Cone snails are sometimes considered monotypic (genus *Conus*), but some researchers have proposed dozens of genera. A recent classification found four logical generic splits: *Californiconus*, *Conasprella*, *Conus*, and *Profundiconus*, but this is not universally accepted. WoRMS and MolluscaBase recognize eight extant genera (accessed January 2019): those four above, plus *Kenyonia*, *Lilliconus*, *Malagasyconus*, and *Pygmaeconus*. To keep things simple, I will consider *Californiconus californicus* as distinct, but all others as being in *Conus*. Interested readers should dig deeper into the literature and decide taxonomy for themselves. There are about 800 species generally recognized worldwide. The turrids are known to be polyphyletic and now some thirteen distinct families are often recognized. The estimated number of species of Conoidea ranges from 3,000 to over 11,000 worldwide. Although other members of Conoidea are venomous, emphasis is placed on *Conus*, as some of the species are known to be dangerously venomous to humans. This includes the dangerously venomous species *C. textile* (Textile Cone), *C. marmoreus* (Marbled Cone), *C. bandanus* (Banded Cone), and *C. striatus* (Striated Cone). *Conus ermineus* (Turtle Cone) has been implicated in delivering painful stings to humans. There are several other species that have been implicated in stings in Hawaii, including *C. obscurus* (Obscure Cone), *C. pennaceus* (Feathered Cone), and *C. nanus* (Dwarf Cone), at least.

Identification: Neogastropods includes the familiar marine snails, but there are also some freshwater taxa. The venomous marine cone snails have external shells, usually in the namesake shape, with the narrower end, known as the base, being anterior. Some species are more elongate and less tapered to the point of being an elongated oval, while others have a sharp taper posteriorly to form a spire. Although narrow, the base has the largest opening of the whorl, known as the aperture, to accommodate the soft body parts. As the snail grows, whorls are added and the shell is enlarged. When the snails are actively foraging, the large, fleshy foot is extended. Anteriorly, the head, antennae, and siphon may be visible, although the mouthparts, which contain the venom apparatus, are usually hidden from view. The siphon is sometimes mistaken as the proboscis, but it is a breathing structure and not used for envenomation. The beautiful shell typically has a disruptive pattern, rendering it cryptic on a complex and dappled background. Over the shell is a covering of tissue called the periostracum, which may be encrusted with algae, making the animal even more difficult to detect.

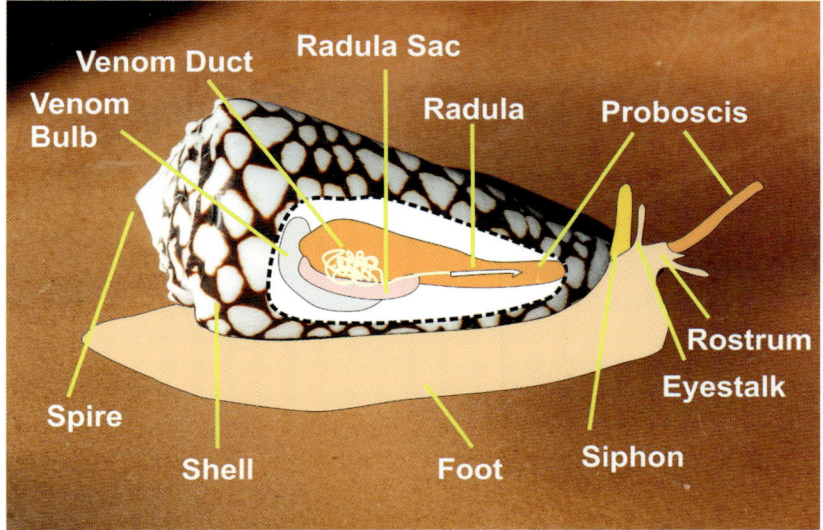

Cone snails have a very advanced venom apparatus, and a treasure trove of conopeptides. *L.L.C. Jones*

The California Cone Snail (*C. californicus*) is only a few millimeters in length, but it is not the smallest, as there are some tiny micro-cones. There are many larger species in the western, southeastern, and central Pacific, as well as the Atlantic, including the Gulf of Mexico. These can be several cm long in shell length. Hawaii's Leopard Cone (*C. leopardus*) grows to an impressive 14 cm in shell length, and 6 cm at the posterior end. Cone snail shells are often marked with exquisite designs, so are favorites among collectors, and some rare and beautiful species may fetch high prices. The potentially deadly Textile Cone (*C. textile*) is among those species sought after as a collectible. It has an intricate pattern of small triangles of white on golden brown, resembling woven fabric. It grows to about 10 and possibly 15 cm. The other highly venomous species also have a pattern of triangles or mottling. The Striated Cone (*C. striatus*) has more conspicuous dark brown banding beneath the white triangles. The Banded Marble Cone (*C. bandanus*) has a beautiful pattern of large white triangles in a dark brown reticulum. It is similar to other marble cones found elsewhere. There are field guides and online resources to help identify cones to species. The Turtle Cone (*C. ermineus*) is highly variable in appearance, ranging from light and monocolored to dark and white mottled. Common names may be more stable than scientific names, as they are popular with shell collectors.

Three cone snails that have claimed human life (left to right): *C. marmoreus*, *C. textile*, and *C. geographus*. L.L.C. Jones

The related, but presumably less-venomous turrids and terebrids are auger-shaped, and the posterior part is narrower. The shells of these snails are also popular among collectors. I know of no significant human envenomations from these families of marine snails.

Distribution: Cone snails are best represented in the tropical Indo-Pacific. There are at least 34 species in Hawaii. The Geographer Cone (*C. geographus*) of the Indo-West Pacific is often regarded as the most venomous invertebrate on Earth, and rivals the most venomous snakes for toxicity, but it does not reach Hawaii. However, it is discussed in more detail in the chapter on American Territories, as it does occur in some U.S. holdings, where fatal envenomations have occurred. There are a few notably venomous species of *Conus* in Hawaii, including *C. textile*, *C. bandanus*, and *C. striatus* (*C. s. oahuensis* is a Hawaiian endemic). There are several other species that have been implicated in human envenomation in Hawaii, including *C. obscurus* (Obscure Cone), *C. pennaceus* (Feathered Cone), and *C. nanus* (Dwarf Cone), at least. Cone snails are only represented in the temperate eastern Pacific of the U.S. by the California Cone, which occurs as far north as the Farallon Islands of San Francisco. It is the only representative of the genus, having diverged from all other groups. It is a familiar snail to southern California

Conus dalli of the Gulf of California looks very similar to *C. textile*, and is thought to be a dangerously venomous species. *"Shellnut"/CCA/Wikimedia Commons*

beachcombers, but as far as I know, is not known to envenomate humans. A large fish-eating cone of the Gulf of California is *C. purpurascens* (Purple Cone). It is said to deliver a painful bite. Another species found throughout the Gulf is *C. dalli* (Dall's Cone). It is very similar in size and appearance to *C. textile* and is also reported to deliver a very painful bite.

According to a 2014 book on cone snails of the southeastern United States and Caribbean, there were 53 valid species, although other authorities consider this conservative. A number of these species have been reported to be venomous to humans. This includes the fish-eating Turtle Cone, plus the Alphabet Cone (*C. spurius*), Jasper Cone (*jaspideus*), Royal Cone (*Conus regius*), and Sozon's Cone (*C. sozoni*). Any large cone found in the Atlantic and Gulf of Mexico should be considered unsafe to handle. None of these species is considered to be dangerously venomous to humans.

Natural History: All of the venomous gastropods inhabit the marine environment. While the various cones can range from the intertidal zone to the deep sea, most discussed here occur in shallow coastal waters, or at least within reach of SCUBA divers. As a group, cone snails occur on sand and mud flats, rocky or coral reefs, or the transition of these habitats, depending on the species. The Textile Cone is usually associated with the interface of coral reefs and sand, often buried in sand or coral rubble, where they seek refuge during the day. *Conus bandanus* and *C. striatus oahuensis* are uncommon taxa also found on hard substrates or buried in sand or rubble near rock or coral reefs. In the western Atlantic *C. ermineus* is often found on rocky rubble, while the *C. spurius* complex is usually found on sandy substrates. Cones are often buried in the substrate during the day but become surface active at night.

Cone snails have separate sexes, and fertilization is internal. After mating, the female lays a variable number of egg capsules, often several hundred. The hatchling larvae, known as veligers, are free-swimming planktonic organisms. The veliger uses ciliated flaps, known as vela, to propel itself in the water column. Some species, however, have velichoncha larvae, which are essentially baby snails. The veliger metamorphoses into a young snail, and then grows into an adult, increasing the size of the shell as it goes by adding to the whorl. Adults may live for over a decade.

All cone snails are carnivorous and have specific diets. There are three primary prey types, specific to different cone species. Some feed primarily on marine worms, others eat mollusks, and some eat fishes. It is often the fish-eating species that tend to be most venomous to humans, presumably because their prey is vertebrate. Fish eaters include *C. ermineus*, the only fish-eating species in the Atlantic. The dangerous Textile and Banded Marble Cones feed on other marine snails, while the Striated Cone is a fish eater. The Purple Cone is also a large fish-eating species. Cones can be ambush predators or they can actively hunt. They often remain partially buried in the substrate to avoid detection while hunting. Cone snails are very slow when surface active, but their bite is extremely fast. When a fish is detected within striking range, it will quickly capture the fish. There are two basic methods of hunting. Some are "hook-and-line" hunters that shoot a "harpoon" (see venom apparatus, below) into the prey, while others are "net hunters," which may catch fish in their broad mouth, then envenomate the prey in the oral "net" using their harpoon. Snail hunters, such as Textile Cones, have a bit of a different strategy. When they happen upon a snail, they harpoon its mantle, then put their head inside the aperture of the victim's own shell to consume it.

Cone snails are fed upon by a variety of animals, including other snails, cephalopods, and some fishes. Some cone snails specialize in eating other cone snails. Humans also consume cone snails. Their venom apparatus is removed during preparation.

Conus ermineus occurs in the Atlantic, but this one is in an aquarium in the Olivera Cone Snail Lab. It is eating a "feeder" Goldfish. *L.L.C. Jones*

Encounters: To the casual beachcomber or tide pooler, some species of cone snails may be encountered in intertidal areas such as tide pools or along beaches. There are also encounters by snorkelers, but SCUBA divers are more likely to encounter a higher diversity of live individuals, because they can reach greater depths and may linger at the bottom and near reefs. Cone snails are sometimes kept in domestic aquaria. They make interesting pets, but when you see what they can do to a fish, it should get your attention, and unless you are prepared to feed them accordingly, they can be a challenge to keep. Cones are largely nocturnal. By day, they usually remain partially buried in the substrate, where they can ambush prey undetected. However, they can and will bite defensively, even when hiding during the day.

People are normally envenomated when handling cone snails, although it is possible to step on them on a beach or in shallow water. They are appealing to shell collectors. Shells on a beach might be dead or empty, but they might also harbor living organisms. Live snails may be washed ashore, especially after a storm. The occupant may be withdrawn inside the shell to avoid desiccation. Cones should never be picked up and put into a swimsuit or carried in the hand. Unless the shell is obviously empty (e.g., worn and bleached), they should be carefully pushed into a bucket, and when you are sure there is no living animal inside, they can be safely handled. They can bite through gloves and wetsuits. The safest way to enjoy their beauty is by buying their cleaned shells from a responsible supplier. Because they are important as human food in the Philippines and elsewhere, these areas can be good sources for the shell trade.

Venom Apparatus, Venom, and Symptoms: Cone snails have an advanced and rather complicated venom apparatus. A snail detects predator and prey using its proboscis, a hollow chemosensory tubule that is usually hidden within the rostrum. The venom is manufactured in a long venom duct in the abdomen, then pumped into radulae in the radular sac. A radula is essentially a slender, pointed hollow tooth that acts as a harpoon. Different species have different radular designs, but there are often one or more recurved barbs anteriorly to hook into prey. The proboscis is also used to aim the radula, much as a blowgun aims a dart. When predator or prey are bitten, a large muscular bulb forcefully ejects a single venom-filled radula through the proboscis into the victim. Each radula is only used once. This action is somewhat analagous to the firing of a nematocyst, which is why most authors consider cone snails to sting, rather than bite, but the envenoming components consist of modified mouthparts. There may be 15–20 radulae in various stages of development at any given time in the radular sac. When threatened, a cone will harpoon the victim with the radula in the same manner in which they procure prey. However, it has been shown that at least some species may selectively inject defensive toxins into predators. These toxins are produced in a specific part of the venom duct. Similarly, if the target is a prey species, they will select the toxins best needed for prey procurement, which are produced in another part of the venom duct. There are also ligaments and salivary glands associated with the venom apparatus. The prey is engulfed in the rostrum and moved through the pharynx into the stomach.

Cone snail venom of the most dangerous species has been relatively well studied, along with numerous other species. Worldwide, there are about 15–30 species thought to have caused serious envenomation in humans, although many others probably cause localized pain. The undisputed champion of lethality is the Geography or Geographer Cone. It is responsible for nearly all cone snail fatalities. The Textile Cone, Striated Cone, and Marbled Cone are three Hawaiian species not to be trifled with, and some fatalities have occurred with those or closely related species outside our region, although accurate reports are difficult to track. There are no species from the mainland U.S. that are known to cause serious envenomation, but case reports of bites from most species are lacking. However, it seems wise to consider all cone snails as potentially dangerous, as they all have complex venoms.

The many species of cone snails have specific peptides that bind and react with specific organs of specific prey, be they worms, mollusks, or fishes, or when used in defense. Cone snail venoms are largely neurotoxic, and fish-eating species evolved to quickly immobilize their vertebrate prey with paralytic neurotoxins. The various peptides interfere with sodium, potassium, or calcium ion channels. It has been estimated that within the genus (assuming all are *Conus*), there are an estimated 75,000 or more bioactive peptides in the venom of species worldwide, and each species of cone snail may have 50–200 or more specific toxic peptides, with specificity for targeting their usual prey items and predators. Many specific peptides have their own particular name, often with the "cono-" prefix. Peptides in cone snails are known as conotoxins, including α-, μ-, ω- conotoxins, conantokins, conopressins, and others. Some fish eaters use specialized forms of insulin to put their prey into a hypoglycemic shock. The venom of some species contains serotonin. The rapid evolution of venoms of cone snails has led to numerous toxinological and evolutionary studies of the Conoidea, which includes the related turrids and terebrids. Those taxa have their own sets of toxins, such as the teretoxins in the terebrids.

Symptoms of less-serious cone snail envenomation include a bee-sting-like pain, with numbness following. The area may remain sensitive for an extended period. Sometimes there is no pain, but the numbness can last weeks. Serious cone snail envenomation from Indo-West Pacific and Hawaiian species includes pain, numbness, weakness, tingling in the mouth and extremities, tremors, vision maladies, nausea, vomiting, dizziness, fainting, difficulty swallowing, slurred speech, respiratory distress, cardiac distress, rapid heartbeat, paralysis, and possibly coma and death. Paralysis of the diaphragm results in the inability of the victim to breathe. There are no records of cone snail fatalities in Hawaii, despite the number of dangerously venomous species occurring there.

To my knowledge, other members of the Conoidea have not been implicated in human envenomation, but researchers have found many toxic peptides in their venom.

Remarks: There are many potential medical applications of conotoxins that can help treat a variety of human maladies, including pain, epilepsy, and Parkinson's disease. One of the two FDA-approved drugs from cone snails is said to be 1,000 times as effective as morphine as an analgesic, and non-habit forming.

Cones are a diverse group of venomous mollusks that come in a variety of shapes (but mostly conical) and colors. The large, slender snail in the middle is a terebrid.
L.L.C. Jones

References and Resources: Al-Sabi et al. (2006), Barbier (2003), Biggs et al. (2010), Bouchet et al. (2011), Chun et al. (2012), Cruz and White (1995), Dutertre et al. (2014a, b), Halstead (1978), Hillyard et al. (1989), Holford et al. (2009), Keen (1971), Kohn (2014), McIntosh and Jones (2001), McIntosh et al. (1995), Morris et al. (1980), Nelson (2004), Olivera (1999, 2002), Olivera and Cruz (2001), Olivera et al. (1988, 2012, 2014), Puillandre and Holford (2010), Puillandre et al. (2012, 2014), Rosenberg et al. (2009), Ruppert et al. (2004), Safavi-Hemami et al. (2015), Schwartz and Meinking (1997), Thomas and Scott (1997), Tucker (2012), Zamora-Bustillos et al. (2009), Zugasti-Cruz et al. (2008).

Nudibranchs (Order Nudibranchia)

"There are a number of reports in Australia of kids engaged in 'Bluebottle' fights—where they throw stranded *Physalia* at each other—being badly stung by inadvertently playing with *Glaucus* and *Glaucilla*, both of which, by concentrating the most venomous of *Physalia's* nematocysts, are much more deadly." (Sea Slug Forum online resource)

Classification: Phylum Mollusca (mollusks); Class Gastropoda (gastropods); Order Nudibranchia.

There are over 3,000 species of nudibranchs worldwide, although there are still many yet to be described. Taxonomy of nudibranchs is confusing and there are many interpretations of hierarchy and nomenclature at all levels, so I follow WoRMS, which recognizes MolluscaBase. Both of these are online resources, but they tend to stay current, are hosted by experts, and they include dates of last edits (in this case, 2018). Nudibranchs that are venomous belong

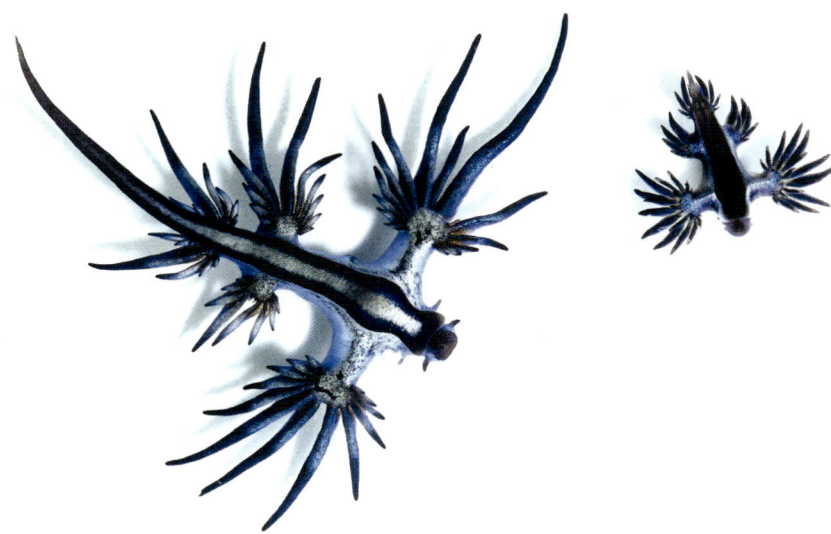

Glaucus atlanticus (the large one) and *Glaucus* (=*Glaucilla*) *marginatus* (the small one). Taro Taylor/CCA/Wikimedia Commons

to the superfamily *Aeolidioidea*, commonly known as aeolid (or eolid) nudibranchs. Worldwide there are eight families of aeolids. Although "aeolid" has the -id suffix of a family, which is retained from previous taxonomy (i.e., all were in the Aeoliidae), the common name is still in widespread use for the superfamily. The aeolids of greatest concern to human health discussed here are the five species in the family Glaucidae. Some authorities consider the family monotypic, with the sole genus being *Glaucus*, while others recognize the genus *Glaucilla* as separate and having four species. To keep things simple, I will refer to the family as monotypic. *Glaucus* nudibranchs have a plethora of common names, including blue slug, sea dragon, sea lizard, blue dragon, sea swallow, blue angel, and blue glaucus nudibranch. The moniker of blue dragon is commonly used, but is also frequently used for a Hawaiian nudibranch, *Pteraeolidia ianthina*, so I am proposing the common names below, based on their known ranges and the fact that the Latin origin of *Glaucus* refers to the glaucous coloration (bluish-gray) of the genus. Species in our area include *G. atlanticus* (Circumtropical Glaucous Nudibranch), *G. marginatus* (Indo-Pacific Glaucous Nudibranch), *G. mcfarlanei* (Kona Glaucous Nudibranch), and *G. thompsoni* (North Pacific Glaucous Nudibranch). *Glaucus marginatus* (Indo-Pacific Glaucous Nudibranch) and *G. bennettae* (South Pacific Glaucous Nudibranch) may also occur in some U.S. holdings.

Technically, all aeolids are venomous and some other taxa may affect humans. Many other nudibranchs are poisonous, and often sport aposematic coloration. At least some of these are commonly called "dorid" nudibranchs, often to differentiate them from aeolid nudibranchs.

Identification: Nudibranchs (meaning "naked gill") lack shells as adults. Some aeolids reach about 40 cm in length, but the smallest are microscopic; most are less than 10 cm. Because they superficially resemble terrestrial slugs, they are sometimes known as sea slugs, although there are other animals in the sea with that moniker. Nudibranchs are among the most beautiful animals in the ocean, often sporting brilliant aposematic colors, which helps protect these often-toxic, slow-moving animals from predation. Nudibranchs are generally benthic, but some are capable of swimming gracefully through the water; those that swim are sometimes known as Spanish dancers. A typical aeolid nudibranch has a slug-like body, the venter of which is called the foot. They lack gills, as they breathe through their skin. They possess structures known as cerata (singular ceras), which are extensions of the mantle, and are normally distributed along the dorsum. Internally, they

A Shag Rug Nudibranch off central California about to feed on a Proliferating Anemone. *Ron Wolf*

are evaginations of the digestive system. Cerata are quite variable in shape, number, and distribution among species, but they are frequently cylindrical, giving the animal a feathery appearance. Cerata have a number of functions, depending on species. They help to increase surface area for respiration, but are also used for digestion, envenomation, and may even harbor photosynthetic zooxanthellae, as do corals. At the anterior end, nudibranchs have structures called rhinophores. These look like antennae, but they are chemosensory rather than visual. Below these are cephalic tentacles that are tactile in nature, helping the animal to navigate. The mouthparts include the radula, which is essentially a chitinous ribbon with teeth.

Members of the genus *Glaucus* have the same body parts as other aeolids but they are highly modified due to their neustonic lifestyle and specialized prey. The basic shape of the mantle is slug-like, tapering posteriorly, but they are dorsoventrally flattened. There are pairs of fleshy appendages, sometimes called arms, on which cerata are clustered. The cerata spread from the arm as do fingers from a human hand. The best known species is *G. atlanticus*. It is blue with white to silvery stripes ventrally, gray on the dorsum, and is longer and more slender than other congeners. It reaches about 3 cm in length. The anterior arms are the longest, and they decrease in size posteriorly; the cerata may emerge directly from the main axis of the mantle posteriorly, rather than

The Spanish Shawl is technically venomous, but its food is a hydroid, and its stings do not affect humans (as far as I know). They are popular eye candy for sport divers in southern California. *Jerry Kirkhart/CCA/Flickr*

from the arm-like appendages. Most seem to have 2–3 pairs of arms and 3–4 pairs of clusters of cerata. *Glaucus* are bilaterally symmetrical with arms and cerata laterally aligned, rather than dorsally like most aeolids. Fun fact: *G. atlanticus* may have a penis longer than its body and has a copulatory spine. Members of the *G. marginata* clade (all except *G. atlanticus*) are also stouter and smaller than *G. atlanticus*, but similar in design. Adults usually reach about 1 cm. They are also not as flamboyantly colored and patterned.

Distribution: There are many species of nudibranchs in the eastern and central Pacific. As of 2004 (hence, out of date), 252 species were reported between Alaska and Baja California. The Atlantic has far fewer species. For example, the northwest Gulf of Mexico has only about 14 species, but *G. atlanticus* is one of them. Despite its specific epithet, this species is not limited to the Atlantic, but is circumtropical to subtropical. The distribution of glaucous nudibranchs is imprecise, so it is difficult to know detailed distribution and specific localities, but they are neustonic in the major currents so it seems reasonable they could be found wherever oceanic gyres take them, along with their food sources. *Glaucus marginata* is widespread in the Indo-Pacific. *Glaucus thompsoni* is found in the North Pacific, including the west coast of Mexico. The holotype was collected 120 km west of Baja California Sur. *Glaucus atlanticus* also occurs in the Gulf of California, at least as far north as the Midriff Islands, but not as far north as California. Hawaiian waters have at least three taxa: *G. atlanticus*, *G. mcfarlanei*, and *G. thompsoni*. *Glaucus marginata* occurs in the North Pacific Gyre, so likely occurs there, as well. *Glaucus mcfarlanei*, as best I can tell, has only been recorded from Hawaii, where it was collected off the Kona coast. The only other species of glacous nudibranch is *G. bennettae* (South Pacific Glaucous Nudibranch), which occurs in the South Pacific. I do not know if it has been reported from American holdings (e.g., American Samoa and Jarvis Island are in the South Pacific, and several islands and atolls are nearly equatorial).

Natural History: Nudibranchs in general are widespread along coastal marine waters, often in relatively shallow areas. They are highly sought-after as eye candy and photographic fodder for SCUBA divers and snorkelers, and some species are even common in tide pools. They often occur in rocky and coral reef habitats. However, *Glaucus* spp. are unique among nudibranchs because they are neuston—planktonic surface dwellers that ride ocean currents. More specifically, they are hyponeuston, occurring at the underside of the surface of the water, with their venter facing up. They remain at the underside of the surface via surface tension and floating. They swallow air to form a bubble, to stay bouyant. If pushed below the surface, they will float up. They are poor swimmers, being at the mercy of currents, but they are able to move in order to feed on their prey. Some other taxa are pelagic, but not truly planktonic, as they may be associated with floating vegetation, debris, or their prey. One such species is the amazing Sargassum Nudibranch (*Scyllaea pelagica*), which matches its algal habitat with striking accuracy. *Glaucus* nudibranchs are also cryptic and/or aposematic, depending on the view. When viewed from above, the blue and white stripes of the venter of *G. atlanticus* probably appears aposematic to aerial predators, while the dorsum, when viewed from below by marine predators, is cryptic. All aeolid nudibranchs are carnivorous, and usually prey on cnidarians. Many of the nearshore species feed on anemones, hard and soft corals, hydroids, and jellyfishes, but *Glaucus* feeds on other neuston, primarily siphonophores, including the Portuguese Man-of-War and Pacific Bluebottle, or the By-the-Wind Sailor. It is apparent they can detect their prey with the chemosensory structures, and when in close proximity can swim to the prey and feed on the colony. They can also be cannibalistic.

Nudibranchs often lay eggs in coils. *Elias Levy/CCA/Flickr*

Nudibranchs are hermaphroditic but cannot fertilize themselves. Instead, when one finds another, they copulate with each other and both may inseminate the other and both may become gravid. In some species there may be mating aggregations, with multiple partners. They lay eggs in spiraled or coiled masses, which are sometimes ribbon-like in shape. The egg capsules may contain several eggs, which may multiply in the capsule. *Glaucus atlanticus* lays numerous strings of eggs that are about 1.7 cm long. The eggs are usually deposited on floating vegetation, flotsam, or jetsam, or else on remnants of prey species. With the many species of nudibranchs, there is quite a bit of variation in specifics of their life history traits. The eggs hatch in about a week, and a veliger larva becomes planktonic, feeding on phytoplankton before metamorphosing into an adult. In nudibranchs, the veliger has a shell, which is discarded as it develops into an adult. In *Glaucus*, however, the shell disappears before the larva hatches.

Encounters: For tide-poolers, snorkelers, and SCUBA divers, encounters with nudibranchs can be commonplace. Many are boldly marked and they are often sought out by nature buffs who want to gaze at them or photograph them. There is little information on encounters with *glaucus* nudibranchs, however. Certainly, a handful of researchers seek them out intentionally, and they occur in general plankton trawls and fishing nets. Most encounters with

A Circumtropical Glaucous Nudibranch floating upside down at the surface.
George Schechter/CCA/Flickr

the public likely happen when they are washed ashore with their prey or floating in swimming areas. Most people know to steer clear of a Portuguese Man-of-War but are probably unaware of the little blue nudibranchs that may accompany them. On a beach, out of water, *Glaucus* will collapse somewhat, so is relatively inconspicuous. If one wants to intentionally see them, it is probably best to look for them when men-of-war are stranded on the beach (and they may be on the prey), and without touching them, put them in seawater to view or photograph. Even though they are stunning animals that might be popular in a public aquarium, nudibranchs of all types are difficult to impossible to maintain, even when given the right type of food.

Venom Apparatus, Venom, and Symptoms: Aeolid nudibranchs do not use their sharp radula to deliver a venomous bite. They sting, and it is purely defensive. They typically feed on cnidarians, and in doing so, they use their prey's unfired nematocysts for their own protection. The nudibranch ingests the tentacles and the unfired nematocysts are then transported into structures called cnidosacs at the tips of the cerata, where they are concentrated. Aeolids are not stung when feeding on cnidarians. The prey's nematocysts don't discharge as they are being eaten, probably because the nudibranch's mucus inhibits a chemical cue for discharge. When a potential prey animal brushes up against the cerata, or tries to feed on the nudibranch, the nematocyst threads are discharged through the epidermis of the cerata.

Not without exception, most aeolid nudibranchs do not deliver a sting meeting the bee sting pain threshold required for inclusion in this book, unless it is on a sensitive part of the body (e.g., tongue, lips, inner arm). Most aeolids do not feed on cnidarians that are particularly venomous to humans, although some people may be sensitive to them and can get contact dermatitis. Nudibranchs in the genus *Glaucus*, however, can deliver extremely painful stings because they are transferring the venom from their highly venomous prey. The venom of *Physalia* is discussed in the chapter on those species. Not surprisingly, symptoms of envenomation by *Glaucus* are similar to that of men-of-war, except that the long whip-like lesions are lacking. However, redness, swelling, and intense pain are symptoms of contact with *Glaucus*, and systemic symptoms may develop. I have read that *Glaucus* selectively sequester and concentrate the most venomous nematocysts of *Physalia*, making the sting worse than its prey (for the same surface area), but I cannot find any quantified information to confirm this assertion.

References and Resources: Bieri (1966), Churchill et al. (2014a, b), Conklin and Mariscal (1977), Gosliner (1979), Hattersley (2009), Hernández et al. (2018), Lalli and Gilmer (1989), Landa et al. (2018), Ottuso (2009), Rosenberg et al. (2009), Ruppert et al. (2004), Schlesinger et al. (2009), Schwartz and Meinking (1997), Thompson and McFarlane (1967), Valdés and Campillo (2004).

Polychaete Worms (Class Polychaeta)

"The most deeply sequenced library (*G. dibranchiata*) expresses the greatest diversity of putative venom toxin transcripts, representing 20 toxin classes that have been convergently recruited into animal venoms, as well as 12 putative toxins that are possibly unique for bloodworms." (von Reumont et al. 2014b)

Classification: Phylum Annelida (segmented worms); Class Polychaeta (polychaetes).

There are over a dozen phyla of animals considered worms, some of which may be venomous to some degree. There are several classes, including Polychaeta (polychaetes), Oligochaeta (earthworms), and Hirudinea (leeches), that may have some toxic compounds, but it is the polychaetes that can cause envenomation. Polychaetes are a huge group of mostly marine animals, with at least 8,000–10,000 species worldwide. Some are sessile (e.g., tubeworms), but most are free-living. Biting species include those in the genera *Eunice*, *Onupis*, and *Glycera*, among others. *Glycera* are the bloodworms, in the order Phyllodocida, and the monotypic family Glyceridae. *Glycera* is the only

Scoloplos sp., a marine polychaete from the eastern Pacific. Ron Wolf

genus of polychaete worms that is known to have complex venom capable of quickly subduing small marine prey. Other potentially biting polychaetes are known as clamworms, ragworms, and lugworms, but these are thought to lack venom, or their venom is poorly known.

The bristled polychaetes belong to the order Aciculata, family Amphinomidae. These are variously known as fireworms, bristleworms, or sea mice. The vernacular names are a little confusing, but bristleworms are those with bristles, fireworms are those noted for painful stings, and sea mice are those that have long bristles. Representative genera include *Amphinome, Chloeia, Hermodice, Hipponoe, Eurythoe, Notopygos, Pareurythoe,* and *Pherecardia*. Tropical fireworms are sometimes kept in home aquaria intentionally, but sometimes native or exotic polychaetes show up in aquaria, hitchhiking in rocks or coral.

Identification: The polychaetes are a highly variable group. They range in size from a few millimeters to a reputed 61 m in length! Bloodworms and fireworms typically reach about 10 cm in length, but both have been recorded to about 35 cm. Overall, they are usually elongate and mostly divided into segments. The head-like anterior portion is known as the prostomium, followed by the peristomium. The posterior segment is the pygidium; between these are the segments of the trunk. The prostomium has the eyes, brain, and sensory organs. The peristomium has tentacular cirri and surrounds the mouth. The animal's trunk is composed of repeating true segments. The segments are about as long as wide, and each has a pair of appendages known as parapodia. The parapodia themselves bear secondary appendage-like structures, including cirri (sensory organs), chaetae, and setae. Setae and chaetae are similar and both are made of chitin (α-chitin on setae and β-chitin on chaetae), but are structurally different. Some references refer to chaetae and setae collectively as setae.

Bloodworms are pinkish in color, and the reddish hemocoelic fluid ("blood") can be seen through the body cavity. When the worm is severed, considerable hemocoelic fluid gushes out—hence the common name. Some fireworms are beautiful animals that have long chaetae and may sport bright aposematic colors. They vary from being pinkish animals with small, lateral chaetae, to those covered in long chaetae. Those that have chaetae forming a mat over the entire body are the polychaetes commonly referred to as sea mice (e.g., *C. viridis*). Because they may be short with long bristles, they don't really appear worm-like.

Aquatic Invertebrates 551

Bloodworms, *G. dibranchiata*, are raised commercially in the northwestern Atlantic. *Bloodworm Depot*

A pileworm from the eastern Pacific. *Ron Wolf*

Bearded Fireworm, *Hermodice carunculata*, is an Atlantic species, but there are fireworms in all tropical oceans. *Philippe Guillaume/CCA/Flickr*

Distribution: Polychaetes and other annelids are found globally, in both temperate and tropical waters. There are an estimated 835 polychaete species and subspecies in the Gulf of Mexico. There are at least 42 described species of bloodworms, all of which are thought to be venomous. The bloodworms in the Atlantic include *G. americana*, *G. dibranchiata*, *G. papillosa*, *G. tesselata*, and *Hemipodus roseus*. Species in the eastern Pacific include *G. americana*, *G. capitata*, *G. dibranchiata*, *G. lapidum*, *G. robusta*, *G. tesselata*, and *H. roseus*.

The Bearded Fireworm, *Hermodice carunculata*, is widely distributed. The Gulf of Mexico sea mouse or bristleworm is *C. viridis*, also widely distributed. In southern California, *C. pinnata* is a common bristleworm. Another California species is *Pareurythoe californica*. There are several other genera and species of amphinomids in the Gulf of Mexico, central Pacific, and eastern Pacific, but these are less well known and/or do not have such a well-developed stinging apparatus. Hawaii is well represented by tropical fireworms, and they are commonly encountered on reefs, especially at night. Hawaiian species include *Amphinome rostrata*, *C. flava*, *E. complanata*, *Hipponoe gaudichaudi*, *N. albiseta*, *N. gregoryi*, *N. labiatus*, *N. megalops*, *Pareurythoe* sp., and *Phercardia striata*.

Natural History: Polychaetes are almost entirely marine, although a few species are known from fresh water. They are found in virtually every type of marine habitat, from intertidal zones to the deepest ocean floor, including the areas around hydrothermal vents. Many species are found in sediments, such as mudflats, but others are found on reefs and some are even planktonic. The familiar bloodworms live in sandy or muddy sediments in intertidal or shallow areas. Bloodworms (especially *G. dibranchiata*) are commercially important as fishing bait and aquarium fish food, constituting a multimillion-dollar industry. They are harvested commercially, mostly in Maine and Nova Scotia, but native species occur along all of our coastlines. Bloodworms are dug up at low tide by people using them for bait. Fireworms are mostly found in reef areas, be they rock or coral. By day, they hide in corals and rocks, but come out at night to forage.

Some species are hermaphroditic, while others have distinct sexes. Bloodworms have two sexes, and during mid-summer, they swim from the sand and mud bottoms to the surface of the water where they release sperm and eggs. The larvae are planktonic at first, but develop through instars in the substrate in silken tubes. Fireworms also have two sexes, and swim to the surface to shed sperm and ova; female Bearded Fireworms are bioluminescent, which can attract mates.

Bloodworms and fireworms are voracious predators. Bloodworms are sometimes said to be detritivores, but they use their proboscis to hunt prey. The venom is highly toxic to crustaceans. Fireworms can seize and envelop their prey. They eat a variety of organisms, but reef-dwelling species seem to prefer coral polyps. In turn, they are preyed upon by a variety of marine animals. Some species may also scavenge. One species in Hawaii lives on driftwood and feeds on Gooseneck Barnacles.

Encounters: Bloodworms and their relatives are usually out of harm's way by being under the surface of sand and mud, or may be under surface debris. They are usually encountered when people dig them up for live bait, or when anglers purchase them in bait shops. Those who commercially harvest bloodworms have an occupational hazard. Fishermen sometimes entice bloodworms to extrude the proboscis, then cut off the tip.

Fireworms are commonly encountered by divers, snorkelers, beachcombers, tide poolers, and marine aquarium employees and enthusiasts. They are sometimes caught by anglers. Fireworms should just never be handled, as their chaetae can even penetrate gloves. Divers can be stung when they brush against them, so wearing full wetsuits is advised. Tide poolers are sometimes stung when lifting objects that the worms are hiding under.

Bristleworms and fireworms can also be abundant in public and domestic aquaria, whether by design or by accident. They can not only cause painful stings in humans, but being voracious predators, can sometimes wreak havoc with tank mates, including fishes and corals. In public aquaria and larger domestic aquaria, some species are a menace, and may need to be eradicated from the aquarium. Before acquiring, I suggest you go online and seek the advice of advanced marine aquarists, so as to not upset the balance of the aquarium system—and to learn how to protect yourself from painful stings.

Venom Apparatus, Venom, and Symptoms: Bloodworms and other polychaetes can bite, and their venom is used both for predation and defense. In some larger species of polychaetes, the bite may be painful, even in nonvenomous ones. Their biting mouthparts are housed inside the body and extruded by a proboscis for feeding or defense. There are four jaws on the tip of the proboscis. Each of the jaws has a fang, venom duct, and associated venom gland. Bloodworms are unique because they have copper integrated into their jaws. The metal strengthens the jaws and the copper ions may be involved with envenomation.

The venom of bloodworms is fairly well known. There are five categories of venom recognized in three species studied (*G. dibranchiata* and two European species). These are pore-forming and membrane disrupting, neurotoxins, protease inhibitors, CAP (CRISP, antigen 5, pathogenesis-related) domain toxins, and other enzymes. The enzymes include hyaluronidase, phospholipases, and metalloproteinases. One specific compound is named α-glycerotoxin. The venom targets invertebrates (e.g., crustaceans), but vertebrates are also affected. The venom is homologous to several venoms, especially of marine origins, including those of scorpionfishes, turrid gastropods, and sea anemones. The neurotoxic compounds have also been likened to those in the widow spiders (α-latrotoxin), scorpions, and snakes. The neurotoxins act on calcium ion channels.

Glycera bites have been described as ranging from a pinch to being similar to a bee sting, although in more serious bites, they can cause more severe skin inflammation. The pain can last for hours. Some people have become allergic to the venom. In one online post, a fisherman was trying to be funny and put one in his mouth. After being bitten, he spit it out, but his tongue and lips swelled up and went numb. Mice injected with

Glycera-gland tissue homogenate developed muscle contractions, scoliosis (curvature of the spine), and flaccid paralysis.

Fireworms only use their chaetae for defense. The long bristles can be extended and plunged into the flesh of a human or predator. There has been debate over whether or not fireworms are venomous. Although they do not seem to have venom glands proper, their hollow chaetae can have toxic compounds within. As with stinging caterpillars, the spines cause physical trauma and the toxic compounds cause envenomation. The chaetae of at least one fireworm have been shown to contain an organic amide termed complanine, for its isolation from *E. complanata*. Two additional compounds described shortly thereafter were termed neocomplanines. These compounds have been shown to be the causative agents of observed dermatitis symptoms. Symptoms include severe pain, redness, swelling, and numbness. A rash usually forms, and in some cases, blisters may form and the skin may slough. The pain can last for several hours, and the itching can last much longer. On occasion, dizziness and nausea accompany dermatitis. In one case report, a person handled ten one-foot-long tropical fireworms in an aquarium, presumably not aware they were venomous. After symptoms escalated, he sought medical attention. The initial response was that of an electrical shock. The pain was severe and his hand remained swollen for two weeks. Allergies and secondary infections are always possible.

Remarks: Technically, leeches are venomous annelids because they possess an anticoagulant to allow blood flow (as with mosquitoes), but do not meet the "bee sting" criterion for inclusion in taxa accounts. Virtually all polychaetes have some sort of bristles and biting mouthparts, so it is best to leave them all alone.

References and Resources: Ahrens et al. (2013), Eckert (1985), Fauchald et al. (2009), Hoover (2006), Jones and Thompson (1987), Kay et al. (2009), Kem (1988), Nakamura et al. (2008, 2009), Reuscher and Shirley (2014), Smith (2002), von Reumont et al. (2014a, b).

True Jellyfishes (Class Scyphozoa)

"The victims included [the] Galveston Island Beach Patrol Chief who was stung about 15 times during his morning workout. . . . As a result, lifeguards are now flying blue flags at their towers to alert swimmers to the jellyfish, known locally as sea nettle. 'We have a very scientific method for establishing when we have a blue flag. . . . We all do a workout in the morning before our shift. If a whole lot of us get stung, it's a blue-flag day.'" (*Houston Chronicle*, S. Wood reporting)

Classification: Phylum Cnidaria (cnidarians); Class Scyphozoa (true jellyfishes); Order Semaeostomeae (flag-mouth jellyfishes); Family Pelagiidae (pelagic jellyfishes); Genus *Chrysaora* (sea nettles).

Sea nettles are among the largest and most widespread jellyfishes that pose a risk to swimmers along coastal areas, but there are some other taxa with a significant sting, which are mentioned below.

Pacific Sea Nettle. This species is often seen at public aquaria, and sometimes at public beaches. *L.L.C. Jones*

Worldwide there are fifteen species of *Chrysaora*, with six in our region. These are the Black Sea Nettle (*C. achlyos*), Bay Sea Nettle (*C. chesapeakei*), Purple-striped Sea Nettle (*C. colorata*), Atlantic Sea Nettle (*C. quinquecirrha*), Pacific Sea Nettle (*C. fuscescens*), and Northern Sea Nettle (*C. melanaster*). There is a related species (same family, different genus; *Pelagia noctiluca*) sometimes called the purple-striped jellyfish, which is also the common name for *C. colorata*. Here *P. noctiluca* is referred to as the Mauve Stinger to avoid confusion—in fact, some internet references erroneously synonymize these two very different species. Another noteworthy scyphozoan that may occasionally enter our marine waters is the Lion's Mane Jellyfish (*Cyanea capillata*, family Cyaneidae). This is actually a species complex that has yet to be sorted out taxonomically. Another scyphozoan that causes harm in humans is the Thimble Jellyfish (*Linuche unguiculata*). It is a causative agent in seabathers' eruption, which is a malady different than that delivered by typical scyphozoans, so is discussed in the Other Aquatic Invertebrates chapter.

Many other scyphozoans occur in North America. Most have mild stings, are poorly known, or are rarely encountered, so are not mentioned in the text. Among the more common larger species that may occur inshore are the Cannonball or Cabbage-head Jellyfish (*Stomolophus meleagris*, family Stomolophidae) and Moon Jellyfish (*Aurelia aurita* or *Aurelia* spp., family Ulmaridae). Both of these are generally considered mildly venomous, although some people have had adverse reactions. Moon Jellyfish occur throughout much of the Atlantic and Pacific, and some populations seem more toxic than others. The Upside-down Jellyfish (*Cassiopea* spp.) is an interesting complex of species that can sting swimmers. The Australian Spotted Jellyfish (*Phyllorhiza punctata*; family Mastigiidae) is non-native, but swarms have occurred in Hawaii, California, and the Atlantic. In some areas it is an invasive species and has been detrimental to shrimp fisheries and other socioeconomic values. It also has a mild sting.

Identification: Members of the Semaeostomeae are large jellyfishes that are best known in the medusa stage. Jellyfish medusae range from about 6 cm in bell diameter to more than two meters, depending on species. The Black Jellyfish has a bell that may be 1 m across, with oral arms that may be 5–6 m long and fishing tentacles that may be nearly 8 m long. The bell is opaque and deep purple to nearly black. The Purple-striped Sea Nettle has a bell that is 70 cm across in large individuals. It is white with purple to dark maroon stripes radiating from the center. The Pacific Sea Nettle has a bell that

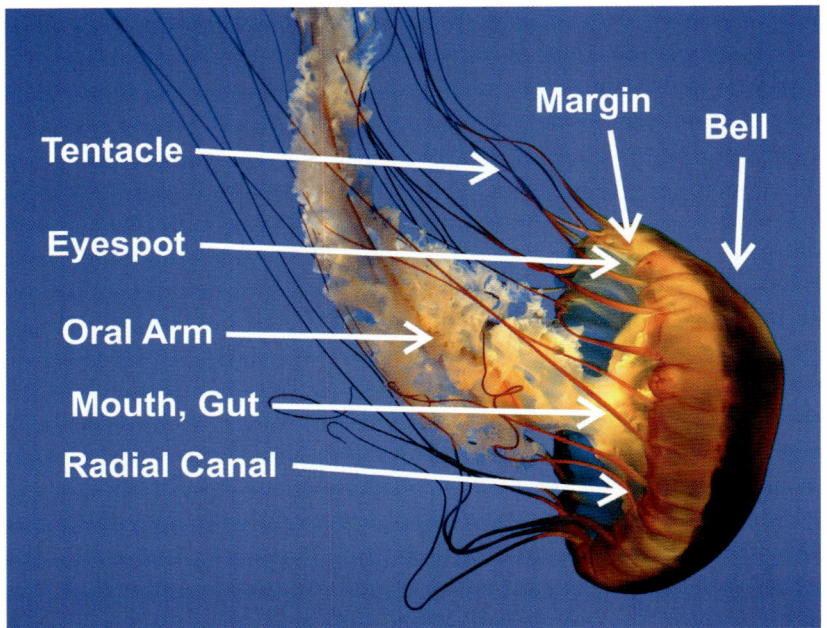

External anatomy of a scyphozoan. *Base photo, Laura Wolf/CCA/Flickr*

may exceed 1 m. It is golden-brown to reddish in color. The oral arms are long and the 24 tentacles may trail 5 m behind. The Atlantic Sea Nettle is similar to the Pacific Sea Nettle, but is smaller, and the bell is typically white to pinkish, often with stripes radiating from the center. Until recently, the Bay Sea Nettle was considered a variant of the Atlantic Sea Nettle, although it was originally described in 1936. However, it has fewer tentacles, fewer lappets, and longer oral arms than the Atlantic Sea Nettle. In Chesapeake Bay, they are often white in color, but may have darker radiating stripes. The Northern Sea Nettle reaches about 60 cm in bell diameter and its 24 tentacles trail about 3 m behind. Like most other sea nettles, it also has linear markings radiating from the top center of the bell. The Mauve Stinger is small, averaging 6.5 cm, but can reach 10 cm. The bell is pink to purple. The eight tentacles can be up to 10 m in length. It is a bioluminescent species and may produce a stunning visual display at night. The undisputed largest jellyfish is the Lion's Mane Jellyfish. There are varied reports of maximum size, but it is often cited as having a bell up to nearly 2.5 m in width and some 800 tentacles reaching at least 15–27 m. Individuals have even been reported to reach 40 m in length—longer than a Blue Whale. Some claim this is the

Lion's Mane Jellyfish.
Top: Lewis Kirkpatrick for Amrae Productions
Bottom: Derek Keats/CCA/ Wikimedia Commons

largest animal on Earth, while others question maximum size claims; it is difficult to gauge the length of an animal that can have tentacles easily broken off. The Upside-down Jellyfish is a favorite in aquariums, because of its unique nature. It generally remains motionless on the seafloor with its bell against the substrate and arms in the water column. Its mouth faces upwards.

Distribution: The Black, Northern, Pacific, and Purple-striped Sea Nettles are all found in the eastern Pacific. The Purple-striped Sea Nettle is common in southern California, but it occasionally occurs north of San Francisco. The Pacific Sea Nettle can reach high densities in northern California northward, and blooms in Oregon and elsewhere have been well publicized. The Northern Sea Nettle is adapted for the cold temperate seas of the far northeastern Pacific, including the Arctic Ocean and Bering Sea. The Atlantic Sea Nettle is found in the Atlantic, while the Bay Sea Nettle is also found along the eastern seaboard and in the Gulf of Mexico. The two species are sympatric in some areas. A population of similar sea nettles is known from the Caribbean, and has been tentatively classified as *Chrysaora* c.f. *chesapeakei*. There have been several studies on the populations of *C. chesapeakei* in Chesapeake Bay, where this species is particularly abundant. The Mauve Stinger is found in the warmer waters of both the Atlantic and Pacific. It is generally pelagic, often in deep water, and usually found in warmer areas, but also occurs in the shallows of Florida and the Caribbean. It occasionally shows up in southern California. The Lion's Mane Jellyfish complex are primarily denizens of cold, boreal waters of both the Atlantic and Pacific in Canada, but can show up in more southerly waters off mainland United States during the winter. In the Gulf of Mexico, they are most often observed in January through March. These animals reach their greatest sizes in cooler waters approaching the poles. Moon Jellyfish are found in both the eastern Pacific and western Atlantic, and the genus is essentially circumglobal, being especially common in temperate areas. Cannonball Jellyfish are common in the eastern Atlantic and Gulf of Mexico, but rarely encountered off California. The Upside-down Jellyfish occurs in Hawaii, the Caribbean, Florida, and the Gulf of Mexico.

Natural History: Depending on conditions, marine jellyfishes may be pelagic, near the surface or in deep water, or found along the coast, including in bays and esturaries. They are often at the mercy of currents in open waters, as they are not strong swimmers. Moon jellyfish are among the

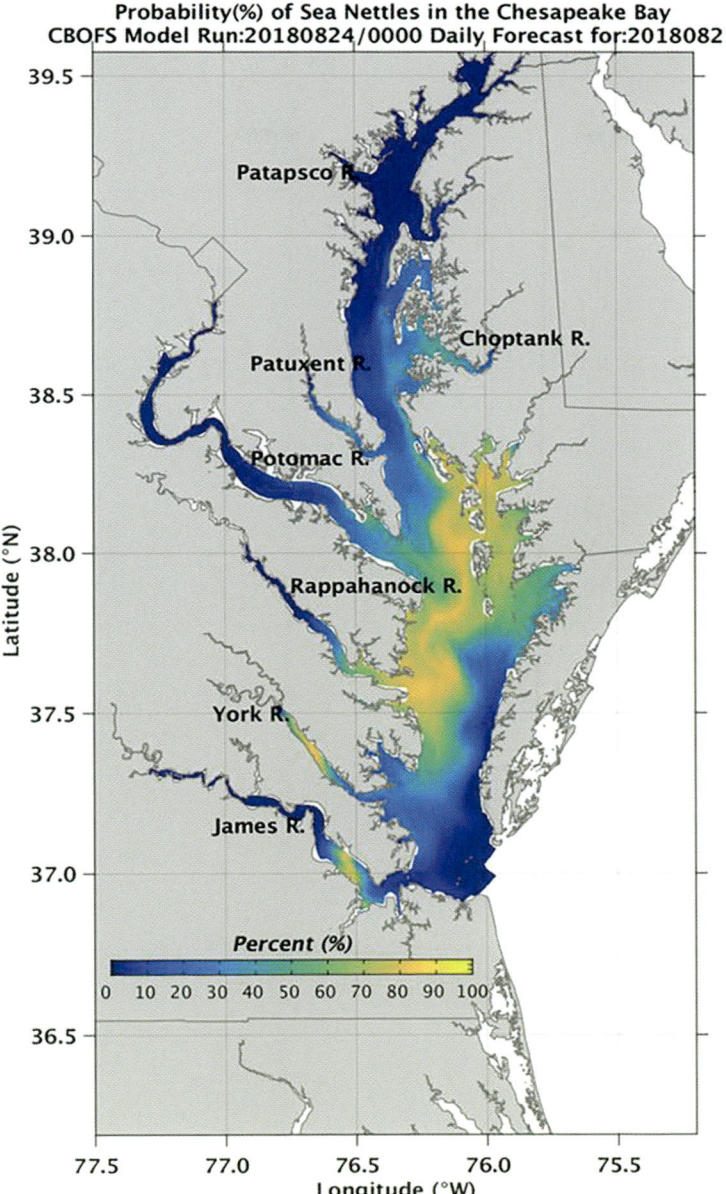

Chesapeake Bay Sea Nettles were recently separated from Atlantic Sea Nettles. They cause tens of thousands of stings, so predictive occurrence models are published. NOAA/Public Domain

most common scyphozoans in inlets, bays, and estuaries, although sea nettles often occur in these coastal features. Black Jellyfish tend to occur along the eastern Pacific inshore waters in large numbers when red-tide events occur. A number of water quality variables, such as temperature, salinity, and pollutants, may influence abundance and distribution. Climate-change factors, such as El Niño/La Niña events, may also influence abundance and distribution patterns. There is evidence that jellyfishes and their relatives are expanding or shifting their ranges globally. There are many articles and books written about jellyfish blooms, why they happen, and what this indicates about our changing environment.

Scyphozoans have a complex and interesting life cycle. The jellyfish starts out life from an egg and sperm, which hatches into a planula larva. The planula settles on the substrate and becomes a sessile polyp, a life-form that is attached by a stalk to a hard surface and has tentacles for feeding in the water column. This life-form is the type seen in a coral polyp or sea anemone. It may spend several years in this stage and produce new individuals asexually through budding. Metamorphosis from polyp to medusa is from a process known as strobilation. A strobila is essentially a stalk derived from a polyp that produces young medusae. The small medusae are stacked onto the distal end and as they mature are released into the water. The young medusae then develop into mature male and female jellyfish. Mature jellyfishes reproduce sexually, producing eggs and sperm. Fertilization can occur in the water column, or in the female who picks up sperm from the water column, depending on species.

In temperate waters, most jellyfishes are usually encountered in the summer, when ocean waters warm up and medusae are fully developed. In southern California, the default ocean temperature is 10–15°C until the summer, due to upwelling of the cold California Current. Southern California waters warm up to at least 20°C in the summer, and El Niño years and Santa Ana wind conditions may result in warmer water and red-tide events and jellyfish blooms. Water temperatures in northern California to the Arctic rarely warm up (except perhaps in some inland waters), but some jellyfishes prefer cold boreal seas. The relatively cool waters and upwelling in California cause the area to be rich in plankton, which is food for jellyfishes. The southern part of the eastern seaboard and Gulf of Mexico is much warmer year-round. March temperatures are typically around 18°C and warm up to over 25°C in the summer. Hawaii is subtropical, so sea temperatures are warm year-round, and subject to jellyfish blooms at any time, as conditions and currents permit.

Aquatic Invertebrates 563

Moon Jellyfishes rarely cause more than an itch, but noteworthy stings have occasionally been documented. *Sonke Johnson/Public Domain/NOAA*

All jellyfishes feed on zooplankton. However, they will also take other small animals, including fishes (especially larvae and their eggs), comb jellies, worms, crustaceans, salps, and other jellyfishes. Not all jellyfishes are opportunistic, as some are specialists. When a small creature comes into contact with the tentacles, stinging cells paralyze the prey. In turn, jellyfishes are eaten by a variety of other animals, including fishes (e.g., Ocean Sunfish, tunas, Butterfish, spiny dogfishes), sea birds, sea turtles, other jellyfishes, and humans. Sea nettles often have crabs of the genus *Cancer* living in them. Unfortunately, jellyfish-eating species often mistake plastics as their prey, and this ecological crisis is helping to cause the demise of many magnificent sea animals.

Encounters: Scyphozoans are a seasonal or periodic safety hazard along all of our coasts. Most jellyfishes have seasonal abundance patterns and there is often a boom or bust occurrence near shore. Storms, strong ocean currents, and favorable environmental conditions can bring jellyfishes into swimming areas. Large numbers are sometimes washed ashore. A good example is the Purple-striped Sea Nettle, which is a periodic menace to southern California bathers. During most of the summer, they are nowhere to be seen, but during blooms, they are everywhere in the surf and on shore. This pattern is mimicked by most sea nettles. There have been several studies of Bay Sea Nettles in Chesapeake Bay, where they are a recurring, seasonal hazard to swimmers. They seem to have highest abundance in certain parts of the bay that provide optimal conditions, especially during warm periods. NOAA has a predictive model so swimmers can determine the best times and areas for recreating. In Oregon, Pacific Sea Nettles periodically reach high densities and can be problematic. The cold water of the northeast Pacific keeps most swimmers out of the water, or at least in thick wetsuits, but jellyfishes are a menace to fishermen in some years. In fact, many (perhaps most?) large species of scyphozoans are problematic to fishermen at various times, not only because they sting but also because they foul fishing nets and get tangled on lines.

Even though it was known for some time, the Black Jellyfish was not scientifically described until 1997. It was considered rare, at least until blooms occurred off southern California in 1989, 1999, 2010, and 2012. Fragments of Black Jellyfishes caused hundreds of stings from Los Angeles to San Diego among swimmers in 1999, and newspaper articles about swarms of these supposedly "rare" jellyfish and their stinging of humans appeared in subsequent years. Similarly, Lion's Mane Jellyfish can cause periodic unwanted appearances in swimming areas. In a well-publicized case, some 50–150 people were stung on one New Hampshire beach in a single day. Even if jellyfishes are actively dodged, the tentacles often break up in the surf and cannot be avoided.

One take-home message of these boom or bust cycles and changes in distribution is that jellyfishes as a group have a very dynamic distribution with occurrence patterns that change with time and environmental conditions.

There is not much left of this Purple-striped Sea Nettle except the bell. The tentacles and oral arms may have been consumed by predators. *Douglas King*

This is the topic of considerable research, as it is happening on a global scale. Not only do ranges shift passively with currents and environmental conditions, scyphozoans may be transported to new areas through ship ballast or as polyps encrusted on substrates such as commercial oysters.

There are undoubtedly tens of thousands of stings every year in North American waters, but they largely go unreported. In one study on stings reported to poison control centers from 2001–2005, there was an annual average of 1,046 exposures nationally, but this includes hydroids and cubozoans. Of these envenomations, 387 were considered minor, 72 moderate, and 1 major. Envenomation from aquatic animals (including stingrays) accounted for only 3 percent of the bite and sting calls. Around the same time period, an average of 724 visits to emergency rooms was reported for jellyfish stings and none were shown to require hospitalization. This was only 0.1 percent of the reported exposures from animal bites and stings. Certainly, these figures grossly underestimate the actual number of stings, as most are adequately treated with first aid without the need for medical intervention; hundreds of stings can occur in a single day during a bloom. For example, on one weekend in May 2015, 475 stings were reported and treated locally on South Padre Island, with *Physalia* and *Chrysaora* the presumed major culprits. One online report claimed that lifeguards estimated 200 jellyfish stings per day by sea nettles in San Diego alone, presumably during summer blooms. Recalling my youth, virtually everyone that went swimming during Purple-striped Sea Nettle blooms (like me) was stung, so there were probably thousands of unreported stings in southern California in a single day.

Venom Apparatus, Venom, and Symptoms: Most sea nettles and their relatives passively sting with their tentacles, both for procuring prey and in defense. Fishing tentacles are those that may target small animals in the water column, while oral tentacles or arms are used to transport food to the mouth. Some species including *Cassiopea* and *Stomolophus* lack tentacles and have only oral arms to catch prey, much like a filter feeder. Fishing tentacles are also more defensive, being longer and having denser nematocysts than oral tentacles. The venom apparatus is a cellular- and subcellular-level mechanism. The cellular organelles and their components responsible for envenomation have a variety of terms (e.g., cnidocyst or cnida), but the most common term for the stinging part of the cell in cnidarians is "nematocyst." When the cnidocil (trigger) of the cnidocyte on the tentacle is touched by an organism, the coiled tubule inside the capsular nematocyst explodes out of the epithelium and injects venom into predator or prey. A variety of spines holds the stinging mechanism in place, so the prey does not swim off. Some cells are adhesive and do not contain toxins.

The venom of sea nettles is not as well studied as the more dangerous box jellyfishes or Portuguese Man-of-War, although there is a fair amount of literature on the venom of the Atlantic and/or Bay Sea Nettle. Some studies simply refer to venom originating from "jellyfish," which can include a host of cnidarians. Sea nettles have venom that affects vertebrates, but the amount of venom delivered is fatal to small fishes; humans rarely receive large enough doses to cause severe systemic reactions or death. However, fishing tentacles of *C. quinquecirrha* are quite toxic, with an intravenous LD_{50} of 0.37 mg/kg. The fishing tentacle of the Black Sea Nettle is less toxic with an intravenous LD_{50} of 6.5 mg/kg.

Pain caused by cnidarian venom is often attributable to serotonin and the release of histamine. Tetramine is also a component in at least some cnidarian venoms; this neurotoxin has a paralyzing effect on vertebrates, not unlike curare. The venom of *C. quinquecirrha* is cardiotoxic, neurotoxic, and hemolytic. Among other compounds, the venom contains amino acids, peptides, polypeptides, and proteins. At least seven enzymes have been identified in Atlantic Sea Nettle venom, including hyaluronidase. The enzymes are similar to those of the Portuguese Man-of-War.

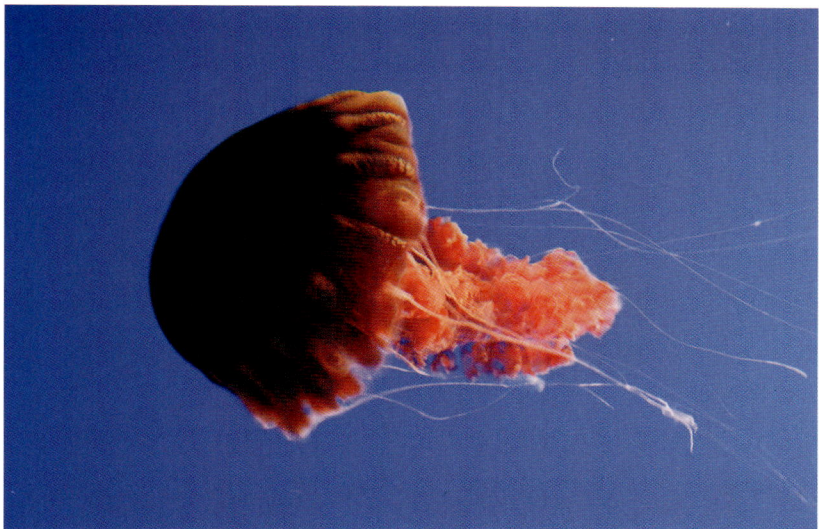
Black Sea Nettle. *"Jim G."/CCA/Flickr*

Symptoms are generally mild to moderate and localized. Typically, a victim will have pain and itching (sometimes severe), accompanied by a rash. The sting site may become swollen, and bleeding may occur. The major effects last for about twenty minutes to an hour, often with lingering effects. In stings with prolonged exposure or to sensitive people, reactions may be more severe. There may be coughing, breathing problems, muscle cramps, light sensitivity, and nausea. Exposure to the eyes can cause redness, burning, and tearing. In the laboratory, test animals have developed dermal necrosis and large doses are lethal. Anaphylaxis has rarely been documented in humans from stings of sea nettles, but anyone with repeated exposure may be at risk. There is at least some cross-reactivity among cnidarians, presumably more so among closely related species.

Lion's Mane Jellyfish apparently have a potent venom. Their huge size and mass of tentacles could be a significant hazard if encountered. Most envenomations probably come from tentacles that break off and drift into the shore. Symptoms are similar to sea nettles. As with all cnidarians, the amount of venom depends on the amount of skin exposure to stinging tentacles.

Pelagia noctiluca is regarded as a formidable stinger. It has a reported LD_{50} of about 20 mg/kg. It is notorious in the Mediterranean, where it can reach high densities and cause huge problems among swimmers and anglers. In Australia, beaches have been closed because of it. Symptoms of stings

The Mauve Stinger is a bioluminescent species. *"Eric + F"/Public Domain/Flickr*

include intense pain, wheals, itching, swelling, and even permanent skin discoloration (presumably this is uncommon). Symptoms may persist up to two weeks. Generally, the symptoms are limited to the skin, but occasionally there can be systemic reactions. The venom can be allergenic and there is a case of a near fatality from anaphylaxis.

Other species of jellyfishes tend to have stings normally considered harmless to mild, but on occasion, can cause distress. For example, *Cassiopea* can exude a slime that will cause itching in swimmers. The Moon Jellyfish was featured in a case study in the Gulf of Mexico wherein the victim had symptoms of pain, itching, ulceration, crusting, and regional lymph gland pain that lasted for days.

Remarks: Jellyfish are an important food in Asia, and Cannonball Jellyfish are harvested commercially in the Gulf of Mexico. The jellyfishes are dried and have 4–5% protein. This animal is rather spherical, resembling a cannonball or head of cabbage, leading to the common names.

References and Resources: Bayha et al. (2017), Birsa et al. (2010), Boulware (2006), Burnett and Calton (1974, 1977, 1987), Burnett et al. (1983), Calder (2009), Calton et al. (1973), Cobbs et al. (1983), Fenner (2005), Forrester (2006), Gershwin (2016), Gershwin and Collins (2002), Graham (2001), Halstead (1971, 1978), Hsieh et al. (2001), Ishikawa et al. (2004), Long and Burnett (1989), Long-Rowe and Burnett (1994), Mariottini et al. (2008), McClain et al. (2015), Mills (2001), Morandini and Marques (2010), Purcell (2012), Purcell et al. (2007), Radwan et al. (2000), Robinson and Graham (2013), Ruppert et al. (2004), Segura-Puertes et al. (2009), Tibballs et al. (2011), Togias et al. (1985), Walker (1988), Wilcox and Yanagihara (2016), Williamson and Burnett (1995).

Box Jellyfishes (Class Cubozoa)

"But as Angel headed back toward shore, she encountered a swarm of box jellies. Her neck, arms, and legs were stung by their thin tentacles. Intense, burning pain coursed through her body as she struggled to swim away from the stinging horde. As she swam, it became harder and harder for her to breathe as fluid filled her lungs, and she was wheezing and gasping for air with every stroke." (Wilcox 2016)

Classification: Phylum Cnidaria (cnidarians); Class Cubozoa (box jellyfishes or sea wasps).

The cubozoans are represented by about 40–50 species that are challenging to assess taxonomically and geographically, although strides have been made in recent years to understand these enigmatic beasts. According to the World Register of Marine Species (WoRMS), there are two orders, Chirodropida (multitentacled box jellyfishes) and Carybdeida (carybdeids). The chirodropids are a smaller group, represented by three families, but in the northwestern Atlantic, only one family, Chiropsalmidae (chiropsalmid box jellyfishes), is represented, with the single species *Chiropsalmus quadrumanus* (Four-handed Box Jellyfish). Taxonomy of the carybdeids is more complex, and some issues still need to be worked out. The WoRMS recognizes five families, four of which have members in U.S. waters: Alatinidae (*Alatina alata*), Carybdeidae (*Carybdea arborifera, C. confusa, C. marsupialis*), Tamoyidae (*Tamoya* cf. *haplonema*), and Tripedaliidae (*Copula sivickisi, Tripedalia cystophora*).

Identification: Superficially, cubozoan medusae are similar to scyphozoans, being gelatinous marine animals with a bell and tentacles armed with nematocysts. However, they are structurally different; the bell is roughly box-shaped and they are usually clear to translucent. Inside the bell of a cubozoan is a mouth-like manubrium that leads to the stomach. At the base of the bell is a ring of tissue called the velarium. The velarium helps with swimming speed and efficiency by increasing water pressure when the animal contracts. At the lower, outer margin of the bell are four sets of invaginations, called rhopalial niches. Inside these are the rhopalia, structures that bear statocysts for balance, and well-developed eyes. The true nature of the eyes is not completely known, but they see well enough to avoid predation and navigate among structures. The tentacle-bearing structures at the base of the bell are called pedalia. There are several morphological and life-history traits used to separate the different taxa, so interested readers should consult the primary literature.

The bell of the Four-handed Box Jellyfish is about 14 cm in width and height, with tentacles that can reach 3 m. The bell of *A. alata* is usually up to about 8 cm high and about half as wide. *Tamoya haplonema* is similar. The other box jellyfishes are small. *Carybdea confusa* has a bell that is about 3 cm wide and 4 cm high. *Carybdea marsupialis* is only about 2.5 cm in bell diameter and *Carybdea arborifera* reaches only about 4 cm in bell width. The tiniest are *T. cystophora* and *C. sivickisi*, which are only about a cm in adult bell width.

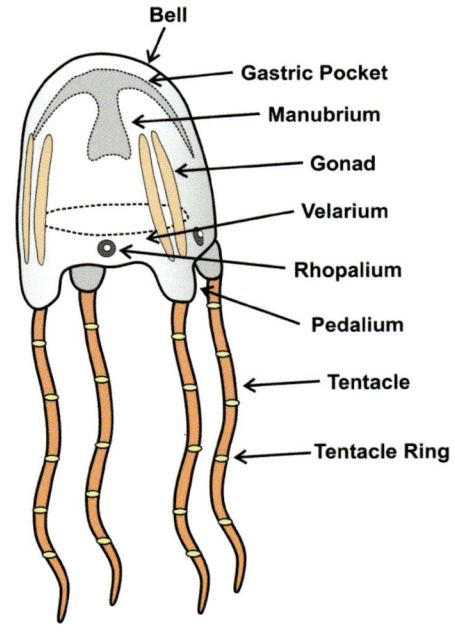

External anatomy of a cubozoan. L.L.C. Jones

Distribution: The Tamoyidae are endemic to the Atlantic, including the Gulf of Mexico. Tripedaliidae occur in all three oceans. Alatinidae and Carybdeidae also seem to occur in all three oceans, but further work is needed on members of these families to determine relationships and refine distribution maps, and there are apparently accidental translocations.

The Four-handed Box Jellyfish is found in tropical and subtropical waters of the Atlantic Ocean, including the Gulf of Mexico and Caribbean Sea. It has been reported off Hawaii, but this record is likely in error. A related form (*Chiropsalmus alipes*) is known from the west coast of Mexico, but to my knowledge has not been reported from California. The cubozoans of southern California have been taxonomically problematic until recently, when it was determined that the eight reported species belonged to a single recently described species, *C. confusa*. This species has mostly been documented from the vicinity of La Jolla and Santa Barbara. *Alatina alata* is

Aquatic Invertebrates 571

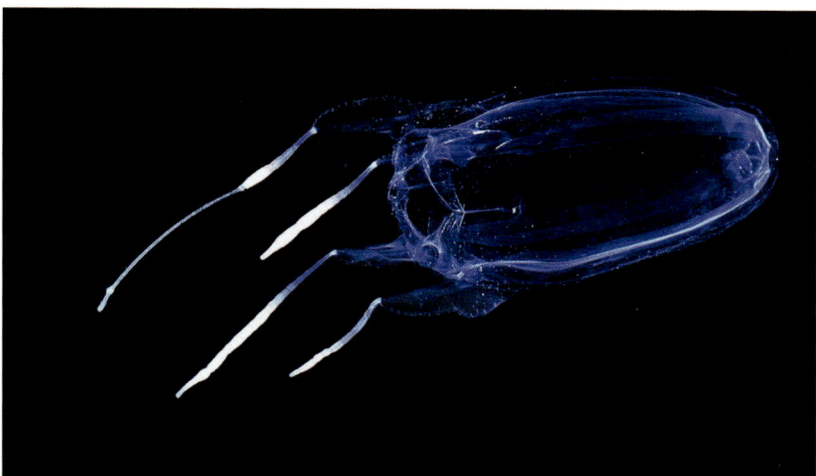

Alatina alata, sometimes called the Hawaiian Box Jellyfish, has a circumtropical distribution. This species has occasionally been documented as causing Irukandji syndrome. *Allen G. Collins*

commonly called the Hawaiian Box Jellyfish in that area, although it is now known to have a nearly circumtropical distribution. It also occurs along the northwestern Atlantic, Caribbean, and the Gulf of Mexico. Even though it was known to occur off Hawaii since the 1800s, it was considered rare. However, in the late 1980s, swarms began showing up with some regularity off Waikiki and other south-shore, lee-side beaches. Other Hawaiian species include *C. arborifera* (an endemic), *C. sivickisi*, and *T. cystophora*. The latter was likely an accidental introduction, and it is not known if it has become established there. *Copula sivickisi* has also been reported from the Bahamas. *Tripedalia cystophora* is apparently native to the Caribbean, and has been reported from both coasts of Florida. *Tamoya haplonema* (or a related, undescribed species) is found in the Gulf of Mexico and southern U.S. Atlantic, into the Caribbean. *Carybdea marsupialis* also occurs in the Gulf of Mexico. The distribution of many species is imprecisely known, hampered by taxonomic issues and translocations.

Natural History: All box jellyfishes live in estuarine or marine waters, but their specific habitat affinities vary by species. They may occur inshore or offshore, and be pelagic, or stay near the bottom. At least some species migrate, as they are strong swimmers, unlike the true jellyfishes and pelagic hydroids that are more at the mercy of the currents or wind. Relatively little is known about the largest species, *C. quadrumanus*, but there was a report of high densities in Matagorda Bay, Texas, in the summers of 1955 and 1956, although according to local fishermen they were known to have been present in good numbers for several years prior to that. In the bay, most were found near muddy bottoms. They were present when bay salinity was high (e.g., during the preceding drought), but disappeared when salinity got low from rain and stream runoff. Similar observations were made from trawls in estuaries in Mississippi. In both cases, Four-handed Box Jellyfishes were found near the bottom. They were found to be the second-most abundant macroplankton in a study of Georgia estuaries during the summer. They appeared to overwinter in those estuaries. They also occur in open waters, as evidenced by reports from other locations and observations off Galveston Island and other Texas barrier islands. There is no reason to believe they are not a regular visitor to the southeastern United States, although they are not always present in swimming areas.

The Hawaiian Box Jellyfish is better known than other U.S. cubozoans. It is a common hazard for beachgoers in Hawaii, so has become a target of research. As with other alatinids, it is a pelagic species. It is usually present near the continental shelf, but it will migrate inshore to reproduce over coral reefs. They occur inshore at night on the 8–12 days following a full moon. In some instances, when the tide recedes, they are left in the shallows, often washing up on the beaches. These occurrences are quite predictable in relation to lunar and tidal cycles, although densities vary tremendously on a monthly and annual basis. Other factors that may influence populations and inshore densities include zooplankton biomass, wind, waves, currents, and topography.

There have been some studies on the habitats used by some of the smaller cubozoans. In southern California, for example, *C. confusa* is often found near the bottom in kelp beds. Most sightings occur from August to November, when ocean temperatures are warm. *Tripedalia cystophora* is sometimes called the Mangrove Box Jellyfish, as it has been found nearshore among mangroves in Florida, the Caribbean, and Hawaii. *Copula sivickisi* is often associated with eelgrass, rocks, and coral, while *C. arborifera* is primarily a coral reef inhabitant in Hawaii. *Carybdea marsupialis* is often associated with sandy nearshore areas of the Gulf of Mexico.

The life history of cubozoans is complex and varies by taxa. The medusae can be either sex and are responsible for sexual reproduction. Some taxa (*Copula* and perhaps *Tripedalia*) are known to court, then pair up, and

Tamoya haplonema, a western Atlantic cubozoan. *Joseph Ryan/CC-SA/University of Florida (UF)*

the male passes a spermatophore to the female. In general, however, males broadcast sperm into the water and females pick them up. Except for chirodropids, fertilization is internal, then the embryos are released. As with other cnidarians, the larval forms are known as planulae. The planula settles on the seabed and metamorphoses into a polyp (= cubopolyp), where it feeds on plankton with its tentacles. Polyps may move around and can bud into other polyps via asexual reproduction or may settle and undergo another metamorphosis. After budding, the cubopolyp may creep around the seabed to find an adequate attachment site. In at least some species, such as *A. alatina*, polyps can form cysts (= podocysts), then continue to develop when conditions are favorable. Some are even known to "hatch" planulae. During final metamorphosis, the polyp tentacles are resorbed and new tentacles are formed for the medusa (= cubomedusa) stage. Unlike scyphozoans, there is no strobilation, as only a single polyp produces a single medusa. The late-stage polyp detaches from the substrate and the young, free-swimming medusa begins life in the water column. There are variations on the life history traits, depending on the taxa.

Box jellyfishes are strong swimmers that can target prey more actively than other cnidarians. They feed on small creatures such as crustaceans and small fishes, and in turn, are fed upon by animals that eat gelatinous marine organisms (see account for True Jellyfishes).

Encounters: In some areas, box jellyfishes can be abundant and are of significant concern when there are blooms or migrations. They generally head to deep water during the day, then come to the surface at night, but it is wise to expect them during the day, as well. Some swimming beaches are closely monitored. The Hawaiian Box Jellyfish along Oahu's south shores has a fairly predictable pattern of spawning migration. Dates of occurrence are posted online to forewarn swimmers, snorkelers, divers, and surfers of when they may be present. Researchers from the University of Hawaii collect beached box jellyfishes during peak occurrence times and have averaged about 270 per month, although during large influxes, they have collected nearly 2,000 in a single day. There is good information on cnidarian sting frequency from Oahu recorded by lifeguards. Although they do not necessarily record the cnidarian culprit, estimates can be made for those belonging to *A. alatina* vs. *Physalia utriculus*, presumably based on relative occurrence by beach and other factors. It is estimated that 90% of the stings on Waikiki, Ala Moana, and Waianae Beaches are due to Hawaiian Box Jellyfish. Stings at Hanauma Bay are about 50% box jellies and 50% bluebottles. On the windward side of Oahu, almost all stings are due to *Physalia*. As an example of an annual count, in 1994 there were about 800 box jellyfish stings recorded on seven Hawaii beaches; this number is probably not far from an annual average, but there is much variation. Of these 800 stings, only about 20–30 required medical care beyond first aid. It has been estimated that *Physalia* and *Alatina* together account for up to about 6,500 stings per year in Hawaii.

In the western Atlantic and Gulf of Mexico, box jellyfishes are a periodic risk, but it is hard to tease out statistics for stings caused by cubozoans among other cnidarians. However, Four-handed Box Jellyfish can be abundant during blooms, when washed into swimming areas. Because of the potential severity of stings, they pose a health risk to swimmers and divers when nearshore populations are high. For example, in 2003 a bloom of *C. quadrumanus* in Florida caused thousands of stings. In 1970, large numbers of *C. quadrumanus* stung bathers near Galveston. This species may be interspersed among scyphozoans and Portuguese Man-of-War. Hundreds of "jellyfish" stings are reported to Texas poison control centers annually. The calls are usually from swimmers near South Padre Island, Galveston, and Corpus Christi. In some published accounts of stings reported to poison control centers, cubozoans may be lumped with other cnidarians and other marine or even freshwater animals, including stingrays and catfishes. To my knowledge, there are never box jellyfish problems in southern California, but *C. confusa* can sting.

Four-handed Box Jellyfish of the western Atlantic and Gulf of Mexico.
Joseph Ryan/CC-SA/UF

Popular swimming beaches usually have warning signs and flags for dangerous marine life, but these vary by region and are not standardized. In the Southeast, the flags are usually purple or blue, while in Hawaii, there is usually a sign specifying "jellyfish" for all types of cnidarians. California has color codes that represent levels of safety, but not specific for marine life. One should never rely on warning flags, anyway; instead, check online, at lifeguard stations, or talk to lifeguards. Box jellyfishes are difficult to see, being almost transparent and relatively small. Because these are tropical to subtropical animals, they are mostly found along the coasts in the late spring through early fall in the Southeast, but year-round in Hawaii.

Venom Apparatus, Venom, and Symptoms: Box Jellyfishes use their venom for both predation and defense. They have cnidae (specialized stinging cells) in their tentacles, and venom is delivered by nematocysts. The nematocysts of some species have been extensively studied and are distinctive, which aids in identification of the culprit in human envenomation. This was done in the case of the fatal Galveston envenomation by *C. quadrumanus*. As a group, box jellyfishes are the most dangerously venomous cnidarians on Earth and rank highly among all marine animals. Some species are reputed to deliver a relatively mild sting, while others, especially the Australian "stingers" (e.g., *Chironex fleckeri* and species in the family Carukiidae) can deliver a potentially lethal sting. *Chironex fleckeri* of the southwestern Pacific is the deadliest known cnidarian, as it attains a large size and has extremely virulent venom, with an LD_{50} estimate of 0.011 mg/kg (intravenous). In the United States, *C. quadrumanus* and *A. alata* are the most commonly encountered cubozoans that have been implicated in moderate to severe stings. There is some evidence that *Tamoya* also has a fairly potent sting. It is possible that some of the other species may deliver painful, or even dangerous, stings.

Much of what we know about box jellyfish venom comes from studies of the Australian "stingers," including *Chironex* and *Chiropsella*. The Four-handed Box Jellyfish is also a potentially dangerous species. The venom of box jellyfishes across the spectrum are variously said to be neurotoxic, hemotoxic, dermonecrotic, cytotoxic, cardiotoxic, and myotoxic; these terms are not mutually exclusive, but this demonstrates how the venom from a stinging incident can affect a host of organ systems. The toxic components of the venom of *C. quadrumanus* and *A. alata* are not completely known, but amino acids, peptides, proteins, and enzymes contribute to the venom's action. Venom of species in our area is probably similar to other box jellyfishes, having compounds that may include histamine or histamine-releasing agents, 5-hydroxytryptamine, catecholamines, quaternary ammonium compounds, proteases, phospholipases, cytolytic enzymes, and hemolytic enzymes. The potent hemolytic venom protein of the Hawaiian Box Jellyfish has been dubbed CAH1 (for *Carybdea alata* hemolysin 1). The venom of carybdeid, *C. marsupialis*, possesses a novel neurotoxic peptide and cytolytic enzymes. *Carybdea rastonii* venom has also been studied, as this species is a major concern for Japanese beachgoers (related forms occur in the United States). The venom contains toxins that were labeled CrTX-A (for *C. rastonii* Toxin A) and CrTX-B. Of these, CrTX-A was found to be present in the nematocysts. It is highly toxic to both crustaceans and mammals, with an LD_{50} of 0.02 mg/kg. Even extremely small doses (0.0001 mg/kg) caused dermatitis in the lab mice. This proteinaceous material was found to be responsible for the derma-

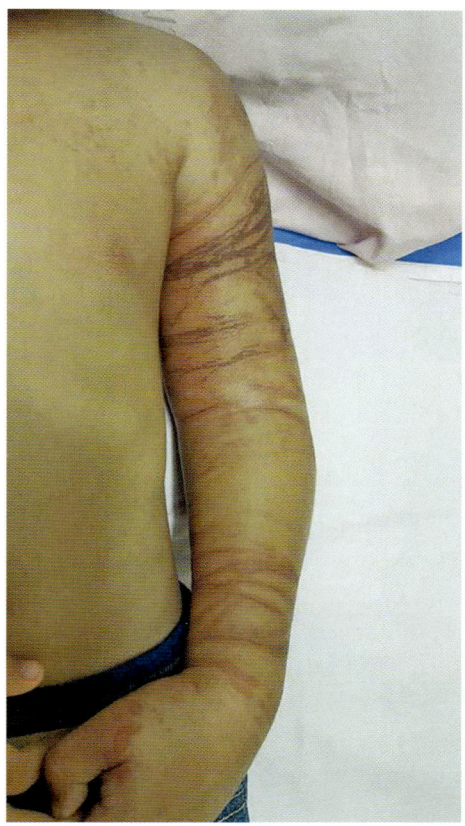
Lesions caused by a cubozoan. *Angel Yanagihara*

titis caused in humans. Note that naming venom components after scientific names is problematic and not intuitive, because taxonomy and nomenclature of the source animal is subject to change, as in all three of the above examples.

The Four-handed Box Jellyfish is generally considered the most dangerous cnidarian in the Atlantic and Gulf of Mexico. Symptoms of stings begin with immediate and intense pain. This is evident at a beach when someone is stung by *Chiropsalmus*, because screaming usually ensues. However, stings by *Physalia* are also extremely painful and these species can co-occur. It has been reported that children will peel the tentacles off the skin, discharging more nematocysts, while adults tend to flee the water, which increases spread of the venom. Other symptoms include pronounced linear, often-raised wheals from the tentacles, discolored skin, blisters, necrosis, and significant swelling. The wheals are similar to those produced by the Portuguese Man-of-War, appearing as if the victim was whipped. However, *Physalia* stings tend to be less severe and have a "beaded" appearance. Cubozoan stings are not beaded, and there may be more extensive contact dermatitis. Wounds also take longer to heal in *Chiropsalmus*. A sting from a cubozoan (depending on species) is often more painful than that of *Physalia* or other cnidarians. The person may rarely be scarred for life by the dermal assault, or it may take months to heal. The wounds can worsen in the hours following exposure and cause severe dermonecrosis and blister formation. The swelling can lead to compartment syndrome. In a study of stings by *C. quadrumanus* in Brazil, 100% of the confirmed stings were extremely painful and resulted in systemic reactions. As with true jellyfishes (but not with *Physalia*), there is usually a rise and then fall in blood pressure. Other symptoms can include nausea, vomiting, fever, lymph node swelling, muscle cramps, weakness, reduced heart rate or palpitations, breathing problems, shock, and (rarely) death.

Stings are rarely fatal from *C. quadrumanus*, but there was one death attributed to this species in the United States. In 1990, off the beach near Galveston, a 4-year-11-month-old boy was stung while swimming. He began screaming, so his mother rushed to his aid and took him out of the water. It was apparent his left arm had been wrapped in tentacles. He was already in grave danger during emergency transport. Twenty minutes post-exposure he was groggy and cyanotic and went into shock. Resuscitation efforts began during transport. He arrived at the hospital thirty-nine minutes later, where the staff tried to revive him, unsuccessfully, for the next forty minutes. Initially, the physicians were not certain which cnidarian killed the boy, because the wounds are similar to Portuguese Man-of-War, but nematocysts left at the wound site confirmed it was *C. quadrumanus*.

Stings from *A. alata* are similar, but not generally as severe. They usually cause immediate and often intense pain, hives and skin rash, and local swelling. The pain usually lasts 2–3 hours with mild to moderate envenomation. Severe stings can also result in muscle weakness, paresthesia, chills, fever, malaise, irregular heartbeat, high or low blood pressure, abdominal pain, nausea, vomiting, difficulties in breathing, and shock. Symptoms can also include those attributable to allergy, treatment, and (later) infection. Some stings cause persistent dermatitis, possibly due to allergy. Lymph glands may also be involved, but this may be an indicator of infection. On rare occasions, there may be nerve damage, and the area may be numb for months. Stings to the eyes can be distressing, resulting in pain, blurred vision, irritation, swelling, and tearing. According to a report of 113 cnidarian stings in Hawaii (most likely *A. alata*), there is always pain, often intense, and/or itching. Of these stings, three had systemic neurological involvement, seven had mild systemic symptoms, eleven had anaphylaxis or anaphylactoid syndrome, twenty-one had persistent or delayed cutaneous syndrome, and six had symptoms consistent with Irukandji syndrome.

In 1952, Hugo Flecker described "Irukandji syndrome," named after the Irukandji tribespeople in northeastern Australia. People there sometimes developed a particular set of symptoms defining the syndrome, but the cause was not known at the time. Then in 1964, one culprit was positively identified as

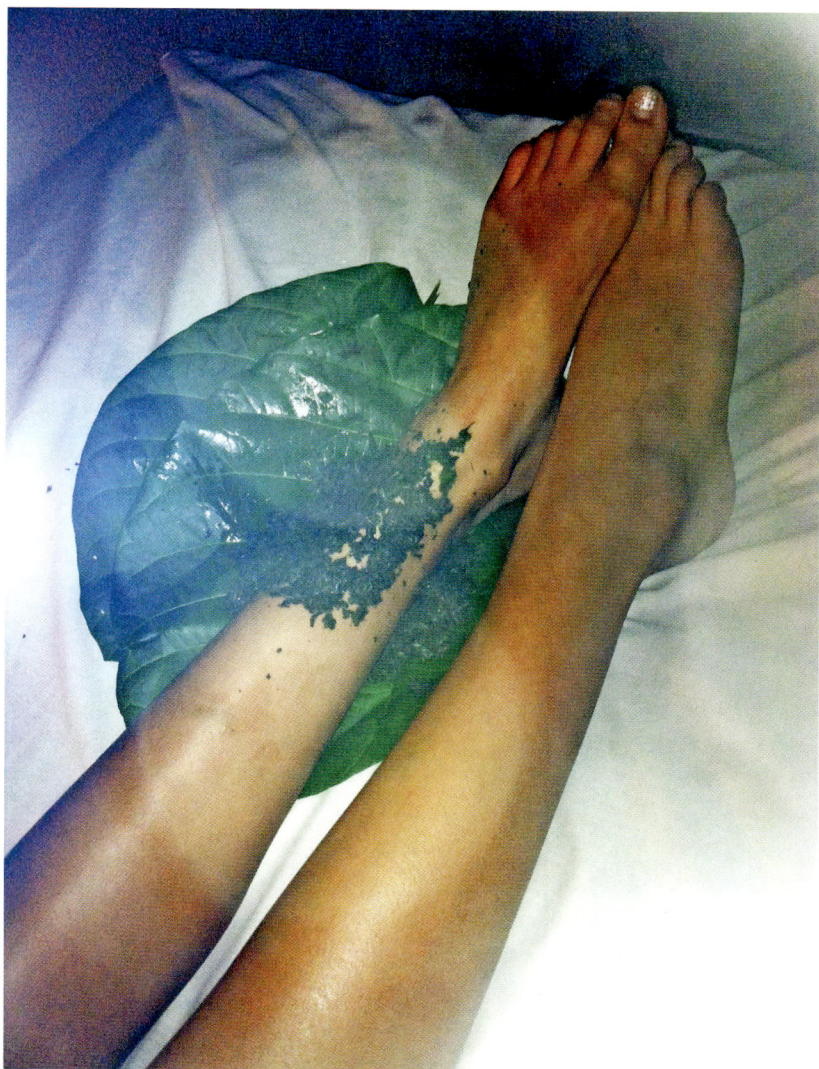

An herbal poultice was applied by a Micronesian shaman to this person stung by a cubozoan. *Giacomo Bernardi*

Carukia barnesi, a diminutive box jellyfish that later bore the common name of Irukandji Jellyfish, and the family Carukiidae was dubbed the Irukandji Box Jellyfish family. Since then, several species of box jellyfishes have been reported to cause the syndrome, or a similar suite of symptoms. At least nine species have been documented as causing this set of symptoms, including *A.*

alata. In 2003, three divers off Key West, Florida, were reported to have had symptoms consistent with Irukandji syndrome. The culprit was not known, but it was suggested to be a carybdeid, which at the time may have included *Alatina*. Several cases of Irukandji syndrome have been attributed to *A. alata* off Hawaii, although the majority of envenomations just show the more classic symptoms. Symptoms of Irukandji syndrome are different than those of classic cubozoan envenomation, beginning with a lack of immediate pain and lack of large linear wheals. However, within five minutes to two hours (usually 20–60 minutes), symptoms arise. These include severe headache, backache, muscle pain, chest and abdominal pain, nausea, vomiting, sweating, anxiety, high blood pressure, rapid heart rate, fluid in the lungs, and a feeling of doom. Irukandji syndrome is not often fatal, but it can be, so it constitutes a medical emergency. Two deaths are attributable to Irukandji syndrome from cubozoans in Australia, but there are probably more, because the culprits are small and colorless, and the lack of immediate pain does not implicate a source. In a study of 116 cases in Australia, only one was fatal, but 64% required hospital care. Worldwide, incidences of Irukandji syndrome are poorly reported.

References and Resources: Baxter and Marr (1974), Bengston et al. (1991), Bentlage and Lewis (2012), Bentlage et al. (2010), Birsa et al. (2010), Boulware (2006), Brinkman and Burnell (2009), Burnett and Calton (1987), Calder (2009), Calder and Peters (1975), Chung et al. (2001), Fenner (2005), Fenner and Hadok (2002), Gershwin (2005, 2006, 2016), Grady and Burnett (2003), Guest (1959), Haddad et al. (2002), Halstead (1971, 1978), Kimball et al. (2004), Kingfords and Mooney (2014), Lawley et al. (2016), Lewis et al. (2013), Little et al. (2006), Nagai et al. (2000), Ohtaki et al. (1990), Ramasamy et al. (2003, 2005), Rottini et al. (1995), Ruppert et al. (2004), Sanchez-Rodriguez et al. (2006), Segura-Puertas et al. (2009), Straehler-Pohl et al. (2017), Tamanaha and Izumi (1996), Tibballs et al. (2011), Wilcox and Yanagihara (2016), Williamson and Burnett (1995), Yanagihara et al. (2002, 2016a, b), Yoshimoto and Yanagihara (2002).

Clinging Jellyfish (*Gonionemus sp.*)

"The clinging jelly in Russian waters packs a powerful punch, with documented mass stinging events. In 1966 a reported 1,000 people were stung one day in June. Symptoms reported can vary and in some cases be severe—such as extreme pain, respiratory difficulty and feelings of paralysis—but the stings are not fatal." (*The Falmouth Enterprise*, 30 May 2017, A. F. Carter reporting)

The stinging form of the Clinging Jellyfish has invaded New England.
Annette F. Govindarajan

Classification: Phylum Cnidaria (cnidarians); Class Hydrozoa (hydrozoans); Order Limnomedusae; Family Olindiidae (= Olindiasidae); Genus and Species *Gonionemus* sp. or *G.* c.f. *vertens* (Clinging Jellyfish).

A few species have been described, but in North America all are generally synonymized with *G. vertens*. However, recent research suggests this taxon probably represents a species complex, and those in our area are probably invasive from more than one introduced population. The stinging varieties are taxonomically distinctive from nonstinging forms, and represent either unnamed or previously named forms. Genetic studies suggest that the northwestern Atlantic clinging jellyfishes have multiple origins and at least some are probably hybrids.

Identification: These are small cnidarians. The bell of a medusa is up to about 2.5 cm in diameter and height. The bell is clear or translucent, except for four radiating canals, which have orange or brownish gonads ventrally, and some other small brownish body parts. Viewed from above, it appears to have an "X" across the bell. It is sometimes called the orange-striped jellyfish because of the distinctive appearance of radiating gonads. Evenly spaced around the margin of the bell are 60–90 tentacles, which bear the stinging cells. There are also adhesive knobs on the tentacles. The polyp of *Gonionemus* is minute and difficult to detect.

Distribution: The natural range of *Gonionemus* sp. is somewhat a mystery, but it is being unraveled. Members of this genus are known from both sides of the Pacific and Atlantic, and in both northern and southern hemispheres. It is likely populations have been moved around in ship ballast, through oyster farming, or by other means. *Gonionemus* cf. *vertens* from North America occurs in the eastern Pacific from southern California to Alaska and into the Arctic Ocean. It is rare south of the Puget Sound, although there have been blooms at least as far south as Santa Barbara, California. In the western Pacific, it is infamous in Russian and Japanese waters, where stings are common. In the western Atlantic, it is primarily known from New England. It was known to occur off Cape Cod as early as 1894, but then "disappeared" with an eelgrass die-off in the 1960s. They started reappearing in the 1990s. Most of the sightings were in the Cape Cod area. Significant human envenomation started about this time in the Northeast, and it has been suggested these were from a more recent invasion from a northwestern Pacific stinging population. *Gonionemus* is currently making headlines as people are being stung in New England and down to New Jersey. Researchers at Woods Hole Oceanographic Institution, among others, are trying to unravel the mysteries of its taxonomy, geographic origins, and sporadic appearances.

Natural History: Clinging Jellyfish occur primarily in eelgrass beds in shallow marine and estuarine waters, but they may also be associated with macroalgae in some areas. Their name derives from their ability to attach to substrates with the adhesive pads on the tentacles. They are also capable of free-swimming by undulation of the bell. They can apparently feed while either swimming or being attached, when small animals come into contact with the tentacles. They are known to feed on copepods and other small planktonic marine animals.

Clinging Jellyfish have both a medusa and a polyp stage. As a medusa, they reproduce sexually. Eggs develop into planula larvae, which are free-swimming. These then settle on substrate to become a miniscule polyp. In the polyp stage, they reproduce asexually via budding. Polyps may produce vermiform frustules, medusae, or cysts. The cysts may then produce more frustules, or may metamorphose into medusae. Frustules are a small vermiform life stage that can creep around the substrate looking for an attachment site. Polyps also have tentacles with stinging cells for feeding and defense. Asexual reproduction in polyps can last for years and medusae may not appear until conditions are conducive to metamorphosis. The polyps are so small they may go unnoticed, so the lack of medusae does not mean the lack of the organism.

Predation is not well known, but an interesting study off Cape Cod showed that some species of crabs readily fed on them, while others did not, and still others did so selectively.

Encounters: In the United States, encounters are either infrequent or rarely noticed—you probably wouldn't notice the small, mostly transparent medusa if you weren't stung. To date, there are no records of human envenomation from the northeastern Pacific, so even if encountered, it is moot. In the Northeast, where some individuals or populations may sting people, they aren't usually encountered in swimming areas, such as open beaches. One might reasonably expect to encounter them in eelgrass beds and estuaries. They can get caught up in nets and other equipment used in shallow water. They may be present around boat docks in stillwater coastal bays.

Venom Apparatus, Venom, and Symptoms: All Clinging Jellyfish have a stinging mechanism to procure prey, but prey is small invertebrate plankton, so it is unclear why some populations or taxa have a venom that is extremely virulent to humans. However, those in the northwestern Pacific are notorious for stinging humans, so it may be a defensive ploy for those populations.

The nematocysts occur in rings on the tentacles. When an animal contacts the nematocyst, it fires a hypodermic tubule into the victim and injects venom; the more nematocyst rings that come into contact with the victim, the more venom injected.

I can find no information on specific properties of the venom, but stings may be fairly serious, and neurotoxic elements are indicated. Much of what we know is from the numerous stings reported from the Sea of Japan and the Russian Coast, but symptoms of some stings in New England are similar. There are local reactions to the sting, but there may also be profound systemic effects. The signature symptoms include immediate burning, with redness and an itchy wheal forming in about 10–15 minutes. This can be followed by excitement, restlessness, delirium, and hallucinations; weakness; chest tightness, labored breathing, and a dry cough; muscle spasms and cramping; sharp muscle and joint pain, especially in the back and legs; profuse sweating; and liver problems. The victim may be so weak they cannot stand or move. Psychological maladies, including a feeling of doom, may ensue. In New England, one person described the pain as being similar to being stabbed by a knife. Another person stung in New Jersey thought he was going to die. A researcher studying New England *Gonionemus* in Cape Cod was stung on the face and said it was like having multiple hypodermic needles in the lip. A media blitz followed some of the stings and at least one person was hospitalized and given morphine for the pain. The effects completely disappear in a few days. I know of no records of anyone being stung in the eastern Pacific and there are no records of fatalities worldwide.

References and Resources: Aznaurian (1964), Carman et al. (2017), Govindarajan and Carmen (2016), Govindarajan et al. (2017), Pigulevsky and Michaleff (1969), Singla (1977), Watson and Govindarajan (2017), Westfall (1970).

Portuguese Man-of-War (*Physalia physalis*) and Pacific Bluebottle (*P. utriculus*)

"Mon, 25 May, 2015, 20:31:08 GMT—Swimmers have reported about 475 jellyfish stings Saturday and Sunday on South Padre Island. The Cameron County Beach Patrol treated about 475 jellyfish and Portuguese man o' war stings during the past two days . . . up from just 200 last week. Beach Patrol put up blue and purple flags to warn swimmers about the jellyfish." (*Valley Central News*, T. Huertas reporting)

Classification: Phylum Cnidaria (anemones and jellyfishes); Class Hydrozoa (hydrozoans); Order Siphonophorae (siphonophores); Family Physaliidae (physaliids); Genus *Physalia* (men-of-war); Species *physalis* (Portuguese Man-of-War) and/or *P. utriculus* (Pacific Bluebottle).

Physalia physalis, Gulf of Mexico. The tentacles were likely eaten by predators or broken off in the surf. *L.L.C. Jones*

The family is monotypic and these are the only two species sometimes recognized, although *P. utriculus* is often regarded as a junior synonym of *P. physalis*. The common name of *P. physalis* is also spelled man-o-war, man-o'-war, and man o' war, alluding to its superficial resemblance to a sailing warship. The name "bluebottle" is commonly used, especially in Australia and Hawaii.

A hydrozoan that is sometimes misidentified by beachgoers as a man-of-war is the By-the-Wind Sailor, *V. velella*. This animal is harmless and easily distinguished from *Physalia*.

Identification: Like other hydrozoans and siphonophores, *Physalia* represents an unusual life-form. It is not actually a single animal, but instead consists of different colonies of multicellular individuals known as zooids. However, the zooids are physiologically connected and cannot live independently, without the other types of zooids.

In outward appearance, the mature form of *Physalia* looks similar to a jellyfish, having a broad upper area and tentacles dangling below in the water. Unlike jellyfishes, they do not have a compressed bell. The upper part of *Physalia*, which floats above the surface, is called the pneumatophore. It is a bluish gas-filled bladder that keeps the colony afloat (hence the common name, bluebottle). Dorsally, the pneumatophore has a thin upper extension called a crest, which functions as a sail; *Physalia* is incapable of moving under its own volition. The pneumatophore can be beautiful; it is often tinged with pink or purple. It is about 10–30 cm long in *P. physalis*, and about half as wide. *Physalia utriculus* is smaller, with a usual pneumatophore size of 5–8 cm and a maximum size of only 15 cm. There are three types of zooids that make up the tentacles. The dactylozooids are used for defense and procuring prey. The gastrozooids mostly function to

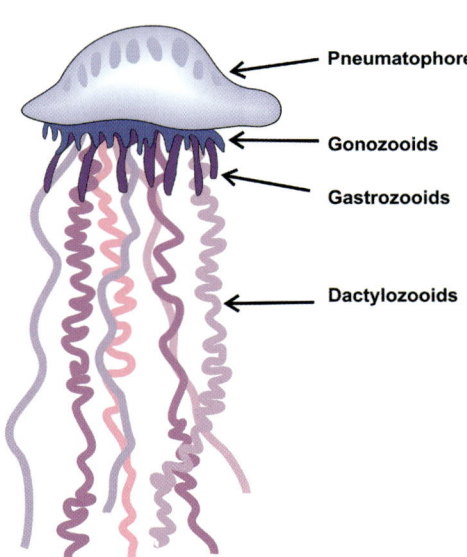

External anatomy of a physaliid. *L.L.C. Jones*

This photograph shows how long the tentacles are and how far from the pneumatophore they can trail. *"Jolt to Horizontal"/CCA/Flickr*

assimilate food. The gonozooids are used for reproduction. The dactylozooid tentacles are coiled and can extend to about 10 m in length in *P. physalis*, but are sometimes reported to be 50 m or more. The tentacles of *P. utriculus* may reach about 12 m. Exact measurements are difficult to assess on gelatinous animals, so reports may be exaggerated.

The By-the-Wind Sailor is sometimes confused with *Physalia*, but it is a small colonial hydrozoan reaching only about 7 cm long. It is has a flat, dark-blue body, which floats on the surface and has a colorless "sail." Although some people may get itchy from contact with the nematocysts, they don't seem to cause any real harm in humans. Sometimes thousands of these are seen at sea or washed up on shore.

Distribution: *Physalia* are primarily tropical to subtropical animals that sometimes venture farther north or south of the Tropics. The Portuguese Man-of-War is found in the Atlantic, while the Pacific Bluebottle is found in the Indian and Pacific Oceans. *Physalia physalis* is a well-known drifter from Texas to Florida, but it occasionally drifts as far north as Nova Scotia, especially after big storms. To my knowledge, there are no confirmed reports of *P. utriculus* from southern California, although there are questionable sightings. However, with climate change and increased El Niño events, never say never. This species does occur in the eastern Pacific and Gulf of California on both sides of the Baja California Peninsula.

Natural History: *Physalia* are completely marine. They are pleuston, drifting around warm oceans at the mercy of the wind, currents, tides, and waves. The crest of the pneumatophore functions as a sail to aid in movement. As it drifts, small fishes known as Portuguese Man-of-War Fish or Shepherd

Pacific Bluebottle, *P. utriculus*. Kyle Hovey/CCA-EQ/Flickr

Fish, *Nomeus gronovii*, may accompany it. These small fish swim around the pneumatophore and nibble on the zooids, but they typically avoid contact with the dactylozooid tentacles. The fish is somewhat resistant to the stings, but it behaviorally avoids the larger stinging tentacles. Other species of fishes may also swim among the tentacles, presumably to avoid predation.

As the pneumatophore floats on the ocean surface, the tentacles dangle below. It does not actively seek prey, so small fish and invertebrates must contact the tentacles to be captured. When prey is captured by the fishing tentacles, it is paralyzed by the nematocysts and moved up to the gastrozooids. Prey is then absorbed by digestive enzymes. Prey of *Physalia* includes many pelagic species, including flying fishes, mackerels, eel larvae, shrimps, chaetognaths, and cephalopods. In one study of *P. physalis* from the Gulf of Mexico, they fed primarily on soft-bodied prey, including planktonic larval fishes and invertebrates. Fish and fish larvae accounted for 94% of the prey, while cephalopods constituted the rest of the diet. However, there are regional differences that likely depend on prey availability.

Predators include the Loggerhead Sea Turtle, Ocean Sunfish, Portuguese-Man-of-War Fish, and Violet Snail. Other creatures that feed on *Physalia* include two other taxa covered in this book that use their tentacles or nematocysts for their own defense: *Glaucus* nudibranchs (see Nudibranch chapter), and blanket octopuses (see Octopuses and Squids chapter).

Physalia are hermaphroditic, having both male and female reproductive organs (gonophores) in each individual, in the gonozooids. Both eggs and sperm are released into the water where the eggs become fertilized. Because physaliids sometimes occur in large groups, up to 1,000 or more, there is genetic exchange between individuals. The eggs then hatch into planulae. The larva develops asexually into a colony by differentiation and budding.

Encounters: Off Florida and the Gulf of Mexico, *P. physalis* can be found at sea or along beaches any time of the year, although they often show up in winter. Wind, currents, and storms can drive large numbers into swimming areas. Huge numbers are sometimes observed on beaches. For example, in February and March 2012, February 2013, and May 2015, there were thousands of Portuguese Men-of-War reported by the media along the coast at South Padre Island. There were similar "invasions" off Florida swimming beaches.

The Pacific counterpart is commonly encountered in Hawaii. While Hawaiian Box Jellyfishes produce most cnidarian envenomation on the lee side of Oahu, stings from the windward side are almost exclusively from *P. utriculus*. It has been estimated that there are 6,500 stings per year from this species, as indicated from data collected on Oahu in 1994. Stings tallied from thirteen beaches showed that nearly 4,000 stings came from Waimanalo Beach alone, and over 1,300 from Bellows Beach (based on holiday and weekend surveys). For Californians venturing south of the border to Baja California, *P. utriculus* is sometimes encountered, especially along open waters or islands in the southern Gulf of California.

A good indicator of the presence of *Physalia* and other cnidarians is seeing them washed ashore. The "bluebottles" can sometimes be seen in the water, but remember that the tentacles may trail far behind or be broken off in the water. There may also be beach warning flags or other information resources for specific beaches (see Cubozoa chapter). They should never be handled or walked on when beached, as the nematocysts remain active. The pneumatophore does not contain stinging cells, but if one picks them up, they can accidentally fling the thread-like tentacles onto their skin. It has been reported that the cnidae may discharge up to a month after the colony has "died."

Not all encounters are during times when animals are washing into shore. There are countless incidences of SCUBA divers, swimmers, and snorkelers swimming into them. They also can get caught up in fishing nets or other equipment used in boating.

Venom Apparatus, Venom, and Symptoms: Men-of-war have a cnidarian stinging mechanism, the nematocyst. These organelles are along the tentacle (primarily the dactylozooid) and discharge when touched or when there is a certain chemical stimulus. *Physalia physalis* dactylozooids can have 750,000 nematocysts each. When discharging, they explosively fire a venom-delivering tubule into the prey or victim. The tentacles of *Physalia* often adhere to the flesh of victims because the stinging cells are well attached to the tentacle and the harpoons and barbs are attached to the flesh. Even upon normal contact, most cells do not discharge—an important consideration when one is stung and tentacles are clinging to them.

There are attributes of neurotoxin, cardiotoxin, myotoxin, and hemotoxin in the venom. Venom composition is fairly well studied in *P. physalis*. It is rather virulent to humans, probably because *Physalia* sometimes feed on vertebrates, and it is also used defensively against vertebrates. The LD_{50} of crude venom has been reported as 0.1 mg/kg (intravenous) to 37 mg/kg (intraperitoneal), based on three studies. In addition to lipids and peptides, it contains several enzymes, including adenosine mono- and tri-phosphatase, ribonuclease, deoxyribonuclease, phospholipase A and B, and a fibrinolysin. "Physalitoxin" is a hemotoxic fraction, and has an LD_{50} of 0.2 mg/kg. Histamine may or may not be released. The venom interferes with calcium ion and sodium ion transport across membranes.

The symptoms of a sting begin with immediate, severe, burning pain and piloerection. The venom causes the rapid formation of skin lesions. The wounds appear as red, linear, open wheals, which look as if the victim has been whipped. There may be a "beaded" look to the surface wheals. The appearance is similar to that of box jellies, so it may be necessary to actually see the assailant or examine the nematocysts under a microscope to know which animal delivered the

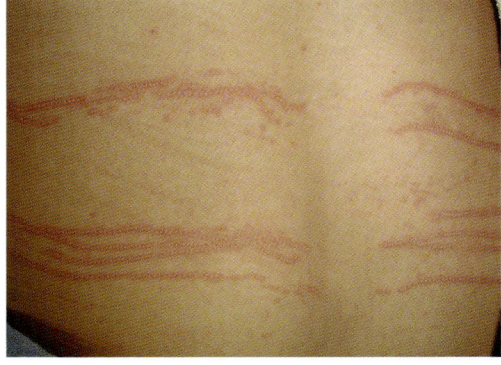

Lesions caused by a Portuguese Man-of-War. Vidal Haddad Jr.

venom. The pain often lasts for hours. The wheals may start to subside in an hour or so, but may last for 2–3 days or sometimes take much longer to heal. The venom can cause extensive dermatitis and dermonecrosis. There can occasionally be permanent scarring.

Typical stings of lower severity from *P. physalis* will not produce systemic reactions, but more severe stings will. In the cases of serious envenomation, symptoms can escalate rapidly. Severe envenomation can include lymph node involvement, muscle pain, spasms, respiratory distress, cardiac dysfunction, renal problems, and loss of consciousness. There are three records of people in the southeastern United States dying from stings inflicted by *P. physalis*, but there are other suspected deaths. Two victims were elderly people with some sort of medical history that may have been a factor in the reaction, but one victim was a 30-year-old SCUBA diver who ascended into the tentacles from below.

The smaller *P. utriculus* has not been implicated in fatal envenomations, but it can also cause serious reactions in humans. Ordinarily, stings from this species are painful, with a burning effect. There is usually a reddish area where the tentacles contacted the skin, but are not as bad as *P. physalis*. In most stings, the red marks disappear in 20–60 minutes. In more severe stings, there may be a more pronounced wound, having raised wheals with a rash and blister formation, as well as lymph node involvement. Sometimes there is a delayed dermatitis caused by the sting. Stings to the eyes can be distressing, and involve pain, itchiness, swelling, tearing, and blurred vision.

Allergies are not uncommon. Antigens from *Physalia* and *Chrysaora* can be cross-reactive, so it is possible that hypersensitivity can occur after being stung by one or both genera. Thus, allergic responses may play a role in some of the symptoms exhibited following recurring envenomation events.

References and Resources: Bardi and Marques (2007), Birsa et al. (2010), Boulware (2006), Burnett (2005), Burnett and Calton (1974, 1977, 1987), Burnett and Gable (1989), Burnett et al. (1983, 1994), Calton et al. (1973), Cormier and Hessinger (1980), Fenner (2005), Fernandez et al. (2011), Forrester (2006), Gershwin (2016), Halstead (1971, 1978), Jenkins (1983), Lane (1960), Månsson et al. (1984), Purcell (1984), Ruppert et al. (2004), Russo et al. (1983), Stein et al. (1989), Stillway and Lane (1971), Tamkun and Hessinger (1981), Tibballs et al. (2011), Walker (1988), Ward et al. (2012), Wilcox and Yanagihara (2016), Williamson and Burnett (1995).

Aquatic Insects (Especially Families Belostomatidae and Naucoridae)

"The painful bites of naucorids are well known, and members of the genus *Pelocoris* are especially noted for their ferocity (Uhler, 1884; Hungerford, 1927). Blatchley (1926) stated from personal experience that *P. femoratus* (Palisot de Beauvois) inflicts a wound more painful for a few seconds than that of a hornet. (Polhemus in Menke, 1979)

Lethocerus medius, one of the giant water bugs, or "toe-biters." L.L.C. Jones

Classification: Phylum Arthropoda (arthropods); Class Insecta (insects); Order Hemiptera (true bugs); Family Belostomatidae (giant water bugs) and Family Naucoridae (creeping water bugs). Some other aquatic insects have a venomous bite.

There are seven genera of belostomatids worldwide, and about sixty species. There are about twenty species north of Mexico, including four *Lethocerus*, one *Benacus*, nine *Belostoma*, and six *Abedus*. Belostomatids have various vernacular names including "toe-biter," "fish killer," and "electric light bug." Worldwide, there are about 400 species of naucorids belonging to forty genera.

There are about five genera and twenty species north of Mexico. Most species belong to the genera *Ambrysus* and *Pelocoris*. They are sometimes called "water bees," because of a bee-sting-like bite. Other families of aquatic true bugs that may have a painful bite include water scorpions (family Nepidae), marsh treaders (family Hydrometridae), backswimmers (family Notonectidae), toad bugs (family Gelastocoridae), and water boatmen (family Corixidae), but I can find little information on the effects of bites to humans. Similarly, there are also a number of aquatic beetles (order Coleoptera) that may give a bite, including predaceous diving beetles (family Dytiscidae), and several others, but there is little information on symptoms or effects on humans.

Identification: Giant water bugs are well named, being large, associated with aquatic habitats, and similar to other true bugs. The largest are the *Lethocerus* and *Benacus*, reaching about 45–65 mm, although in other countries some may be double that size. There are three common *Lethocerus* north of Mexico: *L. americanus*, *L. medius*, and *L. uhleri*. All are similar in appearance. *Abedus* are usually 14–30 mm. Common species include *A. herberti* and *A. indentatus*. The common species of *Belostoma* in most of the United States and part of Canada is *B. flumineum*, at 18–25 mm.

The biting part of a toe-biter (*L. medius*). L.L.C. Jones

Belostomatids are oval when viewed from above, and are dorsoventrally flattened. They tend to be a drab greenish to brownish color. The head, thorax, and abdomen are all streamlined without noticeable constrictions between the major segments. *Lethocerus* and *Benacus* are the longest and narrowest of the group, while *Abedus* is broader. *Belostoma* usually falls somewhere in between. All genera have large, muscular grasping legs for the front pair, while the rear two sets of legs, especially the hindmost, are flattened and have hairs. The front legs are outstretched when the animal is at rest or in a foraging mode. They have sharp tips that are used for pinching and holding prey. The two rear pairs of legs are mostly used for swimming. The setae are paddle-like for swimming but may also gather sensory cues. Legs are also used for terrestrial mobility. They all have large eyes that lack ocelli, so are mostly used for detecting changes in light rather than visual acuity. The antennae are tucked under the head in ventral grooves. Belostomatids breathe through organs called airstraps, which may differ between genera and species. There are a host of other characters that can be used to identify genera and species. For our purposes, the large *Lethocerus* can be differentiated from other genera by size and shape alone. There is no information on the differences in bite or venom among species.

The naucorids are similar to giant water bugs, having swimming legs and large anterior grasping legs. However, they are smaller, tend to be flatter and less elongate, and lack airstraps. They are sometimes called saucer bugs because of their shape. Their head does not reach beyond their thorax. They only reach about 10 mm in length. Species are difficult to differentiate, even under a microscope.

Distribution: Belostomatids are found worldwide in tropical areas, as well as many temperate areas. As a group they are widespread in the United States and southern Canada, but the highest diversity is in the Southwest. Among the more widespread species is *L. americanus*, which is found from California to Florida, north into southern Canada, and south into Mexico. *Lethocerus medius* is found in the southern parts of Arizona, New Mexico, Texas, and south to Costa Rica. *Lethocerus uhleri* and *Benacus griseus* are species of the East, being found from east Texas to Florida. *Belostoma flumineum* is found in most of the United States, while *B. confusum* is found from Arizona to Texas. *Belostoma bakeri* is found from Washington to northern Baja and east to Louisiana and south to Durango, Mexico. *Belostoma lutarium* is a common species in the East. *Abedus herberti* is probably a species complex of Arizona, New Mexico, Utah, and Mexico, while *A. indentatus* is its California counterpart. Two uncommon belostomatids, *L. angustipes* and *B. saratogae*, have an interesting range—they co-occur in some of the same pools in Death Valley, California.

In our area, the greatest diversity of Naucoridae is in the West, especially the Southwest, although there are species throughout the United States and southern Canada, and into Mexico. Most bodies of fresh water that have sufficient habitat have one or more species. *Ambrysus mormon* ranges from California to South Dakota and south to Mexico. *Ambrysus lunatus* is found from New Mexico to Texas, and Mexico. *Pelocoris* is primarily an eastern genus, although all other naucorids are only in the Southwest. *Pelocoris femoralis* is the most widespread species of that genus. Its range extends well into Canada in the north, and south into Latin America.

Natural History: Belostomatids and naucorids are freshwater aquatic to sometimes semiaquatic species. *Lethocerus*, *Benacus*, and *Belostoma* are associated with lentic habitat, while *Abedus* is associated with lotic habitat. Lentic habitat includes marshes, ponds, and other wetlands, including fish hatcheries, where they may be a pest. The lentic species are sometimes found in temporally ephemeral streams. Species may be restricted to pools in streams during low-flow periods. For this reason, the distinction between lentic and lotic is not always clear. Belostomatids overwinter in humid-to-wet protected areas including cracks in mud, in vegetation at the bottom of a pool, and in rocks and logs.

Lethocerus and *Benacus* are capable of flight. *Abedus* does not seem capable of flight. Either way, these insects move between water sources. Overland movement is often associated with the arrival of the North American Monsoon or other rain events. *Abedus* can gauge cues from incoming storms and will evacuate a stream to avoid being swept away by flash floods.

The ability to fly when dispersing may be linked to their reproductive habits. Belostomatids are the only insects known to have parental reversal, where the male cares for the eggs. *Abedus* and *Belostoma* have taken this to an extreme. The female lays eggs on the back of the male, and the male not only protects the eggs, but also keeps them oxygenated. *Lethocerus* and *Benacus* lay eggs on stalks of aquatic vegetation. Male *Lethocerus* also guard their eggs. If males had eggs on their backs, they would not be capable of dispersal flights. Mating occurs in the spring and summer. Belostomatids

Male *Abedus* with eggs on the back. The female is ovipositing on him. *Mike Bogan*

have about 65–250 eggs that take two weeks to two months to hatch. Larvae go through several nymphal stages, all of which are predaceous and presumably venomous.

These animals are fierce aquatic predators. They use ambush tactics, lying in wait to capture edible passersby. They prey on invertebrates and sometimes small vertebrates. *Lethocerus* or *Benacus* have been documented feeding on small snakes, frogs and their tadpoles, fish, and turtles. There is one report of *Lethocerus* feeding on a woodpecker. At the time of this writing, there is an online video of *L. medius* immobilizing a Black-necked Gartersnake, a good demonstration of the power of the venom. One study

A giant water bug can paralyze a snake or frog in fairly short order. *Mike Bogan*

has shown that the bug can liquefy small frogs in relatively short order, which translates to venom that is effective against vertebrates. *Abedus* occasionally feeds on small vertebrates, but more commonly on snails, corixids, and nymphs of their own species, as well as naucorids. Belostomatids sometimes scavenge.

Although large belostomatids are capable of inflicting nasty wounds on potential predators, they tend to use other options in defense first. *Lethocerus* can squirt a foul liquid out of its abdomen that can reach over a meter. They may also feign death when handled. The instars are capable of envenomation, and may be more prone to bite than adults.

Naucorids have a wide range of habitat preferences, depending on the species, and may be indicators of habitat conditions. They may be in still or stagnant to swift water, over mud or rock, and may even crawl along rocks in rapidly flowing water. Some species may even be found in saline water. Naucorids can swim, but they get their name from their habit of creeping over rocks or on the bottom. As with belostomatids, they are voracious predators that feed on invertebrates they can overpower, including terrestrial and aquatic insects and snails.

Encounters: In some areas, large numbers of *Lethocerus* and *Belostoma* may be seen around electric lights; the name "electric light bug," generally refers to these genera. In the wild, these "toe biters" are rather passive, based on input from people who study these animals, but an animal stepped on or brushed against skin can defend itself. Bites typically occur to those with an occupational hazard, especially entomologists and fisheries biologists. However, swimmers may also be bitten, as well as people who do not know what they are and pick them up. One researcher told me he has handled hundreds of *Abedus* without an incident, even when scooped up by the handful, save some bites in the lab by instars, although he is cautious about handling *Lethocerus*. Still, he had no incidences of envenomation by *Lethocerus*. Most fisheries biologists and entomologists I spoke to have never been bitten by belostomatids, although a couple had.

On the other hand, naucorids have a reputation for being quick to bite, and are a nuisance for those having an occupation where encounter is likely. A common cause of bites occurs when biologists are sorting through dip nets or seines. These insects are small, fast, and cryptic, so often go unnoticed when looking through algae for larger items. They often share the same habitat with belostomatids.

Venom Apparatus, Venom, and Symptoms: The venom apparatus consists of salivary glands and the rostrum, which is shorter than in reduviids. There are three types of salivary glands: main, lateral, and accessory. Each of the glands has its own role in producing digestive products. The main and lateral glands possess alveoli that dump into a central canal. The lateral glands are sock-like, and layered with single cells. The main and lateral glands are covered in myoepithelial cells that empty salivary secretions into a central lumen to be injected into the prey or victim with the beak. The prey may be immobilized and is digested extraorally, with the liquefied contents being sucked into the bug through the rostrum.

The venom is used for feeding and defense. There are few studies of the venom of aquatic insects. One study compared the contents of the salivary glands of two North American belostomatids (*L. uhleri* and *B. lutarium*). Venom characteristics reflect their prey preferences. *Lethocerus* and *Benacus* tend to exploit vertebrate prey more than *Belostoma*. Fishes are a common prey item of *Lethocerus*, but not *Belostoma*. The latter also tends to feed on aquatic snails. Not surprisingly, *Lethocerus* has a more potent proteolytic saliva, which reflects the need to immobilize and feed on fishes and frogs.

Lethocerus uhleri produces several digestive enzymes, including three proteases. *Belostoma lutarium* has many of the same digestive enzymes, plus amylase, but the venom lacks one of the proteases and is not as proteolytic. There are some studies of belostomatids outside our borders. *Lethocerus cordofanus* has five proteolytic enzymes. It can liquefy the internal contents of a fish in 10–15 minutes. *Belostoma anuran* immobilizes frogs and toads with phospholipase A_2. The saliva of that species kills small rodents and has been shown to cause cardiac arrhythmia and reduction in contractile force.

There are very few case reports of bites by belostomatids, but in the literature there is frequent reference to their painful bite. It is difficult to assess effects of bites with the paucity of information; most seem to be mild to moderate and rarely require medical intervention. Some state the bite is like a bee sting, while others suggest a more painful or symptomatic outcome. One colleague said the bite of *L. medius* is not as painful as a naucorid, but this is not always the case; in one of the Envenomation Stories in this book, the envenomee was unable to walk without crutches, work, or drive for two days. One reference states that the bite produces "a burning sensation that lasts several hours and causes considerable redness and swelling." In another report I got from a professor, a person was told to not handle *Lethocerus*. However, she did anyway while she was transporting it to another lab; shortly thereafter the professor heard screams echoing in the hallway. Another colleague stated that a bite he got from *L. medius* was extremely painful. One reference states that the bite (of at least some species) is "one of the worst in the insect world." A colleague reported the bite of *A. herberti* to be similar to a paper cut that is resolved in a minute, and definitely not as bad as a bee sting. There are several case reports of bites in Brazil. In all but one case, the offending animals were only identified as belostomatids; in the other, the culprit was known to be a 10 cm *L. delpontei*. Four of the other bites were from belostomatids 3–4 cm long. In all cases, there was a red mark at the bite site, with local, intense, pulsating pain. In four of the cases, there was swelling. One of the bites by a 3 cm belostomatid was to a nine-year-old who was bitten on the hand. The pain spread up the arm and was still intense three hours later. There was pseudoparalysis of the forearm due to pain, with local numbness, followed by paresthesia. The symptoms lasted five hours, then completely resolved. The other bites mostly lasted one to three hours, with no long-lasting effects.

Naucorids are well known for inflicting painful bites, often likened to a bee sting—hence a vernacular name of "water bee." The genus *Pelocoris* is

Naucorids are called water bees because of their bee-like pain, and are quick to bite people who handle or step on them. *L.L.C. Jones*

reported to be particularly aggressive. I have read that symptoms from the bites are short-lived, but also read another account saying the pain lingers. One author said naucorids were much worse than toe-biters, although the author had not been bitten by either of them. There are many naucorid species and cases are rarely reported in the literature, so a good comparison isn't really possible.

Remarks: A bacterial disease thought to be transmitted by biting aquatic insects is known as Buruli ulceration. It is primarily from Africa, but also tropical areas of the New World. In 2014 a report of a Buruli ulcer (a necrotic lesion by *M. ulcerans*) that developed at the bite site of a belostomatid added support to the belostomatid-vector hypothesis.

References and Resources: Choate (2013), Davis (1986, 1996), Haddad et al. (2010), Lauck and Menke (1961), Lytle (1999), Marion et al. (2014), Menke (1979), Merritt and Cummins (1996), Perez-Goodwyn (2006), Rees and Offord (1969), Schmidt (1982), Silva-Cardoso et al. (2010), Smith (1976, 1982), Smith and Larsen (1993), Swart and Felgenhauer (2003), Swart et al. (2006), Velasco and Millan (1998), Weatherston and Percy (1978).

Other Aquatic Invertebrates

"I didn't know what to do. I kept being pushed by the waves into the sea urchins. My husband removed as many spines as possible—some in places where only a husband can look—but a dozen years later, some spines are still embedded under my skin." (personal communication, J. "Urchin Girl" Jones)

Giant Green Anemone of the eastern Pacific. *Ron Wolf*

Classification: In addition to those in taxa in specific accounts, there are some other marine invertebrates that can cause envenomation. These include:

- **Fire Sponge (*Tedania ignis* and related):** Phylum Porifera; Class Demospongiae; Order Poecilosclerida; Family Tedaniidae.

- **Fire Coral (*Millepora* spp.):** Phylum Cnidaria; Class Hydrozoa; Order Anthomedusae; Family Milleporidae. Other hydrozoans called hydroids are known to cause significant skin irritation, termed "stinging hydroids" (below).
- **Stinging Hydroids (various taxa):** Phylum Cnidaria; Class Hydrozoa; Subclass Hydroidolina. Animals referred to here as stinging hydroids are often abundant colonial hydrozoans known mostly from their plant-like colonies of polyps, such as *Pennaria disticha* (Order Anthoathecata) of Hawaii. Hydroids with painful stings include species in the families Aglaopheniidae and Haleciidae.
- **Sea Anemones (Order Actiniaria):** Phylum Cnidaria; Class Anthozoa; Order Actiniaria. While I can find little information on specifics of the venomous nature of the majority of sea anemones in our waters, there is frequent reference to the stings of domestic species causing some degree of harm to humans. Two species that cause painful stings in humans are the appropriately named Stinging Mangrove Anemone (*Bunodeopsis globulifera*; Family Boloceroididae) and Stinging (or Branching) Anemone (*Lebrunia neglecta* (= *danae*); Family Aliciidae).
- ***Linuche unguiculata* and *Edwardsiella lineata*:** Phylum Cnidaria. These are two organisms known to cause seabather's eruption, not to be confused with swimmer's itch. *Linuche* (Thimble Jellyfish) is a scyphozoan (Family Lenuchidae) and *E. lineata* (Striped Sea Anemone) is a sea anemone in the family Edwardsiidae.
- **Long-spined sea urchins (Diadematidae):** Phylum Echinodermata; Order Diadematoida; Family Diadematidae. *Diadema antillarum* and *D. paucispinum* occur in our area. *Echinothrix diadema* is another long-spined species. Other sea urchins may be mildly venomous, but are mostly a concern due to their sharp spines.
- **Crown-of-Thorns Starfish (*Acanthaster planci*):** Phylum Echinodermata; Order Valvatida; Family Acanthasteridae. This is a species complex, and four clades are usually recognized, but nomenclature needs to be sorted out. The Pacific form may be *A. solaris*.

Identification: The list above contains some very different animals; what they have in common is that they are inshore marine invertebrates. Sponges are primitive yet highly developed and diverse organisms. They are sessile, and while some species may simply encrust on rocks, others may be somewhat plant-like. Some can attain impressive sizes. The Fire Sponge has a spongy mass with elevated cones that may grow in a clump. At the top of the cones are oscula, which are about 1 cm in diameter, for outward water flow. Fire Sponges are colored bright orange to red, as a warning to passersby of their venomous nature. Most sponges, including Fire Sponges, have a skeleton of calcium carbonate.

Despite the name, Fire Corals are hydrozoan cnidarians rather than true corals. Hard types of true corals have a solid calcareous structure that is well known as a tropical reef builder, but the skeletal structure of *Millepora* and other hydrozoans is variable, from encrusting to plant-like or similar to a hard coral. Fire Corals are usually yellowish in color, and they have a symbiotic alga that helps gather nutrients from photosynthesis. As with other hydrozoans, the living tissue consists of different types of specialized polyps, including dactylozooids and gastrozooids (see account for *Physalia*). In Fire Corals, the polyps are minute, and mainly situated beneath the skeleton, but the tentacles extend out of the pores (*Millepora* means "thousand pores").

Stinging hydroids are colonies of plant-like hydrozoans similar to Fire Coral but usually without a hard coral-like skeleton. The small stinging polyps often live on a plant-like, flexible base that waves around with the ocean currents. Others essentially coat marine substrates such as rocks, corals, algae, and boat hulls.

Fire Coral isn't a coral; it is a hydroid. *Ryan McMinds/CCA/Flickr*

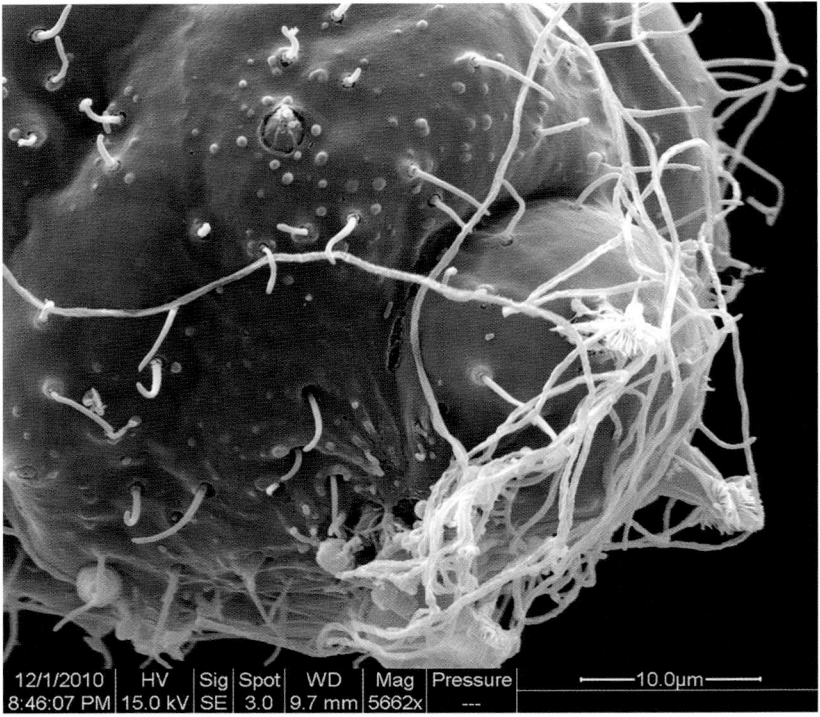

Cnidarians envenomate by discharging nematocysts. In the SEM, the nematocyst tubules of a *Hydra* have been discharged. *Marian Rice and Allison Miller/CCA/Flickr*

Sea anemones are anthozoan cnidarians, like true corals, but lack the hard skeletal support. The adult stages are polyps—often very large polyps. They have a central body, or stalk, and an oral disc on the top, surrounded by tentacles. They attach themselves to rocks or other structures, but are capable of moving to other locations. The abundant and familiar intertidal and shallow-water species of anemone in southern California is the Green Anemone, a large species that gets its color from a symbiotic relationship with photosynthetic algae. In the northern latitudes of both coasts are giant anemones in the genus *Metridium*. The giant of the warmer western Atlantic, Caribbean, and Gulf of Mexico is the Giant Caribbean Sea Anemone. The Stinging Mangrove Anemone is a small nocturnal species typified by translucent tentacles originating from a brown base with white-tipped nodules. The Stinging Anemone is also known as the branching anemone because of its unusual anatomy. During the day it shows its branched pseudotentacles, but at night the true tentacles are extended from the base.

Spines of the Crown-of-Thorns Starfish will deliver a potent epidermal venom into anyone unlucky enough to get jabbed. *Matthew Run/CCA/Flickr*

The cnidarians that cause seabather's eruption are tiny animals. As the name implies, the adult Thimble Jellyfish medusa is the size and shape of a thimble, but the planula larva is about the size of a grain of sand, and it is what causes most known cases of seabather's eruption, although all life stages can cause dermatitis. A species in the eastern and central Pacific is *L. aquila*, but I can find little information about whether or not this animal causes seabather's eruption in California, Hawaii, or beyond. Adult *E. lineata* are small, striped anemones of the subtidal zone, but it is their planulae that apparently cause dermatitis in swimmers. There may be other small organisms that cause the malady. Swimmer's itch is often confused with swimmer's eruption because of similar symptoms, but the former is caused by a parasitic protozoan.

Sea urchins and starfishes are radially symmetrical echinoderms (meaning "spiny skin"). Sea urchins look like giant pincushions, whereas starfishes are dorsoventrally flattened and have arms. The sea urchins that are of most concern to us are the *Diadema* and *Echinothrix*, which have extremely long spines. Depending on the species and locality, the spines are usually black but may be white, or banded. The spines are several times as long as the inner skeleton, known as the test. The spines are movable in a ball-and-socket manner, and not only help protect the animal, but also aid in locomotion. The Crown-of-Thorns Starfish has numerous arms radiating out from the broad central disc, but they too have spines on the aboral surface. While the spines are shorter than those of most sea urchins, they are more pronounced than in other sea stars. They are also different from most other species by their rather flaccid body (spines excluded). They are variously colored, ranging from drab grays and browns to bright pinks, blues, and purples. The clade from the eastern Pacific is distinctive in having a very broad disc, with short arms and spines. Both *Diadema/Echinothrix* and *Acanthaster* are relatively large, conspicuous members of the coral reef communities. Both of these echinoderms have tubular feet for locomotion. They also have anatomical structures known as pedicellariae, discussed in the section on venom apparatus.

Distribution: Sponges, as a group, are worldwide in temperate and tropical waters, but those causing harm are generally restricted to warm waters. Fire Sponges occur in the Caribbean, Atlantic, and the Gulf of Mexico, but primarily in southern Gulf waters and off both coasts of Florida. The fire and red sponges (*Microciona* spp.), and possibly other species in and near Hawaiian waters, are known to envenomate humans. Other genera of sponges have been documented causing some sort of irritation, but the literature is fairly scant and taxonomy is not well refined for these odd animals.

As a whole, cnidarians are found in all temperate and tropical marine waters. Fire Corals are found in many tropical waters. In our area they are primarily known from Florida, especially around the keys. They are also present at Flower Garden Banks, an offshore coral reef off Texas and Louisiana, and probably at offshore drilling rigs as well. They occur in the western Pacific, but are absent from Hawaii and the eastern Pacific. The closely related stinging hydroids in all of our coastal waters can cause mild to moderate human envenomation. There are various species of sea anemones in all of the marine waters of the United States and Canada, but most do not seem to harm humans. The Stinging Mangrove Anemone and Stinging Anemone are found in the tropical and subtropical waters of the western Atlantic, including Florida, the Caribbean, and the Gulf of Mexico. There are more than twenty species of small anemones in the Hawaiian Islands, five of which are endemic. Seabather's eruption caused by Thimble Jellyfish is mostly known from Florida, the Bahamas, and the Caribbean, while that caused by the Striped Sea Anemone planulae occurs in the northwest Atlantic, especially off Long Island, New York.

Sea urchins of various types are found in nearly every nearshore region of the world, in both temperate and tropical seas. The venomous species tend to be tropical. Our primary culprit of envenomation is *D. paucispinum*, a species found in Hawaii, and *D. antillarum*, which is found in south Florida and Flower Garden Banks, and throughout most of the Caribbean. *Echinothrix diadema* also occurs in Hawaii. *Diadema mexicanum* occurs in the eastern Pacific from Mexico to Peru. In the Gulf of California it is fairly widespread and abundant in the southern latitudes, but also occurs in isolated areas as far north as Isla Jorge, in the northern Gulf. It has not been recorded from California, where the water is presumably too cool to allow persistence.

The Crown-of-Thorns Starfish complex is widely distributed in the Red Sea and Indian Ocean, as well as the Pacific. In the Pacific it is found in the Hawaiian Islands to the west coast of Mexico, and south to Central America. It seems particularly abundant in Australia and some other Indo-West Pacific areas. It is absent from the Atlantic, and there is concern it could become an invasive species if introduced.

Natural History: All of these species are marine, or sometimes found in estuaries. The life histories vary by taxon, but in general, marine invertebrates tend to have complicated life histories with larval stages. There is a tendency to have an egg stage, one or more planktonic larval stages, and bottom-dwelling adult stages. The larvae of different groups have different names. Although there are similarities, each group has a distinctive characteristic larval type. Some taxa have asexual reproduction, while others are bisexual; some have both. The summaries below just scratch the surface, so interested readers should consult the many good books and websites on the life cycles of aquatic invertebrates.

Sponges are mostly hermaphroditic, but at any given time they will usually produce just sperm or eggs. Sperm are dispersed into the water and nearby sponges will pick them up while filtering. The sperm then fertilize the eggs within the "female" sponge, where they brood until hatching, at which time they will be released. The larva, sometimes called a parenchymella is ciliated and moves around for 2–3 days, after which it settles on the substrate, or even on another sponge, and grows and develops. Typically, sponge larvae settle near or on the "female" parent. Sponges may also develop asexually, from budding. In this manner, colonies of sponges may expand. The individual cells of sponges dissociated through a sieve will continue to survive, much like an amoeba, but will reassociate with other cells from the same species, to form a new sponge. With this strange life cycle, sponges are sort of "immortal," although they can be killed. Sponges are thought to be a sort of missing link between single- and multicelled organisms.

The life cycle of cnidarians discussed in other sections are similar to Fire Corals, stinging hydroids, and sea anemones, although there are also differences. Fire Corals can reproduce sexually or asexually. Sexual reproduction comes from a medusa stage, unlike true corals. Medusae are formed from specialized structures known as ampullae, then eggs and sperm are released into the sea. Fertilized eggs develop into a planktonic larva that will settle on the seafloor and begin a stage of growth and development, culminating in a small calcareous skeleton and polyps. The polyps can also reproduce asexually by budding, thereby expanding the colony size and skeletal structure. Sea anemones lack a medusa stage. In the sexual phase, eggs are fertilized and develop into a planula larva. The planula settles on the substrate and develops into a polyp. The polyp can then reproduce asexually by budding. Sea anemones are essentially sessile, but they can also move around, albeit in slow motion. It is usually a cnidarian planula of the Thimble Jellyfish that causes seabather's eruption, as even the larvae are able to sting; however, polyps and medusae also sting. The organisms causing seabather's eruption are sometimes referred to as "sea lice," but that is a misnomer. A louse is an insect and a sea louse is a copepod, a type of crustacean.

Echinoderms are bisexual. The males and females release eggs and sperm into the water column. After fertilization, there is a planktonic blastula. This develops into a larva, known as a pluteus or echinopluteus in sea urchins. The pluteus will filter feed for a while (not all do), then will settle on the bottom and metamorphose into young animals, which will continue to grow into adults, at which point they feed on algae or corals. Crown-of-Thorns Starfish are a little different. Spawning is similar, but the young planktonic larvae are known as bipinnariae. Like other larvae of marine invertebrates, they have cilia and swim and feed in the water column. As this early larval stage grows, it metamorphoses into brachiolaria larva. The brachiolaria sink to the bottom and feed on algae. They gain arms as they grow and switch to feeding on coral polyps.

Sponges are filter feeders. They can pump a high volume of water in through their ostia and out through the oscula. In this manner, plankton are brought into the body of the sponge, and filtered out to be absorbed by the sponge. Fire Corals are also filter feeders. Sea anemones are similar; small ones catch small planktonic animals, while larger species catch small fish. It is well known that some species of fishes (especially anemone fishes) use sea anemones for protection by making themselves immune to the toxins in the stinging cells. This concept has led to the development of anti-sting lotions for protection against some cnidarians. The echinoderms tend to be more active feeders. Sea urchins are herbivorous. *Diadema antillarum* is known to feed on many species of algae, as well as diatoms and vascular marine plants. Urchins may also filter feed. In a well-publicized case in southern California, marine pollutants from runoff caused increases in phytoplankton, which caused population explosions in Purple Sea Urchins (*Strongylocentrotus purpuratus*), which then switch to eating holdfasts of giant kelp. This series of events is causing the destruction of kelp beds, which are vital ecosystems for nearshore species in the eastern Pacific. If kept in check, the kelp beds will recover. The Crown-of-Thorns is similarly best known for its destructive capabilities. Population explosions cause harm to coral reefs, because the Crown-of-Thorns feeds on coral polyps. Under the normal situation, the number of starfish is in balance with the ecosystem, and they are just one of the predatory animals that feed on corals. However, in high densities (called outbreaks or plagues) large areas can be stripped of living corals, and only the skeletons remain, which are then colonized by filamentous algae. This converts the ecosystem from a diverse and dynamic one to a simple one. Causes of the outbreaks are thought to be primarily from removal of the starfish's predators. Coral bleaching, however, is generally due to ecological havoc caused by climate change (i.e., rising ocean temperatures) and pollution.

Despite being defensively venomous to vertebrates, all of these taxa are prey for something else. Fire Sponges are eaten by some species of parrotfishes, angelfishes, and sea stars. Fire Corals and sea urchins are eaten by a variety of fishes, perhaps most notably the triggerfishes, which can nip off the sharp spines to get at the test and the goodies inside. Urchins are relished by humans for food in certain areas of the world.

Encounters: These are all shallow-water marine animals that can be encountered by swimmers, snorkelers, divers, surfers, paddlers, anglers, and researchers. Also, there are those with occupational hazards, such as anglers hauling in nets or lines and those cleaning boat hulls, pilings, or other structures that harbor sessile marine life. These creatures are primarily associated with reefs, either of coral or rock, but there are some exceptions. Most of these animals are sessile, or near-sessile organisms, meaning one must brush up against them to be envenomated. Some of them are plant-like, so divers may consider them harmless, or simply ignore them and brush against them as they swim by. Surges of waves and currents can be quite powerful in shallow waters, and many people are stung when they are pushed into the offending creature. Also, people can be injured when they fall off surfboards or paddleboards, or dive into the water. Settling on the bottom or using hands to push off the reef is particularly hazardous, especially around sea urchins.

Venom Apparatus, Venom, and Symptoms: The venom apparatus varies by taxa. Most sponges cause no harm to humans. For example, the ones we use to bathe with do not have hard spines. The spiky sponges can cause mechanical damage, as well as slight to moderate envenomation. The mechanical trauma is caused by spicules, which are sharp calcareous or siliceous spine-like objects in the cells used for support and defense. The slimy skin of venomous sponges has crinotoxins, and when the surface of the victim's skin is penetrated by spicules, the toxins enter the wound. The toxins, which come out of pores, can cause irritation on contact on human skin by themselves, so it is the passive action of the spicules delivering the toxins into the skin or muscle that makes them animals considered to be venomous. Sponge crinotoxins are not particularly well known, but *Tedania* does have presynaptic neurotoxins. There are also fractions that cause a variety of physiological reactions in mammals, including agglutination of human red blood cells. Interestingly, some populations of *Tedania* lack crinotoxins; it is believed they sequester the toxin from planktonic food, so those that feed on poisonous plankton store the toxins. There may be delayed onset of symptoms in humans. Most stings by sponges are considered mild, with itching, burning, and formation of a skin rash. However, it depends on the amount of venom introduced into the victim, plus some people react worse than others. In the rare severe cases, there can be severe pain, itching, redness, swelling, stiffness, blistering, and skin necrosis. In some cases, skin may slough off. There can also be systemic reactions, including fever, chills, dizziness, nausea, and muscle cramps.

Fire Corals are somewhat akin to sedentary jellyfishes. The many pores have tentacles coming out of them, armed with nematocysts. When one brushes against the animal, hypodermic tubules are shot out of pores. Fire Corals and some other species have been shown to contain a phospholipase A_2, termed milleporin-1. Another cytolytic protein has been called MCTx-1 (for *Millepora* cytotoxin 1). *Millepora complanata* venom-purified protein extracts have been shown to have an LD_{50} of 4.62 µg/g. The venom

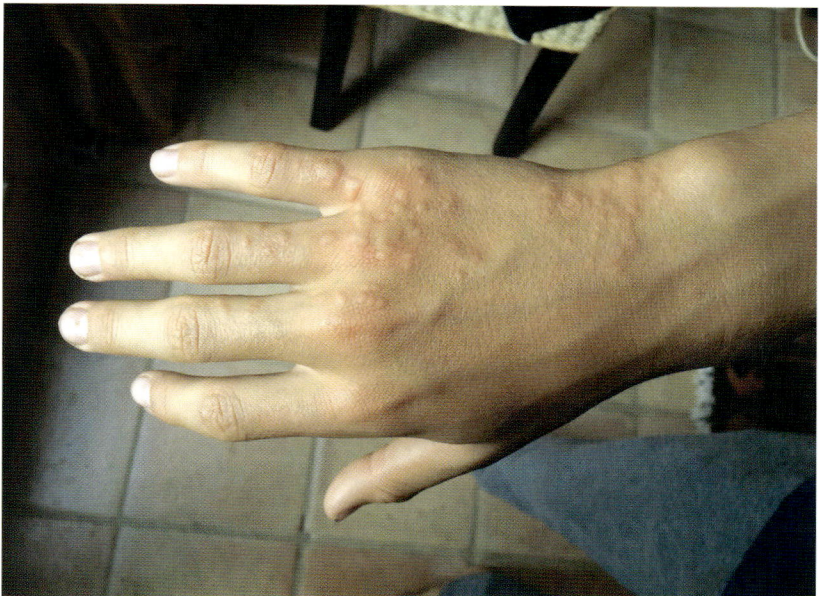

Hydroid-induced welts on a snorkeler. *Giacomo Bernardi*

is hemotoxic and cytotoxic in lab mice. Usual symptoms in humans include rash, burning, redness, and itching (sometimes severe and widespread) that may last up to two weeks. However, there are also severe cases with systemic symptoms, including malaise, fever, and burning pain around the shoulders, muscle paralysis, and neuropathy. Stinging hydroids are problematic in Hawaii, where no Fire Corals exist, and various stinging hydroids also occur along the Pacific and Atlantic Coasts. Stinging hydroids typically cause a red, burning rash that is painful for several hours. There is usually a more severe reaction with greater contact, as when one brushes part of the body against a large colony. Parts of the skin that are hairless are most sensitive to hydroid stings. More severe stings include a large, painful rash that lasts several days, swelling, light sensitivity, blurred vision, and tearing.

Sea anemone envenomation is not well-documented for species in our area, as are some species in the Indian Ocean, Red Sea, Adriatic Sea, and western Pacific, but virtually all publications that discuss cnidarian stings include sea anemones. Like other cnidarians, they have nematocysts in their tentacles. My experience with large Green Sea Anemones in southern California was that they sting, but harpoons did not penetrate the skin of fingers. However, it would be foolhardy to allow one to touch an area of thin skin, such as the inner arm or tongue. For stings documented from a few species in Hawaii (where stings are actually uncommon), there is usually a bee-sting-like reaction. A rash forms with a red halo. The area will be swollen and look bruised for up to two days. In severe cases, there may be bleeding, blisters, and ulcers. There may also be permanent scarring. Some species studied also show neurotoxins that affect sodium and potassium gates.

Seabather's eruption is a bit different than typical cnidarian envenomation. Rather than just a sting by the nematocysts causing immediate pain on contact, the onset of symptoms is usually delayed. What typically happens is that the planula larvae or other small life stages get onto areas of swimmers that get compressed, such as under swimsuits, often in the beltline, buttocks, and breasts. The minute cnidarians discharge their nematocysts and after several hours (e.g., after taking a shower, following a swim) a rash begins to form. The rash may be urticarial, or not, and may include raised lesions, or not, depending on species and circumstances. Symptoms are caused by venom but can be exacerbated by allergy. Some cases are more severe and symptoms may last several days to several weeks. In extreme cases, malaise, fatigue, headache, fever, nausea, vomiting, and abdominal cramps may occur. In rare cases, hospitalization may be required.

Some echinoderms have hard spines on the aboral side that may be sharp and pointed. *Diadema* and *Echinothrix* are the ultimate genera for spine defense. There are long primary spines and shorter secondary spines. The primaries may be three times as long as the width of the test, up to about 300 mm, and needle sharp. There is actually some question if *Diadema* and other long-spined sea urchins are venomous, but several researchers claim they must be, by virtue of the pain that is afflicted to humans, which is believed to be more than mechanical injury alone. Some believe the reaction is due to other invading microorganisms or chemicals introduced with a sting. Even without venom, they are formidable weapons. Invariably, a brush with spines of *Diadema* or other sharp-spined sea urchins will cause physical trauma. The long spines are designed to penetrate deep into flesh, and, having minute retrorse spinules, will break off. There are also venomous pedicellariae, which are small structures on the body of the sea urchin. These appendages are much smaller than the spines, and are used to protect the animal from small marine invaders. Pedicellariae have 2–4 (usually 3) jaws at the end of a stalk. There are several types of pedicellariae, but it is the globose type that is venomous. These appendages can seize and bite small animals, causing immediate paralysis. It is unclear

Diadema antillarum is a long-spined urchin found in Florida and the Caribbean. This is a minefield for a swimmer or snorkeler. *NOAA/Public Domain*

if pedicellariae are capable of penetrating human flesh or even causing any medical problems in our native species—the spines usually get someone's attention first—nevertheless, if pedicellariae are present on the skin, they may continue to bite and for an extended period. Spine venom has not been confirmed in our native species, but *Diadema* and other sharp-spined species cause significant problems in Hawaii, where symptoms include serious puncture damage, pain, aching, numbness, and swelling. In the rare severe cases, there have been muscle aches and rash that lasted for several months. In at least two Hawaiian cases, there has been severe systemic symptoms, including nausea, vomiting, numbness, muscle paralysis, abdominal pain, dizziness, low blood pressure, and difficulty in breathing. It should be noted that thin spines often dissolve in the flesh, while thicker spines may not.

In the Crown-of-Thorns Starfish, venom is produced in the epidermal cells covering the spines. It has been reported to have an LD_{50} (intraperitoneally) of 2.7 mg/kg. There are numerous peptides and proteins that have been isolated from the skin toxins, including at least two phospholipase A_2 enzymes (named $AP\text{-}PLA_2\text{-}I$ and -II). Two lethal factors have been named plancitoxin I and II, which are deoxyribonucleases II, each having an LD_{50}

An Upside-down Jellyfish. Technically, this photo should be in with the True Jellyfishes account. When I took this picture, the other snorkelers who ventured near the mangroves in the Florida Keys and I were all stung by the emissions of nematocyst-laden mucous globs produced by the jellyfish. *L.L.C. Jones*

of 0.43 mg/kg (intraperitoneally). A fraction showing anticoagulant activity has been named plancinin. Venom causes hemolysis, cytolysis, liver damage, and shows cardiovascular activity, at least in laboratory animals. In humans, stings are initially noted by small bluish puncture marks, but within minutes a severe burning will ensue. During the next few hours, the area becomes red and swollen, and the intense pain becomes a dull ache. The area may remain sensitive for about two days. In some cases, there may be severe symptoms, either from extensive contact or allergies. Symptoms in those cases may include numbness, tingling, weakness, nausea, vomiting, swollen lymph nodes, and paralysis. Some signs of allergy may include respiratory distress and hives.

References and Resources: Bakker (1980), Burnett et al. (1995), de Laubenfels (1951, 1957), Fusetani and Kem (2009), Halstead (1978), Haszprunar and Spies (2014), Ibarra-Alvarado et al. (2007), Iguchi et al. (2008), Isbister and Hooper (2005), Karasudani et al. (1996), Mebs (1984), Mebs et al. (1985), Moats (1992), Radwan and Aboul-Dahab (2004), Rifkin (1996), Ruppert et al. (2004), Segura-Puertas et al. (2001), Sevcik and Barboza (1983), Shiomi et al. (1985, 1988a), Sims and Irei (1979), Thomas and Scott (1997).

10 Aquatic Vertebrates

Most aquatic vertebrates are fishes. The higher classification of fishes is complicated, and not necessarily agreed upon, so I've tried to simplify it here. By the way, "fish" is both singular and plural when referencing a single species, whereas "fishes" is plural for more than one species. Fishes occur in either the freshwater or marine environments, and sometimes both, or in brackish water, but most venomous species are marine. According to one recent study that attempted to quantify the evolution of venomous fishes worldwide,

(Top) Dorsal aspect of external anatomy of a stingray (a urolophid). *"bato93"/CCA/Flickr.*
(Bottom) Ventral aspect of the external anatomy of a stingray (a dasyatid). *Yu-Chan Chen/CCA/Flickr*

the authors determined that venom evolved independently in 18 different lineages, among some 2,386–2,962 species. The number of venomous terrestrial vertebrates pales in comparison, with only 450 reptiles, less than a dozen mammals, a couple venomous amphibians, and zero birds.

Cartilaginous fishes include chimaeras, sharks, skates, rays, and thornbacks. The chimaeras (class Holocephali) are an odd group of mostly deepwater marine animals; in Latin, *chimaera* means "monster." There are 3 families, 6 genera, and about 55 species of chimaeras recognized. Of about 500 species of sharks (class Elasmobranchii; infraclass Selachii), few are venomous, and most are in the order Squaliformes, the group that includes spiny dogfishes. While there is great diversity of sharks in terms of shape, size, and habitat use, most Squaliformes are pelagic and typically found in deep water or well offshore. There are only nine species of the distantly related horn sharks (order Heterodontiformes), one of which occurs north of Mexico. They are primarily benthic, and often encountered by divers in southern California. None of the elasmobranchs are believed to be dangerously venomous, but they can cause painful wounds with their sharp, stout, venomous spines.

Among cartilaginous fishes, one group of mostly marine animals rises to the top in terms of risk from envenomation: stingrays. There are more than 630 species of rays (class Elasmobranchii; infraclass Batoidea) worldwide, belonging to 26 families. Eleven of the families have members that include sting-

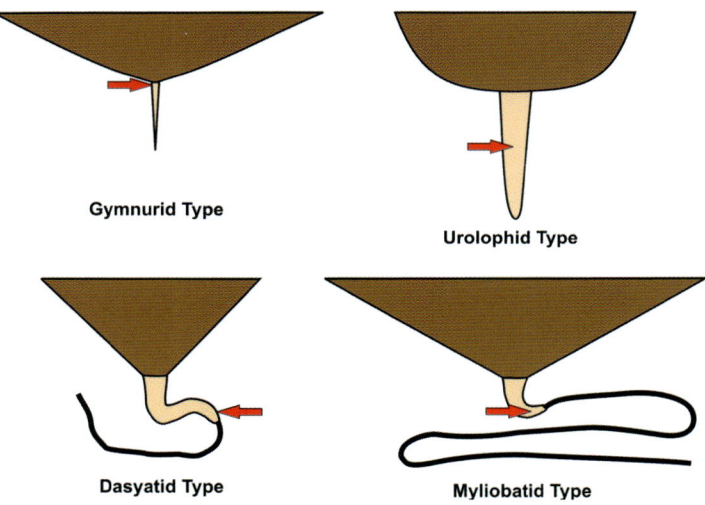

Diagrammatic representation of stingray apparatus types. Dark brown is rear of disc, light brown is caudal appendage, heavy line is caudal filament, and red arrow is location of stinger. *L.L.C. Jones*

Stinger of a small stingray from Gulf of Mexico. *L.L.C. Jones*

rays, a name given to those with at least one venomous caudal spine. Nine of the families occur in North America north of Mexico, and Hawaii. Rays are dorsoventrally flattened, and the head, body, and fins form the disc. They have some sort of caudal appendage, which has the venom apparatus.

The remaining venomous fishes are all bony fishes, having calcified rather than the cartilaginous skeletons. The vast majority (99%) of fishes are the bony fishes, which includes the ray-finned fishes (class Actinopterygii). These are the vertebrate rulers of both freshwater and marine habitats. There are more than 30,000 species worldwide. This is where the great evolution of venom development evolved into numerous lines. The ones that rise to the top in North America are sea catfishes (Ariidae), North American catfishes (Ictaluridae), and a large variety of Scorpaeniformes. Scorpaeniformes is a diverse group with many venomous and nonvenomous species. I divided the venomous taxa into functional groups with similar characteristics and phenotypes: those resembling typical scorpionfishes, lionfishes, and rockfishes plus thornyheads. They are all related and tend to have a similar venom apparatus. I did not attempt to tackle all of the species, or even families and genera, but I tried to address the major players, particularly those in the nearshore environment. Some of the groups have members that are nonvenomous or mildly venomous (rockfishes), to highly venomous (typical scorpionfishes and lionfishes). There are upwards of 350 species of Scorpaeniformes that are known to be venomous.

One taxon account is a hodge-podge of other venomous fishes belonging to several species that have evolved venom. They are put into this "other" category because they are poorly known or have venomous relatives in nearby waters (generally more tropical). However, it is always good to be aware of them. These include toadfishes (Batrachoididae), stargazers (Uranoscopidae), jacks

External anatomy of a bony fish (Starry Rockfish). *John Butler/CCA/NOAA*

Scorpaeniformes are venomous, but many are important food fishes. Pictured are of Redfish from the Northwest Atlantic. *NOAA/Public Domain*

(Carangidae), tangs or surgeonfishes (Acanthuridae), squirrelfishes (Holocentridae), and some less familiar forms. The reader should also refer to the chapter on American Territories, as there are additional families of venomous fishes found in the United States that do not occur in North America.

Lastly, there is a single species of aquatic, venomous reptile in North America, the Yellow-bellied Seasnake. It is but one of more than 60 species of seasnake in the family Elapidae, which includes coralsnakes; the rest are in the Indo-West Pacific.

Chimaeras (Order Chimaeriformes)

Encounter in the Puget Sound, Washington: "I [was] . . . completely surrounded by [spotted] ratfish (hundreds!). We were innocently (and fearfully) swimming under the docks which were completely loaded with them, it was a really cool experience right up to the point when I felt a sharp sting in the palm of my hand somewhat like a bee. I noticed that they seemed to be ramming into us on occasion. Two weeks later I'm . . . with some random divers and they show me pictures of one of their hands swollen pretty large . . . [from] the week before, attempting to grab a ratfish when one of them got 'spined' so deep (between the knuckles into the hand several inches) they were unable to remove it and had to go to the E.R. to have it removed via minor surgery." (Monk, Northwest Divers online forum)

The Spotted Ratfish is the only chimaera that divers are likely to see, and only if they dive in frigid Northwest waters. *Brian Gratwicke/CCA/Flickr*

Classification: Phylum Chordata (chordates); Class Holocephali (holocephalins); Order Chimaeriformes (chimaeras); Families Chimaeridae (shortnose chimaeras) and Rhinochimaeridae (longnose chimaeras).

Chimaeras are the least known of the cartilaginous fishes. Chimaeras are also called ratfishes, ghost sharks (erroneously), spookfishes, or rabbitfishes (erroneously). Worldwide, there are more than 55 species in three families, two families of which occur off North America. The most likely to be encountered are discussed in the Distribution section. Only the Spotted Ratfish (*Hydrolagus colliei*) is likely to be encountered by someone other than ichthyologists or deepwater commercial fishermen.

Identification: Chimaeras are primitive cartilaginous fishes, with some features similar to other elasmobranchs. These strange creatures are easily differentiated from those more familiar animals by their peculiar shape. They are fairly elongate with large heads and often long, tapering tails. They are soft-bodied with smooth skin. As expected, the shortnose chimaeras have short snouts and longnose chimaeras have elongate snouts. There are two dorsal fins, a pair of large pectoral fins, paired pelvic fins, and caudal fins. The first dorsal fin is large and triangular, while the second is slender and long. The caudal fin may be all but vestigial. There may or may not be an anal fin. The intromittent organs, which are located behind the pelvic fins, are known as claspers. One unusual feature that is not shared with sharks is the presence of sexual organs known as tentacula; there is one on the forehead and a pair in front of the pelvic fins. There is a single gill opening

The spine of the anterior dorsal fin of a chimaera is a long and robust weapon.
Ratha Grimes/CCA/Flickr

on each side. There is a well-developed lateral line system. The only species somewhat likely to be encountered by recreational fishermen and divers is the Spotted Ratfish. It has a distinctively marked pattern of white spots over its entire head, body, and tail. It reaches about 1 m in length.

Distribution: Chimaeras are basically deepwater animals, typically found offshore, near the bottom, on the continental shelf or the edge of the shelf, usually deeper than 300 m and often deeper than 1,000 m. In the northeastern Pacific two chimaerids (genus *Hydrolagus*) include the Spotted Ratfish and Black Ratfish (*H. melanophasma*). The Spotted Ratfish is found from Baja California to southeast Alaska. It can occur inshore from northern California northward. There is a deepwater chimaerid, Purple Ratfish (*H. purpurescens*), in Hawaiian waters. Western Atlantic species not found in the Gulf of Mexico include Small-eyed Chimaera (*H. affinis*) and Pale Chimaera (*H. pallidus*). Gulf of Mexico species include Gulf Chimaera (*Hydrolagus alberti*), Large-eyed Chimaera (*H. mirabilis*), and Cuban Chimaera (*Chimaera cubana*). The family Rhinochimaeridae is represented by Longnose Chimaera (*Herriotta raleighana*) in the eastern Pacific, in the western Atlantic, and Gulf of Mexico; Smallspine Chimaera (*H. haeckeli*) in the Atlantic, but not the Gulf; and Atlantic Spearnose Chimaera (*Rhinochimaera atlantica*) in the Atlantic and the Gulf. In Hawaii, one longnose chimaerid is *R. pacifica*.

The longnose chimaera are among the stranger fishes in the deep sea. *NOAA/Public Domain*

Natural History: The natural history of chimaeras is poorly known, no doubt because of their deepwater habitats and sampling issues. The Spotted Ratfish is better known than the others because it is relatively common and widespread, sometimes found in shallow water, and can be maintained in captivity. The other chimaeras are usually found in water between about 300 and 3,000 m deep. Depending on the species, they can be found over rocky substrates or muddy substrates.

They presumably feed on small benthic animals. They do not have sharp teeth; instead, they have three pairs of plates for grinding and cutting prey. Given their mouthparts, it has been assumed they feed primarily on hard prey. They have an extremely strong bite force, amongst the highest of all tested fishes, consistent with the crushing of hard prey. They can inflict a nasty bite.

They have internal fertilization, with males using the claspers to inseminate females. They lay eggs in leathery cases, as do some sharks, skates, and rays. Spotted Ratfish eggs are sometimes visible to SCUBA divers on muddy or sandy bottoms.

Encounters: Most people will never be lucky enough to see one of these interesting animals, unless it is a Spotted Ratfish in a public aquarium or if they are SCUBA divers in the boreal eastern Pacific. In areas like the Puget Sound or off the western Canada coast, they may be very abundant during night dives. Farther south, they tend to stay in deeper water. Recreational fishermen sometimes pull them up when angling for other fishes. Commercial fishermen can encounter them in nets, traps, and probably longlines. Even deepwater species are sometimes encountered by commercial fishermen. For example, even before they were scientifically described, Black Ratfishes were sometimes caught in commercial Sablefish traps. For commercial fishermen, chimaeras of any type are considered an unwanted bycatch, but at least some species are known to be edible.

SCUBA divers may admire ratfishes during their dives, but they should be aware that ratfishes can be territorial. Spotted Ratfishes have been known to bump into divers and occasionally wound them with their spine.

Venom Apparatus, Venom, and Symptoms: Not all species of chimaera have venomous spines, but many do, and the species most likely to be encountered do. The spine is assumed to be strictly for defense. Venom of the Spotted Ratfish, and presumably other spined species, is delivered by a spine in front of the first dorsal fin. The spine is stout, heavy, tapered, and sharp-tipped, so makes for a formidable lancet. It is slightly recurved and has a serrated posterior margin, with the points facing downward to form a hooked barb on contact. It has a core of cartilage surrounded by vasodentine—calcareous tissue that is denser than bone, but infiltrated by blood vessels. It is believed that the structure is a modified dermal denticle. In cross section, it is roughly triangular, with an anterior keel. The spine is covered in a membranous sheath of dermis and epidermis. The venom is produced in the epidermis.

The venom of chimaeras is not well known. Stings are rarely reported and case studies are few. Shallow stings have been compared to a bee sting, but deeper stings can be more severe. One researcher described stings of chimaeras as being fairly serious, but not as bad as those of stingrays. In addition to the venom, the spine itself is capable of causing physical damage, and can break off inside the victim. In some cases, spines need to be surgically removed. Another species, *H. monstrosa*, presents an occupational hazard for fishermen on deep-sea trawlers in the northeast Atlantic. Reports of stings indicate that symptoms can be serious, often requiring medical attention. In one case study, a person stung in the calf felt a burning pain, accompanied by bluish skin coloration and swelling. The victim remained on crutches for three weeks, and symptoms lasted nine weeks. Others stung by this species experienced severe pain and some needed medical attention to relieve pain and/or remove the spine. One victim reported chronic pain from the injury. All of the fishermen reported wearing protective gear, so spines had penetrated gloves and boots. They reported one chimaera per day in their hauls.

References and Resources: Barnett et al. (2009, 2012), Church and Hodgson (2002a), Didier and Rosenberger (2002), Didier et al. (2012), Fields and Lange (1980), Halstead (1978), Halstead and Bunker (1952), Hayes and Sim (2011), Huber et al. (2008), James et al. (2009), Magerøy and Bærheim (1991), McEachran and Fechhelm (1998), Tinker and DeLuca (1973).

Horn Shark (*Heterodontus francisci*)

"Divers can easily approach these sharks and they are generally harmless unless harassed, in which case they may bite their antagonist. Care should be exercised with their spines, which can inflict a painful sting." (Ebert 2003)

The Horn Shark is a docile species that usually allows close approach by divers, but they can bite and sting if grabbed. *Douglas King*

Classification: Phylum Chordata (chordates); Class Elasmobranchii (elasmobranchs); Order Heterodontiformes (bullhead sharks); Family Heterodontidae (bullhead sharks); Genus and Species *Heterodontus francisci* (Horn Shark).

The order and family are monotypic. There are only nine species of these primitive sharks worldwide, all within the genus *Heterodontus*. Most species are known as bullhead sharks. The only species in the United States is the Horn Shark. South in Mexico is a similar species, the Mexican Horn Shark, *H. mexicanus*.

Identification: Horn Sharks are easy to recognize. The have short, blunt heads with ridges over the eyes, a pig-like snout, two dorsal fins, and an anal fin. The presence of an anal fin alone separates them from the other sharks with spines. The dorsal fins each have a spine in front. Horn Sharks can reach over 1 m in length. The overall coloration is mottled tan and gray to brown, with darker brown spots. The pectoral and pelvic fins are wide, using them for support as they rest or move about sluggishly on the seafloor. They have a feature more common in rays than sharks—a spiracle. Because they are not in constant motion, they must use their spiracle to pump water over their gills. Their egg cases are also easy to recognize. They are dark brown with a spiral flange along their oval axes. The developing embryo can be seen within the egg case.

Distribution: This species is found off California and Baja California. It is rare north of Point Conception. It occurs in both the eastern Pacific Ocean and the Gulf of California. It is abundant from southern California to Cabo San Lucas. It has been suggested the species occurs as far south as Peru, but evidence is lacking. The Mexican Horn Shark is found as far north as Magdalena Bay on the Pacific, and into the Gulf of California, the two areas where both species may be sympatric.

Natural History: The natural history of this species is well known, as it is a common inshore denizen of waters near a very populated area that is popular with divers: southern California. These small sharks are frequently encountered by divers in and around rocky areas and kelp beds. It is an inshore species, generally found between about 2–12 m deep, although they have been recorded from the intertidal zone to below 200 m. They migrate seasonally, being found in shallower waters during the summer and deeper waters in the winter. Juveniles usually stay in deeper water until they mature. These young animals are usually found on sandy bottoms, while the adults prefer rocky areas having crevices and caves, often near algal beds.

During the day, they are usually seen resting on rocky reefs or in caves, but are sometimes active. However, they are mostly crepuscular or active at night. While they are graceful swimmers, they spend much time on the bottom, and can "walk" with their pelvic and pectoral fins.

The generic name *Heterodon* refers to the different types (*hetero*) of teeth (*don*). Horn Shark teeth are in multiple rows that function to hold, crush, and grind prey. The teeth closest to the center are long and narrow, to hold prey. The teeth get broader and flatter away from the center, to crush and grind the prey. These different types of teeth are commensurate with their

The unusual spiral egg case of a Horn Shark. *Kevin Stolzenbach/CCA-EQ/Flickr*

feeding lifestyle. They have an extremely strong bite force, especially when stabilized on the substrate, and use suction to help secure prey. They mostly prey on benthic organisms, particularly hard-shelled animals such as bivalves, gastropods, crustaceans, and echinoderms. They also feed on soft-bodied animals such as cephalopods, marine worms, and small fishes.

Horn Sharks are oviparous. They usually lay two eggs at a time, from February through April. They oviposit between 2–4 weeks after mating. Eggs are typically laid in crevices in shallow water. The distinctive spiral egg cases are about 13 cm long. They are often wedged into nooks and crannies, but are sometimes seen washed up on beaches after being dislodged by storms. The eggs hatch in about a month. Young are about 15 cm

The spines are just anterior to the two dorsal fins. *Ed Bierman/CCA/Flickr*

long at hatching. They are known to live at least twelve years in aquaria. One unusual predator of Horn Sharks is the Bald Eagle, which has been reported to catch them in the shallows off Santa Catalina Island.

Encounters: Divers in southern California and Baja California frequently see these animals. They are occasionally caught on hook and line. They are commonly encountered off the Channel Islands, as well as the mainland, even occurring in San Diego Bay. However, they tend to be more abundant off the Channel Islands. In one study, they ranked number 1 in abundance and biomass of any nearshore species caught in traps at Santa Catalina Island, while on the adjacent mainland, they ranked number 24 in abundance and 8 in biomass (out of 98 species trapped).

These little sharks are completely harmless if left unmolested. When encountered during the day they are sluggish, and divers are often tempted to grab them by the tail. This is when accidents occur. Even if a tail-grasped Horn Shark is at arm's length it can maneuver its spines into the attacker and can wrap around to bite.

Venom Apparatus, Venom, and Symptoms: The venom apparatus is completely defensive. It is similar in structure to the other venomous sharks and chimaeras, although the spine is not movable, as it is in chimaeras. There is a single spine in front of the first and second dorsal fins. The spines are

Profile of a Horn Shark head. *Ed Bierman/CCA/Flickr*

stout and sharp, and can inflict a nasty wound. The spine tip is well below the tip of the dorsal fins. The venom is produced in the cutaneous sheath surrounding the spine, lacking a distinctive venom gland. There is more information on shark spines in the section on spiny dogfishes, as those species are better known and the spine structure is very similar. The venom is poorly known, as stings are rare. The few references I could find suggested that deep wounds caused by a Horn Shark can be extremely painful.

Remarks: Horn Sharks and spiny dogfishes can be relatively aged using their spine morphology. Horn Sharks are popular in public aquaria and do well in captivity. Their eggs can also be hatched in captivity, and the embryo can be displayed for the public by backlighting the eggshell. Because they do well in captivity, Horn Sharks are frequently used in research on physiology, brain function, feeding mechanics, and behavior.

References and Resources: Castro (2011), Clarke and Irvine (2006), Ebert (2003), Edmonds et al. (2001), Eschmeyer et al. (1983), Finstad and Nelson (1975), Halstead (1978), Huber et al. (2005), Maisey (1979), Meyer and Seegers (2012), Nelson and Johnson (1970), Pondella and Allen (2000), Segura-Zarzosa et al. (1997).

Dogfish Sharks (Family Squalidae) and Their Relatives

"I must have unhooked a dozen 'dogs' for my clients before one thrashed its tail and spine deep into my knee. It felt like white-hot fire. My lower leg began to swell almost immediately, and in a short time I couldn't put my weight on it. It was excruciating. That night, I couldn't sleep. It actually took two days to disappear. " (Telegram.com, Worcester, Massachusetts, 2013, Mark Blazis reporting)

Atlantic Spiny Dogfish. *Douglas Costa/Public Domain/NOAA*

Classification: Phylum Chordata (chordates); Class Elashmobranchii (elasmobranchs); Order Squaliformes (dogfish, lantern, sleeper, and kitefin sharks); Family Squalidae (dogfish sharks); Genera *Squalus* and *Cirrhigaleus* (dogfish sharks).

North American species include *S. acanthias* (Common Spiny Dogfish), *S. cubensis* (Cuban Dogfish), *S. clarkae* (Genie's Dogfish), *S. suckleyi* (Pacific Spiny Dogfish), and *C. asper* (Roughskin Dogfish). There are numerous species of *Squalus* and *Cirrhigaleus* that are either outside our area or are poorly known deepwater species. The vernacular name of Roughskin Dogfish is also used for a sleeper shark, *Centroscymnus owstonii* (family Somniosidae), but this species does not occur off North America.

There are at least four additional families (or subfamilies) of deepwater squaliform sharks that are known or believed to be venomous in our area. Most of these are poorly known taxonomically and geographically. They include the Etmopteridae (lantern sharks), of the genera *Centroscyllium* (*C. nigrum*, Combtooth or Pacific Black Dogfish, and *C. fabricii*, Black Dogfish), *Etmopterus* (many species), and *Trigonognathus* (*T. kabeyi*, Viper Dogfish); Dalatiidae (pygmy and kitefin sharks), with two species, *Dalatias licha* (Kitefin Shark) and *Squaliolus laticaudus* (Spined Pygmy Shark); Oxynotidae (rough-skin sharks), represented by *Oxynotus caribbaeus* (Caribbean Roughskin Shark); and Centrophoridae (gulper sharks), with *Centrophorus granulosus*, *C. tessellatus*, *C. uyato*, and *Deania profundorum* (Arrowhead Dogfish) in our area. Somniosidae generally do not have spines, or if present, are small. One wide-ranged, spined somniosid is the Velvet Dogfish, *Zameus squamulosus*.

The vernacular, "dogfish," is used for members of several families of sharks, especially squaliforms, but even some bony fishes. In this case, it reflects common usage, rather than taxonomic relatedness. Smooth dogfishes or smoothhounds (family Triakidae) are superficially similar in appearance to spiny dogfishes but lack venomous spines.

Identification: The squaliforms are a group of sharks that includes the familiar spiny dogfishes, also known as spurdogs, or colloquially just "dogs." They are generally shaped like typical pelagic sharks, or cigar-shaped. The moniker "squalus" means "shark" in Latin. They are distinguished from other sharks by this suite of characters: two dorsal fins, sometimes preceded with spines; anal fins absent; five gills; spiracles present; and no nictitating lower eyelid. Only the spiny dogfishes are familiar to anglers and fishermen; the other taxa are small and found in deep water.

The Pacific Spiny Dogfish is about 0.6–1.2 m in length (record 1.6 m). It is gray to light brown above and white on the venter. This color pattern is known as countershading. The animal blends in with the bottom or depths below when viewed from above, but blends in with the skylight when viewed from below. There are white spots on the body. The snout is long and pointed. The caudal peduncle is narrow and has a keel. The Common Spiny Dogfish is very similar, but their ranges do not overlap. *Squalus suckleyi* also has a shorter snout and there are differences in relative fin placement. The Cuban Dogfish is also a slim gray shark with a light venter, but has no spots

Newly described dogfish from the western Atlantic, *S. clarkae*. Ivy Baremore/Mar Alliance

on the body. The dorsal fins have black tips, and the pelvic, pectoral fins, and caudal fin are edged with white. The Roughskin Dogfish is a stouter animal than other species. It has a short snout and the second anal fin is higher than that of the other dogfishes. All of its fins are edged in white, but it lacks the black tips of the Cuban Dogfish. The Shortspine Dogfish is similar to the Cuban Dogfish, but differs slightly in pectoral fin shape and morphology of the dermal denticles. Genie's Dogfish was recently differentiated from *S. mitsukurii* as the species from the western Atlantic. It may take an expert to differentiate some of the squalids, but they all have the same basic shape, arrangement of fins and spines, and all are presumed venomous.

Lantern sharks are structurally similar to squalids with regards to the presence and placement of fins and the presence of spines, but they have one additional feature: photophores. They have large eyes and are small in size, from about 20–90 cm. They are dark in color, so are sometimes called black dogfishes. Unlike pelagic or nearshore species, there is limited need for countershading in the abyssal realm, although some use bioluminescence to produce counterillumination. Instead of having lightly colored venters, they tend to be black, and may have photophores on the ventral surface. The Pacific Black Dogfish is similar to its Atlantic counterpart, the Black Dogfish. Both species resemble spiny dogfishes and are relatively large for etmopterids. *Centroscyllium fabricii* is the largest species, nearly a meter in

The lantern sharks are bioluminescent deep-sea fishes that have venomous spines like dogfishes. *Vince Guida/Public Domain/NOAA*

length, while *C. nigrum* is usually less than 0.5 m long. The bioluminescent organs are scattered about the skin. At least some species of lantern sharks (e.g., *E. spinax*, from the eastern Atlantic) use their illumination to display the dorsal fin spines in the black of deep water, a phenomenon known as aposematic bioluminescence.

The gulper sharks are similar in appearance to other squaliforms, but the lower teeth are larger than the upper teeth. The dalatiids are also similar in appearance, except that kitefin sharks are distinctive. They have the basic squaliform pattern, but both dorsal fins are large and triangular. *Dalatias* has been variously reported as having venomous spines and lacking spines, but this may be an artifact of different taxonomic approaches. *Squaliolus* lacks the enlarged dorsal fins, superficially looks like an etmopterid, and indeed has spines.

Distribution: Spiny dogfishes are found both in temperate and tropical waters. *Squalus acanthias* occurs off the western Atlantic from Canada to Florida and the Caribbean, but apparently does not occur in the northwestern Gulf of Mexico. Its Pacific counterpart, *S. suckleyi* occurs from Baja California north to British Columbia and Alaska. Hawaii has one species of spiny dogfish, *S. mitsukurii*. The Cuban Dogfish is found throughout the northern Gulf of Mexico, as well as the Caribbean and western Atlantic north to North Carolina. The Roughskin Dogfish has a similar distribution off the United States, but is also found in scattered localities in the Pacific, Atlantic, and Indian Oceans. This species was not described until 1973 and is not well known.

The lantern sharks are mostly known from tropical regions. The one species in the eastern Pacific that has been documented off southern California is

C. nigrum. Hawaiian waters have quite a few species, including *C. nigrum*, Blurred Lanternshark (*E. bigelowi*), Blackbelly Lantern Shark (*E. lucifer*), Smooth Lantern Shark (*E. pusillu*s), Hawaiian Lantern Shark (*E. villosus*), and Viper Dogfish. Those from the Atlantic and/or the northern Gulf of Mexico include *E. bigelowi, E. gracilispinis* (Broadband Lantern Shark), *E. hillianus, E. schultzi* (Fringefin Lantern Shark), and *E. virens* (Green Lantern Shark). The Black Dogfish occurs from Virginia to the North Atlantic, and the eastern Atlantic. There are several other species of Etmopteridae in North America.

Two *Centrophorus* are found in the western Atlantic (*C. granulosus* and *C. uyato*), while another (*C. tessellatus*) possibly occurs off Hawaii. The Arrowhead Dogfish has been reported off North Carolina. No gulper sharks are known from the northeastern Pacific. Of the dalatiids, the Kitefin Shark is widely distributed in the Gulf of Mexico, eastern Pacific, and Hawaii. The Spined Pygmy Shark is widespread in the eastern Atlantic and Gulf of Mexico. *Oxynotus caribbaeus* occurs off Florida, the Caribbean, and the southern Gulf of Mexico. The ranges and biology of all deep-sea species are poorly known.

Natural History: The Common and Pacific Dogfish are very similar, and until fairly recently, were not considered separate species, so much of the literature refers only to a single species, *S. acanthias*. They are common species that are mostly found offshore in the south to more commonly inshore to the north. They may also occur in inland marine waters, especially in the north. While they may be found inshore and at the surface, they have also been encountered far offshore and more than 1,220 m deep. Older animals tend to forage near the bottom. Males tend to be found in shallower water than females, except during pupping season. The Shortspine Dogfish is common in deeper Hawaiian waters, usually between 100–500 m.

Squalus acanthias and *S. suckleyi* are sometimes seen in large schools feeding at the surface or near the bottom. Schools can be in the hundreds to over 1,000. Spiny dogfishes are probably the most abundant sharks in our waters. The name dogfish relates to the appearance of their hunting in packs. They are voracious feeders, eating a wide variety of prey items. They mostly feed on schooling prey, such as herring, hake, or smelt, or aggregations of invertebrates, such as shrimp (krill) and squid. In turn, dogfishes are eaten by other sharks, including their own species, larger fishes, seals, and Orcas.

While these animals may be small by pelagic shark standards, they are long-lived and have a low reproductive output. They can live in excess of eighty years and usually give birth to 5–15 pups per season. Males do not reach sexual maturity until about twenty years of age, and females mature

Cuban Dogfish. *Brenda Bowling/Texas Parks and Wildlife Department (TPWD)*

when about 35 years old. Pups are typically born in inshore areas and measure between 20 and 30 cm.

The Cuban Dogfish is generally found over the continental shelf and along the upper slope in the Gulf of Mexico, at depths of 60–380 m. They forage close to the bottom in large, dense schools, presumably feeding on schooling fishes near the bottom, as well as benthic invertebrates. They have about 10 pups per season, which tend to live in shallower water than adults. The other dogfishes are deepwater animals that are not well understood and seldom encountered by people, except perhaps for some specialized fishermen.

Lantern sharks and other deepwater species are inhabitants of the deep sea. Very little is known about their biology because records from deepwater trawls or traps are uncommon and sporadic. Even basic information such as distribution is poorly known. The Pacific Black Dogfish occurs from about 400–1,145 m deep. It is viviparous, with pups measuring 11–13 cm at birth. They are said to feed on deepwater shrimps, cephalopods, and small bony fishes in the mesopelagic realm. The species in the Gulf of Mexico are generally reported from about 200–1,000 m deep, off the slope of the continental shelf.

Pacific Spiny Dogfish are not as important as a commercial fishery as the Atlantic counterpart. In the Pacific, there had been an extensive commercial fishery for the liver oil, but that demand has dropped off. They are not commonly harvested for seafood like the Atlantic species. In 2011, for example, some 11.5 million pounds were harvested by 200 commercial anglers in New England. However, in many fisheries they are considered a disdained bycatch and unwanted nuisance, and in the eastern Pacific, they are normally treated as bycatch and released on site. However, on both coasts, dogfishes have been

The Northwest Atlantic Spiny Dogfish fishery is managed for sustainability, but throughout most of its range the species is in decline from overfishing.
Public Domain/NOAA

overharvested to meet demand at the time, and populations have declined. Conversely, some claim that numbers may have increased due to overharvesting of certain species. Common Dogfish populations are regulated. There is a commercial fishery for Cuban Dogfish, although the fish are not commonly used for human consumption. Deep-sea sharks account for about one-half of the number of shark species worldwide, and with improved technology, commercial fishermen can now harvest deepwater species, either for the liver oil or as a bycatch for other species. For example, in Hawaiian waters, there is a deepwater fishery for lutjanids (snappers) in which Shortspine Dogfish are a common, but unwanted, bycatch. Worldwide, there is a cruel and wasteful fishery for fins of nearly all pelagic species of sharks, and global shark populations are plummeting. In a practice known as finning, sharks have their fins cut off, then they are thrown back in the ocean where they slowly die.

Encounters: Spiny dogfishes are mostly encountered by anglers and commercial fishermen, especially in the North. For fishermen, dogfishes constitute an occupational hazard. The spines are an obvious feature to avoid, but these sharks can also deliver serious bites, and their sandpaper-like skin is extremely rough, due to the dermal denticles. As an example, the aptly named Roughskin Dogfish has a keel and three sharp backwards-pointing teeth on each scale. Lantern sharks and other deepwater species are normally only taken by deep-sea trawls or traps and by researchers.

Venom Apparatus, Venom, and Symptoms: There are two spines on venomous squaliforms, one in front of each dorsal fin. The first dorsal fin is the larger, although in the Roughskin Dogfish the second dorsal fin is nearly as high. The spines are shorter than the top of the flexible, cartilaginous fin, reaching one-third of the way up the fin to approximately the top of the fin, depending on the fin and the species. The crown of the spine projects out in front of the fin, rather than being within tissue. The spine is roughly triangular in cross section, recurved, and the exposed portion is grooved. The spine is supported by cartilage of the fin and articulates with the vertebral column. The spine is composed primarily of vasodentine covered by enamel. It is filled with a cartilaginous core. There is no distinct venom gland. The venom is produced in epithelial cells. A superficial sting may result in no envenomation, but can still cause mechanical damage. A deep wound would result in envenomation. Venom is not injected; rather, a deep wound would cause pressure on the soft tissues, causing venom to flow into the wound. The deepwater species are poorly known, and published records of stings are virtually unknown.

From what is known, it appears the venom apparatus is defensive, although the spine may also aid in locomotion. There is no information on the properties of the venom. It is generally regarded as being mild, although the few case studies indicate stings of spiny dogfishes may be extremely painful, and the pain is greater than would be expected from the mechanical injury alone. Symptoms of envenomation include intense pain that may last for several hours, followed by redness and swelling. The symptoms may spread to the axilla and groin and cause muscular weakness. There is an unconfirmed, but published report of a fatality due to the sting of a spiny dogfish, but details are not known. Secondary infection is presumably common, and could potentially result in a fatal outcome.

References and Resources: Castro (2011), Chen et al. (1979), Claes et al. (2013), Cotton et al. (2011), Ebert et al. (2010), Eschmeyer et al. (1983), Haddad and Gadig (2005), Halstead (1978), Jones and Geen (1976), Kells and Carpenter (2011), Maisey (1979), McEachran (2009), Meyer and Seegers (2012), Parsons (2006), Pfleger et al. (2018), Smith and Wheeler (2006), Smith et al. (2016), Tinker and DeLuca (1973).

Round Stingrays and Similar Taxa (*Genera Urotrygon, Hexatrygon,* and *Plesiobatis*)

"Several years ago, a census of rays on this beach found that about 16,000 rays live along just a few hundred yards of shoreline. And, not surprisingly, the area reports more stingray injuries than any place on Earth—about 400 a year. . . . An instructor at M&M Surfing School in Seal Beach says he's been stung 21 times in the past 30 years. "'I'm known as the king of stingray stings,' he says." (National Public Radio 21 January 2014, Jon Hamilton reporting)

Yellow Round Stingray. *Laszlo Ilyes/CCA/Flickr*

Classification: Phylum Chordata (chordates); Class Elasmobranchii (elasmobranchs); Order Myliobatiformes (thornbacks and rays); Family Urotrygonidae (American round stingrays); Genus and Species *Urobatis halleri* (Haller's Round Stingray) and *U. jamaicensis* (Yellow Round Stingray). Deepwater species with a similar stinging mechanism include two species from monotypic families, *Hexatrygon bickelli* (Family Hexatrygonidae) and *Plesiobatis daviesi* (Family Plesiobatidae).

Before a recent revision, the nomenclature of shallow-water stingrays with rounded discs was confusing, especially since several genera started with a "U": *Urolophis*, *Urobatis*, and *Urotrygon*. These genera were used somewhat interchangeably in the literature, but now there seems good consensus on taxonomic arrangement. *Urolophis* is now placed in its own family (Urolophidae), and are mostly found in Australia, while all North American species are in the family Urotrygonidae. To add to the confusion of common names, some stingrays in different families with a rounded disc are also called round stingrays.

Other species nearby in the Gulf of California include *U. concentricus* (Bullseye Round Stingray), *U. maculatus* (Spotted Round Stingray), plus three *Urotrygon* species: Spinytail Round Stingray (*U. aspidura*), Chilean Round Stingray (*U. chilensis*), and Rogers' Round Stingray (*U. rogersi*).

Identification: Like other stingrays, urotrygonids are darker above than below, a phenomenon known as countershading. The eyes and spiracle are on top, with mouth and gills below. Urotrygonids are aptly named round stingrays. Both of our native species are small, only to about 30 cm in disc width in females. The tail is slightly shorter than the disc length and relatively broad, but not whip-like. The total length is up to about 70 cm in female *U. jamaicensis* and 55 cm in female *U. halleri*. Males are smaller than females. The young are about 6–8 cm disc width at birth. The pectoral fins are broad, rounded, and contiguous with the head and body, forming the disc. The pelvic fins are short but distinct. There is a small caudal fin present at the terminus of the caudal appendage. The spine is located along the mid- to posterior-dorsum of the tail, in front of the caudal fin. Both species are also similar in appearance, having a tan or yellowish to grayish ground color, with shaded markings and spots or reticulations. The Yellow Round Stingray may be more yellow to greenish, and can change tones to better match the substrate. They blend in quite well with sandy and gravelly bottoms. The two species can be differentiated by their range, but there are anatomical features. Yellow Round Stingrays have larger eyes and a narrower disc than Haller's Round Stingray, and they possess dorsal dermal denticles, which are lacking in *U. halleri*.

The Giant Stingaree has a relatively round disc, although it tapers anteriorly, and it is decidedly larger than round rays at 2.7 m in length and 1.5 m in disc width. The caudal appendage is similar to urotrygonids. They are born at about 50 cm disc width. This species is grayish, purplish, or reddish in dorsal coloration.

The unusual deep-sea hexatrygonid, a Sixgill Stingray. *NOAA/CCA-SA/Flickr*

The unusual Sixgill Stingray has a very different appearance from the others. It has a long, wide, flexible wedge-shaped snout and a tapering disc that is broad in front. It has two pairs of six gills, rather than the usual five in all other stingrays. The body is almost gelatinous, and it lacks dermal dentacles. This species reaches about 1.7 m in length and young are around 45–48 cm at birth. The body is pinkish to reddish with some light markings. The snout appears somewhat translucent. The caudal appendage is similar to urotrygonids.

Distribution: The Yellow Round Stingray is primarily a tropical animal found from North Carolina to Trinidad. It is most common in the Greater Caribbean and off southern Florida. It occurs throughout the Gulf of Mexico, but is not common in coastal waters of the northern Gulf. Haller's Round Stingray is reported from Humboldt Bay, California, to Panama, but is most common off southern California and adjacent Mexico. It is found throughout the Gulf of California, where it may co-occur with several species mentioned in the Classification section. Both of the deepwater species are found in Hawaii and the tropical Indo-West Pacific.

Natural History: The two round stingrays are inshore marine species. *Urobatis halleri* is found in about 1–90 m of water, generally over sand and mud bottoms, along beaches, and in estuaries and sloughs. They are often seen (or not seen) partially buried in the substrate. They may migrate in

Haller's Round Stingray. *Ingrid Taylor/CCA/Flickr*

response to environmental conditions, including water temperature, season, currents, tides, and photoperiod. In the spring through fall, they are usually in shallow waters to about 15 m deep. They seek slightly deeper waters in the winter. *Urobatis jamaicensis* is also found in shallow water, usually from about 1–45 m deep. They are mostly found on sandy substrate, often near reefs, but have also been noted over muddy substrate, in eelgrass beds, and in estuaries and sloughs. They seem to have more site fidelity than *U. halleri*, and do not tend to migrate to deeper waters seasonally, but more research is needed on this topic.

Seal Beach, California, has an infamously dense number of *U. halleri* compared to neighboring beaches of southern California, although the species is common elsewhere. Because of the high stingray and human population, many studies on its biology have occurred there. Seal Beach is a draw for *U. halleri* because of warm thermal effluent generated by electricity-producing plants. Seawater is used as a coolant and returned near the shore as warm-water discharge. The San Gabriel River dumps into the Pacific Ocean here as well, so with the warm water, influx of fresh water, and deposition of fine substrate material, the area mimics estuarine conditions. There is also a restored estuary occupied by this species, the Seal Beach National Wildlife Refuge.

Haller's Round Stingrays were monitored there and were mostly found in the first few meters of water depth, generally within 30 m of the surf zone. They tended to have limited activity, but were active day or night, depending on conditions. During the cooler months, they disperse to deeper waters. Yellow Round Stingrays also use nearshore marine and estuary habitat and have been recorded entering rivers.

Urobatis halleri females move inshore to mate in March and April to June and give live birth to young in June to October. Female *U. halleri* average 2–3 young (range 1–6), while *U. jamaicensis* is reported to give birth to 3–4 young. In the warmer waters of Mexico, *U. halleri* can have two litters per year, in the spring and in the fall. The young stay in warm, shallow waters, usually less than 2 m deep.

Both species of round stingrays feed on benthic organisms, including marine worms, crustaceans, mollusks, and small fishes. They expose small benthic animals by using their disc to create a small crater. This also gets the attention of small fishes that may come to feed on exposed invertebrates; in turn, the fishes may become stingray food. Like other species of rays, predominate predators are probably sharks, but marine mammals will also feed on them. Tiger Sharks are likely predators of *U. jamaicensis*.

The Hawaiian deepwater species are reported from 275–680 m (*P. daviesi*) and 500–1,120 m (*H. bickelli*) deep. They likely feed on crustaceans, mollusks, small fishes, and other animals easily overpowered on or under the seafloor. The Sixgill Stingray uses its long, flexible snout for searching for small animals in the silty to rocky substrate, as suggested by the rows of ampullae of Lorenzini on the underside of the snout. They use their extendable mouths to suck up small prey. The Sixgill Stingray is viviparous, having 2–6 pups. The Deepwater Stingray is presumed to be viviparous, although there is a lack of detailed life history information. These deepwater species fall prey to deepwater sharks. Cookie-cutter and Kitefin Sharks are known to excise chunks of flesh from these bottom dwellers.

Encounters: Round Stingrays are most often encountered during the summer and early fall, when they may aggregate in the shallows. It has been estimated that the number of stingray envenomations from all species in U.S. waters ranges between 750 and 1,500 per year, although the latter is a more recent figure and seems more likely. Also, it has been estimated that only a fraction (25% is one estimate) of stings is reported or tracked. Some stings do not cause envenomation, as when the integumentary sheath is lost. Even among cases involving envenomation, most are treated on site (especially by lifeguards) and not reported to a poison control center or emergency room.

Haller's Round Stingrays account for the highest percentage of exposures among stingrays in California. In one media report, 60 people were stung in a single day in San Diego, but it was also reported that a 2 m ray was among the culprits, so at least one sting was due to a different species. Each year there are hundreds of reported stings in California, and the highest incidence is often near Seal Beach, which is not only an important rookery, but also a popular recreational area. According to various studies, there is an average of 226 stings per year in Seal Beach alone, higher during El Niño years, and some years reaching about 500 or more, as in 2008. Seal Beach has the distinction of having the greatest number of stingray stings for a similar-sized area anywhere in the world. There are an estimated 16,000–40,000 Round Stingrays at Seal Beach during the summer. The highest density is near the north end of the pier, an area known as "Ray Bay." It is designated as a surfing area, in part to discourage swimmers and waders. In the calm waters of the Gulf of California, densities of the several species of urotrygonids can be daunting. It is not uncommon to find urotrygonids every couple of meters along the bottom, in shallow water. Urotrygonids are most common in areas without pounding surf, including the protected waters of lagoons and estuaries.

Stings by Yellow Round Stingrays are not well tracked. Atlantic and Gulf of Mexico exposures are probably mostly due to other species, but certainly people in Florida to North Carolina are sometimes envenomated by *U. jamaicensis*. Of the 153 cases of stingray envenomation in Texas from 1998–2004, probably few to none were likely attributable to this species. During 3,211 survey dives in the Gulf of Mexico, Yellow Stingrays were reported only 8 times (0.2%); in the northern Gulf of Mexico, zero were reported during 214 dives. Compare this to detections during about 15% of dives off southern Florida. In Florida to North Carolina, they may be common in eelgrass beds, where they breed.

These are sometimes regarded as docile animals that are approachable by cautious divers, but they will defend themselves if they feel threatened. They are also docile in public aquaria, where they fare well in captivity. Stings occur primarily when swimmers step on them. Divers are sometimes stung if they are too close to the bottom. Surfers are also frequently stung. People are also stung while fishing. Surf fishermen are sometimes stung when they trod upon them. Recreational anglers often catch them from piers and boats and may be stung while trying to remove the hook, release them, or when the animal is thrashing onboard. They are also an unwanted bycatch from bottom trawlers. In Mexico, fishermen may cut the tails off before releasing them, which is a death sentence. Yellow Round Stingrays are feared by commercial fishermen in the Caribbean, so it seems likely stings are not uncommon there.

The two species of deepwater stingrays are rarely encountered, except by commercial anglers and researchers. These are large rays that should be considered potentially dangerous when brought on board.

American round stingrays and similar species (e.g., freshwater stingrays [*Potamotrygon* spp., family Potamotrygonidae]) are common in public aquaria, and not too uncommon in larger home aquaria. One popular aquarium species is the attractively patterned Bullseye Round Stingray of the Gulf of California.

Venom Apparatus, Venom, and Symptoms: There are four spine, caudal appendage, and tail arrangement types recognized among stingrays, named for predominant taxa. The form of the round stingrays and these deepwater species is the "urolophid" or "urotrygonid" type, characterized by a relatively short, but muscular caudal appendage, with a moderately sized spine on the middle or distal third of the appendage. The formidable business end is variously known as the caudal spine, barb, stinger, lancet, and sting. Regardless of terminology, it is similar in basic design among all stingray taxa. The hard portion of the spine is bone-like, composed primarily of vasodentine and covered by enamel. It is tapered to a sharp point at the distal end and has numerous recurved teeth that form a dentate margin. The spine is often dorsoventrally flattened. There is a dorsal ridge and ventrolateral grooves along the axis. Spines are shed and replaced, so in some species there is at least one spine, but there may be several spines or replacements. The hard part of the spine is covered by a sheath of integument. The integument covers the dentate margin, giving it a smooth, slick surface. The sheath is composed of dermis, epidermis, and connective tissue. The venom is produced by glandular cells in the integument, particularly in the ventrolateral grooves.

Envenomation occurs when the animal thrusts the spine into the victim, then withdraws it. The action is similar to a scorpion sting. When the spine enters human flesh, the integumentary sheath is ruptured and venom is discharged up the grooves. When the spine is withdrawn, the integumentary sheath may remain in the wound and continue envenomation. Because stingray spines are usually barbed, they often break off in the wound.

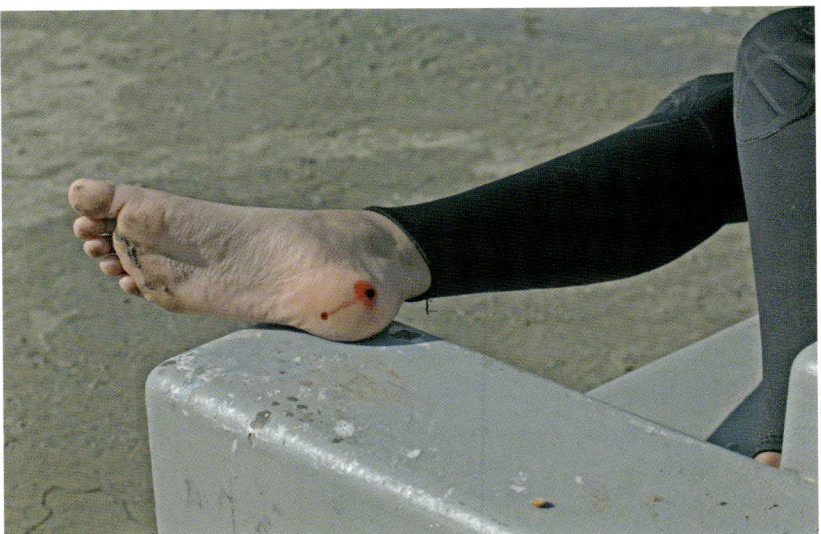

Stingray injuries are common in southern California. Most are due to Haller's Round Stingray, although several species occur there. *Bengt Nyman/CCA/Flickr*

In all stingrays, mechanical injury is caused by the physical properties of the hard part of the spine. Despite their small size, round stingrays have an impressive stinger, about 2.5 to about 4 cm in length. Even though this is smaller than other rays, it is about as long as, and wider than, a small construction nail, and shaped like a small serrated knife. By using powerful blows from the caudal appendage, round stingrays can direct the spine and drive it deeply into the foot or ankle of swimmers.

The venom of stingrays is not well known. This is partly because stingrays do not have distinct venom glands; rather, venom is produced by cells within the integument, so extraction and isolation of toxic fractions is inherently problematic. Much of the scant information we have on venom comes from studies of *U. halleri*. The venom contains serotonin, which helps account for the intense pain, but there are at least five protein fractions, including 5'-nucleotidase and phosphodiesterase. Proteolytic activity has not been observed in *U. halleri*. The LD_{50} is reported to be 28 mg/kg. High doses of its venom are known to cause cardiovascular and respiratory effects in mammals. In laboratory animals, this can cause a precipitous drop in blood pressure and complete cardiovascular collapse. There can also be alterations in the respiratory and central nervous systems, but not on neuromuscular coordination.

Stings from southern California (mostly due to *U. halleri*), in which envenomation was indicated, usually include intense to excruciating pain that increases for about ninety minutes. The area of the wound usually becomes discolored. Swelling is always present. Other common symptoms include weakness, nausea, nervousness, and sweating. Less common symptoms include vomiting, diarrhea, changes in blood pressure, fainting, tremors, cramps, groin and axillary pain, and respiratory distress. Cardiac problems and even spasms or convulsions have occurred. There is occasionally necrosis, especially if untreated. Stings by round stingrays are thought to be less severe than some other larger species. Little is written about the sting of *U. jamaicensis*, but the pain is said to vary from an itch to extreme. The venom is reported to be potent, so symptoms are probably similar to *U. halleri*. Stings from urotrygonids are not known to be fatal.

Freshwater stingrays in home or public aquaria have a venom apparatus similar to urotrygonids, but cause somewhat different symptoms than marine species. The symptoms are often more disconcerting than native species. While the stings are extremely painful, necrosis, skin lesions, and blistering are common effects. There seems to be high enzymatic activity following the sting of a freshwater ray; hyaluronidase is present, but specific enzymes are poorly known. Muscle cramps, spasms, and other neurotoxic symptoms may also occur. Freshwater stingray wounds may take several months to heal.

Because the stinger often breaks off in the muscle tissue, radiographs are useful to find fragments, although they sometimes go undetected. Secondary infection is very common from stings by all species of stingrays, and prophylactic antibiotics are often administered. There is no antivenom for stingrays.

Remarks: One novel approach to reducing the number of exposures among swimmers at "Ray Bay" was conducted when scientists and local authorities joined forces to catch, mark, and remove stingers from Haller's Round Stingrays. Stinger removal or cutting is often done for stingrays in public aquaria and touch tanks, as well, so that accidents won't happen to the public or aquarium staff. Yellow Round Stingray populations are apparently declining in Florida.

References and Resources: Acott and Meier (1995), Ebert (2003), Eschmeyer et al. (1983), Forrester (2005), Halstead (1978), Halstead and Mitchell (1963), Hoese and Moore (1998), Hoisington and Lowe (2005), Jirik and Lowe (2012), Johansson et al. (2004), Kells and Carpenter (2011), Last and McEachran (2016), Lovejoy (1996), Lowe et al. (2007), McEachran and Carvalho (2002), Mull et al. (2008, 2010), Parsons (2006), Plank et al. (2010), Russell (1953, 1959, 1972), Russell and Van Harreveld (1954), Spieler et al. (2013).

Dasyatid Stingrays (Family Dasyatidae)

"Of the 623 cases [of stingray envenomation] reported to the authors in the past four years [all species combined] . . . 112 were seen by a physician at some time . . . 36 victims were hospitalized . . . there were two fatalities . . . [including] eleven year old white male who fell from an inner tube . . . near Galveston . . . he complained of severe pain in his left side . . . [four days later, despite intensive medical intervention] he was pronounced dead." (Russell et al. 1958)

Roughtail Stingray. *Philippe Guillaume/CCA/Wikimedia Commons*

Classification: Phylum Chordata (chordates); Class Elasmobranchii (elasmobranchs); Order Myliobatiformes (thornbacks and rays); Family Dasyatidae (dasyatid stingrays); Genera *Bathytoshia*, *Dasyatis*, and *Hypanus*.

North American dasyatids include *Bathytoshia centroura* (Roughtail Stingray), *B. lata* (Brown Stingray), *H. americana* (Southern Stingray), *H. dipterurus* (Diamond Stingray), *H. sabina* (Atlantic Stingray), and *H. say* (Bluntnose Stingray). Until recently the North American species were usually placed in the single genus *Dasyatis*, and some lists and authorities still use that approach. *Dasyatis brevis* (Whiptail Stingray) is sometimes regarded as a distinct species, but is treated here as a synonym of *H. dipterurus*.

The family Dasyatidae is found worldwide, mostly in tropical and subtropical marine waters, although some occur in fresh water. There are about 90 species currently recognized. One commonly kept marine aquarium species of exotic dasyatid is the Blue-spotted Ribbontail Stingray (*Taeniura lymma*). The Pelagic Stingray belongs to this family, but is covered in a separate account because of its different habits.

Identification: The disc of dasyatids tends to be diamond-shaped, although there is variability between species. The disc is composed of the head, body, and pectoral fins. The eyes and spiracles are on top of the disc and the mouth is below. The small pelvic fins are found behind and underneath the rear of the disc. There are no dorsal, anal, or caudal fins. While dasyatids have long tails that are often whip-like, they are generally not as long and whip-like as those of the myliobatids or aetobatids. There is usually a single caudal spine, but there may be several. They are located on the dorsal surface of the caudal appendage, in an area that marks the approximate transition from the muscular proximal portion to the whip-like distal portion. Our native species are drab in color to blend in with sand. Dasyatids exhibit countershading, being darker on top and light colored below. In the sand, they blend in well,

Southern Stingray. *Rachel Graham/Mar Alliance*

but they do swim in the water column, so the light venter blends with the skylight when viewed from below. Dasyatids can get quite large. One of our species can get over 2 m across the disc, with a maximum total length of about 3 m. Females tend to be larger than males.

The Diamond Stingray is typical of the family, having a distinctly diamond-shaped disc. Females reaching about 1.2 m across the disc to nearly 2 m in total length. It is brown above and whitish below. The Brown Stingray is similar in appearance, but only reaches about 1 m disc width. The Southern Stingray is also similar, but is chocolate brown to olive green or gray above and whitish below. Females reach about 1.4 m in disc width. The Atlantic Stingray is relatively small, with adults measuring 0.4–0.6 m in disc width. It is yellowish to brown above and whitish below. It can also be recognized by its relatively narrow, triangular snout and rounded disc. The Roughtail Stingray also has a less angular disc, and it possesses thorns on its mid-dorsum and tail. It can grow to huge proportions, with large individuals 2.2 m across at the disc, nearly twice that long including tail, and weighing between 270–360 kg. It may be the largest stingray in the world. The Bluntnose Stingray also has a less angular disc, but it lacks thorns of the Roughtail. The Bluntnose varies widely in dorsal color, but those in the southern United States tend to be brown dorsally. It reaches about 1 m in disc width.

Distribution: The Atlantic Stingray is found from Chesapeake Bay to Campeche, Mexico. It is found throughout the Gulf of Mexico. The Southern Stingray occurs from New Jersey to southern Brazil and throughout the Gulf of Mexico and Caribbean. The Roughtail Stingray has a disjunct distribution in the western Atlantic, from Massachusetts to Florida and the Bahamas, plus part of the Gulf of Mexico. The Bluntnose Stingray is found from Massachusetts (rarely), south to Argentina. The Diamond Stingray ranges in the eastern Pacific from southern California to

Atlantic Stingray. *L.L.C. Jones*

northern Chile, including the Galapagos and Hawaii; it has been reported from Canada, but this record is questionable. It is particularly abundant off the coast of Mexico, including the Gulf of California. The Brown Stingray is the common dasyatid of the Hawaiian Islands, hence is often called the Hawaiian Stingray, but it also occurs in the eastern Atlantic and Indo-West Pacific.

Natural History: Most of these dasyatids are found in shallow inshore waters, from near the surf line to about depths of 10–100 m. *Bathytoshia centroura* is more of a cold-water species, so it tends to be found in deeper water. By day, dasyatids commonly bury themselves in sand and mud, often with only their eyes and spiracles above the substrate. They are usually found in shallows of the open ocean or in bays, lagoons, and estuaries. On occasion, some species may enter fresh-water river systems, especially the Atlantic Stingray. They may be found in areas near eelgrass or algae. Atlantic Stingrays are known to cruise along the coast in the surf zone, as they are adept at managing their movements in rough water.

Dasyatids can be active day or night, but in general, they are most active at dusk and at night. Night divers frequently observe them swimming over the bottom in search of prey. Waders and fishermen may see stingrays on the bottom when using powerful lights to look for crabs or flatfishes. In a study of activity patterns of *B. lata*, it was found to be much more active at night while foraging. One interesting observation is a diel reversal in feeding time due to ecotourism. "Stingray City" in Grand Cayman Island is a popular tourist destination, where snorkelers can swim with Southern Stingrays. The rays became adapted to supplemental feeding during the day, so switched from being nocturnal to daytime foragers. Dasyatids eat a variety of animals, but prey mostly consists of benthic invertebrates, including mollusks, crustaceans, worms, and small fishes. They are adept at crushing hard-shelled prey, including clams. They can locate prey with their electrosensory organs, the ampullae of Lorenzini, which are commonly found in sharks and other rays. When feeding, they fan their pectoral fins to expose small animals in the sand or mud bottom; this may also attract small fishes that they may feed on.

During the day, dasyatids are cryptic when buried in the substrate. This behavioral trait is to avoid predation, with the stinger as a back-up defense. Sharks are known to prey on rays. Great Hammerhead Sharks have been observed feeding on Atlantic Stingrays in the wild. In fact, the observation led to a better understanding of the purpose of the strange head shape of the hammerhead. The shark uses its head to pummel and pin the ray so it can be safely

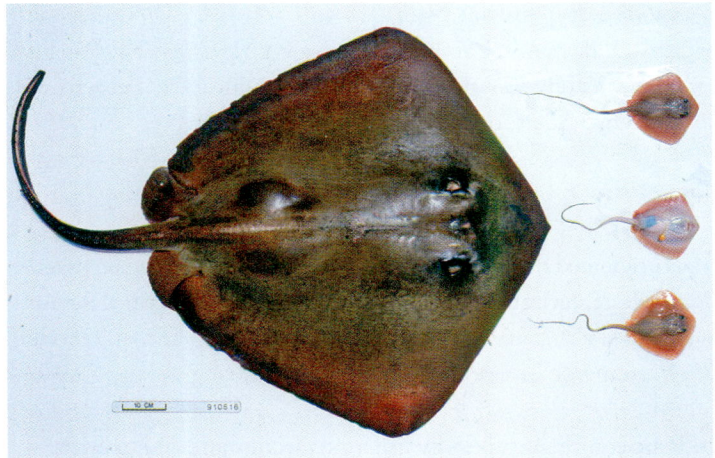

A Bluntnose Stingray and its newborns. *NOAA/Public Domain*

eaten. Bullsharks are also known predators of Bluntnose Stingrays. Aside from sharks and probably some marine mammals, humans are known predators. For example, there is an important fishery for *H. dipterurus* in Mexico. Because of the low reproductive rate of dasyatids, local populations may not be resilient to fishing pressure, so there is concern about unregulated fishing practices.

Dasyatids are sexually dimorphic and the sexes may differ in local migration patterns. Studies on *H. sabina* have shown that the dimorphic tooth patterns of males are temporary—they need long, sharp teeth for holding a mate, as the normal grinding teeth are less than optimal for maintaining a purchase when mating. Stingrays have aplacental viviparity, giving birth to live young without the aid of a placenta. Dasyatids give birth to a low number of young per season. The number of offspring ranges from about 1–10, but is usually 2–4 for most species.

Encounters: For swimmers, waders, snorkelers, SCUBA divers, and anglers in the Atlantic, Gulf of Mexico, southern California, and Hawaii, there is an inherent risk of encountering dasyatids. They can be extremely common in coastal waters, bays, lagoons, and estuaries. Beachgoers in southern California are much more likely to encounter Haller's Round Stingrays than *H. dipterurus*, but the possibility exists, as they are common near rocky reefs. In the Atlantic and the Gulf of Mexico, the Atlantic Stingray is the most common envenomator. For those swimming in the southern Atlantic and Gulf of Mexico, stingrays should be considered omnipresent. Even if a stingray is not buried in the sand or mud, it is difficult to impossible to see. In Hawaii, stings undoubtedly occur from dasyatids, but information is scant. In Kaneohe Bay on Oahu, a radiotelemetry study showed that *B. lata* were typically in deeper water than that occupied by waders, on silt and sand bottoms.

There is little information on regional sting statistics, but in a study of calls to poison control centers in Texas from 1998–2004, 153 stingray envenomations were reported, probably a fraction of the actual number of incidents. Most calls originated during the summer, especially August, when rays are most common inshore. Swimmers and anglers were the usual victims. Seventy-three percent of the victims were males over nineteen years of age, and (fortunately) only 2% were under six years of age. Also not surprisingly, stings mostly took place in the beach areas of Galveston, Corpus Christi, and South Padre Island. Dasyatids are commonly taken by fishermen from shore and in boats.

Dasyatids can get large and they have considerable striking distance, so stings can occur in any number of places on the human body. If stepped on, a dasyatid can usually strike higher on the leg than a urotrygonid. Divers and snorkelers may be stung if swimming near the bottom.

Two other methods of encounters include intentionally diving with stingrays and when they are in aquaria. Travelers may swim with stingrays in a number of places from Florida and the Caribbean. The most famous of these sites is Stingray City, off Grand Cayman Island. These rays are docile, but stings can happen, so caution is advised. Some operators have claimed that they have hosted tens of thousands of snorkelers without an incident, but there are a few records of stinging accidents. Stingrays are common in public aquaria, and there are often touch tanks where the public can feel and feed live stingrays. However, the common practice is to clip off the spines periodically so the public is not at risk. In home marine aquaria, the Blue-spotted Ribbontail Ray is popular for people that have a large enough aquarium with enough sandy bottom space. *Taeniura* is an incredibly beautiful animal that is small for a dasyatid (35 cm maximum disc width). Advanced aquarists can easily maintain them in captivity. Although they are typically quite docile, I personally had a close call with one. I would often feed the rays by hand at the surface of their aquaria. This was normally not a problem until the day someone dropped a large box on the floor. I assume the ray detected the pressure waves and felt threatened. It lashed its tail at me with lightning speed, narrowly missing my arm.

Surfers, inner tubers, and boogie boarders may be stung when they get off their floating devices. When people are suspended on the surface, they may not be as noticeable, and stingrays may not have ample warning of the presence of dangers. Anglers are inherently at risk, and are usually stung when they let their guard down. Dasyatids can be quite large and can thrash their body and tail out wildly. Such an animal on a boat deck with limited space is poised to cause injury.

Venom Apparatus, Venom, and Symptoms: The venom apparatus is similar to round stingrays, but the caudal appendage is usually longer and the caudal spine of large dasyatids is long. The spine is usually situated between one-third and one-half the distance down the tail on the dorsal surface. Appropriately, the stinging apparatus is often referred to as the "dasyatid" type. The spines are among the longest of stingrays, but references to measurements are difficult to find or substantiate in the literature. The largest caudal spine of any stingray is usually attributed to the freshwater dasyatid, *Himantura polylepis*, at 38 cm. A length of 35 cm is commonly cited as a maximum size for the marine Common Stingray, *D. pastinaca*. The Diamond Stingray has a spine about 10 cm in length. The spine of the Atlantic Stingray is about

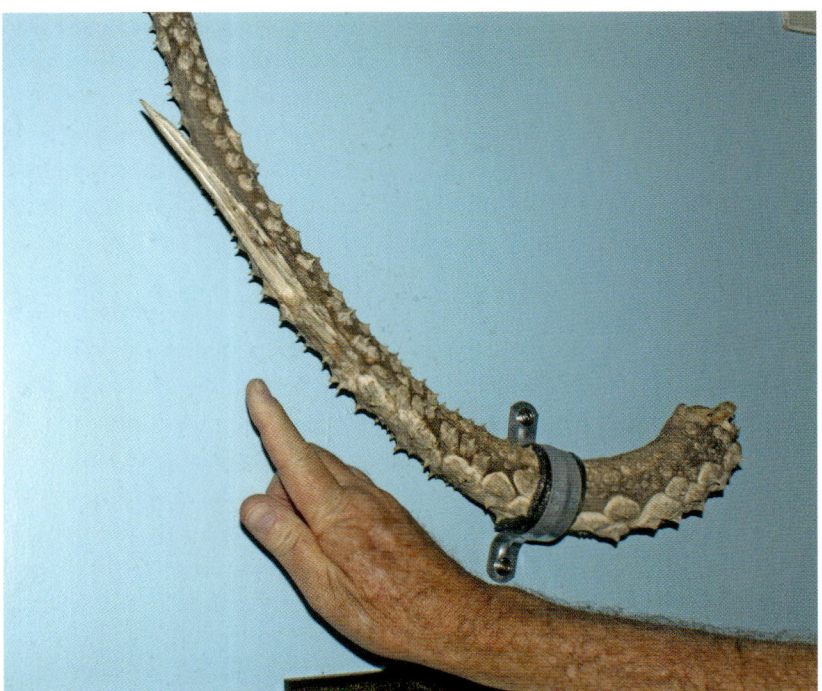

"Selfie" of my hand with the stinger of a large Roughtail Stingray. The visible portion would be about 12 cm x 2 cm—a formidable weapon. *L.L.C. Jones*

25% of the disc width, so perhaps up to 15 cm long. Roughtail Stingrays can have a spine at least 20 cm long. When a dasyatid strikes, it can arch its back high, raise the tail, and give a rapid and precise sting, much like the action of a scorpion, which is why it may strike high on the leg or body.

The venom composition is not well known for the various species, but toxic components are presumed to consist of proteins, including enzymes, and serotonin. The symptoms are similar to those described for the round stingrays, except large dasyatids with intact integumentary sheaths are thought to be more dangerous, if for no other reason than having a longer spine and more venomous tissue. Although rare, there are fatalities recorded for dasyatids, including in the United States. One detailed case study is of the boy in Galveston who was impaled in the abdomen while playing in the surf, reputedly by a Southern Stingray. Another fatality happened to a 12-year-old male in the northern Gulf of California, when a Long-tailed Stingray (*H. longus*, similar to *H. dipterurus*) stung him in the lower abdomen while he was working on a shrimp trawler. In cases where envenomation occurs in the abdomen or thorax by a large stingray, it is difficult to differentiate the medical problems caused by the mechanical injury versus the envenomation, although both play a part. Large doses of stingray venom are fatal to laboratory mammals.

Symptoms of dasyatid envenomation is generally similar to round stingrays. In a study of 153 stingray stings off the Texas Coast, based on poison control center calls, and another previous study, the most common clinical

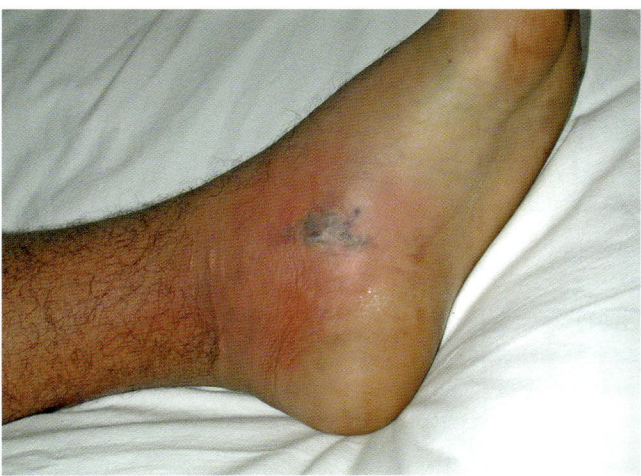

Injury caused by a Southern Stingray. *Vidal Haddad Jr.*

A Southern Stingray in knee-deep water of southern Florida. *L.L.C. Jones*

effects were puncture wounds and dermal irritation or pain. Other symptoms included redness, swelling, bruising, hives, welts, intense itching, bleeding, abdominal pain, hypotension, chest pain, diarrhea, headache, dizziness, muscular weakness, and numbness. In the cases where the outcome was known, about half (53%) of the effects were considered medically minor, 43% were moderate, and 4% were major. Infections are common, and for those seeking medical assistance, physicians usually prescribe prophylactic antibiotics.

References and Resources: Acott and Meier (1995), Bres (1993), Brisset et al. (2006), Cartamil et al. (2003), Corcoran et al. (2013), Dellias et al. (2004), Ebert (2003), Eschmeyer et al. (1983), Forrester (2005), Haddad (2004), Haddad et al. (2004), Halstead (1978), Halstead and Mitchell (1963), Henningsen (2000), Johnson and Snelson (1996), Kajiura et al. (2000), Kells and Carpenter (2011), Last et al. (2016b), McEachran and Carvalho (2002), Meyer (1997), Parsons (2006), Russell (1953, 1959), Russell et al. (1958), Schiera et al. (2002), Schwartz (2005), Snelson et al. (1988), Strong et al. (1990), Tinker and DeLuca (1973).

Pelagic Stingray
(*Pteroplatytrygon violacea*)

"It is believed that the venom of this ray is particularly potent due to the avoidance behavior exhibited by other fish nearby. The danger to humans is minimal due to the pelagic habitat of this species. However, caution must be taken when handling this ray to avoid injury from the venomous spines." (Bester et al., Florida Museum of Natural History website)

Classification: Phylum Chordata (chordates); Class Elasmobranchii (elasmobranchs); Order Myliobatiformes (thornbacks and rays); Family Dasyatidae (dasyatid stingrays); Genus and Species *Pteroplatytrygon violacea* (Pelagic Stingray).

This species is the only member of its genus. In some earlier literature, it was considered to belong to a separate monotypic family, but it is closely related to other dasyatids. However, in other earlier literature, it was not only retained in Dasyatidae, but was considered to belong to the genus *Dasyatis*.

Identification: Adult Pelagic Stingrays are fairly small, reaching about 0.6 m across the disc and 1.2 m long, but captive animals can exceed 1 m in width and weigh 50 kg. They have a broadly curved anterior disc. The disc is thick in cross section. They are dark-purple to blue-green above and purple-gray

The Pelagic Stingray is in the Dasyatidae, but is the only pelagic species. Note the position of the stinger. *"RonW"/CCA-EQ/Flickr*

below. The tail is slender and whip-like. The long caudal spine is about one-third of the way along the tail. It can be differentiated from all other dasyatids by color, having a dark dorsum and venter, the rounded snout, and habitat.

Distribution: This is the most widely distributed dasyatid, being found in all tropical and warm-temperate oceans. In the western Atlantic, they have been recorded from Newfoundland to Uruguay. They occur on both sides of the Pacific and near oceanic islands, including Hawaii. In the cooler waters along California, they were considered uncommon to rare in earlier literature. However, there are now numerous records, especially south of Monterey Bay, suggesting it is actually relatively common. Pelagic Stingrays have been detected off California year-round, but mostly in summer and fall. They follow warm ocean currents and migrate seasonally. During El Niño years, they have been found as far north as British Columbia.

Natural History: *Pteroplatytrygon violacea* is a relatively well-studied species, due to its wide range and abundance, but also because it can be a significant but undesirable bycatch for commercial offshore fishermen targeting pelagic food fishes, including Swordfish, tunas, and sharks. In some fisheries, Pelagic Stingrays may be the most common species caught by longline and drift gill-net fisheries. Bycatch of this species may account for up to 80% of the haul, so constitutes an economic hardship, not to mention safety concerns for deck-hands. These animals are also kept in some public aquaria. Much of what we know about them comes from fisheries research and captive situations.

This is the only pelagic dasyatid. It rarely ventures inshore, and is usually found offshore in the upper 90 m or so of the surface. Pelagic Stingrays need warm water, requiring temperatures above about 19°C. They are well suited for life in the open ocean, swimming with their powerful pectoral fins, rather than undulating as do inshore species. Also, Pelagic Stingrays rely on sight more than electroreception to locate prey, but they do use their arsenal of electroreceptors and other senses to feed. Pelagic Stingrays feed on a variety of vertebrates and invertebrates in the open ocean. Schooling species are frequently taken. Prey includes herring, mackerel, squid, krill, ctenophores, and jellyfishes. To catch their prey, they wrap the pectoral fins around the animal and move it to their mouth. The mouth has sharp teeth to hold onto prey, rather than the crushing and grinding teeth of bottom-dwelling rays.

Pelagic Stingrays migrate seasonally, following warm water currents. In the western Atlantic, they follow the Gulf Stream from December to April and move to the continental shelf during the summer. They are known to aggregate in specific breeding areas. Western Atlantic breeding is thought to occur in the West Indies during the warmer months. There are two migration/breeding populations thought to occur off California: one in the oceanic waters off central California and the other from Japanese and British Columbia waters. Eastern Pacific populations also reportedly breed off Central America November to March.

They have aplacental viviparity. Four to thirteen pups are born alive. The young are about 12–25 cm in disc width at birth. Along with this relatively high number of offspring (for a stingray), they can also grow and reach sexual maturity relatively quickly. Because of this, they have among the highest reproductive rates of elasmobranchs. Captive animals have about five times the growth rate of wild animals. There is uncertainty about the effects of commercial fishing on populations. Some studies suggest the large bycatch may harm populations, especially as bycatch increases with more efficient fishing methods. Other studies suggest that because they are relatively fecund, they are predisposed to recover from heavy fishing pressure. At least one study has shown that populations may actually increase when fishing pressure depresses target species, with which they may compete for food.

Pelagic Stingrays are preyed upon by sharks, including Great White and Oceanic Whitetip Sharks. There has been an amazing online video circulating the Internet that showed an Orca stunning or killing a Pelagic Stingray near the surface with a powerful blow from its flukes.

Encounters: Pelagic Stingrays are mostly encountered by commercial fishermen, researchers, open-water boaters, and public marine aquarists.

Venom Apparatus, Venom, and Symptoms: For general stingray information, refer to the Urotrygonidae, although *P. violacea* has the dasyatid venom apparatus type. The caudal spine of *Pteroplatytrygon* is quite long and located one-third to one-half down the length of the whip-like tail. In one study of 335 specimens, the caudal spine measured 75–156 mm long. Its venom is thought to be very potent. Symptoms are not well-documented, but probably similar to other marine stingrays. Two fatalities have been reported, one to a commercial longliner and one due to tetanus.

References and Resources: Acott and Meier (1995), Ebert (2003), Eschmeyer et al. (1983), Forselledo (2008), Kells and Carpenter (2011), Last et al. (2016), McEachran and Carvalho (2002), Mollet (2002), Mollet et al. (2002), Parsons (2006), Piovano et al. (2010), Ward and Myers (2005), Wilson and Beckett (1970).

Butterfly Rays (*Gymnura altavela* and *G. marmorata*)

"Spiny butterfly rays move into the coastal areas of New Jersey in the summertime and are usually caught by anglers fishing for sharks. They fight hard and are considered gamefish in some parts of the world. While they are massive, maxing out at a width of about 7 to 8 feet [2.1–2.4 m], they are harmless to humans. Their only defense is a tail spine that can cause a painful wound if stepped on." (*On the Water* staff writer, 29 July 2014)

Classification: Phylum Chordata (chordates); Class Elasmobranchii (elasmobranchs); Order Myliobatiformes (thornbacks and rays); Family Gymnuridae (butterfly rays); Genus *Gymnura*; Species *altavela* (Spiny Butterfly Ray) and *marmorata* (California Butterfly Ray). The Smooth Butterfly Ray (*G. micrura*) of the Atlantic and Gulf of Mexico lacks a spine, so is nonvenomous.

Identification: These strange fishes are unmistakable—their appearance is reminiscent of a flying wing aircraft. They have a very broad disc, about twice as wide as long. The tail is short and usually possesses at least one well-developed caudal spine near its base. Females of both species reach a larger size than males. The Spiny Butterfly Ray gets huge. Females usually

This Spiny Butterfly Ray is swimming away from the photographer.
H. Weerman/CCA/Wikimedia Commons

reach about 2 m disc width as adults, while males reach about 1.2 m disc width. The largest confirmed record I am aware of is a 2.6 m female. They often have markings, including marbling, mottling, or spots. Some individuals may be very dark, almost black. California Butterfly Rays are similar, but females only reach about 1.2 m across at the disc, while males reach about 0.9 m disc width. They are olive brown or brown to gray dorsally, sometimes with mottling or spots, and whitish on the venter.

Distribution: *Gymnura altavela* occurs in warm temperate regions of the Atlantic, from Massachusetts to Argentina. It is also found in the eastern Atlantic from the Mediterranean to central Africa. Most references say it has not been recorded from the Gulf of Mexico, while others show it as occurring in the southern Gulf. At any rate, it is absent to rare in the northern Gulf. *Gymnura marmorata* is an eastern Pacific endemic. It is usually said to occur from southern California to Peru. However the taxonomy and identifications of gymnurids off Central America has been questioned.

Natural History: Gymnurids have not been studied as much as other stingrays. The Spiny Butterfly Ray is found in the warm temperate waters to tropical Atlantic Ocean. They occur from about 5–100 m in depth. The California Butterfly Ray is found in tropical to warm-temperate marine waters of the eastern Pacific. Both species are sometimes found inshore, along beaches and lagoons on sandy and muddy substrates. There may be migration patterns following water temperatures, so nearshore activity is usually during the warmer months. With the elongated pectoral fins, they swim by oscillation like a myliobatid, rather than undulation like a dasyatid or urotrygonid. The fin rays are cross-braced for support.

Butterfly rays are viviparous. During parturition, the pups are born rolled up like a cigar, with the spine encased in a protective coating. In this manner, the female can give birth to relatively large young, and not be stung in the process. Spiny Butterfly Rays give birth to 1–8 pups, which are 38–43 cm disc width. California Butterfly Rays give birth to 4–16 young, which are 21–26 cm disc width.

Spiny Butterfly Rays feed on fishes, including small sharks and rays, and several types of invertebrates, including crustaceans, cephalopods, and gastropods. In Mexico, *G. marmorata* was shown to feed on grunts, mullets, and flatfishes. They are also reported to feed on squid occasionally. Butterfly rays stun the prey with their pectoral fins and use the fins to work the food into

Your eye may be drawn to the Roughtail Stingray in the water column, but note the outline of a very large Spiny Butterfly Ray buried in sand that the photographer is eyeing. *Philippe Guillaume/CCA/Flickr*

the mouth. Predators probably include ray-eating species such as sharks and marine mammals.

Encounters: In general, butterfly rays are rare to uncommon but may be locally abundant. The Spiny Butterfly Ray is listed as vulnerable by the International Union for Conservation of Nature and Convention on International Trade of Endangered Species. They are reported to be intrinsically at risk of overfishing and some populations have been extirpated or reduced by 99%, especially in Brazil and the Mediterranean. Fortunately, fishing pressure is lower in the United States, where populations have fared better. They are usually encountered by shore fishermen, boat anglers, and divers when they move to shallow waters of the western Atlantic during the summer. It is possible to see them at some stingray feeding dives, such as Los Gigantes on Tenerife, one of the Canary Islands, although they are not one of the common species.

In a series of surveys that resulted in the capture of more than 20,000 fish from Santa Catalina Island and the adjacent mainland of southern California, only four were *G. marmorata*, compared to 750 *Myliobatis californicus* and 169 *U. halleri*. Surprisingly, no *H. dipterurus* were collected. California Butterfly Rays may be more abundant in warm-water years, as during an El Niño event. They would most likely be encountered during the summer and fall, when ocean temperatures of southern California peak. Spiny Butterfly Rays are more commonly encountered in the southwestern Atlantic, and may be taken by shore anglers.

Butterfly rays are not a significant threat. People may occasionally get stung by stepping on them. Anglers can get stung by not being careful when hauling butterfly rays on board. They are sometimes caught by surf fishermen, so care must be taken when hauling to the beach, and when a curious crowd gathers.

Venom Apparatus, Venom, and Symptoms: The caudal spine is short in length and relatively wide, on a very short caudal appendage. It is referred to as the gymnurid type of venom apparatus. Spines in *G. altavela* from the eastern Atlantic measured about 23 mm for males and 33 mm for females. The spines lacked serrations and the dorsal groove extended 55 percent of the spine length. In a 1 m wide specimen of *G. marmorata* from southern California, the stinger was 17 mm in length and 4 mm wide. The barbs of the dentate margin were deep compared to other stingrays. The stinger is at the base of the tail, which is very short, so it cannot lash out like most other stingrays.

To my knowledge, the venom of butterfly rays has not been researched and I am aware of no case studies of human envenomation, so it is probably best to assume it has an integumentary sheath with the same venomous qualities of other stingrays. It can probably deliver a painful sting, but it is often stated in literature that it is not as dangerous as other stingrays.

Remarks: Much remains to be learned about gymnurid biology, and case studies of envenomation (if they occur), should be published.

References and Resources: Diaz (2008), Ebert (2003), Eschmeyer et al. (1983), Halstead (1978), Halstead and Bunker (1953), Halstead and Mitchell (1963), Kells and Carpenter (2011), Pondella and Allen (2000), Ramirez-Amaro et al. (2013), Russell (1953), Schaefer and Summers (2005), Schwartz (2005), Smith et al. (2002, 2009), Villavicencio-Garayzar (1993), Yokota et al. (2016).

Pelagic Eagle Rays (Genus *Aetobatus*)

"As [he] struggled to get the ray out of the vessel, the animal lashed its ten-foot (three-meter) tail, piercing the man's heart with its venomous barb. . . . Surgeons performed two operations on [him] yesterday and today, ultimately removing the 1-foot (0.3-meter) barb by pulling it through his heart. [He] was listed in critical condition." (Pastino 2006; National Geographic News)

White-spotted Eagle Rays have a very long caudal filament and very long stingers. *"Adam"/CCA/Flickr*

Classification: Phylum Chordata (chordates); Class Elasmobranchii (elasmobranchs); Order Myliobatiformes (thornbacks and rays); Family Aetobatidae (pelagic eagle rays); Genus *Aetobatus*; Species *A. narinari* (White-spotted Eagle Ray) and *A. ocellatus* (Spotted Eagle Ray).

In most earlier literature, aetobatids were considered to belong to the more-inclusive family Myliobatidae, but recent research suggests aetobatids

Profile of a White-spotted Eagle Ray feeding on the bottom. *Tami Warner-Minton/CCA/*

constitute a distinct family. Worldwide, there are five species. The one in the Gulf of California is now classified as *A. laticeps* (Pacific Eagle Ray).

Identification: The White-spotted Eagle Ray is easy to recognize. It is similar in shape to an eagle ray of the family Myliobatidae, but with a distinctive color pattern. It is a very large species, reaching nearly 3 m in disc width and 2.7 m long, excluding the tail. Most adults are probably 1.5–2 m disc width. With the extremely long, whip-like tail it can be nearly 5 m in total length. The whip-like portion of the tail is sometimes called the caudal filament. The pectoral fins are broad and pointed. The posterior margin of the pectoral fins is concave. The small pelvic fins are located at the rear of the disc on either side of the tail. The relatively narrow, wedge-shaped head and snout extends from the front and above the disc. The snout is somewhat dorsoventrally flattened to upturned, giving it a nickname of duck-billed ray. The eyes are on the side of the head, rather than on top. The mouth and gills are on the venter. The spiracles are on the dorsolateral surface of the head near the front of the pectoral fins. The dorsal ground color is black or dark bluish to dark brown, olive, or gray. It has countershading with a whitish venter. The dorsal surface is covered with light spots or ocelli, which may be white, greenish, or yellowish.

The Spotted Eagle Ray is similar in appearance, but the spots or ocelli may be restricted to the dorso-posterior portion of the disc. The Pacific Eagle

Ray is also similar in appearance, but none of these species overlap in range. Until recently, these species were often lumped as the single taxonomic entity, Spotted Eagle Ray (*A. narinari*), with a circumtropical distribution. Because of this, ecological distinction between species is confusing, but they are probably similar in habits.

Distribution: White-spotted Eagle Rays are found on both sides of the Atlantic. In the western Atlantic they are found from North Carolina to Florida and the Caribbean, throughout the Gulf of Mexico, and south to Brazil. They are not uncommon off Florida, but are uncommon in the northwestern Gulf of Mexico and farther north in the Atlantic. *Aetobatus ocellatus* occurs from Hawaii westward throughout the Indo-West Pacific, although it has been suggested the central Pacific populations are distinct. The Pacific Eagle Ray is known from nearby Mexico, but has yet to be reported from California.

Natural History: These are marine species that may occur inshore or offshore. Inshore, it can occur in the surf zone, bays, and estuaries. They are commonly observed in the vicinity of coral reefs, but more often in open water. They are usually near the surface when offshore, but range to about 75 m deep.

They mostly travel in large schools in open water. They socially interact with one another. They are adept at propelling themselves through the water by undulating the pectoral fins. They appear to be flying through the water in slow motion, and with the beak-like snout, it is easy to see why they are called eagle rays. They are well known for their ability to become completely airborne, as when escaping predation, dislodging parasites, or perhaps as a social display.

These rays feed on a variety of animals, including crustaceans, mollusks, echinoderms, and fishes. Their diet includes gastropods, shrimp, octopuses, squids, sea urchins, and some bony fishes. They use their modified snout to root up benthic prey. They have papillae in their mouth that separate mollusks from their shells and will eat only the soft parts. Food is smashed with crushing and grinding teeth.

In turn, they are preyed upon by large sharks, including Silvertip and Great Hammerhead Sharks. In one instance, a Great Hammerhead was seen feeding on a White-spotted Eagle Ray. The attack took place on the surface, but the shark pinned the ray to the bottom with its cephalofoil (hammerhead) to manipulate it for feeding. Also, sea turtles have been found with *A. narinari* spines lodged in them, but they prey on invertebrates and sometimes plant material.

Aetobatids have aplacental viviparity, usually giving birth to four young. They are born at about 15–30 cm disc width. When born, the pups are at risk of predation, as predators have been known to follow adults during the pupping season.

There is a fishery that targets *A. narinari* in the southern Gulf of Mexico.

Encounters: These species are not commonly encountered inshore but they sometimes occur there. More often they are offshore over the continental shelf. Unlike the near-obligate bottom-dwelling rays, they are not as likely to be stepped on, but it can happen if they are resting on the bottom. They are a threat when hauled into a boat after being caught or snagged, and they may occur in nets and lines of commercial fishermen. According to a study in Brazil, most stings by *A. narinari* are to fishermen hauling nets. On rare occasions, these animals have leapt into boats. Because of their large size, they can wreak havoc aboard a small vessel, injuring people by physical trauma and envenomation. At the time of this writing, there is an online video of a White-spotted Eagle Ray that leaped out of the water and landed in a recreational fishing boat in Florida. Some divers are tempted to interact with them, and accidents have happened when engaged in this manner.

Venom Apparatus, Venom, and Symptoms: Aetobatids have the myliobatid venom apparatus, with a very long whip-like tail, and stingers near the disc. *Aetobatus* has a caudal spine similar in design to other stingrays (see account for urotrygonids), although it is not as flattened as some. These species commonly have multiple stingers; there are often up to four stingers, but may be as many as eight. It has been suggested these are replacement stingers, but one spine does not necessarily shed when a new one appears. The spine is long and slender compared to urotrygonids and dasyatids. An adult's caudal spine can be several cm in length, but only a few millimeters wide. In one study, spines of males averaged 60 mm long and in females 75 mm long, with the longest measuring 166 mm. There are short, sharp barbs along the shank of the spine. In some instances, the integumentary sheath, which contains the venom, can be lost, so an unsheathed sting would be a dry sting but it can still cause significant physical trauma. Fortunately, the caudal spine is located near the base of the tail, so they cannot precisely direct a tail-launched sting. Nevertheless, they can thrash and direct bodily movements to protect themselves.

Many readers are probably aware of the well-publicized case in which TV celebrity Steve Irwin, the "Crocodile Hunter," was killed by a large dasyatid when its stinger punctured and broke off in his heart. Only a month after that tragic incident, an 81-year-old man in Florida was struck in the chest by a White-spotted Eagle Ray that leapt into his boat, and its spine also became lodged in the man's heart. The incident occurred when the man attempted to get the thrashing ray out of the boat. He left the spine in place and secured it with cloth. Medical help was immediately sought and they were able to save him in surgery. Another similar situation happened in Florida in 2008 when a ray jumped into a boat and killed a woman, though the cause of death was determined to be blunt-force trauma rather than the effects of the spine or venom.

The venom of this species has not been intensively investigated, but it has been shown to have fibrinogenolytic and anticoagulant activity in the sheath tissue. The wounds may be serious as the stingers are long and there may be up to eight of them. Symptoms are presumed to be similar to that of other stingrays, including, pain, redness, and swelling, and possibly systemic reactions from severe envenomation.

Remarks: It is possible to see these beautiful creatures in some of the larger public aquaria and some tropical travel destinations.

References and Resources: Chapman and Gruber (2002), Cuevas-Zimbrón et al. (2011), Dellias et al. (2004), Haddad et al. (2013), Halstead (1978), Kells and Carpenter (2011), Kumar et al. (2011), McEachran and Carvalho (2002), Ocampo et al. (1953), Parsons (2006), Pedroso et al. (2007), Richards et al. (2009), Sellas et al. (2011), Thorson et al. (1988), Tinker and DeLuca (1973), White et al. (2010, 2013, 2016b).

Giant Mobula (*Mobula mobular*)

"In a group as big as this [a video of hundreds of mobulas in a tightly packed school], the problem is standing out in a crowd. Mobula rays have the most extraordinary way of getting attention. . . . The impact [upon landing on the water after jumping] sends a huge "boom" through the water, the higher they leap, the bigger the bang. It is usually males who do this, signifying their prowess as a mate . . ." (online BBC video)

The Giant Mobula is the only mobulid with a well-developed spine. *Patrik Neckman/CCA/ Wikimedia Commons*

Classification: Phylum Chordata (chordates); Class Elasmobranchii (elasmobranchs); Order Myliobatiformes (thornbacks and rays); Family Mobulidae (devilrays); Genus *Mobula* (mobulas); Species *mobular* (Giant Mobula).

Until recently, mobulids were placed in the large polyphyletic family Myliobatidae. Technically, the Giant Manta (*Mobula birostris*) has a stinger, although it is poorly adapted for envenomation, and stings by this species are unknown, so mantas are not discussed in detail. Mobulids are called devilrays, due to their horn-like projections, or flying rays, because of their aerial acrobatics. A recent revision of *Mobula* indicated *M. japanica* (Spiny Devilray) was a junior synonym of *M. mobular*.

Worldwide, there are eight species of *Mobula*. The two called manta rays were formerly placed in their own genus, *Manta*, and some researchers still recognize that genus. The Giant Manta of the Atlantic, Gulf of Mexico, and Caribbean appears to be distinct from the other mantas (*M. birostris* and *M. alfredi*), but has not been scientifically described. Other North American mobulas lack a caudal spine, including *M. hypostoma* (Atlantic Devilray), *M. munkiana* (Pygmy Devilray, Pacific), *M. tarapacana* (Chilean Devilray, Atlantic and Pacific), and *M. thurstoni* (Bentfin Devilray, mostly circumtropical, except western Atlantic).

Identification: Mobulids differ from other rays by their broad heads that protrude from the anterior margin of the disc, bordered by flap-like structures known as cephalic horns. The mouth is anterior, wide, and has many small teeth. The disc is about twice as wide as long. There is a prominent dorsal fin just in front of the tail. The pectoral fins are concave posteriorly and tapered at the ends. The Giant Mobula is distinguished from other mobulas by the presence of a caudal spine. The Giant Mobula has been reported to reach 3.1 m maximum disc width, but most adults reach about 2.5 cm disc width.

If present, the spine of a manta ray is a nonfunctional, calcified mass.
Bjorn Ognibenl/CCA/Flickr

The Giant Manta, on the other hand, is an enormous animal that may reach 6 m in disc width. The tail is whip-like, but shorter than the body length. A small caudal spine is usually present. The cephalic horns are not as elongate or flap-like as those of *M. birostris*. Giant Mantas are usually dark blue to black above and whitish below, but there are light patches dorsally, especially on the "shoulder," as well as dark patches on the venter. These markings have allowed biologists to recognize individual animals during *in situ* studies. The patterns have also alerted researchers to the presence of two or more species (one as yet undescribed).

Distribution: The Giant Manta is found in most tropical and subtropical oceans and seas. They are uncommon in nearshore waters of the Gulf of Mexico. In southern California, they are usually only present in warm-water years, and are more commonly encountered offshore or near the Channel Islands. They are most common in the tropics, where they may aggregate. They are relatively common off Hawaii.

The Giant Mobula is a wide-ranging species that is nearly circumtropical, although there are no records from the northwestern Atlantic. It is found in the eastern, central, and western Pacific. In the eastern Pacific it occurs from central California to Peru, and is common on both sides of the Baja California Peninsula. It is known from the Hawaiian Islands. Its full distribution is probably not completely known.

Natural History: Scientists are starting to unravel some of the mysteries of mantas because they are large, slow-moving animals that frequent or aggregate in specific areas, and individuals can be recognized by their patterns. In some areas, ecotourists gather to meet the arrival of Giant Mantas. The animals are passive and tolerant of divers and will even allow human passengers to hitch a ride. Giant Mantas can be long-distance migrants that may cross open oceans, but migration patterns (e.g., from satellite telemetry) are just emerging in the literature. The other manta species, the Reef Manta (*M. alfredi*), is more of a local coral reef dweller of tropical waters. Mobulas are less well known. They are relatively fast swimmers (*M. mobular* has been clocked at 8.4 km/hr) that may travel singly or may be found in schools. One large school off San Diego was estimated to contain hundreds of individuals. Much of the knowledge of mobula biology comes from fisheries-related studies.

Mantas and mobulas are epipelagic marine animals, found from offshore to nearshore waters, primarily in the tropics and subtropics, although some mobulas, including *M. mobular*, will enter warm-temperate waters. They are usually found near the surface, but will dive and commonly feed at night. Mantas and mobulas are migratory and will follow ocean currents and changes in water temperatures. For example, *M. mobular* usually migrates out of southern California and Baja California during the winter. Mantas and mobulas are well known for their abilities to aggregate. They are also known for becoming airborne. For example, in the Gulf of California, one does not need to wait long to see an aerial display by a *Mobula* spp. I encourage the reader to go online to find some of the many spectacular videos of leaping mobulas. It is likely that they leap to remove parasites, avoid predation, give social signals, and/or aid in plankton netting.

Mantas and mobulas feed on zooplankton in the water column. Prey includes shrimp, planktonic crabs, and other species. In addition to water temperatures, their migration patterns probably follow that of prey abundance, such as plankton blooms. In the Gulf of California the most abundant prey of *M. mobular* is *Nyctiphanes simplex*, the most common and abundant species of krill in the Gulf. This species is probably a staple off the west coast of Baja California and along southern California, especially during El Niño years, although other planktonic species are consumed. In turn, mobulids are probably preyed upon by sharks and killer whales.

Mating takes place in the spring and summer. Mantas and mobulas have aplacental viviparity. They usually give birth to a single young. Mantas have been documented giving birth by flinging out the young while turning. Young are born every year or two. Young mantas are about a meter disc width and newborn Giant Mobulas are about 70–85 cm disc width.

In some areas of the world, artisanal and commercial fishing has targeted Giant and other mobulas, but they are also a common bycatch of purse seines for tuna and drift nets, as well as recreational fishermen. The meat is sometimes eaten or used as bait, and the gill rakers are popular in Asian markets. In Brazil, the caudal filaments are dried and used as cattle whips. Because of the low reproductive rate, generally small populations, fisheries pressure, a lack of regulatory management, and a host of other threats, mantas and mobulas are of conservation concern.

Encounters: Giant Mobulas are generally present in southern California and Baja California during warm spring to summer months, but may be present year-round if warm temperatures exist, and are year-round residents of Hawaii. These animals are not normally encountered by the general public, except in the Gulf of California, where they can be seen leaping. They may be mixed with non-spined mobulas, so one cannot normally tell which species are jumping. They are also seen when people seek them out, including researchers, recreational divers, and artisanal fishermen. Mobulas may be accidentally caught by recreational anglers, so people should be aware that *M. mobular* has a venomous spine. It is possible that these animals could land in a boat, and with the sheer numbers of leaping mobulas and boaters in the eastern Pacific, it is surprising it does not happen often (or perhaps it is not reported).

Venom Apparatus, Venom, and Symptoms: The venom apparatus is categorized as being of the myliobatid type, although the spine is somewhat different. In *M. birostris*, the spine may be poorly developed or lacking entirely. When present, only the tip is seen emergent from the calcified mass, covered by integument. Of the mobulas, only *M. mobular* has caudal spines. The spine is just behind the relatively high dorsal fin, which is just in front of the tail. The caudal spine mass is about 13 cm long, with the exposed portion being about 3–5 cm long, regardless of the ray's body size. It is sharp and has about 30 lateral serrations on each side. The spine is covered by a heavy black layer of integument. Like the Giant Manta, the spine is embedded in a heavily calcified cartilaginous mass. In a small percentage of individuals the spine is absent.

I can find no case studies of envenomation by *M. mobular*, but it is usually cited as being venomous in many accounts. Until proven otherwise, it is probably best to assume the venom is similar to other stingrays. Therefore, case studies of confirmed stings should be published, even if dry or mild.

Remarks: Before 2016, the species *M. mobular* was assigned to an endangered ray, primarily of the Mediterranean. When *M. japanica* was subsumed, all of a sudden the range of *M. mobular* increased dramatically to become essentially circumtropical, so the species as we know it now is more common. This is a good example of how a name change can alter the conservation status. Of course, local populations and distinct population segments (e.g., Mediterranean) could still be managed as endangered, if management agencies take that approach.

References and Resources: Bizzarro et al. (2009), Camhi et al. (2009), Couturier et al. (2012), Croll et al. (2012), Ebert (2003), Eschmeyer et al. (1983), Gadig and Sampaio (2002), Graham et al. (2012), Marshall et al. (2009), McEachran and Carvalho (2002), Notarbartolo Di Sciara (1987, 1988), Parsons (2006), Sampson et al. (2010), Ward-Paige et al. (2013), White and Last (2016).

Eagle Rays (Genus *Myliobatis*)

". . . felt as though someone had stuck me with a hot electric wire of high voltage . . ." (Halstead and Bunker [1953], regarding a sting to the foot by a Bat Ray)

Classification: Phylum Chordata (chordates); Class Elasmobranchii (elasmobranchs); Order Myliobatiformes (thornbacks and rays); Family Myliobatidae (eagle rays); Genus *Myliobatis;* Species *M. californicus* (Bat Ray), *M. freminvillei* (Bullnose Ray), and *M. goodei* (Southern Eagle Ray). I am using traditional English names here, as the newly suggested Bat Eagle Ray and Bullnose Eagle Ray seem confusing and have not been in common usage.

Identification: Eagle rays have wide diamond-shaped discs, a protruding head with a somewhat duck-bill snout, and a long whip-like caudal filament. There is a small dorsal fin. The eyes are laterally located on the head, in front of the disc. The mouth is relatively small and on the venter of the head. Bat rays have countershading, being brown, olive, gray, or black above, and

Bat Rays are a common site near southern California kelp beds. *Douglas King*

Detail of the head of a Bat Ray on a dock. They are not an uncommon catch, so care must be taken to remove the hook. *"Snapper"/CCA/Flickr*

whitish below. The Bat Ray reaches nearly 2 m in disc width, and a weight of about 90 kg, although they are usually about 1.4 m in disc width and 9–14 kg as adults. The caudal spine is located behind the dorsal fin on the tail.

The Southern Eagle Ray is similar in appearance, and about the same size as the Bat Ray. The Bullnose Ray is also similar, although it may have small white to yellow spots dorsally. It has a longer snout than the other two species and is smaller, averaging less than 1 meter in disc width.

Distribution: Bat Rays are endemic to the eastern Pacific, from northern Oregon (rarely) to central Mexico, and they also occur in the Galapagos Islands. *Myliobatis longirostris* occurs in the eastern Pacific off Mexico, but has not been reported from California. There are no *Myliobatis* in the central Pacific. The Bullnose Ray occurs from Cape Cod (rarely) to northern Argentina. It is found throughout the northern Gulf of Mexico, but is absent from the southern Gulf and parts of the Caribbean. The Southern Eagle Ray is found in the western Atlantic, from North Carolina to south Florida, and Yucatan to Argentina, but has not been reported from the northern Gulf of Mexico or northern Caribbean.

Natural History: The Bat Ray has been extensively studied, as it is commonly encountered in California, even north of Point Conception, both on the mainland coast and around the Channel Islands. In one study of fishes at Santa Catalina Island and the adjacent mainland, it was ranked as the 9th most abundant fish (given sampling design biases) at the island and 7th most abundant off the mainland. It also accounted for the highest biomass of any species of fish off the mainland and 4th highest biomass at the island. It is usually found in relatively shallow coastal waters, from inshore bays to depths of about 50 m, but can occur over deep water in the channel between the mainland and offshore islands.

The Bat Ray is somewhat a habitat generalist, because it is found over sandy and muddy bottoms, as well as rocky bottoms and kelp forests. It can be found along the coast or in bays and estuaries. It is common and often the most abundant large predator in protected waters, such as Tomales Bay, Humboldt Bay, Elkhorn Slough, and Marina Del Rey, among other similar locales. Some bays and sloughs are important feeding and pupping grounds. In California it is present or most abundant in the late spring to early fall. In a study at Tomales Bay, Bat Rays had strong diel travel patterns in relation to the shallow mudflats and more open bay, which was likely a response to thermal change rather than tidal flow.

Bullnose Rays are also common rays in the southeastern U.S. shallows. They often enter estuarine habitat. For example, they are frequently found in Delaware Bay and other estuaries, especially during the summer months. Along the East Coast, they are known to migrate north in the summer and south in the winter. The Southern Eagle Ray is rather poorly known in the northwestern Atlantic, but fairly well researched off the southwestern Atlantic.

Myliobatid rays feed on a variety of benthic organisms. Both *M. californicus* and *M. freminvillei* have been considered key predators of oysters, so have been targeted by fisheries programs to reduce predation on commercial oyster beds. However, research showed there are age-class differences in feeding habits, and they are much more generalized predators. Only adult female Bat Rays are even capable of eating Pacific oysters, due to the small size of the ray's mouth. In one study in Humboldt Bay, clams consumed were usually less than 5 mm in diameter. They also feed on shrimps, crabs (which prey on oysters), polychaete worms, innkeeper worms, and other species, including the occasional small fish. Similarly, in Tomales Bay, only large animals apparently fed on larger prey items. Feeding pits could be seen in the mudflats during low tide. Bullnose Rays have similar habits in the sloughs and estuaries of the eastern United States. They do not target oysters, but some are consumed, and there are also strong age-class differences. Their diet consists of clams, shrimps, crabs, and gastropods. The diet of Southern Eagle Rays is

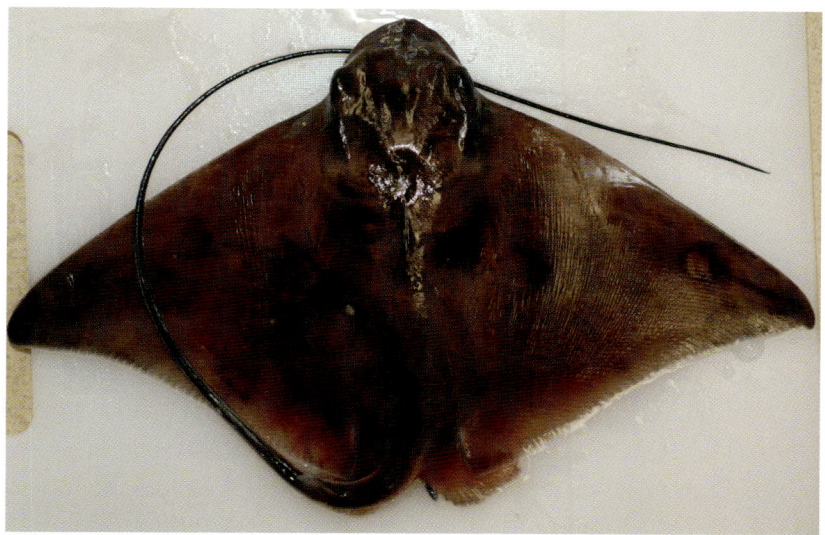

Bullnose Ray, *M. freminvillei*. Robert Agular/CCA/Flickr

more specific, as they target soft-bodied prey, especially polychaetes.

Bat Rays migrate to shallow bays and sloughs during the spring and summer months to feed and reproduce. Mating usually occurs in the summer. After a gestation of 9–12 months, they give birth in the spring and early summer. They have aplacental viviparity. One to twelve (usually 2–5) young are born at about 22–30 cm disc width. Bullnose Rays are similar. They give birth to 4–8 young that are about 25 cm in disc width.

Bullnose Rays are sometimes seen leaping out of the water, especially in the northern part of the range. All species are likely preyed upon by sharks.

Encounters: Bat and Bullnose Rays are commonly encountered by anglers, SCUBA divers, snorkelers, swimmers, and visitors to public aquaria. Bat Rays are sometimes fished commercially, in artisanal fisheries in Mexico, and recreationally. Divers often see one to several Bat Rays in the vicinity of popular kelp forest diving destinations. Bat Rays may be near the surface, in mid-water, or on the bottom. On the bottom they are often partially buried by sand. In the murky waters of bays and estuaries, Bat and Bullnose Rays are difficult to see. Although divers do not use these areas, anglers do. The rays may constitute a relatively common catch of recreational fishermen from shore, piers, and boats.

Venom Apparatus, Venom, and Symptoms: Myliobatid rays have the namesake myliobatid type of venom apparatus. The spine is situated about a third of the way down the whip-like tail on the dorsal surface, at the end of the muscular part of the caudal appendage, before the filament. There are usually 1–2 spines present. The spines from a large series of Bat Rays between 50 and 120 cm disc width ranged between from about 40 × 3 mm to about 80 × 4 mm. The spine is flattened and has dozens of sharp retrorse barbs. The dorsum of the spine has one to several furrows. The dorsoventral furrows are relatively well developed. Bullnose Rays and Southern Stingrays have a similar venom apparatus, capable of causing significant trauma. There is little information on specific venom components of various stingrays, but it is believed the venom in marine species is primarily algogenic, with the intense pain being caused by serotonin.

There are few case studies of specific species of myliobatids, but they are probably similar to those of the eagle rays and round stingrays. The opening quote is from a case study in which a swimmer in northern Mexico accidentally stepped on a Bat Ray and was struck in the heel. The pain was immediate and intense, with profuse bleeding. The victim had a difficult time getting out of the water, as he could not walk on the stung foot. Pain had turned into numbness, and before he could get out of the water, he fainted. Fortunately, he was rescued by a companion. After he got to shore, he fainted again. When he regained consciousness, the severe pain returned and his leg became very swollen and blue. First aid included lancing the wound and an application of cactus juice, a local "remedy." Symptoms continued and he was later admitted to a hospital back in the United States. The wound became infected and necrotic. He was treated with antibiotics and necrotic flesh had been excised.

In another case study of a Bat Ray, a fisherman was struck in the arm, resulting in a 4 cm gash that was 3 cm deep. Symptoms included some pain, swelling, and tenderness. The swelling lasted three days and the wound healed in five days. Interestingly, a website from Florida with information on stingrays lists the Bullnose Ray as harmless, which I assume to be an oversight or an attempt to un-demonize stingrays, in the same way American Cownose Rays are said to cause a sting like a bee. If there is information on mild envenomation due to stingrays, it would be important to have case reports to back up such claims.

References and Resources: Ebert (2003), Eschmeyer et al. (1983), Halstead (1978), Halstead and Bunker (1953), Gray et al. (1997), Hoese and Moore (1998), Kells and Carpenter (2011), Martin and Cailliet (1988a, b), Matern et al. (2000), McEachran (2009), McEachran and Carvalho (2002), Odenweller (1975), Parsons (2006), Russell (1953), Szczepanski and Bengtson (2014), White et al. (2016b).

Cownose Rays (Genus *Rhinoptera*)

"But it chansed our Captaine taking a fish from his sword . . . being much of the fashion of a Thornback. But a long tayle like a ryding rodde, whereon the middest is a most poysoned sting, of two or three inches long, bearded like a saw on each side, which she strucke into the wrest of his arme neere an inch and a halfe: no bloud nor wound was seene, but a little blew spot, but the torment was instantly so extreme, that in foure houres had so swolen his hand, arme and shoulder, we all with much sorrow concluded his funerall, and prepared his grave in an Island by, as himselfe directed: yet it pleased God . . . his tormenting paine was so well asswaged that he eate of the fish to his supper . . . For which we called the island Stingray Isle after the name of the fish." (Russell [1959] from a 1609 account of a stingray envenomation in Chesapeake Bay of Captain John Smith; the species is usually thought to be an American Cownose Ray.)

Classification: Phylum Chordata (chordates); Class Elasmobranchii (elasmobranchs); Order Myliobatiformes (thornbacks and rays); Family Rhinopteridae (Cownose rays); Genus *Rhinoptera*; Species *R. bonasus* (American Cownose

When rhinopterids feed, they extend the cephalic lobes to detect electric field from under the sand. *L.L.C. Jones*

Ray) and *R. brasiliensis* (Ticon Cownose Ray).

Until recently, the cownose rays were usually considered to be a genus within the larger family, Myliobatidae, but were elevated to their own family. Another species, *R. steindachneri* (Pacific Cownose Ray) occurs south of California but is very similar to the Ticon Cownose Ray, and their relationship needs investigating.

Identification: Cownose rays are diamond-shaped, but the disc is wider than long. They reach about 1.1 m in disc width. They can be identified by their strange head shape. The head is elevated and projects beyond the front of the disc. The eyes are on the sides, and spiracles just behind. There is a pair of cephalic lobes below the rostrum. They are distinctive for this family, having evolved from anterior extensions of the pectoral fins. They are sensory appendages having ampullae of Lorenzini for electroreception and neuromasts for mechanoreception. The mouth is ventral to the head. The tail is long and has a caudal filament. The caudal spine is situated near the base on the top of the tail. These rays are relatively smooth-skinned. They are uniformly brownish dorsally with a whitish to yellowish ventral surface. American and Ticon Cownose Rays are very similar, differing in dentition.

Distribution: American Cownose Rays are found in the western Atlantic from New England to Florida, migrating to Trinidad, Venezuela, and Brazil. They occur throughout the Gulf of Mexico. They also range into the eastern Atlantic off the coast of Senegal, Mauritania, and Guinea. The Ticon Cownose Ray is found throughout the Gulf of Mexico, but is apparently absent or rare off the East Coast. The two species co-occur throughout the Gulf of Mexico.

Natural History: Cownose rays are benthic to epipelagic. They are sometimes found offshore in open water, as during seasonal migrations, but they also occur inshore, occupying shallow coastal bays and estuaries. They occur in the water column from the surface down to about 20 m. They are notable for their long-distance migrations, although routes are still not completely known. They travel in large groups called "fevers," which may contain up to 10,000 individuals. This can be an awesome sight when spotted from an aircraft, boat, or when diving with them. In the Gulf of Mexico, they migrate in a clockwise manner from the northern Gulf to the Yucatan. It has been suggested that the western Atlantic population is separate and can migrate as far south as northern South America. There are also inshore to offshore

Fevers of Cownose Rays can number in the thousands. *Kirk Kilfoyle/CCA/Flickr*

migration patterns. There have been studies of American Cownose Rays in Chesapeake Bay, Florida, and elsewhere in the western Atlantic, where there are summer pupping grounds. There has been concern over the economic impact from this species on commercial bivalve beds.

Cownose rays feed primarily on benthic invertebrates, where they may disturb eelgrass beds. They were believed to take a toll on commercial beds of hard clams and Eastern Oysters. However, as with the Bat and Bullnose Rays, the scientific literature shows this is an exaggeration. In fact, oysters are uncommon to nonexistent in stomachs of Cownose Rays in food studies within areas occupied by oysters. While they do feed on hard-shelled prey, they are not specialists and eat a variety of benthic animals. Prey includes clams, gastropods, crabs, lobsters, polychaetes, and fishes. They procure food by cruising along the bottom, and when prey is detected, they suck and blow sand from around the animal. Then they use their cephalic lobes to manipulate and trap prey, as well as to increase suction power. Sand, debris, and shell fragments are then filtered out.

Much of the information on reproduction of *R. bonasus* comes from the Chesapeake Bay studies. Like other stingrays, they have aplacental viviparity. Males mature in 6–7 years, while females mature in 7–8 years. Gestation is about a year. When they arrive at Chesapeake Bay in the late spring, females are pregnant. Usually one pup is born in mid-June to early July. The young are about 25–45 cm disc width at birth. In Chesapeake Bay, the sexes are mixed, but they segregate after breeding in mid-to-late August. Females remain in shallow nearshore habitats until some time in the fall, but it is unclear where males go. It is likely they seek deeper water or forage in areas that are not commercially fished. All American Cownose Rays in the bay that are harvested as bycatch are females, perhaps because that is where commercial fisheries target. Cownose Rays are preyed upon by Bull and Sandbar Sharks, and Cobias.

Encounters: Cownose rays are commonly encountered in the Atlantic and Gulf of Mexico. However, these are active animals that do not usually settle on the bottom near the shore, so are not a high risk to swimmers. The biggest at-risk group is anglers, be they recreational, commercial, or artisanal. When hooked by a recreational angler, they may jump, so caution is required near a boat. Personnel at public aquaria have an occupational hazard, so there should be safety protocols in place. The caudal spines should be cut off and filed down so as to not present a risk to the public. There are cases of aquarium personnel being stung. In a survey of envenomation sent to 216 zoos and aquaria, four stings were reported for *R. bonasus*, which tied for the species with the highest number of exposures. In a study of envenomations in Brazil, 80% of the 137 recorded stingray accidents off the northern coast of São Paulo were from *Rhinoptera*, and almost all victims were fishermen working nets and shrimp trawlers.

Venom Apparatus, Venom, and Symptoms: The venom apparatus is of the myliobatid type, with the caudal spine(s) located near the base of the tail. There may be one or two caudal spines, which are located behind the dorsal fin. If present, the second spine is smaller than the first spine.

Even though there is much information on the natural history and ecology of this animal, there is scant information on venom and envenomation. A recurring comparison online and in publications is that the venom is mild or even harmless and that it has a sting like that of a bee. However, other reports suggest a sting can be decidedly worse. One example is the account of Captain John Smith earlier. A colleague in Brazil observed 12 cases of severe envenomation by this species. He assured me, "To compare stings of cownose rays with bee stings is an erroneous assessment, as the trauma itself is very painful and the venom causes intense pain symptoms." The pain lasts for several hours and the wound site becomes swollen, but there is no necrosis

American Cownose Rays are commonly kept in public aquarium touch pools. Although stingers are periodically cropped to protect visitors, accidents have occurred to aquarium staff. *L.L.C. Jones*

beyond the entry point. One account of a severe sting on an online fishing forum said the sting was "unbearable." A bee-sting-like reaction could probably originate from a superficial poke.

Remarks: The tale of Captain John Smith stung by a ray during his exploration of Chesapeake Bay is one of the earliest case reports of a stingray envenomation in the New World. Captain Smith named the locale Stingray Point, a name that has stood the test of time.

Although seemingly abundant and widely distributed, the IUCN (International Union for Conservation of Nature) lists *R. bonasus* as near-threatened, because of fishing pressure, low reproductive rate, and lack of regulations. *Rhinoptera steindachneri* is also listed as near-threatened, and *R. brasiliensis* is listed as endangered and declining.

References and Resources: Ajemian and Powers (2012), Blaylock (1993), Collins et al. (2007a, b; 2008), Fisher (2010), Forrester (2005), Haddad et al. (2013), Halstead (1978), Hoese and Moore (1998), Kells and Carpenter (2011), Last (2016a), McEachran and de Carvalho (2002), Neer and Thompson (2005), Parsons (2006), Sasko et al. (2006), Smith and Merriner (1985, 1987), Vohra et al. (2008).

Sea Catfishes (Family Ariidae)

". . . approximately 1250–1625+ catfish species should be presumed to be venomous . . . Venom glands have evolved multiple times in catfishes (order Siluriformes), and venomous catfishes may outnumber the combined diversity of all other venomous vertebrates." (Wright [2009])

Gafftopsail Catfish. *Brenda Bowling/TPWD*

Classification: Phylum Chordata (chordates); Class Actinopterygii (ray-finned fishes); Order Siluriformes (catfishes or sheathfishes); Family Ariidae (sea catfishes).

In the United States, there are three species belonging to two genera: *Ariopsis felis* (Hardhead Catfish), *Bagre marinus* (Gafftopsail Catfish), and *B. panamensis* (Chihuil). The family Ariidae is sometimes referred to as fork-tailed catfishes. The family has more than 150 species belonging to about 30 genera. They are much more diverse in the tropics.

Identification: The ariids have the basic shape we associate with a typical catfish, although the caudal fin is deeply forked. Like many other catfishes, they lack scales. There are strong spines on the front edge of the dorsal and

Hardhead Catfish. *Brenda Bowling/TPWD*

pectoral fins. There is an adipose fin dorsally and anal fin ventrally below it. The pelvic fins are abdominal. The barbels are long; those at the corner of the mouth of *Bagre* are particularly long. *Bagre* also has a long extension on the dorsal and pectoral fins. Hardhead Catfish have three pairs of barbels, which are rounded in cross section, while *Bagre* has two pairs, which are flattened in cross section. The Gafftopsail Catfish excretes copious amounts of mucus. The Chihuil is similar to the Gafftopsail Catfish, but the ranges do not overlap.

All of our species are bluish to gray or brown and silvery to whitish below. They grow to about 50–70 cm maximum length. The skull of the Gafftopsail Catfish is said to resemble Christ on a cross, so is sometimes called the crucifixion catfish. These catfishes can produce sound.

Distribution: The family is distributed in both the New and Old Worlds and is mostly tropical and subtropical. The Chihuil is the only species that ranges in the eastern Pacific to California. It has been found from the Santa Ana River in southern California south to Peru. It is rare north of southern Baja California. The Gafftopsail occurs from Maine to Florida (being more common to the south) to Venezuela. It is found throughout the nearshore waters of the Gulf of Mexico. The Hardhead Catfish ranges from North Carolina to the Yucatan Peninsula. It is also found throughout the nearshore waters and lagoons of the Gulf of Mexico.

Natural History: Most Ariidae can tolerate a variety of salinities but are mostly marine- to brackish-water inhabitants. Our three species are mostly estuarine and marine. The Hardhead Catfish also lives in fresh water. The two species in the Gulf of Mexico are among the most commonly caught fishes of the lagoons inside the barrier islands and in estuaries. Salinity in these lagoons varies from hypo- to hypersaline. In the cooler winter months, they are farther offshore, but migrate inshore during the spring and summer, when temperatures and salinities are higher. Most recreational anglers and commercial fishermen consider these species a nuisance, partly because of their venomous spines, but also because they get caught up in nets and exude slime. They are not generally considered good table fare and are usually discarded. However, some people consider them good edibles.

Sea catfishes in the northern Gulf of Mexico breed from May to August, in the shallows of lagoons and in nearshore mudflats. These fish have among the largest eggs of bony fishes. Hardhead Catfish eggs are 12–19 mm in diameter and Gafftopsail Catfish eggs are even larger at 15–26 mm in diameter. Like other ariids, they are mouthbrooders. The eggs are laid by females and transferred to or picked up in the mouth by males. The eggs and larvae develop in the mouth of the males until yolk reserves are used up. The eggs are adhesive, and outside the mouth they pick up sand and sediment, so cannot become oxygenated and will not develop. The eggs and larvae can remain in the Gafftopsail Catfish's mouth for 42–70 days. Even after the yolk sac has been absorbed, juveniles may seek protection in the male's mouth for a while.

Sea catfishes are opportunistic omnivores. The diet includes shrimps, crabs, polychaetes, cnidarians, sea cucumbers, and fishes. Plant material eaten includes algae and sea grasses. They probably use their tactile barbels, and possibly sound, to find food in turbid water. In turn, sea catfishes are preyed upon by larger fishes. They are sometimes used as bait for large Cobia around offshore oil rigs.

Encounters: The Chihuil is rarely encountered in U.S. waters. Hardhead and Gafftopsail Catfishes are frequently caught by nearshore anglers and commercial fishermen. They are commonly caught in the ocean, lagoons, bays, and estuaries by recreational anglers. Anglers are usually aware of the danger of the spines.

Venom Apparatus, Venom, and Symptoms: The venom apparatus of *A. felis* has been described in detail, and is assumed to be similar to *Bagre*. The

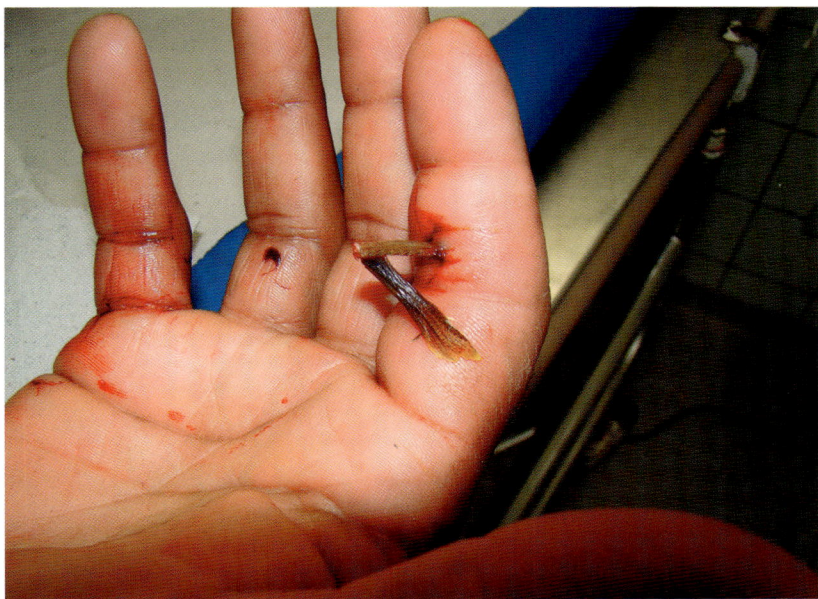

Injury caused by one of the ariid catfishes of the western Atlantic. *Vidal Haddad Jr.*

venom delivery system includes the first dorsal and pectoral fin spines. These spines are heavy and stout, although the newly growing tip is soft. The spines are slightly curved, tapered, and have retrorse denticulations. The bony part of the spine is covered by an integumentary sheath that contains the venom. In these regards, the spines are similar to those of cartilaginous fishes. However, the spines can be erected and locked into position. When the spine enters the flesh, the tissue is lacerated and the venom from the integumentary sheath can enter the wound. A thrashing fish can direct the spine to the perceived threat. There is also an axillary gland and pore near the base of the pectoral fin that presumably contains a toxin, but there is not an effective means for delivering the toxin into the tissues. The venom of ariids is poorly known.

There are some case studies of envenomation by these catfishes. The most common symptoms are profuse bleeding and severe pain. Other symptoms include redness, swelling, itching, a scalding sensation beyond the wound site, muscle tremor and weakness, dim vision, nausea, disorientation, and dry gangrene. Secondary infections are common in catfish stings. There was a case of possible anaphylaxis from an ariid in the Gulf of Mexico. The victim was a fishing guide who had previous exposure; he tested positive to allergy when exposed to extractions from the integument. The Hardhead

This Hardhead Catfish is a long-term resident of the Rockport, Texas, Aquarium.
L.L.C. Jones

Catfish is generally regarded as being more toxic than the Gafftopsail. In Brazil, a fisherman was killed by an ariid catfish sting when he hauled a net out of the water and held it up to his chest. The spine punctured his heart. The spine occasionally breaks off in the victim.

Remarks: The Coral Catfish (*Plotosus lineatus*; family Plotosidae) is a marine species commonly kept in home and public aquaria in North America. It is a seriously venomous species that has caused human fatality. It should never be handled or released into the wild. It is discussed in greater detail in the section on American Territories.

References and Resources: Acero (2002), Acero and Betancur-R (2007), Betancur-R et al. (2007), Birkhead (1972), Eschmeyer et al. (1983), Haddad and Martins (2006), Haddad et al. (2008), Halstead (1978), Halstead et al. (1953), Hoese and Moore (1998), Kells and Carpenter (2011), Mann and Werntz (1991), McKinstry (1993), Muncy and Wingo (1983), Shiomi et al. (1987, 1988b), Wright (2009), Yáñez-Arancibia and Lara-Dominguez (1988).

Freshwater Catfishes (Family Ictaluridae and Others)

Case report of envenomation by a 6 cm Tadpole Madtom: "A 14-year-old boy was injured by a catfish in PA. . . . He had scooped up a small catfish, which he mistook for a tadpole and was quickly wounded in the . . . thumb by one of the pectoral spines. He immediately experienced a burning 'hot pin' pain. Burning pain quickly extended to the entire thumb. The thumb started to swell and became reddened. By 30 min, the swelling and redness had markedly increased while pain remained constant. A rash developed . . . almost to the elbow." (McKinstry, 1993)

Blue Catfish. This researcher knows to wear gloves. *Marshall Smyly*

Classification: Phylum Chordata (chordates); Class Actinopterygii (ray-finned fishes); Order Siluriformes (catfishes); Family Ictaluridae (North American catfishes). There are other families of exotic freshwater catfishes that are established north of Mexico, briefly discussed below.

In the United States, genera (and number of species) of ictalurids include *Ameiurus* (bullheads, seven species), *Ictalurus* (e.g., Blue and Channel Catfishes, four species), *Noturus* (madtoms, 29 species), *Pylodictis* (one species, *P. olivaris*, Flathead Catfish), and Troglobites (a clade of four species).

At least two families of venomous, non-native freshwater catfishes have become established in the United States: Clariidae (air-breathing catfishes) and Callichthyidae (armored catfishes). Other families of venomous catfishes that have been detected in U.S. waters include the Doradiidae (thorny catfishes) and Pimelodidae (long-whiskered catfishes). Additional exotic species of venomous freshwater catfishes may have local populations or could become established in the United States and Canada. There are also non-native, but nonvenomous species in the United States that have become established.

Identification: The ictalurids are familiar to most of us. These are the typical freshwater catfishes commonly encountered in much of the United States and Canada. They vary considerably in size and other features, ranging from the small troglobitic species and madtoms to the large Flathead and Blue Catfishes. Ictalurids have 4 sets of barbels and no scales. There is usually a spine on the dorsal and pectoral fins. There are usually six soft rays in the dorsal fin. The Walking Catfish lacks an adipose fin and a dorsal spine, both of which are present in ictalurids. Armored catfishes are quite different in appearance, having scales resembling bony plates. The general characteristics of the genera (in Ictaluridae, unless otherwise noted) are as follows:

- *Ameiurus*: These include *A. melas* (Black Bullhead) and *A. natalis* (Yellow Bullhead). They differ from other ictalurids by the square, rather than forked, caudal fin. Bullheads are small and stout, usually reaching only about 0.5 kg.
- *Ictalurus*: This genus includes the well known Blue Catfish (*I. furcatus*), Channel Catfish (*I. punctatus*), and Headwater Catfish (*I. lupus*), among others. These have a forked tail. The Blue Catfish is the largest North American catfish, reaching about 165 cm and 68 kg. The Channel Catfish reaches about 18–23 kg. It is variable in color, but often speckled, especially when young. The Headwater Catfish is similar to the Channel Catfish, but smaller, to about 48 cm, and has a deeper caudal peduncle and broader head and snout.
- *Noturus*: This is a large genus of very small catfishes that are superficially similar to bullheads.
- *Pylodictis*: The single species is the second largest of the North American catfishes, reaching 155 cm in length and weighing up to 56 kg. It is mottled and has a broad, flattish head and snout.

- *Troglobites*: This is a clade that includes small troglobitic species in the genera *Satan*, *Trogloglanis*, and *Prietella*.
- *Clarias* (family Clariidae): Two species established in the United States include the Walking Catfish (*C. batrachus*), and Hong Kong Catfish or White-spotted Clarias (*C. fuscus*). The Walking Catfish is usually less than 36 cm in the United States, but in their native range, have been reported up to 61 cm long. In the aquarium trade, they were usually an albino or calico morph, but those in the wild are usually gray to gray-brown with white spots. Most *C. fuscus* are less than 10 cm, but have been reported to be over twice that size. *Clarias* have long barbels.
- *Hoplosternon* (family Callichthyidae): The Brown Hoplo (*H. littorale*) is established in the United States. This is a brown armored catfish having large scales, although there are other nonvenomous large-scaled catfishes in different families that have become established in the United States.
- *Platydoras* (family Doradiidae): The Raphael Catfish (*P. costatus*) has been reported in the United States. It has dorsal armor, which is lacking below the lateral line. It has a color pattern of longitudinal yellow and brown stripes.
- *Perrunichthys* (family Pimelodidae): The Leopard Catfish (*P. perruno*) has been reported in the United States. They lack scales and have three pairs of long barbels. Coloration is brownish with spots and reticulations.

Distribution: The epicenter of North American catfish distribution north of Mexico is in the Midwest and East. However, many of the larger species are also found in other parts of the United States and beyond, where they are non-native. Populations most frequently originated from intentional releases to supplement local sport and food resources. In some cases, these introduced animals have spread and become invasive, threatening native wildlife.

The native range of the Channel Catfish is from New Mexico to Montana and southern Canada and essentially to the East Coast, although they are largely absent east of the Appalachians. They range south into northern Mexico. With introductions, they are now widespread in the Southwest, but largely absent from the Northwest. Blue Catfish is primarily native to the Mississippi River drainage and other rivers, including the Rio Grande and Pecos. It ranges from Nebraska and South Dakota in the north to Belize and

The Margined Madtom, one of the many species of "toms" in the eastern United States. *Jim Atkinson*

Guatemala to the south. It has been introduced to many new areas, including lakes, and is invasive in Chesapeake Bay and other locations. The Flathead Catfish has a similar distribution, occurring as a native in these same major rivers, and is found from the Great Lakes region to northern Mexico. The several species of bullheads are native to the eastern United States. Yellow, Black, and Brown Bullheads are broadly distributed in the East, and many introduced in the West, where they may be invasive. Madtoms are the most speciose and cryptic group of ictalurids. They are well known for their stings. They are mostly found in streams of the eastern United States, with many species in the Appalachians. About half of the known taxa were described in a single journal article. Many species have very small ranges and are of conservation concern. Similarly, members of the Troglobites clade are small catfishes with small, disjunct ranges, aligning with their karst systems.

Among the exotics, the clariids are an Old World family. Walking Catfish were introduced from Asia to the United States in the early 1960s and quickly spread throughout most of Florida through the extensive canal system, and overland. It has been reported from numerous other states, but I know of no large, established populations outside of Florida. The Hong Kong Catfish was intentionally introduced to Hawaii by Chinese immigrants before the 1900s, who raised them in ponds for food. It is now common on the islands of Hawaii, Kauai, Maui, Molakai, and Oahu. The Brown Hoplo is another invasive species that has found a home in Florida waters. The Raphael and Leopard Catfishes have been detected in the Rio Grande. These and other species are likely releases of unwanted aquarium pets, as a variety of catfishes are popular in aquaria. Often, they are intentionally released when they get too large for the tank. This "Releasing Nemo" phenomenon can have disastrous ecological effects (e.g., see the lionfish taxon account). The governments of Canada and the United States have concern over a number of catfish taxa that are likely to wreak ecological havoc, so at least some are banned from importation. This includes all clariids and Stinging Catfish (*Heteropneusta fossilis*, family Heteropneustidae), which can be seriously venomous. Another venomous species, the Wels Catfish (*Silurus glanis*, family Siluridae), is on high alert not because of its sting but because they can grow to over 3 m and could cause ecological upset in the Great Lakes and other freshwater ecosystems. This species has become intentionally (and often illegally) established in many lake systems of Europe and Asia, where it was released for sportfishing.

Natural History: The different taxa have different habitat preferences. The Headwater Catfish occurs over rocky and sandy substrates in riffles, runs, and pools of clear creeks and small rivers. Channel Catfish occur in similar habitat, but in larger bodies of water, and usually below the headwaters. They are also found in lakes. The larger Blue Catfish are found in the main river channels. Flathead Catfish are also found in rivers and lakes. Both of these large species are tolerant of brackish (but not salt) water, so may invade estuarine habitat near river mouths, which is why they have become invasive in areas such as Chesapeake Bay. Bullheads are usually found in the slower parts of rivers (e.g., oxbows, overflow areas), and in ponds and lakes. As a group, the madtoms are often found in small streams, particularly in the mountains, although some occur in shallows of lakes and other sluggish waters. In a sense, they are ecologically similar to some salamanders; they are often found under woody debris, such as logs and rootwads. The troglodytic species are extremely specialized for life in subterranean habitats. For example, *Satan eurystomus* is found from about 300–600 m underground in five springs of the Edwards Aquifer near San Antonio.

North American catfishes are opportunistic carnivores. They feed on a wide variety of animals, the size and type of which is governed by what shares their habitat and what can fit in their mouth. Food includes insects, mollusks, worms, crayfish, amphibians, and smaller fishes. They may also eat plant material. In many areas, large ictalurids and bullheads are invasive species, and often small, rare animals don't stand a chance of surviving or recolonizing invaded native territories. Predators are not well known for most species, but they are likely consumed by a variety of larger fishes, including other catfishes (sometimes their own species), mammals, birds, amphibians, reptiles, and large, predacious invertebrates. Humans are among the more important predators, spending millions of dollars to catch these animals for sport and food. The Channel, Flathead, and Blue Catfish are the most popular species. Channel Catfish are an important aquaculture species. Channel Cats have been introduced into Cuba, Central America, South America, Asia, the Russian Federation, China, and Africa, either for sport, aquaculture, or both. In 1996, 270,000 metric tons of Channel Catfish were farm-raised for food. Ninety percent of the farmed catfish were produced in the Mississippi River Valley.

Given the variety of sizes and habitats of ictalurids, the reproductive histories vary accordingly, but there are commonalities. The smaller species tend to lay fewer eggs than larger species. Madtoms typically lay between dozens and a few hundred eggs. Large individuals of Flathead, Blue, and Channel Catfish can lay tens of thousands of eggs. Ictalurids typically build nests near or under objects such as rock, logs, and overhangs, usually in depressions in the substrate. The males will chase off the female and tend to the eggs by aerating them and guarding them from potential predators, until the young have absorbed their yolk and can fend for themselves. Young catfishes often live in different microhabitats than the adults, to escape predation.

The Clariidae are known as air-breathing catfish because they can respire atmospheric air, and will gulp air from ponds when there are anoxic conditions. Walking Catfish get their name from their ability to leave water and sort-of walk on terra firma during moist conditions—but it is more like controlled wriggling. Because of their ability to breathe air and wriggle across land, clariids can colonize a variety of wetland habitats and have been known to threaten aquaculture facilities. This is ironic, because they were introduced into Florida for aquaculture purposes, being a flavorful fish that is wild and farmed in Asia. Clariids were also common in the aquarium trade, but are now completely banned as "injurious wildlife," except under special permit. While their sting may injure humans, they are deemed injurious by national regulations because of the profound effect they can have on native wildlife. They are ravenous feeders that will eat any type of invertebrate or vertebrate (or their eggs) that they can fit in their mouths.

Similarly, the Brown Hoplo (and some nonvenomous catfishes) is right at home in Florida freshwater systems. It was first detected in 1995, but is already established throughout south and central Florida. It can tolerate a number of freshwater habitats, from lakes and ponds to slow-moving streams and rivers. It is often abundant in areas with muddy bottoms and dense vegetation. Interestingly, it builds a floating foam nest. Eggs are laid below the nest, but after the male fertilizes them, he blows them up into the nest. The Brown Hoplo is considered a traditional delicacy among people with cultural ties to Trinidad and some other areas of South America.

Encounters: Large ictalurids are commonly encountered by most people who go fishing in freshwater habitats, especially in the southern states. Most injuries happen when the angler is attempting to remove the hook. Another method of catfishing that has gained media attention is noodling. With this technique, the fisherman reaches into a catfish nest during the breeding season, and when a male catfish defends its nest by biting the angler, they grab the catfish by the mouth and pull it out. The most commonly targeted species include Blue and Flathead Catfishes, but other species are fair game. Envenomations are a periodic outcome of noodling, either when the animal is grabbed or when the angler tries to control the thrashing prey. In one survey, 59 of 83 noodling injuries were from the fish's spines. Other injuries incurred from noodling include tooth abrasions, cuts from submerged objects, and a well-publicized bite by a cottonmouth. Aquaculturists face an occupational hazard, and there are numerous case reports of catfish stings in workers. Madtoms are too small to be game fishes, so encounters during angling are presumably from accidental exposure to them in the streams or ponds. Researchers and

Researchers seining for madtoms in Tennessee. *Gary Peeples/CCA/Flickr*

fisheries biologists face an occupational hazard when working in freshwater habitats. Species in the Troglobites clade are likely to be encountered only by a few researchers specifically looking for them.

Walking Catfish and Brown Hoplos are usually only encountered in Florida, but they may be extremely abundant in some areas. Encounters are most likely in any body of freshwater, but walking catfish can be encountered on land during wet weather as they are wriggling between aquatic habitats. Hong Kong Catfishes are limited to Hawaii, but are common there, usually encountered in streams.

Venom Apparatus, Venom, and Symptoms: The venom apparatus of catfishes is purely defensive in nature. Not all ictalurids have been studied to determine if they are venomous, but many are, so all should be treated as such. Recent studies have shown that many more catfishes are venomous than previously known, and venom systems have evolved multiple times among the Siluriformes. The commonly encountered genera *Ameiurus*, *Ictalurus*, and *Noturus* all have species known to have caused human envenomation. Among ictalurids, madtoms are often thought to be among the more venomous species, but case reports of stings from other genera suggest many other taxa may be just as virulent, or even more so than some *Noturus*. The Brown Hoplo and Walking Catfish are also known to have venomous spines.

The venom apparatus of ictalurids is the same as that for ariids: the first dorsal and pelvic spines deliver the venom to the victim, and venom is within an integumentary sheath. However, recent studies have shown variability in the spines and venom apparatus. In particular, among madtoms, there were four types of spine apparatuses: serrated spines and venom (the most common case), spines having rear-serration and venom, smooth spines with venom, and smooth spines without venom. Only one species of madtom (*N. gilberti*) lacks venom. The Flathead and Channel Catfishes have serrated spines with venom, the most common of the venom apparatus types. The Brown Hoplo also has dorsal and pectoral spines, but there is also a small spine on the leading edge of the adipose fin. The pectoral spines become red, enlarged, and hooked when they are defending nests. Clariids lack dorsal spines, so their venom apparatus is only on the pectoral spines. These are the spines that allow them to "walk" overland.

Freshwater catfishes have not been particularly well-researched for venom components, but are usually said to contain peptides, some of which have necrotic properties. Researchers have noted that toxic extracts of at least some catfishes not only come from the integumentary sheath of the pectoral and pelvic spines, but are also present elsewhere on the body, including the axillary gland (as in ariids). Those venomous compounds covering the body are proteinaceous crinotoxins. Caudal fin crinotoxins have been shown to be more toxic than the spine glands in some instances. Smaller catfishes may be more prone to envenomate people than larger ones because they are more difficult to handle and the spines are sharper and narrower than those of larger individuals. Physiologists noted that the venom of *C. batrachus* strengthened heart contractions (cardiotonic) in frog muscle.

There are numerous case reports of envenomation in humans by ictalurids. The particular species is not always known or reported, except that sometimes marine species are often separated from freshwater species, and among those, the little madtoms are sometimes fingered as the culprit. In cases of envenomation during aquaculture activities, the species is known, of course—usually Channel Catfish. In large individuals, the spine itself may cause significant traumatic injury. In most reports, symptoms include a painful puncture wound and sometimes profuse bleeding. There is usually swelling and discoloration. The discoloration may be whitish to bluish initially, followed by redness. In mild cases, the symptoms subside within one to a few hours. Some species of catfishes lack venom or have relatively mild venom, while others have more toxic venom.

Symptoms of more severe envenomation include lymph and tendon swelling, trembling, muscle twitching or local spasm, loss of motor control, weakness, nausea, hemorrhage, and necrosis. The pain may last for hours or longer. In some case reports, wound healing and/or swelling lasted for months. Cases specifically identifying madtom envenomations are sometimes said to resemble a bee sting, while in others there can be extreme pain, sometimes resembling an electric shock. The pain may extend into the whole limb. Inflammation, numbness, and stiffness have also been reported.

Perhaps a more important concern than the sting itself is the possibility of secondary infection. Numerous types of pathogenic bacteria have been isolated from catfish sting wounds. The stinger may also break off in the flesh and may cause further trauma and become infected. There are cases of amputation or death due to secondary infection.

References and Resources: Ajmal et al. (2000), Al-Hassan et al. (1985), Arce-H. et al. (2017), Birkhead (1967, 1972), Blomkalns and Otten (1999), Calton and Burnett (1975), Egge and Simons (2006, 2011), Hardman (2004), Hardman and Page (2003), Hubbs (1958), Kaar (2014), Karsky (2009), Langecker and Longley (1993), McKinstry (1993), Murphey et al. (1992), Satora et al. (2008), Scoggin (1975), USFWS (2017, 2018), Whiteside and Burr (1986), Wright (2009, 2012), Zeman (1989).

Typical Scorpionfishes (Subfamily Scorpaeninae)

[Regarding a small sting by a California Scorpionfish] "A few years earlier the patient had been bitten by a rattlesnake. He stated that the pain and other symptoms were quite similar. It is of interest also to note that the patient occasionally works with bees and is rather careless of stings, saying that they cause relatively little discomfort." Halstead (1951)

Spotted Scorpionfish, *Scorpaena plumieri,* a cryptic Atlantic species with intricate beauty. *Richard Bejarano*

Classification: Phylum Chordata (chordates); Class Actinopterygii (ray-finned fishes); Order Scorpaeniformes (mail-cheeked fishes); Family Scorpaenidae (scorpionfishes); Subfamily Scorpaeninae (typical scorpionfishes).

This is a diverse family of venomous fishes with 26 genera and 220 species. There are several North American genera, discussed below. Lionfishes and firefishes (subfamily Pteroinae) and rockfishes and thornyheads (Sebastidae) are discussed in a separate account. Stonefishes (family Synjaceidae) were formerly placed in the Scorpaenidae, but are now in their own family, and are discussed in the section on American Territories.

California Scorpionfish, *S. guttata*, a well-studied scorpaenid.
Kevin Stolzenbach/CCA/Flickr

Identification: The scorpaeniformes are known as the mail-cheeked fishes, because of their characteristic opercular spines. The family Scorpaenidae includes small- to medium-sized fishes with large heads and moderate to large eyes. They have numerous spines and ridges on the head. There is usually a bony ridge called the suborbital stay below the eye and a spiny preoperculum. In addition to the small spines on the head, there is large array of venomous spines on dorsal, pelvic, and anal fins. The pectoral fin is usually broad and the caudal fin is slightly forked to rounded. The pectoral and caudal fins have only soft rays rather than spines.

Most of the Scorpaeninae are benthic, with the general appearance of rocks covered in algae and sponges. They often have fleshy projections on the skin (cirri) that helps them blend in. Typical scorpionfishes are masters of disguise. They may be drab and cryptically colored, but others are brightly colored. Some species have both cryptic and aposematic colors. They are differentiated from lionfishes and firefishes by the lack of very long spines and long, flowing pectoral fins, plus lionfishes are typically banded. Rockfishes tend to be more bass-like, although there are some bass-like scorpaenins, while some thornyheads may resemble bottom-dwelling scorpaenins.

A brief discussion of genera in our area is given below, but there are several excellent references to help identify species, which are often challenging to differentiate. The scorpaenids are well represented in all our waters, but are especially diverse in Hawaii, so the reader is encouraged to use field guides and online resources to appreciate the full array of species.

- *Iracundus*: The single species in our area is *I. signifer*. The anterior part of the dorsal fin mimics a small fish.
- *Neomerinthe*: There are three species in our area. The Spinycheek Scorpionfish (*N. hemingwayi*) is the common Gulf of Mexico and western Atlantic species. It is reddish brown and has a relatively elongate snout. It reaches about 31.5 cm. The Spotwing Scorpionfish (*N. beanorum*) from the Atlantic and the Hawaiian *Neomerinthe rufescens* is a typically deepwater species.
- *Phenacoscorpius*: Represented by the single Hawaiian species *P. megalops*. It is known as the Noline Scorpionfish because it lacks a lateral line. It reaches about 13.5 cm total length.
- *Pontinus*: This includes the Longspine Scorpionfish (*P. longispinis*), which has a distinctively elongate third dorsal spine, and the Highfin Scorpionfish (*P. rathbuni*), which does not. Both are Atlantic species. *Pontinus macrocephalus* is another deepwater species from Hawaii.
- *Rhinopias*: The single species in our area, *R. xenops*, occurs at depths of 59–124 m around Hawaii. It reaches a length of about 15 cm, but is poorly known. Some members of this genus are very unusual algal mimics sporting wild colors and may be popular in the marine aquarium trade.
- *Scorpaena*: This is the most speciose genus of the subfamily. Atlantic species include the familiar *S. plumieri* (Spotted Scorpionfish). It is variously blotched and has white to yellow spots on the underside of the pectoral fin. It reaches about 30 cm. The Barbfish (*S. brasiliensis*) is colorful and usually redder than *S. plumieri*, and has large fleshy cirri over the eye. It has dark brown spots on the pectoral fin. It reaches about 20 cm. The Plumed Scorpionfish (*S. grandicornis*) is about the same size and also has large fleshy cirri, but has white spotting anteroventrally. The Coral (*S. albifimbria*), Goosehead (*S. bergi*), and Mushroom (*S. inermis*) Scorpionfishes are smaller animals at about 8–10 cm. Another species of the inshore shelf and bays is the Smoothhead Scorpionfish (*S. calcarata*). It is a small reddish species with a spot behind the head. It reaches about 10 cm. The Hunchback Scorpionfish (*S. dispar*) is also reddish, but is usually found in deeper water (36–125 m). The Longfin Scorpionfish (*S. agassizii*) is usually reddish, and in deeper water (46–275 m), and has large pectoral fins. In the eastern Pacific, the most common species is the California Scorpionfish (*S. guttata*), which is highly variable in color,

but often brightly banded in reds. Adults have brown to black spots. It reaches about 47 cm. A more drab species is the Stone Scorpionfish (*S. mystes*; not to be confused with the Indo-West Pacific Stonefish), which blends exceptionally well with the background. It can change color rapidly. It has white spots on the underside of the pelvic fin. It reaches about 49 cm. The two species of Hawaiian *Scorpaena* (although sometimes put in other genera) are *S. colorata* and *S. pele*. They are similar to one another and often pinkish. The former is a Hawaii endemic reaching 9.5 cm in length, and the latter reaches 13.5 cm. Both species are generally found below about 100 m.

- *Scorpaenodes*: Hawaiian species include *S. corallinus* (nearshore, 10.5 cm maximum length), *S. evides* (Cheekspot Scorpionfish, nearshore, to about 10.5 cm), *S. hirsutus* (Hairy Scorpionfish, nearshore to about 6 cm), *S. kelloggi* (Dwarf Scorpionfish, nearshore, to about 5 cm), *S. parvipinnis* (Coral Scorpionfish, nearshore, to about 14 cm), and *S. varipinnis* (0–200 m deep, 13 cm maximum length). Species in the western Atlantic are *S. caribbaeus* (Reef Scorpionfish, 25 cm) and *S. tredecimspinosus* (Deepreef Scorpionfish, 6 cm). Both are mottled and spotted. The Rainbow Scorpionfish (*S. xyris*) is a slim, smooth-skinned species of southern California. Overall, it is red to brown with white spots, plus it has a distinctive large, dark spot on the operculum. It reaches 15 cm.

- *Scorpaenopsis*: Five Hawaiian species, including *S. altirostris* (poorly known, about 80–130 m deep, < 5 cm long), *S. brevifrons* (Shortnose Scorpionfish, nearshore, to about 13 cm), *S. cacopsis* (Titan Scorpionfish, nearshore, to 51 cm and 3.5 kg maximum recorded), *S. diabolus* (False Stonefish, nearshore, 30 cm). The Titan Scorpionfish has undergone pressure from harvesting, as it has a large size and mass, and is found near the shore.

- *Sebastapistes*: Six Hawaiian species, including: *S. balleui* (Spotfin Scorpionfish, nearshore, about 12 cm), *S. coniorta* (Speckled Scorpionfish, nearshore, 7.5 cm), *S. fowleri* (Pygmy Scorpionfish, nearshore, < 5 cm), *S. galactacma* (poorly known, collected to 200 m, < 5 cm), *S. mauritiana* (shallow, 9.5 cm), and *S. nuchalis* (poorly known).

- *Taenianotus*: Our species is *T. triacanthus*, the Leaf Scorpionfish, a Hawaiian denizen. This species is notably different than other scorpionfishes. It does resemble the habitat, but is more similar to a frond of macroalga than the bottom décor, per se. It is found close to shore and can be a hazard to swimmers and divers in nearshore reefs.

Leaf Scorpionfishes are perhaps the most unusual scorpaenids in Hawaii. *Bernard DuPont/CCA-SA/Flickr*

Distribution: Scorpaenids are distributed worldwide in temperate to tropical waters, although the diversity is higher in tropical and subtropical regions. Hence, our greatest diversity is from Hawaiian waters (and the Pacific American territories), followed by the Atlantic, and a few in the cool waters of the eastern Pacific. The general distribution of our species is given above in the Identification section. In the eastern Pacific, none of the species are found north of central California. Most of the northwestern Atlantic species are found at various latitudes along the eastern seaboard, and several are in the Gulf of Mexico. Distributions of the nearshore scorpionfishes include: Spinycheek, New Jersey to Florida and the Gulf; Longspine, South Carolina to Gulf to South America; Highfin, Virginia and northeast Gulf to Brazil; Coral, south Florida and Caribbean; Goosehead, New York to Brazil; Barbfish, Virginia to the Gulf to Brazil; Plumed, south Florida to Caribbean; Mushroom, Georgia to Venezuela; Spotted, Maine to Florida and Gulf, and Caribbean; Reef, Florida to Caribbean; Deepreef, North Carolina to Caribbean.

Natural History: Many scorpionfishes occur in the intertidal zone and inshore areas, but others are primarily deepwater animals. Typical scorpionfishes are found in a variety of habitats, including mud or sand, eelgrass beds, rocky reefs, and coral bottoms, depending on the species and its niche. Reef dwellers tend to favor cracks, crevices, and caves, rendering them even more cryptic.

Scorpionfishes tend to remain immobile to go unnoticed by their prey, or perhaps prey is attracted by the cirri. Most species feed primarily on shrimps, crabs, and small fishes, but isopods, cephalopods, and other small animals are also eaten. They protrude their large mouths when prey passes by and rapidly suck in their meal. They may also creep along the bottom to slowly pursue prey. Although venomous and cryptic, scorpionfishes are eaten by larger fishes, including snappers, sharks, rays, and moray eels. Humans are also predators, as scorpionfishes often make good table fare, although ciguatera poisoning has been documented in some scorpaenids. Nearshore scorpionfishes are an important part of the sport fishery.

Scorpaenidae and their relatives have varied reproductive modes, ranging from oviparity to ovoviviparity and external to internal fertilization. Our typical scorpionfishes are oviparous with external fertilization, laying small eggs during the spring to summer. When spawning, the fertilized eggs float to the surface. In the case of *S. guttata*, a well-studied and economically important species of southern California, adults form large breeding aggregations over sandy bottoms with few benthic predators. Spawning peaks in August and occurs around midnight. The eggs are released in the water column and float to the surface where the gelatinous masses float. The eggs hatch in about three days and the planktonic larvae are about 2 mm long. They are fairly nondescript at hatching, except there is a distinct color pattern, and as they grow, the pectoral fins, head,

Unidentified scorpionfish of the Atlantic. It is similar to a Barbfish, but lacks the dark spot over the pectoral fin. *Andrew David/Public Domain/NOAA*

and eyes become larger. The heads become spiny and pectoral fins large and dark by 6 mm. Larvae are encountered from September to November.

Encounters: The most commonly encountered species from California is *S. guttata*. It is an important game fish and commonly taken in recreational and commercial fishing ventures. They are commonly seen by divers and may be particularly abundant around some of the offshore islands. The counterpart in the Gulf of Mexico and the Atlantic is *S. plumieri,* although any of the species mentioned above could be encountered by swimmers, snorkelers, SCUBA divers, and anglers. Nearshore species known to present hazards in Hawaii include *S. diabolus* and *T. triacanthus*. Other species can be encountered also, but they don't seem to be quite as common or cause such a problem from envenomation, but all are probably venomous. Scorpionfishes can be a hazard when they are on a fishing line, in a net, on a spear, or when lying cryptic on the bottom. They are occasionally found in tide pools.

While these animals try to go unnoticed, at least some species (e.g., *S. guttata*) can go on the defensive by erecting their dorsal spines and jabbing at the intruder. This is usually done when the threat is in close proximity. They will also widen their opercula and jab with their head spines.

Venom Apparatus, Venom, and Symptoms: The sting of scorpionfishes is defensive. The venom apparatus consists of bony spines in the dorsal (7–18 spines, depending on species), pelvic (1 spine, each side), and anal fins (2–5 spines), which precede the soft rays. The number of spines and soft rays helps characterize the various genera and species. *Scorpaena* have 12–13 dorsal spines and 3 anal spines. Cephalic spines can deliver crinotoxins during puncture wounds. These spines are also important for identifying genera and species. The dorsal spines are usually responsible for human envenomation. Using *S. guttata* as a well-studied representative, the dorsal spines are relatively long, slender, and sharply pointed at the tip. They are smooth, slightly recurved, and lack serrations. On the anterior portion of the spine, especially distally, are two deep grooves. In these grooves are strips of glandular epithelium that house the venom. The paired venom glands and spine are encased in an integumentary sheath. The pelvic and anal spines are similar, having venom glands in anterolateral grooves. The victim is envenomated when the spine enters the flesh, causing the sheath to push against the venom gland, and squeeze the venom out through the grooves.

There are numerous studies on the venom and symptoms of stings by Scorpaenidae and their allies. However, most studies address lionfishes and stonefishes, rather typical scorpionfishes. The Stone Scorpionfish, a species common in southern California, and False Scorpionfish, a species of Hawaii, should not be confused with the Indo-West Pacific stonefishes.

Even though the venom from many species causes severe pain, it is not entirely algogenic, and can be rather virulent in some species studied. The venom has protein and non-protein factions, but components are difficult to isolate. Venom of both *S. plumieri* and *S. guttata* cause death in laboratory mammals at fairly low doses. One estimated LD_{50} for *S. plumieri* is 0.28 mg/kg (intravenous). That of *S. guttata* was estimated at 0.9 mg/kg (intravenous) for a lethal protein fraction and 2.6 mg/kg for a crude protein extract. In laboratory mammals, hypotension and hemorrhagic, hemolytic, and proteolytic properties were noted. Both of these common scorpionfish species cause changes in cardiovascular and pulmonary function in test animals. Proteolytic enzymes (e.g., Sp-GP), a vasoactive cytolysin enzyme (Sp-CTx), and fish lectin protein (plumieribetin) have been isolated from the venom of *S. plumieri*.

Scorpaena guttata stings result in severe pain, throbbing, swelling, redness or cyanosis, nausea, faintness, and lymph node involvement. High doses of the venom cause pronounced cardiovascular and pulmonary maladies in mammals. In some of the literature, the sting of *S. plumieri* is termed "mild" or "not very poisonous," yet in a study summarizing 14 envenomations by this species in Brazil, symptoms included intense pain, swelling, redness, agitation, and malaise in all cases. In most cases, other symptoms included nausea, vomiting, sweating, and fever. Pain irradiation, diarrhea, enlarged lymph nodes, and rapid or arrhythmic heartbeats were present in some of the victims. The symptoms of *S. brasiliensis* stings (n = 9 cases) were similar, although the Barbfish is sometimes considered more venomous. Pain from scorpionfish stings usually subsides in 1–6 hours.

References and Resources: Abdun-Nur et al. (1981), Andrich et al. (2010), Boletini-Santos et al. (2008), Breder (1963), Brooks (1987), Bulter et al. (2012), Carlson et al. (1971, 1973), Carrijo et al. (2005), Church and Hodgson (2002a), de Santana Evangelista et al. (2009), Diaz (2015), Eschmeyer and Randall (1975), Eschmeyer et al. (1983), Gomes et al. (2010, 2011), Haddad et al. (2003), Halstead (1951, 1978), Halstead et al. (1955), Hoese and Moore (1998), Humann and DeLoach (2014), Kells and Carpenter (2011), Kizer et al. (1985), Love et al. (1987), McEachran and Fechhelm (2010), Schaeffer et al. (1971), Smith and Wheeler (2006), Ziegman and Alewood (2015).

Lionfishes (Subfamily Pteroinae)

"It won't kill you, but it'll make you wish you were dead." [E. Spencer, 2013. National Geographic Explorer online]

Classification: Phylum Chordata (chordates); Class Actinopterygii (ray-finned fishes); Order Scorpaeniformes (scorpionfish-like fishes); Family Scorpaenidae (scorpionfishes); Subfamily Pteroinae (lionfishes); Genera *Pterois* (fire fishes) and *Dendrochirus* (dwarf lionfishes); Species *P. volitans* (Red Lionfish), *P. miles* (Devil Firefish), *P. sphex* (Hawaiian Turkeyfish), and *D. barberi* (Green Lionfish).

The Pteroinae are known by a variety of names, including firefish, zebrafish, turkeyfish, and butterfly cod. Worldwide, there are 12–15 species in the genus *Pterois*. *Pterois volitans* and *P. miles* are non-native invasives in North America that originated from the aquarium trade. Other species commonly seen in pet shops include *P. radiata* (Clearfin Lionfish) and *P. antennata* (Spotfin Lionfish). The most commonly kept species of *Dendrochirus* in aquaria

Antennata Lionfish are popular in aquaria and do not get as large as *other Pterois*. Christian Mehlfuhrer/CCA/Wikimedia Commons

Zebra Dwarf Lionfish. *Silke Baron/CCA/Flickr*

include *D. brachyptera* (Fuzzy Dwarf Lionfish) and *D. zebra* (Zebra Dwarf Lionfish). One other genus and species, *Parapterois heterura* (Bluefin Lionfish), is a beautiful temperate species that is occasionally kept in domestic aquaria.

Identification: Lionfishes are a very distinctive group of scorpionfishes. They are exceptionally beautiful and elegant animals. The dorsal spines of all *Pterois* are very long. The soft rays of the pectoral fins are unbranched, large, and showy. There are spines on the dorsal, anal, and pelvic fins. The spines and pectoral fins are also banded. There are usually cirri above the eyes and around the mouth. The soft rays of the dorsal, caudal, pelvic and anal fins are usually spotted. These animals are popular in domestic marine aquaria, although *P. sphex* is difficult to obtain outside of Hawaii. *Pterois volitans* and *P. miles* are very similar in appearance, with some subtle color and meristic character differences. The North American invasives are usually just referred to as the *P. volitans/miles* complex, as they hybridize. They reach up to 38 cm total length. They are somewhat laterally compressed with narrow aposematic maroon and white banding. *Pterois sphex* has a somewhat different appearance, having broad maroon bands and narrow white bands, with long, white filamentous rays. It is smaller than members of the *volitans/miles* complex, only reaching about 22 cm.

Pterois miles occurs in the Indian Ocean, and its genetic material is in some Atlantic samples. *Andreas Marz/CCA/Flickr*

The Green Lionfish is one of the dwarf lionfish species, so is smaller than *Pterois*, reaching only about 16 cm. It has various amounts of green on the fins, particularly the pectoral fins. It has proportionately very large eyes. The banding on the body is not as distinctive as that of *Pterois*. This species is rarely seen for sale in the domestic aquarium trade, although similar species, *D. brachyptera* and *D. zebra*, are commonly available.

Distribution: In their native habitats, *P. miles* occurs in the Indian Ocean and Red Sea, while *P. volitans* is the tropical western Pacific counterpart. The two species are sympatric in Indonesia. *Pterois volitans* is also native in some U.S. territories in the eastern central Pacific, but are replaced by *P. sphex* in Hawaii, where it is endemic. Most in the aquarium trade are *P. volitans*, because they are commercially collected for U.S. pet stores in the Philippines. *Pterois volitans* is also the most common species in Atlantic and Gulf of Mexico waters (e.g., 93% in one study). Introductions probably originated through release of aquarium stock, but it is possible they came from ship ballast water. Lionfishes were first spotted in the mainland United States off Florida in the 1980s. By 2000, they had spread to North Carolina and southward and westward. Individuals were seen as far north as Rhode Island. As of 2009, they had not become established in the Gulf of Mexico, but were predicted to expand there. Sure enough, by 2012 they were present in most appropriate habitat throughout the Gulf of Mexico, including the entire Texas coastline. In the northern Gulf, most records are from hard-bot-

Hawaiian Turkeyfish, the only native *Pterois* in the United States, outside of Pacific American territories. *Kevin Lino/CCA/NOAA*

tom offshore areas along the continental shelf, but they have even been taken from fishing piers. They are established south into coastal South America.

Pterois sphex and *D. barberi* are both native to Hawaii. The Hawaiian Turkeyfish is endemic to the islands, while the Green Lionfish is also known from the Johnston Islands, 1,360 km southwest of Hawaii. To my knowledge, neither species has become established outside their native range. It is possible that lionfishes other than *P. volitans/miles* complex could also become established off North America.

Natural History: Lionfishes are usually found in marine waters less than 40 m deep, but have been reported to 300 m. In their native habitat, they are most commonly encountered on coral reefs. The Hawaiian species are typically found to about 30 m deep. The coral reef habitat protects them from surge, gives them plenty of crevices to hide in, and provides a bounty of small fishes for prey. *Dendrochirus barberi* is also found in deeper water. In one study of small reef fishes, *D. barberi* was uncommon, and usually found in patch reefs of coral or rock, and was seen at night in areas of rubble. In the same study, six species of other scorpaenids were very common (especially *S. kelloggi*), although *P. sphex* was not recorded.

Non-native *P. volitans/miles* are well established in the western Atlantic, Caribbean, and the Gulf of Mexico, where they are classified as invasive species because they pose a serious threat to coral reef ecosystems. They are the first invasive fishes in the western Atlantic, and it is thought they may be one of the ecologically worst invasive species on Earth. Densities in North Carolina have been recorded as high as 450/hectare. In the northern Gulf of Mexico, recent studies show them to be at densities of nearly 15/hectare at artificial reefs, and it is believed they have not expanded to their full potential in the Gulf. Compare this to their native habitats, where they are considered uncommon to rare because of low densities. They are successful in their new home for a number of reasons. First, there is a lack of large predators, such as large groupers. This is due in part to the reduced numbers of predators from overfishing, but also because larger fishes are naïve predators (may not recognize them as food)—and lionfishes are venomous to native species. It is also possible their vivid color pattern is recognized by fishes as being something to avoid. Lionfishes are effective predators that capitalize on the bounty of smaller animals—prey that is also naïve about this unfamiliar predator. Lionfishes have been shown to reduce fish recruitment in inhabited areas. They feed on a wide variety of smaller fish species from several families, as well as invertebrates. Common fish prey in the Bahamas includes wrasses, gobies, and basslets, but there are many other species in their diet—whatever fits in their mouths. Because they reduce recruitment of a plethora of native fishes, the whole ecosystem is thrown out of balance. While some groupers have been documented to feed on lionfishes occasionally in the western Atlantic, in trials, hungry groupers will not feed on them, so lionfishes outside their natural range are nearly impervious to predation.

Lionfishes are oviparous. *Pterois volitans/miles* breed year-round. They have a courtship ritual that begins at dusk and goes into the night. The egg clusters are released at the surface. In a study of North Carolina and Bahamas introductions, they have been shown to spawn every four days. This is also part of the reason for their success—in one year a female can lay 2 million eggs! At hatching, larvae are about 1.5 mm in length. Like other scorpaenids, larvae have a large head with spines. During the larval period, they are planktonic and can be carried by the currents, which is how they have dispersed so rapidly.

Encounters: Encounters are mostly by marine aquarists, both domestic and public; anglers; commercial fishermen; and SCUBA and skin divers, although they are often deeper than a snorkeler will dive. They are commonly encountered during invasive lionfish eradication derbies. In 2017 there were 23 lionfish harvest derbies for SCUBA divers in Florida, and divers killed 24,029 lionfishes, which hardly made a dent. In some areas, swimmers may be at risk of an encounter. As best I can tell, most envenomations are to domestic aquarists. My token lionfish sting happened when I was capturing a marine aquarium fish in a net and did not see the small "black morph" volitans lionfish on the dark side of the tank. Every so often, I would hear from other aquarists who were stung. In a study of envenomations in Texas from 1998–2006 (before lionfishes invaded the coastal waters), 188 calls to poison control centers were for lionfish stings via aquaria. However, most stings were probably not reported. Similarly, in a study in Brazil in 2004, the author stated that keeping lionfishes in domestic aquaria was growing at an exponential rate, and lionfish stings accounted for one-third of all aquarium accidents he studied. Studies in Europe indicate the same trend. The Florida Poison Control Center website says their most common calls of marine origin are in regards to stings from catfishes, lionfishes, and scorpionfishes.

Stings also occur to divers and anglers. It is usually from not being careful, as when removing the fish from a spear or hook. People have been stung by reaching into coolers containing fish, or even stung by dead fish while preparing it for dinner. Fishermen can also get stung when hauling up nets, or when the fish are left in a precarious area, as on the deck. Lionfishes will defend themselves if they feel threatened by erecting their dorsal spines and darting with amazing speed at a person who is too close. Some divers and fishermen are at risk by the occupational or recreational hazard of lionfish control.

Venom Apparatus, Venom, and Symptoms: Lionfishes have a venom apparatus similar to other scorpionfishes, and it is very well developed. There are 13 dorsal spines, three anal spines, and two pelvic spines (one on each side) that are all venomous. The venom apparatus is classified as the *Pterois* type. The apparatus is similar to that of *Scorpaena*, except that the dorsal spines are typically longer, more slender, and almost straight. The venom glands are also longer and the anterolateral grooves are deep and longer than in *Scorpaena*. This translates into a more advanced defense system for *Pterois*. The integumentary sheath on the spine of *Pterois* is thin, while that of *S. guttata* is relatively thick.

The venom of *P. volitans/miles* can cause cardiotoxic, neurotoxic, and hemotoxic effects. Venom components include proteins and non-protein toxins. Envenomation causes changes in heart rate and strength of contraction, and changes in blood pressure. A sting also stimulates a massive release of acetylcholine and then depletes it. In laboratory studies on test animals, the venom causes fibrillation, then defibrillation of muscle. An LD_{50} of 1 mg protein/kg has been reported.

Most of the genetic material of invasive Atlantic lionfishes is from *P. volitans*.
Andrew David/CCA/NOAA

My lionfish sting was a superficial envenomation. Probably no more than 1–3 mm of a single dorsal spine punctured the skin of a finger knuckle. The poke was felt immediately and the pain started within a few seconds. Despite the miniscule amount of venom that could have entered my finger, the entire hand turned red, swelled up, and was stiff for the entire day. There was a fairly intense burning sensation, which for some reason I did not find all-too unpleasant. I did nothing to counter the effects. The symptoms tapered off during the evening. The pain from a lionfish sting usually peaks in 60–90 minutes and lasts for twelve hours or so, but it may last several days or even weeks. The pain may be intense, throbbing, or felt elsewhere in the body, and numbness may also occur. Symptoms of moderate envenomation may include the formation of blisters, local hemorrhage, and local necrosis.

A friend of mine experienced a severe envenomation while night diving in Micronesia. She was severely stung while doing close-up photographs of a large *P. volitans* in the water column. Apparently, the flash from the other diver triggered a defensive response. The fish jabbed its dorsal spines deep

A "black volitan" lionfish, like the one that tagged me. This color phase is popular in the aquarium trade, although this individual was photographed in the Caribbean, where it is invasive. *Carole Miron/CCA-EQ/Flickr*

into her face. I do not recall the details, except that it was a severe envenomation with systemic complications, and she remained in the hospital for an extended stay. She said she almost died. There are reports of human fatalities from lionfishes, but these are very rare, and some authors flatly state the stings are not fatal. Symptoms of systemic involvement, which is not common, can include headache, nausea, vomiting, delirium, fainting, abdominal pain or cramping, seizures, limb paralysis, changes in blood pressure, respiratory distress, tremors, muscle weakness, and cardiac distress.

Remarks: Lionfishes are popular as table fare, but there is some concern for the potential for ciguatera poisoning from ingesting them. There are no documented cases in the western Atlantic, but there are cases from other species of reef fishes in the Caribbean.

References and Resources: Albins and Hixon (2008, 2013), Barbour et al. (2011), Church and Hodgson (2002b), Cohen and Olek (1989), Côté et al. (2013), Forrester (2008), Freshwater et al. (2009), Greenfield (2003), Haddad (2004), Halstead (1978), Hamner et al. (2007), Humann and DeLoach (2014), Jud et al. (2011), Kulbicki et al. (2012), Morris and Whitfield (2009), Morris et al. (2009), Nair et al. (1985), Saunders and Lifton (1960), Saunders and Taylor (1959), Schofield (2009, 2010), Schultz (1986), Shiomi et al. (1989), Vetrano et al. (2002), Williams et al. (2004).

Rockfishes and Thornyheads (Family Sebastidae)

[After trying to remove the hook from a Black Rockfish] "The next thing I knew the sharp poisonous dorsal spines had sunk into the fleshy palm of my right hand. The pain was excruciating! I dropped on my knees in the bottom of the boat and vomited over the side it was so intense." [B. Thornton, British Columbia fishing website]

Vermilion Rockfish, one of the most targeted species by commercial and sport fisheries now has conservation measures in place. *NOAA/Public Domain*

Classification: Phylum Chordata (chordates); Class Actinopterygii (ray-finned fishes); Order Scorpaeniformes (scorpionfish-like fishes); Family Sebastidae (rockfishes and thornyheads); Genera *Helicolenus* (rosefishes), *Sebastes* and *Sebastolobus* (rockfishes), and *Trachyscorpia* (thornyheads).

Sebastids are often called "rock cod" by fishermen, and the meat is often marketed that way, or as "snapper." Atlantic species are called "redfish" and may be marked as "ocean perch." There are about 130 species of rockfishes worldwide, with nearly half of them occurring in North America, largely in the northeastern Pacific.

Identification: Rockfishes are similar to scorpionfishes, but are generally more bass-like. They tend to be smoother, have smaller, less spiny heads, lack cirri and fleshy tabs, and are more laterally compressed. Although often found near the bottom, they typically are built more for swimming than *Scorpaena* and similar benthic scorpionfishes. They have 12–15 dorsal spines. Thornyheads typically are deepwater bottom dwellers. They are slender, but have a very large and thorny head with large eyes. They are red in color, as are many other deepwater animals. They also have large pectoral fins, similar to scorpionfishes. Rockfishes and thornyheads are typically 30–60 cm in length, but sometimes up to about a meter, depending on species. The single species of *Helicolenus* in our area is the Blackbelly Rosefish (*H. dactylopterus*) of the Atlantic. It is reddish above and pinkish below. It has a short snout and large eyes. It reaches about 20 cm in length.

It is beyond the scope of this book to describe the 60+ species of rockfishes, and even experts are sometimes challenged to identify them to species, especially because of inherent variation in color pattern and ontogenetic changes. Having said that, color pattern is one of the most important features to differentiate species. Fortunately, these are economically very important fishes, so there are several good field guides to help the reader to identify species. Some are decidedly bass-like, showing little resemblance to scorpionfishes (all are *Sebastes*, unless otherwise noted), such as the Bocaccio (*S. paucispinis*) and Silvergray Rockfish (*S. brevispinis*). Others have large, ominous-looking dorsal spines, such as Quillback Rockfish (*S. maliger*) and Cowcod (*S. levis*). While some species are drab, others sport bold color patterns. Treefish (*S. serriceps*) can have a beautiful contrasting black and yellow barring—and pink lips, while Flag Rockfish (*S. rubrivinctus*) and Redbanded Rockfish (*S. babcocki*) have striking contrasting bars of red and white. Many species are reddish in color. Examples of commonly caught species on rod and reel are (or were) Cowcod, Bocaccio, Black Rockfish (*S. melanops*), Copper Rockfish (*S. caurinus*), Blue Rockfish (*S. mystinus*), Vermillion Rockfish (*S. miniatus*), Chilipepper Rockfish (*S. goodei*), and Yellowtail Rockfish (*S. flavidus*). Alaskan Rockfishes of economic importance include Black Rockfish, Blue Rockfish, Yellowtail Rockfish, Yelloweye Rockfish (*S. ruberrimus*), Dusky Rockfish (*S. variabilis*), Widow Rockfish (*S. entomelas*), Pacific Ocean Perch (*S. alutus*), and Shortspine Thornyhead Rockfish (*Sebastolobus alascanus*). The two reddish species of the Atlantic are Acadian Redfish (*S. fasciatus*) and Golden Redfish (*S. norvegicus*).

Treefish, one of the species with aposematic coloration. *Douglas King*

Flag Rockfish, a favorite among divers and anglers. *L.L.C. Jones*

Distribution: Worldwide, rockfishes and thornyheads are found in the Pacific, Atlantic, and Indian Oceans, but they are particularly well represented in the temperate northeast Pacific. There are only three species in the western Atlantic, and only *H. dactylopterus* occurs in the Gulf of Mexico. A single species of thornyhead is found in the Gulf and western Atlantic, the Atlantic Thornyhead (*Trachyscorpia cristulata*). It is a deepwater species, found from 130–1,100 m.

Some rockfishes are warm-water temperate species. In the eastern Pacific, they typically range from southern California to Mexico or even South America. The other group are cold-water temperate species in central and northern California northward into the Aleutian Islands, the Bering Sea, and the adjacent northwestern Pacific. Some species cover a broad range of latitudes and can occur from the Bering Sea to Mexico. Species that are inshore to the north may be offshore and deepwater at the south end of their range. In the Atlantic, distribution is: Acadian Redfish occur from New Jersey to Iceland; Golden Redfish occur in North Atlantic and North Sea, and Grand Banks; Blackbelly Rosefish occurs from Nova Scotia to Brazil, and is scattered in Gulf of Mexico; Atlantic Thornyhead is found on both sides of the Atlantic, and in the western Atlantic from Massachusetts to the Caribbean and Gulf of Mexico.

Natural History: As a group, sebastids range from the intertidal zone to at least 1,200 m deep, but most we encounter are found from nearshore shallows to about 250 m deep. As a rule, rockfishes are not in the open ocean, preferring coastal areas. Most species can be found over a fairly wide range of depths, partly in response to water temperatures and seasonal preferences and life-history traits. Rockfishes are a dominant part of the fish community everywhere along the eastern Pacific. They are conspicuous in kelp beds, over rocky reefs and wrecks, over or on sand and mud bottoms, or in the water column. The deeper species are often benthic. They tend to not be surface dwellers (except when larval), and those in the water column are termed midwater species. Many are found in a particular stratum of the water column to partition resources with other sebastids. Some species are termed transitional, meaning they occur from the bottom up to several meters above the bottom. Young rockfishes often occupy shallower waters than adults, where small food resources abound.

Rockfishes are among the most important predators where they occur. They eat a variety of marine organisms, including shrimps, copepods, amphipods, crabs, squids, octopuses, and a wide variety of fish species. Young rockfishes are pelagic and feed on smaller fellow plankton, then segue to larger organisms as they grow. Rockfishes are eaten by a wide variety of other marine creatures.

Unlike *Scorpaena* and their allies, sebastids give birth to live young. Eggs develop in the ovaries and when they reach a certain developmental stage, they are extruded. An indicator of this larval stage is when the eyes turn black. Eggs and larvae that are not extruded will be resorbed, then the ovaries will repair for future reproductive events. Studies of egg counts in southern California ranged from 1,240 eggs to 2,680,000, depending on the species. They usually extrude larvae in late winter to early spring (ranging from November to May, depending on species) in southern California. Spawning season is longer in duration in the south than in the north.

These are slow-growing, long-lived species. Adults may reach 15–50 years of age, depending on the species, and a few can live in excess of 100 years. The Rougheye Rockfish (*S. aleutianus*) has a record estimate of 205 years old! However, the dwarf species grow rapidly and have short life spans. Many to most species can have multiple broods in southern California, but usually single broods to the north.

Encounters: Pacific rockfishes are commonly encountered by recreational anglers, commercial fishermen, SCUBA divers, snorkelers, researchers, and personnel of public aquaria. These are economically important species in both the Pacific (especially) and Atlantic. Twenty-eight species are considered important in the sport fishery of southern California and 29 are considered important in commercial fishery. Other species are important in the economics of central and northern California to Alaska. Alaska and British Columbia have about 30 rockfish species making up their groundfish fishery.

Party boats offer both inshore and offshore angling, targeting a variety of species, and commercial fishermen can work a wide range of depths. According to one report, in 1986, 20 billion tons of rockfishes were in sport and commercial hauls. In 1985, recreational fishermen landed 8 million rockfishes. The fishery is worth more than 1 billion dollars annually. As a group, rockfishes are of conservation concern because many have been overfished (see Remarks section).

Envenomation mostly occurs when someone lands a fish on a boat. Rockfishes erect their spines, and the sting comes from a thrashing fish or when one is stepped on, bumped into, or held. Removing the fish from the hook is probably the most common path to being stung. It is likely that many people are stung every year, but rarely reported.

Shortraker Rockfish. Large specimens like this can live to be 190 years old. *NOAA/CCA*

Venom Apparatus, Venom, and Symptoms: Rockfishes have a milder sting than typical scorpionfishes, so they have not received the attention of scorpaenids and their relatives. However, certain researchers decided to help close the knowledge deficit by describing the spine apparatus of one representative species (Brown Rockfish, *S. auriculatus*) and comparing it to others of the genus and among scorpaenid genera. They looked for venom glands in 14 California species, and synthesized information on envenomations from personal communication sources. Rockfishes have 13 dorsal, 3 anal, and 2 pelvic spines (one on each pelvic fin), some to all of which may contain venom glands. Only 2 of the 14 species were shown to possess venom glands in all spines, the Brown Rockfish and the Quillback Rockfish. The main spines of concern are the dorsal spines, but if someone holds the fish by its venter, the pelvic and anal spines can cause a sting. The spines are slender, without serrations, and tapered. The pelvic spines are stout. The strong spines are widest in the middle. They are slightly recurved and bear a very sharp trigonal tip. Anteriorly is a medial ridge, which separates the anterolateral grooves. The venom glands are in the grooves. The grooves are generally shallow, but are deep in the pelvic spines. The bony part of the spine and glands are covered by an integument. If the integument is stripped off, the whitish glands can be seen in the grooves. The various species have

Quillback Rockfish has an impressive array of venomous dorsal spines.
Ratha Grimes/CCA/Flickr

differing degrees of venom gland development (e.g., those with vs. without venom on particular spines, venom gland length and volume, depth of dorsolateral grooves). Compared to other scorpaeniforms, rockfishes have the least developed venom system.

Little is known about the venom components. A species of *Sebastes* from the western Pacific, *S. schlegelii*, has been shown to have an L-amino acid oxidase crinotoxin, a substance also found in rattlesnake venom. It has antimicrobial properties in the fish's skin mucous.

In one published account of envenomation, one author stated, "while working as a deck hand on San Diego sportfishing boats during the summers of 1946–1948, [I] was wounded at one time or another by the spines of practically every species of fish taken on the boat. On several occasions, wounds from the stings of bocaccio and vermilion rockfish, *S. miniatus*, among others, produced deep pain, throbbing, swelling, and chills." These are typical symptoms, along with redness and sometimes axillary pain and nausea. The burning pain is usually immediate and always worse than the puncture of a spine from a nonvenomous species. Fortunately, the symptoms are of short duration. Infections are not uncommon, and most cases treated in hospitals and clinics are likely due to infection.

Atlantic Thornyhead from the deep waters of the Gulf of Mexico. *NOAA/CCA/Flickr*

Remarks: Little is known about thornyheads, as they are usually encountered only by researchers doing deepwater trawls or in deep-sea submersibles.

Some species of rockfishes are prone to overfishing. For example, Yelloweye and Canary Rockfish are United States federally designated as overfished species and must be released. Rockfishes are more closely managed these days, as most species are slow growing, long-lived with relatively low fecundity, and highly sought after by anglers and commercial fishers. The Monterey Bay Aquarium Seafood Watch website is a good resource for knowing which species are sustainable and should be eaten or avoided.

References and Resources: There are many references on the biology and fishery of the various rockfish species, and only a few of those, mostly related to multiple species or venom, are included here. Boehlert and Yoklavich (1984), Butler et al. (2012), Charter and Sandknop (2000), Dorn (2002), Echeverria (1987), Eschmeyer et al. (1983), Gunderson and Sample (1980), Lenarz (1987), Love et al. (1990), Miller and Lea (1972), Parker et al. (2000), Roche and Halstead (1972), Woodby et al. (2005), Wourms (1991).

Yellow-bellied Seasnake (*Pelamis platura*)

"*Pelamis platurus* are found in groups along slicks . . . the aggregations range in size from five to several thousand snakes . . ." [Kropach 1971, during a study of Yellow-bellied Seasnake aggregations in the Gulf of Panama]

Yellow-bellied Seasnakes are helpless on land. *William Flaxington*

Classification: Phylum Chordata (chordates); Class Reptilia (reptiles); Order Squamata (snakes and lizards); Family Elapidae (elapids); Genus and species *Pelamis platura* (Yellow-bellied Seasnake).

Seasnakes are usually placed in the elapid subfamily Hydrophiinae. In much of the literature, the specific epithet is *platurus*, but that is the wrong gender for the Latin genus. This is the only species in the genus, although a change to *Hydrophis platura* has been suggested.

Identification: *Pelamis platura* ranges from about 50–113 cm in total length, although specimens from the eastern Pacific are on the low end of the scale. It is a laterally compressed snake, especially on the posterior half of the body, as the tail functions in propulsion. *Pelamis* lacks ventral scutes, which are not needed for locomotion. It has aposematic coloration. The dorsum is black, which contrasts sharply with the lemon-yellow venter. The demarcation line between the colors is straight to wavy. The light coloration on the tail in typical populations is generally white and the black coloration is wavy and spotted. The head is long and somewhat triangular when viewed from above.

Distribution: The Yellow-bellied Seasnake has the widest range of any snake on Earth, being circumtropical, except for its absence in the Atlantic. It is found in all tropical marine waters of the Indian and Pacific Oceans, including the eastern Pacific, where it is the only species of seasnake. There is concern it may become an invasive species in the Atlantic, via introduction through the Panama Canal or other means. It is well distributed in the Gulf of California, especially in the southern end, but has been recorded as far north as San Felipe. Apparently, it is a summer visitor, presumably following warm water currents north, but probably does not breed off Baja California. It is rarely recorded from southern California, as the water is usually too cool. It appears during El Niño events, when the water is warmer than normal. Historically, there were only three confirmed sightings (beach strandings), from San Clemente Island to San Diego. Then in 2015 another washed up on a beach in Oxnard; it made quite the news splash. At least two others appeared on southern California beaches in late 2015 and early 2016, doubling the number ever documented in the continental United States. *Pelamis platura* is not common in Hawaiian waters, either, and probably does not breed there. They are only occasionally seen, as when stranded on a beach after a storm. One reference says there are about 20 sightings from Hawaii.

Natural History: The Yellow-bellied Seasnake is completely marine. Among the many species of seasnakes, it is the only one that is considered pelagic. North American *Pelamis* are following warm water currents from the south. The Yellow-bellied Seasnake does range well out into the open ocean, but like other seasnakes it is mostly encountered along coastal areas. Genetic diversity is low, worldwide, suggesting genes are exchanged on all coasts of all shores through passive long-range migration. Their movements are passive because they are not overly strong swimmers and they use currents as their most effective mechanism for long-distance travel. They will seek out areas where prey is available, such as slicks, drifts, current interfaces, and under flotsam and jetsam. In tropical waters, thousands of individuals of all age classes may aggregate along slicks. Their body is highly modified for aquatic life—they lack belly scutes and a body form to allow efficient terrestrial movement.

One habit found in *Pelamis* that is not known from terrestrial snakes or even other seasnakes is that of knotting and tight coiling—other snakes do coil and writhe, but it is not really the same as in *Pelamis*. Young animals do this with higher frequency than adults, but all age classes have been observed in aquaria and in the wild. Basically, knotting and coiling is when the snake coils up into a tight, compact ball, including knots, and writhes through the coils and knots. This is apparently done to facilitate shedding, stretching the skin after shedding, ridding parasites, and to avoid predation. This behavior probably evolved because *Pelamis* is pelagic and has no contact with substrate or solidly planted objects so it uses itself as a substrate to shed.

The reproductive habits of *P. platura* are poorly known, but they give birth at sea, although there are records of likely reproduction near shore. They are not known to breed at our latitudes, or even along Baja California or Hawaii. Yellow-bellied Seasnakes are generally considered ovoviviparous. They give birth to 1–10 young. The offspring may be protected by the female. The neonates are more brightly colored than adults, and are 22–29 cm in length. In Panama, young are usually born in September to possibly December, while in Mexico, it has been suggested they breed year-round.

Yellow-bellied Seasnakes feed exclusively on fishes. They are not particularly species-specific, but prey is generally small. Studies on the diet have shown a wide range of prey species, perhaps commensurate with the species inhabiting the environments they dwell in. Their aggregations near slicks and under flotsam and jetsam suggest they are there to feed, as small fishes are abundant in these habitats.

Predators include fishes (e.g., snappers), sea birds (e.g., pelicans, sea raptors), and marine mammals (e.g., sea lions). Invertebrates such as octopuses and crabs have been known to feed on seasnakes. Tiger, Nurse, and Smooth Hammerhead sharks have also been reported to feed on seasnakes. However, during experiments, potential predators would not feed on live Yellow-bellied Seasnakes. They often have bite marks on the tail, so it may act as a decoy, as does a brightly colored lizard tail. There is some evidence that the flesh of this species is poisonous.

Encounters: Only the luckiest of the lucky would likely encounter this magnificent serpent in U.S. waters, except near some U.S. territories in the west-central Pacific. It is possible encounters along California and Hawaii may increase with climate change or El Niño events. The observations to date in southern California were of individuals stranded on beaches, although there are unconfirmed reports of offshore sightings. These animals are virtually helpless on land and they cannot return to sea; when stranded they generally expire from exposure. Where they are a safety concern is in tropical areas where commercial fishermen get them as bycatch in nets. Yellow-bellied Seasnakes are often regarded as mild-tempered, but there are accounts of their ability to bite repeatedly. No bites are known to have occurred in Hawaii or California.

Venom Apparatus, Venom, and Symptoms: *Pelamis* and other seasnakes have a venom apparatus similar to coralsnakes (proteroglyphous), except that the primary and secondary venom glands are separated from one another and both empty into the venom duct. The venom-delivering fangs are not much larger than the other teeth. There is a distinct venom gland, which empties into the slit-like opening at the tip of the fangs. Although the venom apparatus is not as advanced as in some other snakes, it is effective. The fangs are 0.9–2.8 mm long (average 1.7 mm). The venom yield is low, ranging from 0.9–5.0 mg/kg dry weight. The LD_{50} has been reported to be 0.67 mg/kg (subcutaneously) and 0.18–0.44 mg/kg (intravenously). It has been estimated that the lethal dose for a human is 3.7–7.5 mg, depending on the weight of the victim.

The venom is neurotoxic and myotoxic. It is generally considered to be less toxic than other seasnakes. Seasnakes evolved venom to quickly immobilize fish prey, so its use as a defense mechanism is secondary. Several proteins have been isolated from this species, including three-fingered toxins, cysteine-rich secretory protein (CRISP), 5'-nucleotidase, and metalloproteinases. Names for some of the toxins include pelamitoxin, pelamis toxin a (α), and pelamis toxins b and c. The venom contains variants of phospholipase A_2 and a variety of other components.

Along with a generally mild disposition, low venom yield, small fangs, and a small mouth, these snakes are not usually considered a great risk of fatal injury to human adults, but the venom is very potent. A number of accounts of fatalities have been published, mostly anecdotally, but the authenticity of these reports has been questioned. This species may deliver a dry bite; there are numerous accounts of bites being asymptomatic. In typical *Pelamis* bites, pain is not immediate nor intense, and symptoms are latent, from five minutes to eight hours after the bite. When envenomation occurs, reported symptoms include drooping eyelids, swelling, discoloration, blood disorders, muscle tenderness, pain, numbness (tongue, hands, and throat), shortness of breath, muscle paralysis, nausea, and vomiting. If death were to occur, it would likely be due to respiratory distress.

In one case report that *may* represent a typical mild envenomation, a herpetologist bitten on the hand experienced swelling, slight hemorrhage at the bite site, stiffness, and deep pain in the hand and wrist. The bite site was extremely sensitive. The pain was mostly swelling-induced and lasted twenty-eight hours. By fifty-six hours following the bite, the symptoms were gone, save some slight sensitivity.

References and Resources: Campbell et al. (2004a, b), Culotta and Pickwell (1993), Ernst (1992), Ernst and Ernst (2011a), Grismer (2002), Halstead (1978), Kropach (1971), Lemm (2006), Lomonte et al. (2014a), Masunaga et al. (2008), Mori et al. (1989), Pickwell (1971), Pickwell and Culotta (1980), Roly et al. (2014), Sheehy et al. (2012), Shipman and Pickwell (1973), Solórzano (2011), Stebbins (2003), Stebbins and McGinnis (2012), Tu (1987), Tu et al. (1976), Vallarino and Weldon (1996), Vick et al. (1973), Zug (2013).

Other Aquatic Vertebrates (i.e., Other Fishes in Class Actinopterygii)

"Longjaw squirrelfish punctures cause immediate pain far out of proportion to the injury. One angler reported a sting from this fish that was so painful he nearly fainted." (Thomas and Scott 1997)

Classification: Phylum Chordata (chordates); Class Actinopterygii (ray-finned fishes).

Even in the Age of Science, we still do not know how many fish taxa are venomous, but we are gaining on it. Certainly, the major players have been identified elsewhere in this book, but there are some minor players. Some species of these groups of fishes listed below are thought or known to be venomous, although many members of these families are poorly known or may be nonvenomous. In a study across fish taxa from 2016, 50–58 families of fishes have at least some venomous species that had developed independently some 18 times.

Any species of squirrelfish with long opercular spines should be handled carefully by anglers. Pictured is a Longjaw Squirrelfish found in Hawaii and beyond.
Paul Asman and Jill Lenoble/CCA/Flickr

- **Toadfishes (family Batrachoididae):** At least some members of the genera *Thalassophryne* and *Daector* are venomous, but these are not known to reach U.S. waters. *Opsanus tau* of the western Atlantic is known to be venomous. Some species of midshipmen (genus *Porichthys*) have been shown to be venomous, while others lack a venom apparatus.

- **Jacks (family Carangidae):** At least some jacks are venomous, but they are poorly known, and most jacks are probably not venomous. The venomous species includes the leatherjackets (*Oligoplites* spp.) and Leatherback or Spotted Queenfish (*Scomberoides lysan* (= *sanctipetri*)).

- **Tangs or surgeonfishes (family Acanthuridae):** At least some species of tangs are known to be venomous, including the genera in our area, *Acanthurus*, *Chaetodon*, *Naso*, *Paracanthurus*, and *Zebrasoma*.

- **Stargazers (family Uranoscopidae):** At least some genera and species of stargazers are venomous. This includes those of the genus *Astroscopus* in North America: *A. guttatus* (Northern Stargazer), *A. y-graecum* (Southern Stargazer), and *A. zephyrus* (Pacific Stargazer). At least the Atlantic species have been confirmed to be venomous. Other genera and species in U.S. North American waters are likely or known to be venomous, including *Kathetostoma averruncus* (Smooth Stargazer) and *Xenocephalus egregius* (Freckled Stargazer). Rather than trying to figure out which stargazers are venomous and which may not be, it is best to leave them all alone.

- **Squirrelfishes (family Holocentridae):** At least one species of squirrelfishes is reported to be venomous. The Hawaiian Longjaw Squirrelfish (*Sargocentron spiniferum*) is notorious among fishermen for its venomous sting. The Atlantic Longspine Squirrelfish (*Holocentrus rufus*) is also said to be venomous. It seems likely that all species with long pre-opercular spines should be eyed with caution. Other species with short pre-opercular spines are not as well known.

- **Orbicular velvetfishes (family Caracanthidae, or family Scorpaenidae, subfamily Caracanthinae):** The family has at least some species known to be venomous to humans. *Caracanthus typicus* (Hawaiian Orbicular Velvetfish) is, but is considered harmless by FishBase; however, I can find no information that anyone has researched this species. At least one of the species in the western Pacific American territories is venomous to humans (see American Territories chapter).

- **Soapfishes (Family Grammistidae, or family Serranidae, subfamily Grammistinae):** Soapfishes get their name from their thick mucous secretions. This coating contains potent crinotoxins. Although not usually considered venomous, they do have dorsal spines, and a jab could introduce crinotoxins, so technically it could be considered envenomation. Two genera in our area include *Rypticus* and *Suttonia*.

All of the above taxa are known to sting, and there are more families with members known or thought to be venomous, but poorly studied. For example, at least some clingfishes of the genus *Acyrtus* (family Gobiesocidae) from the East Coast were recently determined to be venomous, but I can find no information on human envenomation. There are even some taxa known to cause venomous bites. The combtooth blennies (family Blenniidae) of the genus *Meiacanthus* are venomous-fanged blennies found in the west Pacific and Indian Oceans (see chapter on American Territories). One-jaw eels (family Monognathidae) are strange deep-sea fishes, some of which are found in North America, but are not known to be a risk to humans. As one researcher pointed out, we are much more behind the curve of knowing about venomous fishes than we are about venomous terrestrial animals.

Identification: Each of these families of fishes is fairly distinctive, and in some cases their venom apparatus helps to identify them to family. Toadfishes are superficially similar to stargazers, because they are both bottom-dwelling species that have large heads with a body that may taper posteriorly. Toadfishes tend to be drab in color with fleshy projections. Midshipmen are interesting toadfishes because they have four rows of photophores, which can create visible light (hence the name, midshipman, as if lighting the way). Some toadfishes produce audible sounds. Stargazers are so named because their eyes are usually higher atop of the head. In one recent media blitz, the dorsal aspect of the head caused it to be dubbed the "Homer Simpson Fish" due to the resemblance of the namesake cartoon character. Stargazers not only have venom, but they can produce defensive electrical shocks, something that has evolved a few times among fishes.

Jacks are a speciose group of common fishes that generally range from mackerel or somewhat torpedo-shaped to laterally flattened. The two that are known to be venomous are both of the former type. They also both have a leathery skin. Jacks are typically active swimmers in the water column. They tend to be silvery in color to match their mid-water habitat. Squirrelfishes

A Southern Stargazer can not only envenomate but can also produce a nasty electrical shock. *Kirk Kilfoyle*

are mackerel to perch-shaped fishes. They are usually associated with reefs and often in deeper water than a snorkeler would be, although many occur within SCUBA diving range. They are usually reddish in color, which is cryptic for deeper waters or among reefs.

The tangs, or surgeonfishes, are laterally flattened, as are many other reef dwellers, such as butterflyfishes and angelfishes. They can be identified to family by one of their defensive apparatus, the spines on their caudal peduncle. Many tangs are brightly colored, making them appealing as marine aquarium species. There are numerous field guides available that help identify the many types of tangs from tropical waters.

Caracanthids are strange fish that are ovoid in shape but laterally compressed. They have small tubercles on their skin that give the animals a velvety appearance. Soapfishes are more-or-less bass-shaped fishes with a long tapering head that ends in an upturned mouth. They are extremely slimy due to the copious amounts of mucous covering the body. The different species are variously striped, spotted, or mottled.

Distribution: The toadfishes and stargazers are both temperate to tropical animals. There are several North American species. In the western Atlantic and Gulf of Mexico, the batrichoids include the North Atlantic Midshipman

This Specklefin Midshipman has photophores that bioluminesce. *Douglas King*

(*P. plectodon*), South Atlantic Midshipman (*P. porosissimus*), Leopard Toadfish (*Opsanus pardus*), Oyster Toadfish (*O. tau*), and Gulf Toadfish (*O. beta*). Off California, there are Specklefin Midshipman (*P. myriaster*) and Plainfin Midshipman (*P. notatus*). A habitat model has shown that the decidedly venomous-to-humans toadfish, *T. nattereri*, has potential habitat in the northern Caribbean and Gulf of Mexico, but their current range appears to be from the southern Caribbean to Brazil. All other species in the genus are also to the south.

The Northern Stargazer is found from New York to North Carolina. The Southern Stargazer is found in the western Atlantic from North Carolina and the Gulf of Mexico to northern South America. The Pacific Stargazer is found in the eastern Pacific from California to Peru. The Lancer Stargazer is found from North Carolina to Florida and the Gulf of Mexico. The Smooth Stargazer is found from California (rarely) to Peru and the Galapagos Islands.

The Leatherjacket and Spotted Queenfish are usually considered the venomous jacks, but their congeners, at least, are probably similarly armed. *Oligoplites altus* and *O. refulgens* are found just south of southern California in both the Pacific and Gulf of California, to Peru. *Oligoplites saurus* is found in the northern Gulf of Mexico and from Maine to Uruguay. *Oligoplites palometa* and *O. saliens* are found in the Caribbean and southern Gulf of

I counted five species of tangs in this mixed school on a Hawaiian reef, but these fish are docile, and cautious snorkelers can revel in their beauty when swimming among them. *Steve Jurvetson/CCA/Flickr*

Mexico, and farther to the south. *Scomberoides lysan* is found in the Hawaiian Islands and Indo-West Pacific.

Tangs are well distributed worldwide in all tropical waters. This includes Hawaii, the Gulf of Mexico, the western Atlantic, and the Gulf of California, but they do not reach southern California. There are about twenty species of tangs in Hawaii, including those of the genera *Acanthurus* (about twelve species), *Chaetodon* (two species), *Naso* (four Hawaiian species, commonly called unicorn tangs) and *Zebrasoma* (two Hawaiian species). Tangs of the Atlantic and/or Gulf of Mexico include the Ocean Surgeon (*Acanthurus bahianus*), Atlantic Blue Tang (*A. coeruleus*), Doctorfish (*A. chirurgus*), and Gulf Surgeonfish (*A. randalli*). In the Gulf of California the Yellowtail Surgeonfish (*Prionurus punctatus*) is very abundant, and some other species also occur there, particularly in the southern part of the Gulf.

Squirrelfishes are also common reef dwellers and are speciose in the waters of Hawaii, the Atlantic, and Gulf of California, but seem to be absent from southern California. One species, the Hawaiian Longjaw Squirrelfish, has been reported to cause human envenomation, but little is known about squirrelfish venom apparatus and glands, if present, for nearly all species.

Caracanthus typicus occurs in Hawaii, and there are two other species in the American west-Pacific territories. Atlantic soapfishes are in the genus *Rypticus*. Species include *R. bistrispinus* (Freckled Soapfish), *R. saponicus* (Greater Soapfish), *R. maculatus* (Whitespotted Soapfish), *R. subbifrenatus*

(Spotted Soapfish) in the Atlantic. An eastern Pacific species is *R. bicolor* (Cortez Soapfish), which occurs in the Gulf of California. *Suttonia lineata* (Palestripe Podge) occurs in Hawaii.

Natural History: All of these species are carnivorous, eating smaller fishes and invertebrates, except for tangs, which are grazers. They are bottom dwellers (toadfishes and stargazers), midwater species (jacks), or reef dwellers (most tangs, caracanthids, soapfishes, and squirrelfishes). Given that there are many natural history traits that differ from one taxon to another, I defer the reader to primary literature, or one of the many fine field guides.

Encounters: Encounters range from rare to common. Snorkelers, SCUBA divers, spear-fishers, and anglers may encounter some of these species. The largest at-risk group are anglers who catch carnivorous species, and spear-fishermen. Most of our native species are rarely, if ever, implicated in a significant human envenomation, which is why they are listed in this "other" category. Some can cause significant pain, but systemic reactions are rare. Hawaiian fishermen often catch the Hawaiian Longjaw Squirrelfish and Spotted Queenfish, so care should always be taken when unhooking these animals. Most accidents happen when a fish thrashes. Thrashing, by the way, is not only due to a fish trying to escape, but also a means for them to envenomate an assailant, as they may direct the spines into the human threat. Tangs are not normally caught on rod and reel, but they could be speared or hauled up in a net. Toadfishes and stargazers may also be caught while angling, and they are sometimes seen by divers, although many species are cryptic, somewhat resembling some scorpionfishes or sculpins.

Tangs can be extremely abundant, especially near reefs, but envenomations are presumably rare. They are also popular as tropical marine aquarium pets throughout the United States, and some of the more common species in domestic aquaria including the Indo-West Pacific Blue Tang (*Paracanthurus hepatus*), Yellow Tang (*Z. flavescens*), Sailfin Tang (*Z. velifer*), and "Naso" Tang (*N. literatus*), among a host of others in the genera *Acanthurus*, *Chaetodon*, and *Prionurus*. *Acanthurus* is a very large genus with many species that are popular in aquaria. The Yellow Tang is possibly the most common species for the domestic aquarium, aside from some damsels and clownfishes. Tangs and their more venomous relatives, the rabbitfishes (family Siganidae; see the chapter on American Territories), are docile aquarium fish that flee before trying to defend themselves in aquaria, although one in a net is prone to thrash.

Venom Apparatus, Venom, and Symptoms: The venom apparatus of the vast majority of venomous bony fishes (95%) involves spines, most of which are associated with fins. The use of spines is for defense only, except that it has been reported that some jacks can use their anal spines to envenomate smaller fishes used for food.

Toadfishes have 2–4 spines at the front of the first dorsal fin and opercular spines. The decidedly venomous species (i.e., *Thalassophryne* and *Daector*) have two dorsal venomous spines anteriorly that are nearly hidden by soft tissue. They are sharp and stout, and the venom gland surrounds the spine, and can be seen when the integumentary sheath is removed. The opercular spines originate from the bony operculum. Both of these spine types are venomous, at least in some species. They are hollow and venom flows through the hollow spine and out through an opening at the tip. The venom apparatus is considered to be among the most highly developed of fishes. Whether or not *Porichthys* spp. and other taxa are venomous has been the topic of some debate and limited research. While some researchers failed to find a venom apparatus and associated gland in *Porichthys*, other researchers have recently determined that at least *P. porosissimus* is venomous and there were two case reports of envenomation in humans. There is an axillary gland near the pectoral fin that may contain venom. On the Pacific side, *P. myriaster* is said by some to be nonvenomous, while others state it is probably mildly venomous.

Thalassophryne nattereri is the most studied species of toadfish because human envenomation is well-documented and not uncommon. It is primarily from northeast Brazil where they constitute a health risk. Studies on the venom of *T. nattereri* have isolated numerous proteins and enzymes. One class of toxins novel to these toadfishes are called "natterins," a class of venom fractions with kininogenase enzyme activity. Stings by these fishes can be fairly serious and cause pain, swelling, hemorrhaging, neurotoxic effects, and necrosis. Recent studies of the South Atlantic Midshipman show that more than 80 proteins were found to be in the venom. Effects on lab mice and two human case reports demonstrated that these animals are indeed venomous to mammals. In the human cases, stings from the opercular spines caused immediate and intense pain lasting about two hours. In one case, there was also reddening of the wound site. Studies in lab mice showed a variety of clinical effects of envenomation.

The stargazers are similar in general appearance to toadfishes, and have 3–5 small dorsal spines at the leading edge of the dorsal fin. They also possess spines on the cleithral bones of the shoulder girdle on each side above the large pectoral fins. These are similar in appearance to the cleithral spines of toadfishes, but structurally different. At least some *Astroscopus*, *Kathetostoma*, and *Uroscanopus* are known to possess a cleithral venom apparatus, I can find no information about the venom of our local species, or effects on humans, although the European species *Uranoscopus scaber* is known for envenomating people, and there are cryptic reports from the past of possible fatalities due to its sting. That species is said to have effects on humans similar to *Scorpaena* and *Thalassophryne*. In one study involving four specimens of three genera, all had cleithral spines with an associated venom gland. Stargazers are also capable of causing electrical shock in humans, which may cause an occupational hazard among commercial fishermen and recreational anglers alike.

The jacks are different than most other venomous fishes because they are fast-swimming midwater species; other venomous fishes tend to be bottom or reef dwellers. Jacks have dorsal fins that may be separate in adults, but joined to the soft rays of the dorsal fin in juveniles. Similarly, there are anterior anal fin spines that may also be venomous. There are seven dorsal and two anal spines in Leatherbacks that are venomous. The dorsal spines are not particularly well developed, but the anal spines are about three times as long and stout. Rather than having well-developed venom glands, the venom is produced in the skin covering the spines. The characteristics of the venom of carangids is not known, but enough anglers are stung to know how stings manifest themselves in humans. Young animals cause a bee-sting-like pain that only lasts 30–40 minutes, while the stings from adult Leatherbacks cause fairly intense pain that can last about four hours. There is also swelling and bleeding. The stings from the anal spines are said to produce a more intense pain than the smaller dorsal spines.

Squirrelfishes have an array of spines on both the fins and the head. On the anterior dorsal fin, or anterior part of the single, joined dorsal fin, there are 11–12 stout, sharp spines. The anal fin has four spines in the front, and the pelvic fin has one spine. Depending on species, there are zero to several well-developed spines in the head and gill area. In some species, the preopercular spine (i.e., in front of and lying over the gill cover) is the venomous spine. It is not known how many species are venomous. However, all of the spines can inflict painful wounds, even if not venomous. As best I can tell, nothing is known about the venom of holocentrids and there is very little information on stings. In Hawaii, the large species *S. spiniferum* is a common target of anglers, and they are sometimes stung when the fish thrashes about on the end of a hook or spear. Pain from a sting is said to be immediate and intense, well out of proportion with just the physical trauma associated with the sharp preopercular spine. Similarly, at least one holocentrid in the western Atlantic, *H. rufus*, is said to have a venomous preopercular spine. The Atlantic Longjaw Squirrelfish (*Neoniphon marianus*) also has a long preopercular spine.

Tangs are also adorned with a number of stout, sharp spines of the dorsal, anal, and pelvic fins. All genera and many species seem to be venomous to some degree based on the morphology of the spines (having anterolateral grooves), but venom glands are not always apparent. When present, the glands are covered with connective tissue and are within the spine grooves. As with other fishes having venomous fin spines, the spines can be erected to aid in penetrating a predator or territorial rival. In a few species examined in one study, *Acanthurus*, *Paracanthurus*, *Prionurus*, and *Zebrasoma* all had anterolateral grooves in the dorsal or anal spines. Only *Paracanthurus* and *Prionurus* had conspicuous venom glands, but there was a small sample size of one species in each genus. There are also 1–6 caudal spines (depending on genera and species) on each side of the caudal peduncle. These spines are like scalpels and when at rest they lie along the peduncle in a groove. When the animal is threatened, it can partially erect the spine by contracting muscles at the base of the spine. When erect, the spine faces forward. Some species have permanently erect caudal spines. Among several species of *Acanthurus* examined, all had well-developed caudal spines and venom glands, but it is likely other genera also have venom glands, as well. Some species, such as *N. literatus* and Achilles Tang (*A. achilles*) of Hawaii and Baja California have aposematic coloration on the caudal peduncle where the spines are. The scalpel is different than other spines on fishes because they are freestanding, rather than attached to vertebrae (i.e., fin spines) or head bones (e.g., cleithrum, preoperculum, or operculum).

Virtually nothing is known about the venom of surgeonfishes. People are occasionally stung when they catch them or take them off a spear. Humans can be cut or punctured by any of the fin or caudal spines. These cuts can be fairly deep and the pain from a cut may be out of proportion with the physical trauma. This suggests some species are essentially mildly venomous, while others are more so. When a person is cut or stabbed by a spine, there can be profuse bleeding, and the pain ranges between mild and severe. In severe cases the pain is said to be immediate and intense, and may last up to about twelve hours, although some lingering pain may last for a week. There may be burning and swelling that extends beyond the sting site. Muscle aches may develop around the sting site and there may be nausea.

The caracanthids are often regarded as a subfamily of the Scorpaenidae, and they have a similar venom apparatus. They possess 6–7 dorsal spines, one very small pelvic spine, and two anal spines. Little is known about the Hawaiian species, and I cannot confirm if it is or is not venomous. It is likely that it is rarely encountered, so the potential for stings is minimal, anyway. Other species from the Indo-West Pacific are known to be venomous to humans.

The soapfishes are regarded by some as just being poisonous because they have a crinotoxic coating in the mucous exudate. As the toxins are released into the surrounding water, they can cause irritation in humans swimming nearby. But they can also be considered venomous because the three dorsal spines of *Rypticus* and seven dorsal spines of *Suttonia* could introduce the crinotoxins into human flesh. The toxins are peptides known as grammistins, which are both hemolytic and ichthyotoxic, so have both a poisonous and venomous nature. Grammistins have been shown to cause death in lab mice when injected. There are numerous other crinotoxic fishes, such as moray eels and clingfishes, but they are not included here. Some clingfishes of the western Atlantic have been shown to possess a true venom apparatus, but I can find no information of human envenomation.

Remarks: Tangs and lionfishes are among the popular marine aquarium pets and until recently there was a thriving business in commercial collecting in Hawaii. At the time of this writing, all commercial and private collecting from Hawaii has been banned.

References and Resources: Collette (1966), Eschmeyer et al. (1983), Gosline and Brock (1965), Halstead (1978), Halstead et al. (1972), Hoese and Moore (1998), Humann and DeLoach (2014), Jones (1988), Kells and Carpenter (2011), Lopes-Ferreira et al. (1998), Magalhães et al. (2005), McEachran and Fechhelm (1998, 2010), Randall (1955), Randall et al. (1971), Shiomi et al. (2001), Smith and Wheeler (2006), Smith et al. (2016), Thomas and Scott (1997), Walker and Rosenblatt (1988), Winterbottom (1971), Ziegman and Alewood (2015).

PART III

11 Venomous Animals of American Territories

Taxa accounts in this book address species in North America north of Mexico, but also includes Hawaii, so it would not be complete without mention of other U.S. holdings. The territories under U.S. government oversight are all tropical to subtropical islands and atolls with coral reefs. Because of relative remoteness, information on species distribution is wanting in some areas, so the islands and their taxa are briefly discussed here rather than in the taxa accounts. There are several taxa not found in the mainland United States or Hawaii, so some brief taxa accounts are given below. Although the islands and reefs are spatially small, they are biologically rich.

American territories. The Caribbean territories are similar to Florida, while the west-central Pacific Islands are more diverse and have some taxa of the western Pacific. *Wikimedia/CCA-SA*

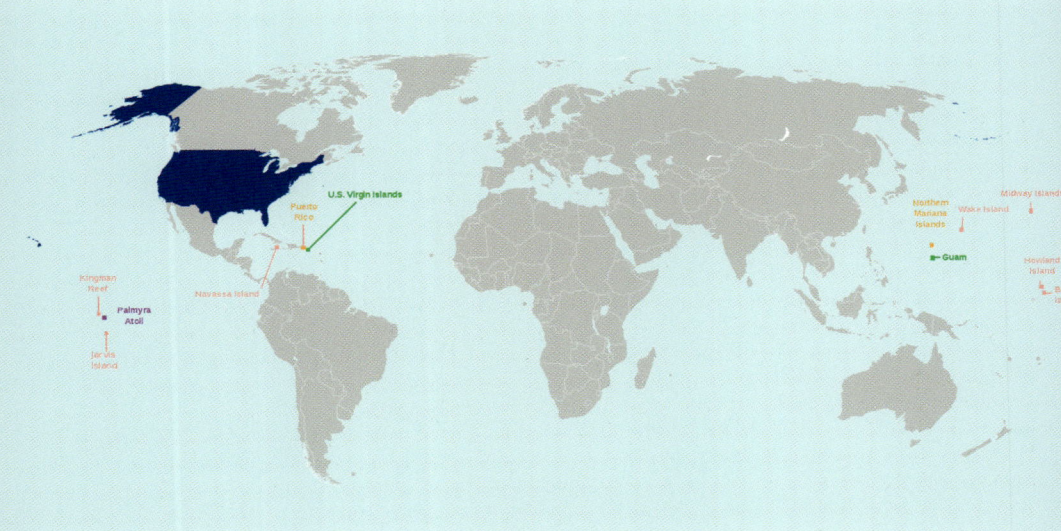

Collectively, the U.S. island territories boast thousands of marine species, eight national parks and monuments, and nineteen national wildlife refuges. Only five of the sixteen islands or island groups are permanently inhabited. Their locations can be divided into these broad marine geographic areas:

- **Caribbean (northwest Atlantic):** Navassa Island, Puerto Rico (inhabited), and U.S. Virgin Islands (inhabited).
- **Central Pacific:** Baker, Howland, Jarvis, and Midway Islands; Johnston and Palmyra Atolls; and Kingman Reef.
- **West-Central Pacific (mostly north Pacific or equatorial):** American Samoa (inhabited; this is the only U.S. territory in southwestern central Pacific), Guam (inhabited), Commonwealth of Northern Mariana Islands (inhabited), and Wake Island.

Not without exception, these islands tend to have few native terrestrial animals, venomous or otherwise, save birds and insects. One major exception is the scorpion fauna of the Caribbean Islands. There are more than one hundred described species in the Caribbean, which is among the best studied areas. Puerto Rico alone has at least nineteen species of scorpions, all but one of which belong to the Buthidae, the family that contains most of the highly venomous scorpions worldwide. These include numerous species of *Centruroides* and *Tityus*, which contain dangerously venomous species. *Centruroides gracilis* (which also occurs in Florida) and *C. griseus* account for many of the envenomations on Puerto Rico and the U.S. Virgin Islands. The effects of the venom of the different species are not well known, but most stings apparently do not require medical attention. Nevertheless, caution is advised, especially if children are stung. There are also several species of centipedes in the Caribbean, the most venomous of which is the notorious *Scolopendra gigantea* and/or *S. alternans* of Puerto Rico and the U.S. Virgin Islands, discussed in greater length in the centipede chapter. The invasive *S. subspinipes* of some Pacific islands is also discussed in that chapter.

There are a few spiders that are venomous to humans on some of the islands that are either native or introduced. There are several native species of *Loxosceles* in the West Indies. The British and U.S. Virgin Islands have an endemic recluse spider, *L. virgo*, while there is a similar species, *L. caribbea* from Puerto Rico. They are assumed to be similar to other recluse spiders discussed in the taxon account. Guam has a native spider, *Cheiracanthium diversum*, similar to other yellow sac spiders covered in the taxon account. Brown

Widows, black widows, and giant crab spiders (*Heteropoda venatoria*) are found on islands in the Pacific and Atlantic. They are excellent stowaways. There are also many native and non-native Hymenoptera on islands of both oceans. Many of those in the Caribbean are known from or similar to taxa in the United States. A recent arrival in Guam is the Great Banded Hornet (*Vespa tropica*), and paper wasps had already become established. European Honey Bees are probably everywhere close to human habitation. Native and non-native ants are everywhere. Puerto Rico has twelve species of fire ants, genus *Solenopsis*, and two species of trapjaw ants, genus *Odontomachus*. Little Fire Ants (*Wasmannia auropunctata*) have become established on several islands.

Puerto Rico and the U.S. Virgin Islands also have a dipsadid snake that is now known to envenomate humans—the Puerto Rican Racer (*Borikenophis* (= *Alsophis*) *portoricensis*). In a few case reports, envenomation occurred when victims were bitten for 1–4 minutes. Symptoms included hemorrhaging and significant swelling (e.g., to the elbow from a finger bite), but symptoms subsided in about a week. The venom has been studied, and it was shown to have high metalloproteinase and gelatinase activity, among other enzymatic activity. In fact, metalloproteinase activity was very high for a dipsadid and comparable to some rattlesnakes. The Brown Treesnake is one notorious invasive species that has overrun Guam and caused the extinction of most of its endemic bird fauna; this species is a potential threat to Hawaii and other Pacific islands. It is discussed in the chapter on colubrids and their relatives. There are an estimated two million Brown Treesnakes on Guam. The rest of Micronesia, Samoan Archipelago, and other Pacific islands are on high alert to avoid the accidental establishment of this species on their islands.

Unlike the terrestrial fauna, the marine waters of these tropical and subtropical isles are replete with native venomous taxa, although there are differences with longitude and latitude. The more southerly and westerly Pacific islands tend to have the highest diversity. However, survey efforts vary widely between island and atolls, so it is difficult to get an accurate count of species. FishBase, an online resource that has reported species counts by region, lists the following numbers of near- and offshore marine fish species among U.S. islands or island groups: Hawaii (1,273), Guam (1,015), Commonwealth of Northern Marianas (800), Puerto Rico (765), American Samoa (745), and U.S. Virgin Islands (569). The smaller island groups are poorly known, so I will emphasize those listed above. The species numbers given do not necessarily align with other studies, and that is likely because of differential survey efforts and emphasis on nearshore or reef fishes. According to one

The Lined Surgeonfish rarely shows up in Hawaii, but is common throughout Micronesia. *L.L.C. Jones*

study, Guam and the Northern Marianas have the highest recorded marine biodiversity of any area of comparable size on Earth, with a whopping 5,640 marine species recorded, including over 1,000 fish species, while it compares that to only about 600 reef species of fishes in Hawaii—less than half of what FishBase reports, which includes offshore species. Numerous species found in Micronesia and Samoa are not found in Hawaii, but Hawaii does have high endemism. Despite the lack of precise occurrence records, it is apparent that all of these waters off these tropical isles are among the most biologically rich places on Earth.

The westernmost Pacific island territories have three additional families of venomous fishes not found in the central or eastern Pacific: stonefishes (family Synanceiidae), coral catfishes (family Plotosidae), and rabbitfishes (family Siganidae), all of which can deliver a moderate to severe envenomation. Also, some fangblennies include venomous members of the combtooth blenny family, Blenniidae. Because these families are not discussed in the taxa accounts, they are briefly described below. According to published surveys and field guides of nearshore species, the Samoan Archipelago is home to about twenty species of typical scorpionfishes and lionfishes, four siganids, one plotosid, thirty-two acanthurids, and dozens of squirrelfishes and carangids (most of which are not venomous), as well as three species

Reef Stonefish are the most venomous fish species known. *Daniel Dietrich/CCA/Flickr*

of nearshore stingrays: Spotted Eagle Ray (*Aetobatus laticeps*), Kuhl's Maskray (*Neotrygon* (= *Dasyatis*) *kuhlii*), and Pink Whipray (*Paleobatis* (= *Himantura*) *fai*). Micronesia, which includes Guam and the Commonwealth of the Northern Marianas, has a similar ichthyofauna. There are thirty typical scorpionfishes and lionfishes, sixteen rabbitfishes, thirty surgeonfishes, one plotosid (not found in northern Micronesia), and two synanceiids across the archipelago. Nearshore stingrays include Kuhl's Maskray (aka, Blue-spotted Stingray), Spotted Eagle Ray, and Blotched Stingray (*Taeniurops* (= *Taeniura*) *meyeri*). Micronesia comprises a large area of Oceania, so Guam and the Northern Mariana Islands have fewer species than the more western and southern islands, which are more contiguous with the western Pacific. Reef Stonefish (*Synanceia verrucosa*), Red Lionfish (see genus account), and Pelagic Stingrays (see species account) are found in all of the U.S. west-central Pacific islands. The Red Lionfish (among other lionfish species) is native here, but it is treated primarily as an invasive species in the taxon account, as it has become a widespread menace in the western Atlantic and Gulf of Mexico. Caribbean fishes are already covered in this book, because most of them also occur off the Florida Coast, although diversity of both venomous and nonvenomous species increases as one goes south.

The Reef Stonefish, which is known from Guam, the Commonwealth of the Northern Mariana Islands, and American Samoa, is usually thought to be the most venomous fish on Earth. It and other synanceiids are responsible for human fatalities, although case reports are surprisingly few and sometimes information is misleading. Stonefishes are related to scorpionfishes and used to be placed in the same family. They are extremely cryptic—even more so than most scorpionfishes. They can easily go undetected in nooks and crannies of coral reefs or when buried in the substrate. Most people are envenomated when they step on them in shallows or reach into a crevice. The venom delivery system is more advanced than scorpionfishes. People are stung by the dorsal spines, which are extremely sharp and have a large, bulbous venom gland. The potent venom is well studied, and a few references indicate an LD_{50} ranging from about 0.02 to 0.8 mg/kg. The most lethal fraction is a protein called stonustoxin, which has an LD_{50} of 0.0017 mg/kg! The venom is a combination of algogenic, cytotoxic, hemotoxic, cardiotoxic, neurotoxic, and myotoxic compounds. Symptoms include extreme pain, hypotension, respiratory distress, anticoagulation, paralysis, and sometimes death. When it occurs, death usually comes from extreme, irreversible hypotension. Early literature seems to indicate a relatively high fatality rate, while later literature suggests early reports may be exaggerated. Everyone agrees, however, that the pain is immediate and intensely agonizing, and it takes several days to recuperate. Hot-water immersion helps with the pain somewhat, but prescription analgesics and prophylactic antibiotics are often prescribed. There is an antivenom produced for stonefish envenomation in Australia.

Rabbitfishes are represented by twenty-eight species in the Indo-West Pacific, and several species range east into Guam, the Northern Marianas, and the Samoan Islands. These are *Siganus argenteus*, *S. fuscescens*, *S. punctatus*, and *S. spinus*. *Siganus vermiculatus* also occurs in Guam. The other species are to the south and west of these island groups, including the Fox-faced Rabbitfish (*S. (Lo) vulpinus*), a popular aquarium pet in North America. They are called rabbitfishes for their habit of grazing on algae along reefs, mangroves, or flats, as they remind one of nibbling rabbits. Many species are colorful, but they can change to drab colors instantaneously to match the background. When alive they can be identified by color and pattern, but when dead they have to be identified by meristics and morphometrics, due to the lack of color. They are laterally flattened to somewhat fusiform in shape, and often swim among corals in pairs. Rabbitfishes are an esteemed food fish. They are not only caught in the wild but are also raised in pens for food. They have an arsenal of thirteen dorsal, two anal, and seven pelvic spines. Many people are envenomated, yet there are few accounts of the venom or symptoms of en-

Rabbitfishes are a family of Indo-West Pacific venomous fishes that do not reach Hawaii. Pictured is *S. doliatus*. L.L.C. Jones

venomation. They are an occupational hazard in fish farms, and fishermen, swimmers, snorkelers, divers, and aquarists are sometimes stung. One recent study showed that *S. fuscescens* venom has properties similar to scorpionfishes and possesses a stonefish-like toxin. Their venom causes pain, hemolysis, and swelling. The stings are usually characterized as being intensely painful. Two online forum reports from aquarists confirmed this when stung by *S. vulpinus*. One victim said a single spine to the hand caused severe pain all the way to the shoulder. Another said it was more painful than a Red Lionfish. There are reports of systemic problems, including nausea, but there is little specific information. Rabbitfishes should never be released into the wild, as some species have become established elsewhere in the world.

The Plotosidae are commonly known as eel catfishes because of their elongate body and tapering tail. Of the handful of species, only one, *Plotosus lineatus*, is found in our area, in the Samoan Islands. It is commonly known as the Coral Catfish, as plotosids are the only group of catfishes found in coral reefs. They are also found in seagrass beds. Juveniles occur in tightly packed schools of over a hundred individuals, while adults are solitary and hide under overhangs during the day. They are also common marine aquarium pets, and their tendency to form tightly packed schools is intriguing to watch. They are sold as juveniles, but they fare well and grow in the aquarium, so should never be kept in small aquaria or released into the wild, even if they outgrow their tanks. Their single dorsal spine and two pelvic spines are retrorsely serrated and venomous. They also possess potent crinotoxins. The Coral Catfish is considered to be among the most venomous fishes in the world. The LD_{50} in mice and rats is about 0.3 mg/kg in two studies (intraperitoneal and intravenous). A lethal factor in the crinotoxin has an LD_{50} of 0.71 mg/kg (intravenous). The venoms are known to be neurotoxic, hemotoxic, and edema-forming. Although there are relatively few case reports, symptoms of human envenomation are well documented. It has been reported that juveniles may deliver a bee-like sting, but large individuals and groups of juveniles are more dangerous. However, even the sting from a single juvenile can cause symptoms far worse than a bee. Stings are accompanied by severe pain and swelling, often with cyanosis that becomes erythematic. Severe stings cause intense pain, swelling (sometimes massive), numbness, paresthesias, paralysis, lymph node maladies, nausea, fever, weakness, and gangrene. It may take several months for symptoms to completely resolve. There are reputedly several deaths attributed to *P. lineatus*, but some authorities question these fatal outcomes, which are reported in older literature. Hot-water immersion helps, but the effects can be long-lasting, and additional analgesics and prophylactic antibiotics are often prescribed.

The west-central Pacific territories also have an unusual group of combtooth blennies, which are small, common denizens of reef communities. Those in the tribe Nemophini (fangblennies), in the genus *Meiacanthus*, are one of only two groups of fishes with a venomous bite (the other group includes one-jaw eels, family Monognathidae; see Other Aquatic Vertebrates chapter). *Meiacanthus* are sometimes called poison-fang blennies, but venom-fang blennies would be more accurate. The Forktail Blenny (*M. atrodorsalis*) is found in Guam, the Northern Marianas, and American Samoa. Related combtooth blennies in other genera may bite, but are not venomous, and some feed on the scales and fins of other fishes. The venomous species also have nonvenomous relatives that are Batesian and aggressive/behavioral mimics. Forktail Blennies grow to about 11 cm and feed on plankton and small benthic organisms, so their bite is thought to be completely defensive. Most fishes do not feed on them, but if one is brought into a predator's mouth, the blenny bites them, and the would-be predator spits them out by distending their jaws and opercula, and quivering the head. *Meiacanthus* has a pair of enlarged canines on the lower jaw. Venom is produced in a gland that surrounds the lower canines and empties into grooves in the fangs. Researchers found *M. grammistes* and *M. atrodorsalis* have three novel venom types

Forktail blennies are one of only two fish types that have a venomous bite. *Bernard DuPont/CCA/Flickr*

not found in other fishes: an X-class phospholipase A_2, proenkephalin, and neuropeptide Y. These chemicals are known to cause inflammation and hypotension. There is also a transient analgesic effect, but the bites are painful. One researcher who studied Forktail Blennies in Guam was bitten through plastic bags by fish he had speared and placed in his swim trunks. The two bites were described as being immediately painful, much like a mild bee sting. The wound bled freely for about ten minutes, and in about two minutes, inflammation started to appear. It peaked at about fifteen minutes at a diameter of 10 cm around each wound. Around the bite site was a 2 cm, raised white ring that remained for about two hours. Local inflammation lasted about twelve hours. The area around the bite site remained hard for several days.

Yellow-bellied Seasnakes are circumtropical in the Indo-Pacific, so they are found throughout all of the islands, but there is at least one confirmed record of another seasnake from American Samoa, the Banded Sea Krait (*Laticauda colubrina*), although the species is rare and most reports are of similarly patterned marine eels. It is a very docile species, but it can bite, and the venom is potent. For many years there was a scare that Australian Tiger Snakes inhabited Suvai'i of the western Samoan Archipelago, although it turned out the reports were of the Pacific Boa, a nonvenomous species.

There are many species of venomous marine invertebrates found in the island territories. Those of the Caribbean are discussed in the text. It is more

There is only one record of the Banded Sea Krait from American waters—American Samoa. *Elias Levy/CCA/Flickr*

difficult to determine precise distributions of marine invertebrates found in the various reefs of the Pacific islands. References may simply cite distributions as "Indo-Pacific." Even "widespread throughout the Indo-Pacific" does not mean much when determining specific localities of islands in the central and west-central Pacific. Cone snails that are venomous to humans have been found in many of the islands, and these are discussed in the taxon account, but (*Conus* (*Gastridium*) *geographus*) is worthy of its own discussion, below. Other venomous invertebrates covered elsewhere in this book that undoubtedly occur in the Pacific include various cephalopods, *Glaucus* nudibranchs, polychaetes, anemones, scyphozoans, cubozoans, Pacific Bluebottle, and stinging hydrozoans. I can find no record of the extremely dangerous box jellyfishes in the genus *Chironex* occurring in U.S. territories, although there is a sighting record for Palau in western Micronesia. Caution is still advised, as jellyfish-like organisms sometimes show up in new places when carried by currents or ship ballast. Also, there are several other genera and species of cubozoans that occur throughout much of the western and central Pacific, including *Alatina alata*, the so-called Hawaiian Box Jellyfish, discussed in the taxa accounts. Some of these cubozoans can cause Irukandji syndrome. I can find no record of any species of the deadly blue-ringed octopuses reaching as far east as the U.S. territories in the central-western Pacific.

The Geography, Geographic, or Geographer Cone is not found in Hawaii or the North American mainland, so it does not have a specific account in the text, but the general information on cone snails does address the venom

Geographer Cone feeding on a goldfish in its tank in a research lab. *L.L.C. Jones*

apparatus and other attributes common to all *Conus*. Although I cannot vouch for which Pacific island territories do or do not have *C. geographus*, it is said to be widespread in the Indo-West Pacific, and there is reference to its presence in Guam, including at least one fatal bite. This species is noteworthy because it is one of the most toxic animals on Earth. The shell of this marine snail is about 7–15 cm long, although a 10 cm specimen is considered large. It is more cylindrical in shape than most cones, has a thinner shell, and has mottled brownish and whitish bands across its axis. The epithet *geographus* is in reference to the pattern resembling a map. The foot is boldly mottled, so it may be aposematic, but it can also be attractive to potential predators; hence, the potent defensive venom evolved to protect this large, thin-shelled, conspicuous snail. By day Geographer Cones sleep in sand to avoid predation, and they are active at night. They are piscivorous, and large individuals can feed on fishes up to about 13 cm in length. They are usually found in areas of coral, coral rubble, and sand near coral reefs. They can occur from the intertidal zone to a depth of about 200 m. Based on two studies, the venom has an estimated LD_{50} in humans of 1–3 μg/g and 0.029–0.038 mg/kg. These studies estimated the values by trying to determine how much venom is injected when they bite, and how much it takes to kill a human, based on studies of venom yield and human mortality from case studies. Either way, there is an overall mortality rate of about 65–70%, even with hospital care. Of the reported human envenomations, 100% of the eight children who were stung died. Even a mere 0.0002–0.0005 mg of venom caused severe paralysis. Based on a small sample, injected venom yield ranged from an average of 21.5 (± 0.5) μl in a 79 mm specimen to 84.0 (± 3.0) in a 99.5 mm specimen. The venom is composed of numerous peptides, many of which have been researched in hopes of finding a new drug for pain and diseases. The venom is neurotoxic and myotoxic. As mentioned in the text for the taxa account of cone snails, this species (at least) has both defensive and predatory venom fractions that are manufactured in different parts of the venom duct. These are selectively used, depending on whether the animal detects predator or prey. When a prey species is bitten, conantokin-G peptides are a conspicuous lethal component, but when a predator/human is bitten, α-GI peptides are a conspicuous lethal component. Most people are stung on the hand when they pick them up. *Conus geographus* is said to be more defensive than other cones when handled. While other species withdraw into their shell, Geographer Cones are quick to use their proboscis to find and harpoon the threat. There may or may not be pain from the sting, followed by weakness and muscular paralysis. The victim may have trouble breathing, so treatment usually involves trying to keep the patient alive long enough for the venom to pass through their system. This means keeping the airway open and providing respiratory and cardiac support. The fatality in Guam was to a spear fisherman who picked one up and rolled it up in his sleeve. He collected the animal around midnight, and within a half hour, he started feeling faint, with numbness and weakness throughout his body. He did not remember feeling a bite. He was admitted to the hospital at 01:14 and was already becoming delirious and unable to speak. He died at 02:55. Unfortunately, there is no antivenom available. The phenotypically similar and closely related *C. tulipa* of the Indo-West Pacific is also probably extremely dangerous, but there is less information on that species and it may be confused with *C. geographus*.

References and Resources: In addition to references cited for various taxa accounts, see: Borges (2015), Borsa et al. (2007), Casewell et al. (2017), Dutertre (2014a, b), Fahim et al. (1996), Garnier et al. (1995), Halstead et al. (1956), Jones (1983), Kiriake et al. (2017), Lee et al. (2004), Losey (1972), McIntosh and Jones (2001), Mundy et al. (2010), Myers (1989), Paulay (2003), Phoon and Alfred (1965), Rice and Halstead (1968), Santiago-Blay (1987, 2009), Shiomi et al. (1986, 1987, 1988b), Teruel et al. (2015), Wass (1984), Weldon and Mackessy (2010), Woodland (1990), Yoshiba (1984).

12 Envenomation Stories: What Is It Like to Be Bitten or Stung?

This section shows what a variety of taxa can do to a human being when they are envenomated. These narratives were rewritten from individual testimonials to fit the format of the book. Nearly all of these accounts were taken from unpublished first-hand testimonials by the victims, although some were from witnesses and attending physicians. Most case reports published elsewhere are written by physicians and medical researchers who emphasize the medical intervention of extreme cases (including fatalities), while I focus more on the circumstances that led to the envenomation, the basic symptoms, and the associated timeline. I do not limit myself to extreme cases, because it is useful to know what happens in typical cases. However, there is no point in demonstrating what happens during a dry bite, or a mild case (e.g., a typical bee sting). In addition to an array of stories of people bitten or stung by common animals (e.g., Western Diamond-backed Rattlesnake, wasps), I attempted to find stories of people who were bitten or stung by species that are rarely reported in the literature, such as centipedes, octopuses, and rear-fanged snakes. All of the transcribed stories were reviewed for accuracy and content by the contributor. As is usually done for case reports, the victims remain anonymous.

One thing to notice is that these stories are skewed to biologists who have an occupational or recreational hazard. This is not by design; being a biologist myself, I am frequently in contact with my peers, some of whom are quick to provide their stories. Researchers often keep good records and are adept at identifying the animal, and some had good recollection of the circumstances and symptoms. Another group of the populace that find themselves in harm's way includes hobbyists and outdoor enthusiasts. The exposure to hobbyists to native and non-native species alike is an important aspect of this book. Keeping venomous animals in a domicile is a surprisingly common pastime, be it legal or otherwise. Pet owners and collectors often become victims to their source of their pleasure.

In each account I categorized the envenomation into the two factors of exposure and a one-line explanation on how the person got tagged.

Categorizing Factors Leading to Envenomation

When reporting a bite or sting, it is important to document which taxon is responsible, identified to the most refined taxonomic unit possible (e.g., Haller's Round Stingray is better than simply "stingray," which could be from any number of species in multiple genera and families). Also, it is important to not guess as to a culprit if not known—we have seen that spiders are often blamed for any number of maladies due to other causes—and this leads to inaccurate and irresponsible reporting.

There are currently only two commonly used categories to explain circumstances that lead to human envenomation: "legitimate" and "illegitimate." Legitimate essentially means a bite or sting happened by accident, whereas illegitimate means that it occurred because of unsafe handling practices. However, these are oversimplifications that don't really explain the circumstances leading up to envenomation. While reviewing case reports, I found there were recurring themes, so I developed an Exposure Code, which is a simple reporting and coding system to better categorize those circumstances. First, most bites and stings happen when a particular activity puts a person in harm's way. This accounts for how the animal is encountered. I refer to this as "Activity," and give it a two-letter category code. Second, when the animal is encountered, there is an explanation of how the person came into physical contact, rather than their getting away unscathed. For example, the animal may have been unseen or was not known to be venomous and was handled without concern. I refer to this as "Contact," and it is also given a two-letter code. Together, these factors ultimately led to envenomation. Activity + Contact is my Exposure Code reporting system. The various two-letter codes in each of the categories are not necessarily mutually exclusive, so I select the most obvious or overriding explanation. For example, if a person is handling a pet snake, the Activity is PT (pet trade), rather than IN (intentional handling) because they would not have been handling the snake if it was not their pet in the first place.

In addition to documenting the species and having a coding system, it is helpful to add a brief written explanation, usually no more than one sentence in length. I put this in parentheses after the Exposure Code. The explanation includes the general type of animal (e.g., rattlesnake or wasp) to set the stage, and a brief account of how the person was tagged. The written explanation puts the codes into context and helps the reader to quickly assess what happened.

The Activity and Contact codes I use in the Envenomation Stories are:

ACTIVITY:

- **Routine (RO).** People going about their daily business. Can be indoor or outdoor. Includes activities such as shopping, walking to the car, gardening, and sleeping.

- **Occupational (OC).** By the nature of their job, people find themselves potentially in harm's way. Includes activities such as research (e.g., entomology, herpetology, fisheries biology, toxinology), fieldwork (e.g., for a logger, rancher, farmer, field biologist, commercial fisherman), and working in zoos and public aquariums.

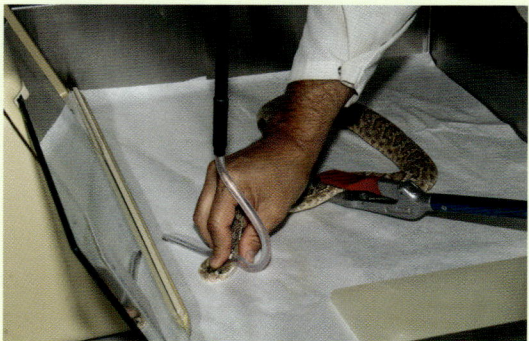

Milking rattlesnakes is but one of many occupational hazards. *L.L.C. Jones*

- **Pet trade (PT).** Activities include the keeping of native and exotic animals in a home or as part of the collecting or shipping of animals for the pet trade.
- **Recreational (RE).** This includes activities primarily in the out-of-doors, such as hunting, angling, photography, bird-watching, snake-watching, and hiking.
- **Interactive (IN).** Activities include intentional interactions with a venomous animal, whether or not the creature was known to be venomous.
- **Other (OT).** Any other activity that does not fit the above categories; this should always have an explanatory qualifier in parentheses.

CONTACT:

- **Failure to detect (FD).** This is a true accident when someone fails to see or hear the animal. Examples are stepping on an unseen and unheard rattlesnake or being bitten by a recluse spider while putting on clothes.
- **Lapse (LA).** This is when one has a lapse in judgment or lets their guard down. Examples include a hobbyist cleaning a terrarium while not paying attention to the location of all tank mates or a herpetologist who gets bitten when the snake bag is too close to their body.
- **Naïve (NA).** This is when someone knows they are interacting with an animal, but they do not believe it is venomous or capable of envenomation. Examples include someone being bitten by an octopus,

not knowing it is venomous, or when someone picks up a coralsnake, thinking it is a kingsnake or incapable of penetrating human flesh.

- **Cavalier or casual (CA).** This is when someone knows an animal is venomous, but is not overly cautious. This may happen when people try to freehand, move, or kill a snake without using proper precautions. It may also include someone who feels bravado from handling dangerous animals.
- **Intentional (IN).** This is when someone knows the animal is venomous, but decides to contact it anyway. This includes altruistic behavior (i.e., "taking a bullet" for a loved one), venom testing, curiosity about effects, and self-immunizing.
- **Malicious (MA).** On rare occasions, venomous animals may be used to harm other humans or themselves. There are cases of homicide and suicide from venomous animals (remember Cleopatra?).

According to popular belief, a famous suicide using a venomous animal as the weapon was Cleopatra and an asp.
Guido Reni, 1640/WikiArt/Public Domain

- **Other (OT):** Any other attitude or behavior that does not fit, but should always have an explanatory qualifier.

Here are some examples taken from the Envenomation Stories. Itemizing and entering exposure cause data in this manner (taxon: categories (brief explanation)) would be very simple and useful at poison control centers and emergency rooms, even when a full explanation is not given. This would yield a better picture of factors leading to exposure for summary reports.

- Western Diamond-backed Rattlesnake (*Crotalus atrox*): RO-FD (accidentally stepped on an unseen rattlesnake)
- Fitch's Octopus (*Octopus fitchi*): RE-NA (bitten while free-handling octopus thought to be harmless)
- Puss caterpillar (*Megalopyge bisessa*): IN-IN: (swiped finger against caterpillar to test venom effect)

Terrestrial Invertebrates

The Pant Leg Prowler: Giant Desert Centipede (*Scolopendra heros*)

At 71 years of age, a herpetologist who had spent his life doing fieldwork in the desert Southwest, Central America, and South America had a remarkable record for avoiding bites and stings. For example, he handled and collected thousands of venomous snakes without being tagged, although he occasionally was stung by some other critters, including Arizona Bark Scorpions and a large Amazonian scorpion in the buthid genus, *Tityus*. He had collected large centipedes over the fifty or so years of fieldwork but had never been stung. That changed one day in September 2016. Although retired to southern Arizona, he is still an avid biologist. He explained the circumstances leading up to the sting, "While not paying close enough attention to what I was doing, a very large *S. heros* went under my pant leg and stung me just above the ankle on the back side of my leg. I thought I had backed into a yucca because the initial pain was more like a series of pin pricks. I ran my hand across my pant leg and the culprit emerged, disappearing into the grass." The "pin pricks" are consistent with the pokes from the many sharp feet, which some believe to be venomous (this has yet to be confirmed), and they pale in comparison to the forcipules, anyway. The sting happened at 14:00 and by 16:00 the skin on the lower leg became hypersensitive, similar to his experience with Arizona Bark Scorpions. His ankle hurt and he noted severe pain in the muscles of his lower leg—which he ranked a 9 out of possible 10. He added that "The pain appeared to be confined to my lower leg, and one of the effects was that it caused my leg to jump at intervals of 16 seconds. It was extremely uncomfortable."

He put on a Lidocaine patch and took 10 mg of morphine at 17:00. At 18:00 he took another 10 mg of morphine. The Lidocaine and morphine did nothing to alleviate the pain, so he added a margarita to the mix of first-aid remedies. By 19:30 the pain started to subside and was gone by 22:00. He had no trouble sleeping. The next morning he experienced some nausea and seemed to have a slight fever. He felt fine by afternoon, although his lower leg and foot were swollen for a couple of days. The dime-sized hemorrhage at the bite mark dissipated after a few days, and there were no long-lasting effects, other than a profound respect for centipedes.

Category: RO-FD (centipede crawled up the pant leg)

How to Not Handle a Centipede: Giant Desert Centipede (*S. heros*)

In 2013, a 40-year-old man whose livelihood is to entertain and educate the public about wildlife, including all things venomous, was envenomated by a 20 cm *S. heros*. After leading a "Learn your Lizards Walk" at Boyce Thompson Arboretum in Superior, Arizona, someone brought him a centipede that they had captured in a jar. The crowd was interested in this impressive creature, so he then attempted to hold the animal behind its head for show-and-tell, as one might try to restrain a snake. He was well aware these animals are venomous, but underestimated their capacity to cause harm, as there is little information on the stings from native *Scolopendra*. It squirmed in his grasp and managed to get a forcipule into his thumb by the nail. He allowed it to remain in contact for about 45 seconds (onlookers love that sort of thing!).

There was no immediate pain, but within about five minutes his thumb was starting to hurt and pain continued to build. Ultimately, it became the most excruciating pain he ever experienced (a most definite 10 on a 0–10 scale). As he put it, "it was worse that my rattlesnake bite (including allergy to the venom), broken ankle, bark scorpion sting, vasectomy, tarantula hawk sting, and broken/dislocated finger." He did not seek medical attention immediately, but called his doctor several times. To combat the excruciating pain, he took two Aleves, two Benadryls, two ibuprofens, and two hydrocodones that were left over from his broken ankle. At one point he feared he may have overdosed, but his doctor assured him he did not, and that he would sleep like a baby. *Au contraire*—he could not sleep from all the pain, despite the medication. He actually went to a football game that night (spectator, not participant), but the pain was so great that he cried and writhed in the car. His wife then drove him to the emergency room of a local hospital. After being lectured at the ER, he received some pain medication and the next day he took two oxycodones. The pain subsided in about twenty-four hours. His thumb and hand were swollen for about nine days, although swelling started to subside on Day 3. The bite site was discolored. Eight days later, he realized the wound was infected, because the thumb was hot, hard, and itchy. He was put on antibiotics for another week. His advice: "Bottom line is that you don't want to get bitten [stung] by a Giant Desert Centipede. Ever, EVER!"

Category: OC-LA (centipede stung him while holding it behind the head)

Twin Bed Stings: Giant Desert Centipede (*S. heros*)

As happened every night, a Tombstone, Arizona, couple drifted off to sleep in their bed. On the first of two eventful nights, the husband was rudely awoken by a sharp pain in his neck. They looked around to see what nailed him. At first he thought it was probably an Arizona Bark Scorpion, but they failed to find the culprit. Then they noticed there were two puncture marks in the back of his neck. The pain was immediate and intense. He described it as being jabbed by a red-hot needle. He rated it as a 7 on a 1–10 scale. The pain radiated from the site onto his back and remained intense for about 3–4 hours. It became red and swollen, but there were no systemic effects. The pain was so intense, he and his wife discussed whether or not to visit the ER but ultimately decided against it. Most of the pain was gone in about twenty-four hours, although it remained swollen and sensitive for several days. He summarized his experience by saying, "it was pretty bad."

The very next night, his wife was also rudely awoken by a sharp pain on her finger. Once again, they tried to find the culprit, and bingo! They saw a 15-cm Giant Desert Centipede slinking under her pillowcase. They captured the offender in a jar for a biologist friend of theirs. Her sting was more superficial, as the centipede apparently did not latch on, but there was pain and swelling in her finger. Her sting wasn't as bad, but wasn't fun, either. Although the centipede was dispatched, it gave its life to science when it was preserved in alcohol for their friend's biology class.

Category: RO-FD (stung by centipede while sleeping in bed)

Size Doesn't Matter: Arizona Giant Hairy Scorpion (*H. arizonensis*)

There is a widespread belief in the Southwest that "large scorpions are harmless, while small scorpions are dangerous." This is completely false, as the algogenic qualities of scorpions are related to specific taxa and certain factors of envenomation, rather than being based on size. Members of the genus

Hadrurus are the largest North American scorpions, and the stings are often said to be similar to a bee sting or less, giving rise to this misinterpretation of size and pain. However, in addition to mild envenomations, I have become aware of numerous instances that show *Hadrurus* can sometimes pack a wallop. For example, a biologist was handling an Arizona Giant Hairy Scorpion and admitted that "he was rushed," so he let his guard down and was solidly tagged in the meat of the hand. At first, he said it was just like a cactus thorn, but it was followed by an intense throbbing pain (6 of 10) for about one hour, with no pain whatsoever after 90 minutes. However, there was a maddening intense itch for about 24 hours, starting the next day. Similarly, a social media report from a hobbyist relayed his story of being tagged by the same species. He picked the scorpion up by the telson (known as "tailing," which is a rather risky approach), and he was stung on the finger, near the nail. This caused him to drop and slap the scorpion onto his solar plexus, where it proceeded to sting him twice. The pain from the thoracic stings was intense. He said it was "literally like being punched in the stomach." He had been stung by *Hadrurus* before but admitted that the pain from stings to the torso is far worse than the hand. He added that this was the only time he experienced abdominal cramps from this species. The wounds left large red welts that were still present at least 14 hours post-sting. He highly recommends not "slapping a scorpion to your chest if it escapes."

Category (First story): OC-LA (zoo personnel stung while handling a scorpion). Second story, PT-CA (hobbyist stung while handling scorpion, then dropped onto chest)

The Dancing Leg: Arizona Bark Scorpion (*Centruroides sculpturatus*)

A young adult colleague related a story she vividly remembers of her scorpion stings in Tucson. One resulted in virtually no reaction, so may have been a vaejovid, but the other was something to write home about. She was in the 6th grade at the time when she walked into her garage barefoot. "It felt like I stepped on a carpet tack." She looked down and saw a tiny scorpion that was only about 25 mm long, but despite its size, it packed a punch. Within minutes she felt the tingling and numbness move up her leg. There was a little swelling. The pain peaked at about 2 hours post-sting, rating a 6 out of 10. Within a few hours her entire leg was affected, and her foot was completely numb and pin-prickly. Then her leg started doing its own thing. She found it rather disconcerting that she had no control over her spasming leg, which she said had "visible fasciculations." She became "worried and a bit panicky," so called the poison control center. She was told that if the effects reached beyond her leg, she should check in to an ER. The paresthesias never extended beyond the leg, so she attempted to sleep, but it was futile. Whenever she started to drift off, her leg twitches kept her from slumber. She got no sleep for twenty-four hours and the next morning she had trouble walking. Her symptoms completely resolved in about two days.

Category: RO-FD (stung by an unseen scorpion that was stepped on)

Venomous Brain Teaser: Arizona Bark Scorpion (*C. sculpturatus*)

A retired scientist in southern Arizona has had years of experience free-handling scorpions without incident. He claims to have handled more than a hundred individuals, including Arizona Bark Scorpions, usually for on-lookers in an educational show-and-tell situation. However, he had received 5–6 stings from species unknown when he was grabbing for snakes or lizards under covered objects. He sometimes teaches classes at Biosphere 2, near Tucson. One day while he was there he met a couple Russian scientists who were interested in scorpions, as they had had little exposure to them. To accommodate their interest, he decided to pick the only Arizona Bark Scorpion present that day. This species often rests with its tail curled up flat against a rock, and that's just what it was doing at the time. Rather than try to coax it into a free-handle position, he pinned its telson down with his finger. It didn't go as planned and he got tagged.

While he has been stung by several scorpions, he said this one was in a class of its own. There was an intense stabbing pain that he rated as a 7 or 8 out of 10. He said it was worse than the other scorpion stings, which were less than that of a wasp. He does not recall some of the other usual symptoms, such as numbness, tingling, or muscle contractions, but one symptom was especially disconcerting—he was having trouble remaining cognizant. He called 911, who recommended he be medevaced out and taken to a nearby hospital, but he refused, since that alone would cost about $3,000. The 911 center patched him through to the Arizona Poison Control Center, who assured him that the symptoms were within the realm of normalcy, but he should monitor his symptoms and be in touch if needed. However, he could not shake the feeling that he wasn't quite right mentally. He was also weak and the intense pain spread to his shoulder. His concern over his situation impelled him to call the Poison Control Center again later. They said he should be better in 1–2 days, but again, to monitor symptoms and call if needed. As I have heard in most cases of *C. sculpturatus* envenomation, he could not sleep the first night. However, about twenty-four hours post-sting, symptoms subsided and his anguish dissipated.

Category: OC-LA (stung by scorpion while tailing it during an educational presentation)

An Altruistic Superhero: Arizona Bark Scorpion (*C. sculpturatus*)

At the Arizona-Sonora Desert Museum, a favorite pastime during summer nights is to look for the abundant *C. sculpturatus* with a black light as they scurry into rock wall crevices. One of the staff entomologists was working that night and saw a small girl reach for one. To protect her he reached his hand forward to block the girl from the scorpion, and in the process, was stung on the tip of his index finger. He was well aware of what bark scorpions were capable of for small children versus adults. As he put it, "She was at the age where she might have to go to the hospital." Immediately, he felt "a little bit of pain" from the tip of the finger, which then spread into the base of the thumb. The whole area became numb within five minutes. In 10 to 15 minutes the area became extremely sensitive. This is when he said his "superpowers began." He had hypersensations of touch. "Even being in an area where people were talking was almost unbearable, as their breath would cause every nerve to fire." He became curious about the new sensations, which spread halfway up his arm, and even started to enjoy the feeling of his new supersenses. In about an hour, he said he "started to feel electrical currents going back and forth from my shoulder to my forearm, and my shoulder became fatigued." When he drove home that night, he left his arm out the window because the sensation of wind on his skin "felt euphoric." He never had any systemic reactions. He was able to sleep that night, and by morning the whole area became numb. By noon the next day (21 hours post-sting), the symptoms resolved.

After his relaying this story to me, one of his coworkers said her experience after being stung on the finger was not nearly so pleasant, as she had widespread pain and paresthesia. At first her fingers went numb, then the hand went numb, then it spread to the other hand, and then to both feet. She also had slurred speech, trouble swallowing, and uncontrollable eye movement. This caused her such concern that a coworker called the poison control center for her, and the staff assured her that at her young-adult age, and being in good health, there is no cause for alarm, unless she was having trouble breathing. She could breathe fine, so she just rode it out and recovered without further incident.

Category (first story): OC-IN (scorpion stung him when he altruistically fended off a sting to a small girl reaching for it)

Hiding in the Laundry Room: Arizona Bark Scorpion (*C. sculpturatus*, presumptive)

A 64-year-old, slight-of-build (147 cm; 45 kg) woman was going about her chores of tidying up her suburban Las Vegas home. As she was straightening bags on the laundry room floor, she was stung by a scorpion on the ring finger of her right hand. She received a single sting, but it clung to her finger before being flung off across the room, where it scampered off. The frightened woman screamed, calling for help. She said the pain was immediate and intense. She ranked it a 20 out of 10! She took a Benadryl, and her neighbor recommended she take another. When she drank the water to wash down the pill, she noticed it had a salty taste. She started getting fasciculations in the eyelids and tongue. These disconcerting symptoms prompted her to have her neighbor drive her to the emergency room. When she arrived, her vision was blurred and her blood pressure was very high. Numbness spread to the entire right arm, then the left, and finally included her knee and nose. She was unable to walk.

She was given a mild sedative to calm her down when she arrived at the ER. The physician administered antivenom (presumably Anascorp). After the first vial failed to make a significant improvement to her symptoms, she was given a second forty minutes later. The antivenom seemed to work and she began to feel relief. The numbness and other symptoms began to disappear in reverse order they appeared. She remained in the ER from 9:00 a.m. until late in the afternoon. It was about a month before her stung finger lost its numbness.

Although we cannot be certain the culprit was an Arizona Bark Scorpion, it is likely. The woman claimed these were the species encountered in her yard, and there is supporting literature about this non-native being established in the Las Vegas Valley. Also, a 2010–2015 study of stings reported to poison control centers in Las Vegas area included more than 4,500 stings during that time. About 170 per 100,000 people in Las Vegas are stung by scorpions annually, and they largely have symptoms consistent with envenomation by *C. sculpturatus* (fasciculations and uncontrolled eye movement).

Category: RO-FD (stung by scorpion that was hiding among bags)

Like Father, Like Son: Striped Bark Scorpion (*C. vittatus*)

A young family was out camping at Mills Canyon on the Canadian River in New Mexico. The father, a biologist, had previously taught his 10-year-old son how to capture scorpions by carefully grabbing the telson with thumb and forefinger (tailing). The child had employed this technique successfully many times to remove scorpions (*Paruroctonus utahensis*) from their home in the Albuquerque area. While looking around camp after dark, the son encountered a Striped Bark Scorpion. One of the problems with tailing is that *C. vittatus* (and *C. sculpturatus*) rest on the substrate (and often on the underside of rocks) with their metasoma curled to the side and flat against the ground or rock, so it is difficult to pick them up in this manner. The dad was aware that *C. vittatus* was not nearly as dangerous to a child as *C. sculpturatus*. The son decided to try it himself (like father, like son!) and managed to get tagged in the thumb. He screamed and cried for about fifteen minutes when the pain was intense. The pain lasted for probably 2–3 hours, then subsided. There was a little swelling in the thumb, but aside from that and the pain, there were no other symptoms.

An interesting side note—I heard this story at a herpetology conference in the Big Bend of Texas, and there I hiked on a field trip with another father-son tag team. We encountered numerous scorpions. Among them were Striped Bark Scorpions (and *Diplocentrus lindo*). Both father and son were attempting to handle both types of scorpions, although they were cautious to not get stung. As the 11-year-old boy stated, "I don't mind if I get stung. I can take a hit. I've been stung by Striped Bark Scorpions before and it is no worse than a hornet sting. Now, Arizona Bark Scorpions—that's a different story—I would never handle one of those!"

Category: RE-CA (stung by scorpion when trying to collect it by tailing)

Stung by a New Species: Gray's Vaejovis (*Vaejovis grayae*)

A professional venom extractor was stung on the index finger while collecting a new species of scorpion. It was a "little brown job" that was later described as *V. grayae*. He had previously been stung by numerous invertebrates in the course of his occupation of collecting spiders and scorpions (including recluse spiders, widow spiders, and bark scorpions), but this one was noteworthy because he said the sting of this new species "hurt like hell!" He described the pain as a 5–6 out of 10 on a pain index. There was no numbness. The pain lasted for several hours and was gone within a day. There was local swelling and inflammation limited to the finger.

This is noteworthy because this species is but one of about twenty species in the *V. vorhiesi* group of small brown scorpions that live in the mountains of Arizona, New Mexico, and Sonora. Stings from this group of scorpions are often considered trivial, rating a 1 or so on the 0–10 pain index, decidedly less than that of a bee sting. Even if stings are not from so-called medically important species, reports of reactions are important for future reference and to establish a baseline. In this case, a very painful sting was received from one species within a group usually considered "harmless."

Category: OC-LA (stung by a scorpion while trying to free-hand collect it)

Stung in Bed: Stripe-tailed Scorpion (*Paravaejovis spinigerus*)

A middle-aged woman was lying in bed one night in her Tucson home when she felt what seemed like a thorn stuck in her pajamas. She looked to see what it was and discovered a Stripe-tailed Scorpion in her bed, which had stung her at least 2–3 times on the hip. At first there was no pain, save the sticker-like pokes. However, after a few minutes, the level of pain escalated until it reached about an 8–9 out of 10. "It was one of the most painful things I ever experienced." The severe pain lasted 1–2 hours, and then gradually subsided. There was some redness and swelling of the local area, but effects did not spread beyond the sting site. She used ice to help alleviate pain, which she said helped some. She also took a hydrocodone tablet, which she said did absolutely nothing. An area of about 50 mm became "strangely numb." This numbness lasted several weeks and the area was tender to the touch. Most symptoms were gone in about a month. However, even after several weeks following the sting, a friend had bumped against her at the sting site and she experienced "sting memory"; she said she "nearly buckled from the pain." All symptoms eventually disappeared and there were no long-lasting effects. She did not seek medical attention.

The victim claimed to know the difference between the three common species of the area, *C. sculpturatus*, *Hadrurus arizonensis*, and *P. spinigerus*, but species identification was not confirmed. Stripe-tails are not as adept at climbing as are bark scorpions, and the latter often cause stings in bed. However, she said she had also been stung by *C. sculpturatus*, and this sting lacked the systemic involvement of a bark scorpion, suggesting *P. spinigerus* was indeed the culprit. It is possible that the location where she was stung caused the intense pain. *Paravaejovis* is common in southern Arizona and undoubtedly accounts for many of the stings reported to poison control centers.

Category: RO-FD (stung by a scorpion that had gotten into the bed)

Hundreds of Scorpion Stings: Many Species

A healthy, young adult male researcher working on scorpions of west Texas (Devil's River to El Paso) was stung by scorpions on more than one occasion. In fact, he was stung by every species of scorpion that occurred there (15–20 species), as recognized in the late 1990s. In total, he was stung hundreds of times. How and why did this happen? It all started because he is color blind and had a difficult time seeing scorpions well enough under a black light to catch them with forceps, which isn't that efficient a method, anyway. He realized that catching them by hand was much more effective. Also, he says he has a high pain threshold, so going to great lengths to avoid a sting "just wasn't that important" when pitted against the need to collect scorpions for his research. Many of his collecting sites were extremely remote, so it was important to not let specimens get away when it took hours to get somewhere. This situation does make him a valuable source of information on stings by a variety of species, since little is known about the stings of most. I asked him to rate scorpion stings, in general, on a scale of 1–10. In his experience, the sting of a *C. sculpturatus* is about an 8, that of *C. vittatus* is a 6, and the rest are 1–3. Those at the lower end of the scale are *Diplocentrus* spp. and *Vaejovis* of the *vorhiesi* complex (but see the Envenomation Story of *V. grayae*, above), which he rates as a 1. *Paruroctonus gracilior* is middle of the road, as a 3. When asked about symptoms, it is usually local pain, redness, and itching that subsides in a short period (except *Centruroides*). One exception was the night when he was stung twenty-seven times. He developed gastrointestinal difficulties the next day, presumably due to the amount of venom he received the night before. Also, both species of *Centruroides* caused "sting memory."

After the pain of the sting had subsided, it reoccurred when the sting area was touched, as if he were being stung all over again. Curiously, he has also been tagged by a Grassland Massasauga while collecting it, and a tarantula hawk and velvet ant, quite by accident.

Category: OC-CA (stung while hand-collecting scorpions for research)

A 106-kg High School Jock Gets Ass Kicked by a 1-g Spider: Western Black Widow (*Latrodectus hesperus*)

A 185 cm, 106 kg senior in high school awoke in bed one morning in Glendale, Arizona, with muscle pain in the front of his right thigh. He was in enough pain to ask to stay home that day, but his mother insisted he go to school. As he was walking to school, the pain got worse. He also noticed a slightly raised bite mark on the underside of his right forearm. Two minutes into his first class, he asked to go to the nurse, as the pain in his leg became increasingly worse. The nurse suggested he go to the hospital. As his grandmother was driving him to the hospital, the muscles in his midsection began to tighten up and his breathing became labored. Once inside the emergency room, the pain spread throughout his body. Every muscle in his body ached. They put him on an IV drip with morphine. Despite the painkillers, he described the pain as 9+ on a 0–10 scale. The ER staff watched and monitored him for several hours. They concluded he had been bitten by a black widow, but at the time there was no "smoking gun" of a culprit. They gave him a prescription for some analgesics and sent him home.

When he arrived home, he went straight to his room to find what caused the malady, and sure enough, there was the crumpled body of a small female black widow. He surmised that he was bitten by the spider and swatted it, but exactly how it happened will never be known. The pain was a dull but severe ache that affected his muscles for three days before subsiding. He could not get comfortable. He relayed this story some forty years later, but he remembered it as if it were yesterday. To his testimonial, he added this snippet, "It is kind of ironic that I was bitten by a spider, as I've kept a collection of rattlesnakes all of my adult life."

Category: RO-FD (bitten by black widow that had gotten into the bed)

A Streamside Hazard: Long-jawed Orbweaver (*Tetragnatha* sp.)

In the spring of 1987, a biologist was doing a research project for a natural history class at a southern California university. The study site was along the litter-riddled lower Santa Ana River. To get to the site, the biologist had to work his way through dense cattails to reach the water. As he and his classmates were carrying buckets and seines, he said he was "shocked by a crazy pain on [his] arm." It took him a moment to determine what was going on, then he looked down to see a large *Tetragnatha* sp. on his arm. He tried to swipe it off, but it "hung on" for a bit before he could finally dislodge it. He said "the pain was very intense and lasted for quite a while. Some swelling developed later and the site stayed red for a while." He added that "Regarding pain, this was much more significant than the black widow bites I had in the late 1990s, and the speed and intensity of the pain was so surprising and shocking, it literally created confusion as to what was going on at first." Tetragnathids are not usually listed as being venomous to humans, probably because bites are presumably rare and case reports even rarer. Fortunately, the biologist was able to identify the spider to genus, which helps our understanding of envenomation by poorly known animals. This bite was fairly significant in terms of both pain and local reaction, so at least some members of this family should be regarded as venomous to humans.

Category: OC-FD (brushed into and bitten by an unseen spider near a wetland)

Please Pass the Tarantula: Iodius Tarantula (*Aphonopelma iodius*)

A woman was hiking with her husband and friends in Mt. Diablo State Park, California, in November 2009. Along the trail they encountered a jogger and a tarantula at about the same time. The jogger gently scooped up the tarantula and discussed its natural history with the hikers. The jogger asked if anyone else wanted to let the docile animal walk on them and only the woman answered in the affirmative. However, while they were chatting, the tarantula was starting to crawl up the jogger's neck, and the woman "was afraid it would get onto his face." She surmised that she "reached out too quickly and must have sort of grabbed it," as it then bit her with both fangs on the base of the thumb near the wrist. She said the bite was immediately painful, so she promptly

Migrating *A. smithi* are a tourist draw. *Diana Stralberg*

This is the culprit that could only take so much of being passed around. *Diana Stralberg*

released it. She described it like "an extremely painful bee sting." She became worried about the effects of the venom, so tracked down one of the local rangers, who assured her she would be fine. She does recall some local redness and swelling but does not remember any itching. The pain subsided in a couple hours. In two days, all effects of the bite had faded away.

Category: RE-CA (bitten by tarantula when passing it around to hold)

Biting the Hand that Kneads You: Rio Grande Gold Tarantula (*A. moderatum*)

A Texas biologist caught more than one hundred tarantulas in spring 2017. One evening in May he came across an adult female *A. moderatum*. As he and others usually do when handling tarantulas, he let it crawl into his hand and gently cupped it around the animal. After he got it back to his truck, he decided to make sure of its sex, so he held it with his fingers against the legs and cephalothorax. As he started to turn the spider over, he said "she partially slipped out of my fingers and I was left grasping her abdomen. She then bent forward, and plunged both of her 1 cm long fangs into the meaty part of the base of my thumb!" He was startled, so dropped her, but couldn't let her loose in the truck, so he attempted to pick her up in the same gentle fashion as he had on the road. However, he noted that "she wasn't having any of that, and she bit my other hand as well."

Rio Grande Blonde Tarantula, *A. moderatum. Troy Hibbitts*

"The bite itself was just big sharp teeth sticking into me, no more painful than a bite from a Coachwhip or getting poked with a sharp thorn. I didn't even feel any 'sting.'" The bite sites were itchy for the next couple of days.

Category: IN-CA (bitten while grasping a tarantula)

That Feeling of Doom: Yellowjackets (*Vespula* spp.), Author's Testimonial

As a young field biologist working in California and the Pacific Northwest, I was stung hundreds of times over several years by yellowjackets and bald-faced hornets. I was stung so frequently, I became quite apprehensive about doing fieldwork, but it was my job. I generally received multiple stings when I or a fellow crew member stepped on a ground or log nest. The stings were individually painful (3–4 on a scale of 0–10), but the more stings, the worse it was. For years I had never had an indication of becoming sensitized, although I saw it happen to fellow field biologists.

That all changed one day when I received a mere two stings to the face. I quickly developed respiratory distress and nearly fainted. Fortunately, I had a rescue inhaler for asthma in my vest, which helped my breathing. I was right next to a road when it happened. While in a very confused state, I waved down a car and garbled to the driver to get me an ambulance. For the next half hour I slumped over, unable to move, drooling on my vest, and fighting to maintain consciousness. I had an overwhelming urge to sleep, but I thought that if I fell asleep I would never wake up again. I experienced that infamous feeling of impending doom associated with anaphylaxis. The ambulance arrived, I got to a hospital, I was treated, and I fully recovered.

After the encounter I went to an allergist. A skin test confirmed my allergy to yellowjackets and "other wasps," but not to European Honey Bees. After this, I received five years of immunotherapy. I reacted systemically during the first three injections. After five years of therapy, the insurance company declined to continue paying for treatment, so I was still nervous when going into the field. About a year later, I was stung again fifteen times in the scalp when I hit an aerial nest with my hardhat. I had forgotten to put the injectable epinephrine in my vest. Fortunately, all I got out of the experience was a throbbing scalp that felt sensitive for several days.

Category: OC-FD (stepped on an unseen yellowjacket nest)

Don't Go for the Gold: Golden Paper Wasp (*Polistes aurifer*)

A biologist at a southwestern desert zoo was tasked with, among other things, removing paper wasp nests from public areas. Normally, he would gently capture the nest in a container to relocate it without incident, but he now has trepidations about moving the nests of *P. aurifer*. On one occasion he was removing a nest in a bent-over grass clump. The Golden Paper Wasps seemed more agitated and belligerent than the similar Yellow Paper Wasp (*P. flavus*) and other species on the museum grounds, as he was promptly stung. He said the pain reached an 8 of 10 for about five excruciating minutes (I expect this translates to a 3.5 of 4 on the Hymenopteran Sting Pain Index), followed by "a merciful numbness for 15–20 minutes." For the next two hours there was painful throbbing (5 of 10). The numbness gradually tapered off and completely resolved the next day, without itching. He said this was easily the worst *Polistes* envenomation he has had, and that the common *P. major castaneicolor* differs by having a less-intense pain, which he described as being "a 6 of 10 for 10 minutes, followed by a throbbing ache for about 30 minutes; tired and a bit disoriented for a few hours after that; and local itching for about 24 hours, starting the next day." He stated that he can deal with *P. major*, but he now has a new mantra, "fear the yellow!" In the following category, "OT" was used because he did not have a lapse in judgment and was not cavalier—he was trying his best to not get stung.

Category: OC-OT (zoo personnel stung while relocating paper wasp nest)

Shot with a Nail Gun: Paper Wasps (*Polistes* spp.), Author's Testimonial

While writing this book, I was trying hard to not be envenomated while doing fieldwork or taking photos. I thought I was going to make it sting- or bite-free. Then in summer of 2016, I was looking for lizards and snakes to photograph in the Big Bend of Texas, and it happened. A subterranean nest of paper wasps was under the rocks I was turning and I received several stings from what appeared to be *P. dominula*. Having been allergic to yellowjacket and wasp stings in the past (see Envenomation Story, above), I hopped in my car, got my epinephrine pen ready, and waited for a reaction. Nothing happened. Even the pain was not that great (I would rate it a 1.5 of 4 on the Hymenopteran Sting

Photo taken of the culprits (presumably *P. rubiginosus*) at their nest inside a kiosk. *L.L.C. Jones*

Pain Index). However, it was interesting to see paper wasps nesting under the surface, rather than in the usual elevated umbrella paper nest.

Fast forward a couple weeks on the same trip. I was still in Texas but at the far eastern side of the state, in the Big Thicket. I was just coming off a wetland trail, and wham! It felt like I was shot in the shoulder by a nail gun! I ran like mad, dropping stuff along the way, but did notice a flying insect out of the corner of my eye. I got into my car and drove back to see what "nailed" me. There I saw a nest of red wasps. The nest was in a crevice of a kiosk at the trailhead, suggesting the inhabitants were *P. rubiginosus*; the similar *P. carolina* usually nests in the open, in a proper umbrella nest. I snapped a few pics from the safety of my car to document the culprits. I did nothing more than walk near the kiosk to warrant the attack. Fortunately, I only received a single sting, because it was a doozy. There was an immediate, powerful, intense, stabbing, searing pain. The intense part only lasted about 10–15 minutes, but the area was inflamed and hot for three days. I would rate it a 3 on the Hymenopteran Sting Pain Scale or 8 on a 10 scale, with a European Honey Bee being 3 or 4 on a 10 scale. It continued to itch, ache, and was sensitive to the touch for two weeks afterward, with a relapse at Day 10. The sting site later became a small, elevated welt, which periodically itched like crazy. The area remained slightly red, swollen, and sensitive for about two months.

In both of these cases, the typical "umbrella nest" was nowhere to be seen, as they were nesting in crevices. Unlike yellowjackets, which have aposematic buzzing, these things are quiet when they fly and when you disturb their nest, so I didn't notice them until they stung me. I used to think *Polistes* were kinder and gentler than yellowjackets, but I may have to revise that assessment.

Category: RE-FD (stung by wasps when walking near their nest)

Caught by Ballooning Shorts: Tarantula Hawk (*Pepsis* spp.)

A researcher in Texas was a passenger in a car while his partner was driving them to a study site at Big Bend Ranch State Park, Texas. They were in a Jeep with the doors off. He was wearing shorts that were ballooning out because of the wind. You probably know where this is going. Yup, a large tarantula hawk (probably *P. grossa*) was captured by the balloon shorts and the wasp stung him on the thigh. Although he claims to have a high pain threshold, having been stung by all manner of scorpions, bees, and even bitten by a rattlesnake, the wasp exceeded that threshold. On a scale of 0–10, he gave it a 10+. He described it as a hot poker stabbing him in the leg, with an intense burning that would just not let up. He said a comparison is like "having boiling water poured on you." They immediately stopped the car, and full of adrenaline he tried to make the pain dissipate by stomping his feet as he paced around the desert. Stomping did nothing and he had no ice or analgesics or other methods to attempt to alleviate the extreme pain. The pain was intense for 1–2 hours. It remained red and swollen for two days. He stated that a tarantula hawk is "a force to be reckoned with." It was far worse than any scorpion sting (he has been stung by two species of *Centruroides*) and worse than the sting from a large red and black velvet ant that got into his pack. That was only an 8+ on the pain index, but his finger did swell to 1.5 times its normal size. When asked how he would rate the sting of a European Honey Bee, he indicated that bees don't really register on his pain index.

Category: OC-FD (stung when tarantula hawk was blown into clothing)

I Didn't Sign Up for This! Velvet Ant (*Dasymutilla* spp.)

At Big Bend Ranch State Park in west Texas, a crew was collecting invertebrates for research at a Texas university. One of the crew members was an undergrad who picked up a species of white velvet ant with forceps, but the animal was able to wriggle free. Velvet ants (actually wasps) are notoriously tough, having a rigid exoskeleton, so the animal could not be pinched well by the forceps. Unfortunately, its captor was holding it over his leg, and it dropped onto his thigh where it proceeded to sting him. The pain was immediate and intense. He then "screamed bloody murder." The pain remained intense for at least 2–3 hours and the undergrad decided he had had enough and was driven five hours back to the airport in El Paso, where he was flown home. The only first aid he received was a topical cream to help alleviate the burning.

The species was unreported, but there are several white velvet ants, including the Thistledown Velvet Ant (*D. gloriosa*), which is thought to be a creosote bush fruit mimic. These are common dwellers of the three hot deserts, from California to Texas. As one insect identification website states, it delivers "a far more painful sting than its cute, fluffy appearance implies."

Category: OC-LA (stung by wasp when it was dropped from collecting tongs)

Sit-ups on an Ant Mound: Red Imported Fire Ant (*Solenopsis invicta*)

For those living in Texas, Red Imported Fire Ants (RIFAs) constitute a well-known hazard. For one junior in high school, they were particularly problematic on one occasion. The high school football team was always told to beware of and avoid RIFA mounds. Our victim knew that there were numerous mounds in the area where they did calisthenics. He told me, "That day, I turned around to see a huge fire ant mound, probably 2 feet wide by 1 foot long. So I moved over, and dropped down to do my sit ups." What he didn't see was the other mound where he moved to. He started doing his sit-ups when he noticed his back started burning. Of course, he immediately knew what the problem was. He leaped to his feet and a friend helped brush the ants off him. He had to "drop trou" then and there to get them off. He explains, "I probably got . . . about 500 or 600 stings, but it was impossible to count because the right lumbar area of my back was one huge welt with stings one on top of the other." The swelling and redness lasted a couple of hours. He iced the area, and the swelling subsided, but he did develop the tell-tale pustules that most people get. He had to miss practice the rest of that day, but being a trooper, was back at it the next.

Category: RO-FD (stung en masse while doing calisthenics on ant nest)

Pogos by the Mouthful: California Harvester Ants (*Pogonomyrmex californicus*)

California Harvester Ants were used in Native American rituals in south-central California. Groark (1996) stated, "The ants were ingested alive in massive quantities in order to induce prolonged catatonic states, during which hallucinogenic visions were reported to manifest." The basic method of ingestion was by swallowing a small ball of 4–5 ants clinging to a wet down feather. Up to fifty to ninety such balls would be swallowed over a period of days, if the participants were willing.

A miner, and self-proclaimed stuntman–scientist living in the same area as the Native American tribes decided to take a chance and emulate the activity, albeit in a less time-consuming method. He attempted to swallow some two hundred live *P. californicus* poured into his mouth at once. It was videotaped, so the reaction was well-documented. The effect was immediate. He said "the ants all stung me at once." He immediately spat them out, and scraped them out with his finger, having been stung countless times in the mouth. No ants got into his throat or were actually swallowed. His mouth, lips, and face rapidly swelled. The pain was described as excruciating, although he said he has a high threshold for pain. After two hours he experienced respiratory difficulty and was taken to the hospital upon the urging of others. Unfortunately, the hospital was an hour away, so they went to the local fire station first for medical assistance. When EMTs arrived, they feared for his life, so they administered epinephrine three times and he was transported to the hospital. His condition stabilized and after several hours of observation and counsel he released himself.

Why did he do it? After reading the article by Groark, he thought he "just had to try it," being a fan of Hymenoptera and native cultures. When asked if he had a spiritual or hallucinatory experience, he confirmed he did, more intense than any he had ever had before. However, he sums it up with this: "It was a life-changing experience and I am glad I did it, but I cannot recommend anyone try this."

The LD_{50} of the California Harvester Ant is 0.60 (intravenous) mg/kg, making it one of the most venomous insects on Earth.

Category: IN-IN (stung by ants while attempting to ingest for psychedelic effect)

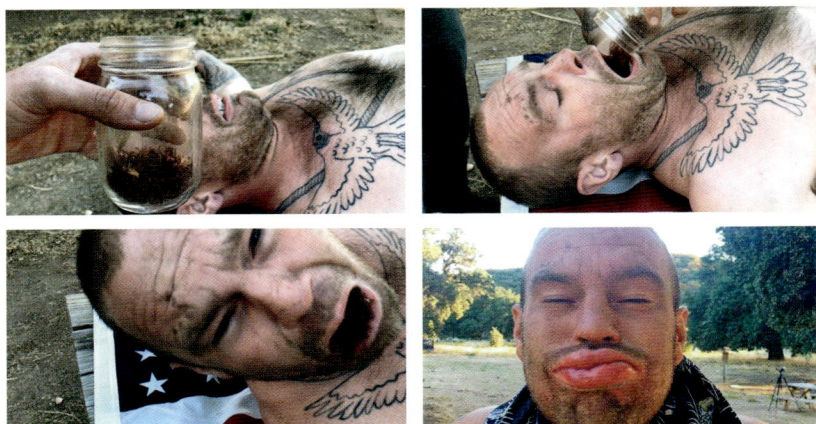

(Top Left) The jar of harvester ants just before the attempted ingestion. (Top Right) Pogo pouring. (Bottom Left) The harvester ants were immediately spat out, but the ants were quick to sting, so damage done. (Bottom Right) Swelling caused by the stings, but the systemic reactions were probably more noteworthy. *Jerod Blazer*

More Trou-Dropping from Ants in the Pants: Harvester Ant (*P. barbatus*)

A young wildlife photographer was stung by an unidentified species of harvester ant near Willcox, Arizona. It was in a grassy area, so the Red Harvester Ant (*P. barbatus*) comes to mind, but there are other species in that general area including the Maricopa Harvester Ant (*P. maricopa*). When first asked to contribute a testimonial, she responded with "here's my testimonial . . . IT HURT LIKE HELL!" Having been stung before, she knew better than to stand near harvester ant nests and always takes great care to avoid them. It was but a single "rogue ant" that stung her. She felt it sting under her pant leg on the upper thigh. She immediately knew what it was and tried to kill it by pounding on her leg. Then it stung her a second time. She then "dropped trou" behind a tree, and its crumpled corpse dropped to the ground. There was immediate, intense pain registering an 8 on a 10 scale. She described it as feeling like a red-hot needle being stabbed into the skin. This pain remained intense and her thigh throbbed for at least another eight hours. The stings raised two red welts about the size of a dime that hurt quite a bit for about two days. The welts increased in size to about that of a half-dollar over about the next three days. The wound site itched (not intensely) for about 2–3 days and it was tender for 4–5 days. Symptoms completely resolved after about two weeks.

Category: RE-FD (stung by a single ant that ventured up a pant leg)

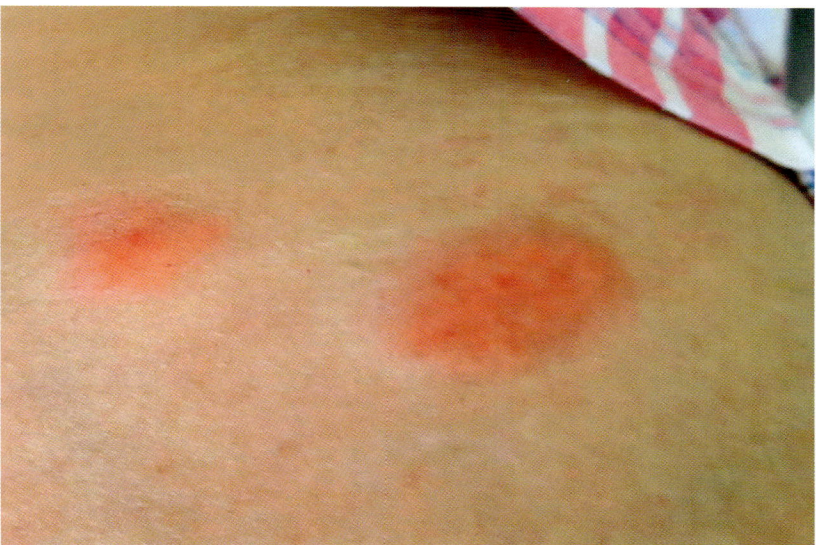

The two harvester ant stings that were under trousers. *R. C. Clark*

Undercover Bees: "Killer" European Honey Bees (*Apis mellifera*)

Africanized Honey Bees have pretty much displaced pure strains of the European Honey Bee along the southern U.S. borderlands. This story involves a herpetology class field trip, being led by the instructor and his father, plus a colleague; the students were fortunate to be led by three respected Texas herpetologists. However, venomous reptiles were not the problem on this particular day. The twelve or so students and their leaders were looking for reptiles in Willacy County, in the Tamaulipan thorn scrub of south Texas, an area renowned for its diversity of reptiles and amphibians. The class was spread out over several hectares while they searched for the resident herpetofauna to see and photograph. Some of the participants, including the father and two others, were concentrating on the fallen debris of an old barn. One particularly promising item was a side of the barn. It took four people to lift the old wall, but as it was being hoisted, someone yelled "BEES!" There was a very large bee hive under the wall. They dropped the wall and everyone scattered. Those that were far enough away (200–250 m) were able to get to their cars or far enough away without being stung. The rest of the people were not so lucky and most received multiple stings before reaching safety. Because the father was one of the four lifting the wall, he was in harm's way and was almost immediately covered in bees on the head and face. As he ran toward his son and a pond, he tripped, so the son helped get him to safety while trying to fend off bees, all the while being stung. They headed for the nearby pond and dove in, staying in the mud and water until the bees mostly disappeared. Each of them sustained 50+ stings, mostly to the head, face, and neck. Most of the father's stings were to the back and top of his head. He was removing stingers for several days. This experience gave him an intense headache, plus pain at the many small sting sites. The son sustained more stings to the face, and it became completely swollen, even with an oral antihistamine. He also complained of weakness and nausea. The swelling decreased after 8–12 hours, but for several days, when he awoke, his face would be swollen again, especially near the eyes. That decreased each day as he became active.

While it may sound like panic and chaos, they actually made several good choices that helped to keep this experience from turning into even more of a nightmare. First, people who rapidly "bee-lined" it to their cars got out of harm's way and avoided more stings. In general, diving into water does not necessarily dissuade an attack from Africanized bees (or yellowjackets). However, the two who headed for the pond attributed their good fortune to having the bees' pheromones stifled by the smelly mud bath.

Category: OC-FD (stung en masse when bee nest under coverboard was disturbed)

Stabbed with a Soda Straw: Wheel Bug (*Arilus cristatus*)

My biologist friend relating this story claims to have been "interested in bugs since about 2 years of age." By the mid-1980s, when he was ten years old, he distinctly recalled being bitten by a Wheel Bug, but some of the details (localized symptoms besides pain, and time to complete recovery) are a little fuzzy now.

He was out walking along a creek in southern Kansas, armed with his butterfly net, following some mammal tracks and looking for butterflies to nab. As he was walking along the shoreline, he dipped below some overhanging branches, but in doing so picked up a hitchhiker. He glanced at it and assumed it was a leaf-footed bug, a group of true bugs that are superficially similar to assassin bugs. Being into bugs, he didn't mind having it along for the ride and after several minutes had forgotten it was there. Then he felt "a sudden, excruciating pain." He quickly wiped the insect off, but then he

An adult Wheel Bug. *Danny Martin*

looked closer at it and noticed the tell-tale cog-like thorax, an unmistakable trait of the Wheel Bug. He said the thorax "reminded me of the ancient synapsid *Dimetrodon*." Obviously, this kid was destined to become a biologist. He looked it up later and knew he was tagged by a Wheel Bug. He described the pain as being "immediate and sharp . . . like someone taking a plastic drinking straw and shoving it through my arm." That analogy isn't too far off considering the mouthparts of assassin bugs. He added that it was more of a physical pain than a wasp or bee sting. The pain subsided relatively quickly, but the bite site was "a bit sore for a while afterwards."

While researching this book, I heard several stories of bites from reduviids and they seem to fall into two categories: those like this that hurt like heck (usually worse than a bee or wasp), and those that actually cause some localized dermonecrosis, and take some time to heal. My friend got off easy, but it still left a mark in his memory banks.

Category: RE-NA (bitten by bug that got on him while hiking)

A Black-lighting Biting: Western Corsair (*Rasahus thoracicus*)

During an evening of blacklighting for insects at Boyce Thompson Arboretum near Phoenix, an entomologist felt an insect land on her shoulder. Reflexively, she swiped at it with her hand, and it proceeded to bite her.

She described the bite as being about as painful as a yellowjacket sting. However, it was different than a typical wasp sting because the "bite site stayed hard and a little swollen for weeks." Eventually a hard piece of tissue, like a scab, broke loose and left a small hole in her finger. She summarized the experience as "Nothing exciting, but different from other stings and bites I have had."

Category: OC-FD (bitten by bug that landed on her during night-lighting collection)

Toddler Kissed by a Bug: Eastern Bloodsucking Conenose (*Triatoma sanguisuga*)

Kissing or conenose bugs are the only parasitic species covered in this book, with an otherwise "no parasites allowed" rule. This exception is made because kissing bugs are a member of the venomous assassin bug family, Reduviidae. While kissing bugs are parasites, this account shows that they will also bite defensively, and do have venom that causes symptoms consistent with some other venomous species.

In 1997, an entomologist and his family were on vacation. When they stopped for gas, they collected insects under streetlights as a family activity. He had been teaching his three-year-old son about animals that were OK to touch and those that were not. Since the toddler was only three, he hadn't yet grasped all of the nuances of safe versus unsafe critters, although the child knew about snakes, bees, wasps, spiders, and scorpions. Unfortunately for the boy, he hadn't been taught about reduviids yet, and they look like a pretty normal bug that dad might want to have, so the toddler decided to help. When he tried to pick it up, the kissing bug bit him on the pad of his right index finger. As judged by the child's reaction, it was probably a painful bite. As his father explained, "the intensity and duration of the crying indicated to me that the pain must have been pretty significant, as this was, and still is, a child that never cried." His dad added that "I can only assume it was akin to the sting of a honey bee, which for me is about a 6 on a scale of 10. . . . His 'boo-boo,' or the discomfort felt as a result of the bite, persisted for about seven days." The toddler is now thirteen years old and he still enjoys collecting insects and reptiles with his dad; the conenose bite was a sort of "rite of passage."

Category: IN-NA (toddler picked up bug out of curiosity)

Beautiful but Itchy: Iris Eyed Silk Moth Larva (*Automeris iris hesselorum*)

Silk moths are beautiful as both adults and larvae, so many are sought after and maintained by lepidopterists. In the insect husbandry department of a Southwest zoo, one of the keepers reached into a terrarium housing ova-reared larvae when one of them got lodged in his armpit. He said there was an "immediate burning sensation, ranking a 6 of 10 on the pain scale." He washed and rinsed the area in cold water to alleviate the pain, but it made no difference, so he then iced the area for 10 minutes. This did offer relief by numbing his skin, but when the numbness wore off, the pain resumed. The pain was mostly gone in about 30 minutes, but it itched for several hours afterward.

Category: OC-LA (zoo personnel stung while maintaining a captive brood of silkmoth larvae)

Look Before You Swipe: Puss Caterpillar (Asp) (*Megalopyge opercularis*)

During the summer of 1969, a student was working for Texas Parks and Wildlife Department in east Texas. He was clearing deer census trails to help TPWD estimate the deer population for hunting management. Being as it was hot and steamy work, he would wipe the sweat off his brow with his shirt sleeve. OK, we know where this is headed. During the course of clearing vegetation, he managed to pick up a puss caterpillar on his sleeve. As he wiped the sweat off his face, he smeared the caterpillar from behind his eye to the corner of his mouth. It left a red streak in its wake. He rated the pain a 6 on a 0–10 scale, but noted he has never experienced a 10, which he would consider unbearable. He said "It felt like someone had just placed a branding iron to my face," but then he added, "I don't know how a branding would really feel, but anyway like very hot metal to the skin." His boss then dropped everything and they headed to town. He said there was a lot of pulsating pain and it gave him one of the worst headaches he ever had. When they arrived at a small town about thirty minutes later, the pain was subsiding, so he did not go to the doctor, but he added "I at least got the rest of the day off."

The larva of Southern Flannel Moth and its congers are the most notorious of the stinging caterpillars in the United States. Based on one study of 2,130 stings in 1958, our student suffered the most common symptom, "severe local pain" (98.2%), but also had a severe headache (29.4%), indicating a typical, moderate envenomation. The sting was a long time ago and he did not remember lymph node involvement or even swelling.

The week before I photographed the *M. opercularis* for this book at Palmetto State Park, the Park Ranger told me a mother and her daughter were both envenomated in the same area and required an ambulance ride to a nearby hospital. I did not know the outcome of those stings, but severe cases can cause systemic reactions and require hospitalization.

Category: OC-FD (accidentally stung by caterpillar while wiping sweat off forehead)

Just Wanted to See What It Could Do: Puss Caterpillar (*M. bissesa*)

Flannel moth caterpillars, commonly known as puss caterpillars or tree asps, are well-known hazards in the East, and especially Southeast, where five species often make headlines, but they are poorly known in Arizona. There are three species of *Megalopyge* in Arizona: *M. bissesa*, *M. lapena*, and *M. opercularis*, although the latter is primarily a southeastern species. A field naturalist who knew members of the genus were venomous happened upon a puss caterpillar feeding on a Soapberry bush while hiking in Arizona's Cochise Stronghold, Dragoon Mountains, in the mid-90s. On a curious whim, and knowing of the caterpillars only from books, he decided to find out how much of an effect the sting would have when lightly brushed against the back tip of his middle finger. The sting site turned only a bit red, and there was no swelling, but there was pain and lots of it. Initially there was just a local burning and tingling, but pain soon increased in intensity and crept up his arm—to a 10 out of 10. It remained intense for about six hours, then fairly quickly subsided; there were no residual effects. However, for those six hours, he described it like "putting your finger, hand, and arm in a vice and applying heavy pressure." He had muscle cramping and spasms from fingertip to shoulder. He said it was the worst pain he ever felt, very similar to but worse than his Pacific Bluebottle and Portuguese Man-of-War stings, and less localized but much more intense and longer-lasting than either yellowjacket and paper wasp stings. He described the pain as comparable to that of his knee replacement surgery if he got way behind in his medication, and the opioids had completely worn off. Two 200 mg Ibuprofen were taken about 2–3 hours after the sting but were ineffectual.

Category: IN-IN (swiped finger against caterpillar to test venom effect)

Bit by a Mega-Fly: Giant Robber Fly (*Promachus aldrichii*)

A field biologist was tasked with conducting surveys for the federally listed Delhi Sands Flower-loving Fly (*Rhaphiomidas terminatus abdominalis*), in its native habitat of the Colton Dunes, California. As part of his fieldwork, he often collected specimens of non-federally listed species that co-occurred with *R. terminatus*. On August 1999, he managed to sweep up a member of the giant robber fly genus, *Promachus*, in his butterfly net. Not only was it a giant robber fly, it was a giant among giants—the largest he had ever seen. The species was *P. aldrichii*, and by his estimation was about 50 mm long. As he was securing the fly to be transferred to a kill jar, he grasped the body through the net, and it managed to bite him on the tip of his finger. As he put it, "I was quite surprised that this fly could puncture my thick calloused fingertip, as I was an avid guitar and bass player with many years of thick callous build-up, so much so that I would routinely extinguish lit cigarettes on them as a "party trick" without pain." Anyway, he said the bee-sting-like pain was immediate and intense, about a 6 out of 10. The intense part of the pain lasted about two hours, and then gradually diminished over time. The tip of his finger had a small, red puncture wound, and his finger remained slightly swollen and tender to the touch for the next two days. He had no other symptoms or long-lasting effects.

Robber flies are good examples of venomous animals for which there are virtually no case reports, since bites are uncommon and they are not considered medically significant. While there are many "biting" flies, most are parasitic. Robber flies are truly venomous and use their bite primarily for predation on other insects. But as we can see, it can be used for defense.

Category: OC-CA (bitten by fly while transferring it from net to collecting jar)

Mud Daubers Can Sting? Blue Mud Dauber (*Chalybion californicum*)

Mud daubers are fairly well-known wasps that build mud tubes, especially under bridges, awnings, and overhangs. Most people consider them harmless, as they are either very disinclined to sting, or the sting is nothing to write home about. However, this is not completely true.

This story comes from an entomologist who distinctly remembers a run-in with Blue Mud Daubers in the 5th grade. He was walking home

Blue Mud Daubers are best known as predators of black widows, but they can also deliver a painful sting to humans. *Sam Droege/Public Domain*

from school in Redlands, California, along some railroad tracks, where he would catch lizards. One day, he noticed an intact cardboard box in a small, sandy-bottomed wash. That was too tempting, so he had to look under it for lizards. He doesn't actually recall if there were any lizards, because upon lifting the box he was immediately attacked by Blue Mud Daubers. They began stinging him on his skin and through his clothes. He ran away and didn't stop running until there were no more wasps about. He was stung three times on the arm and shoulder blade. He vividly remembers the pain being pretty intense, similar to a Honey Bee, and would rate it a 6 out of 10. As best he can recall, the pain lasted about four hours before starting to decrease. He clearly remembers significant swelling in the area of the stings, plus redness and hotness. In fact, each of the sting sites was swollen to the diameter of a baseball. He says that "the areas surrounding each sting location turned bright strawberry red in color and was hot to the touch. The swelling, heat, and angry red coloration remained for about five days before gradually returning to normal." Our school child is now a fully grown entomologist working on (among other things) stinging insects; to this day, he said this was his worst reaction from hymenopteran stings he has had.

I had to play devil's advocate and asked if he was sure about the identity of the culprit; I really didn't doubt it, but needed to hear him say it. He assured me that he was already savvy about such things, saw their diagnostic mud tubes, and was certain they were not one of his other favorite metallic blue wasps—the Steel Blue Cricket Hunter (*Chlorion aerarium*). He added that he has always been fascinated by mud daubers, which are spider hunters. Their booty includes black widows, which they stuff into their tubes. He added that a good educational experience is to cut open one of the tube nests and watching the spiders fall out.

Category: RE-FD (stung by wasps when nest was disturbed under cover object)

A Man of a Thousand Stings: Many, Many Insects

Justin O. Schmidt ("The King of Sting") has always been a celebrity because of his Hymenopteran Sting Pain Index, his flowery prose when describing the pain of a sting (see his book, "Sting of the Wild"), and, of course, the many stings he has received over the years. Justin got his experience not because he generally allows himself to be stung (contrary to popular belief), but because his research on hymenopterans puts him in harm's way. What we may not realize is that there are numerous other biologists who also have been stung or bitten repeatedly in the course of their research, due to their occupational hazard. This story centers on one such individual I dubbed the "Man of a Thousand Stings."

To start with, he is an entomologist specializing in Hymenoptera, as is Justin Schmidt, although he also shared information with me about the stings and bites from other insects he has accrued over the years. As he put it, "What sets me apart is not that I've received super-nasty stings, but that I've been subjected to a particularly large number and diversity of stings and bites, all told—several hundred different species, with about 1,000 total stings or bites. When what you do for both a living, as well as personal enjoyment, is to catch insects (especially bees and wasps), you set yourself up for a variety of experiences." His list includes many species of wasps and bees, including the usual assailants featured in this book (e.g., yellowjackets, paper wasps, tarantula hawks, velvet ants, "killer" bees, bumble bees, and fire ants), but also cicada killers, ichneumons, and stingless bees. It doesn't stop there; he has also been stung or bitten by a variety of other insects, including assassin bugs, stinging caterpillars, and aquatic hemipterans and beetles.

In general, he rates the tarantula hawks and social wasps and bees as being the worst of the Hymenoptera. As he put it, "Mercifully, I survived my few run-ins with killer bees." He is not remiss for avoiding the experience of the top-rated sting from the Neotropical Bullet Ant. Among the bees, honey bees and bumble bees are the worst, as they have large colonies to defend. He mentioned that carpenter bees, which do not live in large colonies, are nothing to write home about. However, he also stated that a few stingless bees have mandibular secretions that cause blisters when they bite, so technically they are venomous. They also live in colonies and are quite capable of defending themselves. He remarked that "if you want a fun time, imagine a few hundred [stingless] bees flying around your hair, your ears, your nostrils, and your eyebrows, all biting and pulling your hair, while giving off a chemical that smells like rancid butter." The ichneumon wasps are generally regarded as not capable of stinging, but he has been stung several times by those in the genus *Ophion*, and likens their sting to a hot light bulb placed against your skin. He has also been stung by male cicada killers (*Sphecius speciosus*), which, being males, are not venomous, but he adds that being forcefully stabbed by their very sharp pseudo-stinger is indeed quite painful. He mentions that among ants, Red Imported Fire Ants are probably his least favorite, as he has been attacked by swarms, which he dubbed a "THOROUGHLY unpleasant" experience.

Then there are the non-hymenopterans. Of these, he states that the stinging caterpillars are the worst of the lot, but assassin bugs are no picnic, either. He said that "the sheer level of discomfort, both initial and persistent, involved in a good slug caterpillar (*Parasa*) venom, is more unpleasant to contend with than any bee or wasp or ant sting." The same is true for other stinging caterpillars. He described the pain as being like a hot iron being pressed against the skin for several seconds. The intensity is short-lived, but pain covers a large area and takes a long time to completely fade away. He has been bitten by a variety of assassin bugs. He said the worst of those were *Rasahus* (corsairs) and *Sirthenea* (only *S. stria* occurs in the U.S.; most are primarily tropical). He says the pain is "akin to having a red-hot needle jabbed into one's flesh, and considerably slower to wear off than any bee or wasp sting; hot, intense, and persistent."

An interesting side note: As part of his research, he marked thousands of sweat bees, which are not regarded as particularly painful (a 1 on the Hymenopteran Sting Pain Index). However, despite precautions, he was stung hundreds of times in the same spot, over several years, due to how he had to hold the animals to mark them. With so many venomous stings in such a restricted area, that part of his thumb became a gangrenous green callus that eventually fell off. However, within a few years, the tissue grew back, and his thumb appears normal now.

Category: OC-CA (routinely stung while free-handling insects to collect)

Terrestrial Vertebrates

Show-and-Yell: Gila Monster (*Heloderma suspectum*)

In 1989, a herpetologist with the Arizona Game and Fish Department was demonstrating the marvels of Gila Monsters and their venom to an audience at the grand opening of an animal rehabilitation center. As he was exposing the teeth on the lower jaw, the animal twisted and managed to bite him on the left index finger. The Gila Monster latched on with tenacity and ended up biting him five times, each time through to the bone. He managed to extricate the lizard by putting a leather glove in its mouth and pushing it in a little farther every time the lizard relaxed. He finally got his finger out in about two minutes. He described the pain as being immediate and one of the worst he had ever felt; he gave it a 9 out of 10 ranking. "My finger felt like it was on fire and this wave of fire kept moving up my arm." He got nauseous and had difficulty breathing. He was quickly transported to a hospital, but was in shock in about ten minutes. His blood pressure was so low that they could not administer pain killers (opiates) until about three hours after the bite. There was pain in his kidneys and blood in his urine. He remained at the hospital for thirty-six hours. He went back to work in about a week, but complete recovery was not for several weeks. He still has a small white discoloration and very slight movement constraint of the left index finger. The only other permanent damage he claimed was to his legacy: "this is what I will be remembered for . . . but at least I admit to being bitten (unlike some other herpetologists I know)."

Category: OC-LA (bitten by lizard when showing a group its venom apparatus)

Bit through the Bag: Gila Monster (*H. suspectum*)

In August 1984, a herp enthusiast was road-cruising southern Arizona. He found a particularly nice Gila Monster and decided to illegally collect it. He put it into a large pillow case, tied the end, and then placed it into a cooler with two Tiger Rattlesnakes (which can be legally collected with a hunting license). A couple days later he started to head back home to southern

California, but noticed there was a strong stench coming from the cooler, because the Gila Monster had defecated in the bag. He knew he had to clean up the mess and pulled into a self-serve car wash, out of the way of the viewing public. As he pulled the bag out of the cooler, a police officer pulled up nearby and left the car running to do some paperwork. As he put it, "Never taking my eye off of the officer, I reached for the knot-end of the pillowcase to switch it out with a clean one, only to realize too late that I had not grabbed the knot, but the lizard's head, as it locked on to my left index finger through the case."

He explained what followed next. "Immediate pain and then panic was felt when I realized the lizard wasn't going to let go . . . 15–20 long seconds of 'clamp-relax-clamp' chewing (soaking the case with venom) went on before I decided I would have to yank the lizard off and hope for the best." It worked but the wound began bleeding profusely. He grabbed a towel, wrapped his hand in it, and went to the other side of the car, hoping the officer would not notice these strange goings-on. He chose to not seek medical attention and drove to the nearest hotel. As he was checking in, he was very light headed, sweating profusely, and almost passed out. Although he was in bad shape and had a blood-soaked towel around his hand, the hotel clerk gave him a room. He got to his room, unwrapped his hand, and soaked it in ice water in the sink. This offered a little relief, but was also causing pain, so he took a cold shower. As he described it, "The pain had become unbearable, much like being severely burned while being hit with a hammer. . . . I flopped around, rolling back and forth on the bed trying to cope with the pain." He described it as the worst pain he had ever experienced. The pain started to subside after about six hours.

When the pain became bearable, he cleaned up the mess (and the lizard) and headed home. Residual pain and sensitivity lasted about a week. There was some loss of mobility in the hand for a while longer, and then he fully recovered. After a few weeks he drove back out to Arizona, and released the Gila Monster where he had found it.

Although I cannot condone poaching, it is a fact that some herp enthusiasts collect or keep venomous animals illegally (though many venomous animals are kept legally). There are numerous case reports of people being envenomated by illegally collected or kept animals. One problem, as demonstrated in this case, is that illegal collectors may not seek medical attention for fear of retribution or the loss of their booty. Self-remedies, such as cryotherapy, may also cause a great deal of physical harm.

Category: PT-LA (bitten by lizard when he grabbed the collection bag)

Unlucky Number 13: Banded Gila Monster (*H. s. suspectum*)

A colleague studying Gila Monsters in Nevada relayed an interesting story about a nearly fatal Gila Monster envenomation by one of his wild-roaming tagged study animals in a national conservation area. A seemingly physically fit, but homeless rock climber from a nearby "hobo camp" had seen three Banded Gila Monsters in the wild over several years and he always felt obligated to handle them. On the last encounter things went awry. When he attempted to grab the monster as it was trying to escape, it turned around and bit him on the back of the hand. It latched on with the proverbial "tenacity of a bulldog." Try as he and his friend might, the Gila held on for 10 to 15 minutes, allowing a large amount of venom to enter his system. Eventually, they got it off by prying its jaws open with a flat rock. When my friend heard about the encounter, he proceeded to check on the well-being of the resident Gila and found that Number 13 only had five remaining teeth—the others were in the victim or the ground. Fortunately, when they finally got the lizard off, the victim stopped his friend from killing it. The victim stated that because he was having such trouble breathing ("10 percent of normal") he considered giving himself a tracheotomy. Two first responders were rock climbers who attempted to stabilize him. He was then put on a board by a park ranger and transported to the nearby hospital, where he went to the ER. From there he remained in the ICU for four days. On day five he was moved to the general care area, and then he checked out. The severe bite caused extensive swelling to his hand, arm, and neck. One of the symptoms he recalled vividly was long-term, uncontrollable diarrhea, which continued well after being released from the hospital. The victim claimed, "it took several weeks to be O.K. again." He also stated that handling Gila Monsters would now be a thing of the past.

Category: RE-CA (person was bitten for 10 to 15 minutes after grabbing a wild Gila Monster)

Biting the Hand that Feeds You: Beaded Lizard (*Heloderma horridum*)

A law enforcement officer in California not only served and protected, but he also researched venomous animals and gave educational seminars about their biology. One of his display animals was a young captive-hatched Beaded Lizard. Although not found north of Mexico, this species can be encountered in Sonora, less than a day's drive from southern Arizona, where it co-occurs

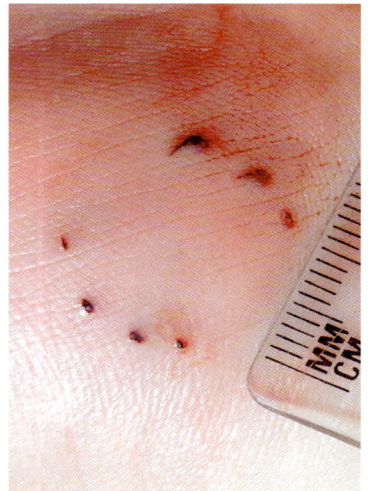 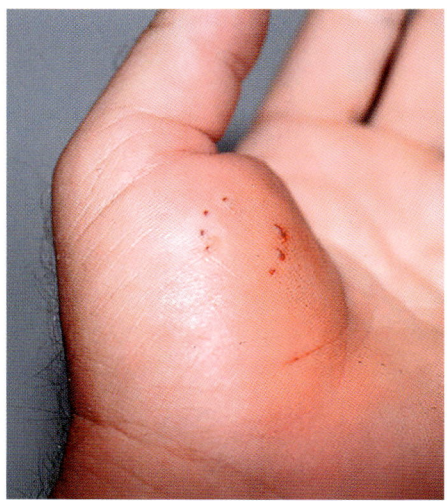

(Left) The bite mark. (Right) Swelling and a burning pain followed the exposure.
Michael Cardwell

with the only other congener, the Gila Monster. Unlike *H. suspectum*, bites from *H. horridum* are poorly documented. This story is an example of one of the thousands of envenomations in North America from exotic animals in captivity.

The 49-year-old man in good health was reaching into the cage of his Beaded Lizard to clean it. Unfortunately, the Beaded Lizard assumed it was feeding time and grabbed the man by the palm of the hand as he reached over it. The lizard hung on for about three seconds. It left a large circular pattern of tooth marks and bled freely. There was an immediate and significant burning sensation. He allowed it to bleed under the faucet for about ten minutes before applying direct pressure to slow bleeding. He disinfected the bite with rubbing alcohol. After a few minutes, he said "pain had intensified dramatically, feeling like my hand was on a hot stove." He took three Excedrin. Within a half hour the hand and thumb were swollen and red, and pain was radiating from the bite site. The armpit lymph glands became tender. Swelling continued to spread up into the forearm. After about 1.5 hours post-bite, he noted that pain was diminishing and nearly gone after 2.5 hours. After 96 hours, the swelling was completely gone, but there was still tenderness on the forearm and hand that resolved very slowly over the next several weeks.

Category: PT-LA (bitten by pet lizard in terrarium)

Should Have Used a Longer Snake Stick: Trans-Pecos Copperhead (a variant of *A. laticinctus*, formerly known as *A. contortrix pictigaster*)

In 2003, a herpetologist who was well versed in aspects of first aid for snakebite was tagged by one of the two Trans-Pecos Copperheads he collected in west Texas. They were captured at night and retained to be photographed the next day. The larger one, which bit him, was a good-sized adult at about 75 cm. He admits to the error of using tongs that were "WAY too short for the job at hand." The snake was uncooperative and would not pose for a photo. While he was repositioning the animal, he was bitten on the index finger. One fang hit a knuckle, while the other met its mark in the meaty part of the finger. He thought to himself, "I hope it's a dry bite . . . I hope it's a dry bite . . ." Almost immediately, he felt the sensation of a wasp-like sting and thought, "Nope, not a dry bite." He quickly gathered his things and drove to the nearest phone booth about fifteen minutes away. He described the escalating symptoms during those first fifteen minutes. "The pain was pretty intense (maybe a 7 or 8 on a scale of 0–10), and my hand felt like it was getting slammed in a car door with every heartbeat." By the time he reached the phone, his finger had already become substantially swollen and was oozing fluid. He called 911 and met the ambulance as he was heading for the nearest town with a hospital. This is where his familiarity with first-aid procedure of snakebite came into play. Apparently, the emergency room physician who was instructing the EMTs on what to do was locked into an "old-school" first-aid regimen. The EMTs were instructed to incise the bite marks and suck out the venom. Our victim declined, saying this was no longer an acceptable method. Then they were instructed to use a tourniquet on the arm to slow the spread of venom, but he pointed out this could be a dangerous practice and declined. Then they were instructed to wrap his arm with an elastic bandage. He pointed out that he was not bitten by an Australian elapid, and this was not a normal first-aid measure for copperhead bites, but he figured it would do no harm and it might satisfy the physician and EMTs, so allowed it. When they arrived at the hospital, ninety minutes post-bite, one of the EMTs commented that their patient was doing well and did not even get nauseous—perhaps it was the power of suggestion, but just then the herpetologist lost his breakfast.

Following proper first-aid procedure, he had been marking the progress of the swelling on his arm with a pen every fifteen minutes. He pointed

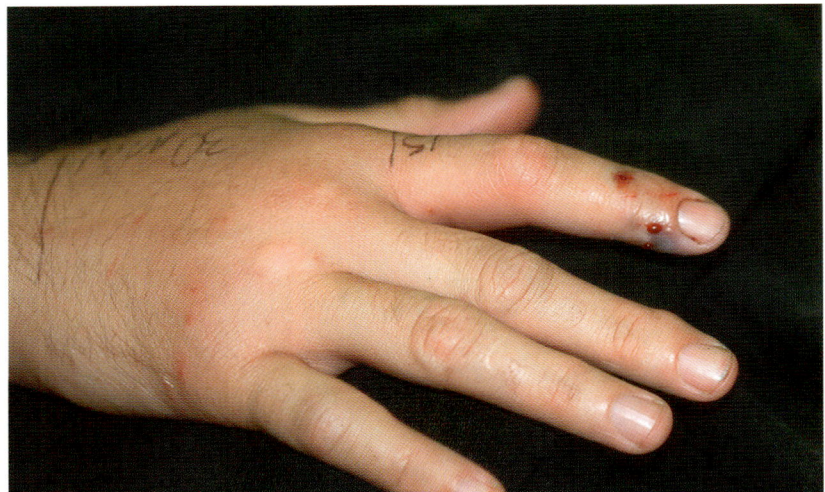

Early photo of the wound caused by the "Trans-Pecos" variant of the Western Copperhead. *Troy Hibbitts*

This is the other individual of the "Trans-Pecos" variant of the Western Copperhead that didn't bite the victim. *Troy Hibbitts*

out that the swelling had not progressed past the forearm after ninety minutes, so he would not need antivenom (this could have been CroFab or the Wyeth polyvalent at this time). The ER staff was pleased to hear that, as they had just used up their antivenom stores treating a serious bite from a Western Diamond-backed Rattlesnake. (That victim, by the way, died later that week.) The herpetologist was given anti-nausea drugs and painkillers, including morphine, and was admitted to the hospital for overnight observation. He said he had no intention of spending the night in the hospital, but since he was high on morphine, he went along with it. He was released the next day. The swelling extended to the elbow and lasted 4–5 days. His finger had a single black bleb. The digit remained stiff for about a month.

The take-home message is that it is very important for people who are at risk of envenomation to know proper first aid. Herpetologists dealing with venomous snakes often know what to do and what not to do, but ER physicians may not be so well versed, as they are responsible for countless maladies and may not be tuned in to the latest and greatest information on envenomation. Physicians who regularly see snakebite cases, or are venom specialists, become well versed in both first-aid measures and medical treatment. These people are a rare and valuable asset. Also, it would have been advisable for the ER physician to call a poison control center to be instructed on proper procedure. After all, poison control centers were primarily created to assist physicians with cases of poisoning and envenomation.

Category: RE-LA (bitten while manipulating to photograph snake with tongs that were too short)

Not in the Tube Yet: Northern Cottonmouth (*A. piscivorus*)

A university professor in biology and herpetology in South Carolina was conducting research on snakes at the Pee Dee Research and Education Center, a facility managed by Clemson State University, in the northeastern corner of the state. He used the practice known as tubing to restrain venomous snakes. After the snake is in the correctly sized clear plastic tube, it can be safely measured and sexed, and there is essentially zero chance of being bitten. However, the tube needs to be of the appropriate size, and you need to safely coax the snake into the tube. While the herpetologist was trying to determine which tube would best fit a 50–60 cm Northern Cottonmouth, he let his guard down, and the snake managed to sink a fang each into his middle and index finger. He knew the drill, so put the snake carefully back in the bag, took his ring off, then raced off to the nearest hospital, which was about a half hour away.

There was immediate, intense pain, which he rated a 10 out of 10. He said it was not just localized, but included "widespread burning and throbbing."

He noted another unpleasant symptom on the way to the hospital: nausea, although he did not vomit. At the ER, they informed him it was a "pretty bad bite," and administered an unknown number of vials of CroFab, plus an IV drip with morphine. He believed the combination worked well to alleviate the pain and stop the progression of swelling and tissue damage that could be wreaked by this highly hemorrhagic species. The swelling was abated partway up the forearm, and there was little discoloration. The swelling subsided in about 7–10 days. There were no long-lasting effects.

Category: OC-LA (bitten by snake while trying to get it into a handling tube)

French Kiss: Eastern Diamond-backed Rattlesnake (*Crotalus adamanteus*)

A physician in South Florida relayed this story about an unusual circumstance leading to envenomation by a highly dangerous snake. An individual who was intoxicated by alcohol brought a juvenile Eastern Diamond-backed Rattlesnake to show his also-intoxicated friend. The friend thought it would be funny to stick his tongue out at the snake. Apparently, the snake was not amused, and bit him on the tongue. Despite the small size of the snake, the bite was severe. The tongue is highly vascularized and in the entrance to the airway, so it is decidedly a bad place to be bitten. The patient almost died very quickly. His air supply was being compromised, venom had spread quickly, he was "bleeding out from everywhere," and his platelets were through the roof. He was initially given twelve vials of CroFab, and later received another eight vials. The ER physicians were able to save his life and stabilize him. The man completely recovered and is doing fine now.

Category: IN-CA (thrill seeker was bitten while toying with a rattlesnake)

Trick or Treat: Western Diamond-backed Rattlesnake (*C. atrox*)

On Halloween night 2009, a 58-year-old male was bitten when he stepped barefoot onto his front porch. Thankfully, the snake did not bite a trick-or-treater, because the bite was no treat. A 45 to 60 cm Western Diamond-backed Rattlesnake bit his toe on the left side of his foot. On a scale of 1–10, he ranked the early pain as a 5. He was immediately driven to the emergency room of a Tucson hospital, where he remained for forty-eight hours. The bite area swelled and turned gray, then spread to about 25 cm. He was given pain medication and eighteen vials of CroFab over the next ten hours. After twenty-four hours, when he lowered his foot off the bed, he reported the pain index as "10++++++++, like a thousand daggers in my left foot." That lasted a week before it began to subside. All symptoms were gone in about two months. He returned to the hospital when tests showed his blood was not coagulating normally. He remained there for several days until he stabilized. The only long-lasting effect is that his wife claims "he still hisses occasionally."

Category: RO-FD (accidentally stepped on an unseen rattlesnake)

The Show Must Go On: Western Diamond-backed Rattlesnake (*C. atrox*), with Allergy

In 1998, a professional wildlife educator gave a lecture on local wildlife to an audience in central Arizona, using live animals. During his presentation, he was bitten by a 147 cm Western Diamond-backed Rattlesnake. He was using tongs to handle it, but swung it around too close to his body, when the animal struck him on the right calf with both fangs. He said he let his guard down by breaking a cardinal rule of snake handling—he was a little groggy from some cold medicine—so probably had a momentary lapse of reason. The audience was unaware of the incident and he continued to finish the show without letting on what had happened—he didn't want to alarm the audience, and the show must go on, right? He did not seek medical attention right away, but remained calm. About forty minutes after the bite (and after the show) he used a snakebite venom extractor. Even though some time had elapsed, he saw clear amber liquid come out of the fang holes. He then monitored the progress of his symptoms. He began to swell up and got covered in hives, so about six hours after the bite he went to a local area emergency room. He was then helicoptered to a Phoenix hospital with expertise on snakebite.

The effect of the venom was significant, but additionally, the doctors informed him he was also having an allergic reaction to the venom, making for a wide range of symptoms. He said pain was not a significant symptom. The bite itself was a 3 on a 0–10 scale, and there were some shooting pains for about an hour from the bite site. His entire leg and foot became swollen and purple, but the swelling was never really painful. His face was swollen and one eye was swollen shut. Although he had never been bitten by a rattlesnake

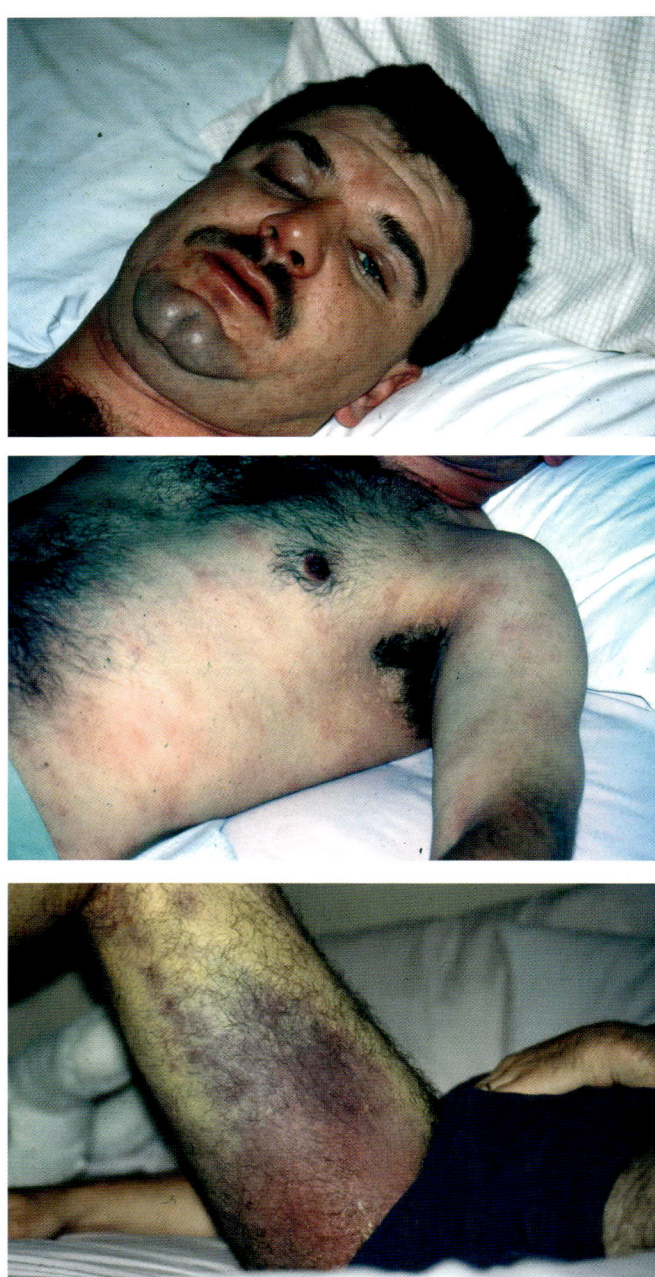

(Top) Two hours post-bite, the victim's face is swollen and the right eye is shut. (Middle) Two hours post-bite, the victim is covered in hives. (Bottom) Seven days post-bite, there is still bruising on the thigh. *Courtesy of Phil Rakoci*

before, he had been tagged by a copperhead in 1995, and had plenty of previous encounters from rattlesnakes in the field and in captivity. At the Phoenix hospital he was administered twenty vials of antivenom, which in 1998 would have been the equine polyvalent from Wyeth. We assume he was tested for allergy to antivenom and the tests would have been negative. He also received antihistamines and steroids. After spending two nights in the hospital, including the intensive care unit, he was released.

He was on crutches for two weeks. The hemorrhaging started to fade about Day 9 and was gone by about Day 15. Most of the swelling was gone in about three weeks. All symptoms resolved in about one month. There were no long-lasting effects, except that twice in several months following the bite, when he skinned and ate road-killed rattlesnakes for a Southwest cooking class, he broke out in hives. Later, when he ate the meat, but did not do the skinning, he did not have an allergic reaction.

The use of an extractor is controversial, but probably never hurts if one does not use a tourniquet, incise the wound, or use cryotherapy. He and his doctor both felt that getting out additional venom may have helped save a part of his leg, although it was likely the fluid was serum rather than venom.

Category: OC-LA (bitten by a rattlesnake while holding it with tongs)

It's Complicated: Western Diamond-backed Rattlesnake (*C. atrox*)

A male hobbyist first rescued a Western Diamond-backed Rattlesnake from a construction site when it was a small juvenile, but by 2009 it grew in captivity to about 1.5 m long, much larger than most *C. atrox* encountered in the wild. He had fed the animal a live rat when the temperatures were high in late October and thought it had eaten, but it hadn't. Then the temperature was to drop drastically and he saw the rat moving in the cage, so he thought it best to get the rat out of the large terrarium, which had two front doors. As he was attempting to remove the rat with tongs in his right hand through the left door, the snake struck at the tongs, overshot them, glanced off the door, and struck the middle finger of his left hand. As he put it, "It put about a centimeter gash in my finger with lots of bleeding. The other fang left a small mark on the ring finger. I was envenomed by only one fang. So starts the long story." That was at 22:00 of Day 1.

He and his wife responded immediately. She called 911, but to save time, drove him directly to the nearest hospital in Tucson, with the 911 personnel on the line the whole time. They were told the hospital had antivenom. When

794 VENOMOUS ANIMALS OF NORTH AMERICA

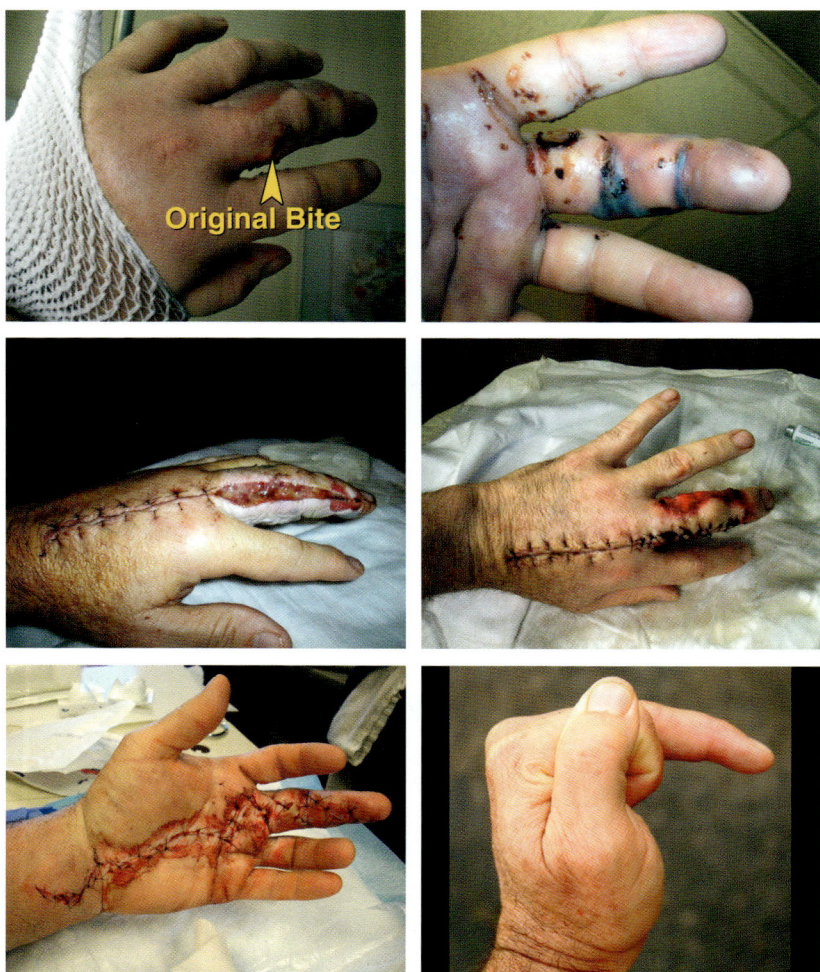

(Top Left) Bite site. (Top Right) Before fasciotomy. (Middle Left) After fasciotomy but before wound closure. (Middle Right) After fasciotomy and wound closure. (Bottom Left) Palm view. (Bottom Right) Permanent disfigurement. *Paul Condon*

they arrived at the hospital twenty minutes later, he was shaking badly. He was given morphine for the pain, but the antivenom was not provided quickly, although he was told the pharmacy had it and it was on its way. It was not until 02:45 (Day 2, 4.75 hours post-bite) when he received a single vial of CroFab. Five more vials were started at 04:15, about six hours post-bite. They continued to give him antivenom into Day 3, for a total of sixteen vials.

On Day 3, it was obvious his hand was getting much worse. The swelling in the arm was massive and there was compartmentalization in the hand and considerable tissue damage. Although one physician told him it would get better, he asked for the opinion of a vascular surgeon, who took one look at his hand, and upon seeing its condition, recommended an emergency fasciotomy. Three incisions were made. One from the middle of the palm to the tip of the middle finger and two more along the sides of the top of the finger, one extending down to the wrist. As he put it, "pain afterwards was an understatement." They gave him IV injections about every two hours for pain for the entire ten days he remained in the hospital, three in ICU. After the original fasciotomy, the incisions were left open to drain but wrapped. The wrap became hard as a rock from the fluids. Two additional surgeries were required, one to partially close the incisions and the last to complete the closure.

After discharge from the hospital, pain management continued and extensive hand therapy began. Therapy continued for six months but extensive scar tissue was forming, affecting the hand and limiting use of the middle finger. The first of what turned out to be five surgeries to remove scar tissue along the tendons and in one, replace a tendon with one from his leg, was at the end of April 2010. A Methicillin-resistant Staphylococcus aureus (MRSA) infection began about a week post-surgery. This required the placement of a peripherally inserted central catheter (PICC line) in his left arm for home administration of antibiotics. Intravenous antibiotics were also required as a prophylaxis after subsequent surgeries, for a total of about eighteen weeks with PICC line. During one of the IV placements, he had a bad reaction to the antibiotics. He said, "the room was spinning rapidly 360°." The final surgery was in February of 2011. All the surgeries were ultimately unsuccessful, leaving him with permanent hand damage. The middle finger to this day does not bend, which could cause some unique societal problems. As he says, "I worry about police seeing me approach and think I'm giving them the finger while gripping the steering wheel." He summarized his experience with this statement: "I tell this because I think it is important for people to know how serious and arduous a bite can be. Also, the cost is very high, not only emotionally and physically, but also financially. I had insurance. Without insurance, my bill alone for the ten days, three of which were in ICU, was approximately $325,000.00. This does not include the hand therapy and additional surgeries." The sixteen vials of antivenom alone were $5,000 each. It also shows how critical it is for medical facilities in snakebite-prone areas to have antivenom on hand. It is also important for the medical staff and envenomees alike to understand the urgency of receiving an adequate dosage early on, before the venom seriously damages tissues.

Category: PT-LA (bitten while feeding a pet rattlesnake in a terrarium)

Hobbyist Using Old-school First Aid: Eastern Black-tailed Rattlesnake (*C. ornatus*)

A man in his early twenties was changing water in the terrarium of his pet meter-long Eastern Black-tailed Rattlesnake. He kept his eye on the snake, which was at the other end of the terrarium. After changing the water, the lid was not securely attached, and the snake struck him in the bony part of the thumb with both fangs from the other side of the cage. Both species of black-tails are considered mild in temperament. The owner thought the snake bit him because it was expecting to be contacting food, rather than its being a defensive bite. This was some forty years ago, and the first aid of the day was to use a tourniquet, incise the wound, and use suction to remove the venom, before using antivenom. The idea is that this would remove most of the venom before it gets into the system, and slow the spread of the remaining venom. He applied the tourniquet above the bite on the arm and made four incisions. He had several suction devices on the thumb. He said "every time my hand and arm turned white, I would loosen the tourniquet and let the blood flow freely into the sink." He kept this up for some time. He said he lost about a pint and a half of blood into the sink. He had seen oozing of yellow liquid at the wound site. He swelled up to just above the wrist. He actually had some antivenom on hand and called a pharmacist and doctor to ask if they could administer it if need be (the pharmacist said he was not allowed). In addition to the local symptoms, he got an intense headache and remembers fasciculations and piloerection. He watched as the swelling increased, kept in contact with his doctor, and waited for symptoms to get worse before administering antivenom. He was aware of allergies to horse serum, so did not want to use antivenom unless need be. Since the symptoms did not escalate much, he just waited it out. He said his thumb was twice as large and described it as looking like "a rotten banana." The swelling subsided and there were no long-lasting effects.

This bite is a good example of how someone used the recommended first-aid procedure of the day. Fortunately, no permanent harm came from the experience. As a keeper of "hot" reptiles, he did have the foresight to have polyvalent antivenom on hand, but apparently that did not do him any good in this situation. When I met this individual, he still maintained a large, legal collection of rattlesnakes and Gila Monsters, some of which had been confiscated from poachers and "loaned" to him by the state Game and Fish agency, under his care.

Category: PT-LA (bitten while feeding a pet rattlesnake in a terrarium)

Payback for Probing: Southwestern Speckled Rattlesnake (*C. pyrrhus*)

A 22-year-old hobbyist captured a 76 cm Southwestern Speckled Rattlesnake and transferred it to a snake bag. He then worked its tail out of the open end of the bag and put a wire probe in the cloaca to determine the sex. The snake crawled back along its body, exited through the opening, and bit him on a finger with both fangs. At first he felt no pain, save that of a puncture, but within ten minutes he felt tingling and numbness. The effects spread and swelling became extensive. The symptoms moved onto his arm, then eventually to his chest. Over time, his arm and upper body, including the rib cage, became discolored from hemorrhaging. He vomited and became disoriented. He faded in and out of consciousness. After the bite, he was quickly rushed to a hospital (in about ten minutes). He was calm, but felt embarrassed for having been bitten. He was fairly cognizant of the medical treatment he received, but that became overridden by the effects of the bite.

He does not remember exactly what happened in the hospital. He became completely covered in hives, suggesting a possible allergic reaction. This makes sense since he was exposed to venomous snakes since the age of eleven, when he started keeping them as pets. He received thirty-five vials of CroFab and was in intensive care for seven days. He needed surgery and a skin graft to help heal the finger. Some symptoms lingered for about a year. Permanent damage was done to the bitten finger, which became twisted. The medical bill was $207,000.

Category: PT-LA (bitten by rattlesnake in a snake bag)

Sprayed in the Eyes: Southwestern Speckled Rattlesnake (*C. pyrrhus*)

In 1986, a young professional herpetologist tasked with maintaining a collection at a zoo was accidentally envenomated in the eye. The culprit was a 0.7 m Speckled Rattlesnake. He said the snake was nervous and would often strike at the cage when staff serviced the exhibit. On one particular day when he was changing the snake's water, it struck at the screen top and a drop or two of venom sprayed him in the left eye. He was wearing contacts. He immediately began to irrigate the eye with saline. He was quickly driven to the nearest hospital where they also irrigated his eye with saline for an extended period. The area of the eye was painful, swollen, and exuding pink tears, suggesting blood in the exudate. At its peak, he ranked the pain as a 5–6 out of 10 (his reference for a perfect 10 was when a heavy tool was dropped on his toe and the toenail shot out!). At the hospital he was kept under observation to determine if symptoms were escalating and becoming systemic. After two hours or so, he was released with a bandaged eye. Being a trooper, he decided to keep an appointment for dancing that evening, but as he sweated, the bandage started to come off. When he got home, he decided to have a look at the damage and said his eye "looked like raw hamburger." The pain and swelling lasted for about ten days and there was no permanent damage.

Category: OC-LA (sprayed in the eye by a misplaced strike from a captive rattlesnake)

Nicked by a Dangerous U.S. Rattlesnake: Mojave Rattlesnake (*C. scutulatus*)

In 2001, a man whose livelihood included collecting rattlesnakes for the production of antivenom was nicked by a 60–75 cm Mojave Rattlesnake in southeastern Arizona. The word "nicked" refers to the fact that a single fang scraped against his middle finger, so he did not receive a well-delivered bite. Nevertheless, his skin was pricked and venom got into the wound. He used a common first-aid treatment of the time of incising the wound area and attempting to suck the venom out with a commercially available venom extractor—this practice is not common today and almost universally advised against. However, in this case, it probably just made little difference. The immediate sensation was a stinging to the finger, but it was "not bad." He sat down and waited for EMTs to arrive. Within 10–15 minutes he started feeling tingling around his lips and became nauseous. He had an overwhelming sensation to sleep, due to falling blood pressure, and the EMTs repeatedly slapped him to keep him awake. The EMTs monitored his status while trying to dispatch a helicopter to transport him to a hospital, but this was during the monsoon and inclement weather was hampering efforts from Tucson. They were able to locate a helicopter from Safford, Arizona.

The last thing he remembered was being led on a gurney down a gravelly path to the helicopter, probably about forty minutes after the bite. He was told the EMTs had to cut off his brand new shirt and "use the paddles" to restart his heart during transport. After the incident, he had seen some of the medical records that showed a rapidly decreasing blood pressure. On one of the records an EMT had written "he is going down." He reached the hospital about one hour after the bite.

At the Safford hospital, they gave him twelve vials of Antivenin Crotalidae Polyvalent (Equine) from Wyeth (now Pfizer), the commercially available antivenom of the time. Since most populations of Mojave Rattlesnakes (including the one he was bitten by) have a notoriously potent neurotoxic venom, the serum from this antivenom included that of the Tropical Rattlesnake (*C. durissus*), a species or species complex that has an extremely virulent neurotoxin. He awoke in the hospital the next day. He was in extreme pain and they administered very high doses of morphine; he said he received "a LOT" of morphine over the next four days. His hand was very swollen and the swelling extended to the elbow, and even (somewhat) to the arm pit. The swollen area was discolored, but not to the degree of a serious bite by a strongly hemotoxic species. His EKG was monitored and considered unusual. They considered doing a bypass, but he refused. He went in on a Monday and was discharged on a Friday. He did not completely recover for several months. He said his lingering symptoms were hard to pinpoint, but he just didn't feel well and he knew it was because of the envenomation. He thought that perhaps he had a sensitivity or anaphylactic reaction to the venom, given the rapid decline in blood pressure, nausea, and loss of consciousness. He also has a history of exposure to venomous snakes, including a bite by a juvenile Cantil (*Agkistrodon bilineatus*), a highly venomous pit viper from Mexico.

One long-term effect of the bite was his renewed sense of extreme caution. He rarely collects rattlesnakes anymore, and never gets close to the business end. He was very fortunate that he did not receive a well-planted bite, or he probably would not be around to tell the story. Although this bite had classic symptoms of neurotoxic envenomation, it was also apparent that tissue-destructive enzymes contributed to his symptoms. This is why pigeonholing a snake species or population as neurotoxic versus hemotoxic is dangerous and physicians always need to treat symptoms as they arise.

Category: OC-LA (bitten by a captive snake used to make antivenom)

No Previous Bites, but Down in Minutes: Prairie Rattlesnake (*C. viridis*) with Anaphylaxis

A zookeeper who had worked with venomous snakes for seventeen years had a perfect record of not being bitten. In addition to his job, from 2001–2005 he also had a research project on the ecology of Prairie Rattlesnakes in Colorado. For his independent study, when snakes were collected, they were measured in restraining tubes and then transferred to cloth bags to determine mass with a spring scale. He had performed this task hundreds of times and had marked some five hundred snakes. However, in 2005, his perfect record was tarnished. During his independent study in the Colorado prairie, he had tubed and weighed a 1 m adult, but as he was carrying the bag he somehow managed to get bitten through the bag in the hand—even though his hand was above the perfectly tied knot. It wasn't so much a bite as it was a scrape of a single fang on the edge of his hand. On that particular day, there was a small crowd gathered to watch him, and he may have been a bit distracted or in a bit of a hurry to get done. He does remember that the individual snake was particularly lively. He felt immediate pain much akin to a bee sting. After he realized he had been bitten, he looked at his hand and noted there was "an angry red inflammation moving across my hand."

Someone in the group immediately called 911, who recommended he sit down. That was the last thing he remembered, as he became unconscious within four minutes of being bitten. He was not far from a fire station when this happened, so paramedics were no more than ten minutes out. When they arrived, they thought he was having an anaphylactic reaction from a bee or wasp sting. They were half right, but those present indicated anaphylaxis was due to a rattlesnake. It was fortuitous that one of the nearby hospital's snakebite experts showed up on the scene to check out his research project. When he arrived, he saw the paramedics present and quickly figured out what happened. He got on the phone to the ER staff, and they prepared for the arrival of a snakebite victim in anaphylaxis. The paramedics noticed his breathing had stopped. They inserted a tracheal tube to get oxygen into him. They dispatched a helicopter, which arrived a few minutes later, which transported him to the nearby hospital. In transit, he was given intravenous drugs, including high doses of Benadryl to counteract the allergic response. In about fifteen minutes he was at the hospital, "in the company of snakebite treatment experts." He drifted in and out of consciousness, so doesn't remember all the details of his care. Despite being in the care of experts, he said "I stayed at the hospital for three days, even though

I wasn't expected to live for 24 hours." Even with the seemingly superficial nature of the bite, he had swelling up to the shoulder. During his hospital stay, he received twenty vials of CroFab "and many other drugs." While he was recovering, his arm became covered in hives, so he was put on a regimen of prednisone. He was also given prophylactic antibiotics. He recovered completely, save one qualifier. As he explained it, "No debility or other physical anomaly lingered after the bite, save the psychological effect of a near-death experience." For a couple of years after the bite he was fearful of going into the field. While he occasionally still works with snakes in the wild, he is especially cautious and vigilant. He no longer works with venomous snakes at the zoo.

In order to explain his hyper-reaction to the snake bite, it was suggested that his long-term battle with allergies and asthma made him at risk of an onslaught by his immune system, but it seems more likely that it was his long-term exposure to dried snake venom that sensitized him. He was around the many different species of snakes at the zoo, so undoubtedly inhaled lots of aerosolized venom, but he was also exposed to Prairie Rattlesnake venom (e.g., on the tubes and bags) while processing hundreds of snakes. He has been stung by bees and wasps, but has never had an allergic reaction to those venoms, so the "hyperactive immune system" explanation doesn't seem as likely.

Although some of the stories in this section suggest less-than-ideal situation management by victims, nearby companions, or medical professionals, in this case, everything was well-orchestrated, from the 911 call, to the emergency response team, to the medivac, and ultimately to the treatment by snake-bite specialists. In this case, our victim was in grave danger and the outcome could have been fatal if it were not so well orchestrated. After this experience, the hospital staff modified their snakebite protocol to include potential complications from allergy and anaphylaxis.

Category: OC-LA (anaphylactic reaction from a rattlesnake following a bite through a snake bag)

Too Small for the Tube: New Mexico Ridge-nosed Rattlesnake (*C. willardi obscurus*), with Serum Sickness

On July 15, 1995, a herpetologist was leading a crew of 4–5 people for an ecological study of a montane snake community in the Animas Mountains of New Mexico. It had been a dry summer, but on the day before, the first significant monsoon rains arrived. It brought out the snakes, and his crew collected four New Mexico Ridge-nosed Rattlesnakes and two Banded Rock Rattlesnakes. The team was tasked with marking some snakes and implanting radio transmitters in others. At mid-day, the herpetologist started processing his booty by "tubing" a neonate *C. w. obscurus* with his smallest tube in preparation to mark the snake and get a blood sample. Although tubing is among the safest ways to handle a venomous snake, there was apparently enough room for this tiny snake to turn around in the tube and bite his left pinky finger twice. He noticed three puncture marks. He immediately knew it was not a dry bite, as his finger started stinging and burning. Being well versed in post-bite procedure, he took off his ring in anticipation of swelling and starting marking the progression of the swelling with a waterproof marker. He also knew that as rattlesnakes go, this species has a small venom yield (especially a juvenile) and a "weak" venom. Rather than aborting the mission and leaving his hard-earned specimens unprocessed, he decided to proceed with implanting the larger snakes with radio transmitters while monitoring his symptoms. As he was working, he noticed pain was intensifying and he was having difficulty with manual dexterity, so decided to seek medical attention. It was about a 330 m drop and 1.5 km back to the cars, so they began hiking back. He elevated his arm but noted that "even slight pressure from a glancing twig or leaf would cause severe pain." He ranked the baseline pain after the start of swelling as only a 2–3 on a 1–10 scale, but when something contacted his skin, such as a twig, it shot up to a 6–7. He was driven to the nearest medical clinic, arriving about twilight.

There was a physician's assistant on duty at the clinic who was concerned by the symptoms and recommended a fasciotomy. He was alarmed by this notion, as capillary blood return to his fingers was still rapid and there was no discoloration of the bite site. He refused the treatment, but was put on an IV drip. The PA insisted on having him flown to a hospital in Tucson, rather than being driven. There were a couple failed attempts to get him in the air because of the storms, so he was instead shuttled by ambulance to Tucson, where he arrived in the wee hours of the morning. By this time the swelling had peaked between the shoulder and the elbow. He added, "I started getting periodic spasms of shooting pains in my hand/arm and I'd rate those at a 6." He was met at the Tucson hospital by "a toxicologist, ER attending physician, and two residents who took one look at the annotated arm and asked, 'herpetologist?'" Thinking his bite was not severe, and because he had a previous experience with serum sickness, he declined antivenom treatment. However, the ER staff was insistent and told him to sign a waiver that he was refusing to follow their advice. He gave in. They administered morphine, which caused him to drift off. When he awoke he recalls being told he had received thirty vials of antivenom, which he stated "was grossly excessive." His symptoms were completely resolved in about forty-eight hours post-exposure. The next afternoon he was released.

A few days later, he decided to go back to work on his study. Shortly after arriving, he started to develop delayed symptoms of serum sickness. Hives started to develop on his buttocks, then spread to his back, and eventually his entire body. He said "the itching was deep and intense. I had blurred vision at times, hot and cold sweats, diarrhea and vomiting (sometimes simultaneously), and mild delirium." He sought medical attention again, and he was given Benadryl, which did not seem to alleviate symptoms—in fact they increased when he received the antihistamine, but that seems likely due to escalating symptoms. Three days after they started, the symptoms disappeared and he finally returned to the field site.

One good thing that came out of the experience was that he rewrote his safety protocols to better avoid accidents and to have more efficient treatment should accidents occur.

Category: OC-LA (bitten by rattlesnake that turned around in handling tube)

It Wasn't a Lizard: Arizona Ridge-nosed Rattlesnake (*C. w. willardi*)

July 5, 1981. A snake aficionado from Phoenix who had never seen or photographed a Ridge-nosed Rattlesnake in his own state figured he might have better luck if he traveled south a few miles into Sonora. After a few days of exploring the vicinity of Cananea, he was plumb tuckered out and ready to take a nap at his campsite. However, before his nap, he noticed a small wood-and-tin structure leaning against some rocks near a stream and decided to take a look. He noticed a piece of cardboard in the structure that was begging to be looked under. When he investigated it, he saw a Madrean Alligator Lizard dive under the oak leaves, so he plunged his hand into the leaves to capture it . . . but of course, it wasn't a lizard. He was bitten on the middle segment of his left index finger by a 25 cm *C. w. willardi*. His finger started to swell immediately. It quickly spread to his other fingers and his hand. As he explained it, "My buddy had trouble driving a standard transmission so I had been doing all of the driving and was now forced to attempt to drive that

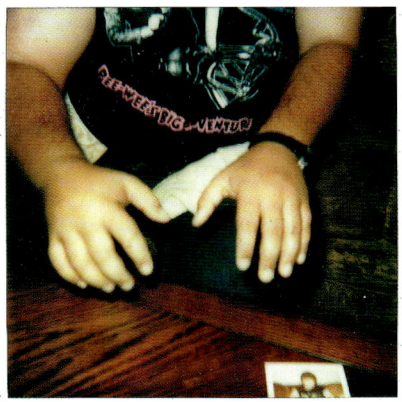

Bite injury from the Arizona Ridge-nosed Rattlesnake. *Courtesy of Kerry Crowther*

winding highway all the way back to Nogales [Arizona], snake-bit, and going on 48 hours with no sleep." He said the pain was annoying, but bearable. He safely arrived at a hospital in Nogales, where he was given antivenom 3–4 hours post-bite. They kept him under observation that night, then released him the next day. He drove himself back to Phoenix. At its peak, the swelling extended to his elbow. After two weeks the swelling was gone, and by the third week he was back to normal.

Category: RE-LA (grabbed a rattlesnake thought to be a lizard under a cover object)

Fast-forward five years to July 5, 1986. Our snake aficionado went back to Sonora once again to look for reptiles to photograph. He added, "We parked in exactly the same spot and I showed my buddy where I had been bitten before heading up." They walked up the canyon until it came to a fork, so the two split up. They agreed on a time to meet back at the car, so they could rest up and head back home. He headed to a nearby rock pile and found a large adult *C. w. willardi* in front of the first rock he came to. The snake was warm and active, but he managed to get it to settle into a nice pose using a stick and took some photographs. Then, as he put it, "I got closer and took a few more. Then I got closer. The snake was holding quite still, so I got closer . . ." But then it started to saunter off, so he used the camera lens to keep it from leaving, and it struck him directly in the middle segment of his right index finger. The bite could have just as easily been to the forehead. In his own words, "SON OF A BITCH did I feel STUPID!" He was

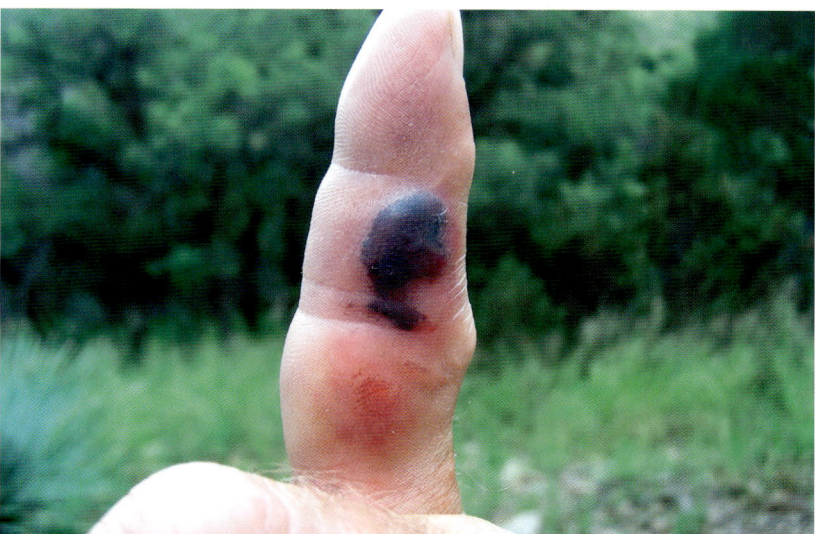

Local effects of a bite by Arizona Ridge-nosed Rattlesnake. *Kerry Crowther*

three-quarters of a mile from the car and walked back slowly, so as to not exert himself. He arrived forty-five minutes before the agreed-upon time, so blared on the horn in an S-O-S pattern to alert his friend. An hour after the meeting time elapsed, he was getting worried because his symptoms were getting more serious: the pain was getting severe, the swelling had extended to below his shoulder, and he was getting nauseous and extremely light-headed. He considered leaving his buddy and driving by himself to a hospital. He pinned a note to a tree explaining what had happened in case he lost consciousness. A young nurse from Cananea happened to be on a picnic that day and when she arrived saw the victim was in distress and tried to convince him to go to a hospital in town, but he declined. His finger became discolored and blisters were developing. After five hours his friend finally came strolling back to camp. He wrapped his arm in a sleeping bag to buffer it from jarring on the road, and they headed home. At this point, since he was about a half-day into the bite, he figured there was no longer a benefit for medical attention (probably too late for antivenom to be effective), so he decided to ride it out. The swelling reached his neck and there were fluid-filled blisters on his fingers, hand, and arm. The swelling subsided within three weeks. It was about six weeks before he completely recovered, but there was local permanent damage. He has arthritis in the finger and joint pain when he tries to close his hand. His final statement about this latest experience: "Now . . . if I could just choose better, more trustworthy herping buddies . . ."

Category: RE-LA (bitten by rattlesnake when trying to photograph it)

The Hat Trick: Multiple Rattlesnake Bites over Time (*C. cerastes*, *C. lepidus*, and *C. willardi*)

A cabinetmaker from southern California, now living in southern Arizona, has the unusual pastime of documenting snakes and their progeny over time. As a hobbyist with a penchant for venomous snakes, and in the course of his field projects, he has sustained fifteen rattlesnake bites from four different species. His bite history started in 1969 with a captive neonate Sidewinder, about 17 cm long. He was trying to force-feed it when it wriggled around and sunk a fang into his finger. He described the pain as a hot-burning sensation, probably a 5 on a 0–10 scale (he gives a European Honey Bee a 4). Being unsure what to expect, he went to a San Diego area hospital. He told the attending physician what had bitten him and the doctor promptly sent him home. The bite self-resolved in about three days. He received two other bites from Sidewinders, another neonate and one about 30 cm in length. There was some swelling, but no discoloration from either of these bites and they resolved in less than a week. He said the Sidewinder bites weren't that bad, just a bit painful and uncomfortable.

Next, enter the small mountain rattlesnakes from southeastern Arizona. He was first bitten by a large (60 cm) Banded Rock Rattlesnake he had in captivity, as he was "hand training it to make it docile." The serpent didn't take well to handling and it bit him three times in succession. The fangs met their marks on his fingers three times, while the other three hit the nail. The bites were more significant than that of his previous experience with Sidewinders. He described the pain as being deep-seated down to his bones. He said it felt like having his hand "repeatedly being hit hard by a hammer." The pain continued to grow over time and ultimately reached a 10 on a 0–10 scale. He said it was pure agony when the hand was touched. He tried to determine what a "10" might feel like, then explained he had never given birth, although he had broken his wrist, and had some other painful experiences, but this was the worst, and the pain was relentless. Swelling reached his shoulder and he could not move his digits. At first aspirins helped, and he dared not put his hand below heart level because it brought about intense throbbing pain. This time, the pain was so great he sought medical attention, although he only wanted to receive stronger pain killers. However, he ended up being taken on a gurney and was taken to the intensive care unit. They gave him five vials of the crotalid polyvalent from Wyeth, the antivenom available at the time. After being administered antivenom, he said, "the episode became surreal and I experienced terror." They reduced the rate of antivenom administration, thinking he might be experiencing the mental anguish from receiving antivenom, and his mental state gradually improved. They also gave him Benadryl and prednisone. He was moved from the ICU, but remained under observation another night at the hospital. He was released about forty-eight hours after being admitted. On the third and fourth days he once again experienced mental anguish, which he described as a "deep depression." In two weeks he was back to work, and fully recovered some time after that.

As if this experience weren't enough, he was subsequently bitten nine more times by Banded Rock Rattlesnakes, although the snakes were smaller and none of the bites as serious. These bites were received while he was conducting an independent study. These bites came from his handling animals in the field. His method was to "tail" them with one hand and use a hat in the other to keep the business end away, and ultimately calm the snake so it could be photographed and marked. Most of the bites actually came when he was marking the animals with spray paint, since he was preoccupied with that task. Although these bites were less severe, he did experience the same feeling of having his hand "smashed by a hammer." The pain was less intense, however, probably a 6 on a 0–10 scale for the neonates. He explained that bites by neonates became swollen to the wrist, while larger animals caused swelling on the forearm, and in one case (a 46 cm male), to the shoulder. This particular bite also had additional symptoms of blisters, prickly sensation on various parts of the body, a roaring sound in the ears, and a buildup of fluids in his side. He also had permanent damage to the bitten finger, which became contorted. In all of these instances the bites were painful, but did not affect his mental state.

To wrap up, he was also bitten by two Arizona Ridge-nosed Rattlesnakes, one neonate and one 30 cm individual, while employing the "tail and hat" method. The pain was actually less than that of a bee sting, registering only a 2–4 (depending on snake). He did get some blebs at the bite site from the larger specimen. He recovered completely in a few days from both bites. But then, there is the report of the envenomation by the fourth species, the Red Diamond Rattlesnake, but that is a whole new story.

Category: RE-CA (bitten by rattlesnakes while using unsafe handling practices)

More on the Hat Trick: Red Diamond Rattlesnake (*C. ruber*) with Secondary Infection

Our friend the cabinetmaker (see "The Hat Trick" Envenomation Story, above) did not limit his independent study to small rattlesnakes, and on one occasion, he was tagged by a Red Diamond Rattlesnake from the west side of the Coast Range of southern California. His study area was on a boulder- and talus-strewn hillside, where the rocks were covered by a deep tangle of vines, so not the easiest area to work in. Red Diamond Rattlesnakes are known for their docile temperament and relatively mild venom, but there are large quantities of it, and they lose their good nature when being handled. In March of 2011, he was tailing a 122 cm male, which was attempting to seek cover in the rocks. When he got it out, he used a snake hook to keep the business end at bay, but snakes are strong and can move their body forward on the hook, giving the neck and head more striking distance. To keep the head at a safe distance, he kept repositioning the snake, but at one point it got too much elbow room and tagged him on the hook-holding hand, on the middle finger. He dropped the snake, which then went into the rocks with a female Red Diamond. There was immediate pain from the bite. Although he was feeling a little agitated by the affair, he continued to handle a couple more snakes, but the pain increased as his hand was getting swollen, so 2–3 hours post-bite, he left. In fact, the pain increased for the next twelve hours and the swelling increased over the next twenty-four hours. Pain was constant for the next 2–3 days. His hand became discolored. This bite left him psychologically affected and for the next two weeks he had no desire whatsoever to mess with venomous animals.

Then things took a turn for the worse, as symptoms of envenomation segued into symptoms of infection. Blebs had formed on his finger and these were weeping clear fluid, which continued for weeks. The wound wasn't healing, so he finally sought medical attention on Day 14. He was given two types of antibiotics and was instructed to clean the wound twice daily and rewrap it each time with new wet-dry gauze. New pink skin was forming under the old sloughing skin, so things were looking up, and he was given a flap graft. The surgery seemed to work, and he began a regimen of exercises to get movement back in his hand and fingers. However, blisters subsequently formed and he was informed he had yet another bacterial infection. Despite five weeks of three types of antibiotics, the tissue damage was just too great, and the finger could not be saved. Eventually, everything healed, and he is still studying his snakes, but primarily with a video camera rather than fingers.

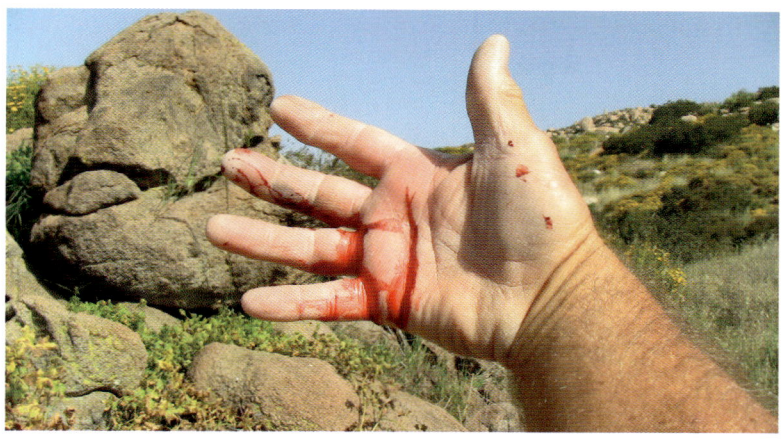
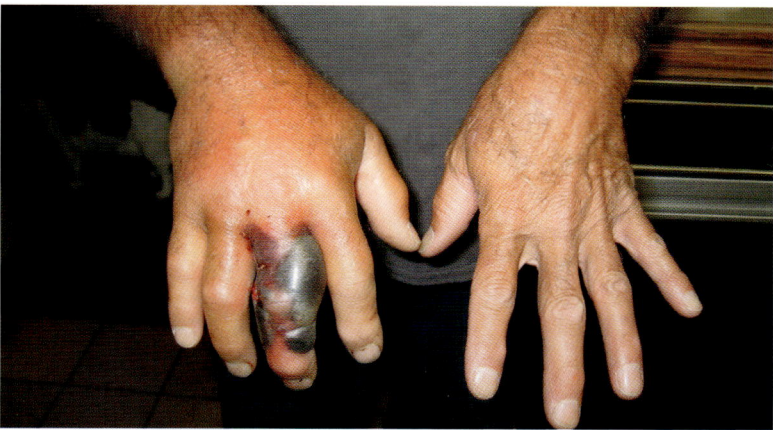
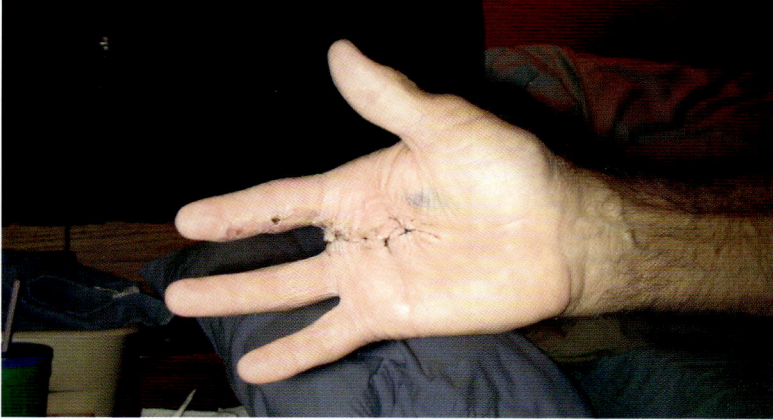

(Top) Day one of bite by Red Diamond Rattlesnake. (Middle) Day five. (Bottom) Day ninety-five. *Courtesy of John Porter*

There is an old saying among biologists, "How do you tell if someone at a conference is a herpetologist?" The answer is, "he or she is the one missing a finger."

Category: RE-CA (bitten by rattlesnakes while using unsafe handling practices)

Small Snake, Big Bite: Pygmy Rattlesnake (*Sistrurus miliarius*)

In 1995, a university professor in biology and herpetology in South Carolina was conducting road-cruising research on snakes of the Carolina Sandhills National Wildlife Refuge. After capturing snakes, he brought them back to the lab to be weighed and measured. Pygmy Rattlesnakes were the most common species encountered at night, so he frequently had to handle them to be measured. To determine length, he held them behind the head with one hand, and stretched them out with the other. On one afternoon, his hand slipped and the serpent managed to sink two fangs on the underside of his thumb.

The effect was immediate, marked by intense pain, which he rated a 9 out of 10. He described it like being "smacked with a hammer really hard." He immediately headed for the ER, arriving in about twenty minutes. The attending physicians decided not to use antivenom but did administer a morphine IV drip. The herpetologist noted that the morphine worked very well to alleviate pain, and at times he almost forgot he had been bitten. The finger and hand started to swell soon after he was tagged, and eventually swelling extended up the arm. That night he said his hand and arm to the elbow were "about as swollen as they could get without bursting." The thumb was black, blue, and throbbing. He noted he could see no lines in his hand, and his palm was convex. By the next morning the swelling had included his entire arm and even part of his upper torso. His friends gave him a new nickname: "Popeye." He ended up spending three nights in the ICU. He couldn't bend his arm for a week. About two weeks later the swelling subsided, although the underside of his thumb remained "squishy," suggesting the tissue never really healed completely. He also has periodic pain at the base of the bitten thumb.

Category: OC-LA (bitten while pinning a rattlesnake behind the head)

Systemic Immediately: Harlequin Coralsnake (*Micrurus fulvius*)

In rural north Florida, a 7-year-old boy mistook a 50–60 cm Harlequin Coralsnake for a Scarlet Snake, which he proceeded to catch. When he retrieved it from the net, his mother said "the snake was faster than Google." It bit him twice on the thick of the hand above the wrist. His mother immediately loaded him in the car, and they sped off to the nearest ER, where they arrived about 1.5 hours later. A snakebite specialist met him at the hospital and became the attending physician. The bite was a life-threatening emergency, being described as "immediately systemic." The physician said the boy "fainted and was retching and vomiting, plus his eyelids were drooping and they had to physically close his jaw." By hour 9, the victim had central paralysis. The antivenom that had been at the hospital had been sent to another, and there was an unfortunate delay in receiving more, so it was not administered until eight hours post-exposure. He was given more than five vials, rather than the usual three, partially because of the severity of the bite, but also because of the delay in receiving antivenom. The patient had a slow recovery. After several months he complained of falling down while running, due to his double-vision caused by nerve damage. Fortunately, because the victim was young, his body eventually healed and his vision came back so that he was able to run normally again. The patient and physician stayed in contact long after the bite, talking on the phone every couple of weeks to check progress. This 7-year-old kid was a real trooper, and thankfully, not only has he completely recovered, but he went on to receive a herpetological and medical education.

Category: IN-NA (picked up a coralsnake, thinking it was a harmless species)

Treating Fifty Bites in Five Years: Texas Coralsnake (*M. tener*)

Dr. Spencer Greene is a medical toxicologist who specializes in snakebites. He started his career in Phoenix, then Tucson, and finally Houston, at the Baylor College of Medicine. In Arizona, he encountered patients with envenomation by the Arizona Coralsnake, and added "they were nothing, and patients were not hospitalized." However, the bite of the Texas and Harlequin Coralsnakes are much worse. In Houston, he has treated about fifty coralsnake bites in about five years. They ranged from mild to (rarely) severe.

One thing that stood out about all envenomations, however, was the "horrible pain accompanying a bite." Pain usually lasts 12–24 hours. Patients invariably "needed to be given high doses of opioids in the first 24 hours to help ameliorate pain." The pain often involved the whole extremity. Other symptoms often included slight redness and swelling, nausea, headache, and sometimes feeling of electric shock. Although there are documented cases of skeletal muscle paralysis, he had only witnessed droopy eyelids, double vision, and difficulty swallowing as systemic symptoms in their most extreme cases. Although most cases were resolved within twenty-four hours, some patients experienced tingling for a week or more. He only used antivenom once, as their facility doesn't "use antivenom just for pain control when it comes to coralsnakes." Patients are usually under observation for at least eighteen hours, and longer if symptoms persist.

When asked about any specific cases that stood out, he noted there were many, but one recent example from 2017 was when an 11-year-old boy was bitten on the finger by *M. tener*. His friends told him the snake was a Scarlet Kingsnake, so he picked it up. The excruciating pain indicated otherwise. Dr. Greene said this was not an unusual type of encounter, as he noted that only two of his fifty patients were bitten completely by accident: one stepped on the snake, and the other while picking up the keys he dropped. The rest were from intentional handling. About half of the people handling the snakes did not believe or know them to be venomous, while the other half knew they were handling highly venomous serpents. He said this is contrary to exposures by pit vipers that he and his colleagues treated, where most were unintentional.

Category: Not Applicable (a compilation of exposure causes)

They Never Bite! Ring-necked Snake (*Diadophis punctatus*)

One researcher distinctly remembers a surprising envenomation that occurred some thirty-five years earlier because it was so uncharacteristic for such a docile species to bite. And the bite was surprisingly painful. He was looking for animals under cover objects in Oak Mountain State Park, Alabama, when he saw a huge Ring-necked Snake trying to dart into a crevice. Fearing it would escape, he quickly grabbed it mid-body . . . and then it quickly grabbed him near the index finger. After all these years, he could not recall how long the animal hung on, but when he felt the stinging pain, he pulled the snake's head off his hand. He had never been bitten by any of the other hundreds of other ring-necks he had handled over the years, so this was unexpected. He said, "The bite was accompanied immediately by a sharp stinging sensation with pain approximating that of a paper wasp (*Polistes* spp.) from that region." He added that the pain also lasted longer than that of a paper wasp. These many years later, he recalls the intense pain lasted about an hour, but did not completely go away for some time. There was redness of the bite site, but no swelling of the hand or forearm. He adds that he is now more cautious when handling large ring-necks and holds them behind the head to avoid being bitten.

Ring-necked Snakes occur over most of the country and about a dozen subspecies are currently recognized, although it seems likely that it is a species complex that has yet to be completely understood. As currently recognized, the taxon occurring in most of Alabama is *D. p. stictogenys* (Mississippi Ring-necked Snake).

Category: OC-CA (bitten by a snake while free-handling)

Didn't Know It Was Venomous: "Western" Hog-nosed Snake (*Heterodon* sp.)

In 2000, a woman who worked in a pet shop was feeding a "Western" Hog-nosed Snake (a former taxonomic designation, often still used in the pet trade, but could be one of three western species) when the captive-bred snake bit her. She did not know these rear-fanged snakes were venomous, and not wanting to hurt the snake, did not try to extricate it from her hand immediately. It hung on "for an extended period," although that length of time is unclear, as it was not reported to the physician who attended her. However, her husband was very good at keeping records and clearly documented her symptoms in a series of notes, which are transcribed here. The husband also took photos, shown here by permission of the physician, who monitored her condition for several days.

She was bitten on Day 1 (Friday) on the right hand at about 14:00–14:30, on the web between the little and ring fingers. Within ten seconds the swelling started, first in the hand, and within an hour had spread to the forearm. The hand and arm were very painful and the arm was extremely itchy. The affected area had become red from inflammation. Within two hours she had a headache and was nauseous. She became dizzy. She took Tylenol for pain. Her husband noted "she did not feel fingers by the end of the day." On Day 2 the swelling and inflammation spread to her upper arm. The

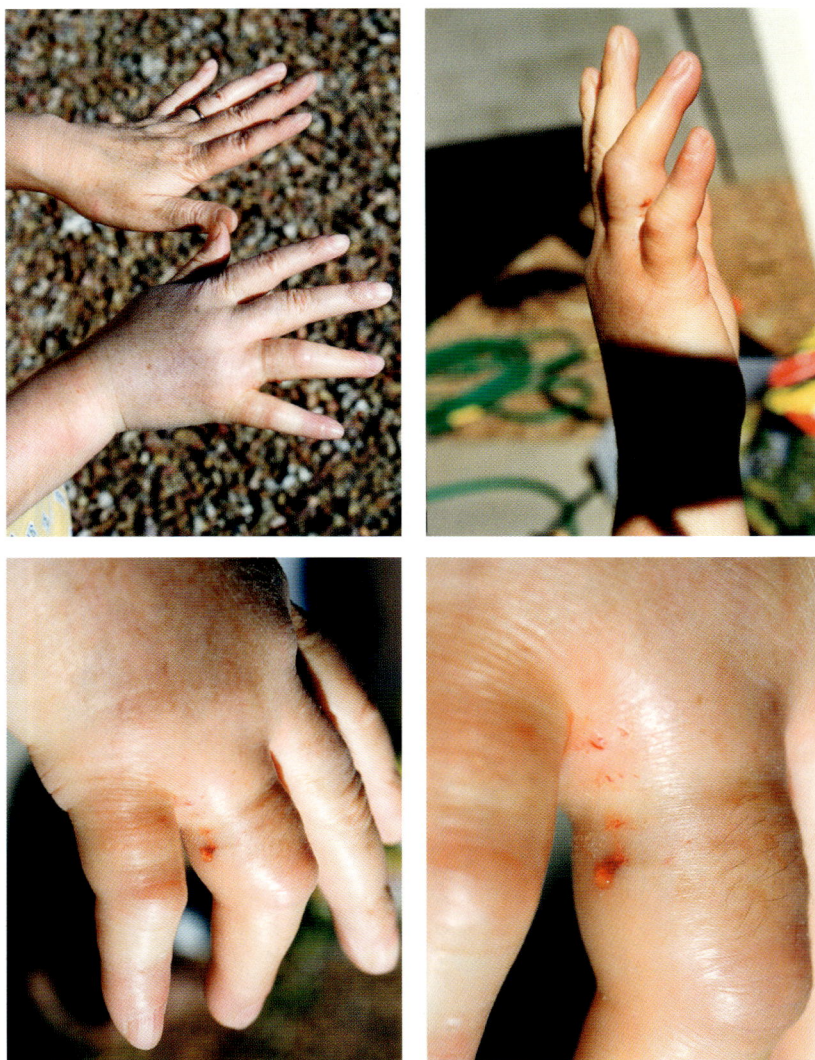

Here and next page: Lesions caused by a "western" Hog-nosed Snake. *Photos courtesy of the attending physician, Mark Wolfson.*

pain had become even more intense and the nausea and dizziness continued. She still could not feel her fingers. On Day 3 the swelling increased to the shoulder. The area became covered in small yellowish blisters. Again there was headache, nausea, and dizziness, plus if the area was touched, it felt like "sharp spines" were being jabbed into her flesh. On Day 4, the swelling started to subside at the two fingers. The arm was still very red and she still had the sys-

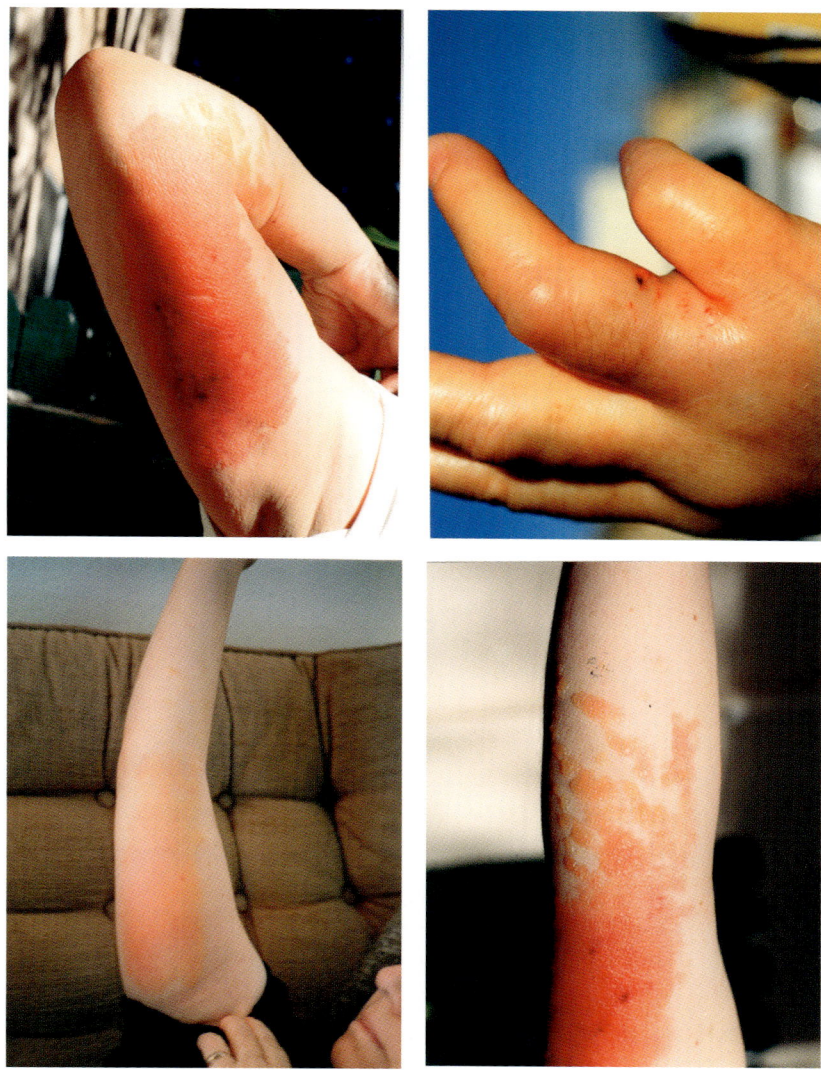

temic symptoms, including vomiting. There were thick orange blisters on the arm. By Day 5 swelling receded, but some new blisters formed on the upper arm. While the pain, nausea, and headache remained, the dizziness started to subside. However, she developed a deep pain "to her bones." The notes said, "Bones felt hot from finger to upper arm." On Day 6 all symptoms began to diminish, but her arm became "deep red" and there were still blisters. Then on Day 7 some symptoms subsided while others worsened. The notes state, "Upper arm all intense red with deep red areas, colors very intense about 10 inches long on upper arm and 3–4 inches wide. Pain in bones more intense; nausea and dizziness were worse. Forearm is yellow and less swollen, almost normal." On Day 8 the husband wrote, "healing was well under way." She was back to normal in four to five weeks.

Category: PT-NA (bitten by pet snake in terrarium)

A Little Experiment: "Western" Hog-nosed Snake (*Heterodon* spp.)

This report involves a 30–40 cm (snout–vent length) captive "western" hog-nosed snake, a species that has since been split into three distinct species. A professional herpetologist was showing the snake to two colleagues when it started to hiss and strike. He assumed the behavior was a bluff, typical of hog-nosed snakes, until the animal bit him on the left hand just below the thumb and started chewing. One of the fellow herpetologists with him laughingly asked if it hurt, to which he replied, "It did." In fact, he stated that "it was the most painful bite I have experienced from a snake (and I have been bitten by some larger colubrids and even a meter-long *Boiga dendrophila* [another rear-fanged colubrid, commonly known as the Mangrove Snake, which can deliver a decidedly nasty venomous bite])." Expecting the snake would eventually realize he was not food and the bite would not result in a serious envenomation, he allowed the animal to chew on his hand. To his surprise, the chewing continued for about five minutes. Then he tried to encourage the snake to release its grip by immersing it in cold running water, but the snake only continued to chew, biting even harder. He eventually removed it by prying its mouth off with a blunt probe. During this drawn-out removal process, the animal kept biting and penetrating deeper for approximately ten minutes. Each time the snake chewed harder, he yowled in pain.

He said the pain started out as a 3 on a 0–10 scale, but became a 7 when the snake bit harder as he was trying to get it off. There was bleeding commensurate with a normal puncture wound, but it stopped soon after the snake was removed. The pain lasted less than two hours, but by then the swelling had started, first in his hand. After about two days, the swelling reached his elbow. During the peak, his hand was so swollen that he could not touch his fingers to his thumb, but there was no pain whatsoever. After Day 4, the swelling began to subside, and there were no long-lasting effects or permanent damage.

This is one of those cases where it was fortunate to have a scientist on the receiving end of the fangs to record the envenomation and symptoms of the bite of a poorly known (in terms of envenomation) species.

Category: OC-CA (bitten while free-handling a captive snake)

Eye of the Cat: Northern Cat-eyed Snake (*Leptodeira septentrionalis*)

Although the Northern Cat-eyed Snake ranges from south Texas to South America, it is rare in the United States, and there is little information on the effects of envenomation by this rear-fanged species. The species is considered monotypic, so the account here from Ecuador adds to the only other known human envenomation, which happened to a zoo worker.

Although *L. septentrionalis* is considered quite docile, a herpetologist conducting an inventory of reptiles and amphibians in Ecuador in 2007 was bitten. The snake had been showing no signs of aggression or stress, so it was being loosely held when collected. After a few minutes of handling, the snake bit him between the pinkie and ring finger. He did not try to pry it off, but after about five minutes, he tried to coax it off his hand by dabbing ethanol around its mouth. There was no immediate reaction or pain, but the following day he said his hand "swelled from about my middle finger down about 4–5 cm down the pinkie-side of my palm. I had no internal hemorrhaging, just swelling to the point that I couldn't move my middle finger, ring finger, or pinkie much." The pain only reached about a 3–4 on a scale of 10, about the same as a bee sting. Swelling lasted for about two days and a dull pain lingered for a few days more.

Category: OC-CA (bitten while free-hand collecting snakes)

But They're Harmless! Gophersnake (*Pituophis catenifer*)

When a future restoration ecologist moved to southern Arizona in 1983, he was a kid in a candy store. He had an appreciation for all living things and the incredible diversity of the area and was fascinated by venomous animals. He went on to write a book on venomous animals of the area and worked in a poison control center. Being a snake aficionado, he did not hesitate to pick up and handle wild Gophersnakes or other "harmless" snakes. Over the years he claims to have handled hundreds of Gophersnakes, which are often the most commonly encountered snakes within their range. Gophersnakes have a variable temperament, ranging from docile to nippy, but usually closer to the former. On one occasion, when he first arrived in Arizona, his experience with a Gophersnake was a little different.

As he had done so many times, he picked up a 1.3 m Gophersnake around its midsection. This particular snake was not one of the docile ones, and actually gave him a solid bite on the middle part of the forearm. Unlike a typical Gophersnake bite, it was not a strike-and-release. It held on for "a little while." He was surprised by its behavior, but did not want to rip it off his arm, so as to not harm the snake. He cannot remember just how long it hung on, but it wasn't an extended period. He didn't think much of it at first, but was again surprised by what happened to his arm over the next two days. The entire forearm continued to slowly swell up and became "slightly discolored." The area of the bite was rather painful, but nothing to write home about. By Day 3 the swelling had subsided.

As snake bites go, this was not bad, but for one of the most common snakes invariably dubbed "harmless," this was an eye opener. Gophersnakes have only rarely been documented as causing venomous reactions, and they do not even have rear fangs. Gophersnakes are superficially similar to rattlesnakes, and sometimes they even rattle their rattle-less tail. This sometimes concerns people who find them in their yards, and thousands of Gophersnakes are removed from yards each year. However, I certainly don't want to cause alarm about Gophersnakes (which include the subspecies known as the Bullsnake) and their close relatives the pinesnakes. They almost never cause harm to humans, and bites are never serious, but you may want to think twice about allowing a biting Gophersnake to hang on if it decides to defend itself. These animals are a valuable part of the environment because they do humankind a great favor by being terribly effective natural rodenticides.

Category: RE-NA (bitten while free-handling snake generally considered harmless)

Envenomation Stories: What Is It Like to Be Bitten or Stung? 819

The Thanks She Got For Saving Its Life: Terrestrial Gartersnake (*Thamnophis elegans*)

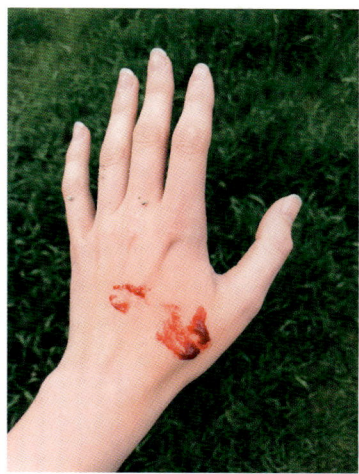
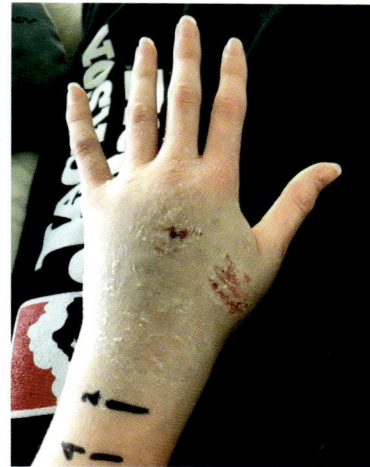
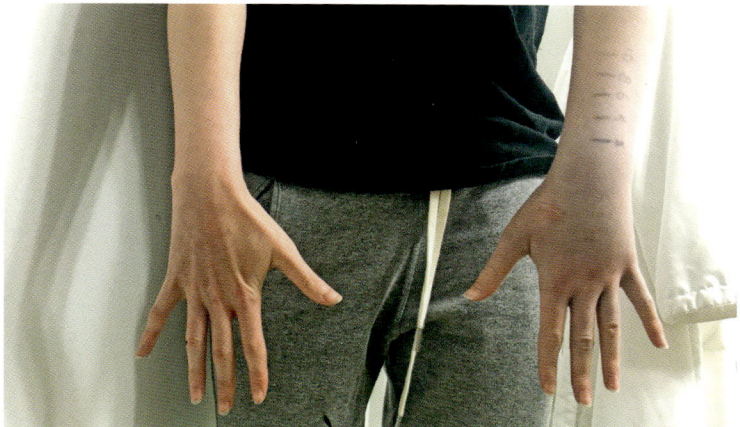

(Top Left) The injury soon after the bite. (Bottom) The hand and arm swelled considerably. Note the markings denoting advancement of swelling. (Top Right) The hand and arm turned gray and became swollen before skin began sloughing. Complete resolution took several weeks. *Courtesy of Savannah Burgess*

In May 2018, a healthy 22-year-old wildlife photographer found a large Wandering Gartersnake (*T. e. vagrans*) on the road near her home in Wyoming. As she picked it up to move it off the road, it "latched on with a strong grip." She had never heard that gartersnakes were venomous, so she didn't bother to disengage its grip while she walked it over to her house. After about a minute she pried it off her hand and released it into her yard. She noted that she "immediately felt a strong tingling and burning sensation . . . which began to spread quickly." Within thirty minutes, the area of the bite began to swell. It was itchy and the whole hand burned, so she applied Benadryl cream. About three hours after the bite her entire hand was swollen and "filled with cold fluid." It turned blue from hemorrhaging, and nerve pain began in the arm. She couldn't close her hand or move her fingers well. At about 4–5 hours post-bite, her hand was very cold, discolored, and numb; she lost the use of her hand and fingers. Using a marker, she tracked the swelling, which spread up the arm about an inch every two hours. The nerve pain reached the elbow. She took some Ibuprofen to alleviate pain.

The next day she noted "the nerve pain in my whole arm was enough to make it hard to sleep, but still not unbearable. When I woke up the swelling and edema had spread all the way down to my elbow during the night. My skin was also dry and flaky." At its peak, the pain was 7 out of 10. Then the itching and burning disappeared, but her whole arm was sore. She applied more Benadryl lotion and scrubbed off her flaky skin. By that evening the arm was still swollen, nerve pain persisted, and her hand was tender. On Day 3, it was still tender, but the swelling was beginning to subside, and the discoloration became greenish, as the bruising healed. She started regaining use of her hand. By Day 12 she was mostly healed, save some scarring at the bite site, but she still has some lingering tenderness and discoloration several weeks post-bite.

Category: IN-NA (bitten while free-handling snake thought to be harmless)

Lyre, Lyre, Thumb on Fire: California Lyresnake (*Trimorphodon lyrophanes*)

In June 1991, a recreational snake enthusiast and budding biologist was out road cruising in the vicinity of San Gorgonio Pass, California. He was thrilled to come across one of his favorite snakes—a California Lyresnake—a rear-fanged colubrid found primarily in that state. As any snake aficionado would do, he safely pulled over, leaped out of his car, and grabbed the yearling snake. To his surprise, the snake didn't seem to like the idea, so bit him on the left thumb, below the knuckle. Also to his surprise, the snake didn't let go. He had many years of experience with the species, having collected them in the past and having two in his personal collection. He had never

A California Lyresnake, similar to the one that bit the victim. *M. D. Wilson*

experienced this defensive behavior, although they are good at bluffing, which includes tail rattling and a characteristic defensive display. The snake quickly worked its fangs into position, which he distinctly felt. It was not inordinately painful, as he ranked it a 3 of 10 on the pain scale. He knew that these animals were one of the rear-fanged, mildly venomous snakes, and at this point, he became curious to know what the effects would be (hence, he became a passive tester). The snake continued to chew on him for about three minutes. He then extricated the snake from his thumb when he felt pins and needles and tingling in his thumb. There were two small puncture wounds that bled slightly for about fifteen minutes. The tingling sensation lasted about thirty minutes. Then he experienced a localized burning sensation that lasted about an hour, which he ranked as a 5 on the pain scale. The snake had earned its freedom, and he released it where he found it. He had no other symptoms that evening, so headed home and went to bed.

When he awoke the next morning, symptoms had not dissipated. As he put it, "The skin immediately surrounding the two puncture marks had turned a dark purple, almost black color, while my entire thumb, from the very tip to where the heel of the palm meets the wrist was a light blue color (cyanotic) and slightly swollen." The two joints of his thumb were stiff and the thumb was tender, which he ranked a 4 on the pain scale. Biologists often make for good victims, as they may keep a record or track the progress of symptoms. Discoloration of his thumb lasted about seven days, while the bite area was discolored for about fourteen days. Tenderness lasted three days, swelling lasted seven days, and joint stiffness lasted thirty days. He had no permanent damage from the bite.

Category: RE-CA (bitten while free-handling snake)

Aquatic Invertebrates

Tiny Octopus, Big Bite: Probably Fitch's Octopus (*Octopus fitchi*)

In May 1995, a 29-year-old woman found an octopus among clams in a net bag in the northern Gulf of California. She was unaware that octopuses are venomous. She described it as being uniformly tan, having a mantle the size of a ping-pong ball, and an arm span of about 10 cm. She picked up the octopus and placed it on the flat palm of her hand. The animal then moved onto her fingers, constricted them, and bit her. She said the pain was immediate and extreme, ranking a 10 out of a possible 10. She described it as being, literally, a "red-blinding pain;" everything turned red and she temporarily lost her vision. While she screamed from the pain, a friend with her removed the octopus within a minute, but she had no recollection of this. The next thing she remembered was that her vision returned and her middle finger was bleeding. She took four Ibuprofens and drank a beer. Once her vision returned, the pain lessened to a 7, and later to a 5, after the drugs took effect. When she returned to Tucson two days later, her hand was swollen to the wrist. The Arizona Poison Control Center recommended she use a hot water bath and massage the affected area three times per day. She did not seek other medical attention. The swelling and pain lasted about ten days. There was slight numbness of the bite site for about two weeks. There was no permanent damage.

The species was not positively identified, but Fitch's (aka Lilliput Octopus) is the likely candidate. It matches the description, is reputedly quick to bite, and has been called "dangerously venomous," though published case reports are wanting. Several other species of octopus occur in the Gulf of California, including two other small, unmarked, intertidal species: Hubb's Octopus (*O. hubbsorum*) and Mexican Pygmy Octopus (*Paroctopus* [= *Octopus*] *digueti*). See the next account for a bite by the latter.

Category: RE-NA (bitten while free-handling octopus thought to be harmless)

But It's So Cute: Mexican Pygmy Octopus (*Paroctopus digueti*)

In November of 2016, a small octopus was found in a dead clam shell along a beach off Sonora, Mexico. It was collected in less than 3 m of water. Photographs confirmed the species was *P. digueti*. It was within a group of clams that were collected for eating. Two people held the octopus prior to the envenomation. The third person to hold the animal was an 11-year-old female, but the octopus had had enough handling. It bit her on the wrist. She described it as "the octopus sucking down on me and the wrist feeling pressure then something like a fingernail pushing into the skin, then a needle, then two razor blades pinching into the wrist," which is when the real pain started. Pain increased from the physical trauma to the point of an 8.5 on a scale of 1–10. The victim has been stung three times by bees and considers bee sting pain is a 4. There was swelling above the elbow and blotchy skin appeared within a half hour. The pain leveled off, then declined after a couple hours, but by Day 2 she lost the use of her hand. By Day 4 she was able to close the hand, but she had a weak grip. Three weeks later, the wound was still not completely healed.

The *P. digueti* bit this little girl moments after the photo was taken. *Bruce Christman*

She was given Benadryl liquid immediately, then ice and pressure. She was given topical Lidocaine 30–40 minutes post-bite. She was kept on Benadryl at night for ten days after the envenomation to help alleviate the itching. The local Seri Indians recommended hot water as a first aid, which is supported for many types of marine envenomations.

Category: RE-NA (bitten by free-handling octopus thought to be harmless)

Bad Kitty! Blue-ringed Octopus (*Hapalochlaena* sp.)

In the account of cephalopods, I mentioned how blue-ringed octopuses used to be sporadic but occasionally available pets for domestic aquaria. A university researcher gave a friend three newly hatched blue-ringed octopuses. She was intending to raise them to adulthood in her home aquarium. They were given to her in a Styrofoam cup, covered in a plastic bag. Given that octopuses are escape artists, this was probably not the best way to secure them. She set the cup down on the counter in her apartment, and proceeded to get cleaned up before making dinner. When she returned to the kitchen, one of her cats had knocked the cup over and only two of the three blue-rings could be located. Apparently, the cat had eaten the third specimen. The cat started showing signs of respiratory distress, so she called a vet, but admitted she had no money to pay for treatment. Nevertheless, the vet made a house call and gave the cat an injection of 4-aminopyridine. The cat survived.

There are several species in the genus, which is considered one of the most venomous animals on Earth. While an LD_{50} has not been determined, they are said to carry enough of their potent venom to kill twenty-six adult humans, and have been responsible for many human fatalities. Make a note: these incredibly venomous animals make terrible pets! And I hope the market for domestic aquaria has dried up. This would be a horrible animal to have established in the wild. This story is a little atypical since it was a cat that was envenomated, but I just had to include it.

Category: PT-OT (cat presumably bitten while eating a hatchling octopus)

Two Halves Made a Hole (Author's

Testimonial): Purple-striped Sea Nettle (*Chrysaora colorata*)

Living near the beach in southern California was a great place for me to grow up, and I spent most summer days "catching rays (sun, not sting)" and bodysurfing. Although it hadn't even been scientifically described yet, the Purple-striped Sea Nettle was regarded by me and other beachgoers as a nemesis. These animals are more-or-less a southern California specialty. Every so often, there would be blooms when these animals would be driven inshore. They were often encountered in the surf, and their bodies on the beach were always evidence of when it was a bad time to enter the water. Even though I would loathe swimming with them, it was hard to keep a young beach bum out of the water on a warm day.

The general scenario is that I would continue my watersports, but tried to avoid swimming into them. However, it was not uncommon to be stung by tentacles that had broken off. The one event I distinctly remember was one day when I was bodysurfing and before I could change my course, I was pushed by a wave straight into one of these large jellyfish. I thought, "that can't be good." I got my footing and looked behind me. There were two halves of a sea nettle that I neatly severed with my torso. That got me out of the water. Pretty much my entire torso had turned into a red welt. Typical of stings, this encounter caused a maddening, burning rash. The local, old-school "first aid" was to rub sand on the wound and use language that our mothers would never approve of. This was like rubbing sandpaper onto a rash. Hot water or vinegar (or urine) was thankfully never even considered among us beach bums—we merely waited for the symptoms to pass.

The other thing that was also disconcerting during jellyfish blooms was that even if one could successfully avoid being stung, just being in the water caused one to itch all day, even if one could not see tentacles, suggesting very small remnants of tentacles or loose cnidae were stinging us, giving us something akin to swimmer's itch.

Category: RE-CA (stung while bodysurfing during jellyfish bloom)

A Bum (Numb) Leg: Portuguese Man-of-War (*Physalia physalis*)

Back in 1984, a twenty-year-old male was swimming for exercise just beyond the breakers on the Gulf of Mexico side of Florida, about 20 m offshore. He was well aware that the Portuguese Man-of-War was a creature to be avoided, but accidents happen. As he was swimming, he felt a sharp stinging sensation over his entire right leg. He immediately knew what had happened. He lifted his head out of the water and there he saw the characteristic blue pneumatophore of a large *P. physalis*, about 3 m from him. He was situated between the siphonophore and the shore, so the tentacles had drifted into his path with the tidal currents. Knowing where the tentacles were, he attempted to "back out" of the tentacles before swimming back to shore. His leg became numb and paralyzed, but he was able to make it back to shore while swimming with only one leg. When he got to shore, he examined his leg and noticed the dotted, whip-like pattern left by a tentacle or tentacles wrapped around his leg. The wound started from the crotch of his swimming trunks, down onto his calf, and on the tops of his feet (from trying to kick the tentacles off). Onshore, the pain turned to numbness and tingling with muscle spasms. The symptoms included alternating hot and cold sensations and profuse sweating of the right leg. His leg was cold to the touch. He said the pain was not as disconcerting as the other symptoms. He did not apply any first aid and did not recall taking any medication for the pain, except perhaps a couple aspirin. The symptoms slowly dissipated over the next few hours. He went to bed about seven hours post-sting and slept "just fine." The next morning the symptoms had disappeared except for a deep itch at the sting site, which lasted several more hours. He regained use of his leg in the morning and there were no long-lasting effects of the sting.

He said, "My strongest memory about the sting was that it was less pain than I would have expected and more weird sensations like I have never felt; i.e., the hot and cold sensations, sweating, and muscle spasms on the afflicted leg. Granted, I would not want this to happen to my face and head, which can easily happen while swimming into jelly tentacles."

Category: RE-FD (swam into Portuguese Man-of-War)

Wrapped up with the Ladies: Pacific Bluebottle (*P. utriculus*)

While vacationing in Hawaii, a woman went out for a dip in the ocean. An unseen Pacific Bluebottle had managed to find its way into the top of her swimming suit, and as she put it, "the tentacles wrapped around 'the ladies.'" This was not her first rodeo, having been bitten or stung by a number of other critters. Compared to some of her worse stings and bites, and other painful experiences, she said this envenomation was tolerable. She said it stung and burned for about thirty minutes. In its wake it left some temporary red streaks.

Category: RE-FD (swimmer stung by bluebottle that drifted into her)

A Bug Put Him on Crutches: Giant Water Bug (*Lethocerus americanus*)

In the late 1960s a fledgling herpetologist was snorkeling in a tributary of Hayfork Creek in northwestern California for his research on Western Pond Turtles. His brother was helping him with the work. He admits that some of the details were lost over time, but this is what he does remember. After snorkeling, when he stood up in the shallows, he felt what seemed to be a "hornet sting" on his leg behind his knee. He swatted at it and discovered a giant water bug had bitten him. Based on the location and size of the bug, it was almost certainly *Lethocerus americanus*, a species widespread in much of the United States. He said the bite "hurt like hell!" On a scale of 0–10, he gave it a 5, but added that he has a higher pain threshold than normal people. He likened the experience to being stabbed by an ice pick. Although the intense pain only lasted about an hour, he had to discontinue work and was unable to drive, so his brother drove him to his house. The leg became swollen and he had to use crutches for a day or two because it hurt too much to put weight on the leg.

Category: OC-FD (bitten by giant water bug while snorkeling)

Toe-biter Bites Finger: Giant Water Bug (*L. medius*)

A filmmaker in Virginia remembers a bug bite from his youth with great clarity. He grew up in Fort Huachuca, Arizona, where he had a fondness for living things, some of which inhabited his above-ground pool. When he was eleven or twelve, their pool was in need of maintenance, as it was being

Lethocerus medius. Professionals can usually handle water bugs safely, but if they do bite, it can be extremely painful. *L.L.C. Jones*

overtaken by algae and aquatic life. One day he saw a giant water bug in the pool and the temptation was too great, so he picked it up. The bug then proceeded to bite him on the webbing between his index and forefinger as it ejected a foul-smelling fluid. He described it as being immediately painful, worse than a wasp sting. As a kid, he said he would have given it a perfect 10 on the pain scale, but now, he said it would still be a solid 8. He screamed and cried. His grandmother applied an unusual first-aid ointment—toothpaste. The bite site swelled up and formed a pustule a couple days later. It throbbed for 2–3 days and was still sore for about a week. Even though he was only a child at the time, he knew what a giant water bug was, and had a field guide to confirm it.

Category: IN-CA (picked up giant water bug)

Aquatic Vertebrates

The Dive Must Go On: Round Stingray (*Urobatis/Urolophus/Urotrygon* sp.)

On a trip to a beach in Sonora, Mexico, in November 2016, several friends had the dubious honor of being envenomated by a host of terrestrial and marine creatures, two of which appear as other stories in this section. The victim thought the culprit was probably *Urobatis halleri*, but several species of round stingrays occur in the Gulf of California.

In November, the water in the Gulf of California can get fairly nippy, so to be comfortable, one of the group donned a full wetsuit before going snorkeling. He managed to get the suit on and was shuffling backward into the water, when wham! He was only in about 30 cm of water when he felt "a sharp pain on my ankle and something moving under my foot." He lifted his foot and saw a small (20–25 cm-long) round stingray swim off. He said "the initial envenomation felt like a jolt of electricity and ranked as a 6 on a pain scale of 1–10." However, within 30–45 seconds, the pain quickly dissipated to a 3–4. The stinger went through the booty and caused a puncture wound that was probably 15–20 mm deep, in his estimation. The stinger did not break off inside. Because it was such a hassle to put on the wetsuit, he decided to grin and bear it and continue with a short dive before returning to shore to assess the situation.

After snorkeling for 20–30 minutes, he ended his dive. When he took his booty off, the wound continued to bleed for about forty-five minutes post-sting. The pain at this point was about a 5. His group suggested he put the ankle in hot water and upon doing so, he got immediate relief. In the hot water, he ranked the pain as a 2. When he would take it out, it went back to being a 5–6. At this point he said it "was more of a throbbing puncture wound." After about two hours in hot water, he removed his foot and was able to walk on it. He said the pain was bearable at a 2–3 for the rest of the night. His ankle swelled slightly. For the next week, he said the pain was only a 1–2, but was tender to the touch. After about three weeks, the pain was gone, but it was still sensitive when pressure was applied.

Category: RE-FD (stepped on unseen stingray in shallows)

Swingin' and Stingin': Round Stingray (*Urobatis/Urolophus/Urotrygon* sp.), Author's Testimonial

In 1973 Mexico Highway 1 was completed. This paved road ran the entire length of the Baja California Peninsula and opened the door for an onslaught of tourists, particularly from southern California. I was among those who were eager to explore the newly accessible region. My brother and I periodically went down "The Baja," usually on fishing adventures. If we were limited by time, this usually meant we could only make San Felipe at the far north end for a weekend fishing trip. On one occasion, my brother, his girlfriend, and I piled into a car and headed to San Felipe. There we rented a *panga* (open fishing skiff) to do some nearshore fishing for the morning. The area near San Felipe is shallow and has extreme tides, so it was not a grand deep-sea fishing adventure, just something to feel the tug of whatever was in the water below us. We caught numerous shallow-water fishes, such as sand bass. When a fish was brought to the surface, we would allow the *pangera* (captain) to remove it from the hook and toss the animal back into the drink. My brother's girlfriend was a novice angler, but she was having fun. Unfortunately, she didn't really know anything about stingrays. She brought up a dinner-plate sized round stingray and wanting to have it removed and released, swung it over to the *pangera*. He put up his hand to protect his face and whap! It stung him on the index finger.

Three things happened simultaneously. First there was the loud screaming. Then there was the bloodletting. Although this was back in the 1970s, I still vividly remember blood shooting out of his finger. And lastly, I remember my brother scolding his girlfriend for swinging the stingray at the *pangera*. He was obviously in terrible pain and could no longer operate the boat. My brother took control of the motor and we headed back to the shore, which was at least a half hour away. The entire time, the poor man groaned and whined, and was doubled over in pain. My brother's girlfriend spent the whole time apologizing. We finally made it to the beach and the *pangera* sauntered off, clutching his finger in a bloodied rag. He got a really nice tip.

Category: RE-OT (angler swung stingray into boat captain while fishing)

Another Bad Cat: Sea Catfish (*Ariopsis felis* or *Bagre marinus*)

A California biologist was invited by his friend to give a talk at Louisiana State University. While there, they decided to go fishing in the nearshore shallows of the Gulf of Mexico. Although they were targeting Red Drum and other sportfish, they started catching marine catfishes, which are ordinarily considered an unwanted bycatch. However, the biologist loved fishing for freshwater catfishes (Ictaluridae), so he was having a blast catching and releasing them. He knew to avoid the spines, but accidents happen. One of the marine cats got hung up in the net and as he was trying to free it, it managed to puncture the middle of his palm with its dorsal spine. As he put it, "the pain was excruciating!" He tried to get the pain to stop, but it was pointless, as the pain was just too intense. As a comparison, he explained, "I have been poked by bullheads a bunch of times, but never had that intense pain. This was a bit shocking to me and I've learned to respect the catfish spines quite a bit more." Part of the spine got stuck in his palm. The area turned red, and it remained lodged in his hand for several days, before working its way out.

Category: RE-LA (stung while getting catfish out of the fishing net)

As a related story, I was chatting with some old-timers in Rockport, Texas, and they relayed a story to me about a Hardhead Catfish envenomation. Someone fishing on a dock for Speckled Trout kept catching nothing but Hardheads. After he landed about ten of these slimy critters, he showed his disgust by stomping on one. Well, that stout dorsal spine went right through his shoe and planted deep into the sole of his foot. They said the color drained right out of him and he was obviously in extreme pain. Fishing was done that night. There is a moral to this story.

Category: RE-IN (stung while intentionally stomping on catfish)

Wiggle-Frenzy on a Kayak: Sea Catfish (Ariidae)

A beach camper in the northern Gulf of California was with a group of friends angling for their usual table fare: fish for tacos. He was surprised when he kept catching catfish, as he had never seen nor heard of that marine species. They were too small to keep, so he released them over the side of his sit-on-top kayak. He explained that "they had an annoying habitat of exploding into a wiggle-frenzy" when brought to the surface. He grew weary of getting splashed incessantly while trying to release them over the side, so he brought one on board to unhook it. It thrashed right into his calf. The pectoral fin went deep into the flesh and twisted, and he knew he had better yank it out before it could do more damage. As he put it, "the pain was immediate and excruciating." He then said "I groaned uncontrollably as I paddled back to camp." Except for his extreme muscle spasms, he said it was the worst pain he had ever felt, and gave it a 7 out of 10, compared to a bee or wasp sting at a 4. The pain lasted about forty-five minutes. He described it as a "deep, constant pain, somewhat like a cramp." There was slight redness and some swelling. Although he was concerned for his well-being in such a remote area, the pain was completely gone in about two hours. He attributed the pain to both the venom and the physical trauma of yanking out the retrorse-barbed spine.

The culprit was one of two ariid species, the Cominate Sea Catfish (*Occidentarius platypogon*) or Chihuil (*Bagre panamensis*). Both are common throughout the Gulf of California, and the latter occasionally ranges into southern California.

Category: RE-LA (stung by catfish while trying to remove hook)

Keep a Spittoon Handy: Madtom Catfishes (*Noturus* spp.)

A Florida man has spent his entire life working in the outdoors on plants and animals, and his work in the aquatic environment included certain occupational hazards. His various jobs included the legal harvest of watersnakes (at night) and freshwater plants, as well as turtle farming. In the Lake Okeechobee area, he was occasionally stung by Tadpole Madtoms (*N. gyrinus*), and some unidentified *Noturus* poked him in the turbid waters of the Apalachicola River system. His immediate thought upon being stung was invariably that he was tagged by a Florida Cottonmouth, but he would soon realize the culprit was instead one of these little catfish. There was a sigh of relief, but the pain was immediate and lasted for about fifteen minutes. However, he claims to have a sure-fire way to alleviate the pain: chewing tobacco spit. At least it worked before he gave up chew. He says it also worked on Florida scorpions, bees, and wasps. He said madtom stings were "the least of his worries, as broken bottles, fish hooks, and other human debris could be more dangerous."

Category: OC-FD (accidentally stepped on madtom catfishes)

Florida Beach Land Mine: Hardhead Catfish (*Ariopsis felis*)

Red tides are caused by toxic dinoflagellates, and the periodic outbreaks can wreak havoc on marine organisms. Red tides can cause fish die-offs due to the lack of oxygen, as dinoflagellate blooms consume most of the oxygen in the water. A man out for a swim said that after a Florida beach was re-opened, following such a die-off, "nose-pick do-gooders buried dead fish at the water's edge." While running on the beach to take a swim, he landed where a Hardhead Catfish was buried just below the surface. Dead or not, its spine was just as sharp and heavy as in life, and the venom still worked. The spine broke off into his foot. He was in significant pain and had to "drive a stick-shift VW bug through traffic to the hospital." There he received morphine to alleviate the pain, and the medical staff had to remove the barbed spine from his foot. He said the foot was swollen for a few days, but it eventually healed completely.

Category: RO-FD (stepped on a dead catfish buried in the sand)

Stabbed by an Alien Invader: Red Lionfish (*Pterois volitans*, possibly *P. miles* or hybrid)

A recreational SCUBA diver was down about 40 feet on her dive at the south shore of Little Cayman Island in the Caribbean. She admitted she was a little bored with the dive. A lionfish got her attention and she decided that she would touch the "flowery parts," which I assume were the pectoral fins. The next thing she knew, she was poked by a single dorsal spine in the finger. She was not certain if she got tagged accidentally while touching the animal or if it intentionally jabbed her in defense (they do that). She said the puncture was not very deep. As soon as she realized she was stung, she pulled away. There was immediate pain, and blood flowed freely from her finger, appearing as a small green cloud at that depth. She said the pain was "More than a 10 [on a 0–10 scale] . . . it was the kind of pain I have never experienced . . . goes very deep inside, kind of a pulsating pain." The site of the sting turned red, and it swelled rapidly. The swelling included all of her fingers and most of her hand. She said the wound site also turned blue (cyanotic). Later, the sting site turned darkish red to purplish. Within two hours, the finger got blistered.

She and her dive master knew the best first aid was to "apply heat, as much as you can stand." They took this to heart. When she first boarded, she placed her finger on the motor, but it was too hot. The captain then heated some towels and wrapped her finger in it. When they got back to their resort, she immersed it in water as hot as she could stand, assisted by her dive partner. The partner placed his hand in the water next to hers, to regulate the temperature, since her hand was so sensitive. They kept filling up the water basin with hot tea kettle water at the resort, probably for about forty-five minutes to an hour. She claimed "the heat made the pain go away."

This event only happened two weeks before this writing, and her finger is now sloughing skin.

Category: RE-IN (touched lionfish out of curiosity while SCUBA diving)

When Is a Fish Not a Fish? When It Is a Stone: Reef Stonefish (*Synanceia verrucosa*)

A California professor related a story from about a decade ago when his son was stung by a Reef Stonefish in French Polynesia. While this was a while back, he remembers it well, and the effects of the sting are still manifesting themselves today. Although this story is from Moorea, the Reef Stonefish occurs throughout Micronesia, including the islands of the westernmost U.S. territories. By the way, he was stung in Opunohu Bay, which translated means "belly (or soul) of the stonefish."

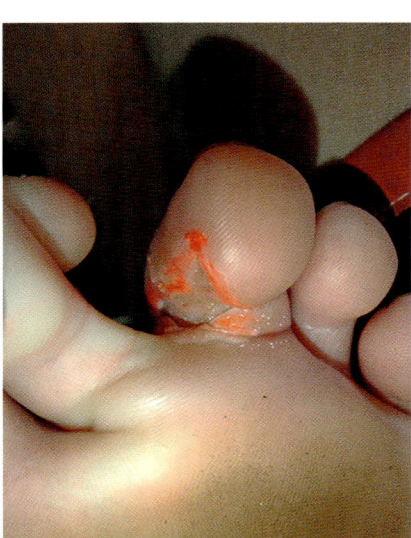

The small wound pales in comparison to the symptoms of envenomation.
Giacomo Bernardi

The professor was teaching a field class to about twenty U.S. students, as he does annually. They had been snorkeling in the bay and were standing in about 30 cm of water discussing what they had observed. His 6-year-old son yelped and said he had stepped on a sharp stone, but it felt like a needle went in his middle toe. In about 30 seconds the intense pain hit him and the boy started screaming. As the father put it, "He screamed for about two hours, the time it took us to reach the hospital. He did not cry because he was screaming too hard for crying." The child wasn't one to complain, so the father said the pain had to be a 10 on a scale of 10. Hot water and urination on the wound had no effect. A friend drove the boy and his dad to the hospital. There was a helicopter waiting to transport them to Tahiti, but about the time they arrived, the pain began to subside, so they did not transport him. At the hospital, the ER staff monitored his heart and administered painkillers and antihistamines, then put him on an IV drip. As far as the father knew, no antivenom was given. When the child was out of danger, he was released. The next day the wound started turning black, and tissue began sloughing off. For the next month, as instructed at the hospital, the father cleaned the wound and irrigated the area with diluted iodine solution and reapplied bandages three times per day.

When asked if the culprit was seen, the answer was "no, but the physicians believed it had to be a stonefish." Fortunately, the injury appeared to be from a single spine. The initial prognosis was of a cone snail envenomation, but the symptoms were consistent with Reef Stonefish—and the evidence mounted when another person was brought in later that day after being stung by a stonefish that he saw in the very same bay. The story didn't end in Polynesia, as the boy developed *Staphylococcus* abscesses after that, which were contagious to other members of the family. Physicians believe that the sting caused the boy's current compromised immune system and contagious staph abscesses.

Category: RE-FD (stepped on unseen stonefish while snorkeling)

No Biggie, but Noteworthy: Shortjaw Leatherjacket (*Oligoplites refulgens*)

A recreational fisherman was angling with a group of friends in the land of the Seri Indians in the fall of 2016, at Bahia Sargento, off Sonora, Mexico. They were fishing from shore with lures and caught and released several Leatherjackets (*O. saurus*) from shore without incident. Then one of the anglers caught a related species, *O. refulgens*. Being a rather savvy biologist, he knew this group of fishes was venomous but the fish managed to stick him with one of its anal spines anyway. He was nabbed in the palm of the hand. There was "minor bleeding, as one would expect from a puncture wound." However, there was also immediate pain indicative of being stung by a venomous animal. The pain increased to about that of a bee sting (which he rates a 3 out of 10). The pain lasted about 30–40 minutes before it subsided. There were no lingering effects.

While this was not a serious envenomation that might be written up in a medical journal, it does show that *O. refulgens* is also venomous like its Gulf of California congener, *O. saurus*.

Category: RE-LA (stung while landing a fish)

This Shortjaw Leatherjacket stung the angler with its anal fin spines. *Bruce Christman*

Rabbitfishes and surgeonfishes, like this Foxface Rabbitfish and Naso Tang, are popular marine aquarium pets, and rarely sting, as they are docile animals. *L.L.C. Jones*

13 Medical and Pharmacological Values of Venom

One of the most exciting aspects of venom research is that of isolating venom compounds and developing drugs that can benefit society. Venom is powerful stuff, and in the complex cocktail of ingredients, across the venomous animal world, are tens or possibly hundreds or thousands of specific chemicals that react specifically on the cells of their predators or prey. When researchers discover which fractions cause particular physiological responses, they can then be isolated, synthesized, and refined into useful products. For example, venom fractions of some snakes variously cause coagulation of blood, anticoagulation of blood, or changes in blood pressure. Neurotoxic peptides alter the way neurons communicate, which, among other things, may translate chemical signals to the brain as pain. Probably the most commonly researched animals are reptiles, cone snails, spiders, and scorpions.

Drugs have been developed that can be used in medical diagnostics, therapeutics, and other uses, such as cosmetics and insecticides. In one 2014 publication, there were twenty-one venom uses and venom-developed drugs that were used by physicians to diagnose various physiological functions, such as Lupus coagulants or anticoagulants, specific protein activities, and fibrinogen levels.

A great deal of research is going into therapeutics. Probably the first use to come along was for the treatment of bites and stings. Antivenoms have saved countless thousands of human lives, worldwide. The same 2014 publication mentioned above lists 16 drugs that are currently used in medical treatment, fourteen of which are FDA-approved and commercially available. One of our local reptilian superheroes is the Gila Monster. Two drugs have been developed that are used in controlling Type II diabetes, one of the biggest health problems in human society: Exenatide (Byetta and Bydureon). A drug from the Pygmy Rattlesnake, Eptifibatide (Intigrilin), is used to treat acute coronary syndrome. There are numerous other venom-derived drugs from animals beyond our geographic area. Several of these drugs have come from vipers, particularly South American pit vipers. One drug derived from cone snails is Ziconotide (Prialt), which is an extremely powerful analgesic for treating severe chronic pain. It is far more effective than opioids and non-habit forming. In fact, the relief of pain is a major focus of animal venom research. Among those animal groups targeted are the cone snails because of their profound effects on neurotransmission, but of course, there are many other neurotoxic taxa that may have as-yet unrecognized potential for becoming the next wonder drug. Opiates are currently the most widely used prescription analgesics, but they can be habit forming, and, as we have seen in several of our stories of people who have been envenomated, they don't always work. In addition to the numerous drugs that are FDA-approved, there are several more being developed and in clinical trials. Some of the venom fractions and drugs are being investigated for their use in the battle against cancer and the location of tumors, as well as for heart disease, multiple sclerosis, appetite suppression, muscular dystrophy, autoimmune disorders, stroke, epilepsy, drug addiction, and reversing paralysis.

Beyond diagnostic and therapeutic, animal venom has other uses. One touched upon earlier is cosmetics. For example, venom extracts from an Old World viper can smooth out skin wrinkles. One economically important use of venom products beyond medicine is as insecticides. Killing insects is what spiders and scorpions do, and they do it very well. Virtually all species of those two groups are venomous, so there is untold hidden potential. Insects have a huge effect on human society. Most of it is good, such as pollination, but some insects also transmit diseases. Malaria alone, which is carried by mosquito vectors, affects more than three billion people in ninety countries and kills 1.2 million people annually—or one person every twenty-six seconds. Others insects wreak havoc on crops. Control of insect pests is a multi-million dollar effort, and, unfortunately, the mechanism to do so is often through control with dangerous chemicals, draining wetlands, and other ecologically unsound practices. Of course, insects can become resistant to chemical agents, so stronger compounds and newer chemical agents are developed in response. Organisms used as biological control agents can cause ecological problems themselves.

The use of bioactive peptides from spiders and scorpions shows great promise for controlling insect pests. These arthropods often have strong neurotoxins that target invertebrates, a mechanism that evolved long ago to

Medical and Pharmacological Uses of Venom   839

Pygmy Rattlesnakes produce the chemical Eptifibatide, an anticoagulant, which is used to treat high blood pressure. It functions as a blood thinner by reducing platelets. *L.L.C. Jones*

quickly disable thrashing insect prey. It has been shown through venomic studies that a single species of spider may have in excess of 1,000 peptides in its venom, and it has been estimated that there may be as many as 20 million spider venom peptides worldwide—we have our work cut out for us! Spiders in our area with insecticidal potential include *Agelenopsis aperta*, *Eratigena agrestis*, and *Calisoga* spp., among others.

Similarly, many invertebrates and vertebrates produce antimicrobial peptides in their venom or crinotoxins. For example, some scorpions will actually spray venom on their carapace, presumably for the antimicrobial properties.

References and Resources: Casewell et al. (2013), Da Silva et al. (2014), Escoubas and King (2009), Fox and Serrano (2007), King (2011), King and Hardy (2013), Lewis and Garcia (2003), Takacs and Nathan (2014). See also references under the Cone Snail chapter and the Institution profile for the Cone Snail Lab in Utah.

14 Some Institutions Associated with Venomous Animals

Arizona Poison and Drug Information Center

The American Association of Poison Control Centers had its crude beginnings in the 1950s. One of its early pioneers was Dr. Albert Picchioni, a professor at the College of Pharmacy, University of Arizona. As a necessity, he spent considerable time answering questions about poisoning to medical staff. His efforts evolved into the current Arizona Poison and Drug Information Center (AzPDIC) in Tucson. By the 1970s, hundreds

Display in the lobby of the APDIC. *L.L.C. Jones*

Responders at poison control centers have computers and reference materials at the ready but are already intensively trained in poisoning and envenomation. *L.L.C. Jones*

of similar centers had popped up across the country, but it wasn't until the 1980s when centers became accredited. Currently, there are fifty-five centers nationally.

The Arizona Poison and Drug Information Center in Tucson serves all of Arizona (about 2.5 million people), except for Maricopa County (the greater Phoenix area), which serves about twice as many people. The Tucson center receives about 40,000 calls per year and has about 60,000 outgoing calls per year. The Phoenix center handles about twice the volume. At the Tucson center, about half of the calls are informational, as when someone wants to know which spiders are venomous. The other half are from callers who have had actual exposures. The AzPDIC area is situated in an area renowned for its diversity of venomous animals, so not surprisingly, about 20% of the calls are regarding venomous animals, as opposed to the national average of about 2%. There are about 4,000 calls per year regarding exposures from bites and stings. In 2014, for example, there were 1,888 calls about scorpion stings, which paled in comparison to Phoenix, which has about 10,000 calls per year! However, AzPDIC has a larger number of rattlesnake bite queries, despite having half the population of the Phoenix area.

When someone calls the AzPDIC, or any other accredited poison control center, the phone will be answered by a nurse, pharmacist, or physician. The staff not only has knowledge of poisonous substances and treatment, but they also have numerous resources at their fingertips, such as computer databases, in-house and online information resources, and a library. The staff members are Specialists in Poison Information. Furthermore, every call is backed by a physician boarded in medical toxicology. About 20% of the calls are from hospitals and other medical facilities, while 70% are covered in home or field situations. Only about 10% are referred to a medical facility by the AzPDIC, where the staff will help with emergency medical needs (e.g., call 911, be in contact with the caller, as well as medical and transport personnel). Pertinent information is tracked on the telephone calls and data are recorded, and then forwarded to the National Poison Data System. These data are summarized in an annual report; specific attributes of the data can be requested by researchers.

Responding to telephone calls is only part of the mission of the poison centers. They also help with healthcare facilities, research, and education. They are well-poised to assist with research that can help advance the medical profession. For example, they may collect information about patients who have recurrent or delayed effects from rattlesnake bites. The educational outreach program is important to keep accidents from happening, and to let the public know how to respond. The poison and drug information centers are a valuable resource. Service is free and the caller remains anonymous, so the national telephone number (1-800-222-1222) should be at the ready in every household and field notebook. Specialized staffers are there to help 24/7/365.25.

The Spider Pharm

If you think your house has a spider infestation, you should see the home of Chuck and Anita Kristensen, proprietors of the Spider Pharm, a 30-year-old family business. At any given time, they have about 50,000 spiders at their residence, including widow, recluse, tarantula, orb-weaving, lynx, jumping, and an assortment of "house" spiders (e.g., theridiids, pholcids, and filistatids), among others. They also have scorpions, including bark and giant hairies, plus centipedes and the odd insects (most of which are food for the arachnids and myriapods).

These animals are milked for their venom and then usually sold to researchers or for education. This is a very specialized service they provide,

A room where hundreds of spiders are kept in constant heat- and humidity-controlled conditions. *L.L.C. Jones*

taking much of the worry of procuring pure venom away from researchers who would rather spend their time analyzing the property and effects of venom. Undoubtedly, much of what we know about venomous arthropods in this book originated from animals that were raised and milked at the Spider Pharm.

To start the process, arthropods must be collected, so the Kristensens gather their own or get them from other field biologists. Then husbandry comes into play. This massive number of arthropods needs to be housed, maintained, and bred in captivity. One room of their home is climate controlled, allowing animals to breed year-round in an area where they would die or brumate in ambient conditions.

As might be expected, milking spiders is a tedious and exacting science, requiring patience and skill. The animals are anesthetized with CO_2 and then held with tweezers that will electrostimulate the venom-gland muscles. As the venom is extruded, it is collected in a micropipette from the fangs via capillary action. This is all done while being viewed through a dissecting stereomicroscope. Purity of the venom is important for clients, so the equipment must be sterilized. The mouthparts of the spider may even need to be

Chuck Kristensen turned his home into a spider farm. *L.L.C. Jones*

cleaned prior to milking. Also, many spiders regurgitate upon electrostimulation; this cannot be mixed with venom, so gastric contents are aspirated away under the microscope to ensure purity of the venom. This work is exacting, but tedious, as it takes up to one hundred milkings of *Parasteatoda* to get 1 µl of pure venom. Widows and recluses are much more accommodating—large individuals only need five milkings to get 1 µl of venom. This is why entomologists speak of µl (or µg) of venom, while vertebrate biologists speak in terms of ml or mg of venom. It takes a lot of spiders and milking to get a usable volume of venom for the researcher.

When I visited, Chuck was in the process of milking a series of *Physocyclus mexicanus*, a cellar spider (family Pholcidae). Apparently, this species is not only important in research, but it is believed by some laypersons to be the most venomous animal on Earth. To disprove the claim, they did some tests for the television program, *Myth Busters*. Venom was injected into mice, and it showed that the venom of these common house spiders was pretty innocuous. However, it is quick and deadly to invertebrates. Chuck has been bitten by more of these than any other venomous animal, and they only leave a slight swelling and inflammation.

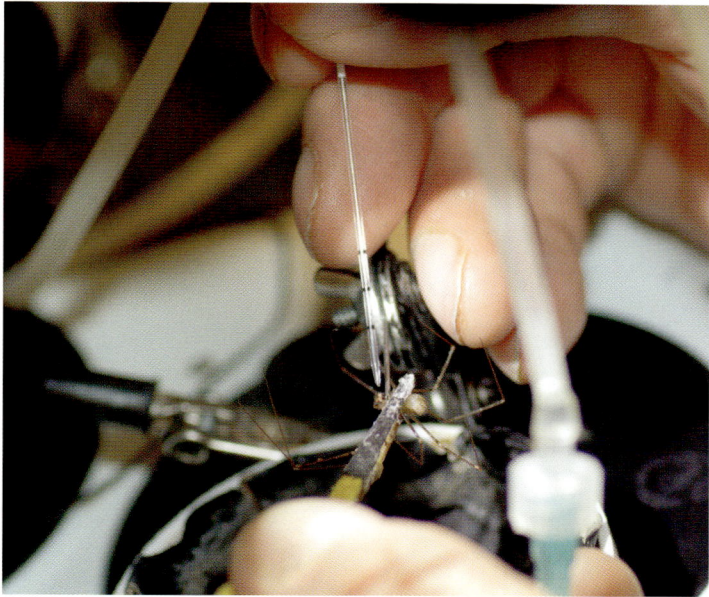

A close-up of Chuck milking a tiny pholcid spider. L.L.C. Jones

His occupation has led to his envenomation by many species, including widow and recluse spiders, Arizona Bark Scorpions, *Vaejovis* scorpions, Giant Desert Centipede, and tarantula hawk. As he was conducting fieldwork, he also received a mass envenomation to his leg by Africanized Honey Bees. Anita was also stung by Arizona Bark Scorpions, and she was once sprayed in the eye by a Green Lynx Spider.

Phoenix Herpetological Sanctuary

The Phoenix Herpetological Sanctuary (PHS) is different than other "herp" societies. Generally, these groups are common-interest clubs, with members who are fascinated with giant snakes and lizards, "designer snakes," other herp pets, or local reptiles and amphibians. What makes PHS different is that they are not actually a club, but instead a nonprofit organization composed of individuals who are committed to saving reptiles. Among their 1,700 reptiles and amphibians at their facility in Scottsdale, Arizona, they also house more than three hundred individuals of more than one hundred species and subspecies of native and exotic venomous reptiles. Most of the venomous reptiles at the PHS facility originated from hobbyists who legally or illegally collected or purchased venomous reptiles for their home menagerie. As we see repeatedly in the pages of this book, hobbyists are frequently among the recipients of bites and stings. Sometimes native, local species are involved, but if from exotic species, finding antivenom is a challenge.

The PHS founders, Russ Johnson, Daniel March, and Debbie Gibson, started a sanctuary for unwanted and confiscated pets until they could find a permanent home, but those humble beginnings grew into something much larger. The PHS has education, husbandry, removal, training, and storage of venomous reptiles as part of their regimen. Probably first and foremost, PHS helps to educate the public about the ecological value of reptiles, including the venomous species we live among. Their educational programs reach out to more than 100,000 people annually.

The PHS provides a service to the greater Phoenix community (more than four million people) with removal of snakes, plus they also train others to conduct safe snake removal. Training starts with nonvenomous species, and then moves onto venomous species, as trainees become accustomed to handling snakes with the proper gear. Those receiving snake removal training include employees of fire departments and others tasked with removing venomous animals, as well as private individuals. Crowd control is always part of the training, as many people are drawn toward the snake out of curiosity, rather than away from it, a situation that can increase the seriousness of a dangerous encounter.

Cody Bartolini is the Curator of Reptiles for the PHS. With the help of his wife, Pia, among others, they must tend to the needs of the three hundred venomous animals in their care. This is a formidable task when one considers that they need to feed, house, and transport animals such as Black and Green Mambas, Boomslangs, Gaboon and Rhinoceros Vipers, Russell's Vipers, King Cobras, spitting cobras, beaded lizards, and a plethora of other dangerously venomous species. They also have native U.S. species such as Gila Monsters, many kinds of rattlesnakes, coralsnakes, cottonmouths, and copperheads. Because exotic herp collectors often seek out the rarest species (most often illegally), the PHS now has some of these animals in their care, so are primed for public display and education. Part of Cody's tasks is gathering animals for public display, such as during reptile shows, where thousands of hobbyists gather. Not only is this educational for hobbyists (past, present, and future), but it gives everyone a chance to see these beautiful and interesting creatures in a public venue, rather than

Some Institutions Associated with Venomous Animals 847

Hundreds of snakes, mostly venomous, are kept securely in terraria. *L.L.C. Jones*

Cody Bartolini is moving a Black Mamba that is twice his length, in order to clean the terrarium. *L.L.C. Jones*

in someone's bedroom. The PHS has an amnesty agreement with Arizona Game and Fish Department. If hobbyists willingly donate their illegal pets, they can avoid prosecution.

Readers can contact the PHS at their website to partake of their many services, such as educating school children, snake removal, donating their illegal "booty," or just getting a tour of their fantastic facilities.

Venom Immunochemistry, Pharmacology, and Emergency Response (VIPER) Institute

The VIPER Institute is located at the University of Arizona's College of Medicine. The Founding Director, Leslie Boyer, MD, a pediatric physician, started the institute to advance the science and treatment of people and animals envenomated by creatures of the American Southwest and beyond. As she puts it, "Tucson is Ground Zero for venomous animals in the United States." The mere handful of permanent and temporary staff supports the work of many. At the time of this writing, there are more than 130 project members, representing scientists and health-care professionals from as close as the same building to as far away as Africa and Australia.

Leslie Boyer, director of the VIPER Institute, with a vial of antivenom. *L.L.C. Jones*

There are currently six primary research programs: (1) coralsnakes, (2) pit vipers, (3) scorpions, (4) venom and antivenom, (5) REGISTER, and (6) pharmecoeconomics. The various projects within these programs are not mutually exclusive, especially as it applies to antivenom, the primary focus at the Tucson center. The VIPER Institute is one of the few organizations to design research projects and lead clinical trials for antivenom efficacy and development. Domestically and abroad, there are inherent challenges associated with antivenom research, development, and production. Research is challenged by the ability to do clinical trials that use the standard experimental- and control-sample approach. The problem is that if someone is bitten or stung by a dangerous animal, they may be in a potentially life-threatening situation (keeping in mind such an outcome is still quite rare), so there may be a legal and ethical imperative to treat each and every patient—in other words, if antivenom is needed, it cannot be withheld. The Institute was the first in the country to find a legal and medically approved "loophole." Some facilities do not use antivenom in treatment, so patients at those hospitals were given placebos. This is how the new antivenom, Anascorp, was tested. Stings from bark scorpions in the United States are of medical importance mostly for children, but in Mexico, bark scorpion stings (where several more

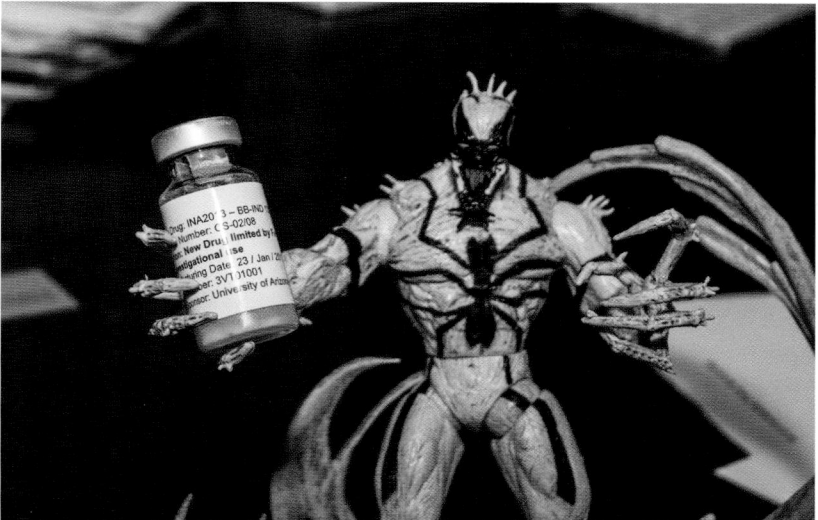

Dr. Boyer's desk features the Marvel Comics character named Anti-Venom who is holding, well, some antivenom. *L.L.C. Jones*

venomous species of *Centruroides* occur) cause thousands of deaths annually. In other clinical trials where a placebo is unnecessary, different types of antivenom (e.g., CroFab, Antivenin (Crotalidae) Polyvalent, and Anavip) have been compared, capitalizing on the cooperation with the extended VIPER Institute network of health-care facilities.

Development of new antivenoms in the United States is a tedious process, as there are lengthy and costly checks and balances in place that can cause significant hurdles. In particular, the Food and Drug Administration has very stringent criteria for a drug to be approved. In the larger context, relatively few people in our country are bitten by dangerously venomous animals, so there is the whole supply and demand problem. Pharmaceutical companies require financial incentive to produce a specialized drug that relatively few hospitals and zoos will purchase. This is a primary concern for the development of coralsnake antivenom, as there are only about 100 coralsnake bites per year in the United States. The REGISTER program is an in-house tracking of clinical trials.

The VIPER Institute has created and maintains an Antivenom Index as a service to the American Association of Poison Control Centers. This is an active inventory of antivenom stocks held in zoos and other facilities. This is particularly useful for zoos where a wide variety of dangerously venomous

A giant thank-you card from children who appreciate the work done by the institute in developing and doing trials on an antivenom for the Arizona Bark Scorpion, a species that is dangerously venomous to children. *L.L.C. Jones*

exotic snakes are kept for display. This information is at the fingertips of herpetologists and curators should someone be bitten at their facility. Interestingly, this problem is not limited to zoos. For example, there are ten to fifteen envenomations annually in the United States by cobras to hobbyists who keep exotic pets, and many more bites due to other types of venomous snakes in home terraria.

Although centered in Arizona, the Institute engages in research with other countries. Mexico, especially, is an important partner. The VIPER Institute has an office at the Institute of Biotechnology for Universidad Nacional Autónoma de México (UNAM) in Cuernavaca, Morelos. As part of a cultural and research exchange program, UNAM has an office at the University of Arizona, where they help sponsor and support fellow venom researchers. A burgeoning partnership is also developing with Africa, where venomous animals pose a much more serious problem than in the United States, and where antivenom stores are severely limited. As Dr. Boyer says, "we have a lot in common with Africa. Arizona also has vipers, elapids, black widows, recluse spiders, and dangerous scorpions."

Some Publications: Boyer et al. (1999, 2001, 2009, 2015), Seifert et al. (1997, 2009).

Arizona-Sonora Desert Museum

The Arizona-Sonora Desert Museum (ASDM) is not a typical museum. Rather, it is a unique zoo and aquarium nestled in ninety-eight acres at the base of the Tucson Mountains. Twenty-one acres are developed for exhibits and trails, while the rest of the facilities are behind-the-scenes. Visitors can view natural displays of live plants and animals along the two miles of trails traversing prime Sonoran Desert habitat. What started as a dream in 1952 grew into one of the most renowned accredited zoos in the nation. What sets it apart from others is that virtually all of the animals are from the Sonoran Desert region, which includes southern Arizona and adjacent Sonora, Mexico, including the Gulf of California and its islands. There are no tigers, elephants, or giraffes, but in their stead you can find White-nosed Coatis, Ocelots, Collared Peccaries (javelina), Mexican Wolves, Harris' Hawks, Thick-billed Parrots, and a host of other seemingly exotic species from Arizona and Sonora.

Since its inception, the directors and staff have always embraced the region's natural biodiversity, venomous and otherwise. One department

One of the husbandry rooms at the ASDM, with Executive Director, Craig Ivanyi, and General Curator, Stephane Poulin. L.L.C. Jones

is replete with venomous animals: Herpetology, Ichthyology, and Invertebrate Zoology (HIIZ). One of their main charges is to give visitors an appreciation of their toxic neighbors and how to live with them. Rattlesnakes, in particular, have been so maligned over the years that it seemed an ethical imperative to educate people about their beauty and ecological value in nature. When people visit or move to the Southwest, they may be naïve about their new surroundings and unwilling to tolerate their serpentine neighbors. Arizona has more species of rattlesnakes than anywhere in the United States, and adjacent Mexico has even more. In any given area of southern Arizona, a homeowner may be surprised to learn they can have 1–5 species of rattlesnakes as neighbors. In addition, there are dozens of venomous invertebrate taxa that can be encountered in southern Arizona and Sonora. For those living in or visiting nearby Mexico, the Gulf of California is a magnificent natural aquarium boasting 822 species of fishes—which is more than northern Caribbean reefs. This includes a diverse array of stingrays, most of which do not reach California. Plus, the gulf harbors a host of tropical marine invertebrates we may not wish to tangle with, such as the Pacific Bluebottle.

In the displays and the behind-the-scenes facilities of the herpetology section of the HIIZ department, there are approximately 130 species of reptiles,

The public reptile and adjoining invertebrate rooms invariably have a plethora of venomous Sonoran Desert denizens on display. L.L.C. Jones

including twenty-five kinds of rattlesnakes, Mexican Beaded Lizards, and Gila Monsters, plus an assortment of mildly venomous colubrids and dipsadids. The ASDM cares for approximately four hundred individual reptiles at any given time. The invertebrate part of HIIZ has members of most of the groups of venomous, ground-dwelling arthropods featured in this book. The main display for these terrestrial taxa is the Reptile and Invertebrate House, where animals are attractively displayed in terraria, accompanied by fact sheets. In these display rooms, most of the taxa are venomous. Displays change over time, but typically one can see numerous species of rattlesnakes, Gila Monster, Arizona Bark Scorpions (all but invisible until you push the black-light button), Arizona Recluse Spiders, giant water bugs, and the impressive Desert Giant Centipedes. Their marine displays include a stingray touch pool with American Cownose Rays. The Atlantic species are displayed as an ambassador species to represent the Golden Cownose Ray of the Gulf of California, a species of conservation concern. They were bred in the Phoenix Zoo and captive-bred rays now thrive in various public aquaria throughout North America. Horn Sharks and other gulf species can also be seen in the Warden Aquarium.

Another display is called Life on the Rocks, where artificial, but shockingly realistic rocks display native reptiles and arthropods in crevices and burrows. The Museum's trails meander through different biotic communities, such as Mojave and Sonoran Desert, desert grasslands, and mountain woodlands. Other displays are interspersed along these trails.

There are educational programs in which people can learn about venomous animals. These include the "Live and (Sort of) on the Loose" and "Fur, Feathers, and Fangs" programs. Staff members safely handle venomous animals with snake hooks from behind partitions. For many visitors, that is close enough. Visitors can also touch and feed the cownose rays, a treat for thousands of school children learning about appreciating nature. The rays are periodically de-barbed so there is no threat from envenomation by these gentle creatures.

Behind the scenes are many more animals than are on display. Some of these are bred in captivity to help conserve the species and supply other zoos with local animals. For example, the small mountain rattlesnakes of Arizona (*Crotalus lepidus*, *C. willardi*, and sometimes *C. pricei*) are highly sought-after by other institutions, and a source such as the ASDM ensures other zoos need not collect them from the wild. Husbandry for terrestrial invertebrates is more challenging than that for terrestrial and aquatic vertebrates, but there are numerous species of venomous centipedes, scorpions, spiders, and insects the museum is either breeding or attempting to breed. The ASDM is always involved in various research programs and they are there to support fellow researchers who need access to venomous organisms. They have an extensive educational outreach program, which includes both classroom and field ventures. For example, in 2011 ASDM hosted the second Biology of the Rattlesnakes Symposium with their partners, the Chiricahua Desert Museum of New Mexico.

Venomous animals are not always confined to the buildings since the grounds are prime wildlife habitat. Due to the natural setting, Gila Monsters and rattlesnakes are commonly encountered on the grounds and are transported away from the area of immediate danger. In the 1970s ASDM started keeping logs of all snakes moved, and this evolved into more formal research, including radiotelemetry of resident and transient snakes. They move about twenty individual snakes out of harm's way annually (some are moved multiple times) and no visitor has ever been bitten. Five species are encountered, but 90% are *C. atrox*. On summer nights, one of the favorite activities of the public is to walk the grounds to view Arizona Bark Scorpions with a black light. When you see how many are in the rock walls, you may think twice about leaning against the walls—but during the day, they are out of harm's way deep in crevices. The museum grounds are also a great place to watch tarantula hawks feeding on nectar or scanning the ground for tarantulas.

Public relations and education are a big part of the staff's job. Approximately 130 staff members and 700 volunteers and docents serve more than 350,000 annual visitors, including 35,000 school children, as well as visiting biologists and television personalities. Among the latter were renowned biologists and entertainers Steve Irwin, Jeff Corwin, and Mark O'Shea. The staff requires proper safety procedures for staff and visitors alike. Safety is of prime importance. There are safety protocols in place as well as information on what to do if bitten. They have an excellent track record with only one envenomation in their more than sixty years of operation, plus an accidental ocular spray while cleaning a cage.

Bottom line: when in Tucson, this place is a must-see. It is rated the number one attraction there, and for those interested in our native venomous fauna, it is their Graceland.

Cone Snail Lab, University of Utah

Although the University of Utah is near a huge hypersaline lake, it is a long way from any ocean that harbors venomous marine animals. Yet the Cone Snail Lab is a premier facility for the study of cone snails (family Conidae) and their relatives (superfamily Conoidea). Baldomero "Toto" Olivera runs the facility, which boasts some 25–30 core staff members at any time, plus there is involvement from countless colleagues worldwide. Dr. Olivera's lab primarily targets the medicinal and pharmacological value of neurotoxic peptides found in the venom of cone snails, collectively known as conotoxins or conopeptides. Cone snail venom contains the ingredients for a treasure trove of potential drugs that show promise for the treatment of maladies of the nervous system, including pain, epilepsy, stroke, and drug addiction. The lab is also involved with complex taxonomic assessments of the Conoidea.

Cone snails are a diverse group of marine mollusks, with more than eight hundred species recognized worldwide. The taxonomy of these snails has been difficult to track over the years, ranging from a single family with one genus (*Conus*) to many families and more than one hundred genera. However, in 2015 a team of experts, including Dr. Olivera, undertook a large-scale molecular phylogenetic analysis of the cone snails. They found that all genera and species should be placed within the single family, Conidae, and that there are

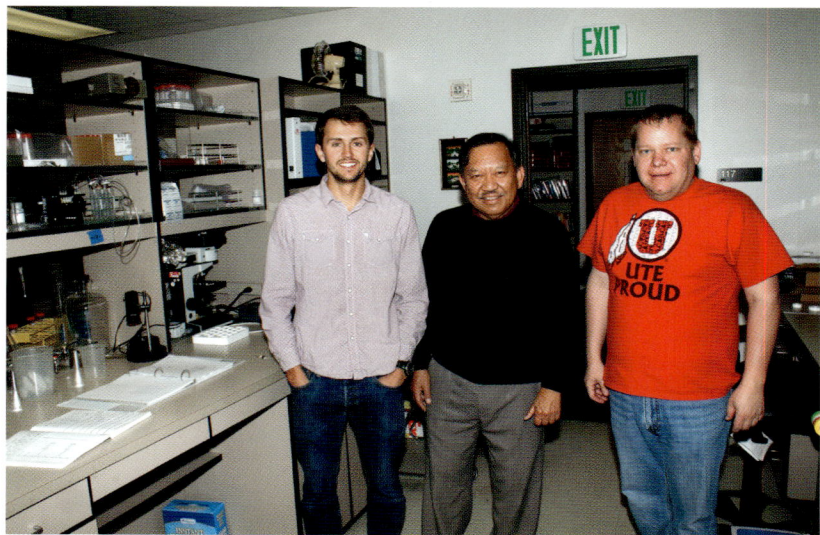

Baldomero Olivera (center) is the principal investigator for the Cone Snail Lab in Utah. *L.L.C. Jones*

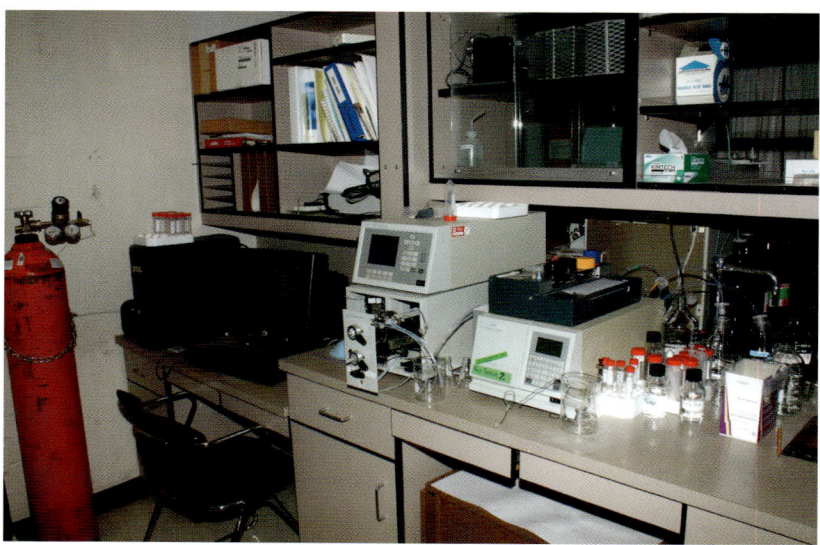

One of the workstations. *L.L.C. Jones*

four natural groupings that align with genera: *Conus, Conasprella, Profundiconus,* and *Californiconus*. Eighty-five percent belong to *Conus*.

It has been estimated that the eight hundred species of cone snails have among them at least 50,000 different conopeptides. Researchers at the Cone

The lab boasts a large collection of cone snail shells. *L.L.C. Jones*

There are numerous aquaria where they maintain live cone snails, such as this *C. purpurascens*. *L.L.C. Jones*

Snail Lab are just scratching the surface of this potential, but several candidate peptides are already being explored biochemically. To obtain venom, the researchers procure specimens primarily from the Philippines, Dr. Olivera's homeland. He explained that not only is the diversity highest in that region, but cone snails are easy to procure because they are an important food source that is commercially harvested. Large numbers of specimens and species can be gathered for research. However, he and his lab mates and colleagues also study species from elsewhere.

Collecting venom can be done via milking through latex glove material—put some fish odor on the glove and present it to captive animals and they will harpoon the latex, leaving venom in a vial on the other side. However, there are inherent problems with this approach because venom yield is low, and the milking of captive cone snails would involve a huge closed salt-water system (which can present its own problems) and would be overly tedious. Also, the venom composition of captive snails changes over time. Hence, wild snails are harvested and their venom glands are extracted. The lab has external suppliers of the venom, so that they can concentrate on analyzing it. Once they have venom, a common starting point is to run samples through HPLC (high performance liquid chromatography) analysis to find peaks in potential peptide activity. They then repeat the process on a finer scale to further isolate venom fractions. These peptide fractions can then be evaluated for their chemical properties and effects on cells, such as dissociated mouse neurons or frog oocytes. For example, they can use potassium chloride infusions and dye to identify voltage-gated ion pumps on mouse neurons. Then they can begin to understand the effects of peptides on ion transfer and the receptor sites. Each of the researchers has their own specific studies, such as effects of peptides on nicotinic acetylcholine receptors.

The final result of some of these studies is the isolation, synthesis, and production of pharmacologically important drugs. A big success to come out of the Cone Snail Lab is the FDA-approved drug, Prialt (Ziconotide), isolated from *C. regius*. It is used for relieving chronic pain in patients who have cancer or who have become tolerant to morphine. In fact, it is the only intrathecal (delivered to the spinal canal) chronic pain medication that is FDA-approved. Currently, there are two other potential conopeptide drugs originating from the lab that are in clinical trials, plus one in pre-clinical trials. Of course, with 50,000 peptides among cone snails, there are undoubtedly many more yet to be discovered.

Probably the highlight of my visit was watching their captive fish-eating species being fed. In their aquaria, among other species, they had *C. ermineus*, the only fish-eating species from the Atlantic; *C. striatus*, a dangerous species from Hawaii; *C. purpurascens*, from the Gulf of California; and *C. geographus*, the most notorious of all cone snail species from the Indo-

Pacific that has caused numerous human fatalities. Cone snails have two basic fishing methods. One method is employed by the Geographer Cone, which includes "netting" one to several fish inside its expansive mouth and then harpooning each of the fish with its radulae. The other method is simply harpooning the fish first, and then engulfing it with its mouth. When a cone snail detects prey, it sends out its siphon to collect chemical cues. When prey is located, the proboscis appears. When the fish is lined up, the harpoon-like radula is shot out into the prey (Goldfish in this case) and venom is injected in hypodermic-needle fashion. The prey is quickly paralyzed by the strong neurotoxins, in order to avoid injury to the snail and its mouthparts. The prey is then engulfed in its large, expandable mouth. One can watch this amazing feat in online videos at their website. The website also has much more information on Dr. Olivera and his staff, their research, and their publications.

Some Publications: Olivera (1999, 2002), Olivera et al. (1988, 2012, 2014).

Venom Analysis Lab, University of Northern Colorado, Greeley

Stephen Mackessy has been the director of the Venom Analysis Lab for over twenty-five years, where he and his graduate and undergraduate students have done research on reptiles, mostly rattlesnakes, to help bridge the gap between natural history attributes and the stuff in the venom glands. According to Dr. Mackessy, their research emphasizes some basic questions about snakes and their venom. What is in the venom? What does the snake do with the venom? And how do different snakes compare in natural history, as it relates to their venom composition? Much of what we know about rattlesnakes and their venom has come from the Mackessy Lab. For example, we now have a better understanding of ontogenetic changes in venom composition as it relates to a rattlesnake's prey. Many or most species of rattlesnakes start out life as primarily lizard-eaters, then as they grow, small mammals become the preferred prey. Although lizards and rodents are both vertebrates, they have a fundamental difference: lizards are ectotherms and mammals are endotherms. An ectothermic prey animal cannot be easily tracked because it does not have an inherent heat signature. Thus, a lizard that is bitten needs to be held onto until it quits squirming, and to avoid injury, the lizard needs to be quickly immobilized. Mammals are bitten, then released, and tracked.

Another day on the job for Steve Mackessy. *L.L.C. Jones*

In general, this means that juveniles tend to have more potent and possibly more neurotoxic venom than adults. Dr. Mackessy has also given us a better understanding of the differences in venom components and relative toxicity among snakes in the western rattlesnake complex. These snakes have been challenging to categorize taxonomically, but differences in the venom between taxa has helped us to get a better understanding of their interrelationships, as well as effects of the toxin. In fact, this is part of the reason I chose to recognize most taxa in the complex as being distinct species—in part it is a bookkeeping exercise, since all of the forms vary in venom attributes.

The Venom Analysis Lab is not large, but it is productive. At any given time, Dr. Mackessy has 5–6 graduate students, about half of whom are master's students and half are doctoral candidates. He typically also has 2–6 undergrads working on various projects. Projects range from a behavioral study of spiny-tailed iguanas in Costa Rica and a long-term ecological study of Prairie Rattlesnakes in the nearby grasslands to detailed biochemical analyses of snake venom. At the time of this writing, the lab is even involved in a forensics case involving human envenomation. Among venomous snakes, Dr. Mackessy and his colleagues have also been increasing our knowledge on characteristics of the venom of dipsadid snakes and their relatives, such as hog-nosed snakes, lyresnakes, and Brown Treesnakes. The researchers have also been expanding their focus to include more studies on potential pharmacological uses of snake venom. One of the targets is cancer cells.

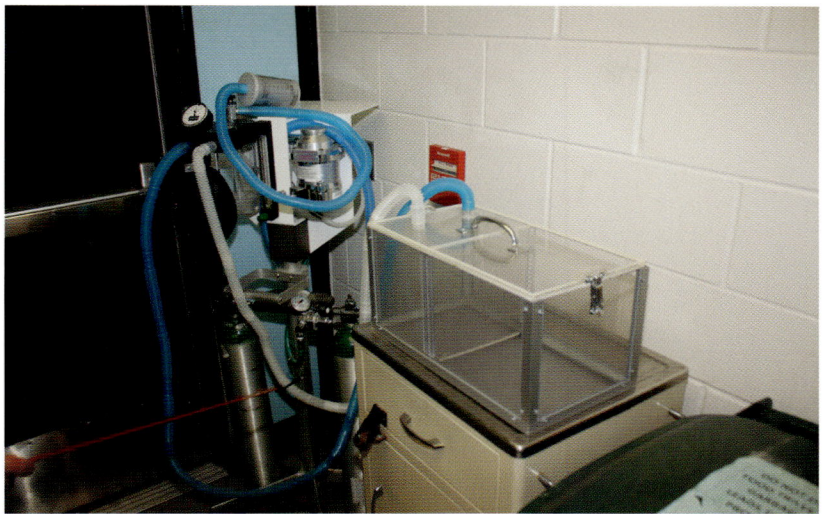
Snakes are anesthetized before being milked, a job that only Steve does. *L.L.C. Jones*

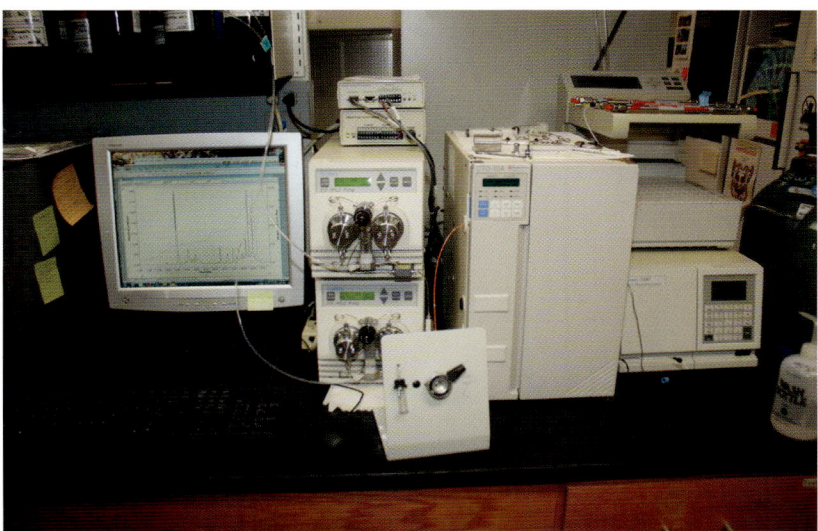
A workstation with a venom protein composition display on the monitor. *L.L.C. Jones*

Their field studies of Prairie Rattlesnakes are quite interesting, where they learn more about ecology than venom. The lab is ideally situated at the edge of the Great Plains in northern Colorado, where they have been marking and monitoring snakes at two dens for about twelve years. To date they have more than 2,000 rattlesnakes pit-tagged. They are getting interesting and valuable information on movement, growth, and site fidelity. On the latter topic, snakes from the two different south-facing dens, approximately 1 km apart, are amazingly dedicated to their same dens, year after year. Long-term studies such as this are few and far between.

Of course, the emphasis of the lab is analyzing venom composition. To acquire venom, they collect snakes in the field and then house them at the university, in a campus-approved animal-care facility. They have about 110 individuals of forty species of snakes in their care. The cages are color-coded with labels that identify the species as venomous (red), rear-fanged/mildly venomous (orange), or nonvenomous (yellow). When venom is extracted, they typically analyze it with HPLC. This takes the venom through a high-pressure pump, which sends venom to a column. The venom is then analyzed by the detector, a spectrophotometer, and finally is divided into components in the fraction collector. The whole process is computer-driven, using software such as Empower to analyze the fractions. They also hope to acquire a mass spectrophotometer, but price is an issue, although it would allow them to better analyze small amounts (micrograms) of venom inexpensively.

Personal safety is of paramount importance at the lab. Dr. Mackessy admits he was envenomated once by a small *C. helleri* when he was a teenager. He insisted I use the adjective "foolishly" when referring to his experience. At his lab now, he takes precautions seriously. The snake room is behind double-locked doors with limited key and card access. Anyone entering the room must wear eye protection, because a snake striking at the screen top can spray venom into the eyes. Dr. Mackessy is the only one allowed to extract venom. He usually anesthetizes snakes first and then extracts venom using the typical milking method. In the twenty-five years of handling thousands of snakes, no one has been bitten—a record he plans to keep.

Some Publications: Hill and Mackessy (2000), Mackessy (1988, 1991, 1993a, b, 1996, 2010a, b, c), Mackessy et al. (2003), Saviola et al. (2013, 2015).

Venom Research Labs and Hospital, Loma Linda University

Loma Linda University is but one of some 450 colleges and universities in the State of California, yet it stands out as being among the top universities for venom research in the world, and has a long history to show it. Two of the world's top toxinologists were among its faculty when the science was in

The laboratories and hospital of Loma Linda University have been graced by numerous researchers, including Findlay Russell, Bruce Halstead, Sean Bush, and Bill Hayes. *L.L.C. Jones*

its early stages: Findlay Russell and Bruce Halstead. They were a rare breed, being both biologists and physicians. They produced hundreds of papers and some books that formed the basis of much of what we know about venomous animals and their effects on humans. Dr. Russell was best known for his work on serpents and stingrays. Dr. Halstead gained a reputation from research on venomous and poisonous marine animals of the world, as well as his broad knowledge of marine animals and their care in captivity. While they were both giants in their field, the legacy of Loma Linda continued, inspired by their contributions.

Sean Bush was another of the medical practitioner/biologist breed, who had gained prominence from his studies on snakebite, with an emphasis on the medical side and patient treatment. He was the featured physician in a television series, *Venom ER*. Dr. Bush had the distinction of bringing envenomation into the media limelight. The program was factually based on cases that came to the Loma Linda Emergency Room, and much of our medical knowledge of envenomation is from Dr. Bush's interactions with patients. Although he has never been bitten himself, he was inspired by the envenomation of his own two-year-old child by a rattlesnake. Sean Bush continues his research at East Carolina University, but he is one of many who have left their mark at Loma Linda.

Dr. Hayes with the infamous Southern Pacific Rattlesnake, the species that causes many serious snakebites in Southern California. It may be hard to see in this photo, but the anterior end of the snake is kept away from his body by a snake hook. *Courtesy of Loma Linda University*

William K. Hayes, who has done a number of studies on venom and venomous animals, has been with Loma Linda University for twenty years. When I asked what his lab should be dubbed, he said "not the Hayes Lab," and that previously it had been known by a variety of monikers that usually included the word "biodiversity." The Biodiversity Lab includes two animal rooms, a procedures room, and two research labs. The research labs are also used by others at the university—after all, this is a world-class research facility and others need to utilize the equipment. Bill is probably best known for his research on venomous snakes of the Great Plains and southwestern California, but his interests and research take him well beyond. While the properties of snake venom have been the focus of much of his work, he has also done behavioral and ecological studies of snakes . . . and much more. He has diverse research interests, and his 5–7 graduate students per year focus on a variety of topics, from venomous animals to bird and lizard ecology. Some of the highlights of Bill's research on venomous animals include: behavior of foraging rattlesnakes; venom composition of different species of rattlesnakes; venom variability within rattlesnake species; and how venomous animals (snakes to widow spiders) regulate venom discharge, or metering. His students have also targeted

Snake terraria at Loma Linda provide a consistently available source of venom that may be difficult to obtain from the wild. *L.L.C. Jones*

venom and venom discharge by centipedes and scorpions. The Biodiversity Lab has a large assortment of venom samples on hand for research from a variety of arthropods and reptiles. There is a large collection of venomous snakes in two animal rooms, from Pygmy Rattlesnakes to mambas. Interaction with these animals is highly regulated among staff. Bill is a firm believer in safety and does extractions himself. He anesthetizes invertebrates, but not snakes. However, he does not press the venom glands, so as to not injure the snake; he lets them do the work.

Loma Linda University is a religious institution that embraces both a spiritual and scientific side of the study of plants and animals. One of Bill's passions is to break down the walls separating religion and science, to allow these worlds to embrace in a common understanding of the wonders of the natural world. Every year, he helps lead two events where the students and the community can experience nature first hand. These wildlife appreciation events are attended by up to 1,500 people. The event features show-and-tell with not only venomous and harmless snakes, but also a wide range of wildlife species. Each event has a theme, such as "Footprints of Africa," where they showcased an African Lion, "Creatures of the Night," and "Expedition Asia." He added, "Heck, one year we even had giraffes out on the lawn, for our 'All Creatures Great and Small' theme and hundreds of people got to feed them by hand."

Carrying on the legacy of venomous marine animal research by Halstead is another colleague, David Hessinger. His research focuses on ion transfer through cell interfaces. His study animals include cnidarians, most notably men-of-war (*Physalia*) and sea anemones. One of his achievements is an edited treatise on the anatomy and functioning of nematocysts. His research is broader than venom research per se, and includes satiety, or feeling "full," and how this relates to obesity in humans, as well as aspects of the physiology of human birth. He has found that using cnidarian research subjects is important in the understanding of some aspects of basic human physiology.

Loma Linda University remains a premier institution for research and treatment of venomous bites and stings. The legacy continues as undergraduate and graduate students help our understanding of these interesting animals as they move on to distinguished careers elsewhere.

Some Publications: Bush (2004), Bush et al. (2000, 2001, 2004), Halstead (1978, 1980, 1992), Hayes and Hayes (1985), Hayes and Mackessy (2010), Hayes et al. (1995, 2008), Hessinger and Lenhoff (1988), Lavonas et al. (2011), Russell (1983), Russell and Saunders (1967), Russell and Scharffenberg (1964), Watson and Hessinger (1989).

Pacific Cnidaria Research Lab (PCRL), University of Hawaii at Manoa

Unlike the other institutional profiles, I could not visit the PCRL in person (it is a bit of a drive from Tucson), so I contacted the lab's director, Dr. Angel Yanagihara, by email. She wrote back and explained why it took a while to get back to me:

> *Aloha Lawrence,*
>
> *I am at the tip of Cape York Australia near Papua New Guinea collecting the most lethal box jelly species,* Chironex. *I have been conducting back to back field efforts for the last few months in the Philippines, Indonesia, and now Australia.*
>
> *Best,*
> *Angel*

Do some people have cool jobs, or what? How did she get into this line of work, with some of nature's deadliest invertebrates? One of her graduate

Angel Yanagihara in the lab with a chirodropid cubozoan. *Courtesy of Angel Yanagihara*

students, Dr. Christie Wilcox, explained it in her book, *Venomous* (Wilcox 2016). Without going into the details of Christie's story (I suggest you read the book instead), I will summarize it here. Angel had nearly completed her Ph.D. in biochemistry from the University of Hawaii, and while out for a swim, she encountered a group of Hawaiian Box Jellyfish. She was seriously envenomated by several jellyfish, when tentacles wrapped around her arms, legs, torso, and neck. As a biochemist, she became intensely interested in the power of the venom from these little colorless sea creatures. Three weeks later, she applied for a grant and started studying venomous cnidarians . . . and hasn't looked back.

The PCRL is one of the few research facilities in the world focusing on cnidarians that are harmful to humans. The lab is ideally situated at Oahu, where venomous cnidarians abound, especially Hawaiian Box Jellyfish and Pacific Bluebottle. These animals provide locally abundant research fodder. Both of these animals and scyphozoans cause problems for local residents, as well as the extensive tourist population of Oahu and the other Hawaiian Islands. Much of what we know about the venom and its effects on humans, ecology, precautions, first aid, and treatment of cnidarian stings originated from the lab.

The research of Dr. Yanagihara, her students, and colleagues focuses on many aspects of the venom: diversity among cnidarians, red blood cell

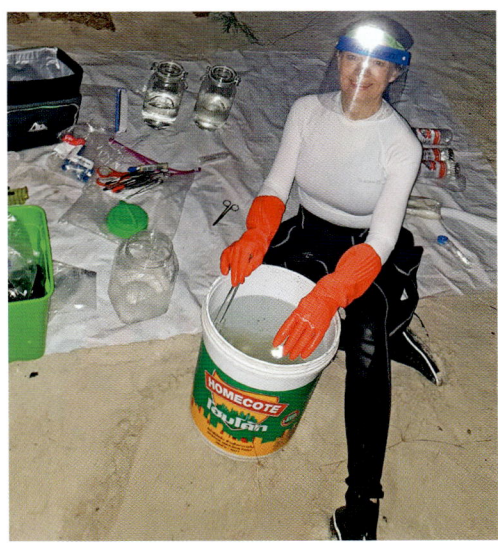

Dr. Yanagihara processing her catch of cubozoans from a night dive. Note that she is wearing a stinger suit under a 1 mm neoprene wetsuit from her dive, and processing with non-porous gauntlets and long tweezers. There is a high price to pay for exposure, so she takes no chances. *Courtesy of Angel Yanagihara*

destruction, effects on pain receptors, and cardiotoxins; structure and functioning of nematocysts; cubozoan phylogeny and taxonomy; population dynamics (especially Hawaiian Box Jellyfish); and fluorescent proteins of *P. utriculus*. She spends much of her time in the field in regions rich with cubozoan diversity. This includes the two species of notoriously lethal chirodropid box jellyfishes, *Chironex fleckeri* and *C. yamaguchii*, from the tropical western Pacific. Some box jellyfishes may also cause Irukandji syndrome, and Dr. Yanagihara and her colleagues are helping to understand more about the mysterious syndrome and which cubozoans stings can cause it.

Perhaps the most pragmatic and useful research topics focuses on precautions and first aid, as cnidarians constitute a health risk in Hawaii and worldwide. The Hawaiian Box Jellyfish (*Alatina alata*) is not unique to Hawaii, and is actually widespread, with colonies in certain areas of Australia, Micronesia, Puerto Rico, and Florida. Researchers in the lab have been tracking their nearshore appearance for many years. *Alatina alata* is unique among the over forty known box jellyfish species in that it comes up from the deep oceans at night and swims actively to coastal shallow reefs to spawn approximately eight to ten nights after each full moon. Coastal disruptions of the food web have led to nuisance numbers of these painful stingers in Mamala

For her night dives to do chirodropid transects in Thailand, Dr. Yanagihara is assisted by a local SCUBA expert and marine oceanographer. *Courtesy of Angel Yanagihara*

Bay, the location of world-famous Waikiki Beach. Various factors including depletion of coastal turtles and reef fish, as well as increased amounts of nutrient rich run-off in the ocean, have led to up to thousands of these stingers appearing during these monthly spawning events since the late 1980s. As part of the long-term monitoring, citizen scientists can assist the lab by helping to collect box jellyfish washed up on shore. In addition to avoiding the nearshore aggregations by knowing peak times, Dr. Yanagihara and her co-workers have tested various methods to reduce pain and lessen other effects of envenomation. She developed a spray that can be applied topically to keep nematocysts from firing and safely remove the unfired "ticking time bomb" nematocysts left on the skin after tentacle contact, appropriately named *Sting-No-More*. She also developed a cream to apply topically to inhibit the venom already injected into the body, something old-fashioned vinegar cannot do. Her discovery has been described in many publications and is the basis for a fully granted U.S. patent.

Some Publications: Bentlage et al. (2010), Chung et al. (2001), Cuypers et al. (2006), Doyle et al. (2017), Wilcox and Yanagihara (2016a, b), Yanagihara (2019), Yanagihara and Shohet (2012), Yanagihara and Wilcox (2017), Yanagihara et al. (2002, 2016a, b, 2017), Yoshimoto and Yanagihara (2002).

National Natural Toxins Research Center, Texas A&M University

When I met Dr. Elda Sánchez, Co-Director of the National Natural Toxins Research Center (NNTRC), I was impressed by the size of the facilities. Then she informed me we were just at the serpentarium; research was conducted in a separate building in the heart of the university campus. Each facility by itself is quite impressive. The NNTRC actually started in the 1970s as a much smaller program but has grown into what it is today. Elda joined the group in the 1990s, and in 2003 they had a major expansion, thanks to a $2.5 million grant they received from the National Institute of Health (NIH). Currently, they are the only NIH-supported snake-venom facility in the United States (hence the "National" in the title).

The serpentarium is in a stand-alone edifice on the outskirts of the campus. This 390 m² facility currently has some 450 snakes, representing about thirty-five species and subspecies, including most taxa from the American Southwest. It has four primary snake rooms where snakes are maintained in captivity for venom extraction. One of the rooms houses snakes from around the Southwest; one is primarily for *Agkistrodon*; one is the *atrox* room; and one has oversized specimens, exotics, and coralsnakes. There are also quarantine rooms for new arrivals. In addition, there are seven large display cases

Elda Sánchez, co-director of NNTRC. *Courtesy of Elda Sánchez*

Dr. Sánchez has ties to her native Venezuela, so maintains *Crotalus vegrandis*, but she also works with colleagues worldwide. *L.L.C. Jones*

for public show-and-tell, with impressive animals such as a massive Eastern Diamond-backed Rattlesnake and a Gaboon Viper. There is a mouse room, where mice are housed and bred for both feeding and research purposes. The focus of the serpentarium is extraction. There are two permanent staff members who tend to the snakes and mice, as well as extract and process venom. Only one of these staffers is allowed to do extraction. He uses the traditional method of snake venom extraction, manipulating the animal with tongs and hooks, and then milking the snake over a beaker. Most extractions are done on the animals maintained in captivity for that purpose. There is a protocol in place for procedure, and what to do if someone is bitten. The two staff members have a perfect record of never having been bitten, but just in case, they are minutes away from a hospital with antivenom and a well-trained staff. Also, the NNTRC has antivenoms on hand for every exotic species they hold in captivity.

Extraction is not just done for on-site research, although some of it is. The NNTRC provides freeze-dried snake venom for other researchers and antivenom production. At the serpentarium, snakes are milked, then their whole venom is pipetted into vials. These are then centrifuged and the purest venom fractions are again pipetted away from the residue (sediments). The purified venom is freeze-dried in a lyophilizer, and stored. The freeze-dried

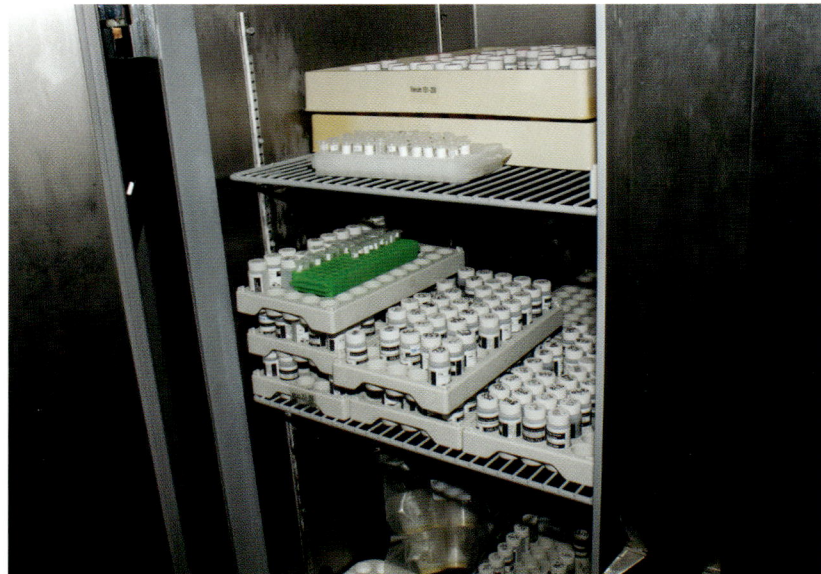

Stores of venom at NNTRC. *L.L.C. Jones*

venom is then sold to other researchers domestically and abroad. One example in our case is that of Texas Coralsnake venom processed at the NNTRC. It is used in the research and production of the experimental Texas/Florida Coralsnake antivenom mentioned elsewhere in this book.

The other part of the facilities on the main campus includes the research labs. There are several primary labs, including general labs, a bloodwork lab, a molecular lab, a tissues lab, and an HPLC lab. The other research labs I visited for my institutional profiles tend to have one or two HPLC units, while the NNTRC has six. The equipment is in constant use by a number of students and researchers. Dr. Sánchez explained the facilities are used by her, other professors, assistant professors, graduate students, visiting graduate students, visiting scientists, and about four undergrads and five graduate students at any given time, as well as an additional 5–10 undergrads working on semester projects for a variety of classes. One such class that Elda teaches covers the biochemistry of venoms and toxins.

The research done here covers a wide gamut of projects. A basic type of research is determining composition of snake venoms, as might be expected from a lab with six HPLC machines. They also test antivenoms. As Elda explained, antivenom research and production has been growing in the past couple of decades, so both private and public organizations are interested in antivenom production and efficacy. They also do work on LD_{50}s, as well as ED_{50}s (Effective Dose in 50% of the drug in those being treated). There are also studies on aptamers—small synthetic molecules (e.g., nucleotides or peptides) that can act as antibodies in cases of envenomation. Researchers at the NNTRC are also cloning venom fractions in bacteria, to produce more useful products for pharmacological uses, as in the treatment of cancers. One of the visiting scientists is engaged in research of allergy and immunity to snake venoms.

The NNTRC has collaborators throughout the world, but especially in the Americas. During my tour, I noticed a number of *Crotalus vegrandis* (= *C. durissus vegrandis*), a species from Venezuela. One of the assistant professors is also from Venezuela. Dr. Sánchez elucidated that she received her Ph.D. in Toxinology from Venezuela, so her venom research understandably crossed borders.

Some Publications: Bohlen et al. (2011), Lomonte et al. (2014), Lucena et al. (2011), McLane et al. (2004), Pérez and Sánchez (1999), Rodriguez-Acosta et al. (2000) Sánchez et al. (2001, 2003, 2005), Vivas et al. (2016).

Venom I (Miami-Dade County) & II (Lake County), Florida: Venom Response Programs

The State of Florida ranks as one of the highest in number of human envenomations per capita, which is not surprising because it is home to some dangerously venomous native animals. In addition, it also harbors the busiest U.S. ports for the shipping and receiving of exotic animals destined for the pet trade, zoos, and aquaria. There are also a number of roadside zoos and reptile ranches in Florida. The state has the highest diversity of non-native, exotic fauna that have become established in the wild, primarily due to escaped and released pets. This includes giant rats, the infamous and massive Burmese Python, and venomous Red Lionfish. The scope of the ecological problem can be seen in the lizard fauna: there are far more non-native lizards established in Florida than native species.

I interviewed Benjamin Abo, a physician and original team member of the famous Venom I program, to learn more about what they do. The program originated in 1998 to help those in need of medical attention after receiving bites and stings in southern Florida. Because of its success, Venom

II was recently added to better facilitate cases in central and northern Florida. The programs respond to all manner of envenomations, but primarily target snakebite emergencies in Florida, and frequently beyond. In addition to the plethora of bites from native species, many people in Florida are tagged by exotics. Dr. Abo stated that "we do whatever is needed to care for people bitten by venomous snakes. Among other things, we put life-saving antivenom in the hands of those that can administer it." They know how and where to locate antivenom stores, have memoranda of understanding with "everyone, including all airlines," and have a helicopter for transport. They also train and assist hospital staff on treatment options and antivenom use. They make themselves available for whatever is needed in case of snakebite. For example, if someone is bitten by their pet cobra, Venom I/II staff will locate the appropriate antivenom, have it delivered to the hospital ASAP, and then have their physicians meet with the ER staff to assist until the victim is in the ICU. Dr. Abo also states that these programs are the "eyes and ears of the poison control centers (three in south Florida)" for cases of envenomation. They maintain a website with information on their emergency response team, facts about venomous animals, and what to do if bitten (e.g., contact us, 24 hours/day).

Dr. Abo says there are about 100–150 snakebites per year in Florida. Bites by Eastern Diamond-backed Rattlesnakes, Florida Cottonmouth, Harlequin Coralsnakes, Pygmy Rattlesnakes, and sometimes Copperheads (in northern Florida) account for the bulk of the domestic species. Antivenom stores are declining and sometimes difficult to find for Harlequin Coralsnakes, but rapid response to this dangerous snake is critical. They also get many calls about snakebite from many exotic species. Cobra bites (and the occasional spitting cobra spray to the eyes), for example, are surprisingly common; according to their website, in 2000–2007 they responded to thirty-seven calls about cobra bites. Other exotics include Green, Black, and Jameson's Mambas; Fer-de-lance; Bushmaster; Puff Adder; Gaboon Viper; Neotropical Rattlesnake; Taipan; Death Adder; and many more. The bites usually occur in the home or an institution. Captives can get loose, putting the public at risk. One caller claimed to have been bitten by a Green Mamba in the wild. While they did not have the culprit in hand to confirm the identity, the symptoms were suggestive of a highly neurotoxic snake, so antivenom for Green Mamba was given, carefully, and successfully. In addition to snake envenomations, they have many interactions with people stung and bitten by many other taxa, including cnidarians, lionfishes, black widows, Brown Recluses, and scorpions. Many of the invertebrates are also exotics that are far more dangerous than our native taxa.

ACKNOWLEDGMENTS

Huge kudos to all of the many people who helped this book come to fruition. First, let me emphasize the many contributors of images listed in the front of the book, many of whom also provided valuable input on venomous animals. I thank the following for reviewing all or parts of the manuscript (exclusive of institutional profiles or their Envenomation Stories): Rich Ayrey, Allen Collins, Aaron Downey, Steve Mackessy, Justin Schmidt, and Steve Werman. Interviews for institutional profiles were conducted with Ben Abo, Cody Bartolini, Keith Boesen, Leslie Boyer, Spencer Greene, William Hayes, Craig Ivanyi, Anita Kristensen, Chuck Kristensen, Baldomero Olivera, Stephane Poulin, Elda Sánchez, Angel Yanagihara, and often various members of their institutional staff.

Many others helped in various aspects of the book, including those who submitted or made available the 1,500 photographs I considered for inclusion but could not be included due to space constraints. Also, everywhere I went during the 4.5 years while preparing this book, I could not help but engage people in discussions of envenomations, symptoms, taxonomy, biology, venom research techniques, and so on. Those who helped in oh-so-many ways include: David Allen, "AnnieH1956," Tom Anton, Bob Ashley, Sheri Ashley, Paul Asman, Brian Aucone, Roy Averill-Murray, Randy Babb, Crystie Baker, Andy Baldwin, Diego Barrales-Alcalá, Paul Barrett, Jeff Barringer, Pia Bartolini, Liz Barta, Heather Bateman, Scott Bauer, Chris Benesh, Alessio Bernardi, Heidi Blasius, Maureen Blasius, Parrish Blasius, Gita Bodner, Jason Boldero, Nick Bonz, Tom Brennan, the Chris Brown not listed in the front, Phil Brown, Richard Brusca, Kathy Burkholder, Bruce Bury, Sean Bush, Dennis Caldwell, Jeff Camper, Mike Cardwell, Judi Carrington, Anne Casey, James Chiucchi, Sean Christensen, June Christman, Sharon Coe, Rob Cole, Kathy Collins, Scott Collins, Polly Conrad, Bill Cooper, Steve Corn, John Corradini, Dennis Crocker, Brian Crother, Kerry Crowther, James Curtiss, Mike Dame, Luke Delano, Dale Denardo,

Jim Dixon, Olliver Dodd, Diane Drobka, Dave Duncan, Doug Duncan, Joe Ehrenberger, Bobby Espinosa, Marty Feldner, Victor Fet, Mark Fisher, Robert Fisher, Lee Fitzgerald, Donald J. Fleischer, Kim Franklin, Denise Garland, James Gatheny, Debbie Gibson, Matt Goode, Matthew Graham, Sean Graham, Anna Gray, Randy Gray, Allan Hack, Bryan Hamilton, Dave Hardy, Trevor Hare, Phil Healy, Ines Hegedus, Toby Hibbitts, Benjamin Hollis, Louis Ira Holmes, Andrew Holycross, Peter Huynh, Robert Johnson, Cristina Jones, Janet Jones, Jason Jones, Josh Jones, Tom Jones, Min Aaron Kang, Steve Killian, Abi King, Allison Kreiss, J. P. Lawrence, Nathan Lawyer, Dan Leavitt, Lee Leavitt, Jeff Lemm, Jill Lenoble, Bill Leonard, Dick Loomis, Tristan Loper, Jeff Lovich, Kim Lovich, Robert Lovich, Bruce Lyon, Dave Lytle, Patti Mahaney, Clark Mahrdt, Don Major, Christopher Maldonado, Stephen Masters, Robert Reed McClure, Kari McWest, Paula Medlock, Terry Merritt, Kacie Miller, Tom Warner Minton, Joe Mitchell, Douglas Moore, Wendy Moore, Amanda Moors, John Morgan, Ray Morgan, Priya Nanjappa, Carl Olson, Charlie Painter, Tony Palmer, John Palting, Kentucky Pete, Lin Piest, Steve Prchal, Dave Prival, Jeremiah Rakoci, Marty Raphael, Chuck Rau, Robin Riechers, Jerry Riepma, Laurie Riepma, Charles Robertson, Phil Rosen, Manny Rubio, Sherby Sanborn, Sean Sartorius, Warren Savary, Li Schmidt, Susan Schuenburg, Gordon Schuett, Cecil Schwalbe, Steven Seifert, Jeffrey Seminoff, Roland Shelly, Bryon Shipley, Robert Siebold, Barry Sinervo, Teresa Singer, Tom Skinner, Michael Smith, Paula Smith, W. Ian Timothy Snyder, Michael E. Soleglad, Stephen Spears, Bernard Sprugg, Mike Sredl, Scott Stockwell, Hanna Strauss, Brian Sullivan, Don Swann, Lauren "Khaleesi" Takerian, Doug Taylor, Emily Taylor, Rebecca Thompson, Barry Thornton, Barney Tomberlin, Ilye Tssuti, Hayley Urbanek, Kevin Urbanek, Ryan Urbanek, James Vanas, Tom Van Devender, Rick Vetter, Laurie Vitt, Kevin Vogt, Bill Voigt, Jim Watt, Burns Weatherby, Barb Webb, Dave Weber, Bill White, Craig Wilcox, Matthew Wilcox, Mike Wilcox, M. D. Wilson, Hunter Winsor, Paul Wolterbeek, Nan Wu, Michael Wunderli, Doug Yanega, and Anne Zones.

Of course, I could not have completed the project without the patience and support of my wife, Janet "Urchin Girl" Jones, plus my stalwart pooches, Cholla the Ebony Lizardhound and Chilipepper Jones the Singing Dalmatian, who were constant companions by my feet while typing, when I should have been taking them for walks. I also thank Aaron Downey and Ross Humphreys of Rio Nuevo Publishers, who probably didn't know what they were getting themselves into when they told me to "go for it!" And to

Kochius colluvius (=K. sonorae) is one of the least common species of scorpion in my study areas. I have no idea how painful the sting may be, and I want to keep it that way. *L.L.C. Jones*

the people I may have forgotten, my humblest apologies. I got so wrapped up in this project that I may have forgotten to jot your names down, but you were all very important in the making of this book, so please email me so that I can include you in a future revision, should it happen.

REFERENCES

There are tens of thousands of potential references, so I focus on books and primary publications that are easily obtainable for free online.

Although the references cited in *Venomous Animals of North America* only represents a small fraction of the available literature on these fascinating creatures, there was still considerable pagination that would have rendered the book even more unwieldy and would have added to the total cost. Not everyone cares to review references, anyway, so they are published in a stand-alone volume. For those who are interested, the References document is available as a separate bound book on Amazon. Go to www.amazon.com and search for Venomous Animals of the United States and Canada References.

GLOSSARY

aboral: In anatomy, the area away from the mouth.

abyss: Generically, this refers to the deep ocean. In the strict sense of marine sciences, the abyssal zone is 4,000–6,000 m deep. In both uses, it refers to an area with no sunlight and having great pressure.

aculeus: The sharp terminal portion of the stinging apparatus of some insects and scorpions, which penetrates the victim and delivers venom. Aculeate species are those that have stingers.

agglutination: Clumping of particles. In the study of effects of envenomation, it usually refers to the binding of antibodies to antigens, or the clumping of blood cells.

aglyphy/aglyphous: A condition in snakes without fangs or long, specialized teeth. This is the normal morphology for snakes considered harmless.

alate: A winged form of social insects (e.g., ants) that are sexually reproductive.

algogenic: Pain-producing.

alveolus (alveoli, plural): A small pit, cavity, or sac. In anatomy, can refer to a number of structures, such as the pulmonary alveoli (air sacs in the lungs), the socket in the jaw for a tooth, or where venom is produced in some glands.

ampulla of Lorenzini: A sensory structure in elasmobranchs (chimaeras, sharks, rays) that detect electrical signals. These organs open through the skin, enabling the animal to detect electrical currents emitted by predators and prey. Some bony fishes also have similar electrosensory receptors.

antibody: A Y-shaped protein, also known as an immunoglobulin (Ig), that will bind with antigens to help neutralize the antigen. For example, specific IgG antibodies can neutralize specific types of venom (to a point).

antigen: A foreign substance that can cause harm if introduced into the human body, such as venom, pollen (if sensitized), virus, or bacteria.

aposematic: Adaptations for warning potential predators, such as vivid coloration, increased sound, or pheromones.

arachnidism: The symptoms following envenomation by an arachnid.

arachnophile: Someone who is fond of arachnids.

arachnophobe: Someone who is afraid of arachnids.

arenicolous: Associated with sandy substrates; synonymous with "psammophilic."

asexual: Reproduction without sexual interaction. Example: a cnidarian polyp, such as a sea anemone, can reproduce by budding.

ballooning: A process by which spiders may disperse through the air. They let out silk from the spinnerets and the differential charge between the ground and sky pulls the silk out and lifts them into the air, where they may be carried by the wind.

Batesian mimicry: A type of aposematic coloration where an animal has a pattern resembling that of another. Example: coralsnake vs. tricolored kingsnake.

benthic: Living on the bottom of a body of water. Example: Typical scorpionfishes and sea anemone.

bioluminescent: Producing biological, rather than reflective, light. Example: A deep-sea lantern shark produces biological light, while a scorpion reflects UV light.

bisexual: Having two sexes, male and female. In human society, this term has a different meaning.

bleb: A small blister on the skin.

bloom: Regarding cnidarians, when large populations appear, usually near the coast. Example: A Portuguese Man-of-War bloom is when thousands may be carried into a swimming beach.

brumation: A hibernation-like state, wherein ectothermic animals have a slowed metabolism, due to cool ambient temperatures. Example: Overwintering rattlesnakes in a den.

buccal: Referring to the area of the mouth.

budding: A process of asexual reproduction, where a single individual can split off another individual.

carapace: A dorsal keratinized structure of the exoskeleton, such as the thorax of a spider.

cardiotoxin: A a type of venom that causes harm to the heart.

caudal: Referring to the tail.

caudal luring: A behavioral trait of certain pit vipers (especially), which wiggle their tail tip to attract prey.

caudal peduncle: The posterior part of a fish that is usually narrower than the body; it anchors and powers the caudal fin.

ceras (cerata, plural): Finger-like projection on the outer surface of aeolid nudibranchs.

chela (chelae, plural): The pincer of a scorpion. Also known as hand or manus.

chelicera (chelicerae, plural): Biting and chewing mouthpart of certain invertebrates, including spiders and scorpions.

chemosensory: Referring to structures that can detect changes in chemical attributes, such as smell or taste.

cheta (chaetae, plural): Chitinous bristle on an invertebrate; it may be a type of seta.

chromatophore: Pigment cell. These cause color changes in the skin of animals, such as squids and octopuses. Examples include melanophore (black pigment cell) and xanthophore (yellow pigment cell).

chrysalis: The pupal stage of a butterfly. It is a keratinized shell that a caterpillar builds; it emerges as a winged adult.

cirrus (cirri, plural): A hair-like or fleshy appendage, such as those on the skin of scorpionfishes.

clade: A group of similar organisms that are believed to have evolved from a common ancestor.

cloaca: The common urogenital opening of some animals, including snakes and birds.

cloacal popping: A defensive display by certain snakes that causes a sound to be emitted from the cloaca; it is often accompanied by excreting foul-smelling substances.

cobweb: A messy web made from spiders of the family Theridiidae.

cocoon: A protective covering made by an animal to protect its young or itself. Examples: silky cover for a moth pupa, sleeping cover for sac spider, and egg sac of a spider.

compartment/compartment syndrome: A compartment in medical terminology is an area of the body that is isolated from nearby areas and held together by tissues such as fascia and bone. The syndrome is when pressure within a compartment builds up to dangerous levels. This causes pain and can threaten a limb or even life.

complex: In taxonomy, a complex is a group of closely related organisms (e.g., species). Members of a complex are often difficult to taxonomically assign or agree upon. Example: The Western/Prairie Rattlesnake species complex, which has caused grief among taxonomists for decades because researchers cannot agree on how to unequivocally differentiate species.

congener/congeneric: Belonging to the same genus.

conspecific: Belonging to the same species.

cosmopolitan: Found across much of the world in appropriate habitat. Example: European Honey Bee.

crepuscular: Active at dusk and/or dawn, during the twilight hours.

cross-reactive: One allergenic substance that is similar to a different substance, which also causes an allergic reaction. Example: A person who is allergic to yellowjackets might have a reaction to Honey Bee venom.

crypsis: Hiding. There are two different uses in biology. (1) Regarding coloration and pattern, it blends in with the background; (2) in taxonomic crypsis, a new taxon is hiding among other similar-looking ones. In the latter, the animals are phenotypically similar, but genetically distinct.

cuticle: A term for a keratinized outer coating, so sometimes this refers to the exoskeleton of arthropods.

cytotoxin: A type of venom that causes harm to cells.

dermal denticle: The scale type of cartilaginous fishes, also known as placoid scale, which is similar in structure to a tooth, having a pulp, nerves, a dentine layer, and an outer enamel-like vitrodentine layer.

dermonecrosis: Death of skin cells. Examples include flesh-eating viruses and an uncommon symptom of recluse spider bites.

desert pavement: In ecology, a dense layer or tightly packed pebbles on the surface of the desert, formed by erosion.

detritivore: An animal that eats detritus, the organic waste from decomposition of dead plants and animals.

diel: Activity on a 24-hour scale, such as sleeping at night and being awake by day or vice versa.

directionality: The tendency to select a specific direction.

diurnal: Active during the day.

Duvernoy's glands: An anatomical feature of colubrid snakes and their relatives that produces proteins. In some species, the secretions may cause a venomous reaction in humans.

ectoparasitoid: A parasite that lives on the outside of an animal (usually arthropods) and eventually kills the host.

Endemism/endemic: Having a geographic range limited to a specific area. Example: The California Butterfly Ray is endemic to the eastern Pacific.

envenomation (= envenoming): Having been bitten or stung by a venomous animal, where venom was introduced. During a dry bite or sting, venom is not introduced.

eusocial: In behavioral studies, this refers to animals that are considered truly social, having a complex system of dependent cooperation. Social insects with different castes (e.g., queens, workers, alates) are eusocial. Examples: European Honey Bees, paper wasps, and harvester ants.

extraoral digestion: Digesting prey outside the predator's body cavity. Chelicerates typically have extraoral digestion, and suck in the prey's liquified organs.

exudates: Secretions.

fossorial: Active under the surface.

foundress: A sexually reproductive queen that establishes a new colony.

fusiform: A body form that is widest in the middle and tapered in the front and back. Many aquatic animals have this general pattern.

genome: The complete set of genetic material making up an organism or a cell.

globalization: Biological globalization is extending a distribution across much of the world into appropriate habitats. Examples: Clinging Jellyfish and Brown Widow Spider.

granivore: An animal that eats grains or seeds.

gravid: Bearing eggs, as opposed to "pregnant," bearing live young.

ground color: An animal's basal or underlying color. Example: A Western Diamond-backed Rattlesnake has a gray or brown ground color, overlain with darker diamond-like markings.

haploid: Having a single set of chromosomes.

hemocoelic fluid: The blood-like fluid that flows between tissues and organs, unconstrained by blood vessels. Many invertebrates have this, including mollusks and arthropods.

hemolytic: Destroying blood cells.

hemotoxic: Causing harm to blood. This term is often misused to categorize certain animals or whole venoms, which may have multiple venom pathways, such as a "hemotoxic" rattlesnake. The toxins can affect much more than blood.

hermaphrodite: A bisexual species that has both male and female reproductive parts. They cannot self-fertilize, so must reproduce in pairs, with each inseminating the other.

hibernaculum (hibernacula, plural): A place where an animal overwinters during brumation or hibernation.

high-grading: Specifically targeting something of high-quality.

hill-topping: A behavioral trait of certain invertebrates to seek out the tops of hills, such as when seeking a mate.

holotype: The individual animal from which the original description of a particular taxon (e.g., species) was scientifically described.

honeydew: The sugary secretion of certain types of insects. When the sap is consumed, it passes through the anus as honeydew and some animals, particularly some hymenopterans, will harvest the substance for larvae. Ants sometimes "milk" honeydew directly from insects, particularly aphids.

hybrid/hybridization: A hybrid is a cross between two species. Hybridize is the verb.

immunoglobulin E (IgE): Antibodies produced by the human immune system to combat antigens. In allergies, IgE is produced in overabundance, causing a systemic reaction very quickly. This can cause anaphylaxis.

immunoglobulin G (IgG): Antibodies produced by the human immune system to combat antigens. IgG can be venom-specific to neutralize some of the effects of venom. IgG can stay in the system for a longer duration than IgE. During immunotherapy, the number of IgG molecules for specific venom increases. There can also be IgG-mediated anaphylaxis.

immunotherapy: A medical treatment administered to build up resistance to venom by increasing the amount of IgG antibodies to a specific venom. People who suspect they are allergic to venom should consider being tested and undergoing immunotherapy.

induration: The hardening of normally soft tissue.

instar: A larval stage of many invertebrates. An insect or arachnid may shed several times, each with distinct instars, before reaching adulthood.

intergrade/intergradation: A cross between two infraspecific taxa, usually a subspecies.

internasal: Between the nares (nostrils).

introgression: The transfer of genetic information from one species to another as a result of hybridization between them and repeated backcrossing.

latrodectism: The symptoms caused by the bite of a widow spider, genus *Latrodectus*, primarily black widows.

lek mating system: When two or more males gather in a communal area to compete for breeding rights.

lentic: Slow to nonmoving water, as in a pond or reservoir.

loreal pit: The heat-sensing pit of pit vipers.

lotic: Flowing water, as in a stream or river.

loxoscelism: The term to describe symptoms derived from the envenomation by a recluse spider, genus *Loxosceles*. Unfortunately, this term is often misused to represent symptoms of unknown origins, or to presume dermonecrosis, so that it has little functional value.

lumen: The interior of a tubular anatomical structure, such as a vein, artery, or venom gland. Venom flows through the lumen. May be synonymous with "duct" in some applications.

Madrean: Referring to the Sierra Madre Occidental of Mexico, and the associated "sky islands" of southeastern Arizona and adjacent New Mexico.

medusa (medusae, plural): A free-living life stage of cnidarians, having an umbrella-like shape. Free-swimming or floating jellyfishes are medusae. The other common form of many cnidarians is the polyp.

melanophore: A black-pigmented skin cell.

mesic: Having moisture, such as a rainforest or riparian area.

midden: A trash heap. Some animals deposit remains of food or other items in a specific area, which may build up their nesting area.

monogynous: A mating system where a male will only breed with a single female during his lifetime, but the female can breed with any males. Example: In eusocial ant colonies, males may only mate with the queen.

monotypic: Literally "one type," referring to a single taxonomic entity. Example: *Micruroides* is a monotypic genus because it only has one species, *M. euryxanthus*.

morphometrics: Body part measurements, including relative lengths (e.g., tail to body length).

myoepithelium/myoepithelial: The former is a layer of outer cells lining an organ, and the latter is the adjective. In venomous animals, myoepithelial cells line venom glands or ducts and squeeze secretions into the lumen.

myotoxin: A type of venom that causes harm to muscles.

nasorostral: Referring to the area in the front of the head, near the nostrils.

necrosis/necrotic: Death of large numbers of cells. The result is a lesion of dead tissue, as seen in some serious rattlesnake bites or a serious recluse bite. In skin, it is called dermonecrosis.

necrotic arachnidism: Necrosis caused by the bite or sting of an arachnid. However, this term is often used specifically for spiders, especially recluse spiders, and is generally a misdiagnosis unless the culprit was confirmed to have caused the envenomation.

nematocyst: A stinging cell of cnidarians. Similar terms include cnidocyte, cnidoblast, and nematocyte.

neonate: A hatchling or recently born individual.

neurotoxin: A type of venom that affects the nervous system.

nocturnal: Active during the nighttime.

ocellus (plural, ocelli): Refers to eyes (especially in invertebrates) or eye-like markings in both invertebrates and vertebrates.

olfaction: The sense of smell and ability to detect smells.

ophidiophobia: A fear of snakes.

opisthoglyphy/opisthoglyphous: The condition of some snakes that have grooved, rear-pointing fangs on the back of the maxilla bone in the upper jaw. Opisthoglyphous species are commonly known as rear-fanged, mildly venomous snakes.

oral arms: Appendages around the mouth of certain cnidarian medusae, including true jellyfishes.

osculum (oscula, plural): The large opening in a sponge.

osteoderm: Literally, "bone skin," referring to bone embedded in skin, as in the beaded skin of a Gila Monster.

ostium (ostia, plural): Hole or cavity in the body. Example: The hole in the side of a sponge's body.

overwinter: To spend the winter, usually in an inactive state, such as rattlesnakes in a hibernaculum in northern latitudes, or a foundress wasp before starting a new colony.

oviparous: Egg-laying.

oviposit: To lay eggs.

ovoviviparous: Retaining eggs in the body, but giving birth to live young.

parapodium (parapodia, plural): Foot-like appendage, as in polychaetes and nudibranchs.

paresthesia: An unusual sensation in the skin that can include tingling, numbness, and burning, among other symptoms. It is often remote from the bite or sting site. Example: Tingling lips from a bite to the hand.

parturition: The act of giving birth to live young.

pedalium (pedalia, plural): The fleshy pad at the four "corners" of cube jellyfishes.

pelagic: Living in the open ocean.

peristomium: The anteriormost true segment of a segmented worm (e.g., polychaete worms).

petiole: A slender structure joining two body parts. The petiole of a hymenopteran joins the thorax and abdomen.

phenotype/phenotypic: Referring to the visual appearance of an animal. This is often opposed to genotype, which is what it is like genetically, regardless of what an animal looks like.

pile: Dense setae that give a hairy or velvety appearance, as on velvet ants, velveteen tarantulas, and bumble bees.

piloerection: A symptom wherein hairs become erect.

planula: A free-swimming larval stage of cnidarians, which is small, usually flattened, and swims by using cilia.

pleuston: Organisms that live at the surface of the water, or at the water/air interface. Examples: Portuguese Man-of-War and Glaucous nudibranchs.

pollen basket: A structure of social bees that is composed of the outer part of the tibia of hind legs and the associated setae, which may be long and stiff. As the bee visits flowers to drink nectar, it gathers pollen in the basket. Also known as the corbicula.

polygynous: A mating system in which a single male may mate with multiple females, but the females mate with only a single male.

polymorphic: Having many forms. Example: Red Imported Fire Ants can have different castes and be of different sizes.

polyp: A structural form of a cnidarian which is generally cylindrical in shape, with the aboral end anchored to the substrate, and the oral end having the mouth and tentacles. Examples: Sea anemone and hydra. The other common form of cnidarians is the medusa. Some cnidarians have both forms.

polyphyletic: In taxonomy, this refers to a group of organisms not derived from a common ancestor, so are not closely related to one another.

population: A geographically defined area for a specific group of plants or animals sharing similar characteristics. For example, in one geographic area of Arizona, Mojave Rattlesnakes predominately have tissue-destroying venom.

procurved: Curved anteriorly.

proleg: Fleshy appendage that is similar in appearance to a walking leg; found on caterpillars and some other insects.

prostomium: The first segment of a segmented worm's body.

proteome: The complete protein composition of an organism, cell, or other entity.

proteroglyphy/proteroglyphous: The condition of some snakes with a hollow, fixed fang on the front of each maxilla bone in the upper jaw. This is the condition of elapids (e.g., coralsnakes).

pruritis: Itching, often intense, and sometimes widespread and remote from the site of a bite or sting.

psammophilic: Associated with sand. This term is commonly used for plants and scorpions, but is synonymous with "arenicolous," a term used for most other animals.

pygidium: The terminal part of certain invertebrates, particularly polychaetes and other annelid worms.

radula: A feature of the feeding anatomy of a mollusk. It is often a rasping, drilling, or harpoon-like structure that is somewhat analogous to the tongue in other animals.

rain harvesting: A behavioral adaptation to gather drinking water from rain. Example: A rattlesnake may drink the water it has captured in its coils.

recurved: Curved posteriorly.

refugium (refugia, plural): An area where an animal seeks relief from inclement weather, such as a hibernaculum to escape cold weather.

rhopalium (rhopalia, plural): A small sensory organ in Scyphozoa and Cubozoa that serves to detect light, maintain balance, and help with swimming.

royal jelly: A secretion of honey bees that workers feed to larvae and the queen. When a queen is needed, large amounts of royal jelly can induce metamorphosis into the queen form.

rugose: Having a rough or wrinkled surface.

saddle: A color pattern feature that roughly resembles a horse's saddle, being broad dorsally, and narrow laterally (think of stirrups). This pattern is commonly seen in snakes, such as lyresnakes and black-tailed rattlesnakes.

saxicolous: Having an affinity for rocky substrates.

sclerite: A thick, hardened segment or section of an arthropod. Similar to tergum or tergite.

scute/scutum: From Latin, "shield," it is usually a large, broad, hard anatomical feature. Examples: the belly scales of snakes and a dorsal feature of ticks and flies.

sensilla (sensillae, plural): A sensory device of one to several cells in the epidermis of certain arthropods. It may be hair-like, peg-like, or a pit.

serous cells: Cells that secrete a watery fluid, which may contain toxic salivary excretions.

sessile: Adhering to the surface, and generally not moving, or moving slowly over time. Examples: sponges and sea anemones.

seta (setae, plural): A hair- or bristle-like anatomical feature of many types of invertebrates. They are sometimes called "hairs," but that term has been widely used for mammals. Setae may or may not be sensory.

sexual dimorphism: Displaying an apparent difference between male and female. Example: Female widow spiders are much larger and robust than males, are often differently patterned, and they lack the pedipalps of males.

site fidelity: Referring to a preference for a particular area, and often returning to it. Example: Rattlesnakes that return to the same hibernaculum year after year have denning-site fidelity.

solenoglyphy/solenoglyphous: A condition of snakes with large, folding fangs in the front of the mouth. This is a dominant feature of vipers. It is considered the most advanced system in snakes.

spermatheca (spermathecae, plural): A sperm-storing anatomical feature of female or hermaphroditic organisms. Sperm storage is beneficial when mates are few and far between.

spicule: A small, sharp siliceous or calcareous structure found in certain invertebrates. In sponges, they deliver crinotoxins into any animal that rubs against them.

spinneret: The highly specialized and advanced anatomical feature of spiders that produces and manipulates silk.

stabilimentum (stabilimenta, plural): A structural feature of the web of some orb-weaving spiders that is often intricate in design, or at least different than the rest of the web. It is often called a web design. The purpose of stabilimenta is not known or agreed upon.

stridulation: The act of producing sound by rubbing together body parts. Examples: crickets, cicadas, and velvet ants.

strobila/strobilation: The strobila is a specialized cnidarian polyp that produces young medusae by splitting transversely. The act of producing medusae is called strobilation.

stylet: In biology, any number of small, slender, and hard structures. This includes the piercing mouthpart of certain insects, the stinging apparatus of hymenopterans, and vestigial shells in octopods.

supercolony: An extremely large colony that is formed when many adjacent colonies of ants (or sometimes other animals) connect. The different colonies not only tolerate one another but work together. Example: The "California Large" is a supercolony of Argentine Ants that stretches some 560 miles.

supraocular: Above the eye, as in the supraocular scale of snakes, which protects the eye.

sympatry/sympatric: When two or more species coexist in the same area. They may frequently encounter each other.

tag/tagged: In vernacular among venomous animal aficionados, this refers to being bitten or stung, regardless if envenomation occurs.

tailing: Vernacular for picking up an animal by the tail, especially to keep from being bitten or stung. Many people are tagged using this generally unsafe handling practice.

tergum (terga, plural)/tergite: A hard, thickened segment of certain arthropods. Similar to sclerite.

thermoregulation: An adaptation wherein an ectotherm maintains body temperature behaviorally, as when a snake basks, then moves into shade.

toxicity: The degree at which a substance can cause harm to an organism. In our use, the LD_{50} is a commonly employed measure of toxicity: the lower the LD_{50}, the higher the toxicity to mice, and presumably, humans. Synonymous with venom "virulence."

trichobothrium (trichobothria, plural): Elongate sensory seta found in arachnids, insects, and myriapods.

troglobite/troglobitic: Living in underground areas, such as caves.

type locality: The location where the holotype was found.

understory: The layer of subdominant plants, like grasses under ponderosa pines.

urticaria/urticarial/urticating: Referring to itching and rash caused by small sharp hairs, setae, spicules, and spines.

velarium: A flap of membrane found in medusae of certain cnidarians, including box jellyfishes.

velichoncha: A larval stage of gastropods.

veliger: A larval stage of gastropods.

velum (vela, plural): A membranous structure having a veil-like appearance.

venom: If you bought this book, I am surprised you are looking up this word.

venome: The complete make up of venom, especially the protein components.

venter: The underside of an organism.

vermiform: Having a worm-like shape.

virulent/virulence: Potency. In our use, it is synonymous with toxicity.

viviparous: Giving birth to live young.

water column: The vertical area of water from the bottom to the interface with air. Most aquatic organisms have a preferred distribution in the water column.

wheal: A small, red, raised area on the skin.

xeric: Dry.

zooid: A single animal that is living in a colony with other similar animals; reproduction is via budding. Example: Pacific Blue Bottle is a colonial organism composed of different zooid types that function differently.

Venom delivery mechanism of Arizona Giant Hairy Scorpion. Not surprisingly, there are many technical terms in this book, and if terms are not in the glossary, they may be in anatomical drawings or descriptions in the text. This image includes the distal metasoma ("tail") and telson, which includes the venom bulb (housing the glands), with aculeus (stinger), and sensory setae ("hairs"). *L.L.C. Jones*

Yellow Paper Wasp. Beauty or the Beast? Both, methinks . . . but mostly the former! *L.L.C. Jones*

INDEX

*page numbers in bold refer to photographs, at species and subspecies level, when known. Photos in page ranges shown in parentheses.

Abedus, 592–598
Abedus herberti, 593, 594, **596**, 599
Abedus indentatus, 593, 595
Acadian Redfish, **618,** 713, 715
Acanthaster, 606
Acanthaster planci, 602, **605**
Acanthaster solaris, 602
Acanthasteridae, 602
Acanthuridae, 618, 726, **730**
acanthurids, 739
Acanthurus, 726, 730, 731, 734
Acanthurus achilles, 734
Acanthurus bahianus, 730
Acanthurus chirurgus, 730
Acanthurus coeruleus, 730
Acanthurus randalli, 730
Acharia stimulea, 294, **298**
Achilles Tang, 734
Aciculata, 550
Acronicta, 295
Actiniaria, 602
Actinopterygii, 617, 682, 687, 696, 704, 712, 725
Acyrtus, 727
aeolid nudibranchs, 17, 542–548
Aeolidioidea, 542
aerial yellowjackets, 211, 212, **217,** 225
Aetobatidae, 662
aetobatids, 647, 662–665
Aetobatus, 662–666
Aetobatus laticeps, 663, 740
Aetobatus narinari, 662–665 (**662, 663**)
Aetobatus ocellatus, 662, 664

Africanized Honey Bees, 46, 52–53, **265,** 266, 268–271, 781
Aglaopheniidae, 602
Agelenidae, 166–167
agelenids, 46, 83, 166–174, 190, 200
Agelena opulenta, 172
Agelenopsis, 166–174 (**168**)
Agelenopsis actuosa, 169
Agelenopsis aperta, 166, 167, 170, 172, 173, 839
Agelenopsis naevia, 169
Agelenopsis oregonensis, 167, 169, 174
Agelenopsis pennsylvanica, 167, 174
Agelenopsis potteri, 169
Agelenopsis utahana, 169
Agkistrodon, 321, 335–336, 342–343, 394, 870
Agkistrodon bilineatus, **348,** 799
Agkistrodon contortrix, 335–341 (**339, 340**), 357
Agkistrodon contortrix contortrix, 335, **339**
Agkistrodon contortrix laticinctus, 336, **337**
Agkistrodon contortrix mokasen, 335, **340**
Agkistrodon contortrix phaeogaster, 335
Agkistrodon contortrix pictigaster, **335,** 336, 787, **788**
Agkistrodon conanti, **65,** 342–348 (**344, 345**), 357
Agkistrodon laticinctus, 335–341

(**337**), 787, **788**
Agkistrodon piscivorus, **324,** 342–348 (**343, 346**), 789–790
air-breathing catfishes, 688, 692
Alacran, 136
Alatina alata, 73, 569–570, **571,** 576–580, 745, 868
Alatinidae, 569, 570
Aliatypus torridus, **208**
Aliciidae, 602
Alphabet Cone, 535
Amaurobiidae, 193
Amaurobius, 202
Ambrysus, 593
Ambrysus lunatus, 595
Ambrysus mormon, 595
ambush bugs, 284
Ameiurus, 687, 688, 694
Ameiurus melas, 688
Ameiurus natalis, 688
American Bumblebee, 274
American coralsnakes, 486, 490
American Cownose Ray, 677–681 (**677, 679, 681**), 853
American round stingrays, 637–645
American Short-tailed Shrew, 509
American short-tailed shrews, 509–514
American Yellow Sac Spider, **186,** 187, 188
Amphioctopus burryi, 521
Amphinome, 550,
Amphinome rostrata, 552
Amphinomidae, 550

anemones, 545, 585, 604, 606, 745
angled orb weavers, **197**
ants, 13, 17, 21, 22, 51, 68, 88, **89**, 98, 148, 153, 238–263, 309, 738
Androctonus, 108
Annelida, 515, 549
Antennata Lionfish, **704**
Anthoathecata, 602
Anthomedusae, 602
Anthophora curta, **317**
Anthophorini, 308
Anthozoa, 516, 602
anthozoan cnidarians, 604
Antrodiaetidae, 193, 202, 203
Antrodiaetus, 193, 194, 202
Antrodiaetus apachecus, 194, 202
Antrodiaetus californicus, **195**
Antrodiaetus gertschi, 194, 202
Antrodiaetus montanus, 194, 202
Antrodiaetus pacificus, 194
Antrodiaetus riversi, 194, 202
Anuroctonus, 136–142
Anuroctonus pococki, **135**, 136, 138, 141–143
Anuroctonus pococki bajae, 136, 138
Anuroctonus pococki pococki, 136, 138
Anuroctonus phaiodactylus, 136, 138, 140, 142
Apache Harvester Ant, 256
Apache Recluse Spider, 154
Aphonopelma, 175–185
Aphonopelma anax, 176
Aphonopelma chalcodes, **97**, **175**, 176, **233**, 234
Aphonopelma chiricahua, 176
Aphonopelma eutylum, 234
Aphonopelma hentzi, 176, **177**, 178, 179, 234
Aphonopelma iodius, 176, 178, 763–765
Aphonopelma joshua, 176
Aphonopelma madera, 176
Aphonopelma marxi, 176, 178
Aphonopelma moderatum, 178, 179, **765**–766
Aphonopelma mojave, 176
Aphonopelma paloma, 176, 178, **179**
Aphonopelma peloncillo, 176
Aphonopelma prentecei, 176

Aphonopelma reversum, 176, 234
Aphonopelma saguaro, 176
Aphonopelma smithi, **764**
Aphonopelma steindachneri, 176, 178, 234
Apidae, 265, 272, 278, 308, 310
Apiomerus, 285, 286
Apiomerus flaviventris, 286, **287**
Apis, 265, 264–271, 773–774
Apis mellifera, **264**–271 (**265**, **267**), 773–774
Apis mellifera carnica, 265
Apis mellifera caucasia, 265
Apis mellifera linguica, 265
Apis mellifera mellifera, 265
Apis mellifera scutullata, 265
aquatic beetles, 593
aquatic insects, 17, 98, 519, 592–600
Arachnida, 42, 108, 119, 127, 135, 143, 154, 166, 176, 186, 193
arachnids, 60, 93, 95, 108–209, 842–845
Araneae, 143, 154, 166, 176, 186, 193
Araneidae, 193
Araneus, 193, 197, 205, 206
Araneus bicentenarius, 197
Araneus cavaticus, 197, 205, 209
Araneus diadematus, **197**, 207
Araneus gemmoides, 197
Araneus occultus, 207
Araneus saevus, 207
Araneus trifolium, 197
Arboreal Orbweaver, 198
Architeuthis, 522
Arctia, 295
Arctic Yellowjacket, 212
Arctiinae, 295
Argentine Ant, 260
Argiope, 193, 197, 206
Argiope aurantia, 197, **206**
Argiope florida, **96**
Argiope trifasciata, 197
Argiopes, 193
Ariidae, 617, 682–686
ariids, 682–686, 694–695
Arilus, 285
Arilus cristatus, **284**, 285, **774**–**775**
Ariopsis felis, **682**, 684, 831, 833
Arizona Bark Scorpion, 18, 34,

46, 69, 76, 108–118 (**109**, **110**), 752, 755–759, 845, **850**, 853, 854
Arizona Black Rattlesnake, 372–377 (**372**, **373**, **375**, **377**)
Arizona Blonde Tarantula, **97**, **175**, 176, **233**, **234**
Arizona Coralsnake, 3, 8, 18, 46, 77, 449, 479–485 (**479**, **480**, **485**), 491, 493, 498, 811
Arizona Giant Desert Centipede, **15**, **100**, 101, **103**
Arizona Giant Hairy Scorpion, **47**, **92**, **106**, 120, **122**, **123**, 754–755, **890**
Arizona Recluse, 46, 154, **156**, 853
Arizona Ridge-nosed Rattlesnake, 457–463 (**457**, **458**, **460**, **463**), 803–806, 807
armored catfishes, 688
Arrow Squid, 521
Arrowhead Dogfish, 630, 633
Arthropoda, 100, 108, 119, 127, 135, 143, 154, 166, 176, 186, 193, 210, 224, 230, 239, 246, 255, 265, 272, 278, 284, 293, 307–317, 515, 592
arthropods, 47, 50, 68–69, 91–98, 100–306, 307, 519, 592, 838, 843, 853, 865
Asian Giant Hornet, **211**
Asian Needle Ant, 309, 311–317 (**316**)
Asilidae, 309
asps. *See* puss caterpillars
assassin bugs, 15, 17, 25, 41, 46, 52, 81, 268, 284–292, 774–775, 776, 781, 782
Astroscopus, 726, 733
Astroscopus guttatus, 726
Astroscopus y-graecum, 726, **728**
Astroscopus zephyrus, 726
Atlantic Blue Tang, 730
Atlantic Brief Squid, 522
Atlantic Devilray, 668
Atlantic Longjaw Squirrelfish, 734
Atlantic Longspine Squirrelfish, 726

Atlantic Pygmy Octopus, 521, 523
Atlantic Sea Nettle, 30, 557, 558, 560, 561, 566
atlantic soapfishes, 730
Atlantic Spearnose Chimaera, 621
Atlantic Spiny Dogfish, **629**, **635**
Atlantic Stingray, 646–654 (**648**)
Atlantic Thornyhead, 715, **719**
Atlantic White-spotted Octopus, 521
auger snails, 531
Aurelia aurita, 557, **563**
Australian Spotted Jellyfish, 557
Australian Tiger Snake, 744
Automeris, 294, 298
Automeris cecrops, 296, 299
Automeris io, 296, **297**
Automeris iris, 296, 299
Automeris iris hesselorum, 777
Automeris louisiana, 299
Automeris patagoniensis, 296, 299
Automeris randa, 296, 299
Automeris zephyria, 296, 299

baboon spiders, 82, 181, 184
backswimmers, 593
Bagre, 683, 684
Bagre marinus, 682, **683**, **686**, 831
Bagre panamensis, 682, 832
Baja California Recluse, 154
Bald-faced Hornet, 212, **215**, 218, 270
bald-faced hornets, 211
Banded Cone, 532
Banded Garden Spider, 197
Banded Gila Monster, **327**, 329, 785
Banded Marble Cone, 533, 536
Banded Rock Rattlesnake, 396–402, 423, 802, 806, 807
Banded Sea-Krait, 744, **745**
Barbfish, 698, 700, 703
bark scorpions, 47, 49, 108–118, 753, 760, 761, 849
Barn Funnel Weaver, 169
Barn Orbweaver, 197
Barronopsis, 168

Bat Ray, 672–676 (**672**, **673**)
Bathytoshia, 646
Bathytoshia centroura, **646**, 649, **652**
Bathytoshia lata, 646, 649, 651
Batoidea, 616
Batrachoididae, 617, 726
batrichoids, 728
Bay Sea Nettle, 557–568
Beaded Lizard, 785–786, 846
beaded lizards, 60, 319, 322, 327, 334
Bearded Fireworm, **552**, 553
Bee Assassin, 268
bee assassins, 286, 289
bees, 8, 13, 17, 18, 20, 21, 22, 23, 35, 40, 46, 53, 54, 68, 88, 98, 243, 264–283, 287, 308, 309, 315, 769, 781–782, 801, 823, 832
Belostoma, 592–600
Belostoma anuran, 599
Belostoma bakeri, 594
Belostoma confusum, 594
Belostoma flumineum, 593, 594
Belostoma lutarium, 594, 598, 599
Belostoma saratogae, 595
Belostomatidae, 592–600
belostomatids, 592–600
Benacus, 592–600
Benacus griseus, 594
Bentfin Devilray, 668
Big Bend Chestnut Scorpion, 136
Big Bend Recluse, 154
Black Bullhead, 688
Black Corsair, 286, 288, 289, 291
Black Dogfish, 630, 631, 633
Black Imported Fire Ant, 247, 249
Black Mamba, **847**
Black Ratfish, 621, 622
Black Rockfish, 712, 713
Black Sea Nettle, 31, 557, 566, **567**
black widow spiders, xvi, 29, 143–153, 313
Black and Yellow Garden Spider, 197, **206**
Black-backed Scorpion, 120
Blackbelly Rosefish, 713, 715

black-clawed scorpions, 136, 138
Black-Footed Spider. *See* American Yellow Sac Spider
black-headed snakes, 483, 497, 499, 503
Blackjacket, 211
black-tailed rattlesnakes, 60, 81, 374, 409–415, 498
black-winged tarantula hawks. *See* tarantula hawks
blanket octopuses, 521–24, 589
Blarina, 509–514 (**513**)
Blarina brevicauda, **509**, 512, 513, 514
Blarina carolinensis, 509, 510, **511**
Blarina hylophaga, 509, 510, 511
Blarina hylophaga plumbea, 510
Blarina peninsulae, 509, 510, 511
Blarina shermani, 510, 512
Blenniidae, 727, 739
blister beetles, **8**, 13, 308
blood-sucking conenose bugs. *See* kissing bugs
bloodworms, 74, 549–555 (**551**)
Blotched Stingray, 740
Blue Catfish, **687**, 688, 689, 691, 692
Blue Mud Dauber, 307, 308, 313, 314, 316, **779**–781
Blue Rockfish, 713
Bluefin Lionfish, 705
blue-ringed octopuses, 62, **64**, 521, 526, 528, 745, 824
Blue-spotted Ribbontail Stingray, 647
Blue-spotted Stingray. *See* Kuhl's Maskray
Bluntnose Stingray, 648, **650**
Bocaccio, 713, 718
bony fishes, **618,** 682–712, 725–735
Boiga irregularis, 502
Boiga dendrophila, 816
Bold Jumping Spider, 199
Boloceroididae, 602
Bombus, 272–277
Bombus bifarius, 274
Bombus bimaculatus, 274

Bombus borealis, 274
Bombus fervidus, 274
Bombus flavifrons, 274
Bombus franklini, 274
Bombus frigidus, 274
Bombus griseocollis, 274
Bombus huntii, 274
Bombus hyperboreus, 274, 275
Bombus impatiens, 274, **275**, 277
Bombus mixtus, 274
Bombus nevadensis, 274
Bombus occidentalis, **272**, **273**, 274
Bombus pensylvanicus, 274, 276–277
Bombus perplexus, 274
Bombus polaris, 274, 275
Bombus sandersoni, 274
Bombus sylivcola, 274
Bombus ternarius, 274
Boomslang, 496, 846
Boreal Bumblebee, 274
Borikenophis portoricensis, 738
Bothriocyrtum, 193
Bothriocyrtum californicum, 194, 202
box jellyfishes, 56, 64, 73, 77, 516, 566, 569–580, 745, 866–869
Brachygastra, 225
Brachygastra mellifica, 212, 225
Brachyponera chinensis, 309, **316**
Branching Anemone. *See* Stinging Anemone
Brazilian Wandering Spiders, 182
bristled polychaetes, 550
bristleworms, 550, 552, 554
Broad-banded Copperhead, 336, **337**
Brown Recluse Spider, 29, 52, **154**, 158, 160, 164
Brown Rockfish, 717
Brown Hoplo, 689–694
brown spiders. *See* recluse spiders
Brown Stingray, 646, 648, 649
Brown-striped Octopus, 521, 523
Brown Treesnake, 502–6, 738, 860
Brown Vinesnake, 496, 499, **500**, 506, 507

Brown Widow, 143–153 (**148**)
Brown-belted Bumblebee, 274
brush-footed butterflies, 295
Buck Moth Caterpillar, 299, **300**, 305
buck moths, 295, 296, 305
bullhead sharks, 624
bullheads, 687, 688, 690, 691, 831
Bullnose Ray, 672–676 (**675**), 679
Bullseye Round Stingray, 638, 643
Bullsnake, 497, 499, 501, 505, 818
bumblebees, 83, 84, 272–277, 279, 281, 312
Bunodeopsis globulifera, 602
Burmese Python, 873
burrowing scorpions, 136, 140, 141
Bushmaster, 64, 874
Buthidae, 108, 737
buthids, 108, 112
Buthus, 108
Buthotus, 108
butterflies, 189, 293, 295, 300, 301
butterfly rays, 658–661
By-the-Wind Sailor, 545, 586, 587

Cabbage-head Jellyfish. *See* Cannonball Jellyfish
Cactus Bee, 308
California Butterfly Ray, 658, 659, 661
California Cone Snail, 533
California Harvester Ant, 256, 771–772
California Lilliput Octopus, **521**
California Lyresnake, 507, 820–21
California Scorpion, **135**
California Scorpionfish, 696, **697**, 698
California Trapdoor Spider, 194, 208
California Turret Spider, 194
California Two-spotted Octopus, 523
California Yellowjacket, 211
Californiconus, 532, 856

Californiconus californicus, 532, 533
Calilena, 168, 170
Calisoga, **176**, 177, 205, 839
Callichthyidae, 688, 689
Callobius, 193, 201, 202, 203, 207
Callistoctopus macropus, 521
Canary Rockfish, 719
Cane Spider, 202
Canebrake Rattlesnake. *See* Timber Rattlesnake
Cannonball Jellyfish, 560, 568
Cantil, **348**, 799
Carabactonidae, 119
caraboctonids, 119, 120
Caracanthidae, 726
caracanthids, 728, 731, 735
Caracanthus typicus, 726, 730
Carangidae, 618, 726
carangids, 733, 739
Cardinal Jumping Spider, 199
Caribbean Reef Squid, 524
Caribbean Roughskin Shark, 630
Carolina Paper Wasp, **228**
Carolina Pygmy Rattlesnake, 468, **469**, **470**
Carolina Wolf Spider, 201, 208
carpenter ants, **310**
Carphophis, 497, 503
cartilaginous fishes, 616, 620, 685
Carukia barnesi, 579
Carukiidae, 576, 579
Carybdeidae, 569, 570
Carybdea arborifera, 569–572
Carybdea confusa, 569–574
Carybdea marsupialis, 569–576
Carybdea rastonii, 576
carybdeids, 569
Cassiopea, 557, 566, 568, **614**
Catalania, 127
caterpillars, 13, 17, 20, 21, 22, 36, 41, 45, 46, 50, 52, 69, 205, 226, 288, 293–306, 313, 555, 777, 781, 782
catfishes, 574, 617, 682–95, 709, 742, 831, 832
cellar spiders, 155
centipedes, 13, 15, 17, 20, 21, 22, 46, 49, 52, 60, 69, 91, 93, 100–107, 113, 122, 400, 401, 461, 476, 737,

748, 752–753, 842, 854, 865
Centrophoridae, 630
Centrophorus granulosus, 630, 633
Centrophorus tessellatus, 630, 633
Centrophorus uyato, 630, 633
Centroscyllium, 630
Centroscyllium fabricii, 630, 631
Centroscyllium nigrum, 630, 632, 633
Centroscymnus owstonii, 630
Centruroides, 76, 108–118, 134, 141, 737, 761, 769, 850
Centruroides exilicauda, 108
Centruroides gracilis, **iv**, 108, **112**, 114, 115, 117, 737
Centruroides griseus, 737
Centruroides guanensis, 108, 111, 114
Centruroides hentzi, 108, 111, 114
Centruroides keysi. See *Centruroides guanensis*
Centruroides sculpturatus, 76, 100, 108–118 (**109**, **110**)
Centruroides vittatus, 108–118 (**111**), 141, 759, 761
Cephalopoda, 517, 520–30
cephalopods, 17, 20, 21, 22, 520–530, 536, 588, 626, 634, 659, 701, 745, 824
Ceropalinae, 230
Chactidae, 128, 135–142
Chaetodon, 726, 730, 731
Chalybion californicum, 308, **779**–781
Channel Catfish, 687–695
Cheekspot Scorpionfish, 699
Cheiracanthium, 186–192
Cheiracanthium diversum, 737
Cheiracanthium inclusum, 186, 187, 188
Cheiracanthium mildei, **96**, 187, 188, 190, **191**
Cheiracanthium mordax, 187, 188, 190
Chelicerata, 186, 193
chelicerates, 93, 181, 186, 193
Chesapeake Bay Sea Nettles, 561
chestnut scorpions, 140, 142
Chihuahuanus, 127, 130

Chihuahuanus coahuilae, **94**
Chihuahuanus crassimanus, **132**
Chihuil, 682, 683, 684, 832
Chilean Devilray, 668
Chilean Recluse, 64, 155, 157, 158, 163
Chilean Round Stingray, 638
Chilipepper Rockfish, 713
Chilopoda, 100–107
Chimaera cubana, 621
chimaeras, 17, 57, 616, 619–623, 627
Chimaeridae, 619
Chimaeriformes, 619–623
Chionactis palarostris, **481**, 498
Chirodropida, 569
Chironex, 576, 745, 866
Chironex fleckeri, 576, 868
chiropsalmid box jellyfishes, 569
Chiropsalmidae, 569
Chriopsalmus alipes, 570
Chiropsalmus quadrumanus, 569, 572–578 (**577**)
Chiropsella, 576
Chloeia, 550
Chloeia flava, 552
Chloeia pinnata, 552
Chloeia viridis, 552
Chlorion cyaneum, 308
Chordata, 327, 335, 342, 349, 355, 359, 366, 372, 378, 383, 389, 396, 403, 409, 416, 421, 425, 431, 435, 441, 445, 450, 457, 464, 468, 473, 479, 486, 490, 495, 509, 619, 624, 629, 637, 646, 655, 658, 662, 667, 672, 677, 682, 687, 696, 704, 712, 720, 725
chordates. See Chordata
Chrysaora, 556, 557, 565, 591
Chrysaora achlyos, 557, **567**
Chrysaora chesapeakei, 557, 560
Chrysaora colorata, **514**, 557, **568**, 825
Chrysaora fuscescens, **556**, 557, **558**
Chrysaora melanaster, 557
Chrysaora quinquecirrha, 557, 566
cicada killers, 308, **309**, 311, 313, 781, 782
Circumtropical Glaucous

Nudibranch, **Front cover**, 12, 542, **547**
Cirrhigaleus, 629
Cirrhigaleus asper, 629
clams, 518, 649, 674, 679
clamworms, 550
Clarias, 689
Clarias batrachus, 689, 695
Clarias fuscus, 689
Clariidae, 688, 689, 692
Clearfin Lionfish, 704
clingfishes, 727, 735
Clinging Jellyfish, 56, **581**–584
Clonophis, 498, 499
Club-horned Wasp, 308
Clubionidae, 187, 188, 194
clubionids, 188, **189**
Cnidaria, 515, 556, 569, 581, 585, 602, 866–869
cnidarians, 4, 15, 17, 21, 22, 45, 56, 72, 73, 84, **516**, 517, 545, 547, 548, 556, 566, 567, 569, 573–577, 581, 582, 589, **604**, 605, 606, 608, 609, 611, 612, 684, 866–869, 874
Coachwhip, **438**
cobras, 60, 64, 319, 851
cobweb spiders, 143
cobweb weavers, 203
Coleoptera, 308, 593
colonial hydrozoans, 587, 602
Coloradia pandora, 294, 299
Colorado Desert Sidewinder, **366**, **369**
colubrids, 20, 21, 22, 70, 72, 319, 321, 495–508, 816, 820, 853
Colubridae, 319, 495–508
comb-footed spiders. See cobweb spiders
combtooth blennies, 727, 739, 743
Cominate Sea Catfish, 832
Combtooth Dogfish, 630
Common Aerial Yellowjacket, 212
Common Desert Centipede, 100, 102, **106**, **124**
Common Eastern Bumblebee, 274, **275**
Common House Spider, 144, 145, 151

Common Octopus, 521, 523
Common Spiny Dogfish, 629, 630
Common Stingray, 652
Common Vampire Bat, **514**
Conasprella, 532, 856
cone snails, 13, 15, 17, 19, 20, 21, 40, 56, 290, 518, 527, 531–**540**, 745, 747, 837, 838, 855–859
conehead bugs. *See* kissing bugs
conenose bugs. *See* kissing bugs
cone-nosed bugs. *See* kissing bugs
Confusing Bumblebee, 274
Conidae, 531–540, 855–859
Coniophanes imperialis, **497**
Conoidea, 531, 532, 539, 855
Contia, 497, 503
Conus, 532, 534, 539, 746, 855, 856
Conus bandanus, 532, 533, 534, 536
Conus dalli, **535**
Conus ermineus, 532, 533, 536, **537**, 858
Conus geographus, 531, **534**, 745, **746**, 747, 858
Conus jaspideus, 535
Conus leopardus, 533
Conus marmoreus, 532, **533**, **534**
Conus nanus, 532, 534
Conus obscurus, 532, 534
Conus pennaceus, 532, 534
Conus purpurascens, 535, **857**, 858
Conus regius, 535, 858
Conus sozoni, 535
Conus spurius, 535, 536
Conus striatus, 532, 533, 534, 858
Conus striatus oahuensis, 534, 536
Conus textile, **531**, 532, 533, **534**, **535**
Conus tulipa, 747
Copula sivickisi, 569, 570, 571, 572
Copper Rockfish, 713
copperheads, 60, 335–341, 343, 346, 347, 364, 474, 488, 512, 846, 874
Coral Catfish, 686, 742, 743

coral catfishes, 739
Coral de Fitzinger, 490
Coral Potosíno, 490
Coral Scorpionfish, 699
Coral de Tampico, 490
Coralillo Bandas Claras, 481, **482**
Coralillo de Mazatlán, 479
Coralillo de Sonora, 479
coralsnakes, 8, 53, 64, 70, 77, 152, 319, 341, 364, 479–494, 498, 618, 723, 812, 846, 849, 870
Coras, 168
Coreidae, 285
Coriniidae, 194
Corixidae, 593
Coronado Island Rattlesnake, 384, 385
cork-lid trapdoor spiders, 193
corsairs, 286, 287, 782
Cortez Soapfish, 731
cottonmouths, 10, 56, 335, 336, 341, 342–348, 364, 512, 846
cow killer, 238, 239
Cowcod, 713
cownose rays, 677–81, 854
crab spiders, 193, 194, 195, **196**, 203, 205
Crabronidae, 308
creeping water bugs, 592
Crescent Octopus, 521, 523
Cross Orbweaver, **197**
Crotalinae, 80, 319, 321
Crotalus, 321, 347, 349, 355, 359, 366, 372, 378, 383, 389, 396, 403, 409, 416, 421, 425, 431, 435, 441, 445, 450, 457, 465
Crotalus abyssus, **349**–353 (**350**, **352**), 379, 404, 451
Crotalus adamanteus, **354**–358 (**357**), 790
Crotalus angelensis, 425, 427
Crotalus atrox, **5**, **322**, **323**, 355, **359**–365 (**361**, **363**), 374, 411, 432, 435, 436, 437, 439, 446, 751, 790–796, 854, 870
Crotalus cerastes, **11**, **366**–371 (**367**-**369**), 806–807
Crotalus cerastes cerastes, 366, **367**

Crotalus cerastes cercobombus, 366, **368**
Crotalus cerastes laterorepens, 366, 368, **369**, 436
Crotalus cerberus, 351, **372**–**377** (**373**, **375**), 451
Crotalus concolor, 351, 376, **378**–382 (**380**), 404, 439, 451
Crotalus durissus, 799
Crotalus exsul. See *Crotalus ruber*
Crotalus helleri, **383**–388 (**384**, **386**), 416, 417, 427, 432, 433, 439, 440, 862, **864**
Crotalus helleri caliginis, 384
Crotalus helleri helleri, **383**, **384**, **386**
Crotalus horridus, 358, 360, **389**–395 (**390**, **392**)
Crotalus horridus atricaudatus. See *Crotalus horridus*
Crotalus horridus horridus. See *Crotalus horridus*
Crotalus lepidus, 60, 81, **396**–402 (**397**, **400**), 439, 459, 806–807, 854
Crotalus lepidus klauberi, **396**, 397, 398, **400**, 401
Crotalus lepidus lepidus, **392**, 396, 397, 401
Crotalus lepidus maculosus, 397
Crotalus lepidus morulus, 397
Crotalus lutosus, 350, 351, 379, **403**–408 (**404**, **405**, **407**), 417
Crotalus mitchellii, 425, 427, 430, 441, 442, 444; See also *C. pyrrhus* and *C. stephensi*
Crotalus mitchellii muertensis, 425
Crotalus molossus, **409**–416 (**411**)
Crotalus molossus estebanensis, 410
Crotalus molossus molossus, **409**, **411**, 415
Crotalus molossus nigrescens, 410, 412
Crotalus molossus oaxacus, 410
Crotalus ornatus, 409–416 (**412**, **413**), 796–797
Crotalus oreganus, 349, 374, 376, 378, 384, 386, 403,

416–420 (**416**, **418**, **419**, **420**), 436, 451
Crotalus oreganus abyssus. See *C. abyssus*
Crotalus oreganus concolor. See *C. concolor*
Crotalus oreganus helleri. See *C. helleri*
Crotalus oreganus lutosus. See *C. lutosus*
Crotalus oreganus oreganus. See *C. oreganus*
Crotalus pricei, 60, 398, 402, **421**–424 (**422**), 854
Crotalus pricei miquihuanus, 421
Crotalus pricei pricei, **421**, **422**
Crotalus pyrrhus, **9**, 351, 412, 425–430 (**426**, **427**), 441, 442, 444, 446, 797–798
Crotalus ruber, **Front cover, 325**, 385, 427, **431**–434 (**432**), 808–810
Crotalus ruber ruber, **Front cover, 325, 431, 432**
Crotalus scutulatus, **75**, 360, 388, 411, 435–440 (**435, 437, 438, 440**), 446, 448, 451, 798–799
Crotalus scutulatus salvini, 436, 437
Crotalus scutulatus scutulatus, **75, 435**, 436, **437, 438, 440**
Crotalus stephensi, 425, 427, 441–444 (**441, 443, 444**), 446
Crotalus tigris, **vi**, 412, 425, 426, 427, 428, 437, 442, **445**–449 (**447, 448**)
Crotalus vegrandis, **871**, 873
Crotalus viridis, 351, 376, 379, 450–456 (**450, 452, 453, 454, 456**), 550, 552, 800–801
Crotalus viridis abyssus. See *Crotalus abyssus*
Crotalus viridis cerberus. See *Crotalus cerberus*
Crotalus viridis concolor. See *Crotalus concolor*
Crotalus viridis helleri. See *Crotalus helleri*

Crotalus viridis lutosus. See *Crotalus lutosus*
Crotalus viridis nuntius, 379, 450, 451, **454**
Crotalus viridis oreganus. See *Crotalus oreganus*
Crotalus viridis viridis. See *Crotalus viridus*
Crotalus willardi, 60, 402, 423, 457–463 (**457, 458, 460, 462, 463**), 806–807
Crotalus willardi amabilis, 457
Crotalus willardi meridionalis, 457
Crotalus willardi obscurus, 398, 457, 459, 461, **462**, 801–803
Crotalus willardi silus, 457
Crotalus willardi willardi, **457, 458, 460**, 461, 462, **463**, 803–806
Crown-of-Thorns Starfish, 74, 81, 519, 602, **605**, 606–608, 613
crustaceans, 150, 153, 285, 519, 528, 553, 554, 563, 573, 576, 626, 641, 649, 659, 664
Ctenidae, 182, 187
Ctenizidae, 193, 202, 203
Ctenocerinae, 230
Cuban Bark Scorpion, 108
Cuban Chimaera, 621
Cuban Dogfish, 629–635 (**634**)
Cubozoa, 516, 569–580
cubozoans, 25, 64, 565, 569–580, 745, 867–868
Cuckoo Bee, 308
cuckoo bumblebees, 273, 276
Cuckoo Yellowjacket, 211
Cupiennius, 182
cuttlefishes, 517, 520, 522, 524, 527
Cyanea capillata, **5**, 557, **559**
Cyaneidae, 557

Daector, 726, 732
Dalatias licha, 630
Dalatiidae, 630
Dall's Cone, **535**
Darna pallavitta, 295, **306**
dasyatid stingrays, 646–657, 655, 659, 665
Dasyatidae, 646–654, 655

Dasyatis, 646, 655
Dasyatis brevis. See *Hypanus dipterurus*
Dasyatis pastinaca, 652
Dasymutilla, 238–245 (**240, 241**), 769–70
Dasymutilla gloriosa, **238**, 239, 770
Dasymutilla klugii, 239, 245
Dasymutilla magnifica, 239
Dasymutilla nigripes, 239
Dasymutilla occidentalis, 239
Dasymutilla quadriguttata, 239
Dasymutilla sackenii, 239
Dasymutilla satanus, 239
Day Octopus, 521, 523
Deania profundorum, 630
Death Adder, 64, 874
Deathstalker Scorpion, 117, **118**
Deepreef Scorpionfish, 699
Deepwater Stingray, 641, 643
Demospongiae, 601
Dendrochirus, 704
Dendrochirus barberi, 704, 707
Dendrochirus brachyptera, 705, 706
Dendrochirus zebra, 705, 706
Desert Fire Ant, 247, 249
Desert Grass Spider, 167, 170
Desert Harvester Ant, 256
Desert Massasauga, xv, 469, **473**–478
Desert Recluse, 154, **157**
Desert Tarantula, 176
Desmodus rotundus, **514**
Devil Firefish, 704
Devil's Chestnut Scorpion, 136
devilrays. *See* mobulas
Diadasia rinconis, 308
Diadema antillarum, 602, 607, 609, **613**
Diadema mexicanum, 607
Diadema paucispinum, 602, 607
Diadematidae, 602
Diademetoida, 602
Diadophis punctatus, 497, **502**, 507, 688, 812–813
Diadophis punctatus stictogenys, 813
Diamond Stingray, 646, 648, 652
Dieunomia, 308
digger bees, 308, 310, 313, 314, 316

Diplocentridae, 136, 142
Diplocentrus, 136, 137, 140, 141, 142, 761
Diplocentrus bigbendensis, 136
Diplocentrus diablo, 136, 138
Diplocentrus lindo, 136, 138, 140, 759
Diplocentrus peloncillensis, 136, 138, 140, 141
Diplocentrus spitzeri, 136, **137**, 138, 140, 141
Diplocentrus whitei, 136, 137, 138, 140, 141
Dipluridae, 167, 170, 194
Dipsadidae, 319, 495, 496, 497
dipsadids, 18, 496, 499, 501, **502**, 738, 853, 860
dipteran insects, 309
diving beetles, 593
Doctorfish, 730
Dog Tick, **43**
dogfish sharks, 629–636
Dolichovespula, 211, 222
Dolichovespula albida, 211
Dolichovespula alpicola, 211
Dolichovespula arctica, 211
Dolichovespula arenaria, 212, 213
Dolichovespula maculata, 213, **215**, 221, 222
Dolichovespula norvegicoides, 212
Doradiidae, 688, 689
dorid nudibranchs, **13**, 542
Doryteuthis, 521
Doryteuthis gigas, 521, 523
Doryteuthis opalescens, 521
Doryteuthis pealeii, 521
Doryteuthis pleii, 522
Doryteuthis roperi, 521
Dosidicus gigas, **518**, 522, **527**
Downy Yellowjacket, 211
Drassodes, 193, 200, 205
Dune Scorpion, 132
Durango Rock Rattlesnake, 396
Dusky Pygmy Rattlesnake, **468**
Dusky Rockfish, 713
Dwarf Cone, 532, 534
dwarf lionfishes, 704, 706
Dwarf Scorpionfish, 699
dwarf tarantulas, 178, **179**, 194
Dysdera crocata, 193, **209**
Dysderidae, 193

Dytiscidae, 593

eagle rays, 672–76
earthsnakes, 503
earthworms, 512, 549
Eastern Bark Centipede, 100, 102
Eastern Black-tailed Rattlesnake, 81, 409, 410, **412**, **413**, 796–797
Eastern Bloodsucking Conenose, 776
Eastern Carpenter Bee, **278**, **279**, 281, **282**, 283
Eastern Cicada Killer, 308
Eastern Copperhead, 79, 80, 335, 336, **339**, **340**, 358
Eastern Coralsnake, 10, **486**
Eastern Diamond-backed Rattlesnake, xv, 354–358 (**354**, **357**), 364, 391, 395, 790, 871, 874
Eastern Gartersnake, 506
Eastern Massasauga, 56, **464**–467 (**465**, **466**), 473, 476
Eastern Pacific Red Octopus, 521, 523, 528, **529**
Eastern Parson Spider, 200, 207
Eastern Puss Caterpillar, **293**
Eastern Velvet Ant, 239, 242
Eastern Yellowjacket, 211
Echinodermata, 515, 602
echinoderms, 17, 519, 605, 606, 608, 609, 612, 626, 664
Echinothrix, 605, 606, 612
Echinothrix diadema, 602, 607
Edwardsiella lineata, 602, 605
Edwardsiidae, 602
Elapidae, 319, 479, 486, 490, 618, 720
elapids, 20, 21, 22, 64, 71, 87, 319, 321, 477, 479, 484, 486, 489, 490, 504, 720, 851
Elasmobranchii, 616, 624, 637, 646, 655, 658, 662, 667, 672, 677
elasmobranchs, 616, 620, 624, 629, 637, 646, 655, 657, 658, 662, 667, 672, 677
electric light bug. *See* belostomatids

Elliot's Short-tailed Shrew, 510
Elongate Twig Ant, 308
Enteroctopus dofleini, **520**, 521, 528
Ephebomyrmex, 255
Epipompilinae, 230
Eratigena, 167–170, 172, 174
Eratigena agrestis, 169, 839
Eratigena atrica, 169
Eratigena duellica, 167, 169
Ericrocis lata, 308
Etmopteridae, 630, 633
Etmopterus, 630, **632**
Etmopterus bigelowi, 633
Etmopterus gracilispinis, 633
Etmopterus hillianus, 633
Etmopterus pusillus, 633
Etmopterus schultzi, 633
Etmopterus spinax, 632
Etmopterus villosus, 633
Etmopterus virens, 633
Euclea, 294, 298, **301**
Euctenizidae, 193
Eumeninae, 212, 308
Euodynerus crypticus, 308
Eunice, 549
European Honey Bee, 14, 23, 40, 41, 65, 68, 89, **264**–271 (**265**, **267**), 277, 738, 766, 768, 769, 773–774, 806
European Hornet, 212, 213, **216**
European Paper Wasp, **224**
Euproctis, 295
Euprymna scolopes, 522
Eurythoe, 550
Eurythoe complanata, 552, 555
Eutichuridae, 186, 187, 188
eutichurids, 187
Everglades Short-tailed Shrew, 509

false black widows, 144, 145, 151
False Scorpionfish, 703
False Tarantula, **176**, 177
fangblennies, 739, 743
Farancia, 497, 499
Feathered Cone, 532, 534
Fer-de-lance, 874
Ficimia streckeri, 496, 500
fiddleback spiders. *See* recluse

spiders
field ants, 308, 315, 316
Filistidae, 155
fire ants, 14, 50, 51, 246–254, 263, 738, 781
Fire Corals, 516, **603**, 606, 609, 610, 611
fire fishes, 704
Fire Sponges, 518, 603, 606, 609
firefishes, 696, 697
fireworms, 74, 550, 552, 553, 554, 555
fish killer. *See* belostomatids
fishes, xvii, 4, 14, 17, 20, 21, 62, 74, 345, 739, 743, 852
Fitch's Octopus, 529, 751, 822
flag-mouth jellyfishes, 556
Flag Rockfish, 713, **714**
flannel moths, 294, 778
Flathead Catfish, 688, 690, 691, 693
fleas, 42, 43
Florida Bark Scorpion, **iv**, 108, 110, 111, **112**
Florida Blue Centipede, 100, 104
Florida Coralsnake. *See* Harlequin Coralsnake
Florida Cottonmouth, **65**, 342–**345 (344)**, 358, 832, 874
Florida Harvester Ant, 256, **257**
Florida Keys Giant Centipede, **65**, 100, 102
flying rays. *See* mobulas
Flying Squid, 522, 524
folding-door trapdoor spiders, 193
Forest Bumblebee, 274
Forest Yellowjacket, 211
Forktail Blenny, 743, **744**
fork-tailed catfishes. *See* sea catfishes
Formica, 308, 311, 313
Formica rufa, 308
Formica obscuripes, 308
Formicidae, 239, 246, 255, 308, 309, 310, 311
Four-handed Box Jellyfish, 569, 570, 572, 574, **575**, 577
Fox-faced Rabbitfish, 741, **836**
Freckled Soapfish, 730
Freckled Stargazer, 726

freshwater catfishes, 687–696, 831
freshwater stingrays, 643, 645
Frigid Bumblebee, 274
Frilled Giant Pacific Octopus, 523
frogs, 148, 153, 338, 454, 471, 476, 492, 501, 597, 598, 599
funnel weaver spiders, 166–174
funnel-web spiders, 167
Fuzzy Dwarf Lionfish, 705
Fuzzy-headed Bumblebee, 274

Gaboon Viper, 64, 871, 874
Gafftopsail Catfish, **682**, 683, 684
garden spiders, 193, 197, 205
gartersnakes, 495, 496, 498, 499, 500, 501, 504, 506, 507, 819
Gastropoda, 517, 531, 541
gastropods, 518, 524, 531, 535, 541, 554, 659, 664, 674, 679,
Gelastocoridae, 593
Genie's Dogfish, 629, **631**
Geographer Cone, 31, 64, **534**, 538, 745, **746**, 747, 859
German Wasp, **211**
German Yellowjacket, 211
Gertschius, 127, 130
Gertschius agilis, 127
ghost sharks. *See* chimaeras
Giant Black-tailed Centipede, 101
giant crab spiders, 193, 738
Giant Desert Centipede, 102, 103, 752–754
Giant Dunes Scorpion, 132
Giant Green Anemone, **601**
Giant House Spider, 167, 169
giant hairy scorpions, 119–125
Giant Manta, 667, **668**, 669, 671
Giant Mobula, **667**–671
Giant Pacific Octopus, **520**, 521, 522, 523, 525, 526
Giant Red-headed Centipede, 101
Giant Robber Fly, 779
Giant Sand Scorpion, 132
giant silkworm moths, 294

giant squids, 522, 523, 525
Giant Stingaree, 638
giant water bugs, 25, 285, 501, 592–600, 853
Gila Monster, **xvi**, 8, 17, 23, 46, 52, 53, 60, 70, 83, 319, 321, 327–334 (**327, 329, 330**), 430, 449, 783–785, 786, 797, 837, 846, 853, 854
Glaucilla. See *Glaucus*
Glaucus, 541–548, 589, 745
Glaucus atlanticus, **Front cover**, **541**–546, **547**
Glaucus bennettae, 542, 545
Glaucus marginatus, **Front cover**, **541**, 542
Glaucus mcfarlanei, 542, 545
Glaucus thompsoni, 542, 544, 545
Glycera, 549, 554, 555
Glycera americana, 552
Glycera capitata, 552
Glycera dibranchiata, 549, **551**, 552, 553, 554
Glycera lapidum, 552
Glycera papillosa, 552
Glycera robusta, 552
Glycera tesselata, 552
Gnaphosidae, 193, 200, 203
Gobiesocidae, 727
Golden Fire Ant, 247
Golden Huntsman Spider, 195, **196**
Golden Paper Wasp, 767
Golden Redfish, 713, 715
Golden Silk Orbweaver, **197**, 198
Gonionemus, 581–584
Gonionemus vertens, 581, 582
Goode's Horned Lizard, **123**
Goosehead Scorpionfish, 698, 700
Gophersnake, 497, 498, 501, **504**, 507, 508, 817–818
Graemeloweus, 127, 130
Grammistidae, 727
Grand Canyon Rattlesnake, 349–353 (**350, 353**), 379
Grand Canyon Recluse, 154
grass spiders, 166–174 (**168**)
grasshopper mice, **91**, 104, 113, 205, 461

Grassland Massasauga, **473**–478 (**475**), 762
Gray's Vaejovis, 759–760
Great Banded Hornet, 738
Great Basin Rattlesnake, **403**–408 (**404**, **405**, **407**)
Greater Soapfish, 730
Green Centipede, 101, 102
Green Lionfish, 704, 706, 707
Green Lynx Spider, 193, 201, **202**, 206, 845
Green Mamba, 846, 874
ground spiders, 193, 201, 203
Guatemalan Beaded Lizard, 329
Gulf Chimaera, 621
Gulf Coast Tick, **40**
Gulf Surgeonfish, 730
Gulf Toadfish, 729
gulper sharks, 630, 632, 633
Gyalopion, 496, 499
Gymnura, 658
Gymnura altavela, **658**–661 (**660**)
Gymnura marmorata, 658–661
Gymnura micrura, 658
Gymnuridae, 658
gymnurids, 659
Gypsy Moth, 295

hacklemesh spiders, 193, 201
Habronattus ophrys, **199**
Hadrurus, 119, 120, 123, 124, 125, 134, 755
Hadrurus arizonensis, **47**, **92**, 120–125 (**122**-**124**), 754–755, 761, **890**
Hadrurus anzaborrego, 120, 121
Hadrurus aztecus, 124
Hadrurus concolorous, 120
Hadrurus gertschi, 124
Hadrurus hirsutus, 120
Hadrurus pinteri, 120
Hadrurus obscurus, 120, 121
Hadrurus spadix, **Front cover**, **119**, 120, 121, 136
Hag Moth, 295, 296, **298**, 299, 305
Hairy Scorpionfish, 699
Haldea, 498, 499
Haldea striatula, 503
Haleciidae, 602
Halictidae, 308

Haller's Round Stingray, 31, 637–650 (**640**), 749
Hapalochlaena, 521, 526, 824
Hardhead Catfish, 682, **683**, 684
Harlequin Coralsnake, 356, 358, 485, **486**–489, 491, 493, 811, 874
Harpactirinae, 181
harvester ants, 10, 18, 29, 46, 50, 52, 68, 255–264, 311, 771–772
Hawaiian Bobtail Squid, 522, 524
Hawaiian Box Jellyfish, **571**, 572, 574, 576, 589, 745, 867, 868
Hawaiian Flying Squid, 522
Hawaiian Lionfish. *See* Hawaiian Turkeyfish
Hawaiian Longjaw Squirrelfish, **725**, 726, 730, 731
Hawaiian Orbicular Velvetfish, 726
Hawaiian Stingray. *See* Brown Stingray
Hawaiian Turkeyfish, 704, **707**
Headwater Catfish, 688, 691
Hebestatis theveneti, 202
Helicolenus, 712, 713
Helicolenus dactylopterus, 713, 715
Heloderma, 327, 328, 329
Heloderma charlesbogerti, 329
Heloderma horridum, **318**, 329, 785–786
Heloderma suspectum, **xvi**, **327**–334 (**329**, **330**), 783, 783–784, 786
Heloderma suspectum cinctum, **327**, 328
Heloderma suspectum suspectum, **xvi**, 328, **329**, 785
Helodermatidae, 319, 327
helodermatids, 21, 22, 319, 327, **328**
Hemileuca, 295, 298, 299, 301, 302, 303
Hemileuca eglanterina, 299
Hemileuca hera, 299, 301
Hemileuca maia, 299, **300**, 301
Hemileuca nevadensis, 299, 301

Hemileuca tricolor, **303**
Hemipepsis, 230–237
Hemipepsis mexicana, 231, 232
Hemipepsis toussainti, 231, 232
Hemipepsis ustulata, 231, 232
Hemipodus roseus, 552
Hemiptera, 42, 284, 285, 592
hemipterans, 68, 285, 287, 781
Hemiscolopendra, 100
Hemiscolopendra marginata, 100, 102
Hentz Striped Bark Scorpion, 108
Hermodice, 550
Hermodice carunculata, **552**
Herpyllus, 193, 200
Herpyllus ecclesiasticus, 200
Herpyllus propinquus, 200, **201**
Herriotta haeckeli, 621
Herriotta raleighana, 621
Heterodon, 495, 497, 505, 625, 813–817
Heterodon kennerlyi, **495**
Heterodontidae, 624
Heterodontiformes, 616
Heterodontus, 624
Heterodontus francisci, 624–628 (**624**, **626**, **627**, **628**)
Heterodontus mexicanus, 624
Heteropneusta fossilis, 691
Heteropneustidae, 691
Heteropoda venatoria, 195, 202, 205, 206, 738
Heteroptera, 285
Hexathelidae, 167
Hexatrygon, 637–645
Hexatrygon bickelli, 637, 641
hexatrygonid, 639
Hexatrygonidae, 637
Hickory Tussock Moth, 295, 299, **300**
High Arctic Bumblebee, 274
Highfin Scorpionfish, 698
Himantura polylepis, 652
Hipponoe, 550
Hipponoe gaudichaudi, 552
Hirudinea, 549
Hobo Spider, 18, 165, 167–174
Hogna, 201, 206
Hogna carolinensis, 201
hog-nosed snakes, 470, 472, **495**, 497, 498, 499, 505, 506, 816–817, 860
Hoffmannihadrurus, 120

Hoffmannius, 127, 128
Holocentridae, 618, 726
Holocentrus rufus, 726, 734
Holocephali, 616, 619
holocephalins, 619
Hololena, 167, 168, 170, 172, 174
Hololena adnexa, 167
Hololena curta, 172
Hololena nedra, **166**, 167
Hololena pacifica, 167
Hololena sula, 167
Hololena turba, 167
honey bees, 21, 203, 218, 221, 265–269, 271, 273, 275, 276, 277, 782
Hong Kong Catfish, 689, 690, 694
Hopi Rattlesnake, 379, 450, 451, **454**
Hopkin's Rose Nudibranch, **13**
Hoplosternon, 689
Hoplosternon littorale, 689
horned lizards, **11**
Hornet Moth, 8
hornets, 210–23, 224, 225, 226, 227, 229, 269, 316
Horn Shark, xvii, 57, 616, **624**–628 (**627**, **626**, **628**), 853
horsefly larvae, 317
Horsefly-like Carpenter Bee, 279
Huamantlan Rattlesnake, 436
Hubb's Octopus, 822
Humboldt Squid, **518**, 522–528 (**527**)
Hunchback Scorpionfish, 698
Hunt Bumblebee, 274
huntsman spiders, 194, 195
Hydra, 604
Hydroidolina, 602
hydroids, 516, 545, 565, 602
Hydrolagus, 621
Hydrolagus affinis, 621
Hydrolagus alberti, 621
Hydrolagus colliei, **618**, **620**
Hydrolagus melanophasma, 621
Hydrolagus mirabilis, 621
Hydrolagus pallidus, 621
Hydrolagus purpurescens, 621
Hydrometridae, 593
Hydrophiidae, 319
Hydrophiinae, 319, 720

Hydrozoa, 516, 581, 585, 602
hydrozoans, 581, 585, 586, 587, 602, 603, 745
Hymenoptera, 98, 210, 224, 230, 239, 245, 246
hymenopterans, 88, 98, 219, 220, 222, 223, 236, 239, 241, 243, 245
Hypanus, 646
Hypanus americana, **6**, 646, **647**, **654**
Hypanus dipterurus, 646, 647, 650, 653, 661
Hypanus longus, 653
Hypanus sabina, 646, **648**, 650
Hypanus say, 646, **650**
Hyphantria, 295
Hyponera, 309
Hypsiglena, 497
Hypsiglena jani texana, 504

Ichneumonidae, 308
ichneumons, 308, 311, 313, 314, 781, 782
Ictaluridae, 617, 687–695, 831
ictalurids, 687–695
Ictalurus, 687, 688, 694
Ictalurus furcatus, **687**, 688
Ictalurus lupus, 688
Ictalurus punctatus, 688
Indo-Pacific Glaucous Nudibranch, **Front cover**, **541**, 542
Indo-West Pacific Blue Tang, 731
inshore squids, 521
Insecta, 42, 210, 224, 230, 239, 246, 255, 265, 272, 278, 284, 293, 592
insects, xv, 10, 20, 21, 22, 50, 52, 54, 65, 91, 98, 142, 210–317, 219, 224, 230, 239, 246, 255, 265, 272, 278, 284, 285, 293, 592, 775, 781–782, 838, 842, 854
Io Moth, 294, 296, **297**, 299, 302, 305
Iodius Tarantula, 763–765
Iracundus, 698
Iracundus signifier, 698
Iridescent Cockroach Hunter, 308
Iris Eyed Silk Moth Larva, 777

Irukandji Jellyfish, 579
Isometrus maculatus, 108, 110, 114, 117
Iuridae, 119, 136
iurids, 119, 120

jacks, 617, 726, 727, 729, 731, 732, 733
Jameson's Mamba, 874
Japanese Hornet, 212
Jasper Cone, 535
jellyfish-like organisms, 516
jellyfishes, 4, 13, 15, 41, 56, 73, 84, 87, 545, 556–568
Johnson's Jumping Spider, 199
jumping spiders, **42**, 193, 198, **199**, 200, 203, 204, 205

Kathetostoma averruncus, 726
Kenyonia, 532
Keys Bark Scorpion, 108
Killer Bees. *See* Africanized Honey Bees
King Baboon Spider, 181
kingsnakes, 10, **11**, **48**, 346, 362, 400, 501, 751
kissing bugs, 24, 41, 42, 46, 49, 64, 284–92, 307, 776
Kitefin Shark, 630, 633, 641
kitefin sharks, 629, 630, 632
Kochius, 127, 130
Kochius colluvius, **877**
Kochius sonorae. See *Kochius colluvius*
Kona Glaucous Nudibranch, 542
Kovarikia, 127, 130
Kovarikia angelena, 127
Kovarikia bogerti, 127
Kovarikia williamsi, 127
Kuhl's Maskray, 740
Kukulcania, 155

Lancer Stargazer, 729
lantern sharks, 57, 630, 631, **632**, 634, 636
large carpenter bees, 65, 278–283
Large-eyed Chimaera, 621
Lasiocampidae, 295
Lasioglossum, 308
Laticauda colubrina, 744, **745**
Latrodectus, 76, 143–153
Latrodectus bishopi, 143, **149**

Latrodectus geometricus, 143, 144, **148**
Latrodectus hesperus, **143**, **144**, **146**, **153**, 762
Latrodectus mactans, 76, 143, **145**, 152
Latrodectus variolus, 143
Leaf Scorpionfish, 699, **700**
leaf skeletonizer moths, 295
leafcutter bees, 313
leaf-footed bugs, 285
leaf-nosed snakes, 497
Leatherback, 726
Lebrunia neglecta, 602
leeches, 549, 555
Lenuchidae, 602
Leirus, 108
Leopard Catfish, 689, 691
Leopard Cone, 533
Leopard Toadfish, 729
Lepidoptera, 293–306
lepidopterans, 293–306
Leptodeira septentrionalis, 497, **501**, 817
Lesser Banded Hornet, 212
Lesser Brown Scorpion, 108, 112, 113
Lethocerus, 592–600
Lethocerus americanus, 593, 594, 827
Lethocerus cordofanus, 599
Lethocerus medius, **592**, **593**, 594, 596, 599, 827–**828**
Lethocerus uhleri, 593, 594, 598, 599
Lilliconus, 532
Lilliput Octopus. *See* Fitch's Octopus *or* California Lilliput Octopus
Limacodidae, 294, 295
Limnomedusae, 581
Lined Snake, 503
Lined Surgeonfish, **739**
Linepithema humile, 260
Linuche aquila, 605
Linuche unguiculata, 557, 602
Liodytes, 498, 499
lionfishes, xvii, 4, 57, 61,74, 617, 696, 697, 702, 704–711, 735, 739, 740, 874
Lion's Mane Jellyfish, 557, 558, **559**, 560, 564, 567
Little Fire Ant, 247, **248**, 249, 250, 738

Living Textile Cone, **531**
lizards, 60, 80, 318, 319, 327, 330, 331, 845–848, 859, 873
Lochmaeus, 295
Loligo, 521
Lolliguncula brevis, 522
Longfin Inshore Squid, 521, 523
Longfin Scorpionfish, 698
long-jawed orb weavers, 203
Long-Legged Sac Spider, 187
long-legged sac spiders, 186
Longnose Chimaera, **621**
longnose chimaeras, 619, 620
Long-nosed Snake, 480, 491, 499
Longspine Scorpionfish, 698
long-spined sea urchins, 602, 612
Long-tailed Stingray, 653
long-whiskered catfishes, 688
Lonomia, 64, 305
Lophocampa caryae, 295, **300**
Loxosceles, 18, 52, 77, 154–163, 172, 191, 737
Loxosceles apachea, 154, 156, 157
Loxosceles arizonica, 154, **156**, 172
Loxosceles blanda, 154, 155, 157
Loxosceles caribbea, 737
Loxoscseles deserta, 154, 155, 156, **157**, 172
Loxosceles devia, 154, 157, 158
Loxosceles intermedia, 158, 161
Loxosceles kaiba, 154, 156
Loxosceles laeta, 155, 157, 161, 163
Loxosecles martha, 154, 157
Loxosceles palma, 154, 157
Loxosceles reclusa, **154**, 155, 157, 161, 163, 172
Loxosceles rufescens, 155, 157, **159**
Loxosceles russelli, 155, 157
Loxosceles sabina, 155, 156, **159**
Loxosceles unicolor. See *Loxosceles deserta*
Loxosceles virgo, 737
lugworms, 550
Lycosa tarantula, 184
Lycosidae, 167, 184, 193, 200, 203
lycosids, 167, 200, 201, 204
Lymantria dispar, 295

Lymantriidae, 295
lynx spiders, 193, 201, 203
lyresnakes, 497, 498, 499, 502, 503, 507, 508, 860

Maaykuyak, 127, 130
Madrean Chestnut Scorpion, 136, **137**, 140
madtoms, 55, 688, 690–695 (**693**), 832
mafia scorpions, 136
mail-cheeked fishes, 696, 697
Malacosoma, 295
Malagasyconus, 532
Mammalia, 509
mammals, xv, 321, 509, 510, 512
Mangrove Snake, 816
Manta. See *Mobula*
manta rays, 668
Marbled Cone, 532, **533**, **534**, 538
Margined Madtom, **690**
Maricopa Harvester Ant, 18, 89, **255**, 256, 258, **259**, 262, 263, 772
marsh treaders, 593
Martha Recluse, 154
Masked Hunter, 286, 288
Masked Shrew, 514
mason bees, 308, 313
massasaugas, 60, 464, 465, 468–470, 473, 475–478
Mastigiidae, 557
Mauve Stinger, 557, 558, 560, **568**
Mediterranean Recluse, 155, 157, 158, **159**
Medusozoa, 516
Megachilidae, 308
megachilids, 310
Megalopyge, 294, 305, 778
Megalopyge bissesa, 299, 778
Megalopyge crispata, 299
Megalopyge immaculata, 299
Megalopyge lacyi, 299
Megalopyge lapena, 299, 778
Megalopyge opercularis, 293, **294**, 299, 777–778
Megalopyge pyxidifera, 299
Megalopygidae, 294
Meiacanthus, 727, 743
Meiacanthus atrodorsalis, 743, **744**

Meiacanthus grammistes, 743
Melanolestes, 285, 287
Melanolestes picipes, 286
Meloidae, 308
Melpomene, 168, 170
Melpomene rita, **169**, 170
men-of-war, 56, 516, 547, 548, 585, 590, 866
Mesobuthus, 108
Mesonychoteuthis hamiltoni, 522
Mexican Beaded Lizard, **318**, 328, 333, 853
Mexican Gartersnake, 498, 507
Mexican Hog-nosed Snake, **495**
Mexican Honey Wasp, 212, 225, 227
Mexican Horn Shark, 624, 625
Mexican Pygmy Octopus, 822, **823**, 824
Microciona, 606
Microtomus, 285, 286, 287, 289
Micruroides, 319, 479, 484
Micruroides euryxanthus, 479–**485** (**479**, **480**)
Micruroides euryxanthus australis, 479, 481
Micruroides euryxanthus euryxanthus, **479**, **480**, 481, **485**
Micruroides euryxanthus neglectus, 479, 481
Micrurus, 77, 319, 480, 481, 484, 485, 486, 489, 490, 491
Micrurus distans, 481, **482**, 485
Micrurus fulvius, 77, 357, 484, 485, **486**–489, 491, 493, 494, 811
Micrurus tener, 77, 484, 489, **490**–494 (**492**), 811–812
Micrurus tener fitzingeri, 490
Micrurus tener maculatus, 490
Micrurus tener microgalbinius, 490
Micrurus tener tener, 490, **492**
Midget Faded Rattlesnake, 351, **378**–382 (**380**)
midshipmen, 726, 727
Milk Snake, 480
Millepora, 602, 603, 610
Millepora complanata, 610
Milleporidae, 602
Miquihuanan Rattlesnake, 421
Mischocytharus, 225
Mischocytharus flavitarsus, 225

Mississippi Ring-necked Snake, 813
Misumena, 203
Misumenoides, 193, 203,
Misumenoides formosipes, 195
Misumenops, 203
Miturgidae, 187, 188
Mobula, 667, 668, 670
Mobula alfredi, 668, 669
Mobula birostris, 667, **668**, 669, 671
Mobula hypostoma, 668
Mobula japonica. See *Mobula mobular*
Mobula mobular, **667**–671
Mobula munkiana, 668
Mobula tarapacana, 668
Mobula thurstoni, 668
mobulas, 667–671
mobulids, 667, 668, 670
Mobulidae, 667
moccasins, 321, 335, 336, 342
Mojave Desert Sidewinder, 366, **367**
Mojave Rattlesnake, 25, 26, 28, 30, 46, **75**, 77, 351, 360, 361, 385, **435**–340 (**437**, **438**), 798–799
moles, 509
Mollusca, 515, 520, 531, 541
mollusks, 56, 517, 518, 520, 522, 531, 541, 855
Monarch Butterfly, 301
monitor lizards, 319, 327, 328
Monognathidae, 727, 743
Moon Jellyfish, 557, 560, **563**, 568
moth larvae. See caterpillars
moths, 293, 294, 295, 300
Mottled Rock Rattlesnake, 396–402 (**397**)
mound ants, 308
mountain kingsnakes, 480
Mourning Cloak, 295
mud daubers, 308, 311, 313, **314**, 779–781
Murder Hornet. See Asian Giant Hornet
Mushroom Scorpionfish, 698
Mutillidae, 235, 238–45
multitentacled box jellyfishes, 569
mygalomorph spiders, 170, 171, 177, **195**, 204

Mygalomorphae, 194
Myliobatidae, 662, 663, 667, 672–676
Myliobatiformes, 637, 646, 655, 658, 662, 667, 672, 677
Myliobatis, 672–676
Myliobatis californicus, 661, **672**, **673**, 674
Myliobatis freminvillei, 672, 674, **675**
Myliobatis longirostris, 673
Myliobatis goodei, 672

Narvesus, 285, 286
Naso, 726, 730
Naso literatus, 731, 734, **836**
Naso Tang, 731, **836**
Natricidae, 319, 495, 496, 498, **505**
natricids, 496, 500, 502
Naucoridae, 592–**600**
naucorids, 592–**600**
nearshore fishes, 609, 738, 739
nearshore octopuses, 523
nearshore scorpionfishes, 700, 701
nearshore squids, 521
nearshore stingrays, 740
Nemertea, 41, 515, **519**
Nemesiidae, 177
nemesiids, 177, 178
Nemophini, 743
Neogastropoda, 531
neogastropods, 532
Neomerinthe, 698
Neomerinthe beanorum, 698
Neomerinthe hemingwayi, 698
Neomerinthe rufescens, 698
Neon Flying Squid, 522
Neoniphon marianus, 734
Neoprocris, 295
Neoscona, 193, 197, 198, 206
Neoscona crucifera, 198
Neoscona domiciliorum, 198
Neoscona oaxacensis, 198
Neotropical Bullet Ant, 782
Neotropical Rattlesnake, 874
Neotrygon kuhlii, 740
Nephila clavata, 206
Nephila clavipes, 193, **197**, 198
Nepidae, 593
Nerodia, 498
Nevada Bumblebee, 274

New Mexico Ridge-nosed Rattlesnake, 457, **462**, 801–803
News Bee, **8**
Night Octopus, 521, 523
nightsnakes, 483, 497, 498, 499, 505
Noctuidae, 295
Noline Scorpionfish, 698
Norape, 294
North American catfishes, 617, 687, 688, 691
North American groundsnakes, 496, 497, 498, 499, 503
North American tarantulas, 176, 181, 182
North American watersnakes, 498
North Atlantic Midshipman, 728
North Pacific Glaucous Nudibranch, 542
Northern Aerial Yellowjacket, 212
Northern Black Widow, 146
Northern Black-tailed Rattlesnake, 409
Northern Cat-eyed Snake, 497, 498, 499, 500, **501**, 507, 817
Northern Copperhead, 335, **340**
Northern Cottonmouth, **342**–348 (**343**, **346**), 789–790
Northern Pacific Rattlesnake, 38, 39, 377, 385, 387, 416–420 (**416**, **418**–**420**), 436
Northern Red-banded Yellowjacket, 211
Northern Sea Nettle, 557, 558, 560
Northern Short-tailed Shrew, **509**, 510, 513
Northern Stargazer, 726, 729
Northern Yellow Sac Spider, **187**
Notocyphinae, 230
Notodontidae, 295
Notonectidae, 593
Notopygos, 550
Notopygos albiseta, 552
Notopygos gregoryi, 552
Notopygos labiatus, 552

Notopygos megalops, 552
Nototodarus hawaiiensis, 522
Noturus, 688, 694, 832
Noturus gyrinus, 832
Novalena, 168, 170
Nudibranchia, 541–548
nudibranchs, 56, 517, 518, 527, 541–548
Nymphalidae, 295
Nymphalis, 295
Nymphalis antiopa, 295

Obscure Cone, 532, 534
Occidentarius platypogon, 832
Ocean Surgeon, 730
Octopoda, 520
Octopus, 520–530
Octopus bimaculatus, 521, 522, 529, **530**
Octopus bimaculoides, 521, 522, 526, 528
Octopus cyanea, 521, 522, 526, 528
Octopus fitchi, 521, 522, 523, 751, 822
Octopus hawaiiensis, 521
Octopus hubbsorum, 822
Octopus joubini, 521, 522
Octopus ornatus, 521, 522, 528
Octopus micropyrsus, **521**
Octopus rubescens, 521, 528, **529**
Octopus vulgaris, 521, 522, 526, 528
octopuses, xv, xvii, 3, 8, 9, 13, 25, 56, 517, 518, 520–530
Odontomachus, 308, **312**, 318
Odontomachus brunneus, 308
Odontomachus clarus, 308
Odontomachus desertorum, 308
Odontomachus haematodus, 308, 309, 315
Odontomachus relictus, 308
Odontomachus ruginodus, 308
Oligochaeta, 549
Oligoplites, 726
Oligoplites altus, 729
Oligoplites palometa, 729
Oligoplites refulgens, 729, 835–**836**
Oligoplites saliens, 729
Oligoplites saurus, 729, 835–836
Olindiasidae. *See* Olindiidae
Olindiidae, 581

Olios, 202, 205
Olios giganteus, 195, **196**, 207
Ommastrephes bartramii, 522
one-jaw eels, 743
Onupis, 549
Opalescent Inshore Squid, 521, 523
Ophion, 308, 782
Opsanus beta, 729
Opsanus pardus, 729
Opsanus tau, 726, 729
Orange Baboon Spider, 181
orbicular velvetfishes, 726
orb-weavers, 205
Oregon Grass Spider, 167
Organ Pipe Mud Dauber, **314**
Organ Pipe Shovel-nosed Snake, 481, 498
Orgyia, 395
Osage Copperhead, 335, 336
owlet moths, 295
Oxybelis aeneus, 496, **500**
Oxynotidae, 630
Oxynotus caribbaeus, 630, 633
Oxyopidae, 193
Oyster Toadfish, 729

Pacific Black Dogfish. *See* Combtooth Dogfish
Pacific Bluebottle, 545, 585–591 (**588**), 827
Pacific Cicada Killer, 308
Pacific Cownose Ray, 678
Pacific Eagle Ray, 663, 664
Pacific Ocean Perch, 713
Pacific Sea Nettle, **556**, 557, 558, 560, 564
Pacific Spiny Dogfish, 629, 630, 634
Pacific Stargazer, 726, 729
Pale Chimaera, 621
Paleobatis fai, 740
Palestripe Podge, 731
Panamint Rattlesnake, **322**, 425, 427, 428, 441–444 (**441**, **443**, **444**)
Pantropical Huntsman Spider, 195
paper wasps, 36, 46, 52, **53**, 65, 211, 212, 219, 224–229, 738, 767–769
Parabuthus, 108
Paracanthurus, 726, 734
Paracanthurus hepatus, 731

Parapterois heterura, 705
Parasa, 296, 298, 782
Parasa chloris, 295
Parasa indetermina, 295
Parasitic Yellowjacket, 212
Parasteatoda tepidariorum, 144
Paratriatoma, 286
Paravaejovis, 127, 128, 130, 761
Paravaejovis pumilis, 128
Paravaejovis spinigerus, 115, 128, 132, **133**, 760–761
Paravaejovis confusus, 128, 132
Pareurythoe, 550, 552
Pareurythoe californica, 552
Paroctopus digueti, 822, **823**–824
Paruroctonus, 127, 128, 131, 134
Paruroctonus boreus, **130**, 134
Paruroctonus gracilior, 134, 761
Paruroctonus stahnkei, **131**
Paruroctonus sylvestris, 134
Paruroctonus utahensis, 759
Pelagia noctiluca, 557, 567, **568**
pelagic eagle rays, 662–666
pelagic hydroids, 571
pelagic jellyfishes, 556
pelagic sharks, 630, 633, 635
Pelagic Stingray, 647, **655**–657, 740
Pelagiidae, 556
Pelamis platura, **720**–724
Peloncillo Chestnut Scorpion, 136, 140
Pelocoris, 592, 593, 595, 599
Pelocoris femoralis, 595
Pelocoris femoratus, 592
Pennaria disticha, 602
Pepsinae, 230
Pepsis, 230–237 (**230**, **233**), 309, 769
Pepsis chrysothemis, 231, 235
Pepsis formosa. See *Pepsis grossa*
Pepsis grossa, 231, 234, 235, 236, 769
Pepsis marginata, 231
Pepsis mexicana, 231
Pepsis mildei, 231, 234, **235**
Pepsis pallidolimbata, 231, 235
Pepsis ruficornis, **237**
Pepsis saphirus, 231
Pepsis thisbe, 231, 234, 235
Perrunichthys, 689
Perrunichthys perruno, 689
Peucetia, 193

Peucetia viridans, 201, **202**
Phenacoscorpius, 698
Phenacoscorpius megalops, 698
Pherecardia, 550
Pherecardia striata, 552
Phidippus, 193, 199, 206, 207
Phidippus apacheanus, 199
Phidippus ardens, 199
Phidippus audax, 199
Phidippus cardinalis, 199
Phidippus carneus, 199, **200**
Phidippus johnsoni, 199
Phidippus regius, 199
Phobetron pithecium, 295, **298**
Pholcidae, 155, 194, 844
Pholcus phalangoides, 207
Phoneutria, 64, 182, 184, **185**
Phoneutria nigriventer, 184
Phyllorhiza punctata, 557
Phyllorhychus, 497
Phyllorhynchus browni, **503**
Physocyclus mexicanus, 844
Physalia, 528, 541, 548, 565, 574, 577, 578, 585–591, 603, 866
Physalia physalis, **585**–591 (**587**), 826
Physalia utriculus, 73, 574, 585–591 (**588**), 827, 868
Physaliidae, 585
physaliids, 585, 589
pileworms, **551**
Pimelodidae, 688, 689
Pine Woods Littersnake, 499, 503
Pink Whipray, 740
pit vipers, 8, 10, 25, 30, 53, 54, 56, 70, 75, 76, 77, 80, 82, 319–324, 331, 340, 343, 347, 352, 422, 461, 498, 812, 838, 849
Pituophis, 497
Pituophis catenifer, 497, 817–818
Plainfin Midshipman, 729
Plains Orbweaver, 197
Platydoras, 689
Platydoras costatus, 689
Plesiobatidae, 637
Plesiobatis, 637–45
Plesiobatis daviesi, 637, 641
Plotosidae, 686, 739, 742
plotosids, 742
Plotosus lineatus, 686, 742, 743

Plumed Scorpionfish, 698
Poecilosclerida, 601
Poecilotheria, 184
Pogonomyrmex, 255–263
Pogonomyrmex apache, 256, 261
Pogonomyrmex badius, 256, 262
Pogonomyrmex barbatus, 256, **257**, 261, 262, 263, 772
Pogonomyrmex californicus, 256, 261, 771–772
Pogonomyrmex comanche, **257**, 261
Pogonomyrmex cunicularius, 262
Pogonomyrmex desertorum, 256, 261
Pogonomyrmex maricopa, **255**, 256, 261, 263, 772
Pogonomyrmex occidentalis, 256, 261
Pogonomyrmex rugosus, 256, **260**, 261
Pogonomyrmex salinus, 256, 261
Pogonomyrmex subdentatus, 261
Pogonomyrmex subnitidus, 260, 261
Polar Bumblebee, 274
Polistes, 224–229, 767–769, 813
Polistes apachus, 225
Polistes annularis, 225
Polistes arizonensis, 225
Polistes aurifer, 225, 767
Polistes bellicosus, 225
Polistes canadensis, 224
Polistes carolina, 225, **228**, 768
Polistes comanchus, 225, **229**
Polistes comanchus navajoe, 228
Polistes dominula, **224**, 767
Polistes dorsalis, 225
Polistes dorsalis dorsalis, 225
Polistes exclamans, 225
Polistes flavus, 225, **227**, 767
Polistes fuscatus, 225
Polistes major, 225, 767
Polistes major castaneicolor, 767
Polistes metricus, 225
Polistes rubiginosus, 225, 229, **768**
Polistinae, 82, 211, 212, 224–229, 231
Polychaeta, 519, 549–557
polychaetes, 17, 20, 21, 549–557, 745
polychaete worms, 56, 549–557

Pompilidae, 230, 231
pompilids, 232
Pompilinae, 230
Pontinus, 698
Pontinus longispinis, 698
Pontinus macrocephalus, 698
Pontinus rathbuni, 698
Porichthys, 726, 732
Porichthys myriaster, 729, 732
Porichthys notatus, 729
Porichthys plectodon, 729
Porichthys porosissimus, 729, 732
Porifera, 515, 518, 601
Portuguese Man-of-War, 31, 73, 545, 547, 566, 574, 577, 578, **585**–591 (**587**), 778, 826
Potamotrygon, 643
Potamotrygonidae, 643
potter wasps, 212, 308, 311, 313, 314
Prairie Rattlesnake, 400, 436, 450–56 (**450**, **452**, **453**, **454**, **456**), 800–801
Prairie Yellowjacket, 211
Prietella, 689
Prionurus, 731, 734
Prionurus punctatus, 730
Pristheancus plagipennis, 289
Profundiconus, 532, 856
Proliferating Anemone, **543**
Promachus aldrichii, 779
prominent moths, 295
Pselliopus, 285
Psithyrus, 273
Pseudomethoca, 239
Pseudomyrmex, 309
Pseudomyrmex gracilis, 308
Pseudouroctonus, 127, 128
Pseudouroctonus reddellii, 131
Pseudouroctonus williamsi, **126**
Pteraeolidia ianthina, 542
Pteroinae, 696, 704–711
Pterois, 704–711
Pterois antennata, **704**
Pterois miles, 704–709 (**706**), 833–834
Pterois radiata, 704
Pterois sphex, 704–**707**
Pterois volitans, 704–**711** (**710**), 833–834
Pteroplatytrygon violacea, **655**–657
Puerto Rican Racer, 738

Puff Adder, 874
Purple Cone, 535, 536
Purple Ratfish, 621
Purple-striped Sea Nettle, xvi, **515**, 557, 560, 564, **565**, 566, **568**, 825
puss caterpillars, 69, 294, 296, 302, 303, 304, 305, 778
Pygmaeconus, 532
Pygmy Devilray, 668
Pygmy Rattlesnake, 60, 464, 468–472 (**468**–**471**), 473, 498, 810, 837, **839**, 865, 874
Pygmy Scorpionfish, 699
pygmy sharks, 630
Pylodictis, 688
Pylodictis olivaris, 688

Quillback Rockfish, 713, **718**

rabbitfishes, xviii, 62, 620, 731, 739, 740, 741, **742**, **836**
Rabid Wolf Spider, 201
Rabidosa rabida, 201
Rabidosa santrita, **204**
ragworms, 550
Rainbow Scorpionfish, 699
Raphael Catfish, 689
Rasahus, 285, 286, 782
Rasahus thoracius, **44**, 286, **290**, 775–776
ratfishes. *See* chimaeras
rattlesnakes, 3, 4, 5, 10, 14, 28, 41, 49, 52, 60, 64, 77, 84, 87, 321, 341, 347, 349–478, **750**, 793, 797, 798–799, 806–807, 846, 852–854, 859, 861, 864
ray-finned fishes, 617
rays, 616, 617, 622, 637–681
rear-fanged snakes, 495–508, 748, 813
recluse spiders, 8, 20, 22, 25, 31, 69, 77, 81, 84, 154–163, 164, 173, 737, 760, 845, 851
Red Diamond Rattlesnake, **Front cover**, **325**, 360, 431–434 (**432**), 436, 807, 808–810
Red Harvester Ant, 256, 772–773
Red Imported Fire Ant, 36, **246**, 247, 260, 770, 782

Red Lionfish, 704–711 (**710**, **711**), 740, 742, 833–834, 873
Red Paper Wasp, 224
red sponges, 518, 606
red velvet ants, 239
Redbanded Rockfish, 713
redfish, **618**, 712
red-haired velvet ants. *See* red velvet ants
Red-headed Giant Centipede, 100
Red-sided Gartersnake, 504, **505**
Red Widow, 143, 144, 145, 146, 147, 148–150, **149**
red-winged tarantula hawks. *See* tarantula hawks
Reduviidae, 284–293, 776
reduviids, 69, 284–293, 775, 776
Reduvius personatus, 288
reef fishes. *See* nearshore fishes
Reef Manta, 669
Reef Scorpionfish, 699
Reef Stonefish, xviii, **740**, 834–835
Regal Black-striped Snake, **497**, 500, 507
Regal Jumping Spider, 199
Regina, 498, 499
reptiles, 52, 70–72, 318, 321, 325–508, 327, 335, 342, 349, 355, 359, 366, 372, 378, 383, 389, 396, 403, 409, 416, 421, 425, 431, 435, 441, 445, 450, 457, 464, 468, 473, 479, 486, 490, 495, 616, 720, 837, 845–848, 852, 853, 859, 865
Reptilia, 349, 327, 335, 342, 355, 359, 366, 372, 378, 383, 389, 396, 403, 409, 416, 421, 425, 431, 435, 441, 445, 450, 457, 464, 468, 473, 479, 486, 490, 495, 720
Reticulated Gila Monster, **xvi**, 328, **329**
Rhadinaea, 497
Rhadinaea flavilata, 503
Rhiginia, 285, 287
Rhinochimaera atlantica, 621

Rhinochimaera pacifica, 621
Rhinochimaeridae, 619, 621
Rhinopias, 698
Rhinopias xenops, 698
Rhinoptera, 677–681
Rhinoptera bonasus, 677–681 (**677, 679, 681**), 853
Rhinoptera brasiliensis, 678, 681
Rhinoptera steindachneri, 678, 681
Rhinopteridae, 677–681
ribbon worms, 41, **519**
Ridge-nosed Rattlesnake, 398, 457–463 (**457, 458, 460, 462, 463**), 801–806
Ring-necked Snake, 483, 497, 498, 499, **502**, 812–813
Rio Grande Gold Tarantula, **765**–766
robber flies, 43, 68, **307**–317 (**311**), 779
rock cod. See rockfishes or thornyheads
Rock Rattlesnake, 81, 235, **396**–402 (**397, 400**)
rockfishes, 57, 84, 617, 696, 697, 712–719
Rocky Mountains Aerial Yellow-jacket, 212
Rogers' Round Stingray, 638
Roper Inshore Squid, 522, 523
rosefishes, 712
Rough Earthsnake, 503
Rough Harvester Ant, 256
Rougheye Rockfish, 716
Roughskin Dogfish, 629, 631, 632, 636
Roughtail Stingray, **646**, 648, **652**, 653, **660**
round stingrays, 637–45, 829
Royal Cone, 535
Rualena, 168, 170
Russell Recluse, 155
Rypticus, 727, 730, 735
Rypticus bicolor, 731
Rypticus bistrispinus, 730
Rypticus maculatus, 730
Rypticus saponicus, 730
Rypticus subbifrenatus, 730

Sabino Recluse, 155, **159**
Saddleback Caterpillar, **298**
Saddled Leaf-nosed Snake, **503**
Sailfin Tang, 731

Salticidae, 193, 198, **199**, 203
salticids, **199**, 302
sand dollars, 519
Sand Scorpion, 132
Sanderson Bumblebee, 274
Sapyga pumila, 308
Sargassum Nudibranch, 545
Sargocentron spiniferum, **725**, 726, 733
Satan, 689
Satan eurystomus, 691
Satan's Velvet Ant, 239, 241
Saturniidae, 294, 295
sawflies, 210, 224, 230, 239, 246, 255, 265, 272, 278
Scarlet Kingsnake, 488, 812
Scarlet Snake, 488, 491, 811
Scarab Hunter Wasp, 308
Scolopendra, 60, 65, 69, 93, 100, 101, 107, 753
Scolopendra alternans, 737
Scolopendra gigantea, 101, 737
Scolopendra hardwickei, 101
Scolopendra heros, **15, 100**, 101, 102, **103**, 104, **105**, 107, 752–54
Scolopendra heros arizonensis, **15, 100**, 101, **103**
Scolopendra heros castaneiceps, 101, **105**
Scolopendra heros heros, 101
Scolopendra longipes, 100, 101
Scolopendra morsitans, 100, 101, 102
Scolopendra polymorpha, 100, 101, **106**, 122
Scolopendra subspinipes, 100, 101, 102, 104, 106, **107**, 737
Scolopendra viridis, 100, 102
Scolopendridae, 100, 103
scolopendrids, 100, 106
Scolopendromorpha, 100, 104
Scoloplos, **549**
Scomberoides lysan, 726, 730
Scorpaena, **701**, 733, 698, 699, 702, 709, 713, 716
Scorpaena agassizii, 698
Scorpaena albifimbria, 698
Scorpaena bergi, 698
Scorpaena brasiliensis, 698, 703
Scorpaena calcarata, 698
Scorpaena colorata, 699

Scorpaena dispar, 698
Scorpaena grandicornis, 698
Scorpaena guttata, **697**, 698, 701, 702, 703, 709
Scorpaena inermis, 698
Scorpaena mystes, 699
Scorpaena pele, 699
Scorpaena plumieri, **696**, 698, 702, 703
scorpaenids, 22, 696–703, 707, 708, 717
Scorpaenidae, 696, 697, 701, 702, 704, 726, 735
Scorpaeniformes, 617, **618**, 696, 697, 704, 712
Scorpaeninae, 696–703
Scorpaenodes, 699
Scorpaenodes caribbaeus, 699
Scorpaenodes corallines, 699
Scorpaenodes evides, 699
Scorpaenodes hirsutus, 699
Scorpaenodes kelloggi, 699, 707
Scorpaenodes parvipinnis, 699
Scorpaenodes tredecimspinosus, 699
Scorpaenodes varipinnis, 699
Scorpaenodes xyris, 699
Scorpaenopsis, 699
Scorpaenopsis altirostris, 699
Scorpaenopsis brevifrons, 699
Scorpaenopsis cacopsis, 699
Scorpaenopsis diabolus, 699, 702
Scorpiones, 108, 119, 127, 135
Scorpionida, 108, 119, 127
Scorpionidae, 135–142
scorpionfishes, xv, 9, 13, 59, 72, 84, 554, 677–711 (**701**), 696, 700, 701, 702, 704, 705, 709, 713, 731, 741, 742
scorpions, 8, 9, 13, 17, 18, 20, 21, 22, 25, 40, 46, 47, **48**, 49, 54, 60, 65, 69, 75, 76, 84, 91, 93, 95, 103, 108–142, 253, 737, 754–762, 838–839, 842, 849, 854, 865, 874
Scyllaea pelagica, 545
Scyphozoa, 556–568, 569
scyphozoans, 573, 574, 602, 745, 876
sea anemones, xvii, 516, 554, 602, 604, 606, 608, 609, 611, 866

sea catfishes, xv, 57, 617, 682–686
sea mice, 550
sea nettles, 556–568, 825
sea slugs, 518, 542
sea snakes, 319
sea urchins, xvii, 74, 519, 601–614
sea wasps. See box jellyfishes
sebastids, 712, 715, 716
Sebastapistes, 699
Sebastapistes balleui, 699
Sebastapistes coniorta, 699
Sebastapistes fowleri, 699
Sebastapistes galactacma, 699
Sebastapistes mauritiana, 699
Sebastapistes nuchalis, 699
Sebastes, 712, 713, 718
Sebastes aleutianus, 716
Sebastes alutus, 713
Sebastes auriculatus, 717
Sebastes babcocki, 713
Sebastes brevispinis, 713
Sebastes caurinus, 713
Sebastes entomelas, 713
Sebastes fasciatus, **618**, 713
Sebastes flavidus, 713
Sebastes goodei, 713
Sebastes levis, 713
Sebastes maliger, 713, **718**
Sebastes melanops, 713
Sebastes miniatus, **712**, 713, 718
Sebastes mystinus, 713
Sebastes norvegicus, 713
Sebastes paucispinis, 713
Sebastes ruberrimus, 713
Sebastes rubrivinctus, 713, **714**
Sebastes serraceps, 713, **714**
Sebastes variabilis, 713
Sebastidae, 696, 712–19
Sebastolobus, 712
Sebastolobus alascanus, 713
segmented worms, 518, 549
Selachii, 616
Semaeostomeae, 556, 557
Sepiida, 520
Serradigitus, 128
Serradigitus agilis, 127
Serradigitus gertschi, 134
Serranidae, 727
Shag Rug Nudibranch, **543**
Shamrock Orbweaver, 197
sharks, 74, 616, 620, 622, 624–636, 649

sharp-tailed snakes, 500, 503
sheathfishes, 682
sheep moths, 295, 299
shrews, xv, 17, 72, 113, 321, 338, 509–14
Shortjaw Leatherjacket, **835**–**836**
shortnose chimaeras, 619, 620
Shortnose Scorpionfish, 699
Shortraker Rockfish, **717**
Shortspine Dogfish, 631, 633, 635
Shortspine Thornyhead Rockfish, 713
short-tailed ichneumons, 308, 311
short-tailed shrews, 509–514
Sicariidae, 154, 155
sicariid spiders, 154, 172
Sidewinder, **11**, 46, 351, 361, **366**–371 (**367**, **368**, **369**), 806
Siganidae, 731, 739
siganids, 739
Siganus argenteus, 741
Siganus doliatus, **742**
Siganus fuscescens, 741
Siganus punctatus, 741
Siganus spinus, 741
Siganus vermiculatus, 741
Siganus vulpinus, 741
silk moth larvae, 294, 777
silk moths, 294, 777
Siluridae, 691
Siluriformes, 682, 687, 694
Silurus glanis, 691
Silvergray Rockfish, 613
Siphonophorae, 516, 585
siphonophores, 545, 585, 586
Sirthenea, 285, 782
Sirthenea carinata, 286
Sirthenea stria, 782
Sistrurus, 86, 321, 347, 394, 464, 465, 468, 473, 474
Sistrurus catenatus, 464–467 (**465**, **466**), 473, 475, 477
Sistrurus miliarius, 358, **468**–472 (**469**, **470**, **471**), 810
Sistrurus miliarius barbouri, **468**, 472
Sistrurus miliarius miliarius, 468, **469**, **470**
Sistrurus miliarius streckeri, 468,

471,
Sistrurus tergeminus, 464, **473**–478
Sistrurus tergeminus edwardsii, 473–478
Sistrurus tergeminus tergeminus, **473**–478 (**475**)
Sixgill Stingray, **639**, 641
skates, 616, 622
sleeper sharks, 630
Slender Inshore Squid. See Arrow Squid
slug caterpillars, 296
slug caterpillar moths, 294
snails, 517, 518, 531
snakes, xv, 8, 10, 13, 14, 15, 17, 21, 22, 26, 29, 34, 46, 49, 52–54, 60–61, 64, 70, 71, 75, 80–81, 85, 153, 318–325, 326, 327, 335–508, 342, 349, 355, 359, 366, 372, 378, 383, 389, 396, 403, 409, 416, 421, 425, 431, 435, 441, 445, 450, 457, 464, 468, 473, 479, 486, 490, 495, 720, 722, 845–848, 859–862, 864–865, 870–871, 874
Smaller Parasa, 295, 296, 305
Small-eyed Chimaera, 621
Smallspine Chimaera, 621
Smeringurus, 128, 131
Smeringurus mesaensis, **129**, 131
Smeringurus vachonis, 134
Smooth Butterfly Ray, 658
smooth dogfishes, 630
Smooth Earthsnake, 503
Smooth Stargazer, 726, 729
Smoothhead Scorpionfish, 698
smoothounds. See smooth dogfishes
soapfishes, 13, 20, 727, 728, 730, 731, 735
social wasps, 232, 235, 310, 781
Solenopsis, 246, 247, 248, 738
Solenopsis ambychila, 247
Solenopsis aurea, 247
Solenopsis geminata, 246–254 (**248**)
Solenopsis invicta, **246**–254, 770
Solenopsis richteri, 246–254
Solenopsis xylori, 247
Somniosidae, 630

Sonora, 496, 497
Sonora palarostris organica. See *Chionactis palarostris*
Sonoran Coralsnake, xv, **2**, 152, **153**, 324, 479–485 (**479**, **480**, **485**)
Sonoran Lyresnake, 507, **508**
Sonoran Shovel-nosed Snake, 8, **481**
Sonoran Sidewinder, 366, **368**
Sorex cinereus, 514
Soricidae, 509
Soricimorpha, 509
Sotanochactus, 136
South Atlantic Midshipman, 729, 732
South Pacific Glaucous Nudibranch, 542, 545
Southern Black Widow, 143, **145**, 146
Southern Carpenter Bee, 281
Southern Copperhead, 335, **339**
Southern Eagle Ray, 672, 673, 674
Southern Fire Ant, 247, 250
Southern Flannel Moth, **293**, **294**, 777
southern house spiders, 171
Southern Pacific Rattlesnake, 374, **383**–388 (**384**, **386**), 434, **864**
Southern Short-tailed Shrew, 509, **511**
Southern Stargazer, 726, **728**, 729
Southern Stingray, **6**, 646, **647**, 648, 649, 653, **654**, 676
Southern Yellowjacket, 211
Southwestern Speckled Rattlesnake, **9**, 60, **226**, **425**–430 (**427**), 442, 445, 797–798
Sozon's Cone, 535
Spanish dancers, 542
Spanish Shawl, **544**
Sparassidae, 193, 202, 203
Speckled Rattlesnake, 9, 425
Speckled Scorpionfish, 699
Specklefin Midshipman, **729**
Sphaerophthalma, 239
sphecid wasps, 308
Sphecidae, 308
Sphecius convallis, 308
Sphecius grandis, 308

Sphecius speciosus, 308
spider wasps, 181, 230–237, 232, 236
spiders, 4, 13, 14, 17, 18, 21, 22, 46–52, 60, 64, 65, 68–69, 75, 91, 93, **95**, 98, 143–209, 737, 749, 837, 838, 839, 842–845, 854
Spilosoma, 295
Spined Pygmy Shark, 630, 633
Spiny Butterfly Ray, **658**–661 (**660**)
Spiny Devilray. *See* Giant Mobula
spiny dogfishes, 564, 616, 628, 629–636
Spiny Oak Slug, **301**
spiny oak slugs, 294, 296
Spinycheek Scorpionfish, 698
Spinytail Round Stingray, 638
spitting spiders, 155, 206
sponges, 13, 15, 17, 20, 74, 304, 518, 602, 603, 606, 607, 608, 610
spookfishes. *See* chimaeras
Spotfin Lionfish, 704
Spotfin Scorpionfish, 699
Spotted Eagle Ray, 662–66, 740
Spotted Queenfish, 726, 729, 731
Spotted Ratfish, **619**–623 (**620**)
Spotted Round Stingray, 638
Spotted Soapfish, 731
Spotted Scorpionfish, 31, **696**, 698
Spotwing Scorpionfish, 698
Squalidae, 629–36
Squaliformes, 616, 629
Squaliolus, 632
Squaliolus laticaudus, 630
Squalus, 629
Squalus acanthias, **629**, 632, 633, **635**
Squalus clarkae, 629, **631**
Squalus cubensis, 629, **634**
Squalus mitsukurii, 631, 632
Squalus suckleyi, 629, 630, 632, 633
Squamata, 80, 319, 327, 335, 342, 349, 355, 359, 366, 372, 378, 383, 389, 396, 403, 409, 416, 421, 425, 431, 435, 441, 445, 450, 457, 464, 468, 473, 479, 486, 490, 495, 720

squids, 3, 517, 518, 520–530
squirrelfishes, 618, 726, 727, 730, 731, 733, 739
Stahnkeus, 128, 130
starfishes, 519, 605
stargazers, 617, 726–728, 731–733
Starry Rockfish, **618**
Steatoda, 144, 151
Steatoda grossa, 144, 151
Steatoda borealis, 144
Sthenoteuthis oualaniensis, 522
stinging caterpillars, 13, 22, 36, 41, 46, 50, 52, 64, 69, 293–306, 777, 781, 782
Stinging Catfish, 691
stinging hydroids, 74, 602, 603, 606, 608, 611
stinging hydrozoans, 745
Stinging Anemone, 604, 606
Stinging Mangrove Anemone, 602, 604, 606
Stinging Nettle Caterpillar, 295, 296, 299, 305, **306**
Stinging Rose, 295, 296, 305
stingrays, xvii, xviii, 3, 5, 8, 13, 14, 18, 22, 23, 45, 57, 59, 62, 64, 72, 73, 74, 84, 85, 565, 574, **615**, 616, 623, 637–681, 829, 830, 852, 863
Striated Cone, 532, 533, 536, 538
Stripe-tailed Scorpion, **133**, 760–761
Striped Bark Scorpion, 36, **91**, 108–118, **133**, 140, 253, 759
Striped Sea Anemone, 602, 606
Stomolophidae, 557
Stomolophus meleagris, 557
Stone Scorpionfish, 699, 703
stonefishes, 62, 64, 696, 703, 739, 741
Storeria, 498, 499, 503
Superstition Mountains Scorpion, 136, 137, **139**
Superstitionia donensis, 136, 138, **139**, 141
Superstitioniidae, 135–142
surgeonfishes. *See* tangs
Suttonia, 727, 735
Suttonia lineata, 731
sweat bees, **41**, 308, 310, 313, 314, 316, 782

swollenstinger scorpions, 137
Synanceia verrucosa, **740**, 834–835
Synanceiidae, 739
synanceiids, 740, 741
Synjaceidae, 696

Tabanidae, 42, 309
Tadpole Madtom, 687, 832
Taenianotus, 699
Taenianotus triacanthus, 699, **700**, 702
Taeniura lymma, 647
Taeniurops meyeri, 740
Taipan, 64, 874
Tamaulipan Hook-nosed Snake, 500
Tamaulipan Rock Rattlesnake, 396
Tamoya haplonema, 569–571, **573**
Tamoyidae, 569, 570
tangleweb spiders, 143
tangs, 62, 618, 726–735 (**730**)
Tantilla, 497
tarantulas, 46, 60, 65, 69, 82, 84, 85, **97**, 175–185, 231, 232, 765, 855
tarantula hawks, 89, 181, **230**–237 (**233**), 312, 791, 855
Tarantula Wolf Spider, 184
Tedania ignis, 601
Tedaniidae, 601
Tegenaria, 18, 167, 168, 170, 172, 174
Tegenaria agrestis. See *Eratigena agrestis*
Tegenaria chiricahuae, 170
Tegenaria domestica, 169
Tengellidae, 155, 194
tent caterpillars, 227, 295
Terebridae, 531, **540**
terebrids, 534, 539, **540**
Terrestrial Gartersnake, 819–820
Tetragnatha, 193, 198, 763
Tetragnatha montana, **198**
Tetragnathidae, 193, 198
tetragnathids, 198, 205, 763
Tetramorium, 309
Teuthoidea, 520
Texas Coralsnake, 486, **490**–494 (**492**), 811–812

Texas Gulf-Coast Coralsnake, **490**, **492**
Texas Lyresnake, **499**
Texas Nightsnake, 504
Texas Recluse, 154
Texas Red-headed Centipede, 101, **105**
Texas Brown Tarantula, 176
Texas Tan Tarantula, 176
Textile Cone, **531**–540 (**534**)
Thalassophryne, 726, 732, 733
Thalassophryne natterei, 732
Thamnophis, 498
Thamnophis elegans, 498, 819–820
Thamnophis elegans vagrans, 498
Thamnophis eques, 498
Thamnophis sirtalis, 498, **505**
Thamnophis sirtalis parietalis, 504
Thamnophis sirtalis sirtalis, 506
Thaumetopoea wilkinsoni, 295
Theraphosidae, 176
theraphosids, 176, 177, 184
Theridiidae, 49, 143, 144
thick-jawed orb weavers, 193
thief ants, 246
Thimble Jellyfish, 556, 602, 605, 606, 608
Thistledown Velvet Ant, **238**, 241, 770
Thomisidae, 193, 203
thornbacks, 616, 637, 646, 655, 658, 662, 667, 672, 677
thorny catfishes, 688
thornyheads, 57, 617, 696, 697, 712–719
ticks, **40**, 42, 44
Ticon Cownose Ray, 678
Tiger Keelback Snake, 496
tiger moths, 295
Tiger Rattlesnake, **vi**, 6, 81, 425, **426**, 427, **445**–449 (**447**, **448**), 783
Timber Rattlesnake, 60, 389–395
Timulla, 239
Titan Scorpionfish, 699
Titiotus, 155
Tityus, 108, 737, 752
toad bugs, 593
toadfishes, 20, 617, 726–733
toe biters. See belostomatids
toothed scorpions, 136

Tortolena, 168
Toxoglossa, 531
Trachyscorpia, 712
Trachyscorpia cristulata, 715, **719**
trapjaw ants, 308, 311–316 (**312**), 738
Trans-Pecos Chestnut Scorpion, 136
Trans-Pecos Copperhead, **335**, 787–789 (**788**)
Treefish, 713, **714**
Tremoctopus gelatus, 521
Tremoctopus gracilis, 521
Tremoctopus violaceus, 521
Triakidae, 630
Triatoma, 285, 286, 292
Triatoma hirsuta, 286
Triatoma lecticularia, 286
Triatoma protracta, 286
Triatoma recurva, **44**, 286, **288**
Triatoma rubida, 286
Triatoma rubrofasciata, 286
Triatoma sanguisuga, 286, 776
Triatominae, 285
Tri-colored Bumblebee, 274
Trichonephila clavipes. See *Nephila clavipes*
Trigonognathus, 630
Trigonognathus kabeyi, 630
Trimorphodon, 497, 502, 505
Trimorphodon lambda, 505, 507, **508**
Trimorphodon lyrophanes, 507, 820–821
Trimorphodon vilkinsonii, **499**
Tripedalia cystophora, 569–572
Tripedaliidae, 569, 570
Triscolia ardens, 308, **309**
Troglobites, 688, 689, 690, 694
Trogloglanis, 689
tropical centipedes, 100, 101, 104
Tropical Fire Ant, 247, **248**, 249
Tropical Rattlesnake, 77, 799
tropical tarantulas, 184
Tropidoclonion, 498, 499
Tropidoclonion lineatum, 503
true bugs, 13, 189, 284, 285, 519, 592, 593, 774
true corals, 516, 603, 604, 608
true jellyfishes, 516, 556–568, 571, 578
true hornets, 212
true vipers, 319

Trypanosoma cruzi, 292
tubeworms, 549
tussock moths, 295, 297
turret spiders, 193
turrid snails, 531
Turridae, 531
turrids, 532, 534, 539
Turtle Cone, 532, 533, 535, **537**
Twig Snake, 496
Twin-spotted Bumblebee, 274
Twin-spotted Rattlesnake, 81, **421**–425 (**421**, **422**)
Two-form Bumblebee, 274
Two-striped Gartersnake, **235**
Typhlochactidae, 136
Typhlochactus, 136
typical scorpionfishes, 9, 57, 617, 696–703, 717
typical snakes, 319, 498

Ulmaridae, 557
umbrella wasps. *See* paper wasps
Ummidia, 193, 202
Ummidia audouini, 194, 202
unicorn tangs, 730
Upside-down Jellyfish, 81, 557, 560, **614**
Uranoscopidae, 617, 726
Urobatis, 638, 829, 830
Urobatis concentricus, 638
Urobatis halleri, 637–641 (**640**), 644, 645, 661, 829
Urobatis jamaicensis, **637**–642, 645
Urobatis maculatus, 638
Uroctonites, 128
Uroctonus, 128, 131
Uroctonus mordax, **134**
Urolophidae, 638
Urolophis, 638
Urotrygon, 637–645, 829, 830
Urotrygon aspidura, 638
Urotrygon chilensis, 638
Urotrygon rogersi, 638
Urotrygonidae, 637–645, 657
urotrygonids, 637–645, 651, 659, 665

vaejovid scorpions, 126–134, 141, 775
Vaejovidae, 126–134
Vaejovis, 127, 128, 131, 761, 845

Vaejovis carolinianus, 130
Vaejovis electrum, 131
Vaejovis grayae, 759–760, 761
Vaejovis intermedius, 133
Vaejovis mexicanus, 133
Vaejovis vorhiesi, 760, 761
Valley Carpenter Bee, 279, **280**, 281
Valvatida, 602
Varanidae, 319, 327
Varanoidea, 327
Velella velella, 586
velvet ants, 8, 46, 238–245, 769–770, 781
Velvet Dogfish, 630
velveteen tarantulas, **176**, 177, 194
Vermilion Rockfish, **712**, 718
Vespa, 212, 215, 218, 270
Vespa affinis, 212, 215
Vespa crabro, 212, 215, **216**
Vespa mandarinia, **211**
Vespa simillima, 212, 215
Vespa tropica, 738
vespid wasps, 82, 210–29, 231, 232, 235
Vespidae, 82, 210–29, 308, 310
Vespinae, 82, 210–23, 228
Vespula, 211, **213**, **214**, 222, 226, 766
Vespula acadia, 211
Vespula alascensis, 211
Vespula atropilosa, 211
Vespula consobrina, 211
Vespula flavopilosa, 211
Vespula infernalis, 211
Vespula intermedia, 211
Vespula germanica, **211**, 215, 221
Vespula maculata, **99**
Vespula maculifrons, 211, 215, **223**
Vespula pensylvanica, **210**, 211
Vespula rufa intermedia. *See* *Vespula intermedia*
Vespula squamosa, 211, 215, 217
Vespula sulphurea, 211, 215
Vespula vidua, 211
violin spiders. *See* recluse spiders
Viper Dogfish, 630, 633
Viperidae, 319, 335, 342, 349, 355, 359, 366, 372, 378, 383, 389, 396, 403, 409, 416, 421, 425, 431, 435, 441, 445, 450, 457, 464, 468, 473
Viperinae, 319
vipers, 60, 80, 87, 319, 321, 335, 342, 349, 355, 359, 366, 372, 378, 383, 389, 396, 403, 409, 416, 421, 425, 431, 435, 441, 445, 450, 457, 464, 468, 473, 838, 851
Virginia, 498
Virginia valeriae, 503

Wadotes, 169
Wadotes calcaratus, 169
Wadotes hybridus, 169
wafer-lid trapdoor spiders, 193
Walking Catfish, 688, 689, 690, 692, 694
Wandering Gartersnake, 498, 504, 505, 507, 819
Wasmannia auropunctata, 247, **248**, 738
wasps, 2, 4, 8, 13, 17, 18, 20, 21, 22, 23, 35, 40, 46, 49, 53, 54, 68, 82, 84, 88, 98, 210–245, 224, 230, 239, 269, 276, 308, 309
water bees, 593, **600**
water boatmen, 593
water moccasins. *See* cottonmouths
water scorpions, 285, 593
watersnakes, 344, 346, 495, 496, 500, 501
Water-walking Wasp, 308
Wels Catfish, 691
Wernerius, 128, 130
West Mexican Coralsnake. *See* Coralillo Bandas Claras
Western Black Widow, 46, **143**, **144**, **146**, 152, **153**, 164, 762
Western Black-tailed Rattlesnake, 46, **409**, **411**, 412, 436
Western Bumblebee, **272**, **273**, 274
Western Carpenter Bee, 279
Western Cicada Killer, 308
Western Copperhead, **335**, 336, **337**, **788**
Western Corsair, **14**, 286, **290**, 775–776

Western Diamond-backed Rattlesnake, 30, 46, 77, **322**, **323**, 355, 356, **359**–365 (**361**, **363**), 371, 432, 436, 751, 788, 790–796
Western Harvester Ant, 256
"western" hog-nosed snakes, 498, 506, 816–817
Western Honey Bee. *See* European Honey Bee
Western Massasauga, 470, 473, 474, **475**, 477, 478
Western Parson Spider, 200, **201**
Western Pygmy Rattlesnake, 468, **471**
Western Rattlesnake, 349, 416, 451
Western Shovel-nosed Snake, 480
Western Spotted Orbweaver, 198
Western Thatching Ant, 308
Western Twin-spotted Rattlesnake, 421
Western Yellowjacket, **210**, 211
Wheel Bug, **284**–292, 774–**775**
Whiptail Stingray, 647
White-spotted Clarias. *See* Hong Kong Catfish
White-spotted Eagle Ray, **662**–666 (**663**)
Whitespotted Soapfish, 730
Widow Rockfish, 713
widow spiders, 14, 25, 36, 49, 65, 69, 76, 81, 143–153
wolf spiders, 167, 193, 203, 206
wood ants, 308
Woodlouse Spider, 193, 208, **209**
wooly bears, 295, 297
wormsnakes, 499, 503

Xenocephalus egregious, 726
Xylocopa, 278–83,
Xylocopa californica, 278
Xylocopa micans, 278
Xylocopa tabaniformis, 278, 279
Xylocopa varipunctata, 279, **280**
Xylocopa virginica, **278**, **279**, 281, **282**, 283

yellow sac spiders, 186–192, 737
Yellow Bullhead, 688
Yellow Bumblebee, 274
Yellow Paper Wasp, **227**, 767
Yellow Round Stingray, **637**–645
Yellow Tang, 731
Yellow-bellied Bee Assassin, **287**
Yellow-bellied Seasnake, 59, 64, 618, 720–725 (**724**), 744
Yelloweye Rockfish, 713
Yellow-headed Bumblebee, 274
Yellowtail Rockfish, 713
yellowjackets, 36, 46, 52, 63, 82, 208, 210–23, 225, 226, 228, 266, 312, 766–767, 769, 774, 781

Zameus squamulosus, 630
Zebra Dwarf Lionfish, **705**
Zebrasoma, 726, 730, 734
Zebrasoma flavescens, 731
Zebrasoma velifer, 731
Zelus, 285, 286, 288
Zodariidae, 194
Zygaena, 295
Zygaenidae, 295